Elastic and Thermoelastic Problems in Nonlinear Dynamics of Structural Members

Scientific Computation

More information about this series at http://www.springer.com/series/718

Jan Awrejcewicz • Vadim A. Krysko

Elastic and Thermoelastic Problems in Nonlinear Dynamics of Structural Members

Applications of the Bubnov-Galerkin and Finite Difference Methods

Second edition

Springer

Jan Awrejcewicz
Łódź University of Technology
Łódź, Poland

Vadim A. Krysko
Department of Mathematics and Modeling
Saratov State Technical University
Saratov, Russia

ISSN 1434-8322 ISSN 2198-2589 (electronic)
Scientific Computation
ISBN 978-3-030-37662-8 ISBN 978-3-030-37663-5 (eBook)
https://doi.org/10.1007/978-3-030-37663-5

Originally published with the title: Nonclassical Thermoelastic Problems in Nonlinear Dynamics of Shells

This Springer imprint is published by the registered company Springer Nature Switzerland AG.
The registered company address is: Gewerbestrasse 11, 6330 Cham, Switzerland

Preface

This book is an extended version of the first edition of *Nonclassical Thermoelastic Problems in Nonlinear Dynamics of Shells: Application of the Bubnov-Galerkin and Finite Difference Numerical Methods* published in 2003.

The previous book material has been essentially extended, since seven new chapters have been added aimed at integration of modern approaches to dynamical systems theory and in particular to understand, model and obtain reliable results while detecting and analysing chaotic dynamics exhibited by thin structural members (note that the book title has been changed in order to cover investigation of dynamics of beams, plates and shells).

In the present extended and partially revised/updated version, there is an attempt to address the large impact of new directions of nonlinear dynamical phenomena in the context of the overall state of the art expressed in the first edition of the book, and also updated and new perspectives of the unsolved problems of thin structural members are presented.

This revised second edition of the monograph edition describes some approaches to the nonlinear theory of beams, plates and shells with an emphasis on the elastic and thermoelastic features. By nonclassical approaches (following the description of the first edition), we mean the description of problems with mathematical models of different sizes (two- and three-dimensional differential equations) and different types (differential equations of hyperbolic and parabolic type in the spatial coordinates). The nonlinearities investigated are also of various categories: geometrical, physical, elasto-plastic periodicity of material nonlinear properties. Creating such types of mathematical models and their detailed justification allows us to achieve the most accurate description of the real behaviour of thin structural members. In addition, the interaction between the strain and temperature fields and coupling between the displacement field and the external influence of a transonic gas flow are addressed. The mathematical treatment of such models helps us greatly in obtaining reliable results by numerical computation.

It appears that the most dangerous situation for thin structural members is the conjunction of a static load with dynamic interactions. Such combined loads very often cause buckling of shell structures, and in many cases a series of bucklings, which can cause fracture. The failure of a structure usually needs a small amount of time. Therefore, the lifetime of a shell structure depends strongly on nonelastic deflections and it is important to mathematically model shell structures as precisely as possible.

Chapter 1 of this monograph is devoted to analysis of the current literature concerning the problems that the title of the book refers to. We emphasize, among other things, the lack of systematic presentation of these problems and their solutions in the existing literature. This chapter has been extended in comparison to that of the first book edition.

In Chap. 2, a general statement of the coupled thermomechanical problems of flexible shells, with application of three-dimensional (3D) heat transfer equations, is presented, emphasizing a serious problem from the point of view of both the theoretical and the numerical approaches. However, the approach proposed here is one of the most powerful approaches, since it allows one to avoid any a priori hypotheses with respect to the temperature distribution through the shell thickness. In addition, a very important coupled problem of the interaction of shell-type structures with a transonic flow is addressed.

Note that in the case of a transonic gas flow, a special, singular flow behaviour can occur for certain profiles. In such flows, shock waves are born, which then move and vanish (they can be recognized as pressure jumps). In addition, the velocity of the excitations generated by the deformed profile is close to the velocities of the gas particles. These phenomena appear during the action of a flow on a profile and the simultaneous action of the profile on the flow.

In fact the problem of the dynamic interaction of a shell with a transonic flow considered here reduces to a two-dimensional problem of the interaction of a flow with an infinite-length cylindrical panel. A mathematical model of the latter problem is presented and a numerical method for its solution is formulated. The dynamic stability of such an infinite panel while interacting with a transonic ideal gas is analysed.

Research into the existence and uniqueness of solutions of linear coupled problems of the thermomechanics of plates and shells, within the framework of both the Kirchhoff–Love kinematic model and Timoshenko-like models with a three-dimensional parabolic equation of heat transfer, is described in Chap. 3. The problem is reduced to an abstract Cauchy problem of a coupled system of two differential equations in a Hilbert space, which generalizes the set of the above coupled problems of thermoelasticity. The solutions to the problems formulated are obtained by applying the Bubnov–Galerkin method in a Sobolev space. Many theorems on the existence, uniqueness, regularity and continuous dependence of solutions on the initial conditions are formulated and discussed.

For linear coupled problems of shallow shells that are rectangular in plan, prior estimates of the errors of the Bubnov–Galerkin method are presented. The estimates are obtained for the general case of a shell of variable thickness, subjected to time-dependent mechanical and thermal loads. These estimates are of great theoretical importance, since they guarantee strong convergence of a series of approximate solutions to the exact solution with a speed not smaller than that described in this chapter, for a wide class of problems.

In Chap. 4, results of numerical analysis of the estimated errors of the Bubnov–Galerkin method for linear problems in the theory of plates are presented. These results show that the estimates provide high efficiency.

In Chap. 5, a mathematical model of coupled, geometrically nonlinear problems of the thermoelasticity of shallow shells, using a Timoshenko-type kinematic theory and taking into account the rotational inertia of the shell and a three-dimensional thermal field (as in previous chapters), is formulated. The problem of the existence and uniqueness of a solution, as well as the convergence speed of the Bubnov–Galerkin method, is analysed; this generalizes results presented in Chap. 3.

In Chap. 6, a more exact mathematical model of the theory of shells is introduced and analysed. In addition to the hypotheses and assumptions of Chap. 5, nonlinearity of the physical properties of the shell material is included. The kinematic Kirchhoff–Love model is used. The mathematical model is obtained by the Biot variation principle.

Chapter 7 contains numerical methods for solution of the problems introduced in the previous chapter, i.e. the finite difference method of order $O(h^2)$ and the method of "elastic" solutions are applied in the numerical computations.

An analysis of the convergence of the solutions and an investigation of the dynamic stability loss of shells for different values of the parameters governing the geometric, physical and thermal properties are carried out.

The Bubnov–Galerkin method is applied to problems concerning the dynamic stability of shells using Timoshenko-type or Kirchhoff–Love models of coupled thermoelasticity with respect to displacements. In Sects. 7.7–7.11, parametric vibrations of flexible shells are investigated. Solutions are approximated using finite difference methods of order $O(h^2)$ and $O(h^4)$ with respect to the spatial coordinates.

Symmetric and nonsymmetric vibrations are analysed. Plates are treated as systems with an infinite number of degrees of freedom. Soliton phenomena in elastic plates in the presence of parametric vibrations are detected and discussed, among other topics. It is shown that both symmetric and nonsymmetric vibrations are included in the mathematical model.

Chapter 8 presents an even more complicated mathematical model than that applied in Chap. 6. Here, besides geometric nonlinearities, the periodicity of material properties during vibration is taken into account. Numerical algorithms for the solution of such problems are presented. Analysis of the dynamic stability and vibrations of plates and shells for different physical and geometric parameters is performed, and the results are presented in a series of graphs.

Chapter 9 is devoted to derivation of a mathematical model of flexible shallow shells (cylindrical and spherical) with an account of geometric nonlinearity. Based on the introduced assumptions, shell thickness and the Kirchhoff–Love hypotheses, the relations between shell stresses and deformations are defined. Then the Hamilton principle yields the PDEs governing shell motion and deformation compatibility. The obtained equations are transformed to their counterpart non-dimensional forms for the case of a spherical panel with the rectangular planform and for a cylindrical panel. Four most important boundary conditions are considered.

Then both the shell deflection function $w(x, y, t)$ and the Airy's function $F(x, y, t)$ are introduced $w = \sum_{i=0}^{N_1} \sum_{j=0}^{N_2} A_{ij}(t)\phi_{ij}(x, y)$, $F = \sum_{i=0}^{N_1} \sum_{j=0}^{N_2} B_{ij}(t)\psi_{ij}(x, y)$ with the help of the properly chosen function φ_{ij}, ψ_{ij}. The Bubnov–Galerkin method (BGM) yields the system of ODEs with respect to the time-dependent functions $A_{ij}(t)$, $B_{ij}(t)$. The obtained equations are supplemented by the boundary conditions, and the obtained Cauchy problem is solved using the 4th order Runge–Kutta method where the time step of integration is estimated based on the Runge's principle and the associated integrals of the BGM are derived explicitly.

First dynamics of a closed shell simply supported is studied, and then vibrations of the simply supported cylindrical panel and spherical panel with rectangular planform are analysed. Many advantages of the employed BGM are outlined with an emphasis on solving both static and dynamic problems. The first can be solved using the dynamic approach based on the set up method. Reliability of the obtained results is illustrated and discussed on the basis of the higher order BGM and the known analytical solution. The benefits of the employed set up method to solve the static problems of flexible shallow shells based on both hyperbolic and parabolic type PDEs are reported with an emphasis put on the solution convergence estimation.

A few case studies aimed at estimation of a number of iterations required for achieving the reliable stationary solutions are performed. It is illustrated that the worked out algorithms allow to solve the wide class of problems of statics and dynamics of flexible shells with the geometric nonlinearity.

The last part of the chapter is devoted to study the flexible shells subjected to either the uniformly distributed and constant load along the shell surface or the impulse-type dynamic load action. A key role in the study is to determine the critical values of static and dynamic load associated with the loss of stability of the shell.

The state of the art of the existing stability criteria is critically overviewed, exhibiting their benefits and drawbacks. Next the critical loads for the cases of closed cylindrical shells, cylindrical panels and spherical shells with rectangular planforms are estimated and the dependencies of the estimated critical load (q_{cr}) versus the shell curvature (k_x, k_y) are reported for both static and dynamic problems. Influence of the most important parameters of the cylindrical panels and spherical shells on their stability loss based on the critical load estimation is studied, which allows to give prescription for engineer to avoid the harmful dynamic stability loss of both local and global character.

Chapter 10 presents numerical analysis of the scenarios of transition from periodic to chaotic vibrations exhibited by shells. Namely, we have detected and studied the so-called modified Ruelle–Takens–Newhouse (RTN) scenario, the modified Ruelle–Takens–Feigenbaum (RTF) scenario as well as the modified Pomeau–Manneville scenario found in our analysed mechanical continuous systems. We have numerically investigated the Sharkovsky's orders of periodicity for PDEs governing dynamics of shallow shells. In the case of analysis of vibrations of the closed cylindrical shell with the curvature $k_y = 112.5$ and $\lambda = 3$ (relative shell thickness) under transversal uniformly distributed harmonic load $q = q_0 \sin(\omega_p t)$, localized in the shell zone characterized by the angle $\phi_0 = 6$, we have detected windows of periodicity which coincide with the order predicted by Sharkovsky's theorem, i.e. 3; 5; $2^2 \cdot 5$; 9. We have also investigated a cylindrical panel under action of the longitudinal two-parametric load $p_x = p_0 \sin(\omega_p t)$ and $p_y = \alpha \cdot p_x$ for $k_y = 24$ and $\alpha = 3$ where we have found also the periodicity windows following the Sharkovsky's order 3; 2·3; 5; 2·5; 7; 2·7; 11; 13. The carried out analysis of the phase and model portraits constructed for both periodic and chaotic shell vibrations revealed that the space and temporal (timing) chaos appear simultaneously, i.e. in the case of shells it refers to occurrence of the space-temporal chaos.

The mathematical models of chaotic vibrations of closed cylindrical shells with the circle cross sections are derived and studied in Chap. 11. We begin with an important problem of estimation of convergence of BGM while studying the chaotic cylindrical shell vibrations. The number of modes required in getting reliable results of the BGM is estimated using the case study of closed cylindrical and geometrically nonlinear shells. A sectorial shell is defined by the central angle φ_0 and it is subjected to the harmonic load $q = q_0 \sin \omega_p t$. The full dynamic palette of the shell regimes is detected and analysed using the so-called dynamic charts presenting shell vibration behaviour versus the chosen control parameters $\{q_0, \omega_p\}$.

In particular, the used numerical tools allow to distinguish zones with the purely exhibited Feigenbaum scenario as well as zone of the modified Ruelle–Takens–Newhouse scenario.

In order to construct the mentioned plane charts, we have put a mesh consisting of the nodes where the vibrations character was identified. Consequently, the dynamic problem for each of the mesh node should be solved and for each node of the mesh $\{q_0, \omega_p\}$ the associated power spectrum should be also analysed. The carried out computations have shown that the mesh resolution $N \times N \geq 350 \times 350$ is optimal with regard to the computational time in both period and chaotic regimes in the investigated plane $\{q_0, \omega_p\}$.

We aimed at getting reliable results while employing BGM to study vibrations of closed cylindrical and geometrically nonlinear shells and hence we can take $N_1 = 1$, i.e. only one series term along the longitudinal shell coordinate.

Since the charts construction is based on analysis of the frequency power spectra and the Lyapunov exponents, the mentioned charts illustrate process of convergence of the power spectrum and the vibrations character for each of the employed mesh approximation. Therefore, for small values of $N_2 \leq 9$ mainly periodic vibrations are presented and only small zones of bifurcations are visible with a lack of chaotic zones. For $N_2 = 10$ the chart undergoes a qualitative change; the bifurcation zones are vanished for the high frequencies but they appear on the frequencies close to the fundamental shell frequency. Further increase of N_2 up to $N_2 = 11$ generates occurrence of large zones of chaotic vibrations on low frequencies and being close to the fundamental frequency, the previous bifurcation areas observed for $N_2 = 9$ are vanished. Further increase of $N_2 = 12$ implies changes on high frequencies within chaotic zones whereas zones associated with low frequencies are practically unchanged. Finally, the constructed charts for $N_2 = 13$ and $N_2 = 14$ are very similar to each other; hence one may conclude that for $N_2 = 13$ the process of convergence is terminated and the obtained results are reliable.

The case study presents investigation of convergence of the Bubnov–Galerkin method for the cylindrical shells under action of the transverse harmonic local load on an example of a shell with $\lambda = 2;\ 6$ and under the width of the external pressure band measured by the central angle $\varphi_0 = 343°$. We demonstrate results obtained by our simulation package devoted to study chaotic vibrations of closed cylindrical shells with the circle cross section based on the Bubnov–Galerkin method in higher order approximations.

The dependencies $w_{\max}(q_0)$ and the vibration scales of the dynamics regimes of the closed cylindrical shells with $\lambda = 2$ and for $\varphi_0 = 90°;\ 180°;\ 343°$ and with $\lambda = 1;\ 2;\ 3;\ 4;\ 5;\ 6;\ 7$ and for $\varphi_0 = 343°$ are reported. In particular, a novel criterion of the stiff stability loss of the cylindrical shells is formulated and illustrated.

We have detected zones with the classical Feigenbaum scenario on the charts $\{q_0, \omega_p\}$ for the closed cylindrical shells with seven Hopf period doubling bifurcations; the latter allowed to estimate the Feigenbaum constant with the accuracy of 0.01%.

We have also detected a novel scenario of transition of the shell periodic vibrations to chaotic vibration referred by us as the modified Pomeau–Manneville scenario as well as the phenomenon of intermittency in the shell vibrations. The latter scenario has been investigated and illustrated by a case study of the closed cylindrical shell with $\lambda = 2$ and for $\varphi_0 = 90°$.

Influence of the area of the transverse local load acting in direction x (longitudinal) and y (circle) as well as of the coordinates of their localization on the shell vibration character is analysed. Finally, the features of the spatial-temporal chaos are illustrated and discussed.

Chapter 12 presents derivation of the mathematical model of a geometrically non-linear closed cylindrical nano-shell and its numerical study. The Kirchhoff–Love hypothesis of the straight normal is employed, and then key steps of the modified couple stress theory are used to derive the shell deformation energy and the stress-strain formulas. Then the resolving PDEs governing shell motion and deformation compatibility are yielded by the Hamilton variational principle. Four boundary conditions are supplemented to the derived PDEs, and zeroth initial conditions are considered. First an algorithm of BGM of the stated problem is presented and then the numerical results associated with the size-dependent closed cylindrical shell with simply supported edges are reported.

The shell is subjected to action of the transversal bond type sinusoidal load with the amplitude $q_0 = 0.323$, $\omega_p = 20.3$, $\varepsilon = 1$, $\lambda = 2$ and with the same loading angle $\varphi_0 = 5.98$ along the entire length of the cylinder.

The signals obtained using different number of the series terms ($N = 5 - 15$) and the Fourier power spectra in the function development of the Bubnov–Galerkin method for $l = 0$ (size-dependent parameter) are presented. In the beginning, for $N = 5, 7$ the cylinder vibrates with the frequency of the excitation load. However, increase of the number of the series terms implies a qualitative change of vibrations that can be followed by the dynamic characteristics. For $N = 9$ three frequencies are found with the values linearly dependent on the excitation frequency ω_p. Further increase of N up to 11 implies occurrence of more frequencies. Therefore, the Fourier power spectrum consists of the following frequencies: $\omega_p = 20.3, \omega_1 = 8.3$, $\omega_2 = \omega_p - \omega_1 = 12$, $\omega_3 = \omega_p - 2\omega_1 = 3.7$, $\omega_4 = \frac{21}{20}\omega_p - \omega_1 = 13.015$, $\omega_5 = \omega_p - \omega_4 = 7.2$, $\omega_6 = \frac{\omega_p}{4} = 5.07$, $\omega_7 = \frac{\omega_p}{20} = 1.015$. For $N = 13, 15$ the dynamic characteristics coincide.

For the case of $N = 5$ the amplitude and vibration form is essentially different in comparison to other N. The signals for $N = 13$ and $N = 15$ coincide. Therefore, one may conclude that the obtained numerical solutions coincide in full for $N = 13$.

We investigate scenarios of transition from the shell regular vibrations to chaotic one depending on the terms number in the Bubnov–Galerkin method and depending on the size-dependent parameter l.

We have detected one period doubling bifurcation for $N = 5$, and a further increase of the load amplitude was impossible due to the large amplitude of vibrations. For $N = 11, 12$ another scenario of transition from regular to chaotic vibrations is observed. For $q = 0.32$, in spite of the excitation frequency ω_p, there is a birth of linearly non-dependent frequency $\omega_1 = 8.3$ as well as of the linear combination of frequencies $\omega_2 = \omega_p - \omega_1 = 12$ and $\omega_3 = \omega_p - 2\omega_1 = 3.7$. For $q = 0.3245$, new linear combinations of the frequencies ω_p, ω_1 appear. Therefore, a scenario of transition into chaos is in agreement with the classical Ruelle–Takens–Newhouse scenario via two linearly independent frequencies and their linear combinations.

For $l = 0$ we have not observed the subharmonic frequency $\omega_p/2$. Occurrence of the periodic doubling bifurcation has been also exhibited in the phase portrait (for instance $\omega_p/11 = 1.845$). For $q = 0.46$ we have detected one more period doubling bifurcation $\omega_p/4$; $3\omega_p/4$, and a new linear combination of the frequencies.

Hence, it is clear that an account of the size-dependent parameter l implies the change of transition from regular to chaotic shell vibrations. In the case of $l = 0.3$ the existence of two scenarios, i.e. Feigenbaum in the form of the period doubling bifurcation and the Ruelle–Takens–Newhouse scenario associated with two independent frequencies and their linear combinations, is detected.

We increase the size-dependent parameter up to 0.5 and investigate a scenario of transition into chaos for different numbers of terms used in the Bubnov–Galerkin series. In the case of $N = 5, l = 0.5$ a transition from regular to chaotic vibrations is realized for large values of the excitation amplitude q via the period doubling bifurcation. Increase of the parameter l for $N = 11, 13$ yielded increase of the vibration amplitude, which was observed also for $N = 5$. However, the scenario of transition from regular to chaotic vibrations has not been changed.

For $l = 0.3$ first a birth of the frequency $\omega_1 = 8.3$ is observed, and then (for $l = 0.5$) the frequency $\omega_2 = 5.16$ appears being linearly dependent $\omega_p - \omega_2 = 15.1$ for $q = 0.51$. Next, the period doubling bifurcation takes place (for $q = 0.53$), and the system regular vibrations move to chaotic vibration via occurrence of new linear combinations of the frequencies. The phase portrait consists of two rings for $q = 0.4;\ 0.51$, which demonstrates the occurrence of a high frequency. The signal is quasi-periodic with two frequencies.

In addition for $q = 0.7$ we have two-frequency shell vibrations where the second frequency is twice larger than the excitation frequency ω_p, ω_3. Then a linearly independent frequency $\omega_4 = 9.5$ in comparison to ω_p, ω_3 appears, and the linear combination of the two mentioned frequencies takes place. In this case, the scenario of transition into chaotic vibrations follows the Ruelle–Takens–Newhouse scenario.

In Chap. 13 the contact interaction of two rectangular plates made from different materials with an account of physical nonlinearity is analysed. In particular, an approach convenient for analysis of stress-strain state (SSS) of multi-layer plates and beams with welded layers and with the occurrence of separated contacting zones due to the technological and exploitation reasons is proposed. The method offers a possibility of studying the constructions with contacts between layers, when the contact condition may depend on coordinates and all non-homogeneity properties of the studied contact.

Important results yielded by our theoretical investigations concern removal of the function describing the contact interaction between the structural members from a number of unknowns.

The given algorithm allows one to study SSS of the structures, in which the dependencies $\sigma_i^{(i)}(e_i^{(i)})$ can be given either in digital forms or can be formulated analytically. In the case of a two-layer plate-plate package with design nonlinearity, phase synchronization takes place in the frequency interval $\omega \in [4,\ 8]$, where there is lack of synchronization (bright zones in the graphs of the phase difference) in certain time intervals, but amplitude locking phenomena appear instead.

Finally, we present modification of the methods of reduction of elliptic type PDEs to ODEs. The mentioned methods possess many advantages in comparison to the Bubnov–Galerkin and Ritz methods since they do not require an a priori knowledge of the system of approximating functions, and the obtained solution satisfies the governing equations. The so far mentioned methods can be also successively applied to solve hyperbolic and parabolic PDEs including 2D and 3D problems (in the case of nonlinear elliptic equations, the use of the variational iteration method was proved rigorously in [118]).

Due to occurrence of a multiplier, the system of governing PDEs becomes nonlinear, and consequently the entire problem becomes nonlinear. The design nonlinearity governs the possibility of switching on/off of the one-sided constraint. Furthermore, inclusion of the physical nonlinearity in the considered problem makes it impossible to solve the problem via classical analytical, semi-analytical or numerical approaches. Therefore, we employ here the method of successive approximations matched with a simultaneous process of improvements of the borders of the contact zones.

This chapter offers also a brief description of advantages/disadvantages of different methods aimed at reduction of PDEs to ODEs, which include the method of Kantorovich–Vlasov, the method of Vaindiner, the method of variational iterations, the method of Arganovskiy–Baglay–Smirnov, the combined method, the matching of the methods of Kantorovich–Vlasov and Arganovskiy–Baglay–Smirnov, the matching of the methods of Vaindiner and Arganovskiy–Baglay–Smirnov and matching of the methods of Vaindiner with the method of variational iterations.

In the case of the variational iterations method employed to our problem, two theorems devoted to its convergence are formulated, and illustrative examples are added.

In addition, a theorem (without a proof) regarding convergence of our proposed iterational procedure associated with a design of nonlinear systems composed of two plates made from different materials is formulated.

Chaotic vibrations of flexible shallow axially symmetric shells with respect to the different employed boundary conditions are analysed in Chap. 14.

The problem is described by a set of two nonlinear PDEs using the polar co-ordinates system. Four different boundary conditions with respect to two functions $\phi = \frac{\partial F}{\partial r}$ and $w(r, t)$ are taken into account, where w is the shell deflection, F is the stress function, and is a radial co-ordinate, whereas t denotes time. The governing PDEs as well as the boundary and initial conditions are recast to the counterpart finite difference formulas, and then the finite difference method is used to transit the spatial problem to that of solving the ODEs with respect to time. The yielded Cauchy problem is solved using the 4th order Runge–Kutta method, the 2nd order Runge–Kutta method, the 4th order Runge–Kutta-Fehlberg method, the 4th order Cash–Karp method, the 8th order Runge–Kutta–Prince–Dormand method, as well as the 2nd and 4th order implicit Runge–Kutta method. Finally, it has been shown that the 4th order Runge–Kutta method was the most convenient to solve the derived ODEs because it offered the required accuracy and the shortest computational time. The used finite difference method stands as one alternative choice of the Faedo–Galerkin, the Ritz and the finite element method.

In order to solve the problem of reliability of chaotic vibrations while numerically solving nonlinear PDEs governing dynamics of spherical axially symmetric shells subjected to harmonic transverse load, the following steps are carried out:

1. Since PDEs are reduced to a Cauchy problem by means of the second-order FDM, the solution strongly depends on the number of nodes involved in the shell radius $r \in [0; b]$ partition. A convergence of the employed FDM is analysed for different shallowness parameters and boundary conditions.
2. Since the Cauchy problem is also solved numerically, the solution essentially depends on both the used method and the employed time step. Therefore, the following methods are used: 4th- and 2nd-order Runge–Kutta (RK4 and RK2) methods, the 4th-order Runge–Kutta–Fehlberg (rkf45) method, the 4th-order Cash–Karp (RKCK) method, combination of the 8th-order Runge–Kutta and Dormand method (rk8pd) as well as 2nd- (rk2imp) and the 4th-order (rk4imp) implicit Runge–Kutta methods.
3. The numerically obtained data are used to construct the time series (signals), phase portraits, Fourier power spectra (FFT), Morlet wavelets and shell deflection surfaces. The following wavelets are used: Meer wavelets, Shannon–Kotelnikov

and Meier wavelets, Daubechies wavelets from db2 to db4, coiflets, Morlet wavelets, complex Morlet wavelets as well as the wavelets based on a derivative of the Gauss function of higher than the 8th order. It should be emphasized that it was not feasible to investigate the considered problem with Meer and Shannon–Kotelnikov wavelets, i.e. the first one did not identify the frequency zones properly, whereas the second one did not yield the expected localization in time. The carried out analysis of the wavelet spectra obtained by the Daubechies wavelets, coiflets and simlets has shown that an increase in the order of the employed filter improves proper frequency identification. Furthermore, in spite of differences between the employed wavelets and their filters, the wavelet spectra yielded by the Daubechies wavelets, simlets and coiflets of the same order give almost the same results. However, the frequency localization yielded by them is not accurate enough, which makes them unsuitable for analysis of vibration of the investigated mechanical system.

Considering the results obtained by means of derivatives of the Gauss functions, the accuracy of the frequency localizations increased with the increase in the order of the derivative. It should be noted that the frequency spectra obtained by the Meer wavelet, which is a smooth variant of the Shannon–Kotelnikov wavelet, have been localized better than in the case of Morlet wavelets only for a low frequency bands. In what follows we give results for Morlet wavelets that were specially chosen and adopted to study the stated problem. In these wavelets, there is no scale function φ, and ψ is given explicitly, i.e. there is no compact carrier. Therefore, to study complex vibration of continuous mechanical system, like a plate/shell, one can employ either complex or real Morlet wavelets or wavelets based on either real or complex derivatives of the Gauss functions of the order higher than 16.

4. Since the definition of chaos given by Gulick [138] is used in this work, one needs to compute Lyapunov exponents (LEs) and define their signs. Here a spectrum of the LEs for each partition of the interval $[0; b]$ was obtained by three methods: Wolf's [351], Rosenstein's [285] and Kantz's [164]. Furthermore, convergence of the above-mentioned methods was also investigated.

Novel results of the study of chaotic vibration of flexible circular axially symmetric shallow shells subjected to sinusoidal transverse load for four different boundary conditions are presented. The study is conducted with the use of the finite difference method (FDM), which differentiates the study from the majority of research conducted with the finite element method (FEM).

Chaos reliability is verified and validated by solving the implemented mathematical model by partial nonlinear equations (PDEs) governing the dynamics of flexible spherical shells and by estimating the signs of the largest Lyapunov exponents (LLEs) with the help of qualitatively different approaches. It is shown how the scenario of transition of the investigated shells from regular to chaotic vibrations depends on the boundary condition. The following cases are considered: (1) movable and fixed simple supports along the shell contours, taking into account shell stiffness (Feigenbaum scenario) and shell damping (Ruelle–Takens–Newhouse scenario), and (2) movable clamping (regular shell vibrations).

It is detected, illustrated and discussed that the employed boundary conditions significantly influence the obtained solutions. For simple nonmovable and movable supports of the shell, for $\omega_p = \omega_0$ (ω_p stands for the harmonic excitation frequency, and ω_0 refers to the shell fundamental frequency), a transition from regular to chaotic shell vibration was realized following the Feigenbaum scenario, whereas for a rigidly clamped shell, a transition to chaos followed the Ruelle–Takens–Newhouse scenario. Finally, in the case of the movable clamped shell, a series of Sharkovsky's windows was reported.

It is found that the number of ribs exhibited by the shell surface strongly depends on the boundary conditions. The maximum number of ribs was detected for the shell with movable simple support, one rib was found for the shell with non-movable simple support and the shell with movable clamping, whereas three ribs were detected in the case of rigid clamping.

Chapter 15 deals with a study of chaotic vibrations of two Euler–Bernoulli beams with a small clearance. In this chapter a methodology to detect true chaos (in terms of nonlinear dynamics) is developed on an example of a structure composed of two beams with a small clearance. The Euler–Bernoulli hypothesis is employed, and the contact interaction between beams follows the Kantor model. The complex nonlinearity results from the von Kármán geometric nonlinearity as well as the nonlinearity implied by the contact interaction. The governing PDEs are reduced to ODEs by the second-order FDM. The obtained system of equations is solved by Runge–Kutta methods of different accuracy. To purify the signal from errors introduced by numerical methods, principal component analysis is employed and the sign of the first Lyapunov exponent is estimated by the Kantz, Wolf and Rosenstein methods and the method of neural networks. In the latter case a spectrum of the Lyapunov exponents is estimated. It has been illustrated how the number of nodes in the employed FDM influences the numerical results regarding chaotic vibrations. We have also shown how increase of the beams distance implies stronger action of the geometric nonlinearity, and hence influence of convergence of the used numerical algorithm for FDM has been demonstrated. The effect of essential dependence of the initial conditions choice on the numerical results of the studied contact problem is presented and discussed.

The following assumptions and hypotheses are introduced:

(1) Each beam is composed of one layer.
(2) The beams are isotropic, elastic and obey Hook's law
(3) The longitudinal dimensions of beams are larger than their transverse dimensions and the beams have the unit thickness.
(4) The axis of each of the beams is a straight line.
(5) The load acts in the direction of the OZ axis and the external forces do not change their direction during the beam deformation.
(6) Contact pressure is estimated within the Kantor model.

(7) Normal stresses regarding the zones being parallel to the axis are small and negligible.
(8) Geometric nonlinearity is taken in the von Kármán form.

The Hamilton energetic principle yielded equations of motion of beams in the form of the four nonlinear PDEs as well as the boundary and initial conditions.

The principal component analysis working as a filter allowing to remove noisy signal components is used.

We have detected the set of parameters, and we have used the methods which guarantee reliability and truth of the obtained solutions. In particular, we have demonstrated that the increase in the number of beam partitions (nodes) implies regularization of the obtained vibrations. After comparing the results obtained within linear and nonlinear problems and taking into account contact between beams, we have also illustrated that the account of the spatial coordinate x in $u(x, t)$, i.e. the increase in the number of equations, decreases the chaotic components in the analysed signals.

In Chap. 16, some mathematical problems which have been briefly mentioned in this monograph but which have not yet been satisfactorily solved are addressed.

In spite of the inclusion of the new material aimed at integration of modern approaches to dynamical systems theory for detection and understanding chaotic vibrations of structural members in this edition, the unsolved problems are seen from a new perspective in comparison to the first edition of the book.

Both authors greatly appreciate the help and assistance of Professor Maksim Zhigalov (Saratov) and Mr. Marek Kaźmierczak (Lodz) who helped us with the book preparation.

Finally, we would like to acknowledge a support of the Russian Science Foundation RSF No. 16-11-10138П grant.

Lodz, Poland
Saratov, Russia
April 2019

Jan Awrejcewicz
Vadim A. Krysko

Contents

1 **Introduction** 1

2 **Coupled Thermoelasticity and Transonic Gas Flow** 19
2.1 Coupled Linear Thermoelasticity of Shallow Shells 20
2.1.1 Fundamental Assumptions 20
2.1.2 Differential Equations 22
2.1.3 Boundary and Initial Conditions 26
2.1.4 An Abstract Coupled Problem 29
2.1.5 Existence and Uniqueness of Solutions of Thermoelasticity Problems 36
2.2 Cylindrical Panel Within Transonic Gas Flow 49
2.2.1 Statement and Solution of the Problem 49
2.2.1.1 Equations of Motion and Boundary Conditions of a Panel 49
2.2.1.2 Equations of Transonic Ideal-Gas Motion 57
2.2.1.3 Equations and Boundary Conditions of Panel–Flow Interaction 69
2.2.1.4 Analysis of Panel Oscillations 73
2.2.1.5 Method of Solution of the Problem 78
2.2.1.6 Interaction Between a Flow and a Panel 91
2.2.2 Stable Vibrating Panel Within a Transonic Flow 92
2.2.2.1 Shock Wave Motion 92
2.2.2.2 Interaction of a Flow and a Panel 97
2.2.3 Stability Loss of Panel Within Transonic Flow 106
2.2.3.1 Stability Criteria 106
2.2.3.2 Analysis of Stress–Strain State 115

3 **Estimation of the Errors of the Bubnov–Galerkin Method** 121
3.1 An Abstract Coupled Problem 122
3.2 Coupled Thermoelastic Problem Within the Kirchhoff–Love Model 142
3.3 Case of a Simply Supported Plate Within the Kirchhoff Model ... 154

3.4 Coupled Problem of Thermoelasticity Within a Timoshenko-Type Model 169

4 Numerical Investigations of the Errors of the Bubnov–Galerkin Method 175
4.1 Vibration of a Transversely Loaded Plate 175
4.2 Vibration of a Plate with an Imperfection in the Form of a Deflection 183
4.3 Vibration of a Plate with a Given Variable Deflection Change 186

5 Coupled Nonlinear Thermoelastic Problems 191
5.1 Fundamental Relations and Assumptions 191
5.2 Differential Equations 195
5.3 Boundary and Initial Conditions 201
5.4 On the Existence and Uniqueness of a Solution and on the Convergence of the Bubnov–Galerkin Method 202

6 Theory with Physical Nonlinearities and Coupling 233
6.1 Fundamental Assumptions and Relations 233
6.2 Variational Equations of Physically Nonlinear Coupled Problems 237
6.3 Equations in Terms of Displacements 241

7 Nonlinear Problems of Hybrid-Form Equations 245
7.1 Method of Solution for Nonlinear Coupled Problems 246
7.2 Relaxation Method 252
7.3 Numerical Investigations and Reliability of the Results Obtained 259
7.4 Vibration of Isolated Shell Subjected to Impulse 267
7.5 Dynamic Stability of Shells Under Thermal Shock 283
7.6 Influence of Coupling and Rotational Inertia on Stability 306
7.7 Numerical Tests 318
7.8 Influence of Damping ε and Excitation Amplitude A 320
7.9 Spatial–Temporal Symmetric Chaos 324
7.10 Dissipative Nonsymmetric Oscillations 345
7.11 Solitary Waves 347

8 Dynamics of Thin Elasto-Plastic Shells 367
8.1 Fundamental Relations 367
8.2 Method of Solution 374
8.3 Oscillations and Stability of Elasto-Plastic Shells 379

9 Mathematical Model of Cylindrical/Spherical Shell Vibrations 415
9.1 Fundamental Relations and Assumptions 415
9.2 The Bubnov-Galerkin Method 422
9.2.1 Closed Cylindrical Shell 425
9.2.2 Cylindrical Panel 427
9.3 Reliability of the Obtained Results 429

9.4 On the Set Up Method in Theory of Flexible Shallow Shells 432
9.5 Dynamic Stability Loss of the Shells Under the Step-Type Function of Infinite Length 435

10 Scenarios of Transition from Periodic to Chaotic Shell Vibrations ... 439
10.1 Novel Models of Scenarios of Transition from Periodic to Chaotic Orbits 439
10.2 Sharkovsky's Periodicity Exhibited by PDEs Governing Dynamics of Flexible Shells 443
10.3 On the Space-Temporal Chaos 448

11 Mathematical Models of Chaotic Vibrations of Closed Cylindrical Shells with Circle Cross Section 451
11.1 On the Convergence of the Bubnov-Galerkin (BG) Method in the Vase of Chaotic Vibrations of Closed Cylindrical Shells.... 452
11.2 Chaotic Vibrations of Closed Cylindrical Shells Versus Their Geometric Parameters and Area of the External Load Action 462

12 Chaotic Dynamics of Flexible Closed Cylindrical Nano-Shells Under Local Load 479
12.1 Statement of the Problem 479
12.2 Algorithm of the Bubnov–Galerkin Method 482
12.3 Numerical Experiment 488

13 Contact Interaction of Two Rectangular Plates Made from Different Materials with an Account of Physical Non-Linearity 495
13.1 Statement of the Problem 496
13.2 Reduction of PDEs to ODEs 501
13.2.1 The Method of Kantorovich-Vlasov (MKV) 501
13.2.2 Method of Vaindiner (MV) 502
13.2.3 Method of Variational Iteration (MVI) 502
13.2.4 Method of Arganovskiy-Baglay-Smirnov (MABS)....... 503
13.2.5 Combined Method (MC) 504
13.2.6 Matching of the Methods of Kantorovich-Vlasov and Arganovskiy-Baglay-Smirnov (MKV+MABS) 505
13.2.7 Matching of the Methods of Vaindiner and the Arganovskiy-Baglay-Smirnov (MV+MABS) 505
13.2.8 Matching of the Methods of Vaindiner with the Method of Variational Iterations (MV+MVI) 506
13.2.9 Numerical Example 507
13.3 Mathematical Background 508
13.3.1 Theorems on Convergence of MVI 508
13.3.2 Convergence Theorem 510
13.4 Contact Interaction of Two Square Plates 512
13.4.1 Computational Examples 513
13.5 Dynamics of a Contact Interaction 516

14 Chaotic Vibrations of Flexible Shallow Axially Symmetric Shells vs. Different Boundary Conditions 521
14.1 Problem Statement and the Method of Solution 522
14.2 Quantification of True Chaotic Vibrations 526
14.3 Modes of Vibrations (Simple Support) 531
14.4 Modes of Vibrations (Rigid Clamping) 535
14.5 Investigation of Occurrence of Ribs (Simple Nonmovable Shell Support) 542
14.6 Shell Vibration Modes (Movable Clamping) 542

15 Chaotic Vibrations of Two Euler-Bernoulli Beams With a Small Clearance 551
15.1 Mathematical Model 552
15.2 Principal Component Analysis (PCA) 557
15.3 Numerical Experiment 558
15.4 Application of the Principal Component Analysis 567

16 Unsolved Problems in Nonlinear Dynamics of Thin Structural Members 573

References 587

Chapter 1
Introduction

In this chapter, we give a brief discussion of the literature on the nonlinear theory of structural members, paying attention particularly to Eastern references, where many interesting results have been obtained and which are (unfortunately) still not well distributed among the worldwide scientific community.

Interest in coupled thermoelastic problems has developed during last 30–35 years. This was caused by the need for mathematical calculations of structures subjected to the action of a nonuniform thermal load (for example in aerospace and nuclear engineering).

During deformation caused by high-speed mechanical and thermal loads, the "coupling effect" appears, which depends on the interaction of the temperature and deformation fields. The first suggestion of this coupling effects was presented by Duhamel [102] in 1837. He introduced into the thermal-conductivity equations a term representing the change of volume of a body. Investigations during the next 100 years were performed mainly in the frame-work of the classical theory of elasticity and thermal conductivity. The theory of the thermal stresses was based on a simplified assumption that there is no influence of the deformation of an elastic body on the heat transfer process. Various aspects of these theories were presented in monographs by Karsloy and Eger [171], Lykov [232] and Kutateladze [211].

A comparison of methods suitable for solving linear coupled problems of thermoelasticity up to 1970 was presented by Nowacki [263–267].

Kozlov [194, 195] has considered the problem of thermoelastic oscillations of an isotropic rectangular plate subjected to a thermal shock. With some generalizing assumptions, a solution to this problem was obtained analytically using a Fourier series approach. The investigations indicated that damping of the oscillations of the plate occurred as a result of the coupling of the deformation and temperature fields. No additional simplifying hypotheses regarding the distribution of the temperature field through the thickness of the plate were used.

The majority of researchers have assumed a polynomial distribution of the temperature through the thickness of the plate or a shell to decrease the order

J. Awrejcewicz, V. A. Krysko, *Elastic and Thermoelastic Problems in Nonlinear Dynamics of Structural Members*, Scientific Computation,
https://doi.org/10.1007/978-3-030-37663-5_1

of the thermoelasticity equations. Malkin [234] first introduced an infinite-series representation for the temperature. Marguerre [237] obtained equations governing the stationary thermoconductivity of a plate by averaging the temperature through the thickness of the plate. This work made the assumption that the temperature is linearly distributed through the thickness of the plate. Podstrigach [279] has extended this method to the case of a nonstationary thermal load for thin shells, without taking their curvature into account. The generalized case of a polynomial temperature distribution through the thickness of a shell was investigated by Borisyuk and Motovilovitz [75]. A linear temperature distribution through the thickness of the shell was also assumed by Shvetz and Lun' [301] in deriving the equations of coupled thermoelasticity of orthotropic shells in the framework of Timoshenko's theory. Shvetz [298] and other authors used the same assumption to obtain generalized variational principles of dynamic coupled problems for isotropic and anisotropic shells. Shvetz and Flachok [299] presented theorems about uniqueness, mutuality and variational principles. These authors derived those theorems for coupled linear equations for an anisotropic shell with a cubic distribution of the temperature through the thickness. The same authors presented solutions extended to problems of mechanical thermodiffusion, also with a cubic distribution of the temperature through the thickness of the shell. Different variants of a polynomial distribution of the temperature through the thickness of the shell were presented in the monograph [256] by Motovilovitz and Kozlov. One of the most promising methods is undoubtedly the operator method [278], which is extremely well applicable to a certain class of linear problems.

In the above-mentioned references, a complete, detailed description of the physical principles of the coupled linear problems of thermoelasticity for thin shell structures is given. Methods of solution and detailed bibliographies are also presented.

Vakhlaeva and Krys'ko [324] have presented solutions of a certain class of problems on the static loss of stability of a shallow shell in a stationary temperature field. Some problems in the theory of plates with boundary conditions not suitable for two-dimensional cases were solved. An article by Kirichenko et al. [181] presented results of calculations characterizing the influence of the coupling of the temperature and deformation fields on the value of the critical dynamic load for a clamped shallow shell.

In all works listed above concerning coupled thermoelasticity problems, all thermal transfer equations are treated as parabolic equations. This treatment was derived from an assumption of immediate heat transfer inside the body. Lord and Shulman [229] presented a new mathematical model of thermo-mechanics, including the velocity of heat transfer. In this new model, terms with thermal inertia were introduced into the heat transfer equations. This changed the type of these equations to hyperbolic. The thermomechanical theory presented in that work was called a generalized theory, in contrast to the classical theory, which did not take into account thermal inertia.

In the monograph by Podstrigach and Kolyano [277] there was presented a theory of generalized thermomechanics, and also some theoretical results and practical examples concerning the effects of a finite speed of heat transfer. In particular, during motion of a plane harmonic wave in an infinite thermoelastic body, an important influence of the heat transfer speed on the increase of the phase velocity was observed. Also, a detailed bibliography was presented. In paper by Shvetz and Lopat'ev [300], some characteristics of dynamic processes at higher frequencies based on linear, three-dimensional equations of a generalized thermodynamics were presented. In particular, an investigation of a distribution corresponding to a plane harmonic wave showed that, for an oscillation frequency lower than a certain frequency characteristic of the material, parabolic-type equations can be applied. Kolyano and Shter [185] applied the principle of conservation of energy to obtain coupled equations of generalized thermomechanics for anisotropic bodies. Kil'chinskaya [178] presented applications of the small-excitation Gauss method to cases of generalized thermomechanics.

Many theorems related to the existence of solutions in the nonlinear theory of shells have been proved by Destuynder [98], whereas in the case of shallow shells a serious mathematical treatment has been carried out by Bernadou and Oden [67]. We are not going to cite all references concerning rigorous mathematical approaches to problems of plates and shells, but the reader is encouraged to follow up the results included in [76, 86–88, 149, 223, 317, 330]. The general conclusions derived from the investigations of coupled problems of thermoelasticity are the following:

1. The quantitative effect of coupling depends on the type of material. For metals this effect is less important; for polymers the effect is relatively large.
2. Consideration of the coupling effect leads to qualitatively new properties in the distribution of shock waves (damping and dispersion).

Kupradze et al. [210], in their monograph, presented the mathematical principles of the theory of elasticity and thermoelasticity. They discussed in detail the problem of the existence and uniqueness of classical solutions (i.e. solutions that are continuously differentiable the necessary number of times) of linear problems of elasticity and of coupled thermoelasticity problems for three-dimensional bodies. They applied potential methods and the theory of multidimensional, singular integral equations. Fischer [117], in his papers, applied Laplace transformation to the generalized theory of strongly nonlinear elliptic systems and obtained results concerning the uniqueness of the solution of a linear coupled thermoelasticity problem, for the Sobolev class and also for continuously differentiable functions. Mikhaylovskaya and Novik [244] presented theorems concerning the solvability of the Cauchy problem and the stability of its periodic approximation for a system of differential–integral equations. As a special case, a system of linear coupled thermoelasticity equations was investigated. Some mathematical aspects of various problems of elasticity were also presented in papers by Vasil'kovskiy [327], Parton and Perlin [272].

Simultaneously with the development and investigation of mathematical models, the creation of approximate methods of solution for problems in the theory of plates and shells has been of great importance. Obtaining exact solutions requires great effort—and does not always give an adequate return on the investment. Exact solutions are not always possible in technological applications [276]. During the last 20–25 years, numerical computation has become one of the most important methods of solution of problems in the theory of plates and shells. In this review we include only a few of the publications concerning the principles and numerical realizations of the Bubnov–Galerkin method and the Ritz method, which is a special case of the Bubnov–Galerkin method.

It is known [217, 218] that investigations of any mechanical system may be made on the basis of the Hamilton–Ostrogradski stationary principle. In other words, for every mechanical system, a functional of a certain structure can be created in such a way that every point of the existence of the functional maps to a certain state of the system. Also, stationary points of the functional correspond to real, practical situations (states). This leads to the conclusion that it is possible to obtain the state of the system in two ways: either indirectly, by applying the method of variational calculation, searching for stationary points of the functional, or by applying methods of solution of partial differential equations, to search for stationary points as solutions of Euler's equations for the given functional. Numerical realizations of the first method correspond to projection methods, and those of the second one correspond to difference methods. Note that, for thermoelastic problems, different options for using the variational principle are presented by Nickell and Sackman [262] and by Kaczkowski [159] for three-dimensional bodies, and in the references mentioned earlier [298, 299] for thin-walled structures. More recent results can be found in the monographs [132, 133, 136, 153, 154, 158, 165, 182, 224, 236, 289, 309, 310, 321, 322, 326, 331].

The Ritz method and Bubnov–Galerkin method belong to the class of projection methods, where a unique functional, obtainable by the Ritz method, possesses extrema in the form of stationary points. The principle of the methods presented here can be described in the following form. The space of solutions of the investigated system is approximated by subspaces, and a search for a stationary point of functional is performed in these subspaces.

This is usually realized by a transformation of the original system of partial differential equations into a system of algebraic equations (in the case of static problems) or ordinary differential equations (for dynamic problems). Note that the Bubnov–Galerkin method is a projection method and can be applied for solving differential equations which are not Euler's equations for any functional (or if the corresponding functional is unknown). In this case the method relies on the projection of the original equations into a certain subspace of the functional space in which there exists an exact solution of the problem. An approximate solution can be obtained from the projected equations as an element of this subspace.

The first application of this method was presented by Ritz [284]. Further development of the Ritz method and the Bubnov–Galerkin method and of the application of those methods in mechanics was presented by Galerkin [120], Galimov [121] and others. Keldysh [173] presented a proof of the convergence of the Bubnov–Galerkin method for linear problems of ordinary differential equations of even order and for partial differential equations of elliptic type with lower-order elements. Mikhlin [245, 248] extended the results obtained by Keldysh to a generalized case of a linear operator equation containing a nonadjoint operator, which describes a wide class of boundary problems for elliptic equations. Krasnosel'skiy [196] presented general theorems about the convergence of the Bubnov–Galerkin method for nonlinear operator-type equations and about error estimation for this method. For this purpose, the author used the error of the best approximation to a sought element by a combination of coordinate elements.

In papers by Petryshyn [275] and Shalov [295], the construction of a squared functional, analogous to the functional of the Ritz method for adjoint problems, was presented for linear equations with a nonadjoint operator, as well as the convergence of the corresponding variational method was discussed. Nashed [260] extended the above-mentioned results to the case of a nonlinear equation with a nonpotential operator.

Unlike the Ritz method, the Bubnov–Galerkin method can be applied effectively to solving not only stationary problems (boundary-type problems for elliptic equations) but also to evolution-type problems (initial–boundary problems for parabolic and hyperbolic equations). It is generally agreed, that the first investigation of the use of the Bubnov–Galerkin method for nonstationary problems was performed by Faedo [108] in 1949; for this reason, the Bubnov–Galerkin method is also known as the Faedo–Galerkin method. In 1931 Krylov and Bogolyubov [200] applied what was effectively the Bubnov–Galerkin method to an investigation of the Cauchy–Dirichlet problem for hyperbolic-type equations of the second order. The way in which that method was applied was quite similar to the currently used method.

A review and comparison of the development of the Ritz method and Bubnov–Galerkin method and their application to nonlinear problems in the theory of plates and shells was presented in a monograph by Volmir [336] in 1956, where methods using the first- and second-order approximation were presented. Further development of computational methods allowed the use of higher-order approximations of the Bubnov–Galerkin method. Results of the application of the Ritz method and Bubnov–Galerkin method to stability problems in the theory of shells were presented in a monograph [135] by Grigolyuk and Kabanov. These authors pointed out that it is necessary to describe the stability loss mode with great nonhomogeneity, either by taking into account a large number of terms in a series or by choosing an appropriate number of basis functions after making a prior estimate.

A detailed description of the use of the Bubnov–Galerkin method in dynamical, geometrically nonlinear problems in the theory of plates and shells was presented in the monograph by Volmir [338]. This monograph also contains a large bibliography presenting the history of the solutions up to 1972. In [339] there were presented some results of work on the dynamic stability of structures interacting with fluids and gases.

Kirichenko et al. [181] applied the Bubnov–Galerkin method to the solution of the problem of the dynamic stability of a shallow shell clamped on the boundary, both with and without taking into account the coupling of the temperature and deformation fields. Various aspects of the numerical application of the Bubnov–Galerkin method were presented in papers by the following authors: Svirskiy [314], Bacinov [57], Mukhopandhayay [257], Chen and Hwang [81], and Gelos et al. [124]. In a monograph by Krys'ko and Kutsemako [201], the application of the Bubnov–Galerkin method to the problem of the stability and oscillations of nonhomogeneous shallow shells was presented.

The Bubnov–Galerkin method initially developed as a numerical method, is currently widely used for solving a large class of problems in mathematical physics; stationary and evolutionary as well as linear and nonlinear. The development of powerful mathematical schemes for proving the existence of solutions obtained by the Bubnov–Galerkin method, based on fundamental theorems of functional analysis, became possible after Sobolev's [307] research on the generalized differential rational function of the Lebesgue class and the formulation of principles for obtaining generalized solutions of problems of mathematical physics. Research on the convergence of the Ritz method and Bubnov–Galerkin method for linear problems, including the theory of plates and shells, and proof of the solvability of particular problems in a Sobolev space was published by Ladyzhenskaya [213]. The mathematical principles of various approximation methods, including the Bubnov–Galerkin method, were summarized in a monograph by Krasnosel'skiy et al. [197] and by Lions [227]. A detailed review of research on the convergence of Bubnov–Galerkin method for nonlinear problems was presented in a monograph by Lions [226], where a large bibliography is presented. Research on the solvability of non-linear problems in the theory of plates and shells was presented by Vorovich [343]. Fundamental results in this area were obtained by Morozov [252, 253] in publications concerning the solvability of nonlinear problems in the theory of thin plates. Morozov stated (applying Sobolev's theorems of inclusion) the existence and uniqueness of the solution. Some aspects of the solvability of various problems in the mechanics of thin-walled structures by means of the Bubnov–Galerkin method were presented by Vorovich and Lebedev [344], Skrypnik [302] and others. The detailed background of the application of the Bubnov–Galerkin method and Ritz method to static problems in the nonlinear theory of shallow shells was presented by Vorovich [343].

Theorems about the existence, uniqueness, regularity and continuous dependence of the solution on the data have been formulated for solutions obtained by the Bubnov–Galerkin method. Chrzeszczyk [84, 85] presented conclusions about the uniqueness and smoothness of the solution to a problem considered also in [180]. Kowalski and Piskorek [190], Gawinecki [122, 123], and Kowalski et al. [190, 191], by the application of the Bubnov–Galerkin method presented theorems about the existence, uniqueness and regularity of the solutions to problems in the linear, three-dimensional theory of thermal stresses and coupled thermoelasticity for both isotropic and anisotropic bodies, with hyperbolic and parabolic types of heat transfer equation.

In the majority of the papers listed above, the convergence of the Ritz method and Bubnov–Galerkin method appears as an extra result of proving the existence of a solution of the problem under investigation. The standard solution (for an arbitrarily selected basis) appears to show convergence in the energy norm for static problems, and weak convergence in the corresponding "energy" spaces for dynamic problems. For properly selected basis systems the results concerning convergence can be stronger and, besides, effective estimates of the convergence speed can be determined. Important results concerning this problem were presented by Mikhlin [247], in which convergence operators were introduced. Here, the deviation of the Ritz method converged to zero if a basis system of eigenelements of an auxiliary operator was introduced that converged to the equation under investigation. In [249] a more exact result was presented, where the main and auxiliary operators satisfied an extra condition of acute-angle form. Fundamental results concerning converging operators satisfying an acute-angle inequality were presented by Sobolevskiy [308] and Ladyzhenskaya [214]. On the basis of these results Mikhlin [249] presented for a series of examples special bases that ensured convergence to null of the deviation of the Ritz method. Bogaryan [71] extended Mikhlin's results concerning the Ritz method to the Bubnov–Galerkin method. Dzhishkaryani [103] obtained prior estimates of the error of the Ritz method in an energy norm, where eigenvalues of an auxiliary operator converging to the original equation were applied. A similar problem was investigated by Vaynikko [329] and Dzhishkaryani [104]. These authors obtained similar results for the Bubnov–Galerkin method. Vaynikko obtained error estimates for the Bubnov–Galerkin method for stationary problems in cases where the spectrum of the auxiliary operator did not appear to be exact. Dzhishkaryani extended the error estimates for the Ritz method to a case of a quasi-linear equation containing a nonlinear operator, as lower terms of a continuous, potential operator having a positive differential. Zarubin [354] presented quite generalized estimates of the convergence speed of the Galerkin–Petrov method for linear and quasi-linear stationary problems. From these results, error estimates can be derived for the Bubnov–Galerkin method, taking as a basis eigenelements of a self-adjoint operator that converges to an operator of the main part of the original system and projects onto it at an acute angle. A detailed review of work on the theory of the errors of numerical methods, including the Bubnov–Galerkin method for stationary problems, is presented in the monograph by Mikhlin [250].

The main publication that dealt with the convergence of the Bubnov–Galerkin method for evolutionary problems was a work by Sobolevskiy [308]. This publication presented theorems about the strong convergence of approximate solutions to the exact solution and about the convergence of errors to zero with a quasi-linear parabolic equation for an arbitrarily selected basis. Zarubin and Tiunchik [355] applied the Bubnov–Galerkin method to a parabolic equation with a nonlinear operator of small-element type, corresponding to a dissipation condition introduced in that publication. As a basis, a system of eigenelements of an auxiliary operator satisfying the acute-angle inequality was used. In that work, results were presented that showed convergence to zero of the deviation of the Bubnov–Galerkin method and a strong convergence of the series approximating the exact solution. Currently there are no publications concerning investigations of the convergence speed of the Bubnov–Galerkin method for hyperbolic equations or for coupled systems similar to thermoelasticity equations.

Note there are a very large number of books, papers and commercial software packages devoted to other numerical methods for the analysis of plates and shells (see, for instance, [147, 153, 316]).

Since, the finite-element method is one of the most popular of those methods especially in the field of engineering, a treatment of this approach is omitted in this book; descriptions can be found elsewhere [306, 347].

An analysis of the current research on the theory of plates and shells and of the convergence speed of the Bubnov–Galerkin method leads to the following conclusions:

1. The Bubnov–Galerkin method appears to be one of the most effective and widely applied numerical methods for solving static and dynamic problems in the theory of plates and shells. Current computational techniques and computer power allow us to apply the Bubnov–Galerkin method with a higher level of accuracy, i.e. to take into account a greater number of series elements in the approximate solution.
2. The main problem related to numerical realization of the Bubnov–Galerkin method for dynamic problems is the lack of a relatively simple estimate of the errors (currently, the reliability of numerical methods is verified experimentally) and, in many cases, the lack of a mathematical background for the proper choice of basis functions. These problems arise not only for dynamic problems in the theory of elasticity and coupled thermoelasticity, but also for a large class of evolutionary problems containing the second derivative with respect to time of a sought function.

The rapid development of computational methods and applied mathematics has allowed researchers to apply more reliable mathematical models. It has forced us to take into account geometric nonlinearities, elasto-plastic deformations and periodic loads when solving dynamic problems. The approaches presented by Il'yushin [155] and Birger [69] are widely applied.

The research on the dynamic stability of plates and shells is based on fundamental work by Bolotin [73, 74]. Various models for describing the behaviour of a structure have been applied to solve nonlinear problems in the dynamics of plates and shells. Historically, one of the first approaches was the application of a model of a rigid–plastic body [140]. For large plastic deformation, this model allows us to obtain sufficiently accurate solutions. When the plastic deformation is much smaller than the elastic deformation, however, this method is not suitable.

Various approaches can be applied to solving problems of cyclic deformation with a small number of cycles. One such approach is based on differential relations between stresses and strains, presented by Birger [69] and Krys'ko [198].

For solving certain problems, finite relations between stresses and strains are widely applied. Although the spectrum of solvable problems is limited to those with loads that are simple or close to simple, such an approach allows a numerical verification.

Results of research on circular plates and cylindrical shells subjected to dynamic loads, where the material elasto-plastic characteristics, were presented by Filippov [116].

Shallow shells were investigated theoretically and experimentally in [78]. The failure of a shell after elasto-plastic deformation was investigated. The boundary problem for a shell subjected to an arbitrary pressure load, with static nonhomogeneous boundary conditions of generalized form was solved by application of an orthogonal-projection Godunov method [127, 129].

The dynamic behaviour of an elasto-plastic plate clamped at an internal edge and not supported at the outside edge was investigated by Stepanov and Kovalenko [312]. An approximate solution for estimating the elasto-plastic deflection of a plate subjected to a short-time impact pressure was presented, with application of an iterative method. Results of theoretical and experimental investigations of the dynamic deflection of a round plate subjected to a uniformly distributed impact pressure load were also presented.

There is a large number of works concerning analysis of the elasto-plastic (physically nonlinear) strain of rectangular plates and of shallow shells that are rectangular in plan.

Krys'ko [198], in his monograph, presented a wide range of problems in the theory of nonhomogeneous shallow shells taking transverse shear into account. The convergence of the Bubnov, Vlasov–Kantorovich and finite-difference variational methods was numerically investigated for static and dynamic problems. The author analysed the influence of different types of transverse impact loads on the stability of shells and also the influence of the deflection on the critical load.

Bauer [62] investigated bending oscillations of simply supported and clamped plates subjected to impact loads of step-like and exponentially decreasing type. A possibility of separation of the spatial and time-dependent elements of the solution was presented.

The reaction of metal plates to an impact load modelling an under-water explosion was presented in [269]. The dynamic problem was solved in Lagrange coordinates by means of a finite-difference method for a two-dimensional axisymmetric state, with an elasto-plastic approximation. Different forms of pressure impulse, different impulse durations, different distributions of the pressure on the plate surface and variable boundary conditions were investigated, and their influence on the deformation process was analysed.

The loss of stability in bifurcation form of a rectangular plate under combined loads was considered in [106]. The problem was solved by representing the equations of the deformation theory in an incremental form, which allowed the authors to include the load history without investigating local loads. The finite-element method (FEM) was applied to cases of plates subjected to transverse and longitudinal loads.

For solving problems in the dynamics of plates and shells, it is necessary not only to know the material characteristics but also, in many cases, to take into account the deformation speed, acceleration and other parameters characterizing the dynamics of the process. These elements have been taken into account in numerical solutions of various problems in the theory of plates and of spherical and cylindrical shells [174, 219, 349, 350].

An analysis of the current research on the dynamic stability of geometrically and physically nonlinear plates and shells, under nonstationary loads, leads to the following conclusions:

1. Research on the dynamic elasto-plastic deformation of plates and shells with geometric nonlinearity has been presented in a large number of publications. However, the dynamic loss of stability of slender elasto-plastic plates and shells that are rectangular in plan, under impact loads of arbitrary character, is poorly represented.
2. The dynamic behaviour of elasto-plastic structures has mainly been investigated in axisymmetric cases. The solutions to such cases are very complex, i.e. the deformation process has to be represented as a function of time for all points of the structure; therefore an analysis of every point in the deformation chart must be performed.
3. In practically all publications, the following assumption is made: the collapse (failure) of a structure takes place when any characteristic point of the structure fails. Bi-linear charts with linear or incremental reinforcement were used in most of the calculations. There were no comparisons of different deformation charts.

Along with problems of coupled thermoelasticity, other cases, including other physical–chemical effects, magnetic fields and transonic gas flows, have been widely investigated.

In a series of works Day [96, 97] presented some qualitative and quantitative aspects of linear coupled thermoelasticity. The author investigated the behaviour of a beam under a thermal load and on one side investigated the conditions, for certain types of problems, under which an uncoupled thermoelasticity theory can be applied. The results presented the possibility of a negative temperature increase for coupled thermoelasticity under boundary conditions that excluded such a phenomenon in classical thermodynamics.

Of the various methods applied for numerical solution of coupled problems, the finite-difference method (FDM) and the FEM were used extremely often.

The FEM was first applied to coupled problems by Nickell and Sackman [262]. The application of the FEM was analysed in details by Gribanov and Panichkin [134].

Analysis of the scientific literature shows that solutions to certain problems with coupling of the temperature and deformation fields had been obtained for linear models only. In a publication by Krys'ko and Mishnik [202], three-dimensional problems in the theory of plates with physical nonlinearities were presented. For the discretization of the spatial coordinates, a finite-difference method with an $O(h^2)$ approximation was applied, and for integration with respect to time, the fourth-order Runge–Kutta method was applied. Numerical results confirmed the effect of coupling on the stress–strain state in nonlinear plate problems.

An analysis of work on current problems in dynamic thermoelasticity with two types of nonlinearity and coupling of the temperature and deformation fields leads to the following conclusions:

1. Only a small number of coupled problems have been solved with nonlinear models. There are no publications concerning the analysis of the dynamic stability of slender shells with coupling of the temperature and deformation

fields. The problems of the complex oscillations of such structures have not been solved.
2. Numerical solution of doubly nonlinear problems of coupled thermoelasticity with a three-dimensional heat transfer equation is very complicated, but it allows one to apply different hypotheses for the temperature distribution through the thickness of the shell.

Problems of the coupled interaction of a shell-type structure with a transonic flow are a special case and have a special complexity. When a gas flows with transonic velocity around a profile, some particular phenomena appear in the flow. In such a flow shock waves, which are pressure jumps, appear, move and disappear. Also, the velocity of movement of an excitation caused by a deformation of the profile is close to the speed of the gas particles. These phenomena appear when a gas flow interacts with a profile and, vice versa, when a profile interacts with a gas flow.

An important scientific goal is the application of mathematical approaches to transonic-flow problems. As stated by Godunov et al. [129], an aircraft can cover the greatest distance at a transonic speed. Also, an aircraft flying at a transonic speed has the best ability to manoeuvre while still preserving high speed, i.e. the relation of manoeuvrability to speed appears to be optimal.

Despite the theoretical and practical aspects, problems of the interaction of a shell and a transonic flow have not received proper attention in the scientific literature. An investigation of travelling pressure jumps, pulsing with a given frequency, was presented by Croker [91] in a publication about a vertical-take-off aircraft. An analysis of such shock waves acting on an aircraft, performed for linear cases, led to the conclusion that these shock waves appear to be the main reason for the relatively high increase of shell vibrations when an aircraft passes through the transonic speed zone.

Volmir [338] and Skurlatov [303, 304] presented pictures from a high-speed camera showing the failure of a steel shell as a result of interaction with a transonic flow with heat transfer.

Analytic research on cylindrical shells in a wind tunnel was presented by Skurlatov and Solomenko [305]. These authors stated that during the interaction, the incoming flow changes from subsonic to supersonic. Two monographs [22, 198] presented a solution obtained by the Bubnov–Galerkin method to a nonlinear problem of forced vibrations of a shallow, simply supported cylindrical panel. The Kirchhoff–Love model was applied.

Investigations of the longitudinal flow of a transonic gas stream around rigid and elastic profiles of conical shape and around a cylindrical panel were performed. The model and the experimental results showed oscillating movements of pressure jumps, i.e. shock waves and their zone of motion was described. Experiments were performed for incoming flows of Mach number M_∞ varying from 0.8 to 1.2 in steps of 0.1. Shock waves were observed visually with a shadow device.

Some publications [338, 339, 361] have presented some properties of the interaction of a transonic flow with the profiles or shells obtained experimentally. These publications showed that during transonic flow, dynamic parameters reach high values and that pulsing shock waves lead to the appearance of grooves in structures. Measurements of the pressure on the surface of a rigid profile showed that the amplitude of oscillations rapidly increased when the Mach number of the incoming stream M_∞ reached 1. Further increase of M_∞ caused a strong decrease of the oscillation amplitude to a level characteristic of the noise of a turbulent boundary layer.

The number of problems investigated in which a shell passes through a transonic-flow region has increased rapidly. The pressure amplitudes and shell deflections reach very high values.

Some other problems of aeroelasticity of plates and shells have been considered in a monograph by Dowell [101].

Analysis of the current literature shows that at this moment there is no theory of the dynamic interaction of a shell with a transonic gas flow. Further development of solution methods for nonstationary transonic flows and their interaction with slender shells would allow better research on this problem.

The easiest problem of dynamic interaction of a shell with a transonic flow is the case of the two-dimensional interaction of a transonic flow with an infinitely long cylindrical panel. The solution for a shallow shell in this case allows us to investigate transonic flow in an asymptotic nonlinear theory when the flow is close to a homogeneous potential flow.

Section 2.2 of this monograph is devoted to current problems of aeroelasticity and presents a method for the numerical solution of the above problem, together with research on the dynamic stability of a cylindrical panel interacting with a transonic flow of an ideal gas. The characteristics of the interaction of a transonic flow with a shallow elastic shell are deduced.

A novel branch of non-linear bifurcational and chaotic dynamics of thin structural members (beams, plates and shells) has been rapidly developed in the last 40 years.

Han et al. [142] matched the Galerkin approach and the Melnikov method to predict and analyze chaotic vibrations exhibited by the elastic cylindrical shells. The numerical investigations were based on the Poincaré maps, phase portraits and time histories.

Hao et al. [143] investigated nonlinear dynamics of a simply supported functionally graded material rectangular plate embedded into thermal environment under the transversal and in-plane excitations. The Reddy's third-order plate theory was employed while deriving the governing equations of motion based on the Hamilton's principle. Then the problem of infinite dimension was reduced to that of 2 DoFs system with the quadratic and cubic nonlinearity. In particular periodic, quasi-periodic and chaotic motions were detected and analyzed.

Zhang et al. [356] analyzed nonlinear and chaotic vibrations of a simply supported orthotropic functionally graded material rectangular plate in thermal environment and subjected to external/parametric excitations. The plate material properties were graded in its thickness direction. The governing PDEs were yielded by the Hamilton's principle and Reddy's third order deformation plate theory. Then the problem was reduced to study three second order nonlinear ODEs, with particular emphasis on investigation of the 1:2:4 internal resonance and the principal parametric (subharmonic) resonance of order 1/2.

Wang et al. [346] investigated chaotic dynamics of a bimetallic shallow shell of revolution subjected to the time-varying temperature. The nonlinear differential equation in time-modes was derived based on an assumed spatial-mode. The influence of static and time-dependent temperature parameters, height shell parameter and damping parameter were taken into account while analyzing numerically the chaotic orbits and transient chaos, chaos with interior crisis, periodic windows, period-doubling scenario, and jump phenomena.

Touze et al. [319] studied a transition from periodic to chaotic vibrations in free-edge, perfect and imperfect circular plates with an account of the von Kármán geometric non-linearity. Then the Galerkin method based on the eigenmodes was used and resulting ODEs were investigated numerically. The complete bifurcation diagram and the corresponding critical forces associated with chaos occurrence were addressed. The numerical results confirmed a number of experimental results devoted to study the shells nonlinear dynamics.

Yuda and Zhiqiang [152] investigated bifurcation and chaos phenomena of a ceramic/metal functionally graded circular plate under one (two) term(s) transversal excitations in the thermal environment. The material properties of the plate varied continuously through its thickness. The Galerkin method was used to derive the Duffing type nonlinear ODE, and the employed Melnikov's approach yielded estimation of the critical set of parameters responsible for occurrence of the graded plate chaotic vibrations. Effects of the material volume fraction index and temperature on the existence of chaotic vibrations were validated by phase portraits, Poincaré maps and the Lyapunov exponents.

Yao et al. [352] studied the multipulse heteroclinic bifurcations and chaotic dynamics of a laminated and chaotic dynamics of a laminated composite piezoelectric rectangular plate. The von Kármán type PDEs based on the Reddy's third order shear deformation theory were yielded through the Hamilton's principle and they included terms responsible for combined parametric and transverse excitations. The primary parametric and 1:3 internal resonances were investigated, and the extended Melnikov's method was used to analyze the Shilnikov type multipulse heteroclinic bifurcations and chaotic vibrations of the mentioned plate. Both theoretical and numerical investigations suggested that chaos in the sense of the Smale horseshoe is reliable for the laminated composite piezoelectric rectangular plate.

Awrejcewicz et al. [55] studied noisy contact interactions of multi-layer mechanical structures composed of a plate and beams and coupled only by the boundary conditions. The novel nonlinear phenomena of the regular and chaotic contact/no-contact vibrations were detected and analyzed based on the Fourier analysis and the Morlet wavelets. In particular, it was demonstrated how the additive white noise cancelled deterministic chaos close to the boundary of chaotic region in the space of parameters and also the windows of on/off switching of the frequencies during the contact dynamics between the plate and beam where illustrated and discussed.

The recent papers/monographs show that the classical continuum mechanics cannot be employed directly to study structural members of micro- and nano-scales.

The strain gradient theory and modified couple stress theory stand for one of the newly proposed extensions to the classical theories to fit the size-dependent effects of micro nano scales.

Lam et al. [216] developed a new set of higher-order metrics to characterize strain gradient elastic bending theory for plane-strain beams. In order to determine the significance of the size dependence, the micron-sized beams were fabricated with the use of bending test based on nanoindenter.

Kong et al. [186] solved the static and dynamic problems of Bernoulli-Euler beams analytically based on the strain gradient elasticity theory. It was found that beam deflections decreased while natural frequencies increased when the beams thickness become comparable to the material length scale.

Zhao et al. [359] derived the nonlinear governing equation of a microbeam based on matching the strain gradient theory and the Hamilton's principle. It was demonstrated that the size effect was significant when the ratio of characteristic thickness to internal material length scale parameter was approximately equal to one or two.

Rajabi and Ramezani [282] developed and studied a microscale nonlinear Bernoulli-Euler beam model with the use of strain gradient elasticity with surface energy and with an account of the von Kármán nonlinearity. The differential equation associated with the first mode of nonlinear vibration of a simply supported microbeam was yielded by the global Galerkin procedure. It was shown that the strain gradient effect on increasing the beam natural frequency was higher than that regarding the geometric nonlinearity. Increasing the beam thickness implied decrease of the strain gradient effect.

Akgöz and Civalek [2] investigated the buckling of the size dependent microbeams made of functionally graded materials using the Bernoulli-Euler beam model and modified strain gradient theory. Both classical and non-classical boundary conditions were employed while studying the higher-order governing equation for buckling. In particular, the effects of the power of the material property variation functions, slenderness ratio, ratio of additional material length scale parameters for two constituents, and beam thickness-to-additional material scale length parameter ratio on the beam buckling response were reported.

In frame of this trend, there is a need to extend the applicability of the classical continuum mechanics to take into account the size dependent material deformation in micro- and nano-scale in both static and dynamic cases.

On the other hand rapidly developed technology is aimed on fabrication of the micro-electro-mechanical integrated systems composed of micro-electronics, micro-sensors and micro-actuators.The microstructural member can be made from metal, silicon-based materials, polymers and functionally graded materials.

Zhang et al. [357] presented a novel Timoshenko beam based on the strain gradient elasticity theory for analysis of the static bending, free vibrations and buckling effects. Based on the finite element method, the convergence, shear locking and comparison investigations were employed to analyze the reliability and accuracy of the numerical solutions. It was demonstrated that small size effects were significant for the beams with small thickness.

Fathalilon et al. [109] investigated the size dependent behavior exhibited by the micro-electrical mechanical structures. It was shown for a given DS voltage a considerable difference between equilibrium position of the gold and nickel micro-beams using the classic beam theory, modified couple stress theory, and modified strain gradient theory.

Hashemi and Asghari [145] studied the flexural vibrations of 3D micro-beams in the framework of the strain gradient theory. The Euler-Bernoulli beams model was yielded by Hamilton's principle. It was demonstrated that the resonant frequencies were significantly dependent on the length scale parameter.

Ausari et al. [9] investigated the size dependent effects exhibited by microbeams. The modified strain gradient elasticity theory was applied to derive the size-dependent analytical formulas corresponding to the frequency response, the resonance frequencies of cantilevers obtained from the developed non-classical model was higher than those yielded by the classical beam theory.

Vatankhah and Kahrobaiyan [328] analyzed the vibration behavior of micro-resonances based on the strain gradient theory capturing the size effects of the micro-resonators. The normalized natural frequency was evaluated and the influence of various parameters was illustrated.

Ansari et al. [10] employed the Mindlin's strain gradient elasticity and Euler-Bernoulli beam theory to study the vibrations of the size-dependent microbeams. In particular, the bending and free vibration problems of microbeams under various kinds of boundary conditions were addressed.

As it has been recently mentioned, the rapid development of the micro/nano-technology stimulates development of continuum mechanics to get the reliable nanomicroscale effects while constructing the appropriate models based on extension of the classical continuum mechanics. The micro-electro-mechanical integrated systems composed of micro-electronics, micro-sensors and micro-actuators can be made from metals, silicon-based materials, polymers, and functionally graded materials, and they can work under various external force mechanical fields, as well as under the thermal, magnetic and electric fields.

Karličić et al. [167] developed a model to describe the free vibration behavior of a cracked nanobeam with an account of the magnetic and temperature fields. The Euler-Bernoulli 1D nanoscale beam model based on the nonlocal elasticity constitutive relation was developed. The classical Maxwell equation yielded the Lorentz magnetic force and the surrounding elastic medium was represented by the Winkler-type elastic foundation. The authors reported the influences of the nonlocal parameter, stiffness of rotational sprig, temperature change and magnetic field strength on the beam frequencies for two types of the boundary conditions.

Alibeigi et al. [3] investigated the buckling response of nanobeams based on the Euler-Bernoulli beam model with an account of the von Kármán geometric nonlinearity using the modified couple stress theory. The studied nanobeam was subjected to two types of thermal loading, uniform and linear, along thickness direction.

Kerid et al. [175] investigated the magnetic field, thermal loads and the small scale effects of a nanobeam structure vibration based on the Euler-Bernoulli and

Timoshenko beam models. The small scale effect was studied through the Eringen's nonlocal theory. It was demonstrated that the resonance frequency change, the magnetic field intensity, the thermal loads and small scale effects depended on the size and periodic square holes.

Arefi and Aranic [12] carried out the nonlocal magnetoelectrothermic-elastic analysis of functionally graded nanobeam under magnetoelectroelastic loads with the use of the third-order shear deformation theory. The influence of the inhomogeneous parameter, initial electric and magnetic potentials, nonlocal parameter and thermal loads were reported.

Ghadiri et al. [126] studied vibration of a rotating noncantilever based on the Euler-Bernoulli beam model with an account of the temperature, small scale effect and centrifugal force. It was shown that the effect of temperature played most important role in the beam vibration.

Pugmo [280] carried out investigation of 3D NEMS (nanoelectromechanical systems). Equilibrium and stability of a nano-plate based NEMS subjected to an electrical field and van der Waals forces were evaluated. The structural instability yielded by the so-called pull-in voltage was analyzed, and dependencies of the amplitude and frequency of the thermal vibrations of the nano-plate functions of the voltage were presented.

Perk and Han [271] analyzed the magneto-electro-elastic nano-plate with a help of the higher-order shear deformation theory. Both Maxwell equation and Eringen's nonlocal differential constitutive relations were employed, and effects of nonlocal parameters, aspect ratio and in-plate load directions were illustrated and discussed.

Pelević and van der Meer [273] investigated numerically flow and heat transfer over a plate covered with carbon nano-fibers. The stochastic and lattic Boltzmann models were used for simulations.

Hosseini and Jamalpoor [150] presented the dynamic characteristics of a viscoelastic nanoplates system subjected to temperature change and including surface effects. Three cases were considered: out-of-phase vibration, in-phase vibration and one nanoplate fixed. The dependencies of frequency of the nanoplates versus various values of the nonlocal parameter temperature change, aspect ratio and the stiffness coefficients of the elastic Pasternak foundation were reported.

Artati [61] studied magneto-hydro-thermal vibration of doubled-layered nanoplates made of functionally graded materials based on the nonlocal strain gradient theory. The governing PDEs were solved using Galerkin's method. It was demonstrated how the type of vibration, temperature rise, nonlocal parameter, strain gradient parameter and material gradation modify vibrations of nanoscale plates.

Kiami [176] analyzed free in-plane and out of plane vibrations of rectangular nanoplates subjected to unidirectional in-plane steady magnetic fields based on the hypotheses of Kirchhoff, Mindlin and higher-order plate theories. The small-scale parameter was incorporated into the exerted forces due to the applied magnetic field. Effects of the length-to-thickness ratio, length-to-width ratio, small-scale parameter, and magnetic field strength on the frequencies were reported.

Karili et al. [168] examined the influence of in-plane magnetic field on the viscoelastic orthotropic multi-nanoplates systems embedded in a viscoelastic medium using the Eringern's nonlocal elasticity and the Kirchhoff's plate theories. Closed form solutions for complex natural frequencies were derived based on the Navier's and trigonometric method.

Karličić et al.[166] adopted the nonlocal Kirchhoff-Love plate theory to study mechanical behavior of single-layered graphene sheet (orthotropic nanoplate). The Lorentz magnetic force was introduced in the classical Maxwell's equations. Influences of the nonlocal and magnetic field parameter, different mass weights, and damping coefficients on the relative damped frequency shift and relative shift of damping ratio were presented.

Many problems of the vibration of plates and shells have been solved by the authors of this book and their co-workers in a series of monographs [17, 18, 21, 22, 31], and in papers [23–26, 32, 33, 36, 37, 40, 203] and reviewed conference proceedings [19, 20, 27–30, 34, 38, 41–49].

In this book we address problems related to the theory of structural members from the point of view of mathematical and numerical investigations only. This means that all problems related to the very efficient and effective asymptotic approaches are omitted here (perhaps with the exception of Sect. 2.2). However, the reader is encouraged to follow up the recent results in this field presented in the monographs [14, 16, 39], in the review paper [7] and the references there in.

Chapter 2
Coupled Thermoelasticity and Transonic Gas Flow

This chapter includes considerations of the coupled linear thermoelasticity of shallow shells and of a cylindrical panel within a transonic gas flow. First, the fundamental assumptions related to the stress–strain relation of the Timoshenko kinematic model are formulated, and then the differential equations are derived. Both the Timoshenko and the Kirchhoff–Love models are taken into account. Then the boundary and initial conditions are formulated. Next, an abstract Cauchy problem for a coupled system of two differential equations in a Hilbert space is considered. This includes the thermoelastic problems of shallow shells modelled by the Kirchhoff–Love and Timoshenko theories defined earlier. In Sect. 2.1.5, theorems related to the existence and uniqueness of a general, "classical" solution to the coupled abstract program are given, and then the corresponding theorems for coupled thermoelastic problems of shallow shells are formulated.

The second part of this chapter is focused on the dynamics of an elastic, infinite cylindrical panel within a transonic gas flow.

First, the equations of motion and boundary conditions of the panel are formulated. It is shown that the problem of the oscillations of a flexible, elastic, infinite, circular cylinder panel subjected to a transverse load can be reduced to an initial-boundary-value problem of mathematical physics governed by an integral–differential equation, which is derived here.

Further, the equations of the transonic motion of an ideal gas are formulated. Next, the equations and boundary conditions of the panel–flow interaction are discussed and illustrated. Free and forced panel oscillations are analysed using the finite-difference method. A new, modified Godunov method for calculation of a transonic flow is presented. An algorithm for calculation of the interaction between a flow and a panel are given, with results of calculations. A stable vibrating panel in a transonic flow including shock wave motion, and the interaction of a flow and a panel are discussed in some detail. The last subsection of this chapter is focused on the stability loss of a panel within a transonic flow. This part includes stability

J. Awrejcewicz, V. A. Krysko, *Elastic and Thermoelastic Problems in Nonlinear Dynamics of Structural Members*, Scientific Computation,
https://doi.org/10.1007/978-3-030-37663-5_2

criteria, as well as an analysis of the stress–strain state of a panel and its plastic deformation.

2.1 Coupled Linear Thermoelasticity of Shallow Shells

2.1.1 Fundamental Assumptions

A shallow shell with variable thickness $h(x, y)$, covering a plane surface Ω_1 and occupying a three-dimensional spatial volume Ω_2, is considered. We assume the shell material to be isotropic, homogeneous and elastic. We introduce in the usual way the coordinate system x, y, z [338]. The coordinates x, y coincide with the directions of the principal curvatures of the middle surface, whereas z has a normal direction towards the centre of the shell curvature.

The components of the displacement vector of a point (x, y) of the middle surface of the shell at time t are denoted by $u(x, y, t)$, $v(x, y, t)$, $w(x, y, t)$. The angles of rotation of the normal to the middle surface in the xz and yz planes are denoted by $\Psi_x(x, y, t)$ and $\Psi_y(x, y, t)$, respectively. The initial shell curvatures related to the coordinates x and y are denoted by k_x and k_y, respectively.

In the analysis of the stress–strain state, we use the Timoshenko kinematic model including inertial rotation [22, 338]. In accordance with this model, the shell fibres that are normal to the middle surface before deformation remain linear but are no longer normal to the middle surface. The deformations related to the fibres normal to the average surface are defined by a plane stress state. For the shell under consideration, the following relations between deformations and displacements are satisfied on the mean shell surface [338]:

$$\varepsilon_{11}^z = \varepsilon_{11} + z æ_{11}, \quad \varepsilon_{22}^z = \varepsilon_{22} + z æ_{22}, \quad \varepsilon_{12}^z = \varepsilon_{12} + z æ_{12}, \tag{2.1}$$

$$\varepsilon_{13}^z = \varepsilon_{13}, \quad \varepsilon_{23}^z = \varepsilon_{23}, \tag{2.2}$$

where $|z| \leq \frac{1}{2}h(x, y)$; $\varepsilon_{ij}(i, j = 1, 2)$ are the tangential deformations of the middle surface; $\varepsilon_{13}, \varepsilon_{23}$ are the shear deformations; and $æ_{ij}(i, j = 1, 2)$ are the bending deformations. The following relations hold:

$$\varepsilon_{11} = \frac{\partial u}{\partial x} - k_x w, \quad \varepsilon_{22} = \frac{\partial v}{\partial y} - k_y w, \quad \varepsilon_{12} = \frac{\partial u}{\partial y} + \frac{\partial v}{\partial y}, \tag{2.3}$$

$$æ_{11} = \frac{\partial \Psi_x}{\partial x}, \quad æ_{22} = \frac{\partial \Psi_y}{\partial y}, \quad æ_{12} = \frac{\partial \Psi_x}{\partial y} + \frac{\partial \Psi_y}{\partial x}, \tag{2.4}$$

$$\varepsilon_{13} = \Psi_x + \frac{\partial w}{\partial x}, \quad \varepsilon_{23} = \Psi_y + \frac{\partial w}{\partial y}. \tag{2.5}$$

It should be noted that the deformations $\varepsilon_{11}, \varepsilon_{22}, \varepsilon_{12}, \varepsilon_{13}, \varepsilon_{23}$ are small in comparison with one, and that the products of deformations can be neglected.

Suppose that in the nondeformed and nonstressed state the shell has a temperature T_0. The temperature of the shell starts to change owing to the surface load and inertial forces, as well as owing to internal heat sources and heat exchange with the surrounding medium. By $\theta(x, y, z, t)$, we denote the temperature increment at the point (x, y, z) at a time instant t. We assume that $|\theta/T_0| \ll 1$, which leads to the conclusion that the temperature change does not vary the elastic and thermodynamic properties of the shell material. The following isothermal constants are used in what follows: E, Young's modulus; ν, Poisson's ratio; α_T, coefficient of linear thermal expansion; ϱ, density; λ_q, coefficient of thermal conductivity; and c_ε, specific thermal capacity for a constant deformation tensor. Assuming that the shell is in a local quasi-equilibrium condition [334], using the relations of noninvertible dynamic processes and taking into account the small values of the deformations and temperature increase, the following Duhamel–Neumann relations are obtained:

$$\varepsilon_{11}^z = \frac{1}{E}\sigma_{13} - \frac{\nu}{E}\sigma_{22} + \alpha_T\theta, \quad \varepsilon_{22}^z = \frac{1}{E}\sigma_{22} - \frac{\nu}{E}\sigma_{11} + \alpha_T\theta, \tag{2.6}$$

$$\varepsilon_{33}^z = -\frac{\nu}{E}(\sigma_{11} + \sigma_{22}) + \alpha_T\theta, \tag{2.7}$$

$$\varepsilon_{12}^z = \frac{2(1+\nu)}{E}\sigma_{12}, \quad \varepsilon_{13}^z = \frac{2(1+\nu)}{E}\sigma_{13}, \quad \varepsilon_{23}^z = \frac{2(1+\nu)}{E}\sigma_{23}. \tag{2.8}$$

These describe the relations between the components of the stress tensor σ_{ij} and the deformation tensor ε_{ij}^z in a plane stress state ($\sigma_{33} = 0$). Solving the system of equations (2.6)–(2.8) for σ_{ij}, we obtain:

$$\sigma_{11} = \frac{E}{1-\nu^2}(\varepsilon_{11}^z + \nu\varepsilon_{22}^z) - \frac{E}{1-\nu}\alpha_T\theta, \quad (1 \leftrightarrow 2), \tag{2.9}$$

$$\sigma_{12} = \frac{E}{2(1+\nu)}\varepsilon_{12}^z, \quad \sigma_{13} = \frac{E}{2(1+\nu)}\varepsilon_{13}^z, \quad \sigma_{23} = \frac{E}{2(1+\nu)}\varepsilon_{23}^z. \tag{2.10}$$

Integrating the stresses (2.9), (2.10) with respect to z and taking into account (2.1)–(2.5), we find the corresponding normal and transverse forces in the middle plane:

$$N_x = \int_{-\frac{h}{2}}^{\frac{h}{2}} \sigma_{11}\, dz = \frac{Eh}{1-\nu^2}\left[\frac{\partial u}{\partial x} - k_x w + \nu\left(\frac{\partial v}{\partial y} - k_y w\right)\right] - \frac{E\alpha_T}{1-\nu}\int_{-\frac{h}{2}}^{\frac{h}{2}} \theta\, dz$$

$$(x \leftrightarrow y, \quad u \leftrightarrow v, \quad 1 \leftrightarrow 2), \tag{2.11}$$

$$S = \int_{-\frac{1}{2}}^{\frac{1}{2}} \sigma_{12}\, dz = \frac{Eh}{2(1+\nu)} \left(\frac{\partial u}{\partial y} + \frac{\partial v}{\partial x} \right), \tag{2.12}$$

$$Q_x = k^2 \int_{-\frac{1}{2}}^{\frac{1}{2}} \sigma_{32}\, dz = \frac{k^2 Eh}{2(1+\nu)} \left(\Psi_x + \frac{\partial v}{\partial x} \right), \quad (x \leftrightarrow y, \quad 1 \leftrightarrow 2), \tag{2.13}$$

where

$$\frac{1}{k^2} = \frac{1}{h} \int_{-\frac{h}{2}}^{\frac{h}{2}} f^2(\frac{z}{h})\, dz = \int_{-\frac{h}{2}}^{\frac{h}{2}} f^2(\xi)\, d\xi,$$

and $f(\frac{z}{h})$ is a function characterizing the distribution of tangential stresses through the thickness of the shell [338]. The moments are obtained after integration of the stresses $\sigma_{11}, \sigma_{22}, \sigma_{12}$, taking account of (2.1)–(2.4), over the shell thickness:

$$M_x = \int_{-\frac{1}{2}}^{\frac{1}{2}} \sigma_{11} z\, dz = \frac{Eh^3}{12(1-\nu^2)} \left(\frac{\partial \Psi_x}{\partial x} + \nu \frac{\partial \Psi_y}{\partial y} \right) - \frac{E\alpha_T}{1-\nu} \int_{-\frac{1}{2}}^{\frac{1}{2}} \theta z\, dz, \tag{2.14}$$

$$(x \leftrightarrow y, \quad 1 \leftrightarrow 2),$$

$$H = \int_{-\frac{1}{2}}^{\frac{1}{2}} \sigma_{12} z\, dz = \frac{Eh^3}{24(1+\nu)} \left(\frac{\partial \Psi_x}{\partial y} + \nu \frac{\partial \Psi_y}{\partial x} \right). \tag{2.15}$$

2.1.2 Differential Equations

Taking into account the entropy balance and the Fourier relation between the components of the heat flow vector and the thermodynamic forces [266], the following heat transfer equation is obtained:

$$c_\varepsilon \frac{\partial \theta}{\partial t} - \lambda_q \nabla^2 \theta = -\frac{E\alpha_T T_0}{1-2\nu} \underline{\frac{\partial}{\partial t} (\varepsilon_{11}^z + \varepsilon_{22}^z + \varepsilon_{33}^z)} + q_2. \tag{2.16}$$

The underlined term links the temperature increase to the speed of the volume change. In this equation, $q_2(x, y, z, t)$ denotes the specific power of the heat sources located within the volume occupied by a shell.

In order to obtain an equation governing shell vibrations in terms of displacements, the following system of equations is used:

$$\frac{\partial N_x}{\partial x} + \frac{\partial S}{\partial y} + p_1 - \rho h \frac{\partial^2 u}{\partial t^2} = 0 \quad (x \leftrightarrow y, u \leftrightarrow v, 1 \leftrightarrow 2), \tag{2.17}$$

$$\frac{\partial Q_x}{\partial x} + \frac{\partial Q_y}{\partial y} + k_x N_x + k_y N_y + q_1 - \rho h \frac{\partial^2 w}{\partial t^2} = 0, \tag{2.18}$$

$$\frac{\partial M_x}{\partial x} + \frac{\partial H}{\partial y} - Q_x - \underline{\rho \frac{h^3}{12} \frac{\partial^2 \Psi_x}{\partial t^2}} = 0 \quad (x \leftrightarrow y). \tag{2.19}$$

This describes the motion of a shell element, including inertial effects associated with motion along the x, y, z axes as well as the inertia of its rotation about x, y [73] (p_1, p_2 and q_1 correspond to the intensity of the external load along the axes x, y and z, respectively).

Substituting (2.11)–(2.15) into (2.17)–(2.19) and attaching to the system of equations obtained the heat transfer equation (2.16), transformed by use of (2.1), (2.3), (2.4), (2.7) and (2.9), the following full system of equations governing the thermoelastic thermoelasticity behaviour of a shell in terms of to displacements is obtained:

$$\rho h \frac{\partial^2 u}{\partial t^2} - \frac{E}{1-\nu^2} \frac{\partial}{\partial x} \left\{ h \left[\frac{\partial u}{\partial x} - k_x w + \nu \left(\frac{\partial v}{\partial y} - k_y w \right) \right] \right\}$$
$$- \frac{E}{2(1+\nu)} \frac{\partial}{\partial y} \left[h \left(\frac{\partial u}{\partial y} + \frac{\partial v}{\partial x} \right) \right] + \frac{E \alpha_T}{1-\nu} \frac{\partial}{\partial x} \int\limits_{-\frac{h}{2}}^{\frac{h}{2}} \theta \, dz = p_1 \tag{2.20}$$

$$(x \leftrightarrow y, \quad u \leftrightarrow v, \quad p_1 \leftrightarrow p_2),$$

$$\rho h \frac{\partial^2 w}{\partial t^2} - \frac{k^2 E}{2(1+\nu)} \left\{ \frac{\partial}{\partial x} \left[h \left(\Psi_x + \frac{\partial w}{\partial x} \right) \right] + \frac{\partial}{\partial y} \left[h \left(\Psi_y + \frac{\partial w}{\partial y} \right) \right] \right\}$$
$$- \frac{E}{1-\nu^2} \left[(k_x + \nu k_y) \left(\frac{\partial u}{\partial x} - k_x w \right) + (k_y + \nu k_x) \left(\frac{\partial v}{\partial y} - k_y w \right) \right] \tag{2.21}$$
$$+ \frac{E \alpha_T}{1-\nu} \frac{\partial}{\partial x} (k_x + k_y) \int\limits_{-\frac{h}{2}}^{\frac{h}{2}} \theta \, dz = q_1,$$

$$\rho\frac{h^3}{12}\frac{\partial^2\Psi_x}{\partial t^2}-\frac{\partial}{\partial x}\left[D\left(\frac{\partial\Psi_x}{\partial x}+\nu\frac{\partial\Psi_y}{\partial y}\right)\right]$$
$$-\frac{1-\nu}{2}\frac{\partial}{\partial y}\left[D\left(\frac{\partial\Psi_x}{\partial y}+\frac{\partial\Psi_y}{\partial x}\right)\right]-\frac{k^2Eh}{2(1+\nu)}\left(\Psi_x+\frac{\partial w}{\partial x}\right) \tag{2.22}$$
$$+\frac{E\alpha_T}{1-\nu}\frac{\partial}{\partial x}\int\limits_{-\frac{h}{2}}^{\frac{h}{2}}\theta\,dz=0\quad(x\leftrightarrow y),$$

$$c_\varepsilon(1+\varepsilon)\frac{\partial\theta}{\partial t}-\lambda_q\nabla^2\theta+\frac{E\alpha_T T_0}{1-\nu}\frac{\partial}{\partial t}\left[\frac{\partial u}{\partial x}+\frac{\partial v}{\partial y}\right.$$
$$\left.-(k_x+k_y)w+z\left(\frac{\partial\Psi_x}{\partial x}+\frac{\partial\Psi_y}{\partial y}\right)\right]=q_2, \tag{2.23}$$

$$D=\frac{Eh^3}{12(1-\nu^2)},\qquad\varepsilon=\frac{E\alpha_T^2T_0(1+\nu)}{c_\varepsilon(1-\nu)(1-2\nu)}. \tag{2.24}$$

The system of differential equations (2.20)–(2.23) is referred to as the full system of equations of the coupled dynamic problem of linear thermoelasticity. It should be emphasized that we have not introduced any limitations on the temperature distribution through the shell thickness, which could lead to a system of equations with different dimensions. The function θ in the parabolic-type heat transfer equation depends on the three spatial variables x, y, z and the time t. The functions u, v, w, Ψ_x, Ψ_y appearing in the hyperbolic-type equation governing motion of a shell element motion depend on the two spatial coordinates x, y and the time t.

The system of equations obtained above governing the thermoelastic behaviour of a shallow shell in the framework of the Timoshenko model can be transformed to a system of equations in the framework of the Kirchhoff–Love model. To this end, we neglect the effect of transverse shear deformations. The rotation angles of the normal to the middle surface have the following form [338]:

$$\Psi_x=-\frac{\partial w}{\partial x},\qquad\Psi_y=-\frac{\partial w}{\partial y}. \tag{2.25}$$

In addition, rotational inertia is neglected, which leads to omission of the underlined term in (2.19). This is introduced simplification leads to the following equations:

$$M_x=-D\left(\frac{\partial^2 w}{\partial x^2}+\nu\frac{\partial^2 w}{\partial y^2}\right)-\frac{E\alpha_T}{1-\nu}\int\limits_{-\frac{h}{2}}^{\frac{h}{2}}\theta z\,dz=0\qquad(x\leftrightarrow y), \tag{2.26}$$

$$H = -D(1-\nu)\frac{\partial^2 w}{\partial x \partial y}, \tag{2.27}$$

$$\theta_x = \frac{\partial M_x}{\partial x} + \frac{\partial H}{\partial y} \qquad (x \leftrightarrow y). \tag{2.28}$$

Substituting (2.28) into (2.18) and attaching (2.17), the following equations governing the motion of a shell element are obtained:

$$\begin{aligned}
&\frac{\partial N_x}{\partial x} + \frac{\partial S}{\partial y} + p_1 - \rho h \frac{\partial^2 u}{\partial t^2} = 0 \quad (x \leftrightarrow y,\ u \leftrightarrow v,\ p_1 \leftrightarrow p_2),\\
&\frac{\partial^2 M_x}{\partial x^2} + 2\frac{\partial^2 H}{\partial x \partial y} + \frac{\partial^2 M_y}{\partial y^2} + k_x N_x + k_y N_y + q_1 - \rho h \frac{\partial^2 w}{\partial t^2} = 0.
\end{aligned} \tag{2.29}$$

Next, the relations (2.11), (2.12), (2.26) and (2.27) are substituted into (2.29) and, additionally, we attach the general heat transfer equation (2.23) modified by (2.25). As a result, we obtain

$$\begin{aligned}
&\rho h \frac{\partial^2 u}{\partial t^2} - \frac{E}{1-\nu^2}\frac{\partial}{\partial x}\left\{h\left[\frac{\partial u}{\partial x} - k_x w + \nu\left(\frac{\partial v}{\partial y} - k_y w\right)\right]\right\}\\
&\quad - \frac{E}{2(1+\nu)}\frac{\partial}{\partial y}\left[h\left(\frac{\partial u}{\partial y} + \frac{\partial v}{\partial x}\right)\right] + \frac{E\alpha_T}{1-\nu}\frac{\partial}{\partial x}\int_{-\frac{h}{2}}^{\frac{h}{2}} \theta\, dz = p_1
\end{aligned} \tag{2.30}$$

$$(x \leftrightarrow y, \quad u \leftrightarrow v, \quad p_1 \mapsto p_2),$$

$$\begin{aligned}
&\rho h \frac{\partial^2 w}{\partial t^2} + \frac{\partial^2}{\partial x^2}\left[D\left(\frac{\partial^2 w}{\partial x^2} + \nu\frac{\partial^2 w}{\partial y^2}\right)\right] + 2(1-\nu)\frac{\partial^2}{\partial x \partial y}(D)\left(\frac{\partial^2 w}{\partial x \partial y}\right)\\
&+ \frac{\partial^2}{\partial y^2}\left[D\left(\frac{\partial^2 w}{\partial y^2} + \nu\frac{\partial^2 w}{\partial x^2}\right)\right] - \frac{E}{1-\nu^2}\left[(k_x + \nu k_y)\left(\frac{\partial u}{\partial x} - k_x w\right)\right.\\
&\left. + (k_y + \nu k_x)\left(\frac{\partial v}{\partial y} - k_y w\right)\right] + \frac{E\alpha_T}{1-\nu}\left[\nabla^2 \int_{-\frac{h}{2}}^{\frac{h}{2}} \theta z dz + (k_x + k_y)\int_{-\frac{h}{2}}^{\frac{h}{2}} \theta\, dz\right] = q_1,
\end{aligned} \tag{2.31}$$

$$\begin{aligned}
&c_\varepsilon(1+\varepsilon)\frac{\partial \theta}{\partial t} - \lambda_q \nabla^2 \theta\\
&+ \frac{E\alpha_T T_0}{1-\nu}\frac{\partial}{\partial t}\left[\frac{\partial u}{\partial x} + \frac{\partial v}{\partial y} - (k_x + k_y)w - z\nabla^2 w\right] = q_2.
\end{aligned} \tag{2.32}$$

The system of differential equations (2.30)–(2.32) governs linear thermoelastic dynamics within the Kirchhoff–Love model (without the inertial effect of shell element rotation). It is clear that in this case we have a system of equations of different types (hyperbolic and parabolic) with different dimensions.

2.1.3 Boundary and Initial Conditions

In order to describe in full a coupled dynamical problem of thermoelasticity, the boundary and initial conditions need to be attached to the system of equations obtained above. This means that we need to define the initial state of the shell and to formulate the conditions that describe the mechanical and thermal interaction with the surrounding medium on both the edge and the surface of the shell. The boundary and initial conditions given below are obtained from a stationary condition of the action of the corresponding Hamilton functional.

As the initial conditions attached to the system (2.30)–(2.32), we take the following distributions (for the time instant $t = 0$) inside the shell of the displacements u, v, w, their velocities $\frac{\partial u}{\partial t}, \frac{\partial v}{\partial t}, \frac{\partial w}{\partial t}$, and the temperature increase θ (the last requirement is equivalent to the introduction of a temperature field at the initial time instant):

$$u|_{t=0} = u_0(x, y), \quad v|_{t=0} = v_0(x, y), \quad w|_{t=0} = w_0(x, y), \tag{2.33}$$

$$\left.\frac{\partial v}{\partial t}\right|_{t=0} = u_1(x, y), \quad \left.\frac{\partial v}{\partial t}\right|_{t=0} = v_1(x, y), \quad \left.\frac{\partial w}{\partial t}\right|_{t=0} = w_1(x, y), \tag{2.34}$$

$$\theta|_{t=0} = \theta_0(x, y, z), \tag{2.35}$$

where $(x, y) \in \Omega_1, (x, y, z) \in \Omega_2$, and $u_0, v_0, w_0, u_1, v_1, w_1, \theta_0$ are given functions.

In considering Eqs. (2.20)–(2.23), we need to add to the initial conditions (2.33)–(2.35) the distributions (for $t = 0$) of the angles of rotation Ψ_x, Ψ_y of the normal to the middle surface and their derivatives $\frac{\partial \Psi_x}{\partial t}, \frac{\partial \Psi_y}{\partial t}$. These distributions have the form

$$\Psi_x|_{t=0} = \Psi_{x0}(x, y), \quad \Psi_y|_{t=0} = \Psi_{y0}(x, y), \tag{2.36}$$

$$\left.\frac{\partial \Psi_x}{\partial t}\right|_{t=0} = \Psi_{x1}(x, y), \quad \left.\frac{\partial \Psi_y}{\partial t}\right|_{t=0} = \Psi_{y1}(x, y), \tag{2.37}$$

where $(x, y) \in \Omega_1$ and $\Psi_{x0}, \Psi_{y0}, \Psi_{x1}, \Psi_{y1}$ are given functions.

At each point of the shell surface $\partial\Omega_2$, one of the following thermal boundary conditions must be applied [187]:

– the temperature of the shell surface,

$$\theta = \theta^0(x, y, z, t); \tag{2.38}$$

– the density of the heat flow through the shell surface,

$$\frac{\partial \theta}{\partial n} = \theta_n^0(x, y, z, t); \tag{2.39}$$

– a condition of free heat exchange with the surrounding medium,

$$\frac{\partial \theta}{\partial n} + \alpha\theta = \theta_n^\alpha(x, y, z, t), \quad \alpha = const > 0. \tag{2.40}$$

Here $(x, y, z) \in \partial\Omega_2, t \in [0, T]$, and $\theta^0, \theta_n^0, \theta_n^\alpha$ are given functions, and T is the total time of observation of the shell.

In the following investigation of rectangular shells, on the sides (for $x = const$ and $y = const$) only a homogeneous condition of type (2.38) will be considered, while on the shell surfaces $\left(z = \pm\frac{h}{2}\right)$, homogeneous conditions of type (2.38) or (2.39) (condition of thermal isolation) will be considered.

The mechanical conditions on the shell edge $\partial\Omega_1$ for Eqs. (2.30)–(2.32) have the following form (at each point of $\partial\Omega_1$, four boundary conditions are applied):

$$u = u^0(x, y, t) \quad \text{or} \quad N_x n_x + S n_y = A^0(x, y, t), \tag{2.41}$$

$$v = v^0(x, y, t) \quad \text{or} \quad N_y n_y + S n_x = B^0(x, y, t), \tag{2.42}$$

$$\begin{aligned} w = w^0(x, y, t) \quad \text{or} \quad & \left(\frac{\partial M_x}{\partial x} + \frac{\partial H}{\partial y}\right) n_x + \left(\frac{\partial M_y}{\partial y} + \frac{\partial H}{\partial x}\right) n_y \\ & + \frac{\partial}{\partial s}[(M_y n_y + H n_x) n_x - (M_x n_x + H n_y) n_y] = C^0(x, y, t), \end{aligned} \tag{2.43}$$

$$\frac{\partial w}{\partial n} = w_n^0(x, y, t) \quad \text{or} \quad M_x n_x^2 + 2H n_x n_y + M_y n_y^2 = D^0(x, y, t), \tag{2.44}$$

where $(x, y) \in \partial\Omega_2, t \in [0, T]$, and n_x, n_y are the components of the external normal to the edge $\partial\Omega_1$. The symbols $\frac{\partial}{\partial n}, \frac{\partial}{\partial s}$ denote differentiation in the direction of the external normal to the edge $\partial\Omega_1$ and in the direction of the edge $\partial\Omega_1$, respectively, and $u^0, v^0, w^0, w_n^0, A^0, B^0, C^0, D^0$ are given functions.

When the Timoshenko kinematic model is used ((2.20)–(2.23)), we need to formulate five mechanical conditions, i.e. the conditions (2.41), (2.42) and the following additional three:

$$w = w^0(x, y, t) \quad \text{or} \quad Q_x n_x + Q n_y = E^0(x, y, t), \tag{2.45}$$

$$\Psi_x = \Psi_x^0(x, y, t) \quad \text{or} \quad M_x n_x + H n_y = F^0(x, y, t), \tag{2.46}$$

$$\Psi_y = \Psi_y^0(x, y, t) \quad \text{or} \quad M_y n_y + H n_x = G^0(x, y, t), \tag{2.47}$$

where $(x, y) \in \partial\Omega_2, t \in [0, T]$, and $w^0, \Psi_x^0, \Psi_y^0, E^0, F^0, G^0$ are given functions.

In the each pair of alternative boundary conditions (2.41)–(2.47), the first is the main condition (this means that during the search for a stationary point of Hamilton's functional this condition is specified in advance), whereas the second is a physically realistic ("real") condition (it is derived from a stationary condition of Hamilton's functional). In the case of a rectangular shell, these boundary conditions have been formulated in a series of monographs (see, for instance [198, 338, 339]).

The following mechanical boundary conditions for different types of support of the shell edge will be considered (for shells that are rectangular in plan):

- simple support on an elastic rib (no compression or elongation in the tangential plane):

$$w = \frac{\partial u}{\partial x} = v = \frac{\partial \Psi_x}{\partial x} = \Psi_y = 0, \quad x = const, \tag{2.48}$$

$$w = \frac{\partial^2 w}{\partial x^2} = \frac{\partial u}{\partial x} = v = 0, \quad x = const; \tag{2.49}$$

- simply supported edge:

$$w = u = v = \frac{\partial \Psi_x}{\partial x} = \Psi_y = 0, \quad x = const, \tag{2.50}$$

$$w = \frac{\partial^2 w}{\partial x^2} = u = v = 0, \quad x = const; \tag{2.51}$$

- clamped edge:

$$w = u = v = \Psi_x = \Psi_y = 0, \quad x = const, \tag{2.52}$$

$$w = \frac{\partial w}{\partial x} = u = v = 0, \quad x = const, \tag{2.53}$$

$$x \leftrightarrow y, \quad u \leftrightarrow v.$$

The conditions (2.48), (2.50), (2.52) correspond to the Timoshenko-type model, whereas the conditions (2.49), (2.51), (2.53) correspond to the Kirchhoff–Love model. Each of the groups of boundary conditions (when a homogeneous boundary condition of type (2.38) is satisfied on the sides of the shell) refers to a particular case of the general conditions, either (2.41)–(2.44) (Kirchhoff–Love model) or (2.41), (2.42), (2.45)–(2.47) (Timoshenko model). For instance, in the group (2.48) the conditions $w = v = \Psi_y = 0$ are related to the homogeneous main conditions (2.45), (2.42), (2.47), whereas the conditions $\frac{\partial u}{\partial x} = \frac{\partial \Psi_x}{\partial x} = 0$ are equivalent to the homogeneous real conditions (2.41), (2.46), and so on.

2.1.4 An Abstract Coupled Problem

Here we consider an abstract Cauchy problem for a coupled system of two differential equations in a Hilbert space which generalizes a series of coupled thermoelastic problems. This also includes the thermoelastic problems of shallow shells modelled by the Kirchhoff–Love and Timoshenko theories defined earlier and is an important subject of this chapter.

First we introduce the necessary notation. Let H be a certain Hilbert space. By $L_p(0, T; H)$ we denote a space of measurable (in an interval $[0, T]$ of the time axis) functions, which have their values in H and are integrable $[0, T]$ (bounded for $p = \infty$), with a norm of order p $(p < \infty)$

$$\|u\|_{L_p(0,T;H)} = \left[\int_0^T \|u(t)\|_H^p \, dt \right]^{\frac{1}{p}},$$

which for $p = \infty$ takes the form

$$\|u\|_{L_\infty(0,T;H)} = \sup_{0 \le t \le T} ess \|u(t)\|_H.$$

By $W_p^k(0, T; H)$ we denote a space of rational functions $u(t)$, which have values in H, possess in the interval $[0, T]$ generalized derivatives (in the Sobolev sense) $u^{(j)}(t)$ up to the k_Ωth order (inclusive), and are integrable with a norm of power p. The space $W_p^k(0, T; H)$ for $p < \infty$ is defined by the norm

$$\|u\|_{W_p^k(0,T;H)} = \left[\int_0^T \sum_{j=0}^k \left\| u^{(j)}(t) \right\|_H^p dt \right]^{\frac{1}{p}},$$

whereas for $p = \infty$ it is defined by the norm

$$\|u\|_{W_\infty^k(0,T;H)} = \sup_{0 \le t \le T} ess \sum_{j=0}^k \|u(t)\|_H.$$

For $1 \le p \le \infty$, each of the spaces $L_p(0, T; H)$, $W_p^k(0, T; H)$ $(k = 1, 2, \ldots)$ is a Banach space. The spaces $L_2(0, T; H)$, $W_2^k(0, T; H)$ are Hilbert spaces, and a scalar product is defined as follows:

$$(u, v)_{L_2(0,T;H)} = \int_0^T (u(t), v(t))_H \, dt,$$

$$(u, v)_{W_2^k(0,T;H)} = \int_0^T \sum_{j=0}^k (u^{(j)}(t), v^{(j)}(t))_H \, dt.$$

By $W_2^k(\Omega)$ we denote the Sobolev space of measurable functions $u(x)$, defined in the space Ω of n-dimensional arithmetic space R^n, which have all generalized partial derivatives $D^\alpha u(x)$ of order up to k inclusive, and are integrable with a norm of the second power in Ω. We define $\alpha = (\alpha_1 + \ldots + \alpha_n)$ as the integer multi-index of differentiation, $|\alpha| = \alpha_1 + \ldots + \alpha_n$, and $D^\alpha = D_1^{\alpha_1} \ldots D_n^{\alpha_n}$, $D_i = \partial/\partial x_i$, denotes a general differentiation operation in the Sobolev sense. The space $W_k^2(\Omega)$ is a Hilbert space and a scalar product is defined in it by the relation

$$(u, v)_{W_2^k(\Omega)} = \int_\Omega \sum_{|\alpha| \le k} D^\alpha u D^\alpha v \, d\Omega.$$

For $k = 0$, the space $W_k^2(\Omega)$ overlaps with the Lebesgue space $L_2(\Omega)$.

Let $H_1, \ldots, H_m$ be m samples of the Hilbert space. We denote by $H_1 \times \ldots \times H_m$ their Cartesian product, i.e. the Hilbert space created by all possible ordered samples of elements $\overline{u} = \{u_i\}_{i=1}^m$, where $u_i \in H_i$ and where the scalar product has the form $(\overline{u}, \overline{v})_{H_1 \times \ldots \times H_m} = (u_1, v_1)_{H_1} + \ldots + (u_m, v_m)_{H_m}$.

Let $u(x, t)$ be a rational function defined for $x \in \Omega \subset R^n$, $t \in [0, T]$, which has its values in R^m and possesses the following property: for all $t \in [0, T]$, the function $x \to u(x, t)$ belongs to a Hilbert space H of the type $W_2^{k_1}(\Omega) \times \ldots \times W_2^{k_m}(\Omega)$. Each function defined in this way is related to a map $u(t)$, which for an arbitrary number $t \in [0, T]$ defines the function $x \to u(x, t)$ as an element of H. Therefore, we shall use the following notation: $\|u(t)\|_H$, $(u(t), v(t))_H$, $\|u\|_{W_p^k(0,T;H)}$, $(u, v)_{W_2^k(0,T;H)}$, where u, v are functions depending on x, t.

In the case of continuous and continuously differentiable functions, the following established notation is applied: $C(\overline{\Omega})$, $C^k(\overline{\Omega})$, $C([0, T]; H)$.

Let A be a self-adjoint, positive definite and (possibly) unbounded operator acting in the Hilbert space H, which we attach to the space $D(A)$; the action of A creates a Hilbert space with the same number of elements and with a scalar product $(u, v)_A = (Au, Av)_H$. Therefore, the following notation is applied: $L_p(0, T; D(A))$, $W_p^k(0, T; D(A))$.

Let H_1, H_2 be two specimens of the Hilbert space. Let us formulate the Cauchy problem for the following system of two differential equations:

$$I_1 w''(t) + Lw(t) + M\theta(t) = g_1(t), \tag{2.54}$$

$$I_2 \theta'(t) + K\theta(t) + Nw'(t) = g_2(t), \tag{2.55}$$

$$w(0) = w_0, \quad w'(0) = w_1, \quad \theta(0) = \theta_0, \tag{2.56}$$

where w, θ are the sought functions mapping the interval $[0, T]$ into H_1 and H_2, respectively. The given functions g_1, g_2 satisfy the conditions

$$g_1 \in L_2(0, T; H_1), \quad g_2 \in L_2(0, T; H_2); \tag{2.57}$$

I_1 and I_2 are bounded self-adjoint, positive definite operators acting in H_1 and H_2, respectively; and L and K are unbounded, self-adjoint, positive definite operators acting in H_1 and H_2, respectively, defined on dense sets $D(L) \subset H_1$, $D(K) \subset H_2$. The inverse operators L^{-1}, K^{-1} are continuous. M and N are linear, unbounded operators acting from H_2 to H_1 and from H_1 to H_2, respectively, linked by the relation $N \subset -M^*$, i.e. for arbitrary elements $w \in D(N), \theta \in D(M)$, the following "linking condition" holds:

$$(M\theta, w)_{H_1} + (Nw, \theta)_{H_2} = 0. \tag{2.58}$$

Also $D(K) \subset D(M)$, $D(L^{1/2}) \subset D(N)$. The bounded operators MK^{-1} and $NL^{-1/2}$ correspond to the operators $H_2 \to H_1$ and $H_1 \to H_2$. The operator $L^{-1/2}MK^{-1/2}$ can be extended to a bounded operator from H_2 to H_1. Finally, w_0, w_1, θ_0 are given elements satisfying the conditions

$$w_0 \in D(L^{1/2}), \quad w_1 \in H_1, \quad \theta_0 \in H_2. \tag{2.59}$$

Definition 2.1 An ordered pair of functions $\{w, \theta\}$ is called a generalized solution to the Cauchy problem (2.54)–(2.56) if $w \in W^1_\infty(0, T; H_1) \cap L_\infty(0, T; D(L^{1/2}))$, $w(0) = w_0$, $\theta \in L_\infty(0, T; H_2) \cap L_2(0, T; D(K^{1/2}))$ and, for arbitrary functions v, η satisfying the conditions $v \in W^1_1(0, T; H_1) \cap L_1(0, T; D(L^{1/2}))$, $v(T) = 0$, $\eta \in W^1_1(0, T; H_2) \cap L_2(0, T; D(K^{1/2}))$, $\eta(T) = 0$, the following assumptions are satisfied:

$$\int_0^T \Big[-(I_1 w'(t), v'(t))_{H_1} + (L^{1/2}w(t), L^{1/2}v(t))_{H_1}$$

$$+(L^{-1/2}M\theta(t), L^{1/2}v(t))_{H_1}\Big] dt - (I_1 w_1, v(0))_{H_1} = \int_0^T (g_1(t), v(t))_{H_1}\, dt, \tag{2.60}$$

$$\int_0^T \Big[-(I_1\theta(t), \eta'(t))_{H_2} + (K^{1/2}\theta(t), K^{1/2}\eta(t))_{H_2} - (Nw(t)\eta'(t))_{H_2}$$

$$-L^{1/2}v(t))_{H_1}\Big] dt - (I_1 w_1, v(0))_{H_1} = \int_0^T (g_1(t), v(t))_{H_1}\, dt. \tag{2.61}$$

Definition 2.2 An ordered pair of functions $\{w, \theta\}$ is called the classical solution to the Cauchy problem (2.54)–(2.56) if

$$w \in W^2_\infty(0, T; H_1) \cap W^1_\infty(0, T; D(L^{1/2})) \cap L_\infty(0, T; D(L)),$$

$$\theta \in W^1_\infty(0, T; H_2) \cap W^1_2(0, T; D(K^{1/2})) \cap L_\infty(0, T; D(K)),$$

and the functions w, θ satisfy the relations (2.54), (2.55) almost always on the interval $[0, T]$ and satisfy the relations (2.56).

Remark 2.1 In Definitions 2.1 and 2.2, maximal smoothness of the functions w, θ is needed, which can be achieved using Theorems 2.1 and 2.2. The integral relations (2.60), (2.61) can be obtained by a general approach used in solving problems of mathematical physics, namely by a formal multiplication of the differential equations (2.54), (2.55) by the functions $\nu(t)$, $\eta(t)$ and integration by parts, with consideration of the initial condition (2.56). In particular, this means that the "classical" solution to the problem (2.54)–(2.56) is its generalized solution.

Let us check if the formally defined Cauchy problem (2.54)–(2.56) is a real generalization of the coupled thermoelasticity problem of a shallow shell defined in Sects. 2.1.2, 2.1.3 (in the case of homogeneous boundary conditions). To do this, we analyse the initial–boundary problem for the system of differential equations (2.20)–(2.23) with the initial conditions (2.33)–(2.37), the homogeneous mechanical boundary conditions (2.41), (2.42), (2.45)–(2.47) and the homogeneous thermal boundary conditions (2.38) (main condition) or (2.39) (real condition). Let $\Omega_1 \subset R^2$ denote a bounded space with a piecewise smooth boundary of C^1 class. With each member of the pair of alternative boundary conditions we associate either the part Ω_1, of the boundary (for mechanical conditions) or Ω_2 (for heating conditions); these parts do not intersect each other. It should be noted that the measure of the part of the boundary corresponding to the main condition is strictly positive, and on the whole of the side surfaces of the shell the main thermal condition is applied. This means that all mechanical boundary conditions may be described via the displacements u, v, w and rotation angles Ψ_x, Ψ_y only (the function θ does not appear in these conditions). Let us denote by $(\mathring{W}^2_2(\Omega_1))^5$ the space of vector functions $\overline{\omega} = (x, y)$ of the form

$$\overline{\omega} = \{u, v, w, \Psi_x, \Psi_y\}; \tag{2.62}$$

all components of this space belong to the space $W^2_2(\Omega_1)$ and satisfy the main and real mechanical boundary conditions of the problem under consideration. By $(\mathring{W}^1_2(\Omega_1))^5$ we denote the space of vector functions of the form (2.62), all components of which belong to $W^1_2(\Omega_1)$ and satisfy the main mechanical conditions of the problem. By $\mathring{W}^2_2(\Omega_1)$ we denote the space of functions $\theta \in W^2_2(\Omega_1)$ satisfying both the main and the real thermal boundary conditions, and by $\mathring{W}^1_2(\Omega_1)$ we denote the space of the functions $\theta \in W^1_2(\Omega_2)$ satisfying the main thermal condition.

The following conditions are applied to the functions and constants occurring in (2.20)–(2.23), (2.33)–(2.37):

$$h \in C^1(\overline{\Omega}_1), \quad h(x,y) > 0, \quad (x,y) \in \overline{\Omega}_1, \tag{2.63}$$

$$\rho, E, \alpha_T, c_\varepsilon, \lambda_q, T_0, T = const > 0, \quad k = const \neq 0,$$

$$k_x, k_y = const, \quad 0 < \nu = const < \frac{1}{2}, \tag{2.64}$$

$$p_1, p_2, q_1 \in L_2(\Omega_1 \times (0,T)), \quad q_2 \in L_2(\Omega_2 \times (0,T)), \tag{2.65}$$

$$\overline{\omega}_0 = \{u_0, v_0, w_0, \Psi_{x0}, \Psi_{y0}\} \in (\mathring{W}_2^1(\Omega_1))^5, \tag{2.66}$$

$$\overline{\omega}_1 = \{u_1, v_1, w_1, \Psi_{x1}, \Psi_{y1}\} \in (L_2(\Omega_1))^5, \tag{2.67}$$

$$\theta_0 \in L_2(\Omega_1). \tag{2.68}$$

The initial–boundary problem under consideration can be transformed to the form (2.54)–(2.56), if the following assumptions are made:

$$H_1 = (L_2(\Omega_1))^5, \quad H_2 = L_2(\Omega_1), \tag{2.69}$$

$$I_1\overline{\omega} = \left\{\rho h u, \rho h v, \rho h w, \frac{1}{12}\rho h^3 \Psi_x, \frac{1}{12}\rho h^3 \Psi_y\right\}, \tag{2.70}$$

$$I_2\theta = \frac{c_\varepsilon}{T_0}(1+\varepsilon)\theta, \tag{2.71}$$

$$L\overline{\omega} = \Bigg\{ l_x\overline{\omega}, l_y\overline{\omega},$$

$$-\frac{k^2 E}{2(1+\nu)}\left\{\frac{\partial}{\partial x}\left[h\left(\Psi_x + \frac{\partial w}{\partial x}\right)\right] + \frac{\partial}{\partial y}\left[h\left(\Psi_y + \frac{\partial w}{\partial y}\right)\right]\right\}$$

$$-\frac{Eh}{1-\nu^2}\left[(k_x+\nu k_y)\left(\frac{\partial u}{\partial x} - k_x w\right) + (k_x+\nu k_y)\left(\frac{\partial v}{\partial y} - k_y w\right)\right],$$

$$\tilde{l}_x\overline{\omega}, \tilde{l}_y\overline{\omega}\Bigg\}, \tag{2.72}$$

$$D(L) = (\mathring{W}_2^2(\Omega_1))^5,$$

$$l_x\overline{\omega} = -\frac{E}{1-\nu^2}\frac{\partial}{\partial x}h\left[\frac{\partial u}{\partial x} - k_x w + \nu\left(\frac{\partial v}{\partial y} - k_y w\right)\right]$$

$$-\frac{E}{2(1+\nu)}\frac{\partial}{\partial y}\left[h\left(\frac{\partial u}{\partial y} + \frac{\partial v}{\partial x}\right)\right] \quad (x \leftrightarrow y, \quad u \leftrightarrow v), \tag{2.73}$$

$$\tilde{l}_x\overline{\omega} = -\frac{\partial}{\partial x}\left[D\left(\frac{\partial \Psi_x}{\partial x} + \nu\frac{\partial \Psi_y}{\partial y}\right)\right] - \frac{1-\nu}{2}\frac{\partial}{\partial y}\left[D\left(\frac{\partial \Psi_x}{\partial y} + \frac{\partial \Psi_y}{\partial x}\right)\right]$$

$$+\frac{k^2 Eh}{2(1+\nu)}\left(\Psi_x+\frac{\partial w}{\partial x}\right) \quad (x \leftrightarrow y), \tag{2.74}$$

$$K\theta = -\frac{\lambda_q}{T_0}\nabla^2\theta, \quad D(K) = \mathring{W}_2^2(\Omega_2), \tag{2.75}$$

$$M\theta = \left\{ m_x\theta, m_y\theta, \frac{E\alpha_T}{1-\nu}(k_x+k_y)\int\limits_{-\frac{h}{2}}^{\frac{h}{2}}\theta\, dz, \tilde{m}_x\theta, \tilde{m}_y\theta \right\}, \tag{2.76}$$

$$D(M) = \mathring{W}_2^1(\Omega_2),$$

$$m_x\theta = \frac{E\alpha_T}{1-\nu}\frac{\partial}{\partial x}\int\limits_{-\frac{h}{2}}^{\frac{h}{2}}\theta\, dz, \quad \tilde{m}_x\theta = \frac{E\alpha_T}{1-\nu}\frac{\partial}{\partial x}\int\limits_{-\frac{h}{2}}^{\frac{h}{2}}\theta z\, dz, \quad (x \leftrightarrow y), \tag{2.77}$$

$$N\overline{\omega} = \frac{E\alpha_T}{1-\nu}\left[\frac{\partial u}{\partial x}+\frac{\partial v}{\partial y}-(k_x+k_y)w+z\left(\frac{\partial \Psi_x}{\partial x}+\frac{\partial \Psi_y}{\partial y}\right)\right], \tag{2.78}$$

$$D(N) = (\mathring{W}_2^1(\Omega_1))^5,$$

$$\overline{\omega}_0 = \{u_0, v_0, w_0, \Psi_{x0}, \Psi_{y0}\}, \quad \overline{\omega}_1 = \{u_1, v_1, w_1, \Psi_{x1}, \Psi_{y1}\}, \tag{2.79}$$

$$g_1 = \{p_1, p_2, q_1, 0, 0\}, \quad g_2 = \frac{1}{T_0}q_2. \tag{2.80}$$

The vector function $\overline{\omega} = \{u, v, w, \Psi_x, \Psi_y\}$ and the given vector functions, $\overline{\omega}_0, \overline{\omega}_1$ (2.79) play the roles of the sought function $w(t)$ and the initial data w_0, w_1. In addition, the following relations hold:

$$D(L^{1/2}) = (\mathring{W}_2^1(\Omega_1))^5, \quad D(K^{1/2}) = (\mathring{W}_2^1(\Omega_1))^5, \tag{2.81}$$

with the accuracy of the equivalent scalar product.

Let us consider further the initial–boundary value problem for the system of equations (2.30)–(2.32) with the initial conditions (2.33)–(2.35) and homogeneous boundary conditions of the form (2.41)–(2.44), (2.38) (or (2.39)). We retain the earlier assumptions about the smoothness of Ω_1 and about the partitioning of the boundary into Ω_1, Ω_2, corresponding to each member of the pair of alternative boundary conditions. We denote by $((\mathring{W}_2^2(\Omega_1))^2 \times \mathring{W}_2^4(\Omega_1)$ the space of vector functions $\overline{\omega}(x, y)$ of the form

$$\overline{\omega}_0 = \{u, v, w\}, \tag{2.82}$$

the components of which belong to $W_2^2(\Omega_1)$, $W_2^2(\Omega_1)$, $W_2^4(\Omega_1)$, respectively and satisfy the main mechanical boundary conditions of the problem. The conditions (2.64), (2.65), (2.68) are applied to the functions and constants occurring in (2.30)–(2.35), and in addition the following conditions are applied:

$$h \in C^2(\overline{\Omega}_1), \quad h(x, y) > 0, \quad (x, y) \in \overline{\Omega}_1, \tag{2.83}$$

$$\overline{\omega}_0 = \{u_0, v_0, w_0\} \in (\mathring{W}_2^1(\Omega_1))^2 \times \mathring{W}_2^2(\Omega_1), \tag{2.84}$$

$$\overline{\omega}_1 = \{u_1, v_1, w_1\} \in (L_2(\Omega_1))^3. \tag{2.85}$$

The initial–boundary value problem under consideration is reduced to that described by (2.54)–(2.56) if we make the assumptions

$$H_1 = (L_2(\Omega_1))^3, \quad H_2 = L_2(\Omega_2), \tag{2.86}$$

$$I_1\overline{\omega} = \{\rho h u, \rho h v, \rho h w\}, \quad I_2\theta = \frac{c_\varepsilon}{T_0}(1+\varepsilon)\theta, \tag{2.87}$$

$$L\overline{\omega} = \Bigg\{ l_x\overline{\omega}, l_y\overline{\omega}, \frac{\partial^2}{\partial x^2}\left[D\left(\frac{\partial^2 w}{\partial x^2} + \nu\frac{\partial^2 w}{\partial y^2}\right)\right]$$
$$+2(1-\nu)\frac{\partial^2}{\partial x\partial y}\left(D\frac{\partial^2 w}{\partial x\partial y}\right) + \frac{\partial^2}{\partial y^2}\left[D\left(\frac{\partial^2 w}{\partial y^2} + \nu\frac{\partial^2 w}{\partial x^2}\right)\right]$$
$$-\frac{Eh}{1-\nu^2}\left[(k_x+\nu k_y)\left(\frac{\partial u}{\partial x} - k_x w\right) + (k_y+\nu k_x)\left(\frac{\partial v}{\partial y} - k_y w\right)\right]\Bigg\}, \tag{2.88}$$

$$D(L) = (\mathring{W}_2^2(\Omega_1))^2 \times \mathring{W}_2^4(\Omega_1),$$

$$K\theta = -\frac{\lambda_q}{T_0}\nabla^2\theta, \quad D(K) = \mathring{W}_2^2(\Omega_2), \tag{2.89}$$

$$M\theta = \left\{ m_x\theta, m_y\theta, \frac{E\alpha_T}{1-\nu}\left[\nabla^2\int_{-\frac{h}{2}}^{\frac{h}{2}}\theta\,dz + (k_x+k_y)\int_{-\frac{h}{2}}^{\frac{h}{2}}\theta\,dz\right]\right\}, \tag{2.90}$$

$$D(M) = \mathring{W}_2^2(\Omega_2),$$

$$N\overline{\omega} = \frac{E\alpha_T}{1-\nu}\left[\frac{\partial u}{\partial x} + \frac{\partial v}{\partial y} - (k_x+k_y)w + z\,\nabla^2 w\right], \tag{2.91}$$

$$D(N) = (\mathring{W}_2^1(\Omega_1))^2 \times \mathring{W}_2^2(\Omega_1),$$

$$\overline{\omega}_0 = \{u_0, v_0, w_0\}, \quad \overline{\omega}_1 = \{u_1, v_1, w_1\} \tag{2.92}$$

$$g_1 = \{p_1, p_2, q_1\}, \quad g_2 = \frac{1}{T_0}q_2. \tag{2.93}$$

The operators l_x, l_y, m_x, m_y appearing in (2.88), (2.90) are defined by the relations (2.73), (2.77). The sought function and initial data play a being sought vector-function $\overline{\omega} = \{u, v, w\}$ and the given vector functions $\overline{\omega}_0, \overline{\omega}_1$ (2.92) play the roles of the sought function $w(t)$ and the initial data w_0, w_1. It should be noted that following relations hold:

$$D(L^{1/2}) = (\overset{\circ}{W}{}_2^1(\Omega_1))^2 \times \overset{\circ}{W}{}_2^2(\Omega_1), \quad D(K^{1/2}) = \overset{\circ}{W}{}_2^1(\Omega_2). \tag{2.94}$$

Remark 2.2 An action space of the operators L, K may be defined as a self-conjugating extension of the initial differential operators defined by (2.72), (2.75) or (2.88), (2.89) initially on sets of smooth enough functions satisfying the corresponding homogeneous boundary conditions. (This definition is due to Friedrichs.)

Remark 2.3 All of the general properties mentioned above of the operators I_1, I_2, L, K, M, N appearing in (2.54), (2.55) either are explicitly obvious (see (2.69)–(2.78) or (2.86)–(2.91)) or can be obtained from known general results. In particular, full continuity of the operators L^{-1}, K^{-1} can be established using theorems of equivalent normalization of the Sobolev spaces described in [214, 243] or using theorems of compact inclusion [214], satisfied for instance for spaces with a boundary of class C^1.

Remark 2.4 The class of Cauchy problems defined by (2.54)–(2.56) is not limited to the problems of the theory of shallow shells theory presented in Sects. 2.1.2, 2.1.3. It is not difficult to check that a thermoelastic problem of a three-dimensional body can be reduced to a problem governed by (2.54)–(2.56) [266], as well as to various other problems of thermoelasticity.

2.1.5 Existence and Uniqueness of Solutions of Thermoelasticity Problems

In this section, theorems related to the existence and uniqueness of the general "classical" solution to the coupled abstract problem (2.54)–(2.56) will be given, and then the corresponding theorems for coupled thermoelastic problems of shallow shells will be formulated.

In proving results related to the existence of solutions, the known scheme of the compactness method will be used [226], and special attention will be paid to prior estimates of exact and approximate solutions of the problem (2.54)–(2.56), which will then be used in the next chapter in obtaining estimates for the Bubnov–Galerkin method.

Let $\{\nu_i\}_{i=1}^{\infty}$, $\{\eta_j\}_{j=1}^{\infty}$ be linearly independent systems of elements from $D(L)$ and $D(K)$, respectively (fully defined in $D(L^{1/2})$ and $D(K^{1/2})$), and let $i(n)$, $j(n)$ be two subseries of a natural series. Let H_1^n and H_2^n be linear vicinities of finite systems of elements $\{\nu_i\}_{i=1}^{i(n)}$ and $\{\eta_j\}_{j=1}^{j(n)}$. Let P_G denote the operator of orthogonal projection of the Hilbert space H $(= H_1$ or $H_2)$. We are going to solve the initial problem (2.54)–(2.56) using the Bubnov–Galerkin method. We consider the following problem:

$$\begin{cases} P_{H_1^n}(I_1 w_n''(t) + L w_n(t) + M\theta_n(t)) = P_{H_1^n} g_1(t), \\ P_{H_2^n}(I_2\theta_n'(t) + K\theta_n(t) + N w_n'(t)) = P_{H_2^n} g_2(t), \\ w_n(t) \in H_1^n, \quad \theta_n(t) \in H_2^n, \quad \forall t \in [0, T], \\ w_n(0) = w_{0n}, \quad w_n' = w_{1n}, \quad \theta_n(0) = \theta_{0n}, \end{cases} \tag{2.95}$$

where $\{w_{0n}\}_{n=1}^{\infty}$, $\{w_{1n}\}_{n=1}^{\infty}$, $\{\theta_{0n}\}_{n=1}^{\infty}$ are arbitrary subseries of elements taken from H_1^n, H_1^n, H_2^n, respectively, satisfying the conditions

$$w_{0n} \to w_0 \text{ in } D(L^{1/2}), \quad w_{1n} \to w_1 \text{ in } H_1, \quad \theta_{0n} \to \theta_0 \text{ in } H_2. \tag{2.96}$$

The exact solution $\{w_n, \theta_n\}$ to the problem (2.95) will be called the approximate solution to the problem (2.54)–(2.56) obtained using the Bubnov–Galerkin method. Because the problem (2.95) is equivalent to a Cauchy problem for a systems of $2i(n) + J(n)$ ordinary differential equations with $2i(n) + J(n)$ sought functions, this solution exists and is unique for an arbitrary natural number n.

Theorem 2.1 *A general solution $\{w, \theta\}$ to the problem (2.54)–(2.56) exists and is unique when the conditions (2.57)–(2.59) are satisfied.*

Proof Let us begin with the existence of a general solution to the problem (2.54)–(2.56). Multiplying scalarly the first equation of (2.95) by $2w_n'(t)$ and the second equation of (2.95) by $2\theta_n(t)$, we take the sum of them and take into account the condition (2.58). Integrating again the equality obtained over the interval $[0, T]$ (where $t \le T$), we obtain:

$$\left\| I_1^{1/2} w_n'(t) \right\|_{H_1}^2 + \left\| L^{1/2} w_n(t) \right\|_{H_1}^2 + \left\| I_2^{1/2} \theta_n(t) \right\|_{H_2}^2$$

$$+ 2\int_0^t \left\| K_1^{1/2} \theta_n(\tau) \right\|_{H_2}^2 d\tau = \left\| I_1^{1/2} w_{1n} \right\|_{H_2}^2 + \left\| L^{1/2} w_{0n} \right\|_{H_1}^1$$

$$+ \left\| I_2^{1/2} \theta_{0n} \right\|_{H_2}^2 + 2\int_0^t \left[(g_1(\tau), w_n'(\tau))_{H_1} + (g_2(\tau), \theta_n(\tau))_{H_2} \right] d\tau.$$

Applying the Cauchy inequality to the right-hand side of the relation given above, we obtain

$$\left\| I_1^{1/2} w_n'(t) \right\|_{H_1}^2 + \left\| L^{1/2} w_n(t) \right\|_{H_1}^2 + \left\| I_2^{1/2} \theta_n(t) \right\|_{H_2}^2$$

$$+2\int_0^t \left\| K^{1/2} \theta_n(\tau) \right\|_{H_2}^2 d\tau \le w + \frac{1}{\delta} \left\| I^{-1/2} g \right\|_{L_2(0,T;H_1 \times H_2)}^2$$

$$+\delta \int_0^t \left[\left\| I_2^{1/2} w_n'(\tau) \right\|_{H_1}^2 + \left\| I_2^{1/2} \theta_n(\tau) \right\|_{H_2}^2 \right] d\tau, \tag{2.97}$$

where $I^{-1/2} g = \{I_1^{-1/2} g_1, I_2^{-1/2} g_2\}$, δ is an arbitrary positive number, and

$$w = \sup_n \left[\left\| I_1^{1/2} w_{1n} \right\|_{H_1}^2 + \left\| L^{1/2} w_{0n} \right\|_{H_1}^2 + \left\| I_2^{1/2} \theta_{0n} \right\|_{H_2}^2 \right]. \tag{2.98}$$

From (2.96), one can conclude that $w < \infty$. We apply to the inequality (2.97) the Gronwill lemma,[1] and finally obtain

$$\left\| I_1^{1/2} w_n'(t) \right\|_{H_1}^2 + \left\| L^{1/2} w_n(t) \right\|_{H_1}^2 + \left\| I_2^{1/2} \theta_n(t) \right\|_{H_2}^2$$

$$+2 \int_0^t \left\| K^{1/2} \theta_n(\tau) \right\|_{H_2}^2 d\tau \le \left(w + \frac{1}{\delta} \left\| I^{-1/2} g \right\|_{L_2(0,T;H_1 \times H_2)}^2 \right) \exp(\delta T).$$

Minimizing the right-hand side of the last inequality and taking account of the fact that $\delta > 0$, we obtain the following prior estimate of an approximate solution to the problem (2.54)–(2.56):

$$\left\| I_1^{1/2} w_n'(t) \right\|_{H_1}^2 + \left\| L^{1/2} w_n(t) \right\|_{H_1}^2 + \left\| I_2^{1/2} \theta_n(t) \right\|_{H_2}^2$$

$$+2 \int_0^t \left\| K^{1/2} \theta_n(\tau) \right\|_{H_2}^2 d\tau \le m(g, w, T), \tag{2.99}$$

where

$$m(g, w, T) = \min_{\delta > 0} \left[\left(w + \frac{1}{\delta} \left\| I^{-1/2} g \right\|_{L_2(0,T;H_1 \times H_2)}^2 \right) \exp(\delta T) \right]$$

$$\equiv \left\{ w + \frac{1}{2} \left[\left(\left\| I^{-1/2} g \right\|_{L_2(0,T;H_1 \times H_2)}^4 \right) T^2 \right.\right.$$

$$\left.\left. +4 \left(\left\| I^{-1/2} g \right\|_{L_2(0,T;H_1 \times H_2)}^2 wT \right)^{1/2} + \left\| I^{-1/2} g \right\|_{L_2(0,T;H_1 \times H_2)}^2 T \right] \right\}$$

[1] The different variants of the Gronwill lemma are given in a series of monographs (see, for instance, [144, 227]). Here and later the following version of this lemma is applied: if $f \in L([0, T])$ and almost always ($t \in [0, T]$), $0 \le f(t) \le C_1 + C_2 \int_0^t f(\tau) d\tau$, where $C_1, C_2 = const \ge 0$, then almost always ($t \in [0, T]$), $f(t) \le C_1 \exp(C_2 T)$.

$$\times \exp\left\{2\left\|I^{-1/2}g\right\|^2_{L_2(0,T;H_1\times H_2)} T\left[\left(\left\|I^{-1/2}g\right\|^4_{L_2(0,T;H_1\times H_2)}\right)T^2 + 4\left(\left\|I^{-1/2}g\right\|^2_{L_2(0,T;H_1\times H_2)} wT\right)^{1/2} + \left\|I^{-1/2}g\right\|^2_{L_2(0,T;H_1\times H_2)} T\right]^{-1}\right\}. \tag{2.100}$$

The estimate (2.99) holds for all n, almost always ($t \in [0, T]$). The series of function pairs $\{w_n, \theta_n\}$ is bounded in the Hilbert space $[W_1^2(0, T; H_1) \cap L_2(0, T; D(L^{1/2}))] \times L_2(0, T; D(K^{1/2}))$ owing to (2.99) (from now on this space will be referred to as $\tilde{H}$). Because any bounded set in the Hilbert space is weakly compact, there exists a sequence of function pairs $\{w_\nu, \theta_\nu\}$ that converges weakly in $\tilde{H}$. Let us denote its weak limit by $\{w, \theta\}$, and let us derive a prior estimate of type (2.99) for the pair of functions $\{w, \theta\}$.

In order to simplify the notation, we write

$$< w, \theta; \nu, \eta; t > = (I_1^{1/2} w'(t), (I_1^{1/2} \nu'(t))_{H_1} + (L^{1/2} w(t), L^{1/2} \nu(t))_{H_1} + (I_2^{1/2} \theta(t), I_2^{1/2} \eta(t))_{H_2} + 2\int_0^t (K^{1/2}\theta(\tau), K^{1/2}\eta(\tau))_{H_2} d\tau,$$

$$\|w, \theta; t\| =< w, \theta; w, \theta; t >^{1/2}.$$

Let C be an arbitrary number greater than $m(g, w, T)$. Let M be the set of such $t \in [0, T]$ for which the following inequality holds:

$$\|w, \theta; t\| \geq C^{1/2}. \tag{2.101}$$

Let us consider the linear bounded functional Φ over $\tilde{H}$ defined by the relation

$$\Phi(\{\nu, \eta\}) = \int_M \frac{< w, \theta; \nu, \eta; t >}{\|w, \theta; t\|} dt, \quad \{\nu, \eta\} \in \tilde{H}.$$

From (2.99), we obtain

$$\Phi(\{\nu, \eta\}) = \lim_{\nu\to\infty} \Phi(\{w_\nu, \theta_\nu\})$$

$$= \lim_{\nu\to\infty} \int_M \frac{< w, \theta; w_\nu, \theta_\nu; t >}{||w, \theta; t||} dt \leq \overline{\lim_{\nu\to\infty}} \int_M \|w_\nu, \theta_\nu; t\| \, dt \leq$$

$$\leq (m(g, w, T))^{1/2} mes\, M, \tag{2.102}$$

and also we get

$$\Phi(\{w,\theta\}) = \int_M \|w,\theta;t\|\,dt \geq C^{1/2} mes\, M. \tag{2.103}$$

From (2.102), (2.103), owing to the inequality $C > m(g,w,T)$, one obtains $mes\ M = 0$. Therefore, for an arbitrary $C > m(g,w,T)$, the measure of the set of $t \in [0,T]$ for which the inequality (2.101) is satisfied equals zero. In what follows, almost always ($t \in [0,T]$), $\|w,\theta,t\| \leq (m(g,w,T))^{1/2}$, i.e. for the pair of functions $\{w,\theta\}$ a prior estimate of type (2.99) can be used:

$$\left\| I_1^{1/2} w_n'(t) \right\|_{H_1}^2 + \left\| L^{1/2} w_n(t) \right\|_{H_1}^2 + \left\| I_2^{1/2} \theta_n(t) \right\|_{H_2}^2$$
$$+2\int_0^t \left\| K^{1/2}\theta(\tau) \right\|_{H_2}^2 d\tau \leq m(g,w,T), \tag{2.104}$$

(almost always ($t \in [0,T]$)).

We now show that $\{w,\theta\}$ is a general solution to the problem (2.54)–(2.56). From (2.104), we obtain

$$w \in W_\infty^2(0,T;H_1) \cap L_\infty(0,T;D(L^{1/2})),$$
$$\theta \in L_\infty(0,T;H_2) \cap L_2(0,T;D(K^{1/2})).$$

The relation $w(0) = w_0$ and the integral identities (2.60), (2.61) are obtained on the basis of (2.95) using a limiting transition similar to that described in [214]. To sum up, the pair of functions $\{w,\theta\}$ satisfies all requirements of Definition 2.1. Therefore, the existence of a generalized solution to the problem (2.54)–(2.56) has been proved.

Now we are going to prove the uniqueness of the generalized solution to the problem (2.54)–(2.56). For this purpose it is sufficient to show that the corresponding homogeneous problem possesses only a trivial solution. Let us make the following substitutions in the integral identities (2.60), (2.61), corresponding to the case of homogeneous data:

$$g_1(t) \equiv 0,\ g_2(t) \equiv 0,\ w_0 = w_1 = 0,\ \theta_0 = 0,\ \nu = 2\nu_\tau,\ \eta = 2\eta_\tau,$$

where $0 \leq \tau \leq T$, and

$$\nu_\tau(t) = \begin{cases} \int_t^\tau w(\xi)\,d\xi, & t \leq \tau, \\ 0, & t > \tau, \end{cases}$$

$$\eta_\tau(t) = \begin{cases} \int\limits_t^\tau \int\limits_0^\xi \theta(\sigma)\, d\sigma\, d\xi, & t \le \tau, \\ 0, & t > \tau. \end{cases}$$

We add together the relations obtained, perform integration by parts and use the condition (2.58) again. As a result, we obtain

$$\left\| I_1^{1/2} w_n(\tau) \right\|_{H_1}^2 + \left\| L^{1/2} \int\limits_0^\tau w(\xi)\, d\xi \right\|_{H_1}^2$$

$$+ \left\| I_1^{1/2} \int\limits_0^\tau \theta(\sigma)\, d\sigma \right\|_{H_2}^2 + 2 \int\limits_0^\tau \left\| K^{1/2} \int\limits_0^t \theta(\sigma)\, d\sigma \right\|_{H_2}^2 dt = 0.$$

This implies that $w(\tau) \equiv 0$, $\theta(\tau) \equiv 0$, $0 \le \tau \le T$, i.e. a homogeneous problem of the form (2.54)–(2.56) possesses only a trivial solution. Therefore, the uniqueness of the generalized solution to the problem (2.54)–(2.56) has been established. Therefore, Theorem 2.1 has been fully proved.

Remark 2.5 In Chap. 3, in deriving a prior estimate for the Bubnov–Galerkin method, the case of homogeneous initial conditions (2.56) will be considered: $w(0) = w'(0) = 0$, $\theta(0) = 0$. Therefore, we take $w_{0n} = w_{1n} = 0$, $\theta_{0n} = 0$ in (2.95) also. It should be noted that if w as defined by (2.98) is constant, then

$$m(g, w, T) = eT \left\| I^{-1/2} g \right\|_{L_2(0,T,H_1 \times H_2)}^2,$$

and the estimates (2.99), (2.104) are reduced to the form

$$\left\| I_1^{1/2} w_n'(t) \right\|_{H_1}^2 + \left\| L^{1/2} w_n(t) \right\|_{H_1}^2 + \left\| I_2^{1/2} \theta_n(t) \right\|_{H_2}^2$$

$$+ 2 \int\limits_0^t \left\| K^{1/2} \theta_n(\tau) \right\|_{H_2}^2 d\tau \le eT \left\| I^{-1/2} g \right\|_{L_2(0,T;H_1 \times H_2)}^2, \tag{2.105}$$

$$\left\| I_1^{1/2} w_n'(t) \right\|_{H_1}^2 + \left\| L^{1/2} w(t) \right\|_{H_1}^2 + \left\| I_2^{1/2} \theta(t) \right\|_{H_2}^2$$

$$+ 2 \int\limits_0^t \left\| K^{1/2} \theta(\tau) \right\|_{H_2}^2 d\tau \le eT \left\| I^{-1/2} g \right\|_{L_2(0,T;H_1 \times H_2)}^2, \tag{2.106}$$

almost always ($t \in [0, T]$).

In our further considerations, special systems composed of eigenelements of certain operators (convergent and making an acute angle with the operators L, K) will be used as the bases for the Bubnov–Galerkin method. Let us introduce the definitions needed.

Definition 2.3 ([360]) The self-adjoint positive definite operators A and A_0 acting in the same Hilbert space H are called convergent if $D(A) = D(A_0)$.

It is known (see [360]) that if the operators A and A_0 are convergent, then $D(A^{1/2}) = D(A_0^{1/2})$, and all of the operators AA_0^{-1}, $A^{-1}A_0$, $A^{1/2}A_0^{-1/2}$, $A^{-1/2}A_0^{1/2}$ are bounded in H, together with their inverses.

Definition 2.4 ([192]) Let A, A_0 be self-adjoint positive definite operators acting in the same Hilbert space. The operators A and A_0 are said to make an acute angle if

$$\gamma(A, A_0) \equiv \inf_{u \in D(A) \cap D(A_0)} \frac{(Au, A_0 u)_H}{\|A_0 u\|_H^2} > 0. \tag{2.107}$$

Let L_0, K_0 be self-adjoint positive definite operators that are convergent and make an acute angle with the operators L and K, respectively. Let $\{v_i\}_{i=1}^{\infty}$ be a basis in H_1 composed of eigenelements of the operator L_0, and let $\{\eta_j\}_{j=1}^{\infty}$ be a basis in H_2 composed of eigenelements of the operator K_0. The eigenelements of the operators L_0 and K_0 will be denoted by λ_i and $æ_j$, respectively: $L_0 v_i = \lambda_i v_i$, $K_0 \eta_j = æ_j \eta_j$, $i, j = 1, 2, \ldots$. It should be noted that because the operators L^{-1}, K^{-1} are fully continuous, the operators L_0^{-1}, K_0^{-1} are also fully continuous, which implies that

$$\lambda_i \to \infty, \quad æ_j \to \infty \quad (i, j \to \infty). \tag{2.108}$$

The systems of elements $\{v_i\}_{i=1}^{\infty}$, $\{\eta_j\}_{j=1}^{\infty}$ will be used in what follows as the basis for the equations of the Bubnov–Galerkin method (2.95). As the initial data for the problem (2.95), we use the projections of the initial data (2.56) onto the finitely measurable subspaces:

$$w_{0n} = P_{H_1^n} w_0, \quad w_{1n} = P_{H_1^n} w_1, \quad \theta_{0n} = P_{H_1^n} \theta_0. \tag{2.109}$$

The following stronger conditions are applied to the given functions g_1, g_2 and the elements w_0, w_1, θ_0 (instead of (2.57) and (2.59)):

$$g_1 \in W_2^1(0, T; H_1), \quad g_2 \in W_2^1(0, T; H_2), \tag{2.110}$$

$$w_0 \in D(L), \quad w_1 \in D(L^{1/2}), \quad \theta_0 \in D(K). \tag{2.111}$$

Theorem 2.2 *If the conditions (2.58), (2.110), (2.111) are satisfied then the "classical" solution to the problem (2.54)–(2.56) exists and is unique.*

Proof Let us define the existence of a "classical" solution to the problem (2.54)–(2.56). We differentiate the first two relations of (2.95), perform scalar multiplication by $2w_n''(t)$ and $2\theta_n'(t)$, and sum the resulting expression with the use of (2.58). We integrate expression obtained over the interval $[0, T]$, where $t \leq T$, and finally obtain

$$\left\|I_1^{1/2} w_n''(t)\right\|_{H_1}^2 + \left\|L^{1/2} w_n'(t)\right\|_{H_1}^2 + \left\|I_2^{1/2} \theta_n(t)\right\|_{H_2}^2$$

$$+ 2\int_0^t \left\|K^{1/2}\theta_n'(\tau)\right\|_{H_2}^2 d\tau = \left\|I_1^{1/2} w_n''(0)\right\|_{H_1}^2 + \left\|L^{1/2} w_{1n}\right\|_{H_1}^2$$

$$+ \left\|I_1^{1/2}\theta_n'(0)\right\|_{H_1}^2 + 2\int_0^t \left[(g_1'(\tau), w_n''(\tau))_{H_1} + (g_2'(\tau), \theta_n''(\tau))_{H_2}\right] d\tau.$$

After applying the Cauchy inequality to the right-hand side of the equation obtained, we obtain

$$\left\|I_1^{1/2} w_n''(t)\right\|_{H_1}^2 + \left\|L^{1/2} w_n'(t)\right\|_{H_1}^2 + \left\|I_2^{1/2} \theta_n'(t)\right\|_{H_2}^2$$

$$+ 2\int_0^t \left\|K^{1/2}\theta_n'(\tau)\right\|_{H_2}^2 d\tau \leq \chi + \frac{1}{\delta}\left\|I^{-1/2} q'\right\|_{L_2(0,T;H_1\times H_2)}^2 \tag{2.112}$$

$$+ \delta\int_0^t \left[\left\|I_1^{1/2} w_n''(\tau)\right\|_{H_1}^2 + \left\|I_1^{1/2}\theta_n'(\tau)\right\|_{H_2}^2\right] d\tau,$$

where $I^{1/2}g' = \{I_1^{1/2}g_1',\ I_2^{1/2}g_n'\}$, δ is an arbitrary positive number, and

$$\chi = \sup_n \left[\left\|I_1^{1/2} w_n''(0)\right\|_{H_1}^2 + \left\|L^{1/2} w_{1n}\right\|_{H_1}^2 + \left\|I_1^{1/2}\theta_n'(0)\right\|_{H_2}^2\right]. \tag{2.113}$$

From the first two relations of (2.95), taking into account the initial data $w_n(0) = w_{0n}$, $w_n'(0) = w_{1n}$, $\theta_n(0) = \theta_{0n}$, we obtain the following inequalities:

$$\left\|I_1^{1/2} w_n''(0)\right\|_{H_1} \leq \left\|I_1^{-1/2}(g_1(0) - Lw_{0n} - M\theta_{0n})\right\|_{H_1}, \tag{2.114}$$

$$\left\|I_1^{1/2}\theta_n'(0)\right\|_{H_2} \leq \left\|I_1^{-1/2}(g_2(0) - K\theta_{0n} - Nw_{1n})\right\|_{H_2}. \tag{2.115}$$

Owing to the special choice of basis systems $\{v_i\}_{i=1}^{\infty}$, $\{\eta_j\}_{j=1}^{\infty}$, (2.109) and (2.111) imply that

$$Lw_{0n} \to Lw_0, \quad L^{1/2}w_{1n} \to L^{1/2}w_1 \text{ in } H_1, \quad K\theta_{0n} \to K\theta \text{ in } H_2. \tag{2.116}$$

From (2.113)–(2.116) and the boundedness conditions on the operators MK^{-1}, $NL^{-1/2}$, the following estimate is obtained:

$$\begin{aligned} \chi \le \sup_n \Bigg[& \left\| I_1^{-1/2}(g_1(0) - Lw_{0n} - M\theta_{0n}) \right\|_{H_1}^2 + \left\| L^{1/2}w_{1n} \right\|_{H_1}^2 \\ & + \left\| I_1^{-1/2}(g_2(0) - K\theta_{0n} - Nw_{1n}) \right\|_{H_2}^2 \Bigg] < \infty. \end{aligned} \tag{2.117}$$

Applying the Gronwill lemma to the inequality (2.112) leads to the result

$$\begin{aligned} & \left\| I_1^{1/2} w_n''(t) \right\|_{H_1}^2 + \left\| L^{1/2} w_n'(t) \right\|_{H_1}^2 + \left\| I_2^{1/2} \theta_n(t) \right\|_{H_2}^2 \\ & + 2\int_0^t \left\| K^{1/2}\theta_n'(\tau) \right\|_{H_2}^2 d\tau \le \chi + \frac{1}{\delta} \left(\left\| I^{-1/2} q' \right\|_{L_2(0,T;H_1\times H_2)}^2 \right) \exp(\delta T). \end{aligned}$$

Minimizing the right-hand side of the last inequality and taking account of the fact that $\delta > 0$, we obtain the following prior estimate of the approximate solution $\{w_n, \theta_n\}$ to the problem (2.54)–(2.56):

$$\begin{aligned} & \left\| I_1^{1/2} w_n''(t) \right\|_{H_1}^2 + \left\| L^{1/2} w_n'(t) \right\|_{H_1}^2 + \left\| I_2^{1/2} \theta_n'(t) \right\|_{H_2}^2 \\ & \qquad + 2\int_0^t \left\| K^{1/2}\theta_n'(\tau) \right\|_{H_2}^2 d\tau \le m(g_1, \chi, T), \end{aligned} \tag{2.118}$$

where $m(\cdot,\cdot,\cdot)$ is defined by (2.100). The estimate (2.118) holds for every n almost always ($t \in [0, T]$).

Furthermore, owing to the definition (2.107) of the quantities $\gamma(L, L_0)$, $\gamma(K, K_0)$ for the pairs of operators L, L_0 and K, K_0, the following relations are obtained from the first two expressions of (2.95):

$$\begin{aligned} & \gamma(L, L_0)\, \|L_0 w_n(t)\|_{H_1}^2 \le (Lw_n(t), L_0 w_n(t))_{H_1} \\ & = (g_1(t) - I_1 w_n''(t) - M\theta_n(t), L_0 w_n(t))_{H_1} \\ & \le \Bigg[\|g_1(t)\|_{H_1} + \left\| I_1^{1/2} \right\|_{H_1\to H_1} \left\| I_1^{1/2} w_n''(t) \right\|_{H_1} \\ & + \left\| MK^{-1} \right\|_{H_2\to H_1} \|K\theta_n(t)\|_{H_2} \Bigg] \|L_0 w_n(t)\|_{H_1}, \end{aligned} \tag{2.119}$$

$$\gamma(K, K_0)\, \|K_0\theta_n(t)\|^2_{H_2} \le (K\theta_n(t), K_0\theta_n(t))_{H_2}$$
$$= (g_2(t) - I_2\theta_n'(t) - Nw_n'(t), K_0\theta_n(t))_{H_2} \le \left[\left\| g_2(t) \right\|_{H_2} \right. \tag{2.120}$$
$$\left. + b_1 \left(\left\| I_2^{1/2}\theta_n'(t) \right\|^2_{H_2} + \left\| L^{1/2}w_n'(t) \right\|^2_{H_1} \right)^{1/2} \right] \left\| K_0\theta_n(t) \right\|_{H_2},$$

where

$$B_1 = \left(\left\| I_2^{1/2} \right\|^2_{H_2\to H_2} + \left\| NL^{-1/2} \right\|^2_{H_1\to H_2} \right)^{1/2}. \tag{2.121}$$

Applying the inequality of an acute angle (2.107) to the operator pairs L, L_0 and K, K_0 and applying the boundedness condition on the operators LL_0^{-1}, KK_0^{-1}, the following prior estimate of the components of the approximate solution to the problem (2.54)–(2.56) are obtained from (2.119) and (2.120):

$$\|Lw_n(t)\|_{H_1} \le c_L \left[\|g_1(t)\|_{H_1} + \left\| MK^{-1} \right\|_{H_2\to H_1} c_K\, \|g_2(t)\|_{H_2} \right.$$
$$\left. + b_2 \left(\left\| I_2^{1/2}w_n''(t) \right\|^2_{H_1} + \left\| L^{1/2}w_n'(t) \right\|^2_{H_1} + \left\| I_2^{1/2}\theta_n'(t) \right\|^2_{H_2} \right)^{1/2} \right], \tag{2.122}$$

$$\|K\theta_n(t)\|_{H_2} \le c_K\, \|g_2(t)\|_{H_2} + b_1 \left(\left\| I_2^{1/2}\theta_n'(t) \right\|^2_{H_2} + \left\| L^{1/2}w_n'(t) \right\|^2_{H_1} \right)^{1/2}, \tag{2.123}$$

where

$$c_L = \frac{\|LL_0^{-1}\|_{H_1\to H_1}}{\gamma(LL_0)}, \quad c_K = \frac{\|KK_0^{-1}\|_{H_2\to H_2}}{\gamma(KK_0)}, \tag{2.124}$$

$$b_2 = \left(\|MK^{-1}\|^2_{H_2\to H_1} c_k^2 b_1^2 + \|I^{1/2}\|^2_{H_1\to H_1} \right)^{1/2}. \tag{2.125}$$

The estimates (2.122), (2.123) are satisfied for all n, almost always ($t \in [0, T]$).

Let $\tilde{\tilde{H}}$ be the Hilbert space $[W_2^2(0, T; H_1) \cap W_2^1(0, T; D(L^{1/2})) \cap L_2(0, T; D(L))] \times [W_2^1(0, T; D(K^{1/2})) \cap L_2(0, T; D(K))]$. According to the prior estimates (2.118), (2.122), (2.123), the sequence of approximate solutions $\{w_n, \theta_n\}$ is bounded in $\tilde{\tilde{H}}$. Because any bounded set in a Hilbert space is weakly compact, a special subsequence of approximate solutions $\{w_\nu, \theta_\nu\}$ exists which is weakly convergent in $\tilde{\tilde{H}}$. Let us denote its weak limit by $\{w, \theta\}$. We shall show that the pair of functions $\{w, \theta\}$ is the "classical" solution to the problem (2.54)–(2.56).

In a manner analogous to that used in the proof of Theorem 2.1, an estimate of (w, θ) similar to (2.104) can be obtained from (2.99). Taking into account (2.118), (2.122) and (2.123), the following prior estimates are obtained:

$$\left\| I_1^{1/2} w_n''(t) \right\|_{H_1}^2 + \left\| L^{1/2} w_n'(t) \right\|_{H_1}^2 + \left\| I_2^{1/2} \theta_n'(t) \right\|_{H_2}^2 + 2 \int_0^t \left\| K^{1/2} \theta_n'(\tau) \right\|_{H_2}^2 d\tau \le m(g', \chi, T), \tag{2.126}$$

$$\| L w(t) \|_{H_1} \le const < \infty, \quad \| K w(t) \|_{H_2} \le const < \infty, \tag{2.127}$$

valid almost always ($t \in [0, T]$) (the last two estimates were obtained using $g_1 \in L_\infty(0, T; H_1)$, $g_2 \in L_\infty(0, T, H_2)$, using the conditions (2.110)). The estimates (2.126), (2.127) imply

$$w \in W_\infty^2(0, T; H_1) \cap W_\infty^1(0, T; D(L^{1/2}) \cap L_\infty(0, T; D(L)),$$

$$\theta \in W_\infty^1(0, T; H_2) \cap W_2^1(0, T; D(K^{1/2}) \cap L_\infty(0, T; D(K)).$$

The uniqueness of the "classical" solution of the problem (2.54)–(2.56) follows from Theorem 2.1, because if the "classical" solution to the problem (2.54)–(2.56) exists then it is also the general solution. Thus, Theorem 2.2 has been proved.

Remark 2.6 For later convenience, we evaluate separately the variants of the prior estimates (2.118), (2.126) which correspond to the case of homogeneous initial conditions (2.56). In this case we take $w_{0n} = w_{1n} = 0$, $\theta_{0n} = 0$ in (2.95), and therefore the estimate (2.117) of the constant χ defined by (2.113) implies $x \le \| I^{-1/2} g(0) \|_{H_1 \times H_2}^2$. Using the latter inequality, we make an estimate from above of the right-hand sides of the inequalities (2.118), (2.126), and we obtain

$$\left\| I_1^{1/2} w_n''(t) \right\|_{H_1}^2 + \left\| L^{1/2} w_n'(t) \right\|_{H_1}^2 + \left\| I_2^{1/2} \theta_n'(t) \right\|_{H_2}^2 + 2 \int_0^t \left\| K^{1/2} \theta_n'(\tau) \right\|_{H_2}^2 d\tau \le M_g(T), \tag{2.128}$$

$$\left\| I_1^{1/2} w''(t) \right\|_{H_1}^2 + \left\| L^{1/2} w'(t) \right\|_{H_1}^2 + \left\| I_2^{1/2} \theta'(t) \right\|_{H_2}^2 + 2 \int_0^t \left\| K^{1/2} \theta'(\tau) \right\|_{H_2}^2 d\tau \le M_g(T), \tag{2.129}$$

for $t \in [0, T]$, where

$$
\begin{aligned}
M_g(T) &= m\left(g', \left\| I^{-1/2} g(0) \right\|^2_{H_1 \times H_2}, T\right) \\
&\equiv \left\{ \left\| I^{-1/2} g(0) \right\|^2_{H_1 \times H_2} + \frac{1}{2} \left[\left(\left\| I^{-1/2} g' \right\|^4_{L_2(0,T;H_1 \times H_2)} T^2 \right.\right.\right. \\
&\left. + 4 \left\| I^{-1/2} g(0) \right\|^2_{H_1 \times H_2} \left\| I^{-1/2} g' \right\|^2_{L_2(0,T;H_1 \times H_2)} T \right)^{1/2} \\
&\left.\left. + \left\| I^{-1/2} g' \right\|^2_{L_2(0,T;H_1 \times H_2)} T \right] \right\} \exp \left\{ 2 \left\| I^{-1/2} g' \right\|^2_{L_2(0,T;H_1 \times H_2)} \right. \\
&\times T \left[\left(\left\| I^{-1/2} g' \right\|^4_{L_2(0,T;H_1 \times H_2)} T^2 + 4 \left\| I^{-1/2} g' \right\|^2_{L_2(0,T;H_1 \times H_2)} \right. \right. \\
&\left.\left.\left. \times \left\| I^{-1/2} g(0) \right\|^2_{H_1 \times H_2} T \right)^{1/2} + \left\| I^{-1/2} g' \right\|^2_{L_2(0,T;H_1 \times H_2)} T \right]^{-1} \right\}.
\end{aligned}
\tag{2.130}
$$

Finally, a theorem about the existence and uniqueness of solutions to coupled thermoelastic problems of shallow shells derived from Theorems 2.1 and 2.2 will be formulated.

Theorem 2.3 *If the conditions (2.63)–(2.68) are satisfied, then the initial–boundary value problem for the system of differential equations (2.20)–(2.23) with the initial conditions (2.33)–(2.37) and homogeneous boundary conditions has a unique (general) solution $\{u, v, w, \Psi_x, \Psi_y, \theta\}$ satisfying the conditions*

$$
\begin{aligned}
\{u, v, w, \Psi_x, \Psi_y\} &\in W^1_\infty(0, T; (L_2(\Omega_1))^5) \cap L_\infty(0, T; (\overset{\circ}{W}{}^1_2(\Omega_1))^5), \\
\theta &\in L_\infty(0, T; L_2(\Omega_2)) \cap L_2(0, T; \overset{\circ}{W}{}^1_2(\Omega_2)).
\end{aligned}
$$

Theorem 2.4 *If the conditions (2.63), (2.64) and*

$$
\begin{aligned}
p_1, p_2, q_1 \in W^1_2(0, T; L_2(\Omega_1)),\ & q_2 \in W^1_2(0, T; L_2(\Omega_2)), \\
\overline{\omega}_0 = \{u_0, v_0, w_0, \Psi_{x0}, \Psi_{y0}\} &\in (\overset{\circ}{W}{}^2_2(\Omega_1))^5, \\
\overline{\omega}_1 = \{u_1, v_1, w_1, \Psi_{x1}, \Psi_{y1}\} &\in (\overset{\circ}{W}{}^1_2(\Omega_1))^5, \\
\theta &\in \overset{\circ}{W}{}^2_2(\Omega_2),
\end{aligned}
$$

are satisfied, then the initial–boundary value problem for the system of differential equations (2.20)–(2.23) with the initial conditions (2.33)–(2.37) and homogeneous boundary conditions has a unique ("classical") solution $\{u, v, w, \Psi_x, \Psi_y, \theta\}$ *satisfying the conditions*

$$\{u, v, w, \Psi_x, \Psi_y\} \in W_\infty^2(0, T; (L_2(\Omega_1))^5)$$
$$\cap\, W_\infty^1(0, T; (\overset{\circ}{W}{}_2^1(\Omega_1))^5) \cap L_\infty(0, T; (\overset{\circ}{W}{}_2^2(\Omega_1))^5),$$
$$\theta \in W_\infty^1(0, T; L_2(\Omega_2)) \cap W_2^1(0, T; \overset{\circ}{W}{}_2^1(\Omega_2))$$
$$\cap\, L_\infty(0, T; \overset{\circ}{W}{}_2^2(\Omega_2)).$$

Theorem 2.5 *If the conditions (2.64), (2.65), (2.68) and (2.83)–(2.85) are satisfied, then the initial–boundary value problem for the system of differential equations (2.30)–(2.32) with the initial conditions (2.33)–(2.35) and homogeneous boundary conditions has a unique (general) solution* $\{u, v, w, \theta\}$ *satisfying the conditions*

$$\{u, v, w\} \in W_\infty^1(0, T; (L_2(\Omega_1))^3) \cap L_\infty(0, T; (\overset{\circ}{W}{}_2^1(\Omega_2))^2 \times \overset{\circ}{W}{}_2^2(\Omega_1)),$$
$$\theta \in L_\infty(0, T; L_2(\Omega_2)) \cap L_2(0, T; \overset{\circ}{W}{}_2^1(\Omega_2)).$$

Theorem 2.6 *If the conditions (2.64), (2.83) and*

$$p_1, p_2, q_1 \in W_2^1(0, T; L_2(\Omega_1)),\ q_2 \in W_2^1(0, T; L_2(\Omega_2)),$$
$$\overline{\omega}_0 = \{u_0, v_0, w_0\} \in (\overset{\circ}{W}{}_2^2(\Omega_1))^2 \times \overset{\circ}{W}{}_2^4(\Omega_1),$$
$$\overline{\omega}_1 = \{u_1, v_1, w_1\} \in (\overset{\circ}{W}{}_2^1(\Omega_1))^2 \times \overset{\circ}{W}{}_2^2(\Omega_1),$$
$$\theta_0 \in \overset{\circ}{W}{}_2^2(\Omega_2),$$

are satisfied, then the initial–boundary value problem for the system of differential equations (2.30)–(2.32) with the initial conditions (2.33)–(2.35) and homogeneous boundary conditions has a unique ("classical") solution $\{u, v, w, \theta\}$ *satisfying the conditions*

$$\{u, v, w\} \in W_\infty^2(0, T; (L_2(\Omega_1))^3)$$
$$\cap\, W_\infty^1(0, T; (\overset{\circ}{W}{}_2^1(\Omega_1))^2 \times \overset{\circ}{W}{}_2^2(\Omega_1))$$
$$\cap\, L_\infty(0, T; (\overset{\circ}{W}{}_2^2(\Omega_1))^2 \times \overset{\circ}{W}{}_2^4(\Omega_1)),$$

$$\theta \in W_\infty^1(0, T; L_2(\Omega_2)) \cap W_2^1(0, T; \overset{\circ}{W}{}_2^1(\Omega_2)) \cap L_\infty(0, T; L_2\overset{\circ}{W}{}_2^2(\Omega_1)).$$

These results can be obtained directly from Theorems 2.1 and 2.2 when either the relations (2.69)–(2.81) (for (2.20)–(2.23) with the initial conditions (2.33)–(2.37)) or the relations (2.86)–(2.94) (for (2.30)–(2.32) with the initial conditions (2.33)–(2.35)) hold.

2.2 Cylindrical Panel Within Transonic Gas Flow

2.2.1 Statement and Solution of the Problem

2.2.1.1 Equations of Motion and Boundary Conditions of a Panel

We start with the fundamental relations governing the dynamics of an elastic, infinite, cylindrical panel. We assume that the shell has a constant thickness, is shallow and slender, and is made of an isotropic material for which Hooke's law holds. We assume also that the deformations are small and that the Kirchhoff–Love hypothesis hold [338]. In addition, we neglect the stresses normal to the middle surface. We assume that the properties of the shell material do not depend on temperature, and we consider dynamic processes to occur in the shell without propagation of longitudinal elastic waves over a finite distance (i.e. we assume that the velocity of those waves is infinitely high). A right-handed Cartesian coordinate system x, y, z is used. The z axis coincides with the normal to the middle surface of the shell, with its direction pointing towards the initial centre of curvature. The cylindrical panel, infinitely long, is part of a shallow cylinder. The panel has rectangular sides along generators of the cylinder. It is assumed that all physical properties in all perpendicular cross-sections, i.e. cross-sections perpendicular to the generators of the infinite cylindrical shell, are the same. Therefore, all of the quantities are independent of coordinate x. In this case we take

$$u = 0, \quad P_x = 0.$$

The equations governing the motion of an elastic, infinite, cylindrical panel have been cited in many publications. However, those nonlinear equations were first derived by Mushtari and Galimov [259]. Following [338], we briefly sketch the derivation of the equations for such a panel starting from a variational principle.

The Hamilton–Ostrogradskij principle has the form

$$\delta \int_{t_0'}^{t_1'} (K - \Pi)dt' + \delta' \int_{t_0'}^{t_1'} E\, dt' = 0. \tag{2.131}$$

The geometrical relations for the deformations can be described through the displacements of the middle surface of the infinite cylindrical panel in the following form:

$$\gamma_{11}^{z} = 0, \quad \gamma_{12}^{z} = 0, \quad \gamma_{23}^{z} = 0,$$
$$\gamma_{23}^{z} = -\frac{h^2 \partial^3 w'}{6(1-\nu)\partial y^3},$$
$$\gamma_{22}^{z} = \frac{\partial v}{\partial y} - \frac{w'}{R} + \frac{1}{2}\left(\frac{\partial w'}{\partial y}\right)^2 - z\left(\frac{\partial^2 w'}{\partial y^2}\right), \tag{2.132}$$
$$\gamma_{33}^{z} = -\frac{\nu}{1-\nu}\left(\frac{\partial v}{\partial y} - \frac{w'}{R} + \frac{1}{2}\left(\frac{\partial w'}{\partial y}\right)^2 - z\frac{\partial^2 w'}{\partial y^2}\right).$$

By applying Hooke's law to the stresses, the following formulas are obtained:

$$P_{11}^{z} = \frac{E'\nu}{1-\nu^2}\left(\frac{\partial v}{\partial y} - \frac{w'}{R} + \frac{1}{2}\left(\frac{\partial w'}{\partial y}\right)^2 - z\frac{\partial^2 w'}{\partial y^2}\right),$$
$$P_{12}^{z} = 0, \quad P_{13}^{z} = 0, \quad P_{23}^{z} = -\frac{E'h^2}{12(1-\nu^2)}\frac{\partial^3 w'}{\partial y^3}, \quad P_{33}^{z} = 0, \tag{2.133}$$
$$P_{22}^{z} = \frac{E'}{1-\nu^2}\left(\frac{\partial v}{\partial y} - \frac{w'}{R} + \frac{1}{2}\left(\frac{\partial w'}{\partial y}\right)^2 - z\frac{\partial^2 w'}{\partial y^2}\right).$$

We now define the internal forces and stresses averaged over a transverse cross-section, parallel to the y axis, of the infinite cylindrical shell. Substituting the expressions in (2.133) into the formulas for the internal forces and moments, we obtain

$$N_x = \frac{E'h\nu}{1-\nu^2}\left(\frac{\partial v}{\partial y} - \frac{w'}{R} + \frac{1}{2}\left(\frac{\partial w'}{\partial y}\right)^2\right),$$
$$N_y = \frac{E'h}{1-\nu^2}\left(\frac{\partial v}{\partial y} - \frac{w'}{R} + \frac{1}{2}\left(\frac{\partial w'}{\partial y}\right)^2\right),$$
$$Q_x = 0,$$
$$Q_y = -\frac{E'h^3}{12(1-\nu^2)}\frac{\partial^3 w'}{\partial y^3}, \tag{2.134}$$
$$T = 0,$$
$$M_x = -\frac{E'h^3\nu}{12(1-\nu^2)}\frac{\partial^2 w'}{\partial y^2},$$

$$M_y = -\frac{E'h^3}{12(1-\nu^2)}\frac{\partial^2 w'}{\partial y^2},$$

$$M_k = 0.$$

From (2.131), the following variational equation for the infinite circular cylindrical shell in the time interval from t'_0 to t'_1 is obtained:

$$\int_{t'_0}^{t'_1}\int_0^L\int_{-\infty}^{+\infty}\left\{\left(\frac{\partial N_y}{\partial y}+P_y\right)\delta v+\left(\frac{N_y}{R}+P'_z-\rho'_0 h\frac{\partial^2 w'}{\partial t'^2}+\frac{\partial}{\partial y}\left(N_y\frac{\partial w'}{\partial y}\right)\right.\right.$$
$$\left.\left.+\frac{\partial Q_y}{\partial y}\right)\delta w'-\left(\frac{\partial M_y}{\partial y}-Q_y\right)\delta\left(\frac{\partial w'}{\partial y}\right)\right\}dx\,dy\,dt'$$
$$-\int_{t'_0}^{t'_1}\int_{-\infty}^{+\infty}\left\{N_y\delta v+\left(N_y\frac{\partial w'}{\partial y}+Q_y\right)\delta w'-M_y\delta\left(\frac{\partial w'}{\partial y}\right)\right\}\Bigg|_{y=0}^{y=L}dx\,dt' \tag{2.135}$$
$$+\rho'_0 h\int_0^L\int_{-\infty}^{+\infty}\left(\frac{\partial v}{\partial t'}\delta v+\frac{\partial w'}{\partial t'}\delta w'\right)\Bigg|_{t'=t'_0}^{t'=t'_1}dx\,dy=0.$$

In this equation the force N_y, the moment M_y and the perpendicular force Q_y are defined via the displacements of the middle surface through the formulas (2.134). In (2.135), all variations δv, $\delta w'$, $\delta\left(\frac{\partial w'}{\partial y}\right)$ are considered as functions of the time t'. Therefore, from the first integral of (2.135) we obtain the following system of three differential equations governing the motion of the infinite circular cylindrical panel:

$$\frac{\partial N_y}{\partial y}+P_y=0, \tag{2.136}$$

$$\frac{\partial Q_y}{\partial y}+\frac{N_y}{R}+\frac{\partial}{\partial y}\left(N_y\frac{\partial w'}{\partial y}\right)+P'_z-\rho'_0 h\frac{\partial^2 w'}{\partial t'^2}=0, \tag{2.137}$$

$$Q_y-\frac{\partial M_y}{\partial y}=0. \tag{2.138}$$

From the last Eq. (2.138) and the relations (2.134), one can observe that (2.138) is identically satisfied.

From the second integral of the variational equation (2.135), we deduce that the following three boundary conditions on the boundaries $y=0$, $y=L$ of the panel must be satisfied:

1. The value of the external load acting parallel to the y axis or the values of the displacements of the edge points of the middle surface in the direction of y axis are given:

$$N_y = N_y^0(t') \quad \text{or} \quad v = v^0(t'). \tag{2.139}$$

2. The value of the external transverse load or of the normal displacement of the edge points is given:

$$Q_y + \left(N_y \frac{\partial w'}{\partial y} \right) = Q_y^0(t') \quad \text{or} \quad w' = w^0(t'). \tag{2.140}$$

3. The external bending moment or the angle of rotation of the normal to an element of the shell around an axis are given:

$$M_y = M_y^0(t') \quad \text{or} \quad \frac{\partial w'}{\partial y} = \alpha_0(t'), \tag{2.141}$$

where $N_y^0(t')$, $v_y^0(t')$, $Q_y^0(t')$, $M_y^0(t')$ and $\alpha_y^0(t')$ are time-dependent given functions.

The third integral of the variational equation (2.135) requires initial and boundary conditions, i.e. conditions for either $t' = t_0'$ or $t' = t_1'$:

1. The displacements of each point of the panel along the y axis or their velocities are given:

$$v_y = v_0(y) \quad \text{or} \quad \frac{\partial v}{\partial t'} = v_0'(y). \tag{2.142}$$

2. The normal (to the middle surface) displacements of each point of the panel or their velocities are given:

$$w' = w_0'(y) \quad \text{or} \quad \frac{\partial w'}{\partial t'} = w_0''(y), \tag{2.143}$$

where $v_0(y)$, $v_0'(y)$, $w_0'(y)$ and $w_0''(y)$ are given functions dependent on y.

We assume that the final conditions are applied at an infinitely large time $t_1' \to +\infty$. Therefore, the final conditions do not influence the solution and can be neglected.

If a simple support (on balls) along the linear boundaries is applied, such that only a rotation of the panel around a line parallel to the x axis is possible, then the boundary conditions (2.139), (2.140), (2.141) read

$$v = 0, \; w' = 0, \; M_y = 0 \quad (y = 0, \; y = L).$$

Using the formula for M_y from (2.134), the above condition can be rewritten in the following way:

$$v = 0,\ w' = 0,\ \frac{\partial^2 w'}{\partial y^2} = 0 \qquad (y = 0,\ y = L). \tag{2.144}$$

When a rigid panel support (clamp) is applied along the linear edges, the boundary conditions (2.139), (2.140), (2.141) have the form

$$v = 0,\ w' = 0,\ \frac{\partial w'}{\partial y} = 0 \qquad (y = 0,\ y = L). \tag{2.145}$$

Both of the boundary conditions (2.144) and (2.145) will be considered later in a problem of transonic aeroelasticity. Integrating (2.136) along the y variable, we obtain

$$N_y(y, t') = -\int_0^y P_y(y, t')\,dy + N_y^0(t'), \tag{2.146}$$

where $N_y^0(t')$ is an arbitrary function of t', and $N_y(0, t') = N_y^0(t')$.

Substituting the expressions (2.134) and (2.136) into the equation of motion (2.137), we obtain the following differential equation:

$$-\frac{E'h^3}{12(1-\nu^2)}\frac{\partial^4 w'}{\partial y^4} + \frac{N_y}{R} - P_y\frac{\partial w'}{\partial y} + N_y\frac{\partial^2 w'}{\partial y^2} + P_z' - \rho_0' h\frac{\partial^2 w'}{\partial t'^2} = 0. \tag{2.147}$$

Now the expressions for N_y from (2.134) are substituted into (2.136) and (2.137), and we obtain the following system:

$$\frac{\partial^2 v}{\partial y^2} + \frac{\partial w'}{\partial y}\frac{\partial^2 w'}{\partial y^2} - \frac{1}{R}\frac{\partial w'}{\partial y} = \frac{1-\nu^2}{E'h}P_y, \tag{2.148}$$

$$\begin{aligned}\frac{E'h^3}{12(1-\nu^2)}\frac{\partial^4 w'}{\partial y^4} - \frac{E'h}{1-\nu^2}\left(\frac{1}{R} + \frac{\partial^2 w'}{\partial y^2}\right)\left(\frac{\partial v}{\partial y} - \frac{w'}{R} + \left(\frac{\partial w'}{\partial y}\right)^2\right)&\\ + P_y\frac{\partial w'}{\partial y} + \rho_0' h\frac{\partial^2 w'}{\partial t^2} = P_z'.&\end{aligned} \tag{2.149}$$

This governs the motion of the infinite circular cylindrical panel under consideration with a continuously distributed load P_y in the direction of the y axis. Here we consider the interaction of the panel with a flow of an ideal gas. Neglecting tangential damping forces in the gas flow, we obtain

$$P_y = 0. \tag{2.150}$$

Therefore, in accordance with (2.146), the normal force N_y in an arbitrary cross-section $y = const$ is constant at any given time instant t', i.e. the normal internal force N_y depends only on the time t' and is equal to the compressional longitudinal force on the boundary $y = 0$ or to the modulus of the load on the boundary $y = L$ acting in the opposite direction.

Therefore, the assumptions about the infinite length of the cylindrical panel, the stress state discussed above and the infinite velocity of elastic waves along the shell lead to the result that the normal average force N_y at a fixed time moment t' in an arbitrary cross-section $y = const$ is constant. From the second relation of (2.134), by integration from 0 to y, we obtain

$$v = v|_{y=0} + \frac{1-\nu^2}{E'h}\int_0^y N_y\,dy + \int_0^y \frac{w'}{R}dy - \frac{1}{2}\int_0^y \left(\frac{\partial w'}{\partial y}\right)^2 dy. \tag{2.151}$$

According to the earlier discussion, we use only the boundary conditions (2.144) and (2.145). Extracting from them the value of v for $y = 0$ and $y = L$ and taking into account the fact that N_y depends only on t', the relation (2.151) can be rewritten in the following way:

$$N_y = \frac{E'h}{(1-\nu^2)L}\left(\frac{1}{2}\int_0^L \left(\frac{\partial w'}{\partial y}\right)^2 dy - \frac{1}{R}\int_0^L w'\,dy\right). \tag{2.152}$$

The expression (2.151) yields

$$v = v|_{y=0} + \frac{1-\nu^2}{E'h} yN_y + \int_0^y \frac{w'}{R}dy - \frac{1}{2}\int_0^y \left(\frac{\partial w'}{\partial y}\right)^2 dy. \tag{2.153}$$

Therefore, for the model under consideration of an infinite cylindrical panel in the case of a purely transverse load (i.e. when (2.150) holds), the equation of motion (2.136) and also Eq. (2.148) are transformed to identities.

The condition $v = 0$ from the boundary conditions (2.144) and (2.145) is already satisfied, whereas (2.137), i.e. (2.149) is transformed into the following integral–differential equation:

$$\begin{aligned} &\frac{E'h^3}{12(1-\nu^2)}\frac{\partial^4 w'}{\partial y^4} - \frac{E'h}{L(1-\nu^2)}\left(\frac{1}{R} + \frac{\partial^2 w'}{\partial y^2}\right) \\ &\times \left(\frac{1}{2}\int_0^L \left(\frac{\partial w'}{\partial y}\right)^2 dy - \int_0^l \frac{w'}{R}dy\right) + \rho_0' h\frac{\partial^2 w'}{\partial t'^2} = P_z'. \end{aligned} \tag{2.154}$$

This integral–differential equation includes only the normal deflection w' of the middle surface and governs the dynamics of the infinite circular panel in the case under consideration. The following boundary conditions from the relations (2.144) and (2.145) are applied to (2.154):

simple support (on balls):

$$w' = 0, \quad \frac{\partial^2 w'}{\partial y^2} = 0 \quad (y = 0,\ y = L); \tag{2.155}$$

stiff support (clamp):

$$w' = 0, \quad \frac{\partial w'}{\partial y} = 0 \quad (y = 0,\ y = L). \tag{2.156}$$

This is because the equality $v = 0$ from the conditions (2.144) and (2.145) is already satisfied.

Finally, we need to apply to (2.154) the initial conditions (2.142) and (2.143). We need to transform these conditions in order to obtain only one transverse deflection w'.

Substituting N_y from (2.152) into (2.153), we express the displacement v in terms of the transverse displacement w':

$$\begin{aligned} v = v|_{y=0} &+ \frac{y}{L}\left(\frac{1}{2}\int_0^L \left(\frac{\partial w'}{\partial y}\right)^2 dy - \int_0^L \frac{w'}{R}dy\right) \\ &+ \int_0^y \frac{w'}{R}dy - \frac{1}{2}\int_0^y \left(\frac{\partial w'}{\partial y}\right)^2 dy. \end{aligned} \tag{2.157}$$

This means that the first equalities in the initial conditions (2.142) and (2.143) include functions $v_0(y)$ and $w_0'(y)$ that are not completely arbitrary. If $w_0'(y)$ is given, then (2.157) uniquely defines $v_0(y)$ and vice versa. We now differentiate the equality (2.157) with respect to the time t', which results in the following conclusion. Setting the functions $w_0'(y)$ and $w_0''(y)$ is same as setting the following pairs of functions: $w_0'(y)$ and $v_0'(y)$; $v_0'(y)$ and $w_0''(y)$; $v_0(y)$ and $w_0''(y)$. This means that the following functions can be taken as the initial functions:

$$w' = w_0'(y), \quad \frac{\partial w'}{\partial t} = w_0''(y) \quad (t' = t_0'). \tag{2.158}$$

Therefore, the problem of the oscillations of a flexible, elastic, infinite, circular cylindrical panel subjected to a transverse load is reduced to an initial–boundary value problem of mathematical physics governed by the integral–differential equation (2.154) with the boundary conditions (2.155) (or (2.156)) and the initial conditions (2.158). Note that the continuum equation, in the case of the infinite cylindrical panel considered here, is satisfied identically.

It turns out that one of the equations obtained possesses only one unknown variable, w', and therefore we can express the other quantities via w'. Differentiating (2.151) with respect to y, we obtain

$$\frac{\partial v}{\partial y} = \frac{1-\nu^2}{E'h}N_y + \frac{w'}{R} - \frac{1}{2}\left(\frac{\partial w'}{\partial y}\right)^2.$$

Substituting N_y from (2.152) into the above equation, we obtain

$$\frac{\partial v}{\partial y} = \frac{1}{L}\left(\frac{1}{2}\int\limits_0^L \left(\frac{\partial w'}{\partial y}\right)^2 \partial y - \int\limits_0^L \frac{w'}{R}\partial y\right) + \frac{w'}{R} - \frac{1}{2}\left(\frac{\partial w'}{\partial y}\right)^2.$$

Substituting the expression for $\frac{\partial v}{\partial y}$ from the last equation into (2.132), we obtain

$$\begin{aligned}
\gamma_{11}^z &= 0,\\
\gamma_{22}^z &= \frac{1}{L}\left(\frac{1}{2}\int\limits_0^L \left(\frac{\partial w'}{\partial y}\right)^2 \partial y - \int\limits_0^L \frac{w'}{R}\partial y\right) - z\frac{\partial^2 w'}{\partial y^2},\\
\gamma_{12}^z &= 0,\\
\gamma_{13}^z &= 0,\\
\gamma_{23}^z &= -\frac{h^2}{6(1-\nu)}\frac{\partial^3 w'}{\partial y^3},\\
\gamma_{33}^z &= -\frac{\nu}{1-\nu}\left(\frac{1}{L}\left(\frac{1}{2}\int\limits_0^L \left(\frac{\partial w'}{\partial y}\right)^2 dy - \int\limits_0^L \frac{w'}{R}dy - z\frac{\partial^2 w'}{\partial y^2}\right)\right).
\end{aligned} \tag{2.159}$$

Therefore, the following expressions for the stress tensor components described via the transverse deflection w' can be obtained from the relation (2.133):

$$\begin{aligned}
P_{11}^{z} &= \frac{E'\nu}{1-\nu^2}\left(\frac{1}{L}\left(\frac{1}{2}\int_0^L\left(\frac{\partial w'}{\partial y}\right)^2 dy - \int_0^L \frac{w'}{R}dy\right) - z\frac{\partial^2 w'}{\partial y^2}\right),\\
P_{22}^{z} &= \frac{E'}{1-\nu^2}\left(\frac{1}{L}\left(\frac{1}{2}\int_0^L\left(\frac{\partial w'}{\partial y}\right)^2 dy - \int_0^L \frac{w'}{R}dy\right) - z\frac{\partial^2 w'}{\partial y^2}\right),\\
P_{12}^{z} &= 0,\\
P_{13}^{z} &= 0,\\
P_{23}^{z} &= -\frac{E'h^2}{12(1-\nu^2)}\frac{\partial^3 w'}{\partial y^3},\\
P_{33}^{z} &= 0.
\end{aligned} \tag{2.160}$$

These results will be used later.

2.2.1.2 Equations of Transonic Ideal-Gas Motion

Now we investigate the motion of a transonic flow of an ideal gas. Let the gas move in a two-dimensional space with coordinates x_1, x_2. It is assumed that there are no external forces (including volume forces), and that in the volume occupied by the gas there are no sources of gas mass and no input of energy. We consider a gas without internal friction and without heat transfer, i.e. the coefficients of viscosity and thermal conductivity are equal to zero. It is known that gas particles can change their entropy at discontinuities [184, 286, 292]. However, in a first approximation, it can be assumed that transonic flow with discontinuities is isentropic and vortex free and preserves those properties with time [292]. Therefore, such a gas flow remains approximately in a quasi-equilibrium state everywhere.

Derivations of the asymptotic equations of motion of a transonic (and close to homogeneous) flow of an ideal gas have been given before [225, 288, 293]. Our approach slightly differs from the previous ones, because we do not use any asymptotic series with respect to a small perturbation parameter. Instead we use Taylor series of certain dependences in the neighbourhood of critical values of some quantities and use integral estimates. We are going now to present briefly the derivation of the equations.

The integral law of mass conservation of the gas, for the space Ω bounded by the closed contour $\partial\Omega$ on the plane x_1, x_2, in the case under consideration has the following form [286, 292]:

$$\frac{\partial}{\partial t'}\int\limits_{w} \rho'\,dw + \int\limits_{\partial w} \rho'(\overline{q}\,\overline{n})\,d(\partial w) = 0. \tag{2.161}$$

The absence of a vortex in the velocity field implies that there exists a potential $\Phi(x_1, x_2, t')$ for the velocity vector, i.e. [184, 287, 292]

$$\overline{q} = grad\,\Phi. \tag{2.162}$$

As is known, one can derive the following Lagrange–Cauchy integral from the equations of motion [292]:

$$\frac{\partial\Phi}{\partial t'} + \frac{q^2}{2} + \int \frac{dP'}{\rho'(P')} = f_a(t'), \tag{2.163}$$

where $f_a(t')$ is a function of the time t'.

It is assumed that both an ideal and a real gas are governed by the following equation of state equation [184, 286, 292]:

$$P'\,\rho'^{-\chi} = const. \tag{2.164}$$

We consider the flow of the gas stream to be close to a homogeneous supersonic flow, in the direction of the axis of the profile of the shell, the x_1 axis. It is assumed that the homogeneously flowing gas stream is close to a critical condition determined by the speed of sound. The equation of state (2.164) can be rewritten in the form

$$P'\,\rho'^{-\chi} = P_\infty \rho_\infty^{-\chi}, \tag{2.165}$$

and the term $P'\,\rho'$ can be expressed by use of the integral in (2.163). We obtain

$$\int \frac{dP'}{\rho'(P')} = \frac{\chi}{\chi-1}\frac{P_\infty}{\rho_\infty}\left(\frac{\rho'}{\rho_\infty}\right)^{\chi-1}. \tag{2.166}$$

According to the known definitions (see, for instance, [184, 286, 292]), the sound velocity is given by

$$a^2 = \frac{dP'}{d\rho'}, \tag{2.167}$$

for a constant entropy. Using the equation of state of the gas (2.165), the velocity a of sound in the gas stream (2.167) can be represented in the following form:

$$a^2 = \chi \frac{P_\infty}{\rho_\infty} \left(\frac{\rho'}{\rho_\infty} \right)^{\chi - 1}.$$

This formula serves to define the sound velocity in a homogeneously flowing gas stream; i.e. infinitely far from the profile, we have

$$a_\infty^2 = \chi \frac{P_\infty}{\rho_\infty}. \tag{2.168}$$

Using (2.166) and (2.168), the Lagrange–Cauchy integral can be represented in the following form:

$$\frac{\partial \Phi}{\partial t'} + \frac{q^2}{2} + \frac{a_\infty^2}{\chi - 1} \left(\frac{\rho'}{\rho_\infty} \right)^{\chi - 1} = f_1(t'). \tag{2.169}$$

Because on homogeneously flowing stream does not change as it flows, at an infinite distance from the profile one obtains

$$\left. \frac{\partial \Phi}{\partial t} \right|_\infty = 0.$$

Thus, the Lagrange–Cauchy integral (2.169) for the flowing stream possesses the following form:

$$\frac{q_\infty^2}{2} + \frac{a_\infty^2}{\chi - 1} = f_a(t').$$

In the above equation we have a function $f_a(t')$; we substitute this into (2.169), and obtain

$$\frac{\partial \Phi}{\partial t'} + \frac{q^2}{2} + \frac{a_\infty^2}{\chi - 1} \left(\frac{\rho'}{\rho_\infty} \right)^{\chi - 1} = \frac{q_\infty^2}{2} + \frac{a_\infty^2}{\chi - 1}. \tag{2.170}$$

Multiplying (2.170) by $(\chi - 1)/a_\infty^2$, we obtain

$$\frac{\chi - 1}{a_\infty^2} \frac{\partial \Phi}{\partial t'} + \frac{\chi - 1}{2} \frac{q^2}{a_\infty^2} + \left(\frac{\rho'}{\rho_\infty} \right)^{\chi - 1} = 1 + \frac{\chi - 1}{2} M_\infty^2, \tag{2.171}$$

where the Mach number for the incoming stream has been introduced. Using the equation of state of the gas (2.165), the Lagrange–Cauchy integral can be represented in the form

$$\frac{\chi-1}{a_\infty^2}\frac{\partial\Phi}{\partial t'}+\frac{\chi-1}{2}\frac{q^2}{a_\infty^2}+\left(\frac{\rho'}{\rho_\infty}\right)^{\frac{\chi-1}{\chi}}=1+\frac{\chi-1}{2}M_\infty^2. \tag{2.172}$$

We introduce the following nondimensional variables and nondimensional time:

$$\xi=\frac{x_1}{L},\quad \eta=\frac{x_2}{L},\quad t=t'\frac{q_\infty}{L}\mu. \tag{2.173}$$

We also introduce a nondimensional perturbation of the velocity vector with components $U(\xi,\eta,t)$ and $V(\xi,\eta,t)$:

$$q_1=q_\infty(1+V),\quad q_2=q_\infty V. \tag{2.174}$$

The relation of the perturbed stream to the homogeneous stream, i.e. to a uniform stream moving at the velocity of sound, is expressed by

$$|U|\ll 1,\quad |V|\ll 1.$$

We introduce the nondimensional density and pressure within the stream in the form

$$\rho=\frac{\rho'}{\rho_\infty},\quad P=\frac{P'}{P_\infty}. \tag{2.175}$$

Finally, the following potential for the perturbations to the velocity vector $\varphi(\varepsilon,\eta,t)$ is introduced:

$$\Phi(x_1,x_2,t')=q_\infty L(\varepsilon+\varphi(\varepsilon,\eta,t)). \tag{2.176}$$

The relations (2.173) and (2.176) yield

$$\frac{\partial\Phi}{\partial t'}=\mu q_\infty^2\frac{\partial\varphi}{\partial t},$$

where the number μ characterizes the velocity of mechanical changes in the gas stream (note that in the case of harmonic oscillations it corresponds to the frequency).

Therefore, using the last equation and (2.174) and (2.175), the Lagrange–Cauchy integral (2.171) can be expressed in the following nondimensional form:

$$(\chi-1)\mu M_\infty^2\frac{\partial\varphi}{\partial t}+\frac{\chi-1}{2}M_\infty^2\left((1+U)^2+V^2\right)+\rho^{\chi-1}$$
$$=1+\frac{\chi-1}{2}M_\infty^2. \tag{2.177}$$

Equation (2.172) takes the form

$$(\chi-1)\mu M_\infty^2\frac{\partial\varphi}{\partial t}+\frac{\chi-1}{2}M_\infty^2\left((1+U)^2+V^2\right)+\rho^{\frac{\chi-1}{\chi}}$$
$$=1+\frac{\chi-1}{2}M_\infty^2. \tag{2.178}$$

The nondimensional density ρ can be obtained from (2.177), and we obtain

$$\rho=\left(1-\frac{\chi-1}{2}M_\infty^2\left(2U+U^2+V^2+2\mu\frac{\partial\varphi}{\partial t}\right)\right)^{\frac{1}{\chi-1}}. \tag{2.179}$$

Analogously, for the nondimensional pressure, we obtain the following from (2.178):

$$P=\left(1-\frac{\chi-1}{2}M_\infty^2\left(2U+U^2+V^2+2\mu\frac{\partial\varphi}{\partial t}\right)\right)^{\frac{\chi}{\chi-1}}. \tag{2.180}$$

With the definition

$$G=M_\infty^2\left(2U+U^2+V^2+2\mu\frac{\partial\varphi}{\partial t}\right) \tag{2.181}$$

and taking into account the relation (2.181), the Lagrange–Cauchy integral (2.179) has the form

$$\rho=\left(1-\frac{\chi-1}{2}G\right)^{\frac{1}{\chi-1}}. \tag{2.182}$$

In an analogous way, we obtain the following relation from (2.180):

$$P=\left(1-\frac{\chi-1}{2}G\right)^{\frac{\chi}{\chi-1}}. \tag{2.183}$$

Let us find the critical value of the derivative $\left(\frac{\partial\varphi}{\partial t}\right)_*$ from the Lagrange–Cauchy integral (2.177) for the critical values of the relevant quantities:

$$\mu M_\infty^2\left(\frac{\partial\varphi}{\partial t}\right)_*=1+\frac{\chi-1}{2}M_\infty^2-\rho_*^{\chi-1}-\frac{\chi-1}{2}M_\infty^2\left((1+U_*)^2+V_*^2\right). \tag{2.184}$$

If we assume, as has already been mentioned, that the incoming stream infinitely far from the profile is close to a critical condition, i.e. $M_\infty \approx 1$, $\rho_* \approx 1$, $U_* \approx 0$, $V_* \approx 0$, then we obtain approximately from (2.184)

$$\left(\frac{\partial \varphi}{\partial t}\right)_* = 0. \tag{2.185}$$

Therefore, in this case, when the gas particles reach the critical sound velocity, we obtain from (2.181) and (2.185) the result

$$G_* = 0.$$

The relation (2.182) between the density ρ and G for a transonic stream can then be approximated by the following Taylor series in the neighbourhood of $G = 0$:

$$\rho = 1 - \frac{1}{2}G + \frac{2-\chi}{8}G^2 + R_\rho, \tag{2.186}$$

where R_ρ is the rest of the series, which is of order G^3. In a similar way, for a transonic flow, the dependence of P on G (2.183) can be approximated by a series of the form

$$P = 1 - \frac{\chi}{2}G + \frac{\chi}{8}G^2 + R_p, \tag{2.187}$$

where R_p is the rest of the series, of order G^3. Multiplying (2.186) by $(\overline{q}, \overline{n})$ the following equation is obtained:

$$\rho\frac{(\overline{q}\,\overline{n})}{q_\infty} = \left(1 - \frac{G}{2} + \frac{2-\chi}{2}G^2 + R_p\right)((1+U)\cos(n,\xi) + V\cos(n,\eta)). \tag{2.188}$$

Therefore, from the above considerations, one can conclude that the values of U, V and $\frac{\partial \varphi}{\partial t}$ are much smaller than one, i.e. there exist positive numbers ε_1, ε_2 and ε_3 such that everywhere in the stream we have

$$|U| < \varepsilon_1 \ll 1, \quad |V| < \varepsilon_2 \ll 1, \quad \left|\frac{\partial \varphi}{\partial t}\right| < \varepsilon_3 \ll 1. \tag{2.189}$$

In these inequalities ($\varepsilon_i \ll 1$), we understand that each of the ε_i is about ten times smaller than 1. Also, the expression "an order smaller" is understood as meaning approximately ten times smaller. We do not know anything about the comparative of the values of ε_1, ε_2 and ε_3, and for a derivation of the transonic asymptotic equations such a comparison is not needed. Of course, any multiplication of those values results in a value order lower than the quantities multiplied. For instance,

$$|UV| \ll |V|, \quad |UV| \ll |U|, \quad \left|U\frac{\partial\varphi}{\partial t}\right| \ll \left|\frac{\partial\varphi}{\partial t}\right|,$$

because the following inequalities hold:

$$|UV| \le \varepsilon_1\varepsilon_2 \ll \varepsilon_2, \quad |UV| \le \varepsilon_1\varepsilon_2 \ll \varepsilon_1, \quad \left|U\frac{\partial\varphi}{\partial t}\right| \le \varepsilon_1\varepsilon_3 \ll \varepsilon_3.$$

It is assumed that the orders of the quantities and their derivatives in relation to the nondimensional time are similar:

$$\left|\frac{\partial U}{\partial t}\right| \le \varepsilon_1, \quad \left|\frac{\partial U}{\partial t}\right| \le \varepsilon_2, \quad \left|\frac{\partial^2 U}{\partial t^2}\right| \le \varepsilon_3, \tag{2.190}$$

i.e. the order of a derivative with respect to the nondimensional time is μ times larger than the order of the quantity which is differentiated. Differentiating (2.186) with respect to t', we obtain

$$\frac{d\rho}{dt'} = \frac{q_\infty\mu}{L}\left(-\frac{1}{2} + \frac{2-\chi}{4}G + \frac{dR_\rho}{dG}\right)\frac{dG}{dt}. \tag{2.191}$$

We assume the existence of the derivatives and a unique convergence of the series and its derivatives. Taking into account these assumptions, we can conclude that the left-hand side of (2.191) and the right-hand side of (2.167) are equal. Substituting the necessary relations from (2.178) and (2.191) into the law of mass conservation (2.161) we obtain

$$\int\limits_{\Omega} \mu\frac{dG}{dt}\left(-\frac{1}{2} + \frac{2-\chi}{4}G + \frac{dR_\rho}{dG}\right) d\Omega + \int\limits_{\partial\Omega}\Bigg[\left(1 - \frac{G}{2} + \frac{2-\chi}{8}G^2 + R_\rho\right)$$
$$+ (1+V)\cos(n,\xi) + V\cos(n,\eta)\Bigg] d(\partial\Omega) = 0, \tag{2.192}$$

where both the space Ω and its contour $\partial\Omega$ are now expressed in nondimensional form in new nondimensional coordinates ξ and η. If a function g under an integral is bounded, $|g| \le \varepsilon$, then for a limited space Ω the following inequality holds:

$$\int\limits_{w} g\, d\Omega \le \varepsilon \int\limits_{w} d\Omega. \tag{2.193}$$

An analogous inequality can be written for the integral along the closed contour $\partial\Omega$. Integral estimates of type (2.193) will be used for estimation of the terms occurring in (2.192). Terms which are an order smaller in comparison with other terms will be omitted.

If we substitute the expression for G from (2.181) into the relation (2.188) and neglect small terms, we obtain

$$\rho(\overline{q}\,\overline{n}) = q_\infty \left\{ \left[1 - \mu M_\infty^2 \frac{\partial \varphi}{\partial t} + (1 - M_\infty^2) U \right.\right.$$
$$\left.\left. + \frac{1}{2} M_\infty^2 (2M_\infty^2 - \chi M_\infty^2 - 3) U^2 \right] \cos(n, \xi) + V \cos(n, \eta) \right\}. \tag{2.194}$$

In (2.194), the term containing U^2 remains, because for $M_\infty = 1$ the coefficient of U is equal to zero. Because the relative orders of the terms occurring in (2.194) and of the value of the coefficient of U are unknown, we need to consider the nonlinear term U^2. Of course, depending on the value of the frequency μ, the relation (2.191) can be simplified.

Substituting G from (2.181) into (2.191) and taking into account the inequalities (2.189) and (2.190), we can identify several separate intervals of positive values of the frequency μ for which the relation (2.191) has various approximate forms:

1. $\mu = 0$ (stationary flow):

$$\frac{d\rho}{dt'} = 0.$$

2. $0 < \mu \le \varepsilon_5 \ll 1$, $\varepsilon_5 \ll \min\left(\frac{\varepsilon_1}{\varepsilon_3}, \frac{\varepsilon_2^2}{\varepsilon_3}\right)$:

$$\frac{d\varphi}{dt'} = -\frac{M_\infty^2 q_\infty \mu}{L}\left(\frac{\partial U}{\partial t} + V\frac{\partial V}{\partial t}\right).$$

This case is sometimes referred to as the low-frequency case and is denoted by $\mu \to 0$.

3. $\varepsilon_5 < \mu \le \varepsilon_6$, $\varepsilon_6 \gg \max\left(\frac{\varepsilon_1}{\varepsilon_3}, \frac{\varepsilon_2^2}{\varepsilon_3}\right)$:

$$\frac{d\rho}{dt'} = -\frac{M_\infty^2 q_\infty \mu}{L}\left(\mu\frac{\partial^2 \varphi}{\partial t^2} + \frac{\partial U}{\partial t} + V\frac{\partial V}{\partial t}\right).$$

This case is referred to as the medium-frequency case.

4. $\varepsilon_6 \le \mu < \infty$:

$$\frac{d\rho}{dt'} = -\frac{M_\infty^2 q_\infty \mu}{L}\frac{\partial^2 \varphi}{\partial t^2}. \tag{2.195}$$

This case is referred to as the high-frequency case.

It is clear that, depending on the relations between the constants ε_1, ε_2 and ε_3, one can also identify other intervals of the frequency μ. However, we have considered the case where the constants ε_1, ε_2 and ε_3 are arbitrary and much smaller than 1, i.e. they are small and the relations between them are not known. In other words, the frequency intervals have been separated for arbitrarily small values ($\varepsilon_i \ll 1$) of the positive constants ε_1, ε_2 and ε_3.

Our aim is to create an asymptotic equation governing the dynamics of a transonic ideal-gas stream which is valid for all intervals of the frequency μ. The most general formula is that governed by (2.195), which holds for medium frequencies. The cases of low and high frequencies are included in the expression (2.195).

Using (2.194) and (2.195), Eq. (2.192) for a transonic stream reads

$$
\begin{aligned}
&-\mu M_\infty^2 \int\limits_{w} \left(\frac{\partial U}{\partial t} + \mu \frac{\partial^2 \varphi}{\partial t^2} + V \frac{\partial V}{\partial t} \right) d\Omega + \int\limits_{\partial\Omega} \bigg\{ (1 - \mu M_\infty^2 \frac{\partial \varphi}{\partial t} \\
&+ (1 - M_\infty^2) V + \frac{M_\infty^2}{2} (2 M_\infty^2 - \chi M_\infty^2 - 3) U^2) \cos(n, \xi) \\
&+ V \cos(n, \eta) \bigg\} d(\partial\Omega) = 0.
\end{aligned}
\tag{2.196}
$$

We compare the terms in (2.196) occurring in the first and second integrals (we compared before only the terms in the same integrals), which yields

$$
\int\limits_{\Omega} \left| \mu M_\infty^2 V \frac{\partial V}{\partial t} \right| d\Omega \ll \int\limits_{\partial\Omega} |V \cos(n, \eta)| \, d(\partial\Omega). \tag{2.197}
$$

Because of the property (2.193) and the condition (2.189), we obtain

$$
\int\limits_{\Omega} \left| \mu M_\infty^2 V \frac{\partial V}{\partial t} \right| d\Omega \ll \mu M_\infty^2 \varepsilon_2^2 \int\limits_{\Omega} d\Omega,
$$

$$
\int\limits_{\Omega} |V \cos(n, \eta)| \, d(\partial w) \le \varepsilon_2 \int\limits_{\partial\Omega} d(\partial w).
$$

Hence, the inequality (2.197) is true if the following relation holds:

$$
\mu \int\limits_{\Omega} d\Omega \ll \frac{1}{\varepsilon_2 M_\infty^2} \int\limits_{\partial\Omega} d(\partial\Omega). \tag{2.198}
$$

It can be seen that the ratio between the integrals appearing in the inequality (2.198) can give arbitrary positive real number. This property can be understood in the following manner. The larger Ω is, the smaller the ratio of the integrals is, and vice versa. However, this property is not always true, because a bounded space with a given area can have a boundary with an arbitrary large length. But the smallest length of the boundary of a space with a given area is the circumference of a circle, in this case the space corresponds to a circle. Hence, the asymptotic transonic equation can be considered in a space Ω that is a circle, a square or other simple geometrical figure. Hence, taking into account the relation (2.197), the equality (2.196) can be rewritten in the form

$$
\begin{aligned}
&-\mu M_\infty^2 \int\limits_\Omega \left(\frac{\partial U}{\partial t} + \mu \frac{\partial^2 \varphi}{\partial t^2}\right) d\Omega + \int\limits_{\partial\Omega} \left\{\left[(1 - M_\infty^2)U - \mu M_\infty^2 \frac{\partial \varphi}{\partial t}\right.\right. \\
&\left.\left.+\frac{M_\infty^2}{2}(2M_\infty^2 - \chi M_\infty^2 - 3)U^2\right] \cos(n, \xi) + V\cos(n, \eta)\right\} d(\partial\Omega) = 0.
\end{aligned} \tag{2.199}
$$

We now use a property of a contour integral of the form $\int\limits_{\partial\Omega} \cos(n, \varepsilon)\, d(\partial\Omega) = 0$. If we consider a low-frequency supersonic flow, then $\mu \to 0$, and from (2.199) one of the equations of motion can be represented in the form

$$
\begin{aligned}
&-\mu M_\infty^2 \int\limits_\Omega \left(\frac{\partial U}{\partial t}\right) dw + \int\limits_{\partial\Omega} \left\{\left[(1 - M_\infty^2 U - \mu M_\infty^2 \frac{\partial \varphi}{\partial t}\right.\right. \\
&\left.\left.+\frac{M_\infty^2}{2}(2M_\infty^2 - \chi M_\infty^2 - 3)U^2\right] \cos(n, \xi) + V\cos(n, \eta)\right\} d(\partial w) = 0.
\end{aligned} \tag{2.200}
$$

Applying the Osrogradskij–Gauss formula, assuming that U, V, $\frac{\partial \varphi}{\partial t}$ are continuous and differentiable, and transforming from the contour integral to a surface integral in (2.199), and then taking into account the arbitrariness of integration over the space Ω, we obtain the following transonic-stream equation:

$$
\begin{aligned}
&-\mu M_\infty^2 \frac{\partial^2 \varphi}{\partial t^2} - 2\mu M_\infty^2 \frac{\partial U}{\partial t} + (1 - M_\infty^2)\frac{\partial U}{\partial \xi} \\
&+ M_\infty^2(2M_\infty^2 - \chi M_\infty^2 - 3)U\frac{\partial U}{\partial \xi} + \frac{\partial V}{\partial \eta} = 0,
\end{aligned} \tag{2.201}
$$

which holds for low, medium and high frequencies.

For low frequencies, when $\mu \to 0$, the following differential equation for a transonic stream is obtained from (2.200):

$$
-2\mu M_\infty^2 \frac{\partial U}{\partial t} + (1 - M_\infty^2 + M_\infty^2)(2M_\infty^2 - \chi M_\infty^2 - 3)U\frac{\partial U}{\partial \xi} + \frac{\partial V}{\partial \eta} = 0. \tag{2.202}
$$

This equation is almost the same as the Lin–Reissner–Tsien (LRT) equation for a nonstationary, transonic, vortex-free (close to homogeneous) flow of an ideal gas

[225]. If we take $M_\infty = 1$ in the expression $(2M_\infty^2 - \chi M_\infty^2 - 3)$ in (2.202), obtain the LRT equation, because the following relation holds:

$$2M_\infty^2 - \chi M_\infty^2 - 3 = -(\chi + 1).$$

We now add the vortex-free equation (2.162) to (2.201) (or to (2.202) for low frequencies). Equation (2.162), using (2.174), can be represented in the following form:

$$\frac{\partial U}{\partial \eta} - \frac{\partial V}{\partial \xi} = 0. \tag{2.203}$$

A simultaneous solution to (2.201) and (2.203) defines the velocity field of the gas particles in the transonic flow far from the profile, for various frequencies of flow vibrations. In the neighbourhood of the profile, we need to take into account the boundary layer and the influence of viscosity on the flow. The vortex-free equation (2.203) can be represented in the following integral form:

$$\int_{\partial\Omega} (U\cos(n,\eta) - V\cos(n,\xi))\, d(\partial\Omega) = 0. \tag{2.204}$$

Indeed, by applying the Osrogradskij–Gauss formula to the (2.203), one also obtains (2.204). However, that the class of functions $U(\xi, \eta, t)$ and $V(\xi, \eta, t)$ which satisfy (2.204) is wider than the class of the functions satisfying (2.203). The first class includes (unlike the second) discontinuous functions.

The components of the velocity vector satisfy the integral equation (2.204) for an arbitrary closed contour $\partial\Omega$. We call this flow a "generalized" vortex-free flow.

For the pressure P after truncation of the Taylor series (2.187), we obtain the following in the transonic approximation

$$P = 1 - \chi M_\infty^2 U - \chi\mu M_\infty^2 \frac{\partial\varphi}{\partial t} - \frac{\chi}{2} M_\infty^2 V^2. \tag{2.205}$$

In the low-frequency range, this reads

$$P = 1 - \chi M_\infty^2 U - \frac{\chi}{2} M_\infty^2 V^2.$$

Analysis of (2.202) leads to the conclusion that the quantity V^2 is an order smaller than U. Therefore the last relation can be rewritten in the following form:

$$P = 1 - \chi M_\infty^2 U. \tag{2.206}$$

For the dynamic coefficient of pressure [292] in the transonic approximation, the following relation holds:

$$C_p = -2U - 2\mu \frac{\partial \varphi}{\partial t}. \tag{2.207}$$

In the case of the stationary and low-frequency states, this has the form

$$C_p = -2U. \tag{2.208}$$

For a transonic flow around an arbitrary profile of infinite length, the nonpenetration condition for an ideal gas can be written [292]

$$\overline{q}_{ng} = \overline{q}_{nP}, \tag{2.209}$$

where $\overline{q}_{ng}$ is the normal component of the velocity of the gas particles close to the profile surface, and $\overline{q}_{nP}$ is the normal component of the velocity of the points of the profile surface. If the profile surface has the form $x_2 = H(x_1, t')$, then [184]

$$q_{nP} = \frac{\partial H}{\partial t'} \left(1 + \left(\frac{\partial H}{\partial x_1} \right)^2 \right)^{-\frac{1}{2}}. \tag{2.210}$$

From the scalar product of the flow velocity vector and the unit vector normal to the profile surface, we obtain

$$q_{ng} = q_\infty \left(V - \frac{\partial H}{\partial x_1}(1 + U) \right) \left(1 + \left(\frac{\partial H}{\partial x_1} \right)^2 \right)^{-\frac{1}{2}}. \tag{2.211}$$

Substituting (2.210) and (2.211) into the nonpenetration condition, we obtain

$$q_\infty \left(V - \frac{\partial H}{\partial x_1}(1 + U) \right) = \frac{\partial H}{\partial t'}.$$

The last expression is then transformed to nondimensional form by use of (2.173), and the condition of nonpenetration of the profile reads:

$$V = (1 + U)\frac{\partial f}{\partial \xi} + \frac{\partial f}{\partial t} \quad \text{or} \quad V = \frac{\partial f}{\partial \xi} + \frac{\partial f}{\partial t}. \tag{2.212}$$

The conditions infinitely far from the profile read

$$U = 0, \quad V = 0 \quad (t \geq 0, \ \xi^2 + \eta^2 \to +\infty), \tag{2.213}$$

which express the fact that the transonic flow at infinity is nonexcited and homogeneous.

We assume that on a surface (ξ, η), a shock wave is governed by an equation $\eta = \eta(\xi, t)$. Expanding the contour ∂w associated with this shock wave, the integral equations (2.199) and (2.204) give the following relations:

$$2M_\infty^2 \frac{\partial \xi}{\partial t} + \left(\frac{V_- - V_+}{U_- - U_+}\right)^2 = M_\infty^2 - 1 + \frac{1}{2} M_\infty^2 (\chi + 1)(U_+ + U_-), \tag{2.214}$$

$$V_- - V_+ = \frac{\partial \xi}{\partial \eta}(U_+ - U_-), \tag{2.215}$$

which are the Reikin–Giugonio conditions in the transonic low-frequency approximation.

2.2.1.3 Equations and Boundary Conditions of Panel–Flow Interaction

We now consider the interaction between a transonic flow and an elastic panel. We assume that between two impenetrable, absolutely rigid, and immoveable semi-surfaces AB and CD an infinitely long circular cylindrical panel with straight, parallel boundaries B and C is fastened. Let us assume that the half-spaces AB and CD are located on the same level, i.e. if they were extended they would belong to one surface. Let us assume also that the panel has either free (ball) or stiff supports along sides B and C (other boundary conditions can also be taken into consideration). Assume that a homogeneous transonic gas flow runs over this panel in a direction transverse to the axis of the infinite circular cylinder, part of which is occupied by the panel under consideration. The direction of the flow is parallel to the half-planes AB and CD. It may be said that the panel, which can be called a profile panel (because it is an object of interest in terms of the theory of elasticity as well as the theory of gas dynamics), has zero attack angle. The interaction of the flow and the panel considered here is completely equivalent to a symmetric flow around a symmetric profile panel, where the plane ABCD becomes the plane of symmetry. This panel serves as a model for an aircraft wing or any part of the surface of any flying aircraft.

Now, let us assume that the motion of the panel is governed by the theory described in Sect. 2.2.1.1. Let us assume also that the motion of the flow is governed by the asymptotic theory of transonic gas flow described in Sect. 2.2.1.2.

In Fig. 2.1, the interaction described above of a transonic flow with an elastic, infinitely long, circular cylindrical panel is illustrated. According to the assumptions that we have made, the whole flow picture is flat.

In Fig. 2.2, an interaction related to the transverse cross-section of the panel is shown. The dashed curve shows the boundary of the supersonic zone of the flow, with a local Mach number $M > 1$. This zone ends, on moving along the direction of the flow, in an impact wave, which is marked by a double dashed line. Looking in the direction of the flow movement, the pressure before the impact wave is smaller

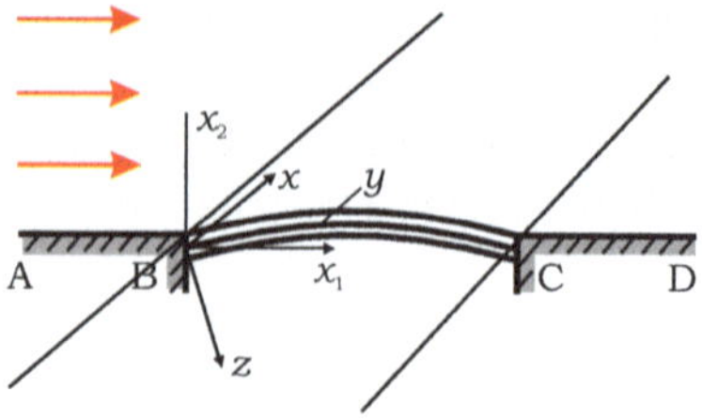

Fig. 2.1 Coordinate system for the panel

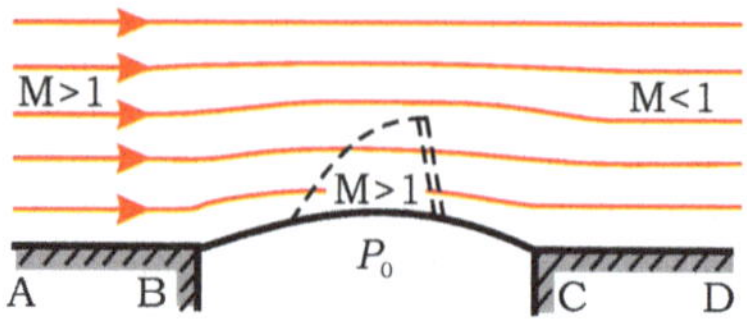

Fig. 2.2 Parameters of flow around panel

than behind the wave, i.e. the impact wave plays the role of a separatrix between regions of different pressure. If the impact wave is in contact with the surface of the profile panel, then the panel is attacked seriously by a nonuniformly distributed load. Hence, the most interesting case of interaction between a flow and a panel seems to be that in which an impact wave moves over the middle part of the panel. Such a motion of an impact wave is possible during flow around a panel in the case of transonic flow, when there are local supersonic zones. This means that for the panel considered here, shock wave motion happens during a flow corresponding to a homogeneous stream with a Mach number M_∞ lying in the interval from 0.75 to 1. Such interactions are analysed in this section.

Nondimensional Quantities Although the panel motion is described here in the Lagrange coordinates x, y, z (given in Sect. 2.2.1.1 and shown in Fig. 2.1) and the motion of the gas flow is described in the Euler coordinates x_1, x_2 (defined in Sect. 2.2.1.2 and shown in Fig. 2.1), the following equalities hold approximately:

$$x_1 = y, \qquad x_2 = -z. \tag{2.216}$$

This is true if we take into account the fact that the panel is shallow and that large displacements do not occur in it. Therefore, the same for rectangular system of coordinates (x_1, x_2), shown in Fig. 2.1, has been chosen for both the flow and the panel.

The problem of interaction between a panel and a transonic flow considered here is governed by the following dimensional and nondimensional parameters:

$$x_1, x_2, t', P', P_\infty, P_0', P_z', \rho_\infty, \rho_0', L, H_*, h, w, q_1, q_2, q_\infty, E', \nu, \chi. \tag{2.217}$$

The radius R has not been included in the system, because there exists the geometrical relation

$$R^2 = \left(\frac{L}{2}\right)^2 + (R - H_*)^2, \tag{2.218}$$

given by Pythagoras' theorem. Of the above system of parameters, only three have independent dimensions. Hence, using Pythagoras' theorem and the theory of dimension, we can choose a system of 16 nondimensional control parameters of the following form [292]:

$$\begin{gathered}
\xi = \frac{x_1}{L}, \ \eta = \frac{x_2}{L}, \ t = t'\frac{q_\infty}{L}, \ P = \frac{P'}{P_\infty}, \ P_0 = \frac{P'_0}{P_\infty}, \ \nu, \\
P_z = \frac{P'_z}{P_\infty}, \ \rho_0 = \frac{\rho'_0}{\rho_\infty}, \ E = \frac{E'}{P_\infty}, \ \delta = \frac{h}{L}, \ w = \frac{w'}{h}, \ \chi, \\
K = \frac{8H_*}{h(1 + 4H_*^2/L^2)}, \ V = \frac{q_1}{q_\infty} - 1, \ V = \frac{q_2}{q_\infty}, \ M_\infty = q_\infty\sqrt{\frac{\rho_\infty}{\chi P_\infty}}.
\end{gathered} \tag{2.219}$$

These 16 parameters define the mathematical model of the interaction under consideration. However, in the analysis we shall introduce some other parameters, which will also be presented in a nondimensional form.

To illustrate our analysis, we introduce the parameter $f_* = H_*/L$, which can be interpreted as an initial relative deflection. It is geometrically linked to the relative thickness δ and to the nondimensional geometrical shell parameter K according to (2.219) via the relation

$$\delta f_* = K\delta(1 + 4f_*^2). \tag{2.220}$$

Hence, the equation of the middle surface has the form $\eta = f(\xi, t)$ instead of $x_2 = H(x_1, t')$. Finally, let us introduce a nondimensional parameter for the membrane stress P_{22} in the middle surface,

$$\sigma_M = \frac{1 - \nu^2}{\delta^2}\frac{P_{22}}{E'}. \tag{2.221}$$

Mathematical Background Returning to the nondimensional quantities (2.173) and (2.174) describing the transonic gas flow, it can be seen that not all of them are included in the system of nondimensional parameters (2.219). Hence, the equations governing the motion of the gas flow in nondimensional form (2.200), (2.204), (2.206) can be used in the statement of the problem of transonic aeroelasticity considered here. In nondimensional form (with the help of the nondimensional quantities (2.219) in the equation of motion of the panel (2.184), and the initial conditions for the transverse deflection w (2.158)), the problem of the interaction between the panel and the transonic gas flow can be stated as the following initial–boundary problem:

$$\frac{\partial^4 w}{\partial \xi^4} - 12\left(K + \frac{\partial^2 w}{\partial \xi^2}\right)\sigma_M + \frac{12(1-\nu^2)}{E\delta^2}\rho_0 \chi M_\infty^2 \frac{\partial^2 w}{\partial t^2} = \frac{12(1-\nu^2)}{E\delta^4} P_z \quad (t > 0,\ 0 < \xi < 1,\ \eta = 0), \tag{2.222}$$

$$\sigma_M = \frac{1}{2}\int_0^1 \left(\frac{\partial w}{\partial \xi}\right)^2 d\xi - K\int_0^1 w\, d\xi \quad (t \geq 0,\ \eta = 0),\ \theta \tag{2.223}$$

$$P_Z = P - P_0 \quad (t \geq 0,\ 0 \leq \xi \leq 1,\ \eta = 0), \tag{2.224}$$

$$P = 1 - \chi M_\infty^2 U \quad (t \geq 0,\ 0 < \xi < 1,\ \eta = 0), \tag{2.225}$$

$$\int_\Omega \left(1 - M_\infty^2 - \frac{1}{2}M_\infty^2(\chi+1)U\right)U\, dt\, d\eta - \int_\Omega 2M_\infty^2 U\, d\xi\, d\eta + \int_\Omega V\, dt\, d\xi = 0 \quad (t > 0,\ -\infty < \xi < +\infty,\ \eta > 0), \tag{2.226}$$

$$\int_\Omega U\, dt\, d\xi - \int_\Omega V\, dt\, d\eta = 0 \quad (t > 0,\ -\infty < \xi < +\infty,\ \eta > 0), \tag{2.227}$$

$$U = 0,\ V = 0 \quad (t \geq 0,\ \xi^2 + \eta^2 \to +\infty), \tag{2.228}$$

$$U = U_0(\xi, \eta),\ V = V_0(\xi, \eta) \quad (t > 0,\ -\infty < \xi < +\infty,\ \eta > 0), \tag{2.229}$$

$$V = 0 \quad (t \geq 0,\ \xi < 0,\ \xi > 1,\ \eta = 0), \tag{2.230}$$

$$V = \frac{\partial f}{\partial \xi}(1 + U) + \frac{\partial f}{\partial t},\quad f(\xi, t) = f_0(\xi) - \delta w(\xi, t) \quad (t > 0,\ 0 < \xi < 1,\ \eta = 0), \tag{2.231}$$

$$w = 0,\ \frac{\partial w}{\partial t} = 0 \quad (t = 0,\ 0 \leq \xi \leq 1,\ \eta = 0), \tag{2.232}$$

$$w = 0,\ \frac{\partial^2 w}{\partial t^2} = 0 \quad (t \geq 0,\ \xi = 0,\ \xi = 1,\ \eta = 0). \tag{2.233}$$

Equation (2.222) governs the motion of the panel. Equation (2.223) defines the membrane stress σ_M, which appears in (2.222) through the transverse deflection w ((2.223) was obtained using (2.221) and (2.160)). The relation (2.224) defines the intensity of the transverse load P_Z through both the pressure of the transonic flow P and the internal shell pressure P_0. The formula (2.206) is transformed into (2.225). The equation of motion of a low-frequency transonic (close to homogeneous) flow of the an ideal gas (2.226) agrees with the equation given earlier for this case (2.200). The vortex-free equation (2.204) in integral form can be rewritten as the integral equation (2.227) using the Ostrogradskij–Gauss theorem. The condition (2.228) holds for the gas flow infinitely far from the panel. The initial flow is defined by the relations (2.229). As specified in the statement of the problem, the front and rear of the panel are fastened to absolutely rigid half-planes AB and CD, and the nonpenetration condition is expressed by (2.230). The nonpenetration condition (2.192) for a deformable panel has been transformed to the condition (2.231), where $\eta = f_0(\xi)$ governs the initial form of the panel. The initial state of the panel is governed by the condition (2.232), which has been obtained from the initial conditions (2.158) by taking $w_0(\xi) = 0$ and $w_0'(\xi) = 0$. The condition (2.233) describes free support (on balls) of the front and rear of the panel, which in nondimensional form is equivalent to the condition (2.155). Of course, instead of a free support along the panel sides we can consider a clamped support; in this case boundary conditions (2.156) have the following nondimensional form:

$$w = 0, \ \frac{\partial w}{\partial \xi} = 0 \quad (t = 0, \ 0 \leq \xi \leq 1, \ \eta = 0). \tag{2.234}$$

We now consider further the problem of the interaction of an elastic, infinite, cylindrical panel with a transonic flow of an ideal gas. The consideration starts from the case where the panel is absolutely rigid, and the flowing stream is stationary. We need to know the initial distribution of the velocity field, i.e. the functions $U_0(\xi, \eta)$ and $V_0(\xi, \eta)$. These functions are found from solutions to Eqs. (2.222)–(2.233), when $U_0(\xi, \eta) = 0$, $V_0(\xi, \eta) = 0$ and the given function $\eta = f(\xi, t)$. The last function is chosen in such a way that for $t = 0$ it describes a plate, whereas at a finite time it describes the given circular panel. Here we propose the following function:

$$f(\xi, t) = \begin{cases} 0, & t = 0, \\ f_0(\xi), & t = 0.02. \end{cases}$$

Therefore, this problem corresponds to the problem of a flow around a panel caused by a nonstationary transonic gas flow, in which a plate becomes a circular shell as a result of bending.

2.2.1.4 Analysis of Panel Oscillations

Now we analyse free and forced panel oscillations. We begin by considering the numerical solution of the problem. First we consider calculations for the panel. In order to check the reliability and correctness of the panel calculations, one first must compare the results of the calculations with known solutions. In this section we consider a method of performing calculations for a panel using the finite-difference method along the spatial coordinates and the Runge–Kutta method.

The equations of motion (2.154) of the panel with the specified initial and boundary conditions can be solved by different methods. The Bubnov–Galerkin method, as applied to this type of equations, is described in [135]. In [340], free vibrations of a panel were analysed using the finite-difference method.

Computational Method This should satisfy the following requirements:

1. The computations must capture high frequencies in the distribution of the transverse deflection w of the transverse cross-section of the infinite cylindrical panel, because the relatively large and sudden nonuniformity in the distributed load caused by a shock wave can lead to a large local curvature of an elastic panel.
2. The algorithm must be sensitive to quick changes with time of the load, and consequently the deflection w in time, because a transonic flow can quickly change the pressure on the surface of the profile panel.
3. The computational method should be stable and must enable further tracing, after stability loss, of the shell dynamics.
4. The method must be economical from the point of view of computational time.

It turns out, that the Bubnov–Galerkin method is not suitable for solving the problem efficiently, because it requires higher-order approximations, which is in conflict to point 4.

The finite-difference method, if it is applied to both the spatial and the time variables, does not posses high stability [325], and therefore does not satisfy point 3.

However, if in the panel calculation one applies the finite-difference technique to the spatial coordinates, and applies one of the methods for solving Cauchy problems of ordinary differential equations (ODEs) with respect to the time variable, one of then [325] the stability of the calculation process increases and one can follow the behaviour of the panel over a long time.

Now we formulate the problem in its general form. In dimensional form, as has been mentioned in Sect. 2.2.2.1, the problem reads

$$\frac{E'h^3}{12(1-\nu^2)}\frac{\partial^4 w'}{\partial y^4} - \frac{E'h}{L(1-\nu^2)}\left(\frac{1}{R}+\frac{\partial^2 w'}{\partial y^2}\right) \times \left(\frac{1}{2}\int_0^L \left(\frac{\partial w'}{\partial y}\right)^2 dy - \int_0^L \frac{w'}{R}dy\right) + \rho_0' h\frac{\partial^2 w'}{\partial t'^2} = P_z', \tag{2.235}$$

$$w' = w_0'(y), \quad \frac{\partial w'}{\partial t'} = w_0''(y) \quad (t'=0,\ 0 \le y \le L), \tag{2.236}$$

$$w' = 0, \quad \frac{\partial^2 w'}{\partial y^2} = 0 \quad (t' \ge 0,\ y=0,\ y=L). \tag{2.237}$$

Note that (2.235) is the same as (2.154). The initial conditions (2.236) are the same as the conditions (2.158). The conditions (2.237) are the simple support conditions (2.155). The panel and the system of coordinates are represented in Fig. 2.1. We introduce the following nondimensional quantities:

$$\xi = \frac{y}{L}, \quad t = At', \quad K = \frac{8H_*}{h(1+4h_*^2/L^2)},$$
$$\delta = \frac{h}{L}, \quad w = \frac{w'}{h}, \quad P_Z = \frac{P'_Z}{E'}, \tag{2.238}$$

and transform the problem (2.235), (2.236) and (2.237) to the following nondimensional form:

$$\frac{\partial^4 w}{\partial \xi^4} - 12\left(K + \frac{\partial^2 w}{\partial \xi^2}\right)\left(\frac{1}{2}\int_0^1 \left(\frac{\partial w}{\partial \xi}\right)^2 d\xi - K\int_0^1 w\, d\xi\right)$$
$$+ 12\frac{(1-\nu^2)\rho'_0 L^2 A^2}{E'\delta^4}\frac{\partial^2 w}{\partial t^2} = \frac{12(1-\nu^2)}{\delta^4}P_Z, \tag{2.239}$$

$$w = w_0(\xi), \quad \frac{\partial w}{\partial t} = w'_0(\xi) \quad (t = 0,\ 0 \le \xi \le 1), \tag{2.240}$$

$$w = 0, \quad \frac{\partial^2 w}{\partial \xi^2} = 0. \tag{2.241}$$

In the system (2.238), the dimensional quantity A appears, which has different values in different problems and characterizes time step. The functions $w_0(\xi)$ and $w'_0(\xi)$ in the initial conditions (2.240) describe the nondimensional form of the corresponding dimensional functions $w'_0(y)$ and $w''_0(y)$ in the initial conditions (2.236). In order to solve (2.239), (2.240) and (2.241) numerically, the central finite differences are used:

$$\frac{\partial w}{\partial \xi} = \frac{1}{2\Delta\xi}(w(\xi+\Delta\xi) - w(\xi-\Delta\xi)),$$
$$\frac{\partial^2 w}{\partial \xi^2} = \frac{1}{(2\Delta\xi)^2}(w(\xi+\Delta\xi) - 2w(\xi) + w(\xi-\Delta\xi)),$$
$$\frac{\partial^4 w}{\partial \xi^4} = \frac{1}{(2\Delta\xi)^4}(w(\xi+2\Delta\xi) - 4w(\xi+\Delta\xi)$$
$$+ 6w(\xi) - 4w(\xi-\Delta\xi) + w(\xi+2\Delta\xi)).$$

These are substituted into (2.239) and into the boundary conditions (2.241). Hence, the problem (2.239), (2.240) and (2.241) reduces to a Cauchy problem for ODEs:

$$\frac{\partial^2 w}{\partial t^2}(\xi, t) = A_1(w(\xi + 2\Delta\xi, t) + w(\xi - 2\Delta\xi, t)) + A_2(w(\xi - \Delta\xi, t) + w(\xi - \Delta\xi, t)) + A_3 w(\xi, t) + A_4 + A_p P_Z(\xi, t), \tag{2.242}$$

$$w = w_0(\xi), \quad \frac{\partial w}{\partial t} = w_0'(\xi) \quad (t = 0,\ 0 \le \xi \le 1), \tag{2.243}$$

$$w(0, t) = 0, \quad w(1, t) = 0 \quad (t \ge 0), \tag{2.244}$$

$$\begin{aligned} w(0 + \Delta\xi, t) &= -w(0 - \Delta\xi, t), \\ w(1 - \Delta\xi, t) &= -w(1 + \Delta\xi, t) \quad (t \ge 0), \end{aligned} \tag{2.245}$$

where

$$A_1 = -\frac{E\delta^2 A^2}{12(1 - \nu^2)\rho_0' L^2 (\Delta\xi)^4}, \tag{2.246}$$

$$A_2 = 4A_1(6(\Delta\xi)^2\sigma_M - 1), \tag{2.247}$$

$$A_3 = 6A_1(4(\Delta\xi)^2\sigma_M + 1), \tag{2.248}$$

$$A_4 = 12A_1\sigma_M, \tag{2.249}$$

$$A_p = -\frac{12(1 - \nu^2)}{\delta^4} A_1, \tag{2.250}$$

and σ_M is defined by (2.223).

Here we refer to the equation system (2.242) because, for every internal point with coordinate ξ, equations of type (2.242) and the initial conditions (2.243) are valid.

To apply central differences to the boundary nodes, we use points outside the contour, i.e. points with $\xi = 0 - \Delta\xi$, $\xi = 1 + \Delta\xi$. The values of the transverse deflection w at these points are derived from the boundary conditions (2.245) and the values of the deflection w on the boundary from (2.244). If values of the transverse deflection w for boundary and off-boundary points are present on the right-hand side of the equation system (2.242), then these values are determined from (2.244) and (2.245).

The dimensionless tensile stress σ_M, following (2.224), after introducing the finite differences, has the following form:

$$\sigma_M = \frac{1}{2}\int_0^1 \frac{1}{(\Delta\xi)^2}\Big(w(\xi + \Delta\xi, t) - w(\xi - \Delta\xi, t)\Big)^2 d\xi - K\int_0^1 w(\xi, t)\, d\xi.$$

The integrals in the last equation were calculated by Simpson's rule with a step $\Delta\xi$ using the standard Fortran IV procedure QSF.

The Cauchy problem (2.243) and (2.242) was solved by a Runge–Kutta method of fourth order, using the standard procedure RGS in Fortran IV. Before applying the Runge–Kutta method, the system of ordinary differential equations of the second order was converted into a system of differential equations of the first order by introducing the derivative $\frac{\partial w}{\partial t}$ as a sought function.

The problem of free vibrations of as infinite, round, cylindrical elastic panel with simply supported edges was investigated in [339]. Following this work, we have assumed

$$K = 10, \quad \delta = 0.0091, \quad P_z' = 0, \quad A^2 = -\frac{\pi^2 E' h^2}{48(1-\nu^2)\rho_0' L^4}\left(1 + 96\frac{K}{\pi^6}\right),$$

$$w_0(\xi) = 0.1\sin(\pi\xi), \quad w_0'(\xi) = 0.$$

Figure 2.3 presents, as solid lines, the results of calculations obtained with the method described above, and the circles show the results presented by Volmir [339].

The problem of forced vibrations of an elastic cylindrical panel of finite length conforming the Kirchhoff–Love model, described by the full nonlinear equation of motion for the displacements, was numerically solved using the finite-difference approach and the Runge–Kutta method for the parameters

$$K = 10, \quad \delta = 0.0091, \quad w_0(\xi) = 0, \quad w_0'(\xi) = 0,$$

$$A^2 = -\frac{\pi^2 E' h^2}{48(1-\nu^2)\rho_0' L^4}\left(1 + 96\frac{K}{\pi^6}\right), \quad P_z' = -\frac{\pi^2 \delta^4 E'}{24}\left(1 + 96\frac{K^2}{\pi^6}\right),$$

and simple support of the edges. The results were obtained by Varygin, the Ph.D student of V. A. Krys'ko.

Figure 2.4 shows by circles some results for points of the central section of a cylindrical panel with edges in the ratio 1:10 obtained by Varygin, and by solid lines, results obtained by the method described in this book. The results of calculations by

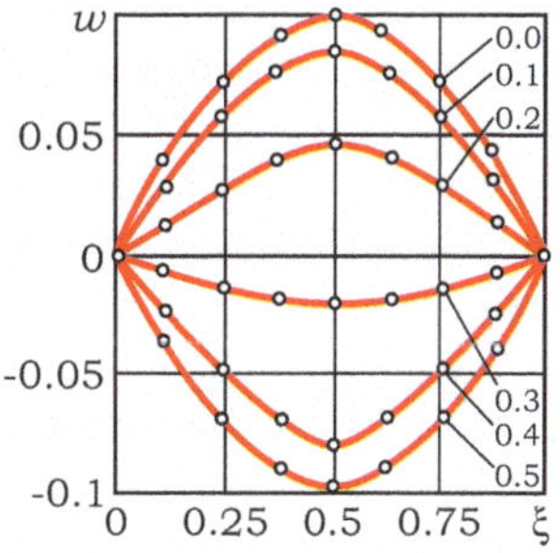

Fig. 2.3 Deflection of panel obtained by different methods, details in text

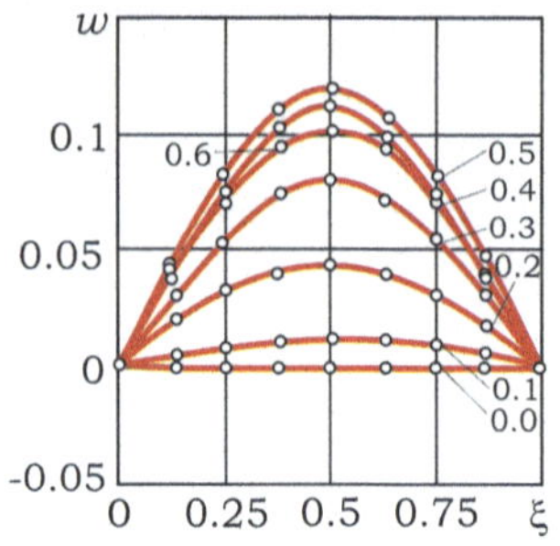

Fig. 2.4 Deflection of panel obtained by different methods; details in text

the method described above agree reasonably with analogous data obtained by other authors.

2.2.1.5 Method of Solution of the Problem

Let us analyse the problem of the stationary and nonstationary flow around an ideally rigid profile. In this subsection, a new method of calculations of a transonic flow is presented. This is the modified Godunov method.

Recently, numerical methods have been applied to calculations of a nonstationary transonic flow of an ideal gas. These methods are based on the Euler equations for gas dynamics or on asymptotic equations of motion of a transonic potential flow. We shall analyse the main trends in the development of numerical methods for calculations of nonstationary transonic ideal-gas flows.

Euler's integral equations have been solved by the Godunov method, based on the theory of failure by planar fracture [127–129]. This method has been widely applied to various cases of transonic flow around absolutely rigid bodies [156, 157, 171]. For solving the Euler equations in the transonic region of flow velocities, a two-step Lacks–Wendorff method [233] was applied, which was used for calculation of the flow in the presence of sinusoidal longitudinal oscillations. The calculations showed that in that case a shock wave excited oscillations while moving along a surface. Calculations by Magnus and Yoshihara [233] showed that an increase of the frequency of the oscillations decreases the amplitude of shock wave motion. Effective numerical solutions of the Euler equations for low-frequency cases using uncommon methods were presented by Beam and Warning [63, 64].

Various schemes for the full and low-frequency equations of the theory of small excitations of a transonic, close to homogeneous, nonstationary flow are presented in [340]. These schemes become the mixed Murman–Cole system [258] in the stationary case.

By extracting nonstationary addends from the solution (by a harmonic approach, applying a relaxation method with finite differences), Ehlers [105] and Weatherill [348], with co-workers, obtained numerical solutions of the full, transonic asymptotic equation. An analogous approach for solving not the full but the low-frequency asymptotic equations was presented by Traci et. al. [320]. As stated in reference

[72], the disadvantage of these methods is the limitations caused by the assumption that the flow is harmonic.

Calculations of nonstationary transonic flows of an ideal gas have been presented in many works, not listed here, where analytical or other solutions were given. For numerical solution, we chose a commonly known and relatively simple method.

As stated by Gorshkov [131], the Godunov method can be used effectively for solving problems of the interaction of flexible bodies with gas flows. To simplify the numerical calculations of the transonic flow, we shall apply idea of the Godunov method to an approximate theory of transonic flow [119].

Let us investigate the flow around an elongated symmetric profile in a two-dimensional nonstationary, low-frequency transonic flow of an ideal gas. The profile is oriented to have zero attack angle relative to the incoming uniform flow. Figure 2.2 presents a map of the flow. The asymptotic transonic equations were derived in Sect. 2.2.1.2. Hence, in dimensionless form, we obtain the following initial–boundary conditions:

$$\begin{aligned}\int\limits_{\Omega} ((1-M_{\infty}^2)U - \frac{1}{2}M_{\infty}^2(\chi+1)U^2)d\eta\, dt - 2\int\limits_{\Omega} M_{\infty}^2 U\, d\xi\, d\eta \\ + \int\limits_{\Omega} V\, d\xi\, dt = 0 \quad (t>0,\ -\infty<\xi<+\infty,\ \eta>0),\end{aligned} \tag{2.251}$$

$$\int\limits_{\Omega} U\, d\xi\, dt - \int\limits_{\Omega} V\, d\eta\, dt = 0 \quad (t>0,\ -\infty<\xi<+\infty,\ \eta>0), \tag{2.252}$$

$$V = \frac{\partial f}{\partial \xi}(1+U) + \frac{\partial f}{\partial t} \quad (t \ge 0,\ 0 \le \xi \le 1,\ \eta = 0), \tag{2.253}$$

$$U = 0,\ V = 0 \quad (t \ge 0,\ \xi < 0,\ \xi > 1,\ \eta = 0), \tag{2.254}$$

$$U = U_0(\xi,\eta),\ V = V_0(\xi,\eta) \quad (t = 0,\ -\infty<\xi<+\infty,\ \eta \ge 0), \tag{2.255}$$

$$V = 0 \quad (t \ge 0,\ \xi < 0,\ \xi > 1,\ \eta = 0). \tag{2.256}$$

The boundary conditions of no intersection (2.253) on the surface of the elongated profile $\eta = f(\xi, t)$ are projected onto the chord $\eta = 0$. At an infinite distance from the profile, the flow is treated as homogeneous and nonexcited—the condition (2.254). The initial velocity field of the gas flow in a general case is random—the condition (2.255).

Following the assumption about the symmetry of the problem, only the upper half-plane $\eta \ge 0$ will be investigated. On the symmetry axis of the flow the condition (2.256) should be satisfied—no mass flow over the symmetry axis. All recalculated boundary and initial conditions should match.

Method of Calculation of the Transonic Flow We assume that the surface is a side of an orthogonal parallelepiped, of which the edges are the axes of an orthogonal

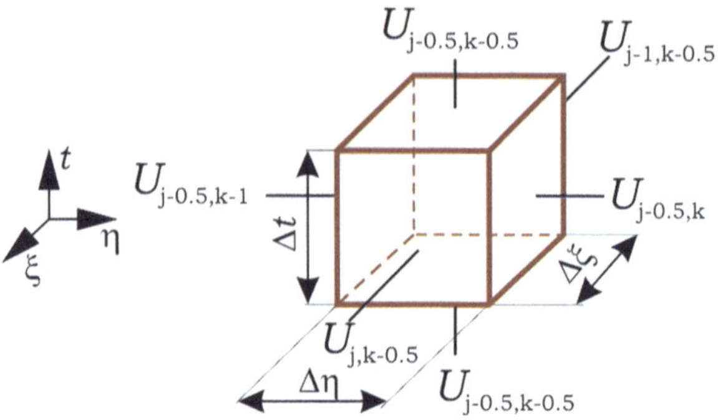

Fig. 2.5 Description of a panel by a parallelepiped

coordinate system ξ, η, t. We assume that the values of U and V are constant on the sides of the parallelepiped. The values of U are indicated in Fig. 2.5; similar indices are used for V. It is assumed also that these elements are constant in the internal volume and are described by $U_{j-1/2,k-1/2}$ and $V_{j-1/2,k-1/2}$, respectively. The nondeformed parallelepiped will be investigated.

Upper indices of similar form correspond to values at the time instant $t + \Delta t$. The integral equations of motion of the transonic gas flow (2.251) and (2.252) have the following form:

$$U^{j-\frac{1}{2},k-\frac{1}{2}} = U_{j-\frac{1}{2},k-\frac{1}{2}} + \frac{\Delta t}{2M_\infty^2 \Delta\xi}\Big(U_{j,k-\frac{1}{2}}\Big(1 - M_\infty^2 - \frac{1}{2}M_\infty^2(\chi+1)U_{j,k-\frac{1}{2}}\Big) - U_{j-\frac{1}{2},k-\frac{1}{2}}\Big(1 - M_\infty^2 - \frac{1}{2}M_\infty^2(\chi+1)U_{j-\frac{1}{2},k-\frac{1}{2}}\Big)\Big) + \frac{\Delta t}{2M_\infty^2 \Delta\eta}(V_{j-\frac{1}{2},k} - V_{j-\frac{1}{2},k-1}), \tag{2.257}$$

$$V_{j,k-\frac{1}{2}} = V_{j-\frac{1}{2},k-\frac{1}{2}} + \frac{\Delta\xi}{\Delta\eta}(U_{j-\frac{1}{2},k} - U_{j-\frac{1}{2},k-\frac{1}{2}}). \tag{2.258}$$

If U and V, which appear in the right-hand side of (2.257), are known at the time instant t, it is possible to calculate the value of $U^{j-\frac{1}{2},k-\frac{1}{2}}$ at the time instant $t + \Delta t$. At this time interval $t + \Delta t$, it is necessary to evaluate all variables in (2.257) and (2.258) before performing further calculations. It is possible to do this using (2.258) and some extra equations which can be found using the idea of the Godunov method for the analysis of a fracture, treating any of the sides of the parallelepiped, shown in Fig. 2.5 as a fracture surface. Here, differently from the Godunov method, the asymptotic problem of a transonic fracture is considered using the full set of the equations of ideal-gas dynamic. The case of a transonic, uniform fracture which is oriented transversely to the flow, treated on the basis of the asymptotic equations is called the Cauchy case. We have

$$\int_{\Omega} \left((1 - M_\infty^2)U - \frac{1}{2} M_\infty^2 (\chi + 1) U^2 \right) dt - 2 \int_{\Omega} M_\infty^2 U \, d\xi = 0, \tag{2.259}$$

$$U(\xi, t_0) = \begin{cases} U_-, & \xi < \xi_0, \\ U_+, & \xi > \xi_0, \end{cases} \tag{2.260}$$

where U_-, U_+, t_0, ξ_0 are constants and Ω is a closed contour on the surface (ξ, t).

In differential form, the integral equations (2.259) have the form

$$-2M_\infty^2 \frac{\partial U}{\partial t} + \left(1 - M_\infty^2 - M_\infty^2(\chi + 1)U\right) \frac{\partial U}{\partial \xi} = 0, \tag{2.261}$$

which, after we make the substitutions

$$\begin{aligned} &\overline{\xi} = -\xi, \quad 2M_\infty^2 \overline{t} = t, \\ &\overline{U}(\overline{\xi}(\xi), \overline{t}(t)) = 1 - M_\infty^2 - M_\infty^2(\chi + 1)U(\xi, t), \end{aligned} \tag{2.262}$$

leads to the known equation [286]

$$\frac{\partial \overline{U}}{\partial \overline{t}} + \overline{U} \frac{\partial \overline{U}}{\partial \overline{\xi}} = 0. \tag{2.263}$$

Solving the characteristic equation, we obtain the equations of the characteristic curves:

$$\begin{aligned} &\xi = \xi_0 - \frac{t - t_0}{2M_\infty^2}(1 - M_\infty^2 - M_\infty^2(\chi + 1)U), \\ &U = U_0. \end{aligned} \tag{2.264}$$

These characteristics, in the coordinate plane (ξ, t), as can be seen from (2.264), have the form of straight semi-lines, starting from the line $t = t_0$ on which the value of the function is constant and equal to the initial value of U_0. Different cases of the characteristics for the initial condition (2.260) are presented in Fig. 2.6. Figure 2.6a, d show the case $U_- < U_+$. Figure 2.6b shows the case $U_- > U_+$ and Fig. 2.6c shows the case $U_- = U_+$. The case $U_- < U_+$ appears in two parts of the figure, Fig. 2.6a, d. This shows that the solution in this case is not unique.

A case similar to (2.259) and (2.260) with discontinuous initial conditions was studied by Riemann, who found a solution containing an arbitrary functional,

Fig. 2.6 (a–d) Characteristics for different cases of the initial condition (2.260)

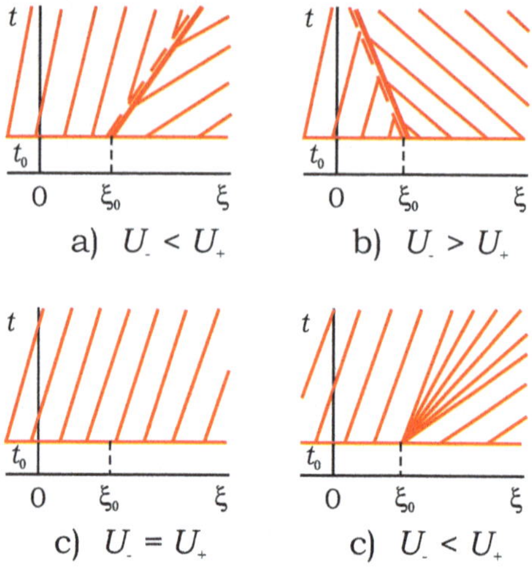

called a simple Riemann wave [93]. The nonautonomous solution to the differential equation (2.261) containing an arbitrary functional has the following form [66]:

$$g(C) = 1 - M_\infty^2 - M_\infty^2(\chi + 1)U,$$

$$C = \xi + \frac{t}{2M_\infty^2}(1 - M_\infty^2 - M_\infty^2(\chi + 1)U), \tag{2.265}$$

where $g(C)$ is any differentiable function of the single variable C. If we differentiate (2.265) with respect to ξ and t separately, we obtain

$$\frac{\partial U}{\partial \xi} = -\frac{2\frac{dg}{dC}}{(\chi + 1)\left(2M_\infty^2 - t\frac{dg}{dC}\right)}, \tag{2.266}$$

$$\frac{\partial U}{\partial t} = \frac{\frac{dg}{dC}(1 - M_\infty^2 - M_\infty^2(\chi + 1)U)}{(\chi + 1)M_\infty^2\left(2M_\infty^2 - t\frac{dg}{dC}\right)}. \tag{2.267}$$

To prove that we can really obtain a solution of (2.259) from (2.266) and (2.267), we observe that (2.265) is always a solution of (2.261), excluding the case when

$$2M_\infty^2 - t\frac{dg}{dC} = 0,$$

and for air $\chi = 1.4$ and $\chi + 1 \neq 0$. The function $g(C)$ can be expressed through the principal values of the function $U(\xi, t)$. For $t = t_0$, we obtain from (2.265)

$$g(C) = 1 - M_\infty^2 - M_\infty^2(\chi + 1)U(\xi, t_0).$$

For the problem (2.259), (2.260) under investigation, the solution of type (2.265), in the case $U_- > U_+$, has the form

$$U(\xi, t) = \begin{cases} U_-, & 2M_\infty^2(\xi - \xi_0) < A(t - t_0), \\ U_+, & 2M_\infty^2(\xi - \xi_0) > A(t - t_0), \end{cases} \tag{2.268}$$

where

$$A = M_\infty^2 - 1 + \frac{1}{2}M_\infty^2(\chi + 1)(U_- + U_+).$$

The solution of (2.268) is presented in Fig. 2.6b. If $U_- < U_+$, then (2.265) gives the following solution for the problem (2.259), (2.260):

$$U(\xi, t) = \begin{cases} U_-, & 2M_\infty^2(\xi - \xi_0) > A(t - t_0), \\ U_+, & 2M_\infty^2(\xi - \xi_0) < A(t - t_0), \end{cases} \tag{2.269}$$

where

$$A = M_\infty^2 - 1 + \frac{1}{2}M_\infty^2(\chi + 1)(U_- + U_+).$$

This case is represented in Fig. 2.6a. If $U_- = U_+$ then (2.265) gives a trivial solution,

$$U(\xi, t) = U_- = U_+, \tag{2.270}$$

which is represented in Fig. 2.6c.

It can be seen that the solution of (2.259) and (2.260) is not a single one. Besides the solution of type (2.265), i.e. the solutions (2.258), (2.259) and (2.270), the following self-modelling solution exists:

$$\begin{aligned} &U(\xi, t) = (1 - M_\infty^2)/(M_\infty^2(\chi + 1)) + 2(\xi - \xi_0)/(t - t_0)(\chi + 1), \\ &\quad M_\infty^2 - 1 + M_\infty^2(\chi + 1)U_- < 2M_\infty^2(\xi - \xi_0)/(t - t_0) \\ &\quad < M_\infty^2 - 1 + M_\infty^2(\chi + 1)U_+. \end{aligned} \tag{2.271}$$

Let us prove it. The self-modelling solution (2.271) corresponds to a simple symmetric expanding wave. This is represented in Fig. 2.6d in the form of a family of straight semi-lines, starting from an initial point of the explosion at $\xi = \xi_0$. Besides this solution (2.271), the solution (2.269) is also shown outside the family of lines.

It is necessary to explain the physical interpretation for a solution of the transverse fracture (2.259), (2.260) in the case of the problem under consideration. In the currently theory of quasi-linear hyperbolic equations, the following criteria for the existence of generalized solutions have been established: criteria based on entropy, smoothness, and conformity of the physical model with a more accurate one [93, 286]. In general, those different criteria are equivalent [93].

The criterion assumed most often is that of conformity of the model of an ideal gas with the more accurate model of a viscous gas. The asymptotic, dimensionless differential equation of a low-frequency, transonic uniform flow of a viscous gas has the form [288]

$$-2M_\infty^2\frac{\partial U_\alpha}{\partial t}+(1-M_\infty^2-M_\infty^2(\chi+1)U_\alpha)\frac{\partial U_\alpha}{\partial \xi}=\alpha\frac{\partial^2 U_\alpha}{\partial \xi^2}. \tag{2.272}$$

Using the transformation (2.262), this equation can be converted into Burger's equation [93, 286]:

$$\frac{\partial \overline{U}_\alpha}{\partial \overline{t}}+\overline{U}_\alpha\frac{\partial \overline{U}_\alpha}{\partial \overline{\xi}}=\alpha\frac{\partial^2 \overline{U}_\alpha}{\partial \overline{\xi}^2}. \tag{2.273}$$

If we analyse the Cauchy problem for (2.272) with the initial conditions (2.260), then we obtain a solution of the equations (2.259), (2.260), which can be obtained from the solution U_α by the following transformation:

$$U(\xi,t)=\lim_{\alpha\to 0}U_\alpha(\xi,t). \tag{2.274}$$

The Cauchy problem governed by Eqs. (2.260), (2.270) with the generalized initial conditions $U(\xi,t_0)=U_0(\xi)$ was presented by Cole [89]. Germain and Bader [125] proved the unique of the solution of (2.259) and (2.260), if the condition $U_- > U_+$ is satisfied at the points of discontinuity of the function $U(\xi,t)$; this yields the existence of a solution from the entropy criterion. Hopf [148] evaluated the solution of (2.272), (2.260) and formulated a strong limiting transformation (2.274). In this way, he proved the existence and unique of the generalized solution of the Cauchy problem (2.259) and (2.260) for bounded and rational initial functions $U_0(\xi)$. Hopf showed, that the generalized solution of (2.259), (2.260) obtained with the transformation (2.274) satisfies (2.261), which is described by the integral equation

$$\iint\limits_{t>0}\left(U\frac{\partial g}{\partial t}+\frac{U^2}{2}\frac{\partial g}{\partial \xi}\right)d\xi\,dt+\int\limits_{-\infty}^{+\infty}g(\xi,0)U_0(\xi)\,d\xi=0,$$

for any smooth finite function $g(\xi,t)$. Detailed bibliographies concerning the cases discussed in this section have been published in the literature [93, 286]. In the monograph [286], it is proved that the generalized solution of the Cauchy problem for (2.263), obtained by means of the limiting transformation (2.274) from the smooth solution of the Cauchy problem for (2.273), satisfies (2.263) in the sense of the integral equality

$$\oint_{\partial\Omega} \overline{U}(\overline{\xi}, \overline{t})\, d\overline{\xi} - \frac{1}{2}\overline{U}^2(\overline{\xi}, \overline{t})\, d\overline{t} = 0, \tag{2.275}$$

for any boundary consisting of a closed contour $\partial\Omega$ on which the measure of the set of points of discontinuity of $U(\xi, t)$ equals zero. The initial condition is satisfied in the sense of a weak convergence:

$$\lim_{\overline{t}\to\overline{t}_0} \int_0^{\overline{\xi}} \overline{U}(\xi, \overline{t})\, d\xi \rightrightarrows \int_0^{\overline{\xi}} U_0(\xi)\, d\xi. \tag{2.276}$$

Taking into account this result, Eqs. (2.262) can be substituted into (2.275):

$$\oint_{\partial\Omega} \left(M_\infty^2 - 1 + M_\infty^2(\chi + 1)U(\xi, t)\right) d\xi \\ - \frac{1}{4M_\infty^2}\left(1 - M_\infty^2 - M_\infty^2(\chi + 1)U(\xi, t)\right)^2 dt = 0.$$

Therefore, keeping in mind that the integral of a constant over a closed contour equals 0, one obtains the integral equation (2.259). Analogously, it can be proved from (2.276), using (2.262), that the initial conditions (2.260) are satisfied in the sense of the following limit:

$$\lim_{\overline{t}\to\overline{t}_0} \int_0^{\xi} U(\xi, t)\, d\xi = \int_0^{\xi} U_0(\xi, t_0)\, d\xi.$$

In that way it was proved that a single, generalized solution of the Cauchy problem (2.259), (2.260) exists. This solution is the limit (2.274) of the solution of the physically real Cauchy problem, with viscosity taken into account, for (2.268). This eigenvalue solution has the following form:

- if $U_- > U_+$, then

$$U(\xi, t) = \begin{cases} U_-, & 2M_\infty^2(\xi - \xi_0) < A(t - t_0), \\ U_+, & 2M_\infty^2(\xi - \xi_0) > A(t - t_0), \end{cases}$$

– if $U_- < U_+$, then

$$U(\xi,t)=\begin{cases} U_-, & 2M_\infty^2(\xi-\xi_0)<(t-t_0)B_-,\\ U_+, & 2M_\infty^2(\xi-\xi_0)>(t-t_0)B_+,\\ \frac{1-M_\infty^2}{M_\infty^2(\chi+1)}+\frac{2(\xi-\xi_0)}{(\chi+1)(t-t_0)}, & B_-<2M_\infty^2\frac{\xi-\xi_0}{t-t_0}<B_+,\end{cases}$$

– if $U_- = U_+$, then

$$U(\xi,t)=U_-=U_+. \tag{2.277}$$

Here,

$$A=M_\infty^2-1+\frac{1}{2}M_\infty^2(\chi+1)(U_-+U_+),$$

$$B_\mp=M_\infty^2-1+M_\infty^2(\chi+1)U_\mp).$$

In this case, the solution (2.269), which is represented in Fig. 2.6a, has no physical interpretation. The solution in the form of a fracture, represented in Fig. 2.6b, is said to be stable and represents a shock wave. Fracture occurs if the velocity components on the left and right sides of an initiating fracture satisfy the condition $U_- > U_+$. The shock wave moves along the flow. Its speed can be calculated from the Reikin–Gugiono condition for a shock wave, which, in the asymptotic transonic mode, has been included in (2.259) and has the form

$$2\,M_\infty^2\frac{\partial\xi}{\partial t}+1-M_\infty^2=\frac{1}{2}M_\infty^2(\chi+1)(U_-+U_+)\,,$$

where $\xi=\xi(t)$ describes the shock wave front.

If the velocity components on the left and right sides of the initial fracture satisfy the condition $U_- < U_+$, then the central straight wave vanishes (solution (2.271), shown in Fig. 2.6d), and the initial fracture disappears.

Following the solution (2.277) of the problem of failure by fracture (2.259), (2.260) and assuming that the initiating side of the fracture is the side of the parallelepiped with indices j and $k-\frac{1}{2}$, we obtain the following formulas for use in calculations:

$$U_{j,k-\frac{1}{2}}=\begin{cases} U_{j-\frac{1}{2},k-\frac{1}{2}}, & (D_{j,k-\frac{1}{2}}>0\wedge A_{j,k-\frac{1}{2}}\geq 0)\\ & \vee(D_{j,k-\frac{1}{2}}<0\wedge B_{j-\frac{1}{2},k-\frac{1}{2}}\geq 0)\\ & \vee D_{j,k-\frac{1}{2}}=0,\\ U_{j+\frac{1}{2},k-\frac{1}{2}}, & (D_{j,k-\frac{1}{2}}>0\wedge A_{j,k-\frac{1}{2}}<0)\\ & \vee(D_{j,k-\frac{1}{2}}<0\wedge B_{j+\frac{1}{2},k-\frac{1}{2}}\leq 0),\\ \frac{1-M_\infty^2}{M_\infty^2(\chi+1)}, & D_{j,k-\frac{1}{2}}<0\wedge B_{j-\frac{1}{2},k-\frac{1}{2}}<0<B_{j+\frac{1}{2},k-\frac{1}{2}},\end{cases} \tag{2.278}$$

where

$$D_{j,k-\frac{1}{2}} = U_{j-\frac{1}{2},k-\frac{1}{2}} - U_{j+\frac{1}{2},k-\frac{1}{2}},$$

$$A_{j,k-\frac{1}{2}} = M_\infty^2 - 1 + \frac{1}{2}M_\infty^2(\chi + 1)(U_{j+\frac{1}{2},k-\frac{1}{2}} + U_{j-\frac{1}{2},k-\frac{1}{2}}),$$

$$B_{j-\frac{1}{2},k-\frac{1}{2}} = M_\infty^2 - 1 + M_\infty^2(\chi + 1)U_{j-\frac{1}{2},k-\frac{1}{2}}.$$

In this equation, for $D_{j,k-\frac{1}{2}} > 0$ and $A_{j,k-\frac{1}{2}} > 0$, the value of $U_{j,k-\frac{1}{2}}$ has to be defined. Also, for logical correctness, it is necessary to mention that for $D_{j,k-\frac{1}{2}} < 0$ the condition $B_{j+\frac{1}{2},k-\frac{1}{2}} > B_{j-\frac{1}{2},k-\frac{1}{2}}$, is always satisfied.

Following the Godunov method [127–130] for the component V tangential to the transverse fracture under consideration (remembering that the component U is an excitation of the longitudinal component of the velocity vector of the gas particles in the transonic flow), we obtain

$$V_{j,k-\frac{1}{2}} = V_{j-\frac{1}{2},k-\frac{1}{2}}. \tag{2.279}$$

The following Cauchy problem describes the destruction of a longitudinal failure by fracture in the asymptotic transonic flow investigated here:

$$\oint V\,dt = 0,$$

$$V(\eta, t_0) = \begin{cases} V_-, & \eta < \eta_0, \\ V_+, & \eta > \eta_0. \end{cases} \tag{2.280}$$

The solution

$$V(\eta, t) = \begin{cases} V_-, & \eta - \eta_0 < t - t_0, \\ V_+, & \eta - \eta_0 > t - t_0, \end{cases}$$

for $\eta = \eta_0$ gives the following calculation formula: if the initial fracture appears on the parallelepiped face with indices $j - \frac{1}{2}$ and k,

$$V_{j-\frac{1}{2},k} = V_{j-\frac{1}{2},k-\frac{1}{2}}. \tag{2.281}$$

In general, it is necessary to mention that the solution corresponding to the case (2.280) may be any of the integrable functions $f_1(t)$ of the variable t which satisfy the following condition:

$$\lim_{t\to t_0} f_1(t) = \begin{cases} V_-, & \eta < \eta_0, \\ V_+, & \eta > \eta_0. \end{cases}$$

The nonuniqueness of the solution to the problem (2.280) results in a formula for estimating the components $V_{j-\frac{1}{2},k}$. To obtain a unique solution, additional equations and limitations are necessary. For example, we can replace (2.281) with the formula

$$V_{j-\frac{1}{2},k} = \frac{1}{2}(V_{j-\frac{1}{2},k-\frac{1}{2}} + V_{j-\frac{1}{2},k+\frac{1}{2}}),$$

which corresponds to an analogous expression known from the theory of fracture in unidimensional acoustics [?]. Following the Godunov method, the component U tangential to a longitudinal fracture on the face of the parallelepiped with indices $j-\frac{1}{2}$ and k can be represented in the form

$$U_{j-\frac{1}{2},k} = \begin{cases} U_{j-\frac{1}{2},k-\frac{1}{2}}, & V_{j-\frac{1}{2},k} \geq 0, \\ U_{j-\frac{1}{2},k+\frac{1}{2}}, & V_{j-\frac{1}{2},k} \geq 0. \end{cases} \tag{2.282}$$

In this way, Eqs. (2.257), (2.259), (2.278), (2.279), (2.281), (2.282) give a set of algebraic equations from which it is possible to evaluate all values necessary for solving (2.251)–(2.256).

It can be seen from these equations that the way presented here of solving a transonic flow around a profile appears to give a real and unique for all regions of the flow, including the vicinity of the fracture.

It should be emphasized that in our case the flow is calculated from the same equations both in the vicinity of the fracture and far from it, because the Reikin–Giugonio conditions for the shock wave (2.214) and (2.215) and Eqs. (2.257), (2.258) are obtained from the same integral equations of motion of a transonic flow (2.251), (2.252).

The values of the components U and V on a boundary can be treated as values on the corresponding sides of bounding parallelepipeds. The values of those elements on the profile surface are not set, because the relationship between them is known (2.253). To calculate the value of the component U on the profile surface and to take into account the variation of the distribution of U, the Newton extrapolation method is applied, here. For two intervals, one of length $\Delta\eta$ and the other of length $\Delta\eta/2$, the following computational formula holds:

$$U_{j-\frac{1}{2},1} = 1.875U_{j-\frac{1}{2},1+\frac{1}{2}} - 1.25U_{j-\frac{1}{2},2+\frac{1}{2}} + 0.375U_{j-\frac{1}{2},3+\frac{1}{2}}. \tag{2.283}$$

Therefore the component $V_{j-\frac{1}{2},k}$ can be calculated from the boundary condition of nonpenetration of the profile (2.283). Of course, it is possible to satisfy the boundary conditions in other ways. It is necessary to mention that calculations according to (2.283) result in no oscillations in the vicinity of the shock wave, but calculations performed taking into account other boundary conditions on the profile, for example in the form

$$U_{j-\frac{1}{2},1} = U_{j-\frac{1}{2},1+\frac{1}{2}},$$

show oscillations.

In the vicinity of the profile, the mesh used in the calculations must be more dense. This means that the parallelepipeds used in the calculations are smaller.

Referring to the characteristic equations (2.264) for the case of uniform transonic destruction of a transverse fracture flow (2.259), (2.260) and assuming that excitations separating two neighbouring fractures cannot meet in the time interval Δt, we obtain a condition for stability of the numerical calculations of the transonic gas flow:

$$\Delta t \leq \min_{U} \frac{2M_{\infty}^{2}}{|M_{\infty}^{2} - 1 + M_{\infty}^{2}(\chi + 1)U|} \Delta\xi,$$

which corresponds to the relationship between Δt and $\Delta\xi$.

Cases of Unstable and Stable Flow Around an Absolutely Rigid Profile The following two problems will be considered. The first, is that of an unstable flow around a rigid profile as a result of a sudden start of the motion of the profile at a transonic speed or, equivalently, a sudden start of the gas flow around the profile. This problem corresponds to the following initial velocity field:

$$U_0(\xi, \eta) = 0, \;\; V_0(\xi, \eta) = 0. \tag{2.284}$$

The second problem is that of the stationary motion of a transonic gas flow around a rigid profile, which can be obtained from any unstable motion (this means that the stationary motion not dependent on the functions $U_0(\xi, \eta)$ and $V_0(\xi, \eta)$), by application of the method of stabilizing in time as $t \to \infty$. The initial conditions (2.255) and boundary condition (2.253) (when the equations (2.284) are satisfied) can coincide at $t = 0$ for points on the profile surface only when the profile is located in a plane oriented at zero angle of attack. This is valid when Eq. (2.284) are satisfied. In other words, we can say that the first problem, concerning a sudden start of the motion of the profile, is equivalent to the problem of high-speed buckling of a plate into the given profile in the gas flow.

Let us investigate the flow around a symmetric profile, modelled as a long arc, oriented at zero attack angle to the incoming flow, with sudden acceleration. For the calculations, we have chosen the following values of the computational parameters: number of parallelepipeds for calculation on the surface of the profile, 30; time interval, $\Delta t = 0.001$. The values of the pressure factors c_p at different time instants for profile with a shape coefficient $f_* = 0.05$ are presented in Fig. 2.7. This profile is placed in flow with a Mach number $M_\infty = 0.785$. In Fig. 2.7, a shock wave is seen as a zone of very high gradient. The shock wave decreases in intensity on moving from the back to the front of the profile. The solid lines shows the results

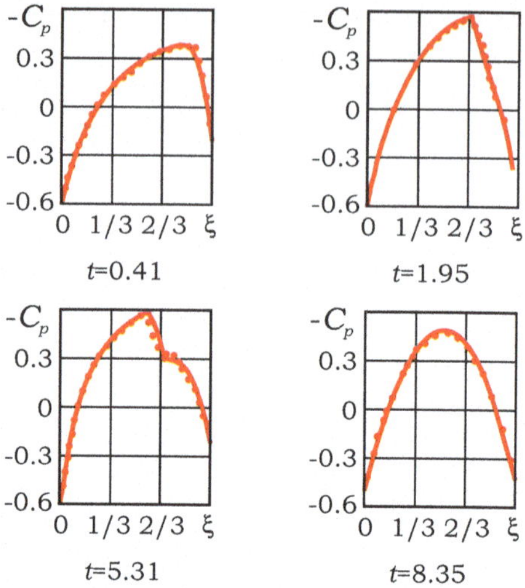

Fig. 2.7 Distribution of the pressure factor c_p at different time instants for $f_* = 0.05$, $M_\infty = 0.785$

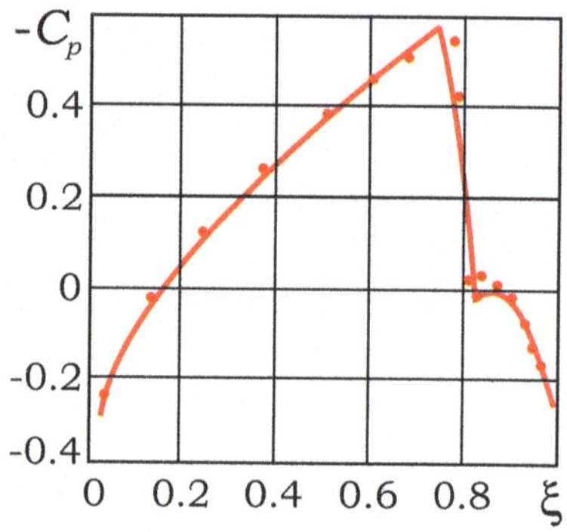

Fig. 2.8 Distribution of the pressure factor c_p for $f_* = 0.03$, $M_\infty = 0.864$

of calculations of the transonic flow obtained with the method described above. The dots show the results given in [72]. Comparison of these results shows reasonable agreement between them.

To obtain better control of the calculation method, another flow around a long, symmetric profile, modelled as a long arc, oriented at zero attack angle to the incoming flow, was considered. This incoming flow was characterized by a Mach number $M_\infty = 0.864$, and the corresponding profile factor was $f_* = 0.03$. Calculations were performed applying the asymptotic method for the time.

In Fig. 2.8 the solid line represents the distribution of the pressure factor c_p on the profile surface, obtained by the calculation method described above. The dots represent analogous results taken from [311]. As can be seen from Fig. 2.8 calculation results show good equivalency of these methods.

We conclude that the numerical method for calculation of transonic flows presented in this section allows us to calculate stable and unstable flows and gives good results.

2.2.1.6 Interaction Between a Flow and a Panel

In this section we shall describe an algorithm for calculation of the interaction between a flow and a panel. The numerical solution of the interaction of a transonic flow and a shallow panel (2.222)–(2.233) consists of two separate problems, of the motion of the panel and of the flow. In solving problems of the interaction of a shell with a flowing medium, it is very important to perform the calculations for the shell and the gas together.

Let us assume that at a time instant t, the values of w, $\partial w/\partial t$, U, V, P_z, f are known. Knowing the values of w, $\partial w/\partial t$ and P_z, from (2.222) and (2.223) (for the time instant t), taking into account the conditions (2.232) and (2.233), it is possible to calculate the values of w and $\partial w/\partial t$ at the next time instant $t + \Delta t$. From the known values of U, V and f obtained from (2.226) and (2.227) at the time instant t, with the conditions (2.228), (2.229), (2.230) and (2.231), the values of U and V at the time instant $t + \Delta t$ can be evaluated. The load at the time instant $t + \Delta t$ is calculated from (2.224) and (2.225). Knowing the value of w at the time instant $t + \Delta t$, the shape of the panel is computed from the second equation of (2.231); this means the value of the function f, at that time instant $t + \Delta t$. Iterating the time step, the calculations described above are performed repeatedly.

At every time step, the results of the calculation of the transonic flow or of the shallow shell must be verified. This is performed by repeating the calculations with a smaller time interval and a greater number of smaller iterations. The results of these calculations are then compared. If that comparison shows that the results agree with the required accuracy, then the calculations of the solution are continued. If the comparison gives a difference between the results, it is necessary to change the time integration interval Δt. For the optimal setting of the time interval Δt, it is possible to perform only one calculation with that value of the time interval for the flow and for the shell. Computations showed that, for the required accuracy, the calculations of the panel were less sensitive and allowed a higher value of the time interval Δt than for the calculation of the transonic flow. The calculations showed that for the chosen rounding of the results, the best values of the time interval Δt were between 0.001 and 0.003. The influence of the value of the time interval Δt will be discussed for the following numbers of elements in the mesh: on the shell, 30 equally spaced nodes were created; in the flow over the panel in the direction of the ξ axis, 30 equally spaced nodes were also created; in the direction of the ξ axis there were approximately 60 nodes in total in the flow, with the distance between them increasing with the distance from the panel. In the direction of the η axis approximately 30 nodes were created, with the distance between them increasing with the distance from the panel.

The stability of the numerical calculations of the problem was checked by comparing the results for different values of the time interval Δt. If the results agreed within the allowed limits then the calculations were treated as convergent.

The calculations of the transonic flow are described in Sect. 2.2.1.5. Here we mention only that for estimating the initial velocity field of the particles of the gas flow, an initial calculation of the flow around a rigid profile was applied.

Calculations of the dynamic motion of the shallow shell were performed according to the method presented in the Sect. 2.2.1.4. It is necessary to estimate the values of some dimensionless parameters. Comparing the sets of dimensionless equations (2.219) and (2.238), one can obtain the following difference in transverse load:

$$A = \frac{q_\infty}{L}.$$

In the calculations, the interval of dimensionless time spent in stabilizing the flow around a rigid initial profile was set to 2.4 in all cases considered. As can be seen from the numerical calculations of a transonic flow around a rigid panel presented in Sect. 2.2.1.5, sufficiently good stability of the flow can be expected after this time interval.

For performing calculations, it is necessary to set the values of the main characteristic parameters. The pressure in the flow, at an infinite distance from the profile was taken as equal to atmospheric pressure, $P_\infty = 101\,330$ Pa. The density of air was taken as $\rho_\infty = 1.2928$ kg/m^3, and the adiabatic Poisson coefficient as æ $= 1.4$. It was also assumed that the panel was made from steel with density $\rho_0' = 7953\,\text{kg/m}^3$, Young's modulus $E' = 20 \times 10^{10}$ MPa and Poisson's ratio $\nu = 0.3$ [177]. Transforming these parameters into dimensionless values, we obtain the following values: $\rho_0 = 6163.8$, $E = 1\,973\,749$. These parameters were not changed during the calculations. The Mach number for the incoming flow varied from 0.75 to 1.0. The relative thickness of the panel δ, the pressure P_0 on the side opposite to the transonic flow, and the geometric dimensionless parameter K of the circular cylindrical panel had varying values. In detail, the relative thickness of the panel δ varied from 0.0005 to 0.01. The initial maximum deflection of the infinite circular cylindrical panel f_* was set to two values, 0.03 and 0.05. For each value of f_*, values of δ were assumed such that the panel started to buckle in the estimated time of calculation. This time interval, during which the interaction between the panel and the transonic flow was investigated, was approximately 6–10 dimensionless time units.

2.2.2 *Stable Vibrating Panel Within a Transonic Flow*

2.2.2.1 Shock Wave Motion

Let us discuss the characteristics of a transonic flow. The complexity of the problems of a transonic flow around a profile is caused by specific characteristics of transonic flows in comparison with subsonic and supersonic flows. The occurrence of shock waves, which can expand and move along the profile (shell) and disappear in the transonic flow, is characteristic of this type of flow. A shock wave is a jump of the values of certain parameters, including the pressure. It is necessary to consider the description of the motion of shock waves.

Classification of the Motion of a Shock Waves To classify the motion of a shock wave during the interaction of a panel and a transonic flow it is necessary first to itemize some simple cases of shock wave motions:

U—motion upstream the flow only,
D—motion downstream the flow only,
S—steady state of a shock wave,
N—no shock wave in a flow.

More complicated forms of shock waves can be created from those simple forms of motion. For example (UD) means that a shock wave moves up and then down, does not disappear and does not stand still (the parentheses mean that this motion repeats many times); (UDN) means that a shock wave moves up, moves down and disappears. This repeats many times. It is sometimes important to put in parentheses (after each letter describing a simple shock wave mode) the coordinates of the time interval and flow space in which such a motion of a shock wave takes place.

Creation, Existence and Destruction of a Shock Wave A shock wave can appear in a transonic flow when a local transonic zone appears. This means that the Mach number must have a value greater than 1 in a certain sub-zone of a flow with $M < 1$. The case of existing local supersonic waves without a shock wave cannot be taken into account. By applying a known estimate of the local Mach number M in a transonic zone [293], we can obtain an approximate condition for the existence of a shock wave in a flow in the following form:

$$M^2 = M_\infty^2 \left(1 + (\chi + 1)U + (\chi - 1)\frac{\partial \varphi}{\partial t}\right) > 1. \tag{2.285}$$

If at a certain point of the flow the velocity has locally reached the sound speed, then $M - 1$ at this point, and, following (2.285), the longitudinal velocity excitation of the gas particles reaches some critical value U_*. This critical value can be calculated from (2.285) after setting $M = 1$:

$$U_* = \frac{1 - M_\infty^2 - (\chi - 1)\frac{\partial \varphi}{\partial t}}{(\chi + 1)M_\infty^2}. \tag{2.286}$$

For a low-frequency motion $\left(\frac{\partial \varphi}{\partial t} = 0\right)$, the critical sound speed depends on the Mach number M_∞ of the incoming flow:

$$U_* = \frac{1 - M_\infty^2}{(\chi + 1)M_\infty^2}. \tag{2.287}$$

The critical values of U_*, calculated from (2.287) for the values of M_∞ used in the computations described in this book, are the following: $U_*(0.785) = 0.2595$; $U_*(0.825) = 0.1955$; $U_*(0.864) = 0.1415$; $U_*(0.9) = 0.098$; $U_*(1) = 0$. Of course, as can be seen from (2.286), the values of the critical parameter U_* in a nonstationary flow differ slightly from the values given above. A shock wave which satisfies the condition for the existence of shock waves (2.285) disappears when this condition is disturbed; this means when, at all points of the flow, the local Mach number satisfies the inequality $M < 1$. In a subsonic flow, a moving shock wave may exist, but the shock wave first appears in a supersonic flow.

Direction of Movement of a Shock Wave From the Reikin–Giugono condition, for a shock wave in the transonic low-frequency approximation (2.215), we obtain a solution for the velocity of motion of a shock wave:

$$\frac{\partial \xi}{\partial t} = \frac{1}{2M_\infty^2}\left(M_\infty^2 - 1 + \frac{1}{2}M_\infty^2(\chi + 1)(U_- + U_+) - \left(\frac{V_- - V_+}{U_+ - U_-}\right)^2\right). \tag{2.288}$$

A shock wave can move up the flow if the following condition is satisfied:

$$M_\infty^2 - 1 + \frac{1}{2}M_\infty^2(\chi + 1)(U_- + U_+) - \left(\frac{V_- - V_+}{U_+ - U_-}\right)^2 < 0. \tag{2.289}$$

This is because the velocity of a shock wave moving up the flow is negative. The inequality (2.289) contains values of parameters on both sides of the shock wave front, and the Mach number of the incoming uniform flow M_∞. Analogously, a shock wave moves down the flow if the following condition is satisfied:

$$M_\infty^2 - 1 + \frac{1}{2}M_\infty^2(\chi + 1)(U_- + U_+) - \left(\frac{V_- - V_+}{U_+ - U_-}\right)^2 > 0. \tag{2.290}$$

It is necessary to mention that the inequalities (2.289) and (2.290) must be considered at a particular instant of time and have a local character. As is known, the velocity component tangential to a shock wave front is continuous on passing through the shock wave. Therefore, treating the shock wave as approximately almost vertical, it can be assumed that

$$V_+ = V_-. \tag{2.291}$$

Taking into account the last Eq. (2.291), the conditions for the direction of motion of a shock wave (2.289) and (2.290) can be converted into the following form:

up:

$$U_- + U_+ < 2U_*; \tag{2.292}$$

down:

$$U_- + U_+ > 2U_*. \tag{2.293}$$

For an existing shock wave, another condition can be derived from (2.285) and (2.287):

$$U_- > U_+. \tag{2.294}$$

From the values of U_* presented above one can conclude that with an increase of the Mach number of the incoming flow M_∞ in the interval under consideration, the relative influence of U_* decreases. This means that for a nonstationary flow around the same profile, motion of a shock wave down the flow may occur earlier for flows with a higher value of the Mach number of the incoming stream M_∞ than for flows with lower values of the Mach number. The value of U_* is the limit of the sum of components U on opposite sides of the shock wave front, $U_- + U_+$, for a certain value of the Mach number of the incoming flow. This provides an estimate of the direction of motion of the shock wave at any given time t. Of course, the value of U_* given here is approximate. As can be seen from (2.287), the value of U_* depends only on the Mach number of the incoming uniform flow M_∞ and appears on the right-hand side of (2.292) and (2.293). The left-hand side of these equations, which means the sum $U_+ + U_-$, depends on many coefficients, including the form of the flow profile.

In a stabilizing stream, the location of a shock wave should also stabilize, and the inequalities (2.289), (2.290), (2.292), (2.293) should become equalities. In such a stabilizing flow, the sum $U_+ + U_-$ can be approximated as $2U_*$. In this case the condition for a shock wave to be stationary is equivalent to the equation

$$U_+ + U_- = 2U_* \tag{2.295}$$

A shock wave, as can be seen from (2.294), can be stationary only in a supersonic flow; this means that the velocity U_- exceeds the local sound speed.

It is obvious that a change of the form of the profile changes the flow around it and the location of any shock waves. The pressure on the panel surface changes correspondingly. The flow and pressure on the panel surface can vary for small changes of the profile. It should be emphasized that for certain transonic flows, very small variations of the profile shape may cause relatively strong changes in the flow and pressure on the panel surface.

The Ability of a Shock Wave to Move up the Stream Is Not Obvious Such a motion of shock waves can be periodic–one shock wave behind another. Let us investigate slightly supersonic flows; this means flows for which the local Mach number M in some bounded sub-regions of the flow is slightly greater than one.

We shall prove that slightly supersonic, almost uniform gas flows exist for which shock waves appear at the rear, move up and disappear. This scheme is repeated. As can be seen from (2.293), the values of the dimensionless longitudinal components U of the velocity excitation vector are close to zero. This means that there exists a positive number ε_U, much less than one, such that $|U_-| < \varepsilon_U$ and $|U_+| < \varepsilon_U$. Thus, for a slightly overcritical transonic flow, we obtain the following from the triangle inequality:

$$|U_- + U_+| \leq |U_-| + |U_+| < 2\varepsilon_U .$$

From this expression, the following inequality can be derived, which is also satisfied:

$$U_+ + U_- < 2\varepsilon_U . \tag{2.296}$$

This yields the result that the inequality (2.292) is also satisfied if

$$\varepsilon_U \leq U_* . \tag{2.297}$$

Let us assume that for any positive constant number ε_U (as mentioned earlier, close to zero), such a transonic flow exists for which the inequalities (2.296) and (2.297) are satisfied. It can be proved that for every U_*, such a flow can be found for which the condition (2.292) is satisfied.

This means that a flow in which a shock wave moves upstream is slightly overcritical and nonstationary. Such a flow appears during flow around a slightly deformed profile panel of a weakly overcritical gas stream with local supersonic zones. In the stabilized state of an overcritical stream, there are no shock waves. Thus, when small deformations of the profile occur, a shock wave is created as in a nonstationary forced flow. A similar situation appears in an explosion of a profile at transonic speed. For a nonstationary flow, the component $\left(\frac{\partial \varphi}{\partial t}\right)$ in (2.285) cannot be neglected. This gives the possibility for the local Mach number to reach a value of 1 earlier than in a stationary flow. This explains the appearance of a shock wave in a nonstationary flow in cases when, for the same parameters, there is no shock wave in a stationary flow. In our investigations, the low-frequency approximation has been investigated, but, as can be seen in Sect. 6.1, $\left(\frac{\partial \varphi}{\partial t}\right)$ is partially taken into account in the equations of motion of a transonic flow element, and it is possible to find shock waves in nonstationary flows when numerical calculations are performed. In that case a shock wave moves upstream and disappears, losing its intensity, while the flow is stabilizing.

Two reasons for shock wave motion are possible: (a) self-excited motion of a shock wave in a forced flow, and (b) motion caused by a change of the geometry of the profile. In slightly overcritical nonstationary flows, the motion of shock waves can be self-excited.

Let us consider some examples of the motion of shock waves. Figure 2.9 shows diagrams of the distribution of shock waves on the surface of a profile panel at certain instants of time t, for the following values of the parameters main flow and panel parameters: (a) $M_\infty = 0.864$, $K = 34.16$, $\delta = 0.007$ $(f_* = 0.03)$, $P_0 = 0$ and fixed support; (b) $M_\infty = 0.9$, $K = 34.16$, $\delta = 0.007$ $(f_* = 0.03)$, $P_0 = 0$ and fixed support; (c) $M_\infty = 0.785$, $K = 36.8$, $\delta = 0.005$ $(f_* = 0.03)$, $P_0 = 0$

Fig. 2.9 (**a–d**) Distribution of shock wave motion for different parameters–details in text

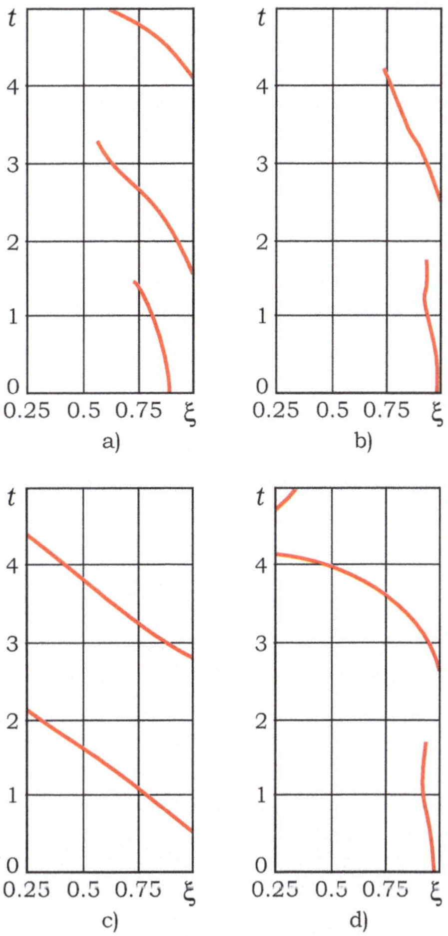

and fixed support; (d) $M_\infty = 0.9$, $K = 36.8$, $\delta = 0.065$ ($f_* = 0.03$), $P_0 = 0$ and fixed support. Figure 2.9c shows a slightly overcritical transonic flow, for which the shock wave motion can be represented as type D2. Figure 2.9d shows the shock wave motion in a case where the panel loses its dynamic stability. This motion of a shock wave can be described as UDNUDN . . .

2.2.2.2 Interaction of a Flow and a Panel

In this section we shall discuss some aspects of the interaction of a flow and a panel. The results presented here illustrate the application of the method described in the previous sections method for solving problems of the interaction of a transonic flow and a shock wave.

As one of the examples, we shall discuss the results for the interaction of a transonic flow with an elastic, infinite, circular cylindrical panel for the following values of the main parameters: $M_\infty = 0.9$, $K = 34.16$, $\delta = 0.007$ ($f_* = 0.03$), $P_0 = 0$. Both edges of the panel are clamped. Figure 2.10 presents the distribution at different time instants of the dimensionless pressure along a section of the panel, calculated from the following formula (the right-hand side of (2.222)):

$$\Delta p = \frac{12(1-\nu^2)}{E'\delta^4}(P' - P_0'). \tag{2.298}$$

This figure shows also the dimensionless transverse deflection w. The pressure Δp represented by a continuous line, and the deflection w by a dashed line. A zone of strong positive gradient on a diagram of the pressure Δp a shock wave. As can be seen from Fig. 2.10, a shock wave exists at the initial time instant $t = 0$. After the start of deformation of the panel, the shock wave moves a short distance upstream and moves down to the rear of the panel, loses its intensity and disappears. A shock wave appears again at $t = 2.8$ at the rear of the panel and starts to move forwards. Such a motion of a shock wave is repeated many times. In the classification of shock wave motion presented in Sect. 2.2.2.1, shock wave has a motion of type (UDN). It can be seen that at the beginning, the deflection w increases over the whole panel. The rear half of the panel has negative deflections; this means that the panel buckles in that region. At the same time, in the front part of the panel, the positive value of the deflection w increases. After the time instant $t = 2.8$, the deflection w decreases in the front part and increases in the rear part.

For better representation of the changes of the transverse deflection w and pressure Δp, diagrams of the variations in time of w and Δp at different points of the panel in the case investigated here are presented in Figs. 2.11 and 2.12. Figure 2.13 presents the variation of the location of the shock wave with time. From Fig. 2.11 it can be seen that the absolute value of the deflection w does not exceed 1, and its value varies from -0.7 to 1. The deflection w at the point $\xi = 0.75$ has a negative value; at the points $\xi = 0.25$ and $\xi = 0.5$ the deflection w is positive. This means that the rear part of the panel buckles. This buckling is caused by the fact that, directly before a shock wave located near the rear of the panel, the pressure in the flow acting on the panel is significantly lower than that on the front part of the panel. From the diagrams presented in Fig. 2.11, it can be seen that in the time interval $(0 - -0.8)$ from the start of the interaction, the whole panel deflects under the pressure caused by the flow. Later, in the next time interval $(0.8 - -5.0)$, the rear part of the panel buckles, but the central and front parts are still deflecting. The frequencies of oscillation of the points of the panel may be considered. At the point with coordinate $\xi = 0.75$, a main low-frequency oscillation and a higher-frequency oscillation with lower amplitude can be observed. Similarly, low and high frequencies, as can be seen in Fig. 2.11, appear at the points with coordinates $\xi = 0.25$ and $\xi = 0.5$. The low frequencies of the points with $\xi = 0.25$ and $\xi = 0.75$ are almost equal, but the oscillations have opposite phases. The high

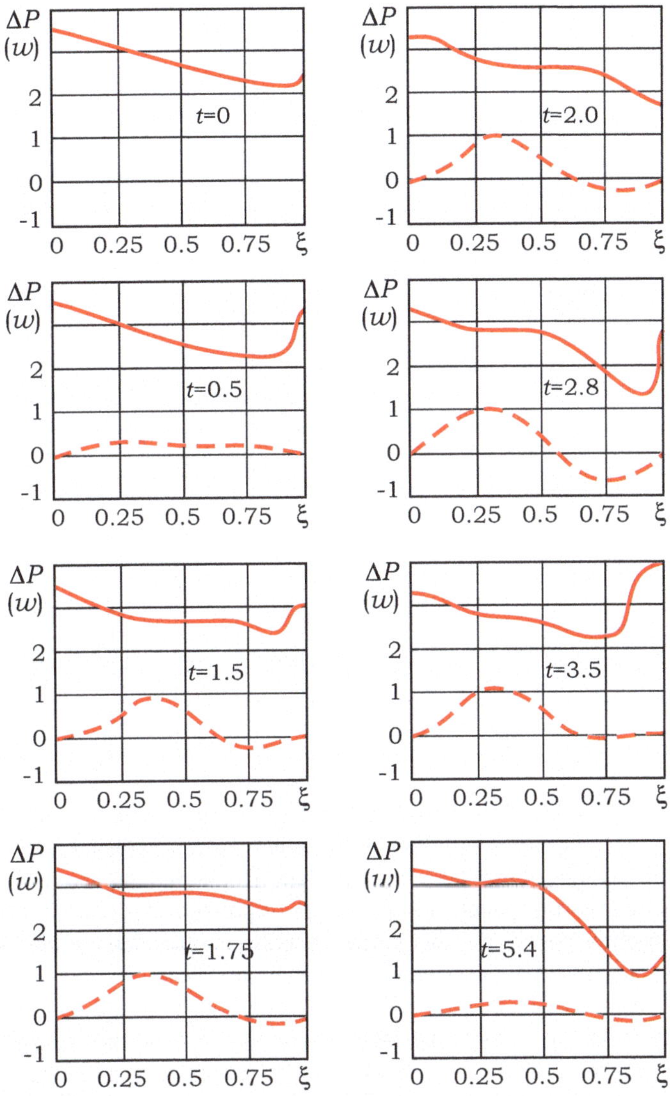

Fig. 2.10 Distribution of dimensionless pressure Δp (*solid line*) and transverse deflection w (*dashed line*) along a section of the panel at different time instants

frequencies of the oscillations of the deflection w at different points of the panel are approximately equal.

It can be seen from the diagrams presented in Fig. 2.12 that the lowest amplitude of the oscillations of the pressure Δp occurs at the point $\xi = 0.25$, which is located close to the front of the panel, and the highest amplitude of the pressure oscillations occurs at the point $\xi = 0.75$, in the rear part of the profile, in the zone of shock

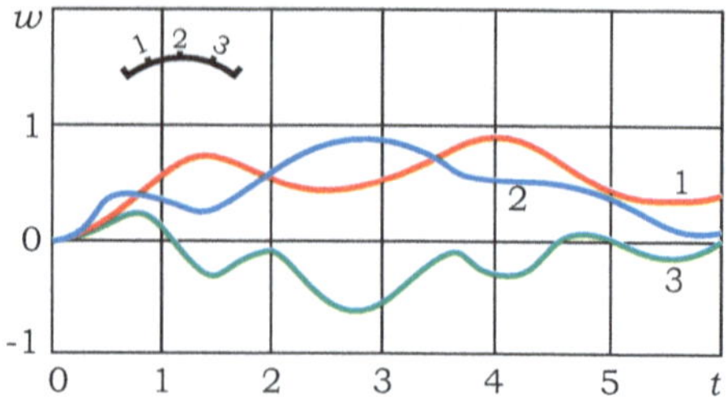

Fig. 2.11 Time variation of transverse deflection w at different points of the panel

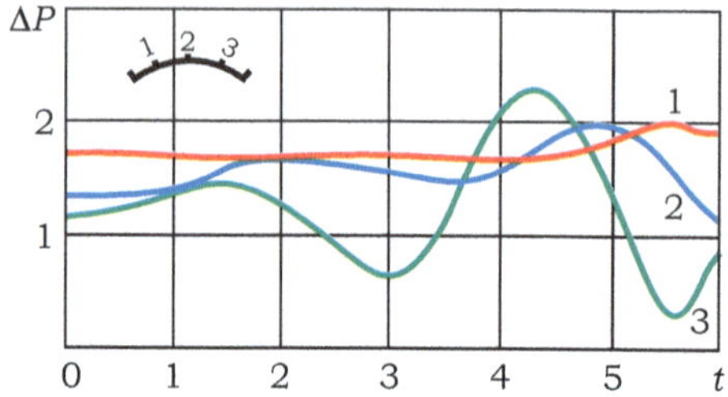

Fig. 2.12 Time variation of pressure Δp at different points of the panel

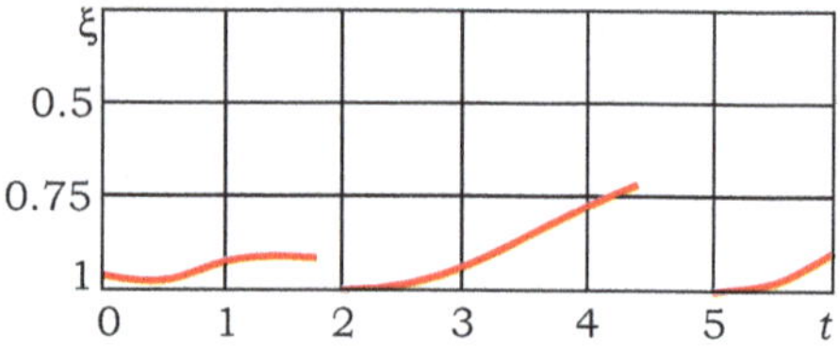

Fig. 2.13 Shock wave motion

wave motion. From these diagrams and Fig. 2.13, it can be seen that the oscillation frequencies of Δp at different points of the panel are approximately the same and match the oscillations of the shock wave motion. This shows that motion of a shock wave in the rear part of panel causes changes of pressure Δp not only in the rear part of the profile, close to the shock wave, but also in other parts of the panel. It is obvious that at greater distances pressure changes are smaller.

Let us consider a second case of the interaction of this panel with a transonic flow. The distributions of the dimensionless pressure Δp and transverse deflection w along a section of the panel at different time instants are presented in Fig. 2.14 for the following parameters: $M_\infty = 0.864$, $K = 34.16$, $\delta = 0.007$ ($f_* = 0.03$), $P_0 = 0$.

The changes of the transverse deflections w with time at various points of the panel are shown in Fig. 2.15. Figure 2.16 presents diagrams of pressure Δp versus time for the same points as those for which deflections are shown in Fig. 2.15. A diagram of the shock wave motion in this case is presented in Fig. 2.17. The case presented in Figs. 2.14, 2.15, 2.16, and 2.17 differs from that presented in Figs. 2.10, 2.11 and 2.13 only in the Mach number M_∞ of the incoming stream. In the first case $M_\infty = 0.9$, and in the second case $M_\infty = 0.864$. In general, the characteristic curves in Fig. 2.14, the transverse deflections w in Fig. 2.15 and the pressure Δp in

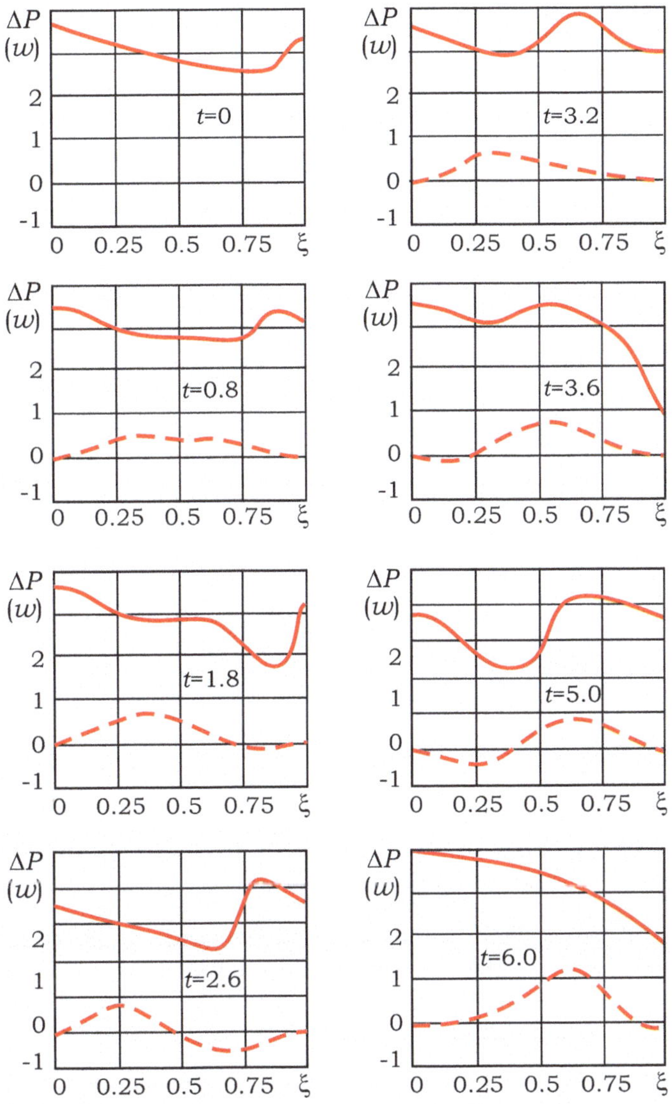

Fig. 2.14 Distribution of dimensionless pressure Δp (*solid line*) and transverse deflection w (*dashed line*) along a section of the panel at different time instants

Fig. 2.16 match the corresponding curves in Figs. 2.10, 2.11 and 2.12. Analysis of the deflection diagrams in Figs. 2.14 and 2.15 leads to the following conclusions: the panel, after first buckling in the rear part buckles also in the front part, and the buckling in the rear part changes into a deflection inside the panel. The high frequencies of the oscillations of the transverse deflection w at the points with coordinates $\xi = 0.25$ and $\xi = 0.75$ approximately match each other. The low

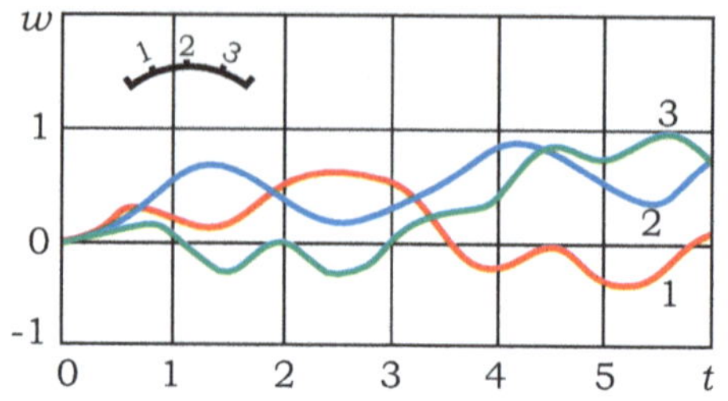

Fig. 2.15 Time variation of transverse deflection w at different points of the panel

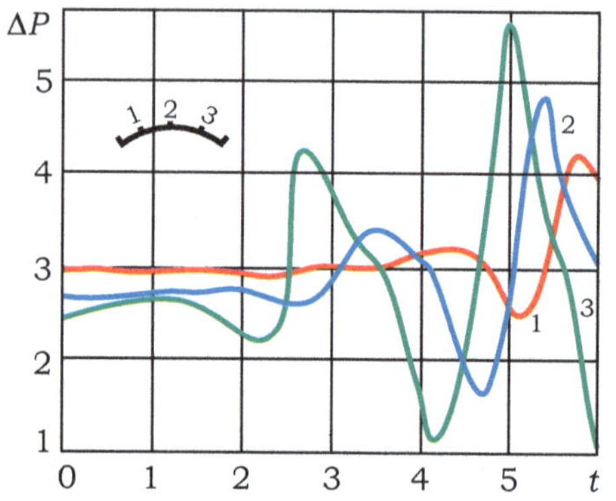

Fig. 2.16 Time vibration of pressure Δp at different points of the panel

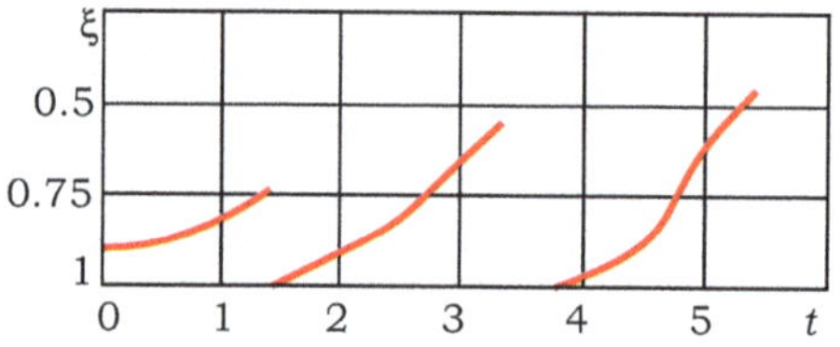

Fig. 2.17 Shock wave motion

frequencies of the oscillations at these points, symmetrically placed around the centre of the panel, also almost match, but the oscillations have opposite phases. The panel buckles successively in the front and rear parts, as was observed earlier in the first case of this interaction.

Dependence on Main Parameters Comparison of the diagrams in Figs. 2.13 and 2.17 leads to the conclusion that the frequency at which shock waves are created is higher for the flow with the lower value of the Mach number for the incoming stream, $M_\infty = 0.864$, than for the flow with the higher value of the Mach number, $M_\infty = 0.9$. Also, for the lower value of the Mach number, shock waves move faster, and further from the rear to the front of the panel, than for the higher value of the Mach number. It is easy to see from a comparison of the shock wave motion presented in Figs. 2.13 and 2.17 and the pressure distribution Δp presented in Figs. 2.12 and 2.16 that the peaks of the pressure Δp at the points of the panel surface investigated correspond to shock waves. A high pressure appears after a low pressure, as at a shock wave front. The peak of the pressure appears at the top of waves of pressure oscillations. For a slightly overcritical gas flow and a small amplitude of the oscillations of the transverse deflection w, moves along the panel surface almost independently of the changes of profile geometry, but with a strong interaction with the flow, which results in peaks of the pressure oscillations. The motion of shock waves described above for a slightly overcritical transonic flow around a profile appears to be the main reason for the strong increase of the amplitude of oscillations of the pressure Δp acting from the stream on the panel. This also happens when the amplitude of oscillations of the transverse panel deflection w is relatively small and the panel does not lose its original dynamic stability.

Figure 2.18 shows the variation with time of the transverse deflection w for the following case: $M_\infty = 0.9$, $K = 47.8$, $\delta = 0.005$ ($f_* = 0.3$), $P_0 = 1$ and simple support on both edges of the panel. Figure 2.19 shows the variations of the pressure Δp at the same points of the panel as those for which the curves in Fig. 2.12 were plotted.

Figure 2.20 represents the motion of the shock wave in this case of interaction of a panel with a flow. This case is characterized by a simple support and a dimensionless pressure under the panel $P_0 = 1$. From Fig. 2.19 it can be seen that Δp is negative at the points $\xi = 0.5$, $\xi = 0.75$, and for $\xi = 0.25$ it has both negative and positive values. That is why the deflections w shown in Fig. 2.18 for the points $\xi = 0.75$ and $\xi = 0.5$ are also sometimes negative. These values oscillate, covering

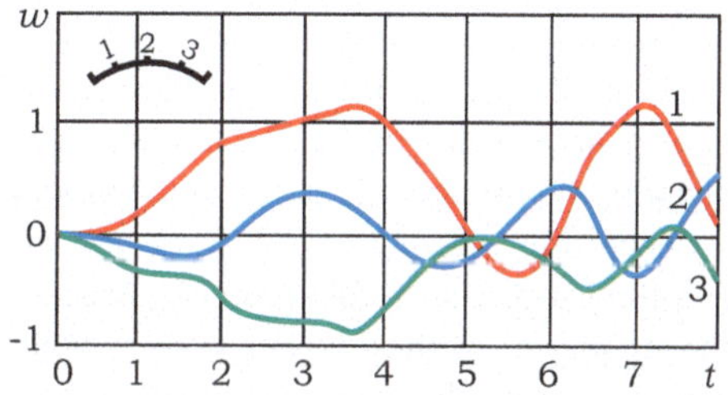

Fig. 2.18 Time variation of transverse deflection w at different points of the panel

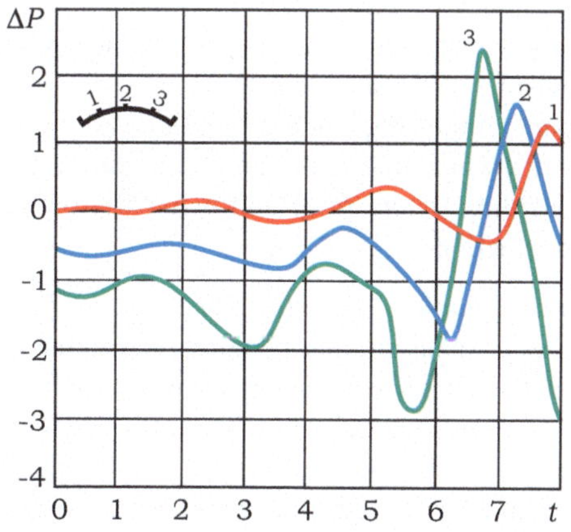

Fig. 2.19 Time variation of pressure Δp at different points of the panel

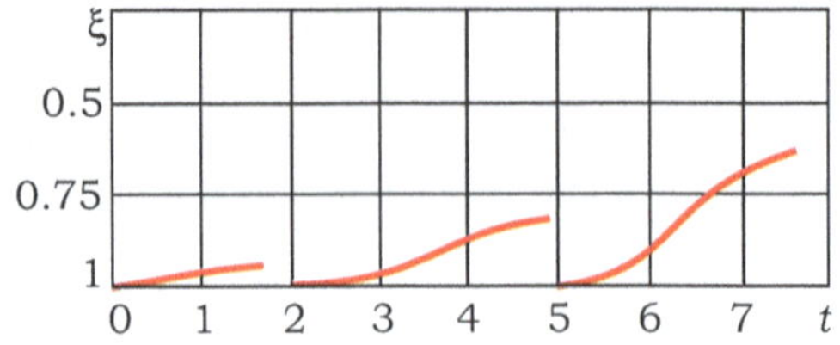

Fig. 2.20 Shock wave motion

both positive and negative values. The front part of the panel deflects only, the other parts of the panel can either slightly buckle or deflect.

In the initial time interval, from 0 to 5.5 dimensionless time units, the front ($\xi = 0.25$) and rear ($\xi = 0.75$) parts oscillate with a low principal frequency in opposite phase. After this time, the oscillations of these parts become more complicated. A high frequency of oscillations, as seen from Fig. 2.18, appears only for $\xi = 0.75$. At the initial time instant, this frequency is similar to the analogous high frequencies obtained in the cases of interaction of a flow and a panel described previously. In general, as can be seen in Fig. 2.18, the oscillations of the panel in the case $P_0 = 1$ are slightly more complicated than in the previous cases with $P_0 = 0$. It is necessary to mention that in the centre of the panel ($\xi = 0.5$), the oscillations of the deflection w and pressure Δp have similar frequencies, but have a small constant phase shift. Comparing Figs. 2.19 and 2.20, it can be seen that the peaks of pressure correspond to shock waves, which periodically appear and move upstream, losing intensity. Comparing Figs. 2.19 and 2.20 leads to the following conclusion: for the same Mach number of the incoming flow $M_\infty = 0.9$ and relative panel camber coefficient $f_* = 0.03$, in all cases of interaction between a flow and a panel investigated, differing mainly in the value of the pressure P_0 under the panel and the method of panel edge support, the principal features of the motion of shock waves are almost identical.

Therefore the motion of shock waves appears to be the reason for the large amplitudes of the oscillations of pressure at the surface points of the panel, and the frequency of these oscillations matches the frequency of shock wave motion. The motion of shock waves up the flow for slightly overcritical flows is repeated periodically. This excites a changing, periodic, nonuniformly distributed transverse load on the panel. A complicated deformation of the panel is caused by this load. Besides the main, low-frequency oscillations of the transverse deflection of the panel, high-frequency oscillations are also observed. Their amplitude is lower, but still comparable to the amplitude of the low-frequency oscillations. An oscillating buckling of the front and rear parts of the panel is characteristic of this case of interaction.

The Change of the Stress State of the Panel has a Complicated Oscillating Form The changes in time of the dimensionless membrane stress on the middle surface of the panel, calculated according to (2.220), are presented in Fig. 2.21. These results were obtained for the following values of the main parameters: $M_\infty = 0.9$, $K = 36.16$, $\delta = 0.007$ ($f_* = 0.03$), $P_0 = 0$. Both edges of the panel were clamped. The figure shows that the membrane stress σ_M is negative. Comparing

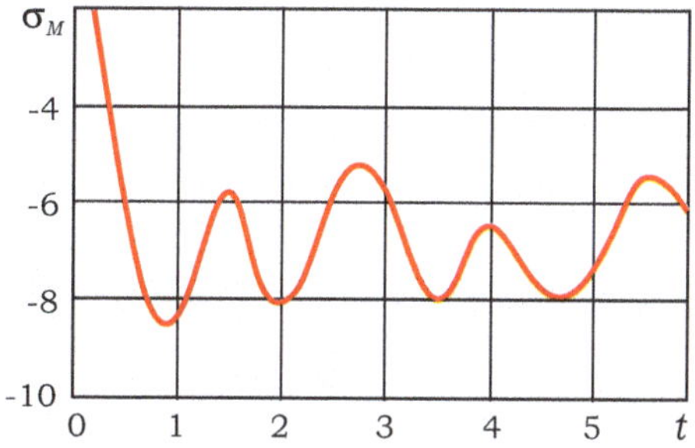

Fig. 2.21 Time variation of membrane stress on the middle surface for $M_\infty = 0.9$, $K = 36.16$, $\delta = 0.007$ ($f_* = 0.03$), $P_0 = 0$, clamped edges

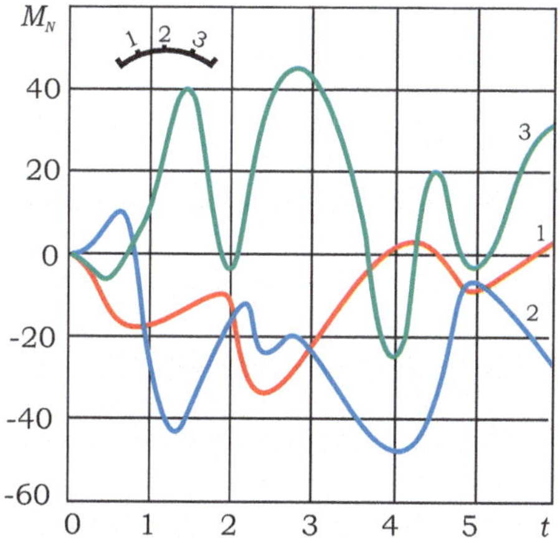

Fig. 2.22 Time variation of bending moment at different points of the panel

the oscillations of the transverse deflection w shown in Fig. 2.11 with the variation of the membrane stress shown in Fig. 2.21 leads to the conclusion that the oscillation frequency of the membrane stress is approximately equal to the high frequency of the oscillations of the transverse deflection. As explained earlier, motion of shock waves is the reason for the high-frequency oscillations of the deflection w, and this causes similar high-frequency oscillations of the membrane stress σ_M. Graphs versus time of the dimensionless bending moment M_u, which can be calculated from the bending moment using

$$M_u = -\frac{12(1-\nu^2)}{E'h^3} M_y, \tag{2.299}$$

are presented in Fig. 2.22.

From this figure it can be seen that the values of the dimensionless bending moment are different at different points of the panel. As can be seen, even for stable oscillations of the panel the bending moment reaches relatively high values, which can lead to plastic deformation. The question of plastic deformation will be discussed in Sect. 2.2.3.2. The nonuniform distribution of the load applied to the panel may result in a large local curvature. As a result, high bending moments M_u

may act inside the panel. In the present case, during stable motion of the panel, the bending moment has different signs at different points of the panel, as can be seen from the graphs in Fig. 2.22.

2.2.3 *Stability Loss of Panel Within Transonic Flow*

2.2.3.1 Stability Criteria

Let us consider the theory of the stability of an elastic panel. The criteria for the dynamic stability of an elastic shell interacting with a transonic flow must be obtained from the allowable load conditions of the panel, treated as a part of a structure. Also, the loss of stability of the structure is important, because after this happens, the structure is unable to satisfy the requirements on it.

We assume that a structure ceases to be functional when it undergoes sufficiently high deformation or when a change of material properties, for instance plastic deformation of an elastic material, occurs. We shall discuss two criteria.

The first criterion says that loss of dynamic stability of a panel happens when the transverse deflection w increases rapidly with small changes in the transverse load; this normally occurs in a nonstationary flow [74, 199].

The second criterion says that for an elastic structure, the appearance of plastic deformation is treated as a loss of dynamic stability [74, 199]. This criterion corresponds to loss of functionality of the structure.

Of course, in investigating the stability of a panel during interaction with a transonic flow, other criteria for dynamic stability can also be taken into account. We shall discuss only the two criteria mentioned above.

The first criterion means that the panel has reached large deflections and, in addition, the transition from one stable motion to another is characterized by a very high velocity. The limits on the deflection and speed of deflection, exceeding of which can be treated as a dynamic loss of stability, can be set arbitrarily.

In the case of interaction of an elastic panel and a transonic flow, investigated here, either a local or a global loss of stability may occur. Local loss of stability is understood as a rapid increase of the transverse deflection w in a small area of the panel. Global loss of stability takes place when the transverse deflection increases rapidly at all points of the panel. Of course, local and global loss of stability are related in that a local loss of stability may start a global loss of stability of the panel. Global loss of stability for a panel with nonzero curvature leads to a global buckling of the shell.

We shall start by investigating the dynamic loss of stability of an elastic cylindrical panel of infinite length during interaction with a transonic flow by applying the first criterion of dynamic stability loss. In Sect. 2.2.3.2 we shall analyse the stability of a panel starting from the second criterion, in which loss of stability occurs after plastic deformation of the panel.

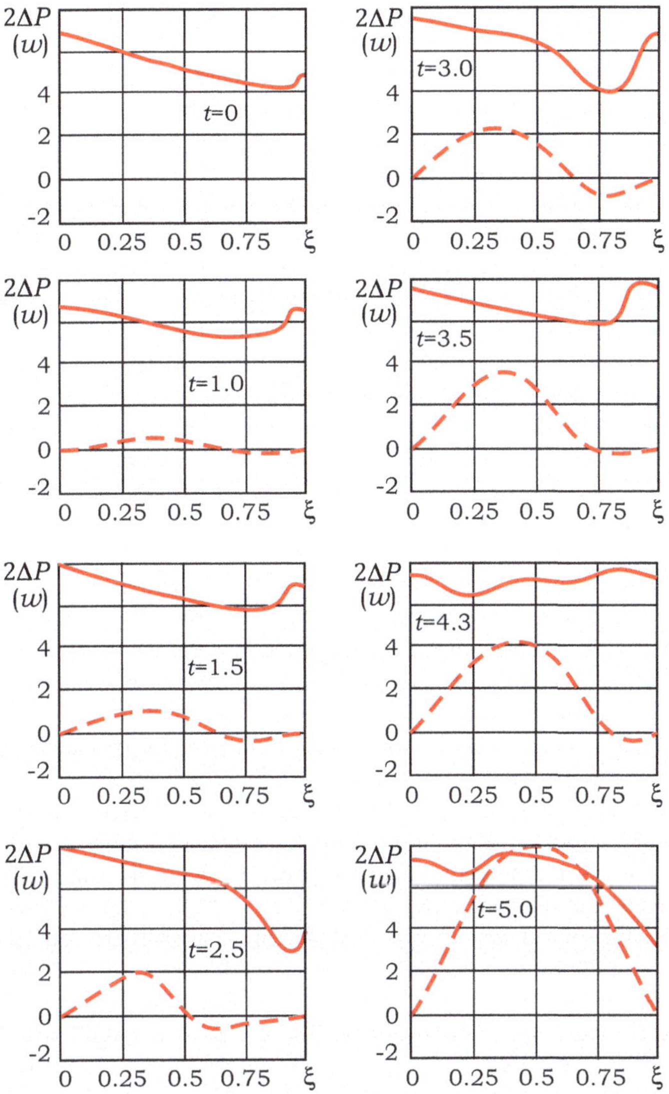

Fig. 2.23 Distribution of dimensionless pressure Δp (*solid line*) and transverse deflection w (*dashed line*) in a section of the panel at different time instants, for $M_\infty = 0.9$, $K = 36.8$, $\delta = 0.0065$ ($f_* = 0.3$), $P_0 = 0$, clamped edges

Influence of Internal Pressure Let us analyse the buckling of the panel for the following parameters: $M_\infty = 0.9$, $K = 36.8$, $\delta = 0.0065$ ($f_* = 0.3$), $P_0 = 0$. Both edges are clamped. Results are presented in Fig. 2.23. The graphs in this figure show, at different time instants, the distribution of the transverse deflection w and the pressure Δp in a transverse section of the infinite-length cylindrical shell. The values of Δp were calculated from (2.298), which characterizes the difference of pressure between the two sides of the shell. As mentioned earlier, a zone of large positive pressure gradient corresponds to a shock wave. A characteristic of the buckling of this panel is that the deflection w in the front part of the panel increases uniformly (see Fig. 2.23) and the deflection in the rear part is negative. This means that the panel buckles in the rear part, and the negative deflection oscillates, increases and decreases. The whole panel has positive deflections, and it buckles. In the case of no internal pressure ($P_0 = 0$) and a sufficiently thin panel, the main influence on buckling is the unidirectional pressure Δp, not the nonuniformly distributed transverse load nor moving shock waves. It is necessary to mention, however, the influence of travelling shock waves (initially moving, regardless of the form of the profile). As mentioned in the Sect. 2.2.2, shock waves undergo excited motion in addition to self-excited motion, depending on the changes of the geometrical form of the profile, which can be seen in Fig. 2.23 for high values of the time t. In other words, during buckling of the panel, the shock wave is located at the front part of the panel. The shock wave can undergo excited motion because of the change of form of the panel, but also self-excited motion if the conditions (2.292) and (2.293) are satisfied.

Let us now discuss a second case of dynamic loss of panel stability, for the following values of the main parameters: $M_\infty = 0.864$, $K = 198$, $\delta = 0.002$ ($f_* = 0.05$), $P_0 = 1$; both edges are clamped. Graphs of the transverse deflection w and pressure Δp at different time instants for this case are presented in Fig. 2.24.

The main difference between the two cases presented in Figs. 2.23 and 2.24 is in the internal pressure P_0. A comparison of the buckling graphs presented in Figs. 2.23 and 2.24 leads to the conclusion that the buckling graph for an internal pressure $P_0 = 1$ becomes complicated when different signs of the pressure Δp appear.

Figure 2.25 presents the variation of the transverse deflection w with time for different values of the pressure under the panel P_0, with the same values of the other parameters of the panel: $M_\infty = 0.785$, $K = 132$, $\delta = 0.003$ ($f_* = 0.05$). Both edges of the panel are clamped. In this figure characteristic signs of dynamic stability loss in the form of buckling are observed. For $P_0 = 0$, buckling happens earlier and faster than for $P_0 = 1$. For $P_0 = 0$, almost the whole panel deflects under a nearly uniformly distributed load, and the loss of stability has an almost symmetric form [198, 338], but is not exactly symmetric. For $P_0 = 1$, as can be seen in Fig. 2.25, during the transition to a new stable oscillating motion, the panel undergoes some transient oscillations that vary between different points of the panel. For $P_0 = 1$, the dynamic loss of stability has a nonsymmetric form [198, 338].

Influence of Geometric Parameters of the Panel Graphs of the transverse deflection w at different points of the panel for $M_\infty = 0.785$, $K = 47.8$, $\delta = 0.005$ ($f_* = 0.03$), $P_0 = 0$ and clamping on both edges are presented in Fig. 2.26. In this diagram, the transition from one stable oscillating motion to another one is easily visible. This means that a dynamic loss of stability takes place. In the time interval from 0 to 2 dimensionless time units, oscillations with a small amplitude in the first stable motion are visible, especially for the point with coordinate $\xi = 0.5$.

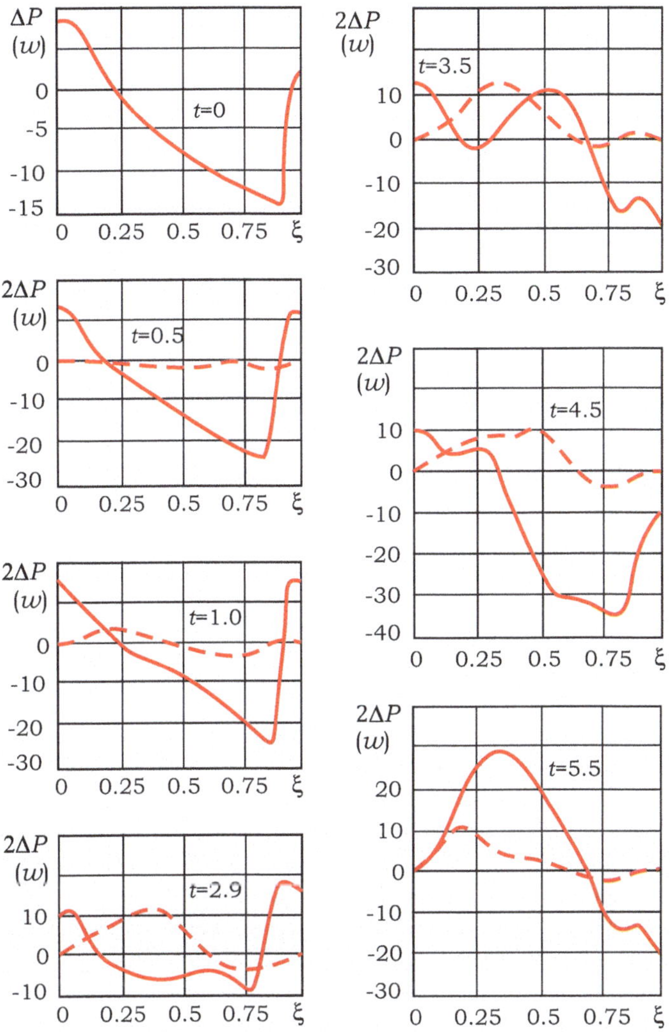

Fig. 2.24 Distribution of dimensionless pressure Δp (*solid line*) and transverse deflection w (*dashed line*) in a section of the panel at different time instants for $M_\infty = 0.864$, $K = 198$, $\delta = 0.002$ ($f_* = 0.05$), $P_0 = 1$, clamped edges

In the time interval after 3.5 units, oscillations with a greater amplitude around a new central location can be observed. This is the new stable motion. The transition from one stable motion to the other in the time interval from 2 to 3.5 time units is characterized by high gradients of the deflection w. It is necessary to mention, however, that not all points of the panel have a high value of the deflection gradient w at the same time. As can be seen in Fig. 2.26, the increase of deflection at the point $\xi = 0.25$ starts earlier than as the points $\xi = 0.5$ and $\xi = 0.75$. This means

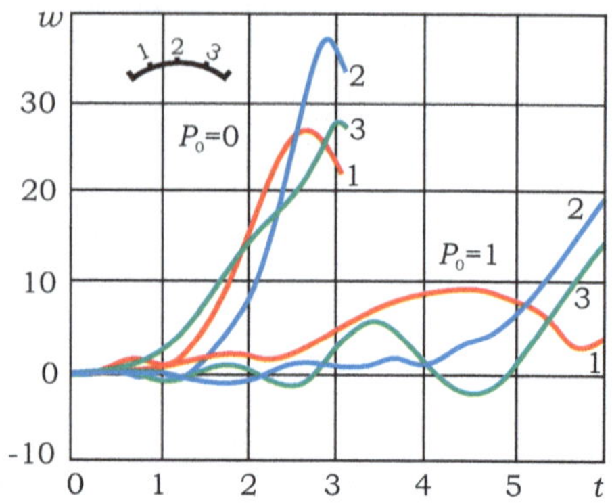

Fig. 2.25 Time variation of transverse deflection w at different points of the panel; $M_\infty = 0.785$, $K = 132$, $\delta = 0.003$ ($f_* = 0.05$), clamped edges

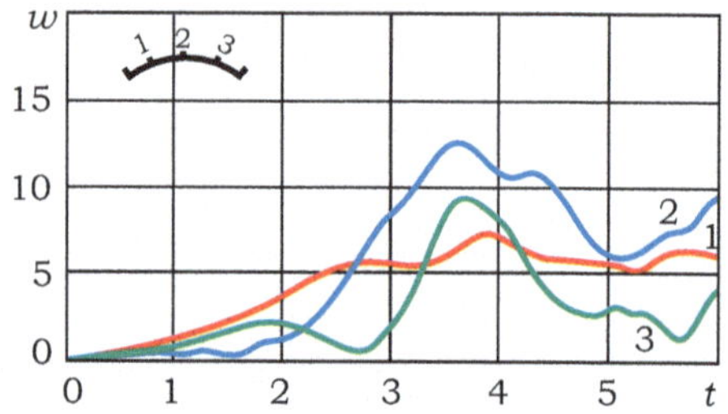

Fig. 2.26 Time variation of transverse deflection w at different points of the panel for $M_\infty = 0.785$, $K = 47.8$, $\delta = 0.005$ ($f_* = 0.03$), $P_0 = 0$, clamped edges

that buckling of the panel occurs in a complicated way: usually the front part of the panel deflects strongly enough to pull the rest of the panel. This effect is seen very well in the graphs of the deflection in Fig. 2.25.

Let us compare the changes of transverse deflection w for the two cases represented in Fig. 2.25 (for $P_0 = 0$) and Fig. 2.26. These two cases differ in the values of the geometric parameters K, δ (or f_*). From these two diagrams it can be seen that for high values of the dimensionless geometric parameter K, the panel buckles faster and the transverse deflections reach higher values. The infinite-length cylindrical panel investigated here is characterized, as stated in Sect. 2.2.1.3, by two independent geometric parameters: K, a dimensionless geometric parameter, and δ, the relative thickness of the panel. As was stated in Sect. 2.2.1.3, one of these calculated parameters can be replaced by the relative camber f_*. The geometry of the panel, which depends on two parameters, makes the investigations more complicated.

We shall now discuss some more examples of the interaction of a panel and a transonic flow. Figure 2.27 presents the time variation of the transverse deflection w at different points of the panel for the following parameters: $M_\infty = 0.9$, $K = 198$, $\delta = 0.002$ ($f_* = 0.05$), $P_0 = 1$. Both edges of the panel are clamped. Figure 2.28 shows correspondingly results for a panel with $K = 119.6$. By comparing these diagrams, it is possible again to conclude that for higher values of the parameter K (or for higher values of the camber factor f_*), the relative deflections w are also higher.

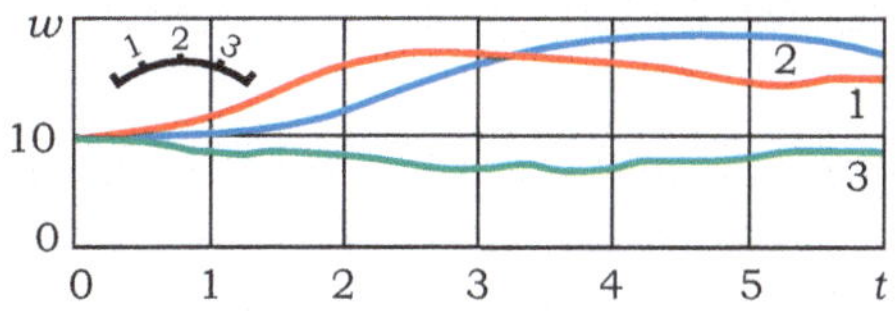

Fig. 2.27 Time variation of transverse deflection w at different points of the panel for $M_\infty = 0.9$, $K = 198$, $\delta = 0.002$ ($f_* = 0.05$), $P_0 = 1$, clamped edges

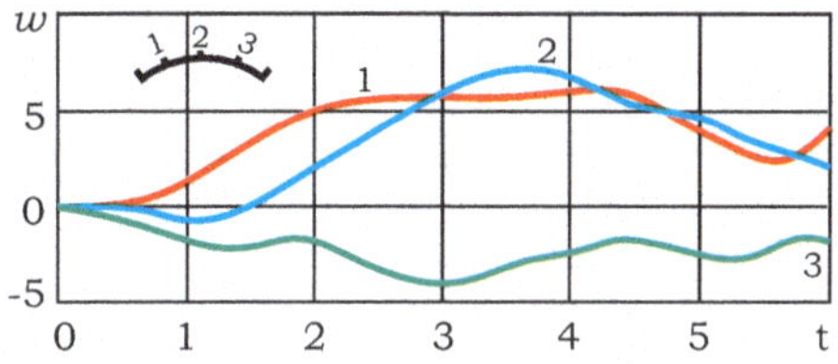

Fig. 2.28 Time variation of transverse deflection w at different points of the panel for $M_\infty = 0.9$, $K = 119.6$, $\delta = 0.002$ ($f_* = 0.3$), $P_0 = 1$

The cases presented in Figs. 2.27 and 2.28 are characterized by the fact that a shock wave is located close to the rear of the panel. This happens when the Mach number of the incoming flow M_∞ is close to one. In these diagrams it can be seen, that the panel does not buckle completely, and has an interesting form. In this case the panel deflects in the front part and buckles in the rear part. It stays for a relatively long period of time in that state and undergoes oscillations with small amplitude.

The motion of shock waves excites high-frequency, forced, oscillating deformation of the panel. In Figs. 2.27 and 2.28 the influence of the shock wave can be seen only for the point with coordinate $\xi = 0.75$ in the rear part of the panel. The graphs of the transverse deflection in Figs. 2.25 and 2.26 show a greater influence of travelling shock waves on the panel deflections than do the graphs in Figs. 2.27 and 2.28. In the case presented in Figs. 2.25 and 2.26, the deflection oscillations have a high frequency for most of the points of the panel. For the case shown in Figs. 2.27 and 2.28, a shock wave moves in the central part of the panel; this happens in an overcritical flow, and takes place in small supersonic zones of the incoming stream for a Mach number $M_\infty = 0.785$. This confirms the high importance of the panel geometry in the interaction of a transonic flow and a panel.

Influence of Mach Number M_∞ of Incoming Flow The influence of the Mach number of the incoming flow on the dynamic stability of a panel appears to be one of the most important influences, because the Mach number influences strongly the location and motion of shock waves. Figures 2.29 and 2.30 present time graphs of the transverse deflection w for three points on the profile for the following parameters: $M_\infty = 0.864$, $K = 119.6$, $\delta = 0.002$ ($f_* = 0.03$), $P_0 = 1$, clamping of edges; and $M_\infty = 0.785$, $K = 119.6$, $\delta = 0.002$ ($f_* = 0.03$), $P_0 = 1$, clamping of edges. Figures 2.28, 2.29 and 2.30 present cases which differ only in the Mach number M_∞ of the incoming flow. Comparing these three diagrams, it is possible to analyse the process of stability loss of the same panel for different velocities of the incoming flow, i.e. different values of the Mach number of the incoming stream. It can be seen that the oscillations of the same point of the panel differ in frequency, especially for the point with coordinate $\xi = 0.75$. This is caused by the difference

Fig. 2.29 Time variation of transverse deflection w at different points of the panel for $M_\infty = 0.864$, $K = 119.6$, $\delta = 0.002$ ($f_* = 0.03$), $P_0 = 1$, clamped edges

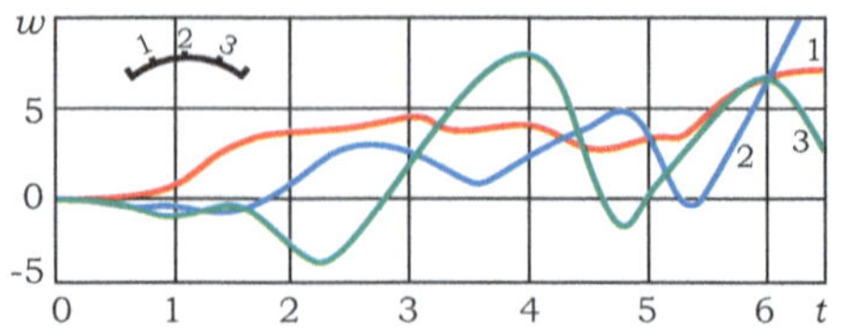

Fig. 2.30 Time variation of transverse deflection w at different points of the panel for $M_\infty = 0.785$, $K = 119.6$, $\delta = 0.002$ ($f_* = 0.03$), $P_0 = 1$, clamped edges

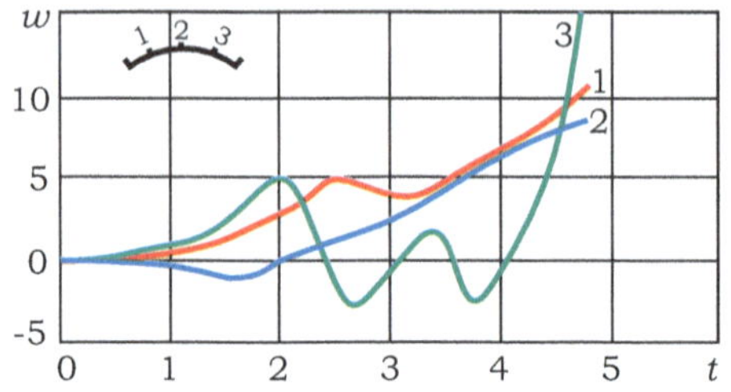

in the shock wave speed for different velocities of the incoming stream, as was described earlier.

A comparison of the graphs of transverse deflection w in Figs. 2.28, 2.29 and 2.30 leads to the conclusion that loss of panel stability happens earlier for lower values of the Mach number M_∞ of the incoming stream, i.e. for $M_\infty = 0.785$ (the case presented in Fig. 2.30). Of course, such rapid buckling in this case occurs during motion of a shock wave in the middle part of the panel, which happens for this value of M_∞. As was mentioned in Sect. 2.2.1, for greater values of the Mach number M_∞ of the incoming flow (values of M_∞ close to one), the zone of shock wave motion is located closer to the rear of the panel, which decreases the influence of the shock wave on buckling of the panel in the present case, for an internal pressure $P_0 = 1$.

In the cases presented in Figs. 2.29 and 2.30, the panel buckles after oscillation around a certain bent state. As can be seen, such oscillation around a bent state of the panel can be treated as stable in a limited sense in the time interval investigated. Following the method presented in [8], we can prove that such oscillations of an elastic panel around a deformed state can be treated as stable. Figure 2.31 presents diagrams of $\left(w, \frac{\partial w}{\partial t}\right)$ for different points of the panel, for the following parameters $M_\infty = 0.864$, $K = 119.6$, $\delta = 0.002$ ($f_* = 0.03$), $P_0 = 1$. Both sides are clamped. According to [8], the motion of a panel point is stable if the absolute values of the deflection w and deflection speed $\frac{\partial w}{\partial t}$ do not exceed preset values ε_w and $\varepsilon_{\dot{w}}$ for given excitations. On the graphs of $\left(w, \frac{\partial w}{\partial t}\right)$ for some points of the elastic panel, values of ε_w and $\varepsilon_{\dot{w}}$ can be chosen, for which w and $\frac{\partial w}{\partial t}$ do not exceed the set limits in the bounded time interval investigated. As can be seen in Fig. 2.31c, the graph of $\left(w, \frac{\partial w}{\partial t}\right)$ for the point $\xi = 0.25$ in the time interval investigated does not exceed the range $0 \leq w < 7.5$, $-3.5 \leq \frac{\partial w}{\partial t} < 9$. Figure 2.31b presents a graph of $\left(w, \frac{\partial w}{\partial t}\right)$ for the point $\xi = 0.5$. At the beginning this graph remains inside the limits $-2 \leq w \leq 5$, $-20 \leq \frac{\partial w}{\partial t} \leq 7.1$ but it then exceeds this range–loss of dynamic stability of the panel has occurred around this point. Figure 2.31a presents

Fig. 2.31 (**a–c**) Diagrams of $\left(w, \frac{\partial w}{\partial t}\right)$ for different points of the panel; $M_\infty = 0.864$, $K = 119.6$, $\delta = 0.002$ $(f_* = 0.03)$, $P_0 = 1$, clamped edges

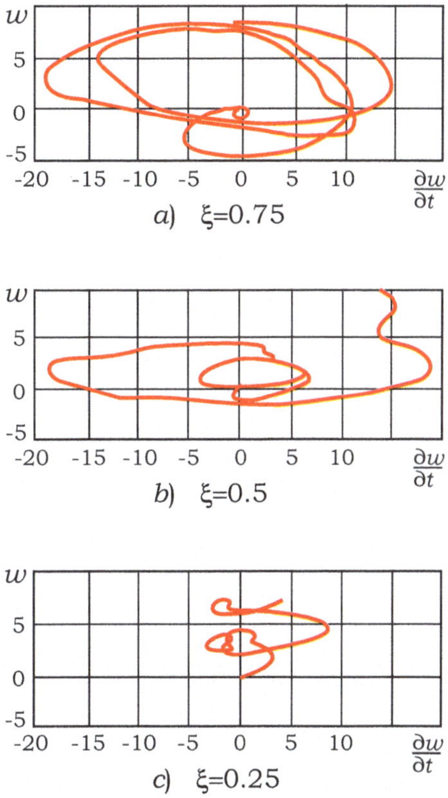

a graph of $\left(w, \frac{\partial w}{\partial t}\right)$ for the point $\xi = 0.75$. This graph does not exceed the range $-5 \leq w \leq 4$, $-20 \leq \frac{\partial w}{\partial t} < 15$. This means that this point does not lose dynamic stability. Loss of dynamic stability begins around random point of the panel and then occurs in other points of the panel. The expanding loss of stability leads to buckling of the panel. After a transition process, which happens at the beginning of the interaction, the panel undergoes oscillations around a bent position.

Critical Values of Parameters The critical values are those values of the parameters for which dynamic loss of stability takes place in the sense described above within the time interval analysed.

For an internal pressure $P_0 = 0$ and clamped edges of an elastic panel, numerical calculations performed according to the method presented above give the values of the critical geometrical parameters K and f_* which are presented in Fig. 2.32. A slight dependence of the results on the Mach number M_∞ of the incoming stream is observed in this case. As mentioned earlier, this results from the unidirectional load. The influence of the Mach number of the incoming flow and, consequently, of shock wave motion on dynamic loss of stability for the case $P_0 = 0$ depends on the

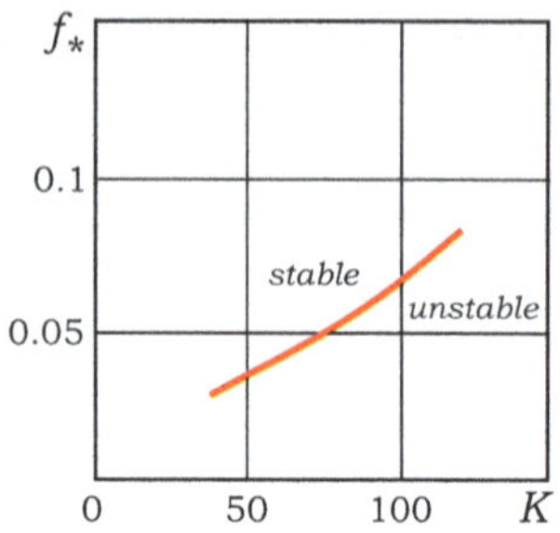

Fig. 2.32 Critical geometrical parameters K and f_* for $P_0 = 0$ and clamped edges

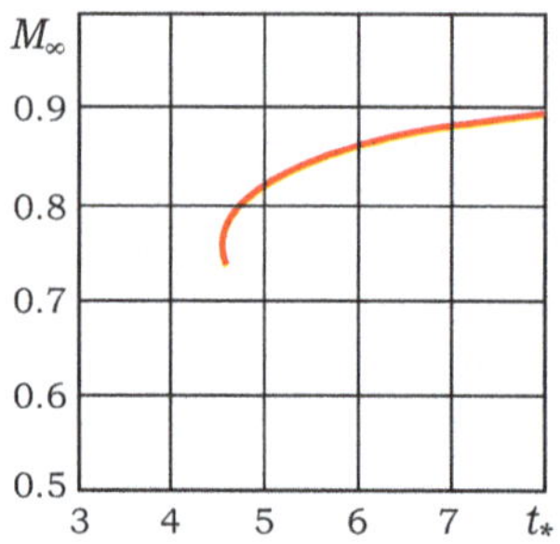

Fig. 2.33 Stability loss time for $P_0 = 1$, $K = 119.6$, $\delta = 0.002$ ($f_* = 0.03$), clamped edges

geometrical form of the panel. Perhaps the importance of the Mach number in the stability loss of a panel for $P_0 = 0$ will be greater if the interaction of a transonic flow and a panel is investigated over a longer period of time.

For the case $P_0 = 1$, an elastic panel can oscillate in a stable manner under excitations originating from the interaction of the panel with the flow (for example, in the case presented in Fig. 2.27).

In this case, during a small time interval, there is no loss of dynamic stability of the panel. As previously mentioned, for some values of the Mach number of the incoming flow M_∞ (see Figs. 2.29 and 2.30), dynamic loss of stability of the panel may happen. This loss of stability occurs at different time instants, depending on M_∞. Figure 2.33 presents a diagram of the stability loss time versus the Mach number M_∞ of the incoming flow for the following parameters of the elastic panel and transonic flow: $P_0 = 1$, $K = 119.6$, $\delta = 0.002$ ($f_* = 0.03$); both edges are clamped.

As has been stated in experimental studies [305, 338, 339], the following characteristic features of the interaction of a transonic flow with a shell can be examined: shock wave motion, and high amplitudes of oscillation of the pressure and transverse deflection. Although the case investigated here is different from those studied in [305, 338, 339], it is possible to talk about qualitative similarities of the behaviour of the panel and flow in the two cases.

2.2.3.2 Analysis of Stress–Strain State

This subsection gives an estimate of the stress–strain state of the panel and an analysis of the occurrence of plastic deformation during interaction of a panel and a flow.

The von Mises yield criterion has the following form:

$$\sigma = \frac{\sigma_S}{\sqrt{3}}, \tag{2.300}$$

where σ_S is the yield point of the shell material, and σ is the shear stress, which can be calculated from the following formula [252]:

$$\begin{aligned}\sigma = \frac{1}{\sqrt{6}}\Big((P_{11}^z - P_{22}^z)^2 + (P_{22}^z - P_{33}^z)^2 + (P_{33}^z - P_{11}^z)^2 \\ +6(P_{12}^{z^2} - P_{23}^{z^2} + P_{13}^{z^2})\Big)^{1/2}.\end{aligned} \tag{2.301}$$

By use of Eqs. (2.160), the right-hand side of (2.301) can be converted to

$$\sigma = \frac{1}{\sqrt{3}}\left((1+\nu^2-\nu)P_{22}^{z^2} + 3P_{23}^{z^2}\right)^{\frac{1}{2}}. \tag{2.302}$$

Substituting the stress from (2.187) into (2.302), we obtain the following formula for calculating the value of the shear stress:

$$\begin{aligned}\sigma = \frac{E'}{\sqrt{3}(1-\nu^2)}(1+\nu^2-\nu)\left(\frac{1}{2L}\int\limits_0^L \left(\frac{\partial w'}{\partial y}\right)^2 dy\right. \\ \left. - \frac{1}{RL}\int\limits_0^L \left(w'\,dy - z\frac{\partial^2 w'}{\partial y^2}\right)^2 + \frac{h^4}{48}\left(\frac{\partial^3 w'}{\partial y^3}\right)^2\right)^{\frac{1}{2}}.\end{aligned} \tag{2.303}$$

Introducing the system of dimensionless parameters (2.216) and (2.219), we obtain the right-hand side of (2.303) in dimensionless form:

$$\begin{aligned}\frac{\sigma}{E'} = \frac{3^{-\frac{1}{2}}}{1-\nu^2}\left((1+\nu^2-\nu)\left(\frac{\delta^2}{2}\int\limits_0^1 \left(\frac{\partial w}{\partial \xi}\right)^2 d\xi - \delta^2 K\int\limits_0^1 w\,d\xi\right.\right. \\ \left.\left. +\eta\delta\frac{\partial^2 w}{\partial \xi^2}\right)^2 + \frac{\delta^4}{48}\left(\frac{\partial^3 w}{\partial \xi^3}\right)^2\right)^{\frac{1}{2}}.\end{aligned} \tag{2.304}$$

Here, the dimensionless coordinate η of the panel may take values in the interval $[-\frac{\delta}{2}, \frac{\delta}{2}]$ only. The von Mises yield condition then has the following form:

$$\frac{\sigma_S}{E'} = \frac{\delta^2}{1-\nu^2}\left(\frac{1+\nu^2-\nu}{4}\left(\int_0^1\left(\frac{\partial w}{\partial \xi}\right)^2 d\xi - 2K\int_0^1 w\,d\xi + 2\frac{\eta}{\delta}\frac{\partial^2 w}{\partial \xi^2}\right)^2 + \frac{\delta^4}{48}\left(\frac{\partial^3 w}{\partial \xi^3}\right)^2\right)^{\frac{1}{2}}. \tag{2.305}$$

If the right-hand side of (2.305) is greater than or equal to the left-hand side of this equation, then we assume that the shell yields at this point. If the right-hand side of (2.305) is less than the left-hand side, then elastic deformation occurs. The von Mises yield criterion (2.305) can also be presented in the following form:

$$\left(\frac{\sigma_S}{E'}\right)^2 \frac{4(1-\nu^2)^2}{(1+\nu^2-\nu)\delta^4} - \frac{\delta^2}{24(1+\nu^2-\nu)}\left(\frac{\partial^3 w}{\partial \xi^3}\right)^2 = \left(\int_0^1\left(\frac{\partial w}{\partial \xi}\right)^2 d\xi - 2K\int_0^1 w\,d\xi + 2\frac{\eta}{\delta}\frac{\partial^2 w}{\partial \xi^2}\right)^2. \tag{2.306}$$

After substituting in (2.306) the values of the parameters used in the studies described in this book, namely $E' = 20 \times 10^{10}$ N/m, $\sigma_S = 24 \times 10^7$ N/m^2, $\nu = 0.3$, the yield condition (2.306) has the form

$$\frac{1}{165\,623.4\,\delta^4} - 0.105\,48\,\delta^4\left(\frac{\partial^3 w}{\partial \xi^3}\right)^2 = \left(\int_0^1\left(\frac{\partial w}{\partial \xi}\right)^2 d\xi - 2K\int_0^1 w\,d\xi + 2\frac{\eta}{\delta}\frac{\partial^2 w}{\partial \xi^2}\right)^2. \tag{2.307}$$

If we assume, as stated in Sect. 2.2.1.6, that the relative thickness of the panel $\delta < 0.01$, the following inequality becomes true:

$$\frac{1}{165\,623.4\,\delta^4} > 603.78. \tag{2.308}$$

Assuming that

$$\left|\frac{\partial^3 w}{\partial \xi^3}\right| < 10^3, \tag{2.309}$$

which can be confirmed by numerical calculations, the von Mises yield criterion can be presented in the following simplified form:

$$165\,623.4\,\delta^4 \left(\int_0^1 \left(\frac{\partial w}{\partial \xi} \right)^2 d\xi - 2K \int_0^1 w\, d\xi + 2 \frac{\eta}{\delta} \frac{\partial^2 w}{\partial \xi^2} \right)^2 = 1. \tag{2.310}$$

In other words, if the conditions (2.308) and (2.309) are satisfied, as is true in the case investigated, the influence of the third derivative $\frac{\partial^3 w}{\partial \xi^3}$ on the appearance of yield zones in the panel is negligible. The second derivative $\frac{\partial^2 w}{\partial \xi^2}$ has, locally, the main influence, and it corresponds, following (2.161), to the bending moment M_y. After integration by parts of the first integral in (2.310) and taking into account the boundary condition (2.234) for simple support or the boundary condition (2.235) for clamping, the von Mises yield criterion can be presented in the following form:

$$165\,623.4\,\delta^4 \left(2 \frac{\eta}{\delta} \frac{\partial^2 w}{\partial \xi^2} - \int_0^1 w \left(2K + \frac{\partial^2 w}{\partial \xi^2} \right) d\xi \right)^2 = 1. \tag{2.311}$$

By solving (2.311) for η at a given time instant t, on the normal to the middle surface of the panel, values of the coordinate η can be found where are boundaries between yielded and elastic zones for the case $\frac{\partial^2 w}{\partial \xi^2} \neq 0$:

$$\begin{aligned} \eta_1 &= \frac{\delta}{2} \left(\frac{\partial^2 w}{\partial \xi^2} \right)^{-1} \left(\int_0^1 w \left(2K + \frac{\partial^2 w}{\partial \xi^2} \right) d\xi + 0.002\,46\,\delta^{-2} \right), \\ \eta_2 &= \frac{\delta}{2} \left(\frac{\partial^2 w}{\partial \xi^2} \right)^{-1} \left(\int_0^1 w \left(2K + \frac{\partial^2 w}{\partial \xi^2} \right) d\xi - 0.002\,46\,\delta^{-2} \right). \end{aligned} \tag{2.312}$$

Of course, only those values of η which are contained in the range $\left(-\frac{\delta}{2}, \frac{\delta}{2}\right)$ are interesting. From (2.311) and (2.312), one can conclude that elastic deformation will appear for the following values of the variable η:

$$\min(\eta_1, \eta_2) < \eta < \max(\eta_1, \eta_2). \tag{2.313}$$

We now present examples showing the occurrence of yield deformation in the cylindrical panel investigated here during interaction with a transonic flow of an ideal gas. Results of computations for the case $M_\infty = 0.9$, $K = 36.8$, $\delta = 0.0065$ ($f_* = 0.03$) and clamping of the edges are presented in Fig. 2.34. The graphs show the distribution of the bending moment M_u in transverse section of an infinite panel, computed at different time instants. The hatched areas in the diagrams above the

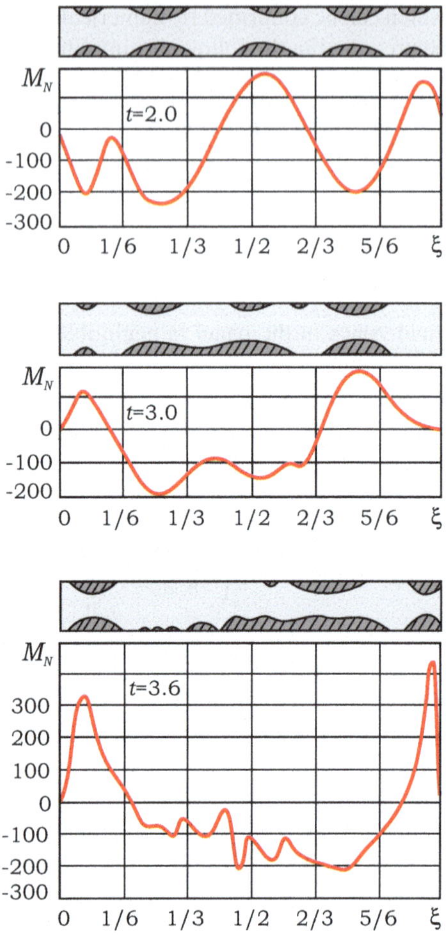

Fig. 2.34 Distribution of bending moment and yield deformation at different time instants t for $M_\infty = 0.9$, $K = 36.8$, $\delta = 0.0065$ $(f_* = 0.03)$, clamped edges

graphs correspond to yield zones in sections of the panel. According to (2.161), M_u is proportional to the second derivative with respect to ξ of the transverse deflection, $\frac{\partial^2 w}{\partial \xi^2}$, and this, as can be seen from the von Mises criterion (2.311), has a strong influence on the occurrence of yield deformations. To visualize the yield regions better, thickness of the panel for which results are presented in Fig. 2.34 was chosen to be abnormally high.

The boundaries of the yield regions were calculated from (2.312), taking into account the inequalities (2.313). In Fig. 2.34 it can be seen that yield deformation occurs, when the bending moment M_u has a sufficiently high value. The yield deformation appears initially on the surface of the panel. In Fig. 2.34 it can be seen that even in the initial stage of buckling of the panel at $t = 2$, relatively large yield regions appear.

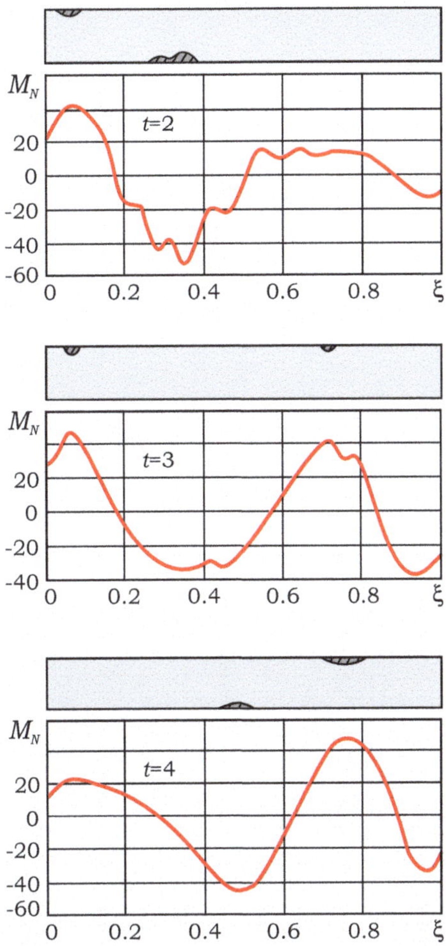

Fig. 2.35 Distribution of bending moment and yield deformation at different time instants t for $M_\infty = 0.9$, $K = 34.16$, $\delta = 0.007$ ($f_* = 0.03$), $P_0 = 0$, clamped edges

The distribution of the bending moment and the areas of yield deformation (hatched) at different time instants t are presented in Fig. 2.35 for the following parameters: $M_\infty = 0.9$, $K = 34.16$, $\delta = 0.007$ ($f_* = 0.03$), $P_0 = 0$. Both edges are clamped. In this case, the elastic panel did not lose its initial, stable form of motion. According to Fig. 2.35, however, yield deformation appeared in the panel. This confirms the influence of yield deformation on the behaviour and stability of a panel during interaction with a transonic gas flow.

Chapter 3
Estimation of the Errors of the Bubnov–Galerkin Method

In Sect. 3.1, an abstract coupled problem is considered and a few theorems related to the estimation of the accuracy of the Bubnov–Galerkin method are formulated and proved. The error estimates hold for a system of differential equations of a rather general form with homogeneous boundary conditions, which corresponds to coupled thermoelastic problems for plates and shallow shells with variable thickness. In addition, a particular case of this problem (with nonhomogeneous initial conditions), where a prior estimate of the errors of the Bubnov–Galerkin method is most effective, is illustrated and discussed. Finally, a prior estimate for the Bubnov–Galerkin method to a problem generalizing a class of dynamical problems of elasticity (without a heat transfer equation) for both three-dimensional and thin-walled elements of structures is given.

In Sect. 3.2, a coupled thermoelastic problem within the Kirchhoff–Love model is considered. We show how to find operators L_0, K_0 that are convergent and form an acute angle with the differential operators L, K, for which full systems of eigenfunctions and the corresponding eigenvalues are known. In Sect. 3.3, a coupled thermoelastic problem for a simply supported plate with constant thickness in the framework of the Kirchhoff–Love model is analysed. It is assumed that the mechanical and heat loads are constant in time and that the plate is square.

In the next section, a coupled problem of thermoelasticity of shallow shells, rectangular in plan with variable thickness is considered in the framework of a kinematic Timoshenko-type model. Some remarks related to estimation of the errors of the Bubnov–Galerkin method are made.

J. Awrejcewicz, V. A. Krysko, *Elastic and Thermoelastic Problems in Nonlinear Dynamics of Structural Members*, Scientific Computation,
https://doi.org/10.1007/978-3-030-37663-5_3

3.1 An Abstract Coupled Problem

We consider the abstract coupled Cauchy problem (2.54)–(2.56) and, in order to simplify our treatment we take $w = w_1 = 0, \theta_0 = 0$ (it is clear that an arbitrary problem can be reduced to one with homogeneous boundary conditions by a simple change of variables). Therefore, the problem analysed has the form

$$\begin{cases} I_1 w''(t) + Lw(t) + M\theta(t) = g_1(t), \\ I_2\theta'(t) + K\theta(t) + Nw'(t) = g_2(t), \\ w(0) = w'(0) = 0, \quad \theta(0) = 0. \end{cases} \tag{3.1}$$

The conditions imposed on the operator coefficients I_1, I_2, L, K, M, N are given in Sect. 2.1.4.

The Bubnov–Galerkin relations appropriate to the problem (3.1) have the form

$$\begin{cases} P_{H_1^n}(I_1 w_n''(t) + Lw_n(t) + M\theta_t(t)) = P_{H_1^n} g_1(t), \\ P_{H_2^n}(I_2\theta_n'(t) + K\theta_n(t) + Nw_n'(t)) = P_{H_2^n} g_2(t), \\ w_n(t) \in H_1^n, \quad \theta_n(t) \in H_2^n, \quad \forall t \in [0, T], \\ w_n(0) = w_n'(0) = 0, \quad \theta_n(0) = 0. \end{cases} \tag{3.2}$$

As the bases $\{\nu_i\}_{i=1}^{\infty}$, $\{\eta_j\}_{j=1}^{\infty}$, we take systems of eigenelements of certain self-adjoint, positive definite operators L_0 and K_0 that are convergent and form an acute angle with the operators L and K, respectively. The eigenvalues of the operators L_0 and K_0 are denoted by λ_i and $æ_i$: $L_0\nu_i = \lambda_i\nu_i$, $K_0\eta_j = æ_j\eta_j$, $i, j = 1, 2, \ldots$ Let us remind ourselves that the H_1^n, H_2^n occurring in (3.2) are composed of a linear vicinity and of finite systems of base elements $\{\nu_i\}_{i=1}^{i(n)}$, $\{\eta_j\}_{j=1}^{j(n)}$, $P_{H_1^n}$ is an orthoprojector of H_1 into H_1^n, and $P_{H_2^n}$ is an orthoprojector of H_2 into H_2^n.

Let us introduce the following notation:

$$\Delta w_n(t) = w_n(t) - w(t), \quad \Delta\theta_n(t) = \theta_n(t) - \theta(t),$$

$$\delta_n^{(1)}(t) = I_1 w_n''(t) + Lw_n(t) + M\theta_n(t) = g_1(t),$$

$$\delta_n^{(2)} = I_2\theta_n'(t) + K\theta_n(t) + Nw_n'(t) = g_2(t),$$

where $\{w, \theta\}$ is the exact solution to the problem (3.1), and $\{w_n, \theta_n\}$ is the approximated solution to the problem (3.1) constructed by the Bubnov–Galerkin method using (3.2).

The first theorem related to the convergence of the Bubnov–Galerkin relations (3.2) to the problem (3.1) is related to a case corresponding to differentiable free terms of (2.110).

Theorem 3.1 *If the conditions (2.58), (2.110) are satisfied then the accuracy of the Bubnov–Galerkin method based on (3.2) applied to the problem (3.1) can be estimated from the following inequality:*

$$\max_{0\le t\le T}\Bigg[\left\|I_1^{1/2}\Delta w_n'(t)\right\|_{H_1}^2+\left\|L^{1/2}\Delta w_n(t)\right\|_{H_1}^2$$

$$+\left\|I_2^{1/2}\Delta\theta_n(t)\right\|_{H_2}^2+\int_0^t\left\|K^{1/2}\Delta\theta_n(\tau)\right\|_{H_2}^2 d\tau\Bigg]$$

$$\le c_1(M_g(T))^{1/2}\Big[\|g_1\|_{L_2(0,T;H_1)}+c_2\|g_2\|_{L_2(0,T;H_2)}$$

$$+c_3(M_g(T))^{1/2}\Big]\tilde{\lambda}_{i(n)+1}^{-1/2}$$

$$+c_4\Big[\|g_2\|_{L_2(0,T;H_2)}+c_5(M_g(T))^{1/2}\Big]^2\tilde{æ}_{j(n)+1}^{-1}$$

$$\equiv 0(\max(\tilde{\lambda}_{i(n)+1}^{-1/2},\tilde{æ}_{j(n)+1}^{-1})), \tag{3.3}$$

where

$$\tilde{\lambda}_k=\min_{i\ge k}\lambda_i,\quad \tilde{æ}_1=\min_{j\ge l}æ_j, \tag{3.4}$$

$$\begin{aligned} c_1&=4\left\|L^{-1/2}L_0^{1/2}\right\|_{H_1\to H_1}T^{1/2}(1+c_L),\quad c_2=\left\|MK^{-1}\right\|_{H_2\to H_1}c_K,\\ c_3&=b_2T^{1/2},\quad c_4=\left\|K^{-1/2}K_0^{1/2}\right\|_{H_2\to H_2}^2(1+c_K)^2,\quad c_5=b_1T^{1/2}, \end{aligned} \tag{3.5}$$

and $M_g(T)$ is the constant defined by the relation (2.130); c_L, c_K are the constants defined by (2.124); and b_1, b_2 are the constants defined by the relations (2.121), (2.125).

Proof Let us introduce the following notation:

$$R_{H_1^n}=E_1-P_{H_1^n},\quad R_{H_2^n}=E_2-P_{H_2^n},$$

where E_i is the identity operator over $H_i (i=1,2)$. The relations (3.1), (3.2) imply

$$\begin{cases} I_1\Delta w_n''(t)+L\Delta w_n(t)+M\Delta\theta_n(t)=R_{H_1^n}\delta_n^{(1)}(t),\\ I_2\Delta\theta_n'(t)+K\Delta\theta_n(t)+N\Delta w_n'(t)=R_{H_2^n}\delta_n^{(2)}(t),\\ \Delta w(0)=\Delta w'(0)=0,\quad \Delta\theta(0)=0. \end{cases} \tag{3.6}$$

We multiply first two equations of (3.6) scalarly by $2\Delta w_n'(t)$ and by $2\Delta\theta_n(t)$ and then sum them. Next we take into account the condition (2.58) and integrate the relation obtained over the interval $[0, T]$, where $t \le T$. As the result, we obtain

$$\begin{aligned}&\left\|I_1^{1/2}\Delta w_n'(t)\right\|_{H_1}^2 + \left\|L^{1/2}\Delta w_n(t)\right\|_{H_1}^2 + \left\|I_2^{1/2}\Delta\theta_n(t)\right\|_{H_2}^2\\ &+2\int_0^t \left\|K^{1/2}\Delta\theta_n'(\tau)\right\|_{H_2}^2 d\tau = 2\int_0^t (R_{H_1^n}\delta_n^{(1)}(\tau), \Delta w_n'(\tau))_{H_1}\, d\tau\\ &+2\int_0^t (R_{H_2^n}\delta_n^{(2)}(\tau), \Delta\theta_n(\tau))_{H_2}\, d\tau.\end{aligned} \tag{3.7}$$

The first term of the right-hand side of (3.7) can be estimated in the following way:

$$\begin{aligned}&2\int_0^t \left(R_{H_1^n}\delta_n^{(1)}(\tau), \Delta w_n'(\tau)\right)_{H_1} d\tau = 2\int_0^t \left(L^{-1/2}L_0^{1/2}L_0^{-1/2}R_{H_1^n}\delta_n^{(1)}(\tau)\,,\right.\\ &\left.\quad L^{1/2}\Delta w_n'(\tau)\right)_{H_1} d\tau \le 2\left\|L^{-1/2}L_0^{1/2}\|_{H_1\to H_1}\|L_0^{1/2}R_{H_1^n}\right\|_{H_1\to H_1}\\ &\quad\times\left\|\delta_n^{(1)}\right\|_{L_2(0,T;H_1)}\left\|L^{1/2}\Delta w_n'\right\|_{L_2(0,T;H_1)}.\end{aligned} \tag{3.8}$$

Owing to the special choice of the basis $\{v_i\}_{i=1}^{\infty}$, we obtain

$$\left\|L_0^{1/2}R_{H_1^n}\right\|_{H_1\to H_1} = \tilde{\lambda}_{i(n)+1}^{-1/2}. \tag{3.9}$$

Using the prior estimate (2.128), (2.129), we obtain

$$\begin{aligned}\left\|L^{1/2}\Delta w_n'\right\|_{L_2(0,T;H_1)} &\le \left\|L^{1/2}w'\right\|_{L_2(0,T;H_1)} + \left\|L^{1/2}w_n'\right\|_{L_2(0,T;H_1)}\\ &\le 2T^{1/2}(M_g(T))^{1/2}.\end{aligned} \tag{3.10}$$

From (3.8)–(3.10) we obtain

$$\begin{aligned}&2\int_0^t \left(R_{H_1^n}\delta_n^{(1)}(\tau), \Delta w_n'(\tau)\right)_{H_1} d\tau\\ &\le 4\left\|L^{-1/2}L_0^{1/2}\right\|_{H_1\to H_1} T^{1/2}(M_g(T))^{1/2}\left\|\delta_n^{(1)}\right\|_{L_2(0,T;H_1)}\tilde{\lambda}_{i(n)+1}^{-1/2}.\end{aligned} \tag{3.11}$$

The second term of the right-hand side of (3.7) can be estimated in the following way:

$$
2\int_0^t \left(R_{H_2^n}\delta_n^{(2)}(\tau),\, \Delta w_n'(\tau)\right)_{H_2} d\tau
$$

$$
= 2\int_0^t \left(K^{-1/2}K_0^{1/2}K_0^{-1/2}R_{H_2^n}\delta_n^{(2)}(\tau),\, K^{1/2}\Delta\theta_n(\tau)\right)_{H_2} d\tau
$$

$$
\leq 2\left\|K^{-1/2}K_0^{-1/2}\right\|^2_{H_2\to H_2}\left\|K_0^{1/2}R_{H_2^n}\right\|^2_{H_2\to H_2}\times\left\|\delta_n^{(2)}\right\|^2_{K_2(0,T;H_1)}
$$

$$
+\int_0^t \left\|K^{1/2}\Delta\theta_n(\tau)\right\|^2_{H_2} d\tau. \tag{3.12}
$$

Thanks to the special choice of the basis $\{\eta_j\}_{j=1}^{\infty}$, we obtain

$$
\left\|K_0^{1/2}R_{H_2^n}\right\|_{H_2\to H_2} = \tilde{\ae}_{j(n+1)}^{-1/2},
$$

and therefore from (3.12) we obtain

$$
2\int_0^t \left(R_{H_2^n}\delta_n^{(2)}(\tau),\, \Delta\theta_n(\tau)\right)_{H_2} d\tau \leq \left\|K^{-1/2}K_0^{1/2}\right\|^2_{H_2\to H_2}
$$

$$
\times\left\|\delta_n^{(2)}\right\|^2_{L_2(0,T;H_1)}\tilde{\ae}_{j(n+1)}^{-1} + \int_0^t \left\|K^{1/2}\Delta\theta_n(\tau)\right\|^2_{H_2} d\tau. \tag{3.13}
$$

Using the inequalities (3.11) and (3.13), we estimate the right-hand side of (3.7) form above as follow:

$$
\left\|I_1^{1/2}\Delta w_n'(t)\right\|^2_{H_1} + \left\|L^{1/2}\Delta w_n(t)\right\|^2_{H_1} + \left\|I_2^{1/2}\Delta\theta_n(t)\right\|^2_{H_2}
$$

$$
+\int_0^t \left\|K^{1/2}\Delta\theta_n(\tau)\right\|^2_{H_2} d\tau \leq 4\left\|L^{-1/2}L_0^{1/2}\right\|_{H_1\to H_1} T^{1/2}(M_g(T))^{1/2}
$$

$$
\times\left\|\delta_n^{(1)}\right\|_{L_2(0,T;H_1)}\tilde{\lambda}_{i(n)+1}^{-1/2} + \left\|K^{-1/2}K_0^{1/2}\right\|^2_{H_2\to H_2}\left\|\delta_n^{(2)}\right\|^2_{L_2(0,T;H_1)}\tilde{\ae}_{j(n+1)}^{-1}. \tag{3.14}
$$

Let us estimate from above the norms of the errors $\left\|\delta_n^{(i)}\right\|_{L_2(0,T;H_i)}$, $i=1,2$. For $\delta_n^{(1)}$, using the prior estimates (2.122), (2.123), (2.128), we obtain

$$\begin{aligned}
&\left\|\delta_n^{(1)}(t)\right\|_{H_1} \le \|g_1(t)\|_{H_1} + \|Lw_n(t)\|_{H_1} + \left\|MK^{-1}\right\|_{H_2\to H_1} \|K\theta_n(t)\|_{H_1}\\
&+\left\|I_1^{1/2}(t)\right\|_{H_1\to H_1}\left\|I_1^{1/2}w_n''(t)\right\|_{H_1}\\
&\le \|g_1(t)\|_{H_1} + c_L\|g_1(t)\|_{H_1} + \left\|MK^{-1}\right\|_{H_2\to H_1} c_K\|g_2(t)\|_{H_2}\\
&+b_2\left(\left\|I_1^{1/2}w_n''(t)\right\|_{H_1}^2+\left\|L^{1/2}w_n'(t)\right\|_{H_1}^2+\left\|I_1^{1/2}\theta_n'(t)\right\|_{H_2}^2\right)^{1/2}\\
&+\left\|MK^{-1}\right\|_{H_2\to H_1} c_K\Bigg[\|g_2(t)\|_{H_2}\\
&+b_1\left(\left\|I_2^{1/2}\theta_n'(t)\right\|_{H_2}^2+\left\|L^{1/2}w_n'(t)\right\|_{H_1}^2\right)^{1/2}\Bigg]+\left\|I_1^{1/2}\right\|_{H_1\to H_1}\left\|I_1^{1/2}w_n''(t)\right\|_{H_1}\\
&\le(1+c_L)\Bigg[\|g_1(t)\|_{H_1}+\left\|MK^{-1}\right\|_{H_2\to H_1} c_K\Bigg[\|g_2(t)\|_{H_2}\\
&+b_2\left(\left\|I_1^{1/2}w_n''(t)\right\|_{H_1}^2+\left\|L^{1/2}w_n'(t)\right\|_{H_1}^2\right)+\left\|I_2^{1/2}\theta_n'(t)\right\|_{H_1}^2\Bigg]^{1/2}\Bigg]\\
&\le(1+c_L)\left[\|g_1(t)\|_{H_1}+\left\|MK^{-1}\right\|_{H_2\to H_1} c_K\|g_2(t)\|_{H_2}+b_2(M_g(T))^{1/2}\right],
\end{aligned} \tag{3.15}$$

and then, by transition to the norm of the space $L_2(0,T;H_1)$, we obtain

$$\begin{aligned}
\left\|\delta_n^{(1)}\right\|_{L_2(0,T;H_1)} \le (1+c_L)\Big[&\|g_1(t)\|_{L_2(0,T;H_1)}+\left\|MK^{-1}\right\|_{H_2\to H_1}\\
&\times c_K\|g_2(t)\|_{L_2(0,T;H_2)}+b_2(M_g(T))^{1/2}\Big].
\end{aligned} \tag{3.16}$$

For the error $\delta_n^{(2)}$, using the prior estimates (2.123) and (2.128), we obtain

$$\begin{aligned}
&\left\|\delta_n^{(2)}(t)\right\|_{H_2}\\
&\le\|g_2(t)\|_{H_2}+\|K\theta_n(t)\|_{H_2}+b_1\left(\left\|I_2^{1/2}\theta_n'(t)\right\|_{H_2}^2+L^{1/2}w'(t)\Big\|_{H_1}^2\right)^{1/2}
\end{aligned}$$

$$\leq (1 + c_k)\left[\|g_2(t)\|_{H_2} + b_1\left(\left\|I_2^{1/2}\theta_n'(t)\right\|_{H_2}^2 + \left\|L^{1/2}w'(t)\right\|_{H_1}^2\right)^{1/2}\right]$$

$$\leq (1 + c_k)\left[\|g_2(t)\|_{H_2} + b_1(M_g(T))^{1/2}\right], \qquad (3.17)$$

and then, by transition to the norm of the space $L_2(0, T;\, H_2)$, we obtain

$$\left\|\delta_n^{(2)}\right\|_{L_2(0,T;H_2)} \leq (1 + c_k)\left[\|g_2(t)\|_{L_2(0,T;H_2)} + b_1T^{1/2}(M_g(T))^{1/2}\right]. \qquad (3.18)$$

Using (3.16) and (3.18), we estimate (from above) the right-hand side of the inequality (3.14), and we obtain

$$\left\|I_1^{1/2}\Delta w_n'(t)\right\|_{H_1}^2 + \left\|L^{1/2}\Delta w_n(t)\right\|_{H_1}^2 + \left\|I_2^{1/2}\Delta\theta_n(t)\right\|_{H_2}^2$$

$$+\int_0^t \left\|K^{1/2}\Delta\theta_n(\tau)\right\|_{H_2}^2 d\tau \leq 4\left\|L^{-1/2}L_0^{1/2}\right\|_{H_1\to H_1} T^{1/2}(M_g(T))^{1/2}$$

$$\times(1 + c_L)\left[\|g_1(t)\|_{L_2(0,T;H_1)} + \left\|MK^{-1}\right\|_{H_2\to H_1} c_K\, \|g_2(t)\|_{L_2(0,T;H_2)}\right.$$

$$\left. + b_2T^{1/2}(M_g(T))^{1/2}\right]\tilde{\lambda}_{i(n)+1}^{-1/2} + \left\|K^{-1/2}K_0^{1/2}\right\|_{H_2\to H_2}^2 (1 + c_K)^2$$

$$\times\left[\|g_2(t)\|_{L_2(0,T;H_2)} + b_1T^{1/2}(M_g(T))^{1/2}\right]^2 \tilde{æ}_{j(n)+1}^{-1},$$

where t is an arbitrary value from $[0, T]$. By introducing the notation (3.5), the last inequality can be transformed to that of (3.3). Therefore Theorem 2.1 has been proved.

Remark 3.1 The estimate (3.3) is true owing to (2.108) in the sense that its right-hand side tends to zero when $n \to \infty$.

Remark 3.2 The estimate (3.3) has been obtained for a bounded operator MK^{-1}. For a series of real coupled thermoelasticity problems (for instance, for shallow shells in the framework of Timoshenko-type models), more restrictive requirements regarding the boundness of the operator $MK^{-1/2}$ are needed. In that case, for the problems (3.1), (3.3), the following prior estimate of the error of the Bubnov–Galerkin method based on (3.2) is valid:

$$\max_{0\leq t\leq T}\left[\left\|I_1^{1/2}\Delta w_n'(t)\right\|_{H_1}^2 + \left\|L^{1/2}\Delta w_n(t)\right\|_{H_1}^2\right.$$

$$\left. + \left\|I_2^{1/2}\Delta\theta_n(t)\right\|_{H_2}^2 + \int_0^t \left\|K^{1/2}\Delta\theta_n(\tau)\right\|_{H_2}^2 d\tau\right]$$

$$\leq c_1(M_g(T))^{1/2}\left[\|g_1\|_{L_2(0,T;H_1)} + c_2^* \left\|I^{-1/2}g\right\|_{L_2(0,T;H_1\times H_2)} + c_3^*(M_g(T))^{1/2}\right]\tilde{\lambda}_{i(n)+1}^{-1/2}$$
$$+c_4\left[\|g_2\|_{L_2(0,T;H_2)} + c_5(M_g(T))^{1/2}\right]^2 \tilde{æ}_{j(n)+1}^{-1}, \tag{3.19}$$

where the constants $c_1, c_4, c_5, M_g(T)$ are the same as in (3.3), and c_2^*, c_3^* are constants defined by the relations

$$c_2^* = \frac{\sqrt{2}}{2}e^{1/2}T^{1/2}\left\|MK^{-1/2}\right\|_{H_2\to H_1}, \quad c_3^* = T^{1/2}\left\|I_1^{1/2}\right\|_{H_1\to H_1}. \tag{3.20}$$

The estimate (3.19) is obtained via the following procedure. Instead of (2.122), we use an analogous but simpler estimate:

$$\|Lw_n(t)\|_{H_1} \leq c_L\left(\|g_1(t)\|_{H_1} + \left\|I_1^{1/2}\right\|_{H_1\to H_1}\left\|I_1^{1/2}w_n''(t)\right\|_{H_1}\right.$$
$$\left.+\left\|MK^{-1/2}\right\|_{H_2\to H_1}\left\|K^{-1/2}\theta_n(t)\right\|_{H_2}\right),$$

and almost always ($t \in [0, T]$) the norm of the error $\left\|\delta_n^{(1)}(t)\right\|_{H_1}$ can be estimated as follows:

$$\left\|\delta_n^{(1)}(t)\right\|_{H_1} \leq \|g_1(t)\|_{H_1} + \|Lw_n(t)\|_{H_1} + \left\|I_1^{1/2}\right\|_{H_1\to H_1}\left\|I_1^{1/2}w_n''(t)\right\|_{H_1}$$
$$+\left\|MK^{-1/2}\right\|_{H_2\to H_1}\left\|K^{-1/2}\theta_n(t)\right\|_{H_2}$$
$$\leq (1+c_L)\left[\|g_1(t)\|_{H_1} + \left\|I_1^{1/2}\right\|_{H_1\to H_1}\left\|I_1^{1/2}w_n''(t)\right\|_{H_1}\right.$$
$$\left.+\left\|MK^{-1/2}\right\|_{H_2\to H_1}\left\|K^{-1/2}\theta_n(t)\right\|_{H_2}\right].$$

Then, using the prior estimates (2.105) and (2.128), we transform to the norm of the space $L_2(0, T; H_1)$:

$$\left\|\delta_n^{(1)}\right\|_{L_2(0,T;H_1)} \leq (1+c_L)\left[\|g_1(t)\|_{L_2(0,T;H_1)} + \frac{\sqrt{2}}{2}e^{1/2}T^{1/2}\left\|MK^{-1/2}\right\|_{H_2\to H_1}\right.$$
$$\left.\times\left\|I^{-1/2}g\right\|_{L_2(0,T;H_1\times H_2)} + T^{1/2}\left\|I_1^{1/2}\right\|_{H_1\to H_1}(M_g(T))^{1/2}\right].$$

Finally, using this newly obtained inequality (instead of (3.16)) and (3.18), we estimate from above the right hand-side of the inequality (3.14).

The estimates (3.3), (3.19) were obtained using an assumption about the non-differentiability of the free forms in (2.110). Now we are going to consider a more general case, (2.57). Let f be one of the functions $g_1, g_2, I^{-1/2}g$. Let us denote by $\tilde{f}$ the following extension of the function f from the interval $[0, T]$ to the whole time axis:

$$\tilde{f}(t) = \begin{cases} 0, & |t| > T, \\ -f(-t), & -T \le t \le 0. \end{cases} \tag{3.21}$$

The integral modulus of continuity of the function $f \in L_2(0, T; H)$ is denoted by $w(f, \delta)$:

$$w(f, \delta) = \sup_{|\xi| \le \delta} \left[\int_0^T \left\| \tilde{f}(t + \xi) - \tilde{f}(t) \right\|_H^2 dt \right]^{1/2}, \tag{3.22}$$

where H is one of the spaces H_1, H_2, $H_1 \times H_2$, and δ is an arbitrary positive number.

Theorem 3.2 *If the conditions (2.57), (2.58) are satisfied, then the error introduced by use of the equations of the Bubnov–Galerkin method (3.2) for the problem (3.1) allows one to obtain the following estimate:*

$$\begin{aligned} &\sup_{0 \le t \le T} ess \Bigg[\left\| I_1^{1/2} \Delta w_n'(t) \right\|_{H_1}^2 + \left\| L^{1/2} \Delta w_n(t) \right\|_{H_1}^2 \\ &\quad + \left\| I_2^{1/2} \Delta \theta_n(t) \right\|_{H_2}^2 + \int_0^t \left\| K^{1/2} \Delta \theta_n(\tau) \right\|_{H_2}^2 d\tau \Bigg]^{1/2} \\ &\le d_1 w(I^{-1/2}g, \delta_n) + \Big[e_1(g, \delta_n) w(I^{-1/2}g, \delta_n) \tilde{\lambda}_{i(n)+1}^{-1/4} \\ &\quad + d_2 (w(I^{-1/2}g, \delta_n))^2 \Big] + c_4 \Big[e_2(g, \delta_n) \tilde{æ}_{j(n+1)}^{-1/2} \\ &\quad + d_3 \left(w(I^{-1/2}g, \delta_n) \right)^2 \Big]^{1/2} \equiv 0 \left(\max(\delta_n, w(I^{-1/2}g, \delta_n)) \right), \end{aligned} \tag{3.23}$$

where

$$\delta_n = \max \left(\tilde{\lambda}_{i(n)+1}^{-1/4} \tilde{æ}_{j(n+1)}^{-1/2} \right); \tag{3.24}$$

d_1, d_2, d_3 are constants defined by the relations (3.36); $e_1(g, \delta)$, $e_2(g, \delta)$ are quantities that are bounded for $\delta \to 0$, defined further by (3.37), (3.38); c_4 is a constant defined by the fourth relation of (3.5); and $\{\tilde{\lambda}_k\}_{k=1}^{\infty}$, $\{\tilde{æ}_l\}_{l=1}^{\infty}$ are the series defined by the relation (3.4).

Proof Let $\varphi(t)$ be an arbitrary kernel of averaging of the class W_2^1 in the interval $[-1, 1]$, i.e. let it be an even function of the Sobolev class $W_2^1([-1, 1])$, having zero values for $t = \pm 1$ and satisfying the equation

$$\int_{-1}^{1} \varphi(t)\, dt = 1. \tag{3.25}$$

Let us introduce the following notation:

$$\varphi_\delta(t) = \frac{1}{\delta}\varphi\left(\frac{t}{\delta}\right) \quad (-\delta \le t \le \delta),$$

$$g_{j\delta}(t) = \int_{t-\delta}^{t+\delta} \tilde{g}_j(\tau)\varphi_\delta(t-\tau)\, d\tau \equiv \int_{-1}^{1} \tilde{g}_j(t+\delta\tau)\varphi(\tau)\, d\tau,$$

$$\Delta g_{j\delta}(t) = g_j(t) g_{j\delta}(t) \quad (0 \le t \le T),$$

where $\delta > 0$, $j = 1, 2$, and $\tilde{g}_j$ are positive definite functions g_j of the form (3.21). In accordance with (3.25), we obtain

$$\Delta g_{j\delta}(t) = \int_{-1}^{1} \left[g_j(t) - \tilde{g}_j(t+\delta\tau)\right] \varphi(\tau)\, d\tau. \tag{3.26}$$

It is clear that, for arbitrary $\delta > 0$, $g_{j\delta} \in W_1^2(0, T; H_j) (j = 1, 2)$, and

$$g'_{j\delta}(t) = \int_{t-\delta}^{t+\delta} \tilde{g}_j(\tau)\varphi'_\delta(t-\tau)\, d\tau \equiv \frac{1}{\delta}\int_{-1}^{1} \tilde{g}_j(t+\delta\tau)\varphi'(\tau)\, d\tau.$$

The last expression, owing to the noneven kernel derivative $\varphi'(t)$, reads

$$g'_{j\delta}(t) = \frac{1}{\delta}\int_{-1}^{1} \left[\tilde{g}_j(t+\delta\tau) - g_j(t)\right]\varphi'(\tau)\, d\tau. \tag{3.27}$$

The relations (3.26), (3.27) imply the following inequalities:

$$\left\| \Delta g_{j\delta} \right\|_{L_2(0,T;H_j)} \le C_{0,\varphi} w(g_j, \delta), \quad j = 1, 2, \tag{3.28}$$

$$\left\| I^{-1/2} \Delta g_\delta \right\|_{L_2(0,T;H_1\times H_2)} \le C_{0,\varphi} w(I^{-1/2} g, \delta), \tag{3.29}$$

$$\left\| I^{-1/2} g_\delta' \right\|_{L_2(0,T;H_1\times H_2)} \leq \frac{1}{\delta} C_{1,\varphi} w(I^{-1/2} g, \delta), \tag{3.30}$$

where

$$C_{0,\varphi} = \sqrt{2} \left[\int_{-1}^{1} (\varphi(\tau))^2 \, d\tau \right]^{1/2}, \quad C_{1,\varphi} = \sqrt{2} \left[\int_{-1}^{1} (\varphi'(\tau))^2 \, d\tau \right]^{1/2}$$

(for instance, for $\varphi(t) = 1 - |t|$, $C_{0,\varphi} = 2\sqrt{3}/3$, $C_{1\varphi} = 2$).

Let us now consider a family of Cauchy-type problems (3.1) with differentiable free terms:

$$\begin{cases} I_1 w_\delta''(t) + L w_\delta(t) + M\theta_\delta(t) = g_{1\delta}(t), \\ I_2 \theta_\delta'(t) + K\theta_\delta(t) + N w_\delta'(t) = g_{2\delta}(t), \\ w_\delta(0) = w_\delta'(0) = 0, \quad \theta_\delta(0) = 0, \end{cases} \tag{3.31}$$

where δ is a positive parameter. Let us denote by $\{w_{\delta,n}, \theta_{\delta,n}\}$ an approximate solution to the problem (3.31) constructed by the Bubnov–Galerkin method. Let us introduce the following notation:

$$\Delta w_\delta(t) = w(t) - w_\delta(t), \qquad \Delta\theta_\delta(t) = \theta(t) - \theta_\delta(t),$$
$$\Delta w_{\delta,n}(t) = w_n(t) - w_{\delta,n}(t), \; \Delta\theta_{\delta,n}(t) = \theta_n(t) - \theta_{\delta,n}(t),$$
$$\Delta w_{n,\delta}(t) = w_{\delta,n}(t) - w_\delta(t), \; \Delta\theta_{n,\delta}(t) = \theta_{\delta,n}(t) - \theta_\delta(t).$$

Let us consider the pair of functions $\{\Delta w_\delta, \Delta\theta_\delta\}$ as the exact solution to the Cauchy problem (3.1) with free terms $\Delta g_{1\delta}$, $\Delta g_{2\delta}$, and the pair of functions $\{\Delta w_{\delta,n}, \Delta\theta_{\delta,n}\}$ as an approximate solution to this problem constructed according to the Bubnov–Galerkin method. The following prior estimates of the form (2.105), (2.106) apply:

$$\left[\left\| I_1^{1/2} \Delta w_{\delta,n}'(t) \right\|_{H_1}^2 + \left\| L^{1/2} \Delta w_{\delta,n}(t) \right\|_{H_1}^2 + \left\| I_2^{1/2} \Delta\theta_{\delta,n}(t) \right\|_{H_2}^2 \right.$$
$$\left. + 2 \int_0^t \left\| K^{1/2} \Delta\theta_{\delta,n}(\tau) \right\|_{H_2}^2 d\tau \right]^{1/2} \leq e^{1/2} T^{1/2} \left\| I^{-1/2} \Delta g_\delta \right\|_{L_2(0,T;H_1\times H_2)}^2, \tag{3.32}$$

$$\left[\left\|I_1^{1/2}\Delta w_\delta'(t)\right\|_{H_1}^2+\left\|L^{1/2}\Delta w_\delta(t)\right\|_{H_1}^2+\left\|I_2^{1/2}\Delta\theta_\delta(t)\right\|_{H_2}^2\right.$$
$$\left.+2\int_0^t\left\|K^{1/2}\Delta\theta_\delta(\tau)\right\|_{H_2}^2d\tau\right]^{1/2}\le e^{1/2}T^{1/2}\left\|I^{-1/2}\Delta g_\delta\right\|_{L_2(0,T;H_1\times H_2)}^2 \tag{3.33}$$

(almost everywhere ($t\in[0,T]$)).

By applying Theorem 2.1 to the problem (3.31), the following prior estimate of type (3.3) of the error introduced by the Bubnov–Galerkin method can be obtained.

$$\left[\left\|I_1^{1/2}\Delta w_{n,\delta}'(t)\right\|_{H_1}^2+\left\|L^{1/2}\Delta w_{n,\delta}(t)\right\|_{H_1}^2+\left\|I_2^{1/2}\Delta\theta_{n,\delta}(t)\right\|_{H_2}^2\right.$$
$$\left.+\int_0^t\left\|K^{1/2}\Delta\theta_{n,\delta}(\tau)\right\|_{H_2}^2d\tau\right]^{1/2}\le\Big\{c_1(M_{g\delta}(T))^{1/2}\big[\|g_{1\delta}\|_{L_2(0,T;H_1)}$$
$$+c_2\|g_{2\delta}\|_{L_2(0,T;H_2)}+c_3(M_{g\delta}(T))^{1/2}\big]\tilde{\lambda}_{i(n)+1}^{-1/2}$$
$$+c_4\left[\|g_{2\delta}\|_{L_2(0,T;H_2)}+c_5(M_{g\delta}(T))^{1/2}\right]^2\text{æ}_{j(n)+1}^{-1}\Big\}^{1/2}, \tag{3.34}$$

where $0\le t\le T$. Because the extension $\tilde{g}_j(0)$ is equal to 0 ($j=1,2,\delta>0$), the following applies in (3.34):

$$M_{g\delta}(T)=eT\left\|I^{-1/2}g_\delta'\right\|_{L_2(0,T;H_1\times H_2)}^2. \tag{3.35}$$

The error of the Bubnov–Galerkin method for the initial problem (3.1) can be estimated in the following way:

$$\left[\left\|I_1^{1/2}\Delta w_n'(t)\right\|_{H_1}^2+\left\|L^{1/2}\Delta w_n(t)\right\|_{H_1}^2+\left\|I_2^{1/2}\Delta\theta_n(t)\right\|_{H_2}^2\right.$$
$$\left.+\int_0^t\left\|K^{1/2}\Delta\theta_n(\tau)\right\|_{H_2}^2d\tau\right]^{1/2}\le\left[\left\|I_1^{1/2}\Delta w_{\delta,n}'(t)\right\|_{H_1}^2+\left\|L^{1/2}\Delta w_{\delta,n}(t)\right\|_{H_1}^2\right.$$
$$\left.+\left\|I_2^{1/2}\Delta\theta_{\delta,n}(t)\right\|_{H_2}^2+\int_0^t\left\|K^{1/2}\Delta\theta_{\delta,n}(\tau)\right\|_{H_2}^2d\tau\right]^{1/2}+\left[\left\|I_1^{1/2}\Delta w_\delta'(t)\right\|_{H_1}^2\right.$$

$$+ \left\| L^{1/2} \Delta w_\delta(t) \right\|_{H_1}^2 + \left\| I_2^{1/2} \Delta\theta_\delta(t) \right\|_{H_2}^2 + \int_0^t \left\| K^{1/2} \Delta\theta_\delta(\tau) \right\|_{H_2}^2 d\tau \Bigg]^{1/2}$$

$$+ \Bigg[\left\| I_1^{1/2} \Delta w'_{n,\delta}(t) \right\|_{H_1}^2 + \left\| L^{1/2} \Delta w_{n,\delta}(t) \right\|_{H_1}^2 + \left\| I_2^{1/2} \Delta\theta_{n,\delta}(t) \right\|_{H_2}^2$$

$$+ \int_0^t \left\| K^{1/2} \Delta\theta_{\delta,n}(\tau) \right\|_{H_2}^2 d\tau \Bigg]^{1/2}.$$

Using (3.32)–(3.35) and (3.28)–(3.30), we obtain

$$\Bigg[\left\| I_1^{1/2} \Delta w'_n(t) \right\|_{H_1}^2 + \left\| L^{1/2} \Delta w_n(t) \right\|_{H_1}^2 + \left\| I_2^{1/2} \Delta\theta_n(t) \right\|_{H_2}^2$$

$$+ \int_0^t \left\| K^{1/2} \Delta\theta_n(\tau) \right\|_{H_2}^2 d\tau \Bigg]^{1/2} \le 2e^{1/2} T^{1/2} \left\| I^{-1/2} \Delta g_\delta \right\|_{L_2(0,T;H_1\times H_2)}$$

$$+ \Bigg\{ c_1 e^{1/2} T^{1/2} \left\| I^{-1/2} g'_\delta \right\|_{L_2(0,T;H_1\times H_2)} \Big[\|g_1\|_{L_2(0,T;H_1)} + c_2 \|g_2\|_{L_2(0,T;H_2)}$$

$$+ \|\Delta g_{1\delta}\|_{L_2(0,T;H_1)} + c_2 \|\Delta g_{2\delta}\|_{L_2(0,T;H_2)}$$

$$+ c_3 e^{1/2} T^{1/2} \left\| I^{-1/2} g'_\delta \right\|_{L_2(0,T;H_1\times H_2)} \Big] \tilde{\lambda}_{i(n)+1}^{-1} + c_4 \Big[\|g_2\|_{L_2(0,T;H_2)}$$

$$+ \|\Delta g_{2\delta}\|_{L_2(0,T;H_2)} + c_5 e^{1/2} T^{1/2} \left\| I^{-1/2} g'_\delta \right\|_{L_2(0,T;H_1\times H_2)} \Big]^2 \tilde{æ}_{j(n)+1}^{-1} \Bigg\}^{1/2}$$

$$\le 2e^{1/2} T^{1/2} C_{0,\varphi} w(I^{-1/2} g, \delta)$$

$$+ \Bigg\{ c_1 e^{1/2} T^{1/2} C_{1,\varphi} \frac{1}{\delta} w(I^{-1/2} g, \delta) \Big[\|g_1\|_{L_2(0,T;H_1)} + c_2 \|\Delta g_2\|_{L_2(0,T;H_2)}$$

$$+ C_{0,\varphi} w(g_1, \delta) + c_2 C_{0,\varphi} w(g_2, \delta) + c_3 e^{1/2} T^{1/2} C_{0,\varphi} \frac{1}{\delta} w(I^{-1/2} g, \delta) \Big] \times \tilde{\lambda}_{i(n)+1}^{-1/2}$$

$$+ c_4 \Big[\|g_2\|_{L_2(0,T;H_1)} + C_{0,\varphi} w(g_2, \delta)$$

$$+ c_5 e^{1/2} T^{1/2} C_{1,\varphi} \times \frac{1}{\delta} w(I^{-1/2} g, \delta) \Big]^2 \tilde{æ}_{j(n)+1}^{-1} \Bigg\}^{1/2}.$$

In the last inequality, δ denotes an arbitrary positive number. Let us assume $\delta = \delta_n$, where δ_n is defined by the relation (3.24). The following inequality is then obtained:

$$\Bigg[\left\|I_1^{1/2}\Delta w_n'(t)\right\|_{H_1}^2 + \left\|L^{1/2}\Delta w_n(t)\right\|_{H_1}^2 + \left\|I_2^{1/2}\Delta\theta_n(t)\right\|_{H_2}^2$$

$$+\int_0^t \left\|K^{1/2}\Delta\theta_n(\tau)\right\|_{H_2}^2 d\tau\Bigg]^{1/2} \le 2e^{1/2}T^{1/2}C_{0,\varphi}\frac{1}{\delta}w(I^{-1/2}g,\delta_n)$$

$$+\Bigg\{c_1 e^{1/2}T^{1/2}C_{1,\varphi}\frac{1}{\delta}w(I^{-1/2}g,\delta_n)\Big[\Big(\|g_1\|_{L_2(0,T;H_1)} + c_2\|g_2\|_{L_2(0,T;H_2)}$$

$$+C_{0,\varphi}w(g_1,\delta_n) + c_2C_{0,\varphi}w(g_2,\delta_n)\Big)\tilde{\lambda}_{i(n)+1}^{-1/4} + c_3e^{1/2}T^{1/2}C_{1,\varphi}$$

$$\times w(I^{-1/2}g,\delta_n)\Big] + c_4\Big[\Big(\|g_2\|_{L_2(0,T;H_2)} + C_{0,\varphi}w(g_2,\delta_n)\Big)\tilde{æ}_{j(n)+1}^{-1/2}$$

$$+c_5e^{1/2}T^{1/2}C_{1,\varphi}\times w(I^{-1/2}g,\delta_n)\Big]^2\Bigg\}^{1/2}$$

(almost always everywhere ($t \in [0, T]$)). By introducing the notation

$$d_1 = 2e^{1/2}T^{1/2}C_{0,\varphi},\quad d_2 = c_1c_3eTC_{1,\varphi}^2,\quad d_3 = c_5e^{1/2}T^{1/2}C_{1,\varphi}, \tag{3.36}$$

$$e_1(g,\delta) = c_1e^{1/2}T^{1/2}C_{1,\varphi}\Big(\|g_1\|_{L_2(0,T;H_1)} + c_2\|g_2\|_{L_2(0,T;H_2)}$$
$$+C_{0,\varphi}w(g_1,\delta) + c_2C_{0,\varphi}w(g_2,\delta)\Big), \tag{3.37}$$

$$e_2(g,\delta) = c_4\|g_2\|_{L_2(0,T;H_2)} + C_{0,\varphi}w(g_2,\delta), \tag{3.38}$$

the last inequality can be transformed to the form (3.23). Theorem 3.2 has been proved.

Remark 3.3 The estimate (3.23) is true because of (2.108), in the sense that for $n \to \infty$ its right-hand side approaches zero.

Remark 3.4 The estimate (3.23) was obtained using an assumption about the bounds of the operator MK^{-1}. If the operator $MK^{-1/2}$ is also bounded, then (together with (3.23)) the following prior estimate of the error of the Bubnov–Galerkin method based on (3.2) for the problem (3.1) holds:

$$\sup_{0\le t\le T} ess\Bigg[\left\|I_1^{1/2}\Delta w_n'(t)\right\|_{H_1}^2 + \left\|L^{1/2}\Delta w_n(t)\right\|_{H_1}^2$$

$$+\left\|I_2^{1/2}\Delta\theta_n(t)\right\|_{H_2}^2 + \int_0^t \left\|K^{1/2}\Delta\theta_n(\tau)\right\|_{H_2}^2 d\tau\Bigg]^{1/2}$$

$$\leq d_1 w(I^{-1/2}g, \delta_n) + \Big[e_1^*(g, \delta_n) w(I^{-1/2}g, \delta_n) \tilde{\lambda}_{i(n)+1}^{-1/4}$$

$$+ d_2^* (w(I^{-1/2}g, \delta_n))^2 + c_4 (e_2(g, \delta_n) \tilde{æ}_{j(n)+1}^{-1/2}$$

$$+ d_3 w(I^{-1/2}g, \delta_n))^2 \Big]^{1/2}, \tag{3.39}$$

where the constants d_1, d_3, d_4 and the quantity $e_2(g, \delta)$ are the same as those appearing in (3.23), d_2^* is a constant defined by the relation

$$d_2^* = c_1 c_3^* e T C_{1,\varphi}^2, \tag{3.40}$$

and $e_1^*(g, \delta)$ is the bounded (for $\delta \to \infty$) quantity defined by the equation

$$e_1^*(g, \delta) = c_1 e^{1/2} T^{1/2} C_{1,\varphi} \left(\|g_1\|_{L_2(0,T;H_1)} + c_2^* \|g_2\|_{L_2(0,T;H_2)} \right). \tag{3.41}$$

The scheme by which the estimate (3.39) is obtained is analogous to the scheme applied to prove Theorem 3.2. Here, instead of (3.34), a prior estimate of type (3.19) of the error introduced by the Bubnov–Galerkin method for the problem (3.31) is used.

All the error estimate for the Bubnov–Galerkin method given above have been obtained for a system of differential equations of a rather general form and with homogeneous boundary conditions, which in the case of its application to a thin-walled structure, corresponds to coupled thermoelastic problems for plates and shallow shells with variable thickness. The free terms g_1, g_2 (corresponding to the vector of mechanical load intensity and to the specific power of the internal heat sources) have been assumed to be time dependent. Let us consider now one of the particular cases of the problem (2.54)–(2.56) (with nonhomogeneous initial conditions), where a prior estimate of the error of the Bubnov–Galerkin method is most effective. This corresponds to a mechanical problem where the thickness of the shell or plate under consideration is constant, the intensity of the mechanical load and the power of the internal heat sources do not depend on time, and the full systems of eigenfunctions of the operators of the static elasticity theory of stationary heat transfer are known. These eigenfunctions are used as a basis. The problem corresponds to the following Cauchy problem:

$$\begin{cases} I_1 w''(t) + L w(t) + M\theta(t) = g_1(t), \\ I_2 \theta'(t) + K\theta(t) + N w'(t) = g_2(t), \\ w(0) = w'(0) = 0, \quad \theta(0) = 0, \end{cases} \tag{3.42}$$

where $w(t), \theta(t)$ are the sought functions; g_1, g_2 are given elements from the Hilbert spaces H_1, H_2; w_0, w_1, θ_0 are given elements satisfying the conditions (2.59); and I_1, I_2, L, K, M, N are operator coefficients satisfying all conditions given in Sect. 2.1.4 and the following conditions:

$$\begin{aligned} &I_1(H_1^n) = H_1^n, \quad L(H_1^n) = H_1^n, \quad M(H_2^n) \subset H_1^n, \\ &I_2(H_2^n) = H_2^n, \quad K(H_2^n) = H_2^n. \end{aligned} \tag{3.43}$$

Here H_1^n, H_2^n are the linear neighbourhoods of finite systems of basis elements $\{v_i\}_{i=1}^{i(n)}, \{\eta_j\}_{i=1}^{i(n)}$, and n is an arbitrary natural number. According to the conditions (3.43), for arbitrary elements $w_n \in H_1^n, \theta_n \in H_2^n$, the following relations are satisfied:

$$P_{H_1^n} I_1 w_n = I_1 w_n, \quad P_{H_2^n} I_2 \theta_n = I_2 \theta_n,$$

$$P_{H_1^n} L w_n = L w_n, \quad P_{H_2^n} K \theta_n = K \theta_n, \quad P_{H_1^n} M \theta_n = M \theta_n.$$

Taking these relations into account, (2.95), (2.109) can be written in the following form:

$$\begin{cases} I_1 w_n''(t) + L w_n(t) + M\theta_t(t) = P_{H_1^n} g_1, \\ I_2 \theta_n'(t) + K\theta_n(t) + P_{H_2^n} N w_t'(t) = P_{H_2^n} g_2, \\ w_n(t) \in H_1^n, \quad \theta_n(t) \in H_2^n, \quad \forall t \in [0, T], \\ w_n(0) = P_{H_1^n} w_0, \quad w_n'(0) = P_{H_1^n} w_1, \quad \theta_n(0) = P_{H_2^n} \theta_0. \end{cases} \tag{3.44}$$

According to the conditions formulated in Sect. 2.1.4 applied to the operators K, L, N, for arbitrary elements $w \in H_1^n, \theta \in D(K^{1/2})$, the following inequality is satisfied:

$$\left| (R_{H_2^n} N w, \theta)_{H_2} \right| \le s_n \left\| L^{1/2} w \right\|_{H_1} \left\| K^{1/2} w \right\|_{H_2}, \tag{3.45}$$

where

$$s_n = \left\| K^{-1/2} R_{H_2^n} \right\|_{H_2 \to H_2} \left\| N L^{-1/2} \right\|_{H_1 \to H_2} \to 0, \quad n \to \infty. \tag{3.46}$$

In the following treatment we mean by s_n an arbitrary numerical series (not necessarily defined by the relation (3.46)). The elements of this series do not depend on $w \in H_1^n, \theta \in D(K^{1/2})$, and then converge to zero as well as satisfy the inequality (3.45).

Theorem 3.3 *If the conditions (2.58), (2.59), (3.43), (3.45) are satisfied then the velocity of convergence of the Bubnov–Galerkin based on (3.44) (applied to the problem (3.42)) is characterized by the following inequality:*

$$\max_{0\le t\le T}\left[\left\|I_1^{1/2}\Delta w_n'(t)\right\|_{H_1}^2+\frac{1}{2}\left\|L^{1/2}\Delta w_n(t)\right\|_{H_1}^2\right.$$
$$\left.+\left\|I_2^{1/2}\Delta\theta_n(t)\right\|_{H_2}^2+\int_0^t\left\|K^{1/2}\Delta\theta_n(\tau)\right\|_{H_2}^2 d\tau\right]^{1/2}$$
$$\le\left\|I_1^{1/2}R_{H_1^n}w_1\right\|_{H_1}^2+\left\|L^{1/2}R_{H_1^n}w_0\right\|_{H_1}^2+\left\|I_1^{1/2}R_{H_2^n}\theta_0\right\|_{H_2}^2$$
$$+2\left\|L^{-1/2}R_{H_1^n}g_1\right\|_{H_1}^2-2(R_{H_1^n}g_1,R_{H_1^n}w_0)_{H_1}$$
$$+\left(\left\|K^{-1/2}R_{H_2^n}g_2\right\|_{H_2}^2 s_nc_n\right)^2 T, \tag{3.47}$$

$$\max_{0\le t\le T}\left[\left\|I_1^{1/2}\Delta w_n'(t)\right\|_{H_1}^2+\frac{1}{2}\left\|L^{1/2}\Delta w_n(t)\right\|_{H_1}^2+\left\|I_2^{1/2}\Delta\theta_n(t)\right\|_{H_2}^2\right]$$
$$\le\left\|I_1^{1/2}R_{H_1^n}w_1\right\|_{H_1}^2+\left\|L^{1/2}R_{H_1^n}w_0\right\|_{H_1}^2$$
$$+\left\|I_1^{1/2}R_{H_2^n}\theta_0\right\|_{H_2}^2+2\left\|L^{-1/2}R_{H_1^n}g_1\right\|_{H_1}^2-2(R_{H_1^n}g_1,R_{H_1^n}w_0)_{H_1} \tag{3.48}$$
$$+\frac{1}{2}\left(\left\|K^{-1/2}R_{H_2^n}g_2\right\|_{H_2}^2 s_nc_n\right)^2 T,$$

where

$$c_n=\left[\left\|I_1^{-1/2}(g_1-Lw_0-MP_{H_2^n}\theta_0)\right\|_{H_1}^2+\left\|L^{1/2}w_1\right\|_{H_1}^2\right.$$
$$\left.+\left\|I_2^{-1/2}(g_2-K\theta_0-NP_{H_1^n}w_1)\right\|_{H_2}^2\right]^{1/2}. \tag{3.49}$$

Proof We differentiate the first two relations of (3.42). The first is then scalarly multiplied by $2w_n''(t)$ and the second by $2\theta_n'(t)$. We then sum the relations, making use of the condition (2.58). As the result, we obtain

$$\frac{d}{dt}\left\|I_1^{1/2}w_n''(t)\right\|_{H_1}^2+\frac{d}{dt}\left\|L^{1/2}\Delta w_n'(t)\right\|_{H_1}^2+\frac{d}{dt}\left\|I_2^{1/2}\theta_n'(t)\right\|_{H_2}^2$$
$$+2\left\|K^{1/2}\theta_n'(t)\right\|_{H_2}^2=0.$$

Integrating this equation over the interval $[0, t]$, where $t \leq T$, we obtain

$$\left\| I_1^{1/2} w_n''(t) \right\|_{H_1}^2 + \frac{1}{2} \left\| L^{1/2} w_n'(t) \right\|_{H_1}^2 + \left\| I_2^{1/2} \theta_n'(t) \right\|_{H_2}^2 + 2 \int_0^t \left\| K^{1/2} \theta_n'(\tau) \right\|_{H_2}^2 d\tau$$

$$= \left\| I_1^{1/2} w_n''(t) \right\|_{H_1}^2 + \frac{1}{2} \left\| L^{1/2} w_n'(0) \right\|_{H_1}^2 + \left\| I_2^{1/2} \theta_n'(0) \right\|_{H_2}^2$$

$$= \left\| I_1^{-1/2} (P_{H_1^n} g_1 - L w_n(0) - M\theta_n(0)) \right\|_{H_1}^2$$

$$+ \left\| L^{1/2} w_n'(0) \right\|_{H_1}^2 + \left\| I_1^{-1/2} (P_{H_2^n} g_2 - K\theta_n(0) - P_{H_2^n} N w_n'(0)) \right\|_{H_2}^2$$

$$= \left\| P_{H_1^n} I_1^{-1/2} (g_1 - L w_0 - M P_{H_2^n} \theta_0) \right\|_{H_1}^2 + \left\| P_{H_1^n} L^{1/2} w_1 \right\|_{H_1}^2$$

$$+ \left\| P_{H_2^n} I_2^{-1/2} (g_2 - K\theta_0 - N P_{H_1^n} w_1) \right\|_{H_2}^2 \leq c_n^2,$$

which means that the following inequality holds:

$$\left\| L^{1/2} w_n'(t) \right\|_{H_1} \leq c_n. \tag{3.50}$$

From (3.45) and (3.50), the following inequality is obtained:

$$\left| (R_{H_2^n} N w_n'(t), \theta)_{H_2} \right| \leq c_n s_n \left\| K^{1/2} \theta \right\|_{H_2}, \tag{3.51}$$

where θ is an arbitrary element from $D(K^{1/2})$.

Taking into account (3.42) and (3.44), we then obtain

$$\begin{cases} I_1 \Delta w_n''(t) + L \Delta w_n(t) + M \Delta \theta_n(t) = -R_{H_1^n} g_1, \\ I_2 \Delta \theta_n'(t) + K \Delta \theta_n(t) + N \Delta w_n'(t) = -R_{H_2^n} g_2 + R_{H_2^n} N w_n'(t), \\ \Delta w_n(0) = -R_{H_1^n} w_0, \ \Delta w_n'(0) = -R_{H_1^n} w_1, \ \Delta \theta_n(0) = -R_{H_2^n} \theta_0. \end{cases} \tag{3.52}$$

We multiply scalarly the first two equations of (3.52) by $2\Delta w_n'(t)$ and $2\Delta\theta_n(t)$, respectively, and sum them. We then use the condition (2.58) and integrate the newly obtained expression over the interval $[0, t]$, where $t \leq T$. As the result we obtain

$$\left\| I_1^{1/2} \Delta w_n'(t) \right\|_{H_1}^2 + \left\| L^{1/2} \Delta w_n(t) \right\|_{H_1}^2 + \left\| I_2^{1/2} \Delta \theta_n(t) \right\|_{H_2}^2$$

$$+2 \int_0^t \left\| K^{1/2} \Delta \theta_n(\tau) \right\|_{H_2}^2 d\tau = \left\| I_1^{1/2} R_{H_1^n} w_1 \right\|_{H_1}^2 + \left\| L^{1/2} R_{H_1^n} w_0' \right\|_{H_1}^2$$

$$+\left\|I_2^{1/2}R_{H_2^n}\theta_0\right\|_{H_2}^2 - 2(R_{H_1^n}g_1, \Delta w_n(t))_{H_1^n} - 2(R_{H_1^n}g_1, R_{H_1^n}w_0)_{H_1}$$
$$+2\int_0^t(-R_{H_2^n}g_2 + R_{H_2^n}Nw_n'(\tau), \Delta\theta_n(\tau))_{H_2}\,d\tau. \tag{3.53}$$

The fourth term of the right-hand side of (3.53) allows us to obtain the estimate:

$$\begin{aligned}&-2(R_{H_1^n}g_1, \Delta w_n(t))_{H_1}\\ &\le \frac{1}{2}\left\|L^{1/2}\Delta w_n(t)\right\|_{H_1}^2 + 2\left\|L^{-1/2}R_{H_1^n}g_1(t)\right\|_{H_1}^2.\end{aligned} \tag{3.54}$$

The last term on the right-hand side of (3.53), owing to (3.51), leads to the estimate

$$\begin{aligned}&2\int_0^t\Big(-R_{H_2^n}g_2 + R_{H_2^n}Nw_n'(\tau), \Delta\theta_n(\tau)\Big)_{H_2}d\tau\\ &\le 2\left(\left\|K^{-1/2}R_{H_2^n}g_2\right\|_{H_2} + c_ns_n\right)\int_0^t\left\|K^{1/2}\Delta\theta_n(\tau)\right\|_{H_2}d\tau.\end{aligned} \tag{3.55}$$

Furthermore, by applying the Cauchy inequality, we obtain

$$\begin{aligned}&2\left(\left\|K^{-1/2}R_{H_2^n}g_2\right\|_{H_2} + c_ns_n\right)\int_0^t\left\|K^{1/2}\Delta\theta_n(\tau)\right\|_{H_2}d\tau\\ &\le \left(\left\|K^{-1/2}R_{H_2^n}g_2\right\|_{H_2} + c_ns_n\right)^2 T + \int_0^t\left\|K^{1/2}\Delta\theta_n(\tau)\right\|_{H_2}d\tau.\end{aligned} \tag{3.56}$$

From (3.53)–(3.56) the estimate (3.47) can be derived.

The right-hand side of the inequality (3.55) can also be evaluated in the following way:

$$\begin{aligned}&2\left(\left\|K^{-1/2}R_{H_2^n}g_2\right\|_{H_2} + c_ns_n\right)\int_0^t\left\|K^{1/2}\Delta\theta_n(\tau)\right\|_{H_2}d\tau\\ &\le \frac{1}{2}\left(\left\|K^{-1/2}R_{H_2^n}g_2\right\|_{H_2} + c_ns_n\right)^2 T + 2\int_0^t\left\|K^{1/2}\Delta\theta_n(\tau)\right\|_{H_2}^2d\tau.\end{aligned} \tag{3.57}$$

From (3.53)–(3.56), we obtain the estimate (3.48). Theorem 3.3 has been proved.

Remark 3.5 During the formulation of Theorem 3.3, two estimates of the accuracy of the Bubnov–Galerkin method based on (3.44) applied to the problem (3.42) have been used. They differ in the following way. Firstly, on the left hand side of the inequality (3.47), the following term appears:

$$\int_0^t \left\| K^{1/2} \Delta\theta_n(\tau) \right\|_{H_2}^2 d\tau, \tag{3.58}$$

which does not occur on the left-hand side of the inequality (3.48). Secondly, the last term of the right-hand side of the inequality (3.48) is half of the last term of the right-hand side of the inequality (3.47). To conclude, the estimate (3.47) should be used only when we need to estimate (3.58) (i.e. when we need to evaluate the convergence velocity $\theta_n \to \theta$ in the norm of the space $L_2(0, T; D(K^{1/2}))$). Otherwise, it is more convenient to apply the estimate *(3.48)* with a smaller right-hand side.

To end this section, we outline the prior estimates of the error of the Bubnov–Galerkin method applied to the Cauchy problem

$$I_1 w''(t) + Lw(t) = g_1(t), \quad w(0) = w_0, \quad w'(0) = w_1, \tag{3.59}$$

which generalizes a class of dynamical problems of theory of the elasticity (without a heat transfer equation) for both three-dimensional and thin-walled elements of structures. The operators I_1 and L occurring in (3.59) satisfy all conditions formulated in Sect. 2.1.4.

If $w_0 = w_1 = 0, g_1 \in W_2^1(0, T; H_1)$ and a system of eigenelements ν_i of a self-adjoint, positive definite, convergent operator L_0 (forming an acute angle with the operator L) is used as a basis, then the error introduced by the Bubnov–Galerkin method for the problem (3.59) reads

$$\begin{aligned} &\max_{0 \le t \le T} \left[\left\| I_1^{1/2} \Delta w_n'(t) \right\|_{H_1}^2 + \left\| L^{1/2} \Delta w_n(t) \right\|_{H_1}^2 \right] \\ &\le c_1 (M_g(T))^{1/2} \left[\|g_1\|_{L_2(0,T;H_1)} + c_3 (M_g(T))^{1/2} \right] \times \tilde{\lambda}_{i(n)+1}^{-1/2} \\ &\equiv 0(\tilde{\lambda}_{i(n)+1}^{-1/2}), \end{aligned} \tag{3.60}$$

where $\Delta w_n(t) = w_n(t) - w(t)$; w_n, w are the approximate and exact solutions to the problem (3.59); $\tilde{\lambda}_k = \min_{i \ge k} \lambda_i$; λ_i are the eigenvalues of the operator L_0; $i(n)$ is the set of basis functions ν_i used in constructing the approximate solutions w_n; c_1 is a constant defined by the first relation of (3.5); c_3 is a constant defined by the relation

$$c_3 = T^{1/2} \left\| I_1^{1/2} \right\|_{H_1 \to H_1}; \tag{3.61}$$

and $M_g(T)$ is a constant defined by the relation (2.130), where we take $g_2(t) \equiv 0$. The estimate (3.60) is analogous to both of the estimates (3.3), (3.19), which hold for the approximate solutions of the corresponding coupled problems.

If $w_0 = w_1 = 0$, $g_1 \in L_2(0, T; H_1)$, then the error introduced by the Bubnov–Galerkin method for the problem (3.59) can be evaluated in the following way, using the basis defined above:

$$\sup_{0 \le t \le T} ess \left[\left\| I_1^{1/2} \Delta w_n'(t) \right\|_{H_1}^2 + \left\| L^{1/2} \Delta w_n(t) \right\|_{H_1}^2 \right]^{1/2}$$
$$\le d_1 w(I^{-1/2} g, \tilde{\lambda}_{i(n)+1}^{-1/4}) + \left[e_1(g_1, \tilde{\lambda}_{i(n)+1}^{-1/4}) w(I^{-1/2} g_1, \tilde{\lambda}_{i(n)+1}^{-1/4}) \tilde{\lambda}_{i(n)+1}^{-1/4} \right.$$
$$\left. + d_2 (w(I_1^{-1/2} g_1, \tilde{\lambda}_{i(n)+1}^{-1/4}))^2 \right]^{1/2}$$
$$\equiv 0 \left(\max(\tilde{\lambda}_{i(n)+1}^{-1/4}, w(I_1^{-1/2} g_1, \tilde{\lambda}_{i(n)+1}^{-1/4})) \right), \tag{3.62}$$

where the notations Δw_n, $\tilde{\lambda}_k$ and $i(n)$ have the same meaning as in (3.60); $w(f, \delta)$ is the integral modulus of continuity defined by the relation (3.22); d_1 is a constant defined by the first relation of (3.36); $e_1(g_1, \delta)$ is a bounded quantity for $\delta \to 0$ defined by the relation (3.37), where we take $g_2(t) \equiv 0$; d_2 is a constant defined by the second relation of (3.36), in which c_3 is a constant defined by the relation (3.61). The estimate (3.62) is analogous to both of the estimates (3.23), (3.39) satisfied for the corresponding coupled problems.

If the linear neighbourhood H_1^n of a finite system of basis elements $\{v_i\}_{i=1}^{i(n)}$ is the eigensubspace for both of the operators I_1 and L for an arbitrary n, $g_1(t) \equiv g_1 = const$ ($g_1 \in H_1$), $w_0 \in D(L^{1/2})$, $w_1 \in H_1$, then the error introduced by the Bubnov–Galerkin method for the problem (3.59) can be estimated as follows:

$$\max_{0 \le t \le T} \left[\left\| I_1^{1/2} \Delta w_n'(t) \right\|_{H_1}^2 + \frac{1}{2} \left\| L^{1/2} \Delta w_n(t) \right\|_{H_1}^2 \right]$$
$$\le \left\| I_1^{1/2} R_{H_1^n} w_1 \right\|_{H_1}^2 + \left\| L^{1/2} R_{H_1^n} w_0 \right\|_{H_1}^2 + 2 \left\| L^{-1/2} R_{H_1^n} g_1 \right\|_{H_1}^2 \tag{3.63}$$
$$- 2(R_{H_1^n} g_1, R_{H_1^n} w_0)_{H_1},$$

where Δw_n possesses the same meaning as in (3.60), $R_{H_1^n} = E_1 - P_{H_1^n}$, $P_{H_1^n}$ is an orthoprojector from H_1 to H_1^n, and E_1 is an identity operator over H_1. The estimates (3.63) are analogous to the estimates (3.47), (3.48) satisfied for the corresponding coupled problems.

The proofs of the estimates (3.60), (3.62), (3.63) are analogous to the proofs of Theorems 3.1–3.3. The estimates (3.60), (3.62), (3.63) can also be formally obtained using the corresponding estimates (3.3) (or (3.19)), (3.23) (or (3.39)), (3.47) (or (3.48)), by deleting all terms the existence of which leads to the occurrence of the second differential equation and the second sought function θ in (2.54)–(2.56).

3.2 Coupled Thermoelastic Problem Within the Kirchhoff–Love Model

A fundamental problem that arises in the application of the various error estimates for the Bubnov–Galerkin method to real thermoelastic problems is the need to find operators L_0, K_0 (convergent and forming an acute angle with the differential operators L, K) for which full systems of eigenfunctions and the corresponding eigenvalues are known. In this section, we show how to find operators L_0, K_0 with the required properties and the required precision of the estimates (3.3), (3.23) for coupled thermoelastic problems of shallow shells, rectangular in plan and with variable thickness, in the framework of the Kirchhoff–Love model. Assuming that the conditions (2.64), (2.65), (2.68), (2.83)–(2.85) are satisfied, we consider the system of differential equations (2.30)–(2.32) with the homogeneous initial conditions (2.33)–(2.35), where u, v, w, θ are the sought functions, defined for $(x, y) \in \Omega_1, |z| \leq \frac{1}{2}h(x, y), t \in [0, T]$. Let us assume that $\Omega_1 = \left(-\frac{a}{2}, \frac{a}{2}\right) \times \left(-\frac{b}{2}, \frac{b}{2}\right)$, where a, b are positive constants, i.e. we are considering a shell that is rectangular in plan. Boundary conditions, corresponding either to simple support of the edges on a flexible rib that cannot be compressed or elongated in the tangential plane (2.49) or to clamping of the edges (2.51), are applied to the shell edge $d\Omega_1$. We apply the homogeneous boundary condition (2.38) ($\theta = 0$) to the shell surface $d\Omega_2$.

The differential operator L for the problem under consideration is defined on $D(L) = (\mathring{W}_2^2(\Omega_1))^2 \times \mathring{W}_2^4(\Omega_1)$ by the relation (2.88). An operator L_0 with a known spectrum will be constructed in a way similar to that presented in [360]. Let us look for the operator L_0 in the form

$$L_0\overline{\omega} = \tilde{L}\overline{\omega} + P_L\overline{\omega}, \tag{3.64}$$

where $\overline{\omega} = \{u, v, w\}$, and $\tilde{L}$ is the self-conjugated extension of a differential operator in the Friedrichs sense, defined initially on $[(\mathring{W}_2^2(\Omega_1))^2 \times \mathring{W}_2^4(\Omega_1)] \cap [(C^2(\overline{\Omega_1}))^2 \times C^4(\overline{\Omega_1})]$ by the relation

$$\tilde{L}\overline{\omega} = \left\{-\frac{\partial^2 u}{\partial x^2} - \frac{1-\nu}{2}\frac{\partial^2 u}{\partial y^2}, -\frac{\partial^2 v}{\partial y^2} - \frac{1-\nu}{2}\frac{\partial^2 v}{\partial x^2}, \nabla^2\nabla^2 w\right\},$$

from which P_L (see (3.64)) is going to be found.

We introduce the following notation:

$$w_k(x) = \begin{cases} \cos \pi k x, & k \text{ odd} \\ \sin \pi k x, & k \text{ even} \end{cases}, \tag{3.65}$$

$$w_k^*(x) = \begin{cases} \cos \pi kx, & k \text{ even} \\ \sin \pi kx, & k \text{ odd} \end{cases}, \tag{3.66}$$

$$w_{kl}^{(1,1)}(x, y) = w_k^*(x) w_l(y), \quad w_{kl}^{(1,2)}(x, y) = w_k(x) w_l^*(y), \tag{3.67}$$

$$w_{kl}^{(1,3)}(x, y) = w_k(x) w_l(y). \tag{3.68}$$

For the boundary conditions (2.49) the vector functions $\overline{v}_{kl}^{(j)} (j = 1, 2, 3)$ of the operator L_0 have the form

$$\overline{v}_{kl}^{(1)}(x, y) = \left\{ w_{kl}^{(1,1)} \left(\frac{x}{a}, \frac{y}{b} \right), 0, 0 \right\}, \quad k = 0, 1, \ldots, l = 1, 2, \ldots, \tag{3.69}$$

$$\overline{v}_{kl}^{(2)}(x, y) = \left\{ 0, w_{kl}^{(1,2)} \left(\frac{x}{a}, \frac{y}{b} \right), 0 \right\}, \quad k = 1, 2, \ldots, l = 0, 1, \ldots, \tag{3.70}$$

$$\overline{v}_{kl}^{(3)}(x, y) = \left\{ 0, 0, w_{kl}^{(1,3)} \left(\frac{x}{a}, \frac{y}{b} \right) \right\}, \quad k, l = 1, 2, \ldots, \tag{3.71}$$

and the corresponding eigenvalues $\lambda_{kl}^{(j)}$ of the operator L_0 are defined by the equations

$$\lambda_{kl}^{(1)} = \pi^2 \left[\frac{k^2}{a^2} + \frac{(1 - \nu) l^2}{2b^2} \right] + P_L, \quad \lambda_{kl}^{(2)} = \pi^2 \left[\frac{(1 - \nu) l^2}{2a^2} + \frac{l^2}{b^2} \right] + P_L, \tag{3.72}$$

$$\lambda_{kl}^{(3)} = \pi^4 \left(\frac{k^2}{a^2} + \frac{l^2}{b^2} \right)^2 + P_L. \tag{3.73}$$

Using the boundary conditions (2.51), the eigenvector functions $\overline{v}_{kl}^{(j)}$ of the operator L_0 have the form

$$\overline{v}_{kl}^{(j)}(x, y) = \left\{ \delta_i^j w_{kl}^{(1,3)} \left(\frac{x}{a}, \frac{y}{b} \right) \right\}_{i=1}^3, \quad j = 1, 2, 3, \quad k, l = 1, 2, \ldots, \tag{3.74}$$

where δ_i^j is the Kronecker delta, i.e. $\delta_i^j = 1, \delta_i^j = 0 \ (i \neq j)$; the corresponding eigenvalues of the operator L_0 are again defined by (3.72), (3.73).

Let us check if, for an appropriate choice of the constant P_L, the operator L_0 is defined by the relation (2.88). Integrating by parts, with use of the homogeneous boundary conditions, leads to the following result:

$$\|L_0 \overline{\omega}\|_{(L_2(\Omega_1))^3}^2 = \int_{\Omega_1} \left[\left(\frac{\partial^4 w}{\partial x^4} \right)^2 + \left(\frac{\partial^4 w}{\partial y^4} \right)^2 + 6 \left(\frac{\partial^4 w}{\partial x^2 \partial y^2} \right)^2 + 4 \left(\frac{\partial^4 w}{\partial x^3 \partial y} \right)^2 \right.$$

$$+ 4 \left(\frac{\partial^4 w}{\partial x \partial y^3} \right)^2 + \left(\frac{\partial^2 w}{\partial x^2} \right)^2 + \left(\frac{\partial^2 v}{\partial y^2} \right)^2 + (1 - \nu) \left(\frac{\partial^2 w}{\partial x \partial y} \right)^2 + (1 - \nu) \left(\frac{\partial^2 v}{\partial x \partial y} \right)^2$$

$$+\frac{(1-\nu)^2}{4}\left(\frac{\partial^2 u}{\partial y^2}\right)^2+\frac{(1-\nu)^2}{4}\left(\frac{\partial^2 v}{\partial x^2}\right)^2\Bigg]\,d\Omega_1+2P_L\int\limits_{\Omega_1}\left[\left(\frac{\partial^2 w}{\partial x^2}\right)^2+\left(\frac{\partial^2 w}{\partial y^2}\right)^2\right.$$

$$\left.+2\left(\frac{\partial^2 w}{\partial x\partial y}\right)^2+\left(\frac{\partial u}{\partial x}\right)^2+\left(\frac{\partial v}{\partial y}\right)^2+\frac{1-\nu}{2}\left(\frac{\partial u}{\partial y}\right)^2+\frac{1-\nu}{2}\left(\frac{\partial v}{\partial x}\right)^2\right]d\Omega_1$$

$$+P_L\int\limits_{\Omega_1}(u^2+v^2+w^2)\,d\Omega_1.$$

From elementary estimates, we obtain the following inequality:

$$\|L\overline{\omega}\|_{(L_2(\Omega_1))^3}\le\alpha_L\,\|L_0\overline{\omega}\|_{(L_2(\Omega_1))^3}\,, \tag{3.75}$$

where $\overline{\omega}=\{u,v,w\}$ is an arbitrary vector function taken from $[(\mathring{W}_2^2(\Omega_1))^2\times\mathring{W}_2^4(\Omega_1)]\cap[(C^2(\overline{\Omega_1}))^2\times C^4(\overline{\Omega_1})]$, and α_L is a positive constant independent of $\overline{\omega}$.

The operator L defined by (2.88) can be transformed to the form

$$\begin{aligned}L\overline{\omega}=\Bigg\{&-\frac{Eh}{1-\nu^2}+\left(\frac{\partial^2 u}{\partial x^2}+\frac{1-\nu}{2}\frac{\partial^2 u}{\partial y^2}+\frac{1-\nu}{2}\frac{\partial^2 v}{\partial x\partial y}\right)+l_1\overline{\omega},\\ &\frac{Eh}{(1-\nu)^2}\left(\frac{\partial^2 v}{\partial y^2}+\frac{1-\nu}{2}\frac{\partial^2 v}{\partial x^2}+\frac{1-\nu}{2}\frac{\partial^2 u}{\partial x\partial y}\right)+l_2\overline{\omega},\\ &D\,\nabla^2\,\nabla^2 w+l_{31}w+l_{32}\overline{\omega}\Bigg\},\end{aligned} \tag{3.76}$$

where $l_1\overline{\omega}$, $l_2\overline{\omega}$ are linear differential operators that contain only first-order derivatives of the functions u, v, w; $l_{31}w$ is a linear operator that depends only on w and contains derivatives with respect to w of third order only; and $l_{32}\overline{\omega}$ is a linear differential operator that contains derivatives of the arbitrary functions u, v of first order only and derivatives of w of second order only. Using (3.76), any of the vector functions

$$\overline{\omega}=\{u,v,w\}\in\left[\left(\mathring{W}_2^2(\Omega_1)\right)^2\times\mathring{W}_2^4(\Omega_1)\right]\cap\left[(C^2(\overline{\omega_1}))^2\times C^4(\overline{\Omega_1})\right]$$

can be represented in the following form:

$$(L\overline{\omega},\tilde{L}\overline{\omega})_{(L_2(\Omega_1))^3}=\int\limits_{\Omega_1}\Bigg\{D(\nabla^2\,\nabla^2\,w)^2+\frac{Eh}{1-\nu^2}\Bigg[\left(\frac{\partial^2 u}{\partial x^2}+\frac{1-\nu}{2}\frac{\partial^2 u}{\partial y^2}\right)^2$$

$$+\left(\frac{\partial^2 v}{\partial y^2}+\frac{1-\nu}{2}\frac{\partial^2 v}{\partial x^2}\right)^2\Bigg]\Bigg\}\,d\Omega_1+\frac{1+\nu}{2}\int\limits_{\Omega_1}\frac{Eh}{1-\nu^2}\left[\frac{\partial^2 v}{\partial x\partial y}\left(\frac{\partial^2 u}{\partial x^2}\right.\right.$$

$$\left.\frac{1-\nu}{2}\frac{\partial^2 u}{\partial y^2}\right) + \frac{\partial^2 u}{\partial x \partial y}\left(\frac{\partial^2 v}{\partial y^2} + \frac{1-\nu}{2}\frac{\partial^2 v}{\partial x^2}\right)^2\Bigg] d\Omega_1 + \int\limits_{\Omega_1}\Bigg[l_1\overline{\omega}\left(\frac{\partial^2 u}{\partial x^2}\right.$$

$$\left.+\frac{1-\nu}{2}\frac{\partial^2 u}{\partial y^2}\right) + l_2\overline{\omega}\left(\frac{\partial^2 v}{\partial y^2} + \frac{1-\nu}{2}\frac{\partial^2 v}{\partial x^2}\right) + (l_{31}w + l_{32}\overline{\omega})\,\nabla^2\,\nabla^2 w\Bigg]\,d\Omega_1. \tag{3.77}$$

Using the Cauchy inequality, integrating by parts and using again the Cauchy inequality, the second term of the right-hand side of the relation (3.77) can be estimated in the following way:

$$\frac{1+\nu}{2}\left|\int\limits_{\Omega_1}\frac{Eh}{1-\nu^2}\left[\frac{\partial^2 v}{\partial x \partial y}\left(\frac{\partial^2 u}{\partial x^2} + \frac{1-\nu}{2}\frac{\partial^2 u}{\partial y^2}\right) + \frac{\partial^2 u}{\partial x \partial y}\left(\frac{\partial^2 v}{\partial y^2} + \frac{1-\nu}{2}\right.\right.\right.$$

$$\left.\left.\left.\times\frac{\partial^2 v}{\partial x^2}\right)^2\right] d\Omega_1\right| \le \frac{(1+\nu)^2}{8}\int\limits_{\Omega_1}\frac{Eh}{1-\nu^2}\left[\left(\frac{\partial^2 u}{\partial x \partial y}\right)^2 + \left(\frac{\partial^2 v}{\partial x \partial y}\right)^2\right] d\Omega_1$$

$$+\frac{1}{2}\int\limits_{\Omega_1}\frac{Eh}{1-\nu^2}\left[\left(\frac{\partial^2 u}{\partial x^2} + \frac{1-\nu}{2}\frac{\partial^2 u}{\partial y^2}\right)^2 + \left(\frac{\partial^2 v}{\partial y^2} + \frac{1-\nu}{2}\frac{\partial^2 v}{\partial x^2}\right)^2\right] d\Omega_1$$

$$= \frac{(1+\nu)^2}{8}\int\limits_{\Omega_1}\frac{Eh}{1-\nu^2}\left(\frac{\partial^2 u}{\partial x^2}\frac{\partial^2 u}{\partial y^2} + \frac{\partial^2 v}{\partial x^2}\frac{\partial^2 v}{\partial y^2}\right) d\Omega_1 + \frac{(1+\nu)^2}{8}$$

$$\times\int\limits_{\Omega_1}\frac{Eh}{1-\nu^2}\left(\frac{\partial h}{\partial x} - \frac{\partial h}{\partial y}\right)\left(\frac{\partial u}{\partial x}\frac{\partial^2 u}{\partial x \partial y} - \frac{\partial v}{\partial y}\frac{\partial^2 v}{\partial x \partial y}\right) d\Omega_1 + \frac{1}{2}\int\limits_{\Omega_1}\frac{Eh}{1-\nu^2}$$

$$\times\left[\left(\frac{\partial^2 u}{\partial x^2} + \frac{1-\nu}{2}\frac{\partial^2 u}{\partial y^2}\right)^2 + \left(\frac{\partial^2 v}{\partial y^2} + \frac{1-\nu}{2}\frac{\partial^2 v}{\partial x^2}\right)^2\right] d\Omega_1 \le \frac{1}{2}\left[1 + \frac{(1+\nu)^2}{8(1-\nu)}\right]$$

$$\times\int\limits_{\Omega_1}\frac{Eh}{1-\nu^2}\left[\left(\frac{\partial^2 u}{\partial x^2} + \frac{1-\nu}{2}\frac{\partial^2 u}{\partial y^2}\right)^2 + \left(\frac{\partial^2 v}{\partial y^2} + \frac{1-\nu}{2}\frac{\partial^2 v}{\partial x^2}\right)^2\right] d\Omega_1$$

$$+\frac{(1+\nu)^2}{8}\int\limits_{\Omega_1}\frac{Eh}{1-\nu^2}\left(\frac{\partial h}{\partial x} - \frac{\partial h}{\partial y}\right)\left(\frac{\partial u}{\partial x}\frac{\partial^2 u}{\partial x \partial y} - \frac{\partial v}{\partial y}\frac{\partial^2 v}{\partial x \partial y}\right) d\Omega_1. \tag{3.78}$$

From (3.77), (3.78), we obtain

$$(L\overline{\omega}, \tilde{L}\overline{\omega})_{(L_2(\Omega_1))^3} \leq \int\limits_{\Omega_1} \left\{ D(\nabla^2 \nabla^2 w)^2 + \gamma(\nu) \frac{Eh}{1-\nu^2} \left[\left(\frac{\partial^2 u}{\partial x^2} \right. \right. \right.$$

$$\left. \left. \left. + \frac{1-\nu}{2} \frac{\partial^2 u}{\partial y^2} \right)^2 + \left(\frac{\partial^2 v}{\partial y^2} + \frac{1-\nu}{2} \frac{\partial^2 v}{\partial x^2} \right)^2 \right] \right\} d\Omega_1 + I, \tag{3.79}$$

where $\gamma(\nu) = \frac{1}{2}\left[1 - \frac{(1+\nu)^2}{8(1-\nu)}\right]$ (it is easy to check that for $0 < \nu < \frac{1}{2}, \gamma(\nu) > 0$), and

$$I = \int\limits_{\Omega_1} \left[l_1 \overline{\omega} \left(\frac{\partial^2 u}{\partial x^2} + \frac{1-\nu}{2} \frac{\partial^2 u}{\partial y^2} \right) + l_2 \overline{\omega} \left(\frac{\partial^2 v}{\partial y^2} + \frac{1-\nu}{2} \frac{\partial^2 v}{\partial x^2} \right) + (l_{31} w \right.$$

$$\left. + l_{32} \overline{\omega}) \nabla^2 \nabla^2 w + \frac{(1+\nu)E}{8(1-\nu)} \left(\frac{\partial h}{\partial x} - \frac{\partial h}{\partial y} \right) \left(\frac{\partial u}{\partial x} \frac{\partial^2 u}{\partial x \partial y} - \frac{\partial v}{\partial y} \frac{\partial^2 v}{\partial x \partial y} \right) \right] d\Omega_1.$$

The inequality (3.79) implies

$$(L\overline{\omega}, \tilde{L}\overline{\omega})_{(L_2(\Omega_1))^3} \geq m_1 \left\| L\overline{\omega} \right\|_{(L_2(\Omega_1))^3} + I, \tag{3.80}$$

where

$$m_1 = \min_{(x,y)\in\overline{\Omega}_1} \left\{ D(x,y), \gamma(\nu) \frac{Eh(x,y)}{(1-\nu^2)} \right\}.$$

Furthermore, using the Cauchy inequality, we obtain

$$\left| \int\limits_{\Omega_1} l_1 \overline{\omega} \left(\frac{\partial^2 u}{\partial x^2} + \frac{1-\nu}{2} \frac{\partial^2 u}{\partial y^2} \right) d\Omega_1 \right| \leq \frac{\varepsilon}{2} \int\limits_{\Omega_1} \left(\frac{\partial^2 u}{\partial x^2} + \frac{1-\nu}{2} \frac{\partial^2 u}{\partial y^2} \right) d\Omega_1 \tag{3.81}$$

$$+ \frac{k_1}{\varepsilon} \left\| \tilde{L}^{1/2} \overline{\omega} \right\|^2_{(L_2(\Omega_1))^3} \qquad (x \leftrightarrow y, \quad u \leftrightarrow v, \quad l_1 \leftrightarrow l_2, \quad k_1 \leftrightarrow k_2),$$

$$\left| \int\limits_{\Omega_1} l_{32} w \nabla^2 \nabla^2 w \, d\Omega_1 \right| \leq \frac{\varepsilon}{4} \int\limits_{\Omega_1} (\nabla^2 \nabla^2 w)^2 \, d\Omega_1 + \frac{1}{\varepsilon} \left\| l_{31} w \right\|^2_{L_2(\Omega_1)}, \tag{3.82}$$

$$\left\| l_{31} w \right\|^2_{L_2(\Omega_1)} \leq \frac{\varepsilon^2}{4} \int\limits_{\Omega_1} (\nabla^2 \nabla^2 w)^2 \, d\Omega_1 + \frac{k_{31}}{\varepsilon^2} \left\| \tilde{L}^{1/2} \overline{\omega} \right\|^2_{(L_2(\Omega_1))^3}, \tag{3.83}$$

$$\left|\int\limits_{\Omega_1} l_{32}\overline{\omega}\,\nabla^2\,\nabla^2 w\,d\Omega_1\right| \le \frac{\varepsilon}{2}\int\limits_{\Omega_1}(\nabla^2\,\nabla^2\,w)^2\,d\Omega_1 + \frac{k_{32}}{\varepsilon}\left\|\tilde{L}^{1/2}\overline{\omega}\right\|^2_{(L_2(\Omega_1))^3}, \tag{3.84}$$

$$\frac{(1+\nu)E}{8(1-\nu)}\left|\int\limits_{\Omega_1}\left[\left(\frac{\partial h}{\partial x}-\frac{\partial h}{\partial y}\right)\left(\frac{\partial u}{\partial x}\frac{\partial^2 u}{\partial x\partial y}-\frac{\partial v}{\partial y}\frac{\partial^2 v}{\partial x\partial y}\right)\right]d\Omega_1\right|$$
$$\le \frac{\varepsilon}{2}\int\limits_{\Omega_1}\left[\left(\frac{\partial^2 u}{\partial x^2}+\frac{1-\nu}{2}\frac{\partial^2 u}{\partial y^2}\right)^2+\left(\frac{\partial^2 v}{\partial y^2}+\frac{1-\nu}{2}\frac{\partial^2 v}{\partial x^2}\right)^2\right]d\Omega_1 \tag{3.85}$$
$$+\frac{k_4}{\varepsilon}\left\|\tilde{L}^{1/2}\overline{\omega}\right\|^2_{(L_2(\Omega_1))^3},$$

where ε is an arbitrary positive number, and $k_1, k_2, k_{31}, k_{32}, k_4$ are positive constants independent of both ε and $\overline{\omega}$. From (3.82) and (3.83), we obtain

$$\left|\int\limits_{\Omega_1} l_{31}\overline{\omega}\,\nabla^2\,\nabla^2 w\,d\Omega_1\right| \le \frac{\varepsilon}{2}\int\limits_{\Omega_1}(\nabla^2\nabla^2 w)^2 d\Omega_1 + \frac{k_{31}}{\varepsilon^3}\left\|\tilde{L}^{1/2}\overline{\omega}\right\|^2_{(L_2(\Omega_1))^3}. \tag{3.86}$$

The union of the estimates (3.81), (3.84)–(3.86) gives us

$$|I| \le \varepsilon\left\|\tilde{L}\overline{\omega}\right\|^2_{(L_2(\Omega_1))^3} + \left(\frac{m_2}{\varepsilon}+\frac{k_{31}}{\varepsilon^3}\right)\left\|\tilde{L}^{1/2}\overline{\omega}\right\|^2_{(L_2(\Omega_1))^3}, \tag{3.87}$$

where $m_2 = k_1 + k_2 + k_{32} + k_4$. Taking into account (3.80), (3.87), we obtain

$$(L\overline{\omega}, \tilde{L}\overline{\omega})_{(L_2(\Omega_1))^3}$$
$$\ge (m_1-\varepsilon)\left\|\tilde{L}\overline{\omega}\right\|^2_{(L_2(\Omega_1))^3} - \left(\frac{m_2}{\varepsilon}+\frac{k_{31}}{\varepsilon^3}\right)\left\|\tilde{L}^{1/2}\overline{\omega}\right\|^2_{(L_2(\Omega_1))^3}, \tag{3.88}$$

where $\overline{\omega}$ is an arbitrary vector function from $[(\mathring{W}_2^2(\Omega_1))^2 \times \mathring{W}_2^4(\Omega_1)] \cap [(C^2(\overline{\Omega_1}))^2 \times C^4(\overline{\Omega_1})]$, and ε is an arbitrary positive number.

With the help of the Korn inequalities [353], the following estimate can be obtained

$$\left\|L^{1/2}\overline{\omega}\right\|^2_{(L_2(\Omega_1))^3} \ge m_L\left\|\tilde{L}^{1/2}\overline{\omega}\right\|^2_{(L_2(\Omega_1))^3}, \tag{3.89}$$

where $\overline{\omega}$ is an arbitrary vector function from $(\mathring{W}_2^1(\Omega_1))^2 \times \mathring{W}_2^2(\Omega_1)$, and m_L is a positive constant independent of $\overline{\omega}$.

We put $\varepsilon = \varepsilon_0 = \frac{m_1}{2}$ into (3.88) and take

$$P_L = \frac{1}{m_L}\left(\frac{m_2}{\varepsilon_0} + \frac{k_{31}}{\varepsilon_0^3}\right); \tag{3.90}$$

then, owing to (3.64), (3.89), we obtain

$$\begin{aligned}
&(L\overline{\omega}, L_0\overline{\omega})_{(L_2(\Omega_1))^3} = (L\overline{\omega}, \tilde{L}\overline{\omega})_{(L_2(\Omega_1))^3} + P_L \left\| \tilde{L}^{1/2}\overline{\omega} \right\|_{(L_2(\Omega_1))^3} \\
&\geq (L\overline{\omega}, \tilde{L}\overline{\omega})_{(L_2(\Omega_1))^3} + m_L P_L \left\| \tilde{L}^{1/2}\overline{\omega} \right\|^2_{(L_2(\Omega_1))^3} \\
&\geq (m_1 - \varepsilon_0) \left\| \tilde{L}\overline{\omega} \right\|^2_{(L_2(\Omega_1))^3} + \left(m_L P_L - \frac{m_2}{\varepsilon_0} - \frac{k_{31}}{\varepsilon_0^3} \right) \left\| \tilde{L}^{1/2}\overline{\omega} \right\|^2_{(L_2(\Omega_1))^3} \\
&= \frac{m_1}{2} \left\| \tilde{L}\overline{\omega} \right\|^2_{(L_2(\Omega_1))^3}.
\end{aligned} \tag{3.91}$$

Because L_0 and $\tilde{L}$ are convergent, the following inequality is obtained from (3.91):

$$(L\overline{\omega}, L_0\overline{\omega})_{(L_2(\Omega_1))^3} \geq \beta_L \left\| \tilde{L}_0\overline{\omega} \right\|^2_{(L_2(\Omega_1))^3}, \tag{3.92}$$

where $\overline{\omega}$ is an arbitrary vector function from $[(\mathring{W}_2^2(\Omega_1))^2 \times \mathring{W}_2^4(\Omega_1)] \cap [(C^2(\overline{\Omega_1}))^2 \times C^4(\overline{\Omega_1})]$, and β_L is a positive constant independent of $\overline{\omega}$. The inequality (3.92) implies that when the constant P_L is chosen to have the value given in (3.90), the operators L and L_0 defined by (2.88), (3.64) form an acute angle, and the constant $\gamma(L, L_0)$ occurring in an inequality of type (2.107) can be estimated from

$$\gamma(L, L_0) \geq \beta_L. \tag{3.93}$$

Taking into account (3.92), we obtain

$$\|L_0\overline{\omega}\|_{(L_2(\Omega_1))^3} \leq \beta_L^{-1} \|L\overline{\omega}\|_{(L_2(\Omega_1))^3}, \tag{3.94}$$

where $\overline{\omega}$ is an arbitrary vector function from $[(\mathring{W}_2^2(\Omega_1))^2 \times \mathring{W}_2^4(\Omega_1)] \cap [(C^2(\overline{\Omega_1}))^2 \times C^4(\overline{\Omega_1})]$. It is clear, by taking into account (3.75) and (3.94), that the norms $\|L\cdot\|_{(L_2(\Omega_1))^3}$ and $\|L_0\cdot\|_{(L_2(\Omega_1))^3}$ are equivalent. Therefore, the spaces of existence of the operators L and L_0 overlap because they have the same linear manifold $[(\mathring{W}_2^2(\Omega_1))^2 \times \mathring{W}_2^4(\Omega_1)] \cap [(C^2(\overline{\Omega_1}))^2 \times C^4(\overline{\Omega_1})]$ in equivalent norms, which means that the operators L and L_0 are convergent. To conclude, it has been shown, that when the constant P_L is chosen in the form (3.90), the operator L_0 defined by (3.64) is convergent and forms an acute angle with the operator L defined by the relation (2.88).

For the problem under consideration the space H_1 overlaps with $(L_2(\Omega_1))^3$, and therefore the inequality (3.75) can be represented in the form

$$\left\| L L_0^{-1} \right\|_{H_1 \to H_1} \leq \alpha_L. \tag{3.95}$$

According to (3.93), (3.95), the constant C_L defined by the first relation of (2.124) can be estimated in the following manner:

$$C_L \leq \alpha_L \beta_L^{-1}. \tag{3.96}$$

An auxiliary operator K_0 with a known spectrum can be constructed in the following way. Let $\overline{\Phi} = \{\Phi_1, \Phi_2, \Phi_3\}$ be a reciprocally unique mapping of the parallelepiped $\tilde{\Omega}_2 = \Omega_1 \times \left(-\frac{1}{2}, \frac{1}{2}\right)$ into the space Ω_2, defined by the relations

$$\Phi_1(\overline{x}) = x, \quad \Phi_2(\overline{x}) = y, \quad \Phi_3(\overline{x}) = zh(x, y), \quad \overline{x} = (x, y, z).$$

We now introduce the differential operator $\tilde{K}$, defined on $\overset{\circ}{W}{}_2^2(\Omega_2)$ by the relation

$$\tilde{K}\theta = -\nabla^2 (\theta \circ \overline{\Phi}) \circ \overline{\Phi}^{-1}, \tag{3.97}$$

and a new scalar product (equivalent to the usual scalar product) in the space $L_2(\Omega_2)$:

$$\begin{aligned} (\theta, \eta)_{L_2(\Omega_2)} &= \int\limits_{\tilde{\Omega}_2} (\theta \circ \overline{\Phi})(\eta \circ \overline{\Phi})\, d\tilde{\Omega}_2 \\ &\equiv \int\limits_{\Omega_2} \theta\eta \left| \det \frac{D(\overline{\Phi}^{-1})}{D\overline{x}} \right| d\Omega_2 \equiv \int\limits_{\Omega_2} \frac{1}{h} \theta\eta\, d\Omega_2. \end{aligned} \tag{3.98}$$

The operator $\tilde{K}$ defined by (3.97) is self-adjoint and positive definite in relation to the scalar product (3.98) (because the operator $-\nabla^2$ has the same properties in relation to the usual scalar product of a space $L_2(\tilde{\Omega}_2)$). We introduce the notation

$$w_{klm}^{(2)}(x, y, z) = w_k(x) w_l(y) w_m(z), \tag{3.99}$$

where $w_k(x)$, $w_l(y)$, $w_m(z)$ are the functions as defined by (3.65). Let us assume that

$$K_0\theta = \tilde{K}\theta + p_k\theta, \tag{3.100}$$

where p_k is a sought positive constant. The eigenfunctions η_{klm} of the operator (3.100), from $\mathring{W}_2^2(\Omega_2)$, are known and have the form

$$\eta_{klm}(x, y, z) = w_{klm}^{(2)}\left(\frac{x}{a}, \frac{y}{b}, \frac{z}{h(x, y)}\right), \quad k, l, m = 1, 2, \ldots . \tag{3.101}$$

The corresponding eigenvalues $æ_{klm}$ of the operator (3.100) are defined by the relations

$$æ_{klm} = \pi^2\left(\frac{k^2}{a^2} + \frac{l^2}{b^2} + m^2\right) + p_k, \quad k, l, m = 1, 2, \ldots . \tag{3.102}$$

For the problem under consideration $H_2 = L_2(\Omega_2)$. We assume that the scalar product (3.98) is defined for the space $L_2(\Omega_2)$. It turns out that the operator K defined on $\mathring{W}_2^2(\Omega_2)$ by (2.89) is not self-adjoint in relation to the new scalar product. Let us multiply the two parts of the differential equation (2.32) by $T_0^{-1}h(x, y)$ and define the operator K again, i.e., instead of (2.89), we take

$$K\theta = -\frac{\lambda_q}{T_0} h \, \nabla^2 \theta, \quad D(K) = \mathring{W}_2^2(\Omega_2). \tag{3.103}$$

Now the operator K is defined as self-adjoint and positive in relation to the scalar product (3.98). Using the method described in [360], the following estimate can be obtained:

$$\left\| K K_0^{-1} \right\|_{H_1 \to H_1} \leq \alpha_K, \quad \gamma(K, K_0) \geq \beta_K, \tag{3.104}$$

for an appropriate choice of p_k (α_K, β_K are positive constants). This implies that the operators K_0 and K defined by the relations (3.100) and (3.103) are convergent and form an acute angle, and the constant c_K defined by the second relation of (2.124) can be estimated in the following way:

$$c_K \leq \alpha_K \beta_K^{-1}. \tag{3.105}$$

The operators I_1, M for the problem under consideration are defined by the relations (2.87) and (2.90), correspondingly. Assuming that the scalar product (3.98) is determined for the space $H_2 = L_2(\Omega_2)$, the operators I_2, N can be defined using a differential equation obtained by multiplication of the two parts of (2.32) by $T_0^{-1}h(x, y)$:

$$I_2\theta = \frac{c_\varepsilon(1 + \varepsilon)h}{T_0}\theta,$$

$$N\overline{\omega} = \frac{E\alpha_T h}{1 - \nu}\left[\frac{\partial u}{\partial x} + \frac{\partial v}{\partial y} - (k_x + k_y)w - z \, \nabla^2 w\right],$$

$$D(N) = (\mathring{W}_2^1(\Omega_1))^2 \times \mathring{W}_2^2(\Omega_1).$$

It is clear that, for an arbitrary vector function $\overline{\omega} \in (L_2(\Omega_1))^3$, the following inequality holds:

$$(I_1\overline{\omega}, \overline{\omega})_{(L_2(\Omega_1))^3} \leq \rho H_0 \|\overline{\omega}\|^2_{(L_2(\Omega_1))^3}, \tag{3.106}$$

where $H_0 = \max_{(x,y)\in\overline{\Omega_1}} h(x, y)$. For an arbitrary function $\theta \in L_2(\Omega_2)$, the following inequality holds:

$$(I_2\theta, \theta)_{L_2(\Omega_1)} \leq c_\varepsilon(1+\varepsilon)T_0^{-1} H_0 \|\theta\|^2_{L_2(\Omega_1)}, \tag{3.107}$$

where the constant H_0 in the inequalities (3.106) and (3.107) cannot be replaced by a smaller one. It follows that for the problem under consideration the norms of the operators $I_1^{1/2}$, $I_2^{1/2}$ are defined by the relations

$$\left\|I_1^{1/2}\right\|_{H_1\to H_1} = \rho^{1/2} H_0^{1/2}, \quad \left\|I_2^{1/2}\right\|_{H_2\to H_2} = c_\varepsilon^{1/2}(1+\varepsilon)^{1/2} T_0^{-1/2} H_0^{1/2}. \tag{3.108}$$

Using some methods due to Korn [246], the Poincaré–Friedrichs inequalities [214] the second fundamental inequality for elliptic operators [214] the and Cauchy–Buniakowski inequalities, the following estimates can be obtained:

$$(L_0\overline{\omega}, \overline{\omega})_{(L_2(\Omega_1))^3} \leq A_L^2 (L\overline{\omega}, \overline{\omega})_{(L_2(\Omega_1))^3},$$

$$(K_0\theta, \theta)_{L_2(\Omega_2)} \leq A_K^2 (K\theta, \theta)_{L_2(\Omega_2)},$$

$$\|M\theta\|_{(L_2(\Omega_1))^3} \leq B_1 \|K\theta\|_{L_2(\Omega_2)},$$

$$\|N\overline{\omega}\|^2_{L_2(\Omega_2)} \leq B_2^2 (L\overline{\omega}, \overline{\omega})_{(L_2(\Omega_1))^3},$$

where $\overline{\omega}$ is an arbitrary vector function from $D(L)$; θ is an arbitrary function from $D(K)$; A_L, B_2 are positive constants independent of $\overline{\omega}$; and A_K, B_1 are positive constants independent of θ. The estimates given above can be represented in the following way:

$$\left\|L_0^{1/2} L^{-1/2}\right\|_{H_1\to H_1} \leq A_L, \quad \left\|K_0^{1/2} K^{-1/2}\right\|_{H_2\to H_2} \leq A_K, \tag{3.109}$$

$$\left\|MK^{-1}\right\|_{H_2\to H_1} \leq B_1, \quad \left\|NL^{-1/2}\right\|_{H_1\to H_2} \leq B_2. \tag{3.110}$$

Because

$$\left\|L^{-1/2} L_0^{1/2}\right\|_{H_1\to H_1} = \left\|(L^{-1/2} L_0^{1/2})^*\right\|_{H_1\to H_1} = \left\|L_0^{1/2} L^{-1/2}\right\|_{H_1\to H_1},$$

and, analogously,

$$\left\|K^{-1/2} K_0^{1/2}\right\|_{H_2\to H_2} = \left\|K_0^{1/2} K^{-1/2}\right\|_{H_2\to H_2},$$

(3.109) implies the following inequalities:

$$\left\| L^{-1/2} L_0^{1/2} \right\|_{H_1 \to H_1} \leq A_L, \quad \left\| K^{-1/2} K_0^{1/2} \right\|_{H_2 \to H_2} \leq A_K. \tag{3.111}$$

The free term g_1 appearing in the first equation of (3.1) is (for the case under consideration) defined by the first relation of (2.93). The free term g_2 of the second equation of (3.1) can be obtained using a differential equation obtained by multiplication of the two sides of (2.32) by $T_0^{-1}h(x, y)$, i.e. we take $g_2 = T_0^{-1}hq_2$.

Using the relations (3.108) and the inequalities (3.96), (3.105), (3.110), (3.111), we can now estimate from above the constants c_i $(i = 1, \ldots, 5)$ defined by the relations (3.5), the constants d_j $(j = 1, 2, 3)$ defined by the relations (3.36) and the varying quantities e_1, e_2 defined by (3.37), (3.38). By substituting these estimates into the right-hand sides of the inequalities (3.3), (3.23), we then obtain prior estimates of the errors of the Bubnov–Galerkin method for the system of differential equations (2.30)–(2.32) with homogeneous initial conditions (2.33)–(2.35) and boundary conditions. Let present, as an example, one of those estimates, obtained from (3.3):

$$\max_{0 \leq t \leq T} \Bigg\{ \int_{\Omega_1} \Bigg[\rho h \left(\left(\frac{d}{dt} \Delta u_n \right)^2 + \left(\frac{d}{dt} \Delta v_n \right)^2 + \left(\frac{d}{dt} \Delta w_n \right)^2 \right)$$

$$+ \frac{Eh}{1+\nu} \left(\left(\frac{d}{dx} \Delta u_n - k_x \Delta w_n \right)^2 + \left(\frac{d}{dy} \Delta v_n - k_y \Delta w_n \right)^2 \right.$$

$$\left. + \frac{1}{2} \left(\frac{d}{dy} \Delta u_n + \frac{d}{dx} \Delta v_n \right)^2 \right) + \frac{Eh\nu}{1-\nu^2} \left(\frac{d}{dx} \Delta u_n + \frac{d}{dy} \Delta v_n \right.$$

$$\left. - (k_x + k_y) \Delta w_n \right)^2 + D(1-\nu) \left(\left(\frac{d^2}{dx^2} \Delta w_n \right)^2 + \left(\frac{d^2}{dy^2} \Delta w_n \right)^2 \right.$$

$$\left. + 2 \left(\frac{d^2}{dx\, dy} \Delta w_n \right)^2 \right) + D\nu (\nabla^2 \Delta w_n)^2 \Bigg] d\Omega_1$$

$$+ \frac{c_\varepsilon (1+\varepsilon)}{T_0} \int_{\Omega_1} (\Delta \theta_n)^2 d\Omega_2 + \frac{\lambda_q}{T_0} \int_0^t \int_{\Omega_1} | \nabla \Delta \theta_n |^2 d\Omega_2 d\tau \Bigg\}$$

$$\leq 4 A_L T^{1/2} (1 + \alpha_L \beta_L^{-1}) (M_g(T))^{1/2} \Bigg\{ \left[\int_0^t \int_{\Omega_1} (p_1^2 + p_2^2 + q_1^2) d\Omega_1 d\tau \right]^{1/2}$$

$$
\begin{aligned}
&+B_1\alpha_k\beta_k^{-1}\left[\int_0^T\int_{\Omega_2} hq_2^2 d\Omega_2 d\tau\right]^{1/2}\\
&+[B_1^2\alpha_k^2\beta_k^{-2}(c_\varepsilon(1+\varepsilon)H_0T_0^{-1}+B_2^2)+\rho H_0]^{1/2}\\
&\times T^{1/2}(M_g(T))^{1/2}\Big\}\tilde{\lambda}_{i(n)+1}^{-1/2}+A_k^2(1+\alpha_k\beta_k^{-1})^2\left\{\left[\int_0^T\int_{\Omega_2} hq_2^2 d\Omega_2 d\tau\right]^{1/2}\right.\\
&\left.+[c_\varepsilon(1+\varepsilon)H_0T_0^{-1}+B_2^2]^{1/2}T^{1/2}(M_g(T))^{1/2}\right\}^2 \tilde{æ}_{j(n)+1}^{-1} \qquad (3.112)\\
&\equiv 0(\max(\tilde{\lambda}_{i(n)+1}^{-1/2}, \tilde{æ}_{j(n)+1}^{-1})),
\end{aligned}
$$

where

$$\Delta u_n = u_n - u,\ \Delta v_n = v_n - v,\ \Delta w_n = w_n - w,\ \Delta\theta_n = \theta_n - \theta,$$

$\{u, v, w, \theta\}$ is the exact ("classical") solution to the problem under consideration, and $\{u_n, v_n, w_n, \theta_n\}$ is its approximate solution found using the Bubnov–Galerkin method. The latter approximate solution is given by

$$
\begin{aligned}
\overline{\omega_n}(x, y, t) &= \{u_n(x, y, t), v_n(x, y, t), w_n(x, y, t)\}\\
&= \sum_{\overline{v}_{kl}^{(j)}\in H_1^n} \alpha_{kl}^{(j)}(t)\overline{v}_{kl}(x, y),\\
\theta(x, y, z, t) &= \sum_{\eta_{klm}\in H_2^n} \beta_{klm}(t)\eta_{klm}(x, y, z),\\
\tilde{\lambda}_{i(n)+1} &= \min\{\lambda_{kl}^{(j)} : \overline{v}_{kl}^{(j)}\overline{\in} H_1^n\},\\
\tilde{æ}_{j(n)+1} &= \min\{æ_{klm} : \eta_{klm}\overline{\in} H_2^n\},
\end{aligned}
$$

where H_1^n, H_2^n are finitely measured subspaces of $(L_2(\Omega_1))^3$ and $L_2(\Omega_2)$, spanned by $i(n)$ basis vector functions $\overline{v}_{kl}^{(j)}$ (defined either by the dependence of the mechanical boundary conditions, by the relations (3.69)–(3.71), or by the relations (3.74)) and on $j(n)$ basis functions η_{klm} defined by the relations (3.101); $\lambda_{kl}^{(j)}$, $æ_{klm}$

are the eigenvalues of the auxiliary operators L_0, K_0, defined by the relations (3.72), (3.73), (3.102); and

$$M_g(T) = M\left[\frac{1}{\rho}\int\limits_{\Omega_1}\frac{1}{h}(p_1^2|_{t=0} + p_2^2|_{t=0} + q_1^2|_{t=0})\,d\Omega_1\right.$$
$$+\frac{T_0}{c_\varepsilon(1+\varepsilon)}\int\limits_{\Omega_2} q_2^2|_{t=0}d\Omega_2,\ \frac{1}{\rho}\int\limits_0^T\int\limits_{\Omega_1}\frac{1}{h}\left[\left(\frac{\partial p_1}{\partial t}\right)^2 + \left(\frac{\partial p_2}{\partial t}\right)^2\right. \tag{3.113}$$
$$\left.\left.+\left(\frac{\partial q_1}{\partial t}\right)^2\right]d\Omega_1\,dt + \frac{T_0}{c_\varepsilon(1+\varepsilon)}\int\limits_0^T\int\limits_{\Omega_2}\left(\frac{\partial q_2}{\partial t}\right)^2 d\Omega_2\,dt,\ T\right],$$

where

$$\begin{aligned} M[w, X, T] &= \min_{\delta>0}\left[\left(w + \frac{1}{\delta}X\right)\exp(\delta T)\right] \\ &\equiv \left\{w + \frac{1}{2}\left[(X^2T^2 + 4wXT)^{1/2} + XT\right]\right\} \\ &\times \exp\left\{2XT\times\right. \\ &\times\left.\left[(X^2T^2 + 4wXT)^{1/2} + XT\right]^{-1}\right\}. \end{aligned} \tag{3.114}$$

Let us remember, that owing to the known theorem about the boundedness of the operation $W_2^2(\Omega) \to C(\overline{\Omega})$, where Ω is a two-dimensional space with a sufficiently smooth boundary [307], an estimate of the convergence $w_n \to w$ in a norm uniform with respect to the variables x, y, t can be easily obtained from (3.112).

3.3 Case of a Simply Supported Plate Within the Kirchhoff Model

Let us consider a coupled thermoelastic problem for a simply supported plate ($k_x = k_y = 0$) with a constant thickness ($h = const$) in the framework of the Kirchhoff–Love kinematic model. This problem is related to our earlier general considerations of the problem described by (2.30)–(2.35), (2.38) and (2.51). In addition, we assume that (1) the mechanical and heat loads are constant in time, and (2) the plate under consideration is square ($a = b$). The last assumption is not essential and is made in order to simplify the prior estimate for the Bubnov–Galerkin method (the considerations below can be easily extended to the more general case of a rectangular plate).

The following nondimensional quantities are introduced:

$$\begin{gathered}
\overline{x}=\frac{x}{a},\quad \overline{y}=\frac{y}{a},\quad \overline{z}=\frac{z}{h},\quad \overline{t}=\frac{\lambda_q}{c_\varepsilon(1+\varepsilon)h^2}t,\\
\lambda=\frac{h}{a},\quad \overline{w}=\frac{w}{h},\quad \overline{\theta}=\frac{\alpha_T a^2}{h^2}\theta,\\
\overline{q_1}=\frac{a^4}{Eh^4}q_1,\quad \overline{q_2}=\frac{\alpha_t a^2}{\lambda_q}q_2,\quad \overline{w_0}=\frac{w_0}{h},\\
\overline{w_1}=\frac{c_\varepsilon(1+\varepsilon)h}{\lambda_q}w_1,\quad \overline{\theta_0}=\frac{\alpha_T a^2}{h^2}\theta_0,\quad \overline{T}=\frac{\lambda_q}{c_\varepsilon(1+\varepsilon)h^2}T,\\
c=\frac{\rho\lambda_q^2}{c_\varepsilon^2(1+\varepsilon)^2Eh^2},\quad \beta=\frac{E\alpha_T^2T_0}{(1-\nu)c_\varepsilon(1+\varepsilon)}.
\end{gathered}\tag{3.115}$$

The initial–boundary value problem considered here has the following form:

$$\frac{c}{\lambda^4}\frac{\partial^2\overline{w}}{\partial\overline{t}^2}+\frac{1}{12(1-\nu^2)}\overline{\nabla}^2\overline{\nabla}^2\overline{w}+\frac{1}{1-\nu}\overline{\nabla}^2\int_{-\frac{1}{2}}^{\frac{1}{2}}\overline{\theta}\overline{z}d\overline{z}=\overline{q_1}(\overline{x},\overline{y}),\tag{3.116}$$

$$\frac{\partial\overline{\theta}}{\partial\overline{t}}-\left(\lambda^2\frac{\partial^2\overline{\theta}}{\partial\overline{x}^2}+\lambda^2\frac{\partial^2\overline{\theta}}{\partial\overline{y}^2}+\frac{\partial^2\overline{\theta}}{\partial\overline{z}^2}\right)-\beta\overline{z}\frac{\partial}{\partial\overline{t}}(\overline{\nabla^2}\overline{w})=\overline{q_2}(\overline{x},\overline{y},\overline{z}),\tag{3.117}$$

$$\begin{gathered}
\overline{w}|_{\overline{x}=\pm\frac{1}{2}}=\overline{w}|_{\overline{y}=\pm\frac{1}{2}}=\left.\frac{\partial^2\overline{w}}{\partial\overline{x}^2}\right|_{\overline{x}=\pm\frac{1}{2}}=\left.\frac{\partial^2\overline{w}}{\partial\overline{y}^2}\right|_{\overline{y}=\pm\frac{1}{2}}=0,\\
\overline{\theta}|_{\overline{x}=\pm\frac{1}{2}}=\overline{\theta}|_{\overline{y}=\pm\frac{1}{2}}=\overline{\theta}|_{\overline{z}=\pm\frac{1}{2}}=0,\quad 0\le\overline{t}\le\overline{T},
\end{gathered}\tag{3.118}$$

$$\begin{gathered}
\overline{w}|_{\overline{t}=0}=\overline{w}_0(\overline{x},\overline{y}),\quad \left.\frac{\partial\overline{w}}{\partial\overline{t}}\right|_{\overline{t}=0}=\overline{w}_1(\overline{x}\overline{y}),\\
\overline{\theta}|_{\overline{t}=0}=\overline{\theta}_0(\overline{x},\overline{y},\overline{z}),\quad -\frac{1}{2}\le\overline{x},\overline{y},\overline{z}\le\frac{1}{2}.
\end{gathered}\tag{3.119}$$

In the differential equations (3.116), (3.117), the following notation has been used:

$$\overline{\nabla^2}=\frac{\partial^2}{\partial\overline{x}^2}+\frac{\partial^2}{\partial\overline{y}^2}.$$

From now on, in this chapter and in Chap. 4, the bars above the nondimensional quantities will be omitted.

Remark 3.6 The thermal-isolation condition $\left.\frac{\partial \theta}{\partial z}\right|_{z=\pm\frac{1}{2}} = 0$ can be used instead of the boundary condition $\theta|_{z=\pm\frac{1}{2}} = 0$. All of the analysis below can be applied in the case of this condition with insignificant changes.

The following notation is introduced:

$$\Omega_1 = \left(-\frac{1}{2}, \frac{1}{2}\right)^2, \quad \Omega_2 = \left(-\frac{1}{2}, \frac{1}{2}\right)^3, \quad H_j = L_2(\Omega_j) \quad (j = 1, 2), \tag{3.120}$$

$$I_1 w = \frac{c}{\lambda^4} w, \quad I_2 \theta = \frac{1}{\beta(1-\nu)} \theta, \tag{3.121}$$

$$Lw = \frac{1}{12(1-\nu^2)} \nabla^2 \nabla^2 w,$$

$$K\theta = -\frac{1}{\beta(1-\nu)} \left(\lambda^2 \frac{\partial^2 \theta}{\partial x^2} + \lambda^2 \frac{\partial^2 \theta}{\partial y^2} + \frac{\partial^2 \theta}{\partial z^2}\right), \tag{3.122}$$

$$M\theta = \frac{1}{1-\nu} \nabla^2 \int_{-\frac{1}{2}}^{\frac{1}{2}} \theta z \, dz, \quad Nw = -\frac{1}{1-\nu} z \nabla^2 w, \tag{3.123}$$

$$g_1 = q_1, \quad g_2 = \frac{1}{\beta(1-\nu)} q_2. \tag{3.124}$$

The space of the functions $w(x, y)$ from $W_2^2(\Omega_1)$ that are equal to zero on the boundary $\partial\Omega_1$ is denoted by $\mathring{W}_2^2(\Omega_1)$, and the space of the functions $w(x, y)$ from $W_2^4(\Omega_1)$ satisfying the mechanical boundary conditions (3.118) is denoted by $\mathring{W}_2^4(\Omega_1)$. The notations $\mathring{W}_2^1(\Omega_2)$, $\mathring{W}_2^2(\Omega_2)$ introduced in Sect. 2.1.4 are used for the manifolds of functions $\theta(x, y, z)$ from $W_2^1(\Omega_2)$ and $W_2^2(\Omega_2)$, respectively, that satisfy the boundary conditions (3.118). Let $D(L) = \mathring{W}_2^4(\Omega_1)$ and $D(K) = \mathring{W}_2^2(\Omega_2)$; we then obtain $D(L^{1/2}) = \mathring{W}_2^2(\Omega_1)$, $D(K^{1/2}) = \mathring{W}_2^1(\Omega_2)$. Using the notations described here, the problem (3.116)–(3.119) can be represented in the form (3.42), where, instead of w_0, w_1, θ_0, the initial data (3.119) appear.

Let n_1, n_2 be natural numbers $n = \{n_1, n_2\}$. Let us denote by H_1^n the linear neighbourhood of the function $\{w_{lkl}^{(1,3)} | 1 \leq k, l \leq n_1\}$, where $w_{kl}^{(1,3)}(x, y)$ are the basis functions defined by (3.68). Similarly, let us denote by H_2^n the linear neighbourhood of the function $\{w_{klm}^{(2)} | 1 \leq k, l \leq n_1, 1 \leq m \leq n_2\}$, where $w_{klm}^2(x, y, z)$ are the basis functions defined by (3.99).

The following conditions are applied to the given functions $g_1, g_2, w_0, w_1, \theta_0$:

$$\begin{aligned} &g_1 \in L_2(\Omega_1), \quad g_2 \in L_2(\Omega_2), \quad w_0 \in \mathring{W}_2^4(\Omega_1), \\ &w_1 \in \mathring{W}_2^2(\Omega_1), \quad \theta_0 \in \mathring{W}_2^2(\Omega_2). \end{aligned} \tag{3.125}$$

Therefore, for the problem (3.116)–(3.119), all conditions required by Theorem 3.3 are satisfied and the convergence velocity of the Bubnov–Galerkin method is characterized by the inequalities (3.47), (3.48). We now show explicitly how the series s_n appears in these inequalities.

For arbitrary functions $w \in H_1^n, \theta \in \overset{\circ}{W}{}_2^2(\Omega_2)$, we have

$$\left|(R_{H_2^n} Nw, \theta)_{L_2(\Omega_2)}\right|$$

$$= \frac{1}{\pi(1-\nu)} \left| \int_{\Omega_2} \sum_{k=[\frac{n_2}{2}]+1}^{\infty} \frac{(-1)^{k+1}}{k} \times \sin 2\pi kz \, \nabla^2 w\theta \, d\Omega_2 \right|$$

$$= \frac{1}{2\pi^2(1-\nu)} \left| \int_{\Omega_2} \sum_{k=\left[\frac{n_2}{2}\right]+1}^{\infty} \frac{(-1)^{k+1}}{k^2} \times \cos 2\pi kz \, \nabla^2 w \frac{\partial \theta}{\partial z} d\Omega_2 \right|$$

$$\leq \frac{1}{2\pi^2(1-\nu)} \left[\int_{-\frac{1}{2}}^{\frac{1}{2}} \left(\sum_{k=\left[\frac{n_2}{2}\right]+1}^{\infty} \frac{(-1)^{k+1}}{k^2} \times \cos 2\pi kz \right)^2 dz \right]^{1/2}$$

$$\left[\int_{\Omega_1} (\nabla^2 w)^2 \, d\Omega_1 \right]^{1/2} \times \left[\int_{\Omega_2} \left(\frac{\partial \theta}{\partial z} \right)^2 d\Omega_2 \right]^{1/2}$$

$$= \frac{1}{\pi^2} \left[\frac{3\beta(1+\nu)}{2} \sigma_{\left[\frac{n_2}{2}\right]+1} \right]^{1/2} \left\| L^{1/2} w \right\|_{L_2(\Omega_1)} \left\| K^{1/2} \theta \right\|_{L_2(\Omega_2)}, \tag{3.126}$$

where

$$\sigma_m^{(4)} = \sum_{k=m}^{\infty} \frac{1}{k^4} \equiv \frac{\pi^4}{90} - \sum_{k=1}^{m-1} \frac{1}{k^4}, \tag{3.127}$$

and $\left[\frac{n^2}{2}\right]$ denotes the largest number not larger than $\frac{n^2}{2}$. In (3.47), (3.48), s_n is an arbitrary series of which the elements approach zero and satisfy the inequality (3.45). From (3.126), we conclude that we can take

$$s_n = \frac{1}{\pi^2} \left[\frac{3\beta(1+\nu)}{2} \sigma^{(4)}_{\left[\frac{n_2}{2}\right]+1} \right]^{1/2}. \tag{3.128}$$

Let us formulate an algorithm for calculating

$$\left\| L^{-1/2} R_{H_1^n} g_1 \right\|_{H_1}^2, \quad \left\| K^{-1/2} R_{H_2^n} g_2 \right\|_{H_2}^2,$$

which occur in the right-hand sides of the inequalities (3.47), (3.48). Taking into account (3.120), (3.122), (3.124), we obtain:

$$\left\| L^{-1/2} R_{H_1^n} g_1 \right\|_{H_1}^2 = 12(1-\nu^2) \left\| (\nabla^2)^{-1} R_{H_1^n} g_1 \right\|_{L_2(\Omega_1)}^2, \tag{3.129}$$

$$\left\| K^{-1/2} R_{H_2^n} g_2 \right\|_{H_2}$$

$$= \frac{1}{\beta^{1/2}(1-\nu)^{1/2}} \left\| \left(-\lambda^2 \frac{\partial^2}{\partial x^2} - \lambda^2 \frac{\partial^2}{\partial y^2} - \frac{\partial^2}{\partial z^2} \right)^{-1/2} R_{H_2^n} q_2 \right\|_{L_2(\Omega_2)}. \tag{3.130}$$

By expanding the function $(\nabla^2)^{-1} R_{H_1^n} q_1$ into a Fourier series with respect to the basis $\{w_{kl}^{(1,3)}\}$ and expanding the functions $\left(-\lambda^2 \frac{\partial^2}{\partial x^2} - \lambda^2 \frac{\partial^2}{\partial y^2} - \frac{\partial^2}{\partial z^2}\right)^{-1} R_{H_2^n} q_2$, $R_{H_2^n} q_2$ into Fourier series with respect to the basis $\{w_{klm}^{(2)}\}$, we obtain

$$\left\| (\nabla^2)^{-1} R_{H_1^n} q_1 \right\|_{L_2(\Omega_1)}^2$$

$$= \frac{4}{\pi^4} \left(\sum_{k,l=1}^{\infty} - \sum_{k,l=1}^{n_1} \right) (k^2 + l^2)^{-2} \left(\int_{\Omega_1} q_1 w_{kl}^{(1,3)} \, d\Omega_1 \right)^2, \tag{3.131}$$

$$\left\| (-\lambda^2 \frac{\partial^2}{\partial x^2} - \lambda^2 \frac{\partial^2}{\partial y^2} - \frac{\partial^2}{\partial z^2})^{-1/2} R_{H_2^n} q_2 \right\|_{L_2(\Omega_2)}$$

$$= \left(\left(-\lambda^2 \frac{\partial^2}{\partial x^2} - \lambda^2 \frac{\partial^2}{\partial y^2} - \frac{\partial^2}{\partial z^2} \right)^{-1} R_{H_2^n} q_2, R_{H_2^n} q_2 \right)_{L_2(\Omega_2)}^{1/2}$$

$$= \frac{2\sqrt{2}}{\pi} \left[\left(\sum_{k,l=1}^{\infty} \sum_{m=1}^{\infty} - \sum_{k,l=1}^{n_1} \sum_{m=1}^{n_2} \right) (\lambda^2 k^2 + \lambda^2 l^2 + m^2)^{-1} \right.$$

$$\left. \times \left(\int_{\Omega_2} q_2 w_{klm}^{(2)} d\Omega_2 \right)^2 \right]^{1/2}. \tag{3.132}$$

Taking into account (3.120)–(3.124), (3.128)–(3.130), we can obtain a more detailed estimate of the error (3.47), (3.48) introduced by the Bubnov–Galerkin method for the problem (3.116)–(3.119). The following is one of the estimates obtained from (3.48):

$$
\begin{aligned}
&\max_{0\le t\le T}\left\{\int\limits_{\Omega_1}\left[\frac{c}{\lambda^4}\left(\frac{\partial}{\partial t}\Delta w_n\right)^2+\frac{1}{24(1-\nu^2)}(\nabla^2\Delta w_n)^2\right]d\Omega_1\right.\\
&\left.+\frac{1}{\beta(1-\nu)}\int\limits_{\Omega_2}(\Delta\theta_n)^2\,d\Omega_2\right\}\\
&\le\int\limits_{\Omega_1}\left[\frac{c}{\lambda^4}(R_{H_1^n}w_1)^2+\frac{1}{12(1-\nu^2)}(R_{H_1^n}\nabla^2 w_0)^2-2R_{H_1^n}q_1R_{H_1^n}w_0\right]d\Omega_1\\
&+\frac{1}{\beta(1-\nu)}\int\limits_{\Omega_2}(R_{H_1^n}\theta_0)^2\,d\Omega_2+24(1-\nu^2)\left\|(\nabla^2)^{-1}R_{H_1^n}q_1\right\|^2_{L_2(\Omega_1)}\\
&+\frac{\beta}{2(1-\nu)}\left\{\frac{1}{\beta}\left\|(-\lambda^2\frac{\partial^2}{\partial x^2}-\lambda^2\frac{\partial^2}{\partial y^2}-\frac{\partial^2}{\partial z^2})^{-1/2}R_{H_2^n}q_2\right\|_{L_2(\Omega_2)}\right.\\
&\left.+\frac{c_n}{\pi^2}\left[\frac{3(1-\nu)}{2}\sigma^{(4)}_{\left[\frac{n_2}{2}\right]+1}\right]^{1/2}\right\}^2 T,
\end{aligned}
\tag{3.133}
$$

where $\Delta w_n=w_n-w$, $\Delta\theta_n=\theta_n-\theta$, $\{w,\theta\}$ is the exact solution to the problem, $\{w_n,\theta_n\}$ is the approximate solution obtained using the Bubnov–Galerkin method, and

$$
\begin{aligned}
c_n=&\left[\frac{\lambda^4}{c}\int\limits_{\Omega_1}\left(q_1-\frac{1}{12(1-\nu^2)}\nabla^2\nabla^2 w_0-\frac{1}{1-\nu}\nabla^2-\int\limits_{-\frac{1}{2}}^{\frac{1}{2}}P_{H_2^n}\theta_0\right.\right.\\
&\left.\times z\,dz\right)^2 d\Omega_1+\frac{1}{12(1-\nu)}\int\limits_{\Omega_1}(\nabla^2 w_1)^2 d\Omega_1+\frac{1}{\beta(1-\nu)}\\
&\left.\times\int\limits_{\Omega_2}\left(q_2+\lambda^2\frac{\partial^2\theta_0}{\partial x^2}+\lambda^2\frac{\partial^2\theta_0}{\partial y^2}+\frac{\partial^2\theta_0}{\partial z^2}+\beta z P_{H_1^n}\nabla^2 w_1\right)^2 d\Omega_2\right]^{1/2}.
\end{aligned}
\tag{3.134}
$$

Furthermore,

$$
\left\|(\nabla^2)^{-1}R_{H_1^n}q_1\right\|^2_{L_2(\Omega_1)},\quad \left\|\left(-\lambda^2\frac{\partial^2}{\partial x^2}-\lambda^2\frac{\partial^2}{\partial y^2}-\frac{\partial^2}{\partial z^2}\right)^{-1/2}R_{H_1^n}q_2\right\|_{L_2(\Omega_2)}
$$

are the quantities defined by (3.131), (3.132); $\sigma_m^{(4)}$ is the series defined by the relation (2.127); and $\left[\frac{n_2}{2}\right]$ denotes the largest number not larger than $\frac{n_2}{2}$.

Let us consider now a problem of the influence of symmetry of the data $g_1, g_2, w_0, w_1, \theta_0$ on the theoretically guaranteed velocity of convergence of the Bubnov–Galerkin method applied to the problem (3.116)–(3.119). Let, for instance, the data of the problem exhibit the following symmetry:

$$g_1(x, y) = g_1(y, x) = g_1(-x, y) = g_1(x, -y), \tag{3.135}$$

$$\begin{aligned} g_2(x, y, z) &= g_2(y, x, z) = g_2(-x, y, z) \\ &= g_2(x, -y, z) = -g_2(x, y, -z), \end{aligned} \tag{3.136}$$

$$w_0(x, y) = w_0(y, x) = w_0(-x, y) = w_0(x, -y), \tag{3.137}$$

$$w_1(x, y) = w_1(y, x) = w_1(-x, y) = w_1(x, -y), \tag{3.138}$$

$$\begin{aligned} \theta_0(x, y, z) &= \theta_0(y, x, z) = \theta_0(-x, y, z) \\ &= \theta_0(x, -y, z) = -\theta_0(x, y, -z), \end{aligned} \tag{3.139}$$

where x, y, z are arbitrary numbers from $\left(-\frac{1}{2}, \frac{1}{2}\right)$. Let us denote by S_1 a set of functions $g_1(x, y)$ defined on Ω_1, satisfying the symmetry conditions (3.135), and let us denote by S_2 a set of functions $g_2(x, y, z)$ defined on Ω_2, satisfying the symmetry conditions (3.136). Let

$$H_1 = L_2(\Omega_1) \cap S_1, \quad H_2 = L_2(\Omega_2) \cap S_2. \tag{3.140}$$

The operators I_1, I_2, L, K, M, N for (3.116)–(3.119) with the symmetric data (3.135)–(3.139) are defined via the relations (3.121)–(3.123). However, we take

$$D(I_1) = L_2(\Omega_1) \cap S_1, \quad D(I_2) = L_2(\Omega_2) \cap S_2, \quad D(L) = \mathring{W}_2^4(\Omega_2) \cap S_2,$$
$$D(M) = \mathring{W}_2^2(\Omega_2) \cap S_2, \quad D(N) = \mathring{W}_2^2(\Omega_1) \cap S_1,$$

and use the notation (3.124). Therefore, the initial–boundary problem (3.116)–(3.119) with the symmetric data (3.135)–(3.139) can be reduced to the form (3.42), where the initial data (3.119) play the role of the elements w_0, w_1, θ_0.

The basis functions to be used in the Bubnov–Galerkin method for the problem (3.116)–(3.119) with the symmetric data (3.135)–(3.139) can be obtained in the following way:

$$\begin{aligned} &\tilde{w}_{kl}^{(1)}(x, y) \\ &= \cos(2k-1)\pi x \cos(2l-1)\pi y + \cos(2l-1)\pi x \times \cos(2k-1)\pi y, \end{aligned} \tag{3.141}$$

$$k, l = 1, 2, \ldots, k \le l,$$

$$\tilde{w}^{(2)}_{klm}(x, y, z) = \tilde{w}^{(1)}_{kl}(x, y) \sin 2\pi m z, \quad k, l, m = 1, 2, \ldots, k \le l. \tag{3.142}$$

It is easy to check that the functions $\tilde{w}^{(1)}_{kl}(x, y)$ defined by the relations (3.141) create a full system in the space $L_2(\Omega_1) \cap S_1$, and that the functions $\tilde{w}^{(2)}_{klm}(x, y, z)$ defined by the relations (3.141) create a full system in the space $L_2(\Omega_2) \cap S_2$. Let n_1, n_2 be natural numbers $n = \{n_1, n_2\}$. Let us denote by H_1^n the linear neighbourhood of the function system $\{\tilde{w}^{(1)}_{kl} | 1 \le k \le l \le n_1\}$, and by H_2^n the linear neighbourhood of the function system $\{\tilde{w}^{(2)}_{klm} | 1 \le k \le l \le n_1, 1 \le m \le n_2\}$. We assume that the data $g_1, g_2, w_0, w_1, \theta_0$ satisfy the symmetry conditions (3.135). All conditions of Theorem 3.3 are then satisfied, and the velocity of convergence of the Bubnov–Galerkin method for the present problem is characterized by the inequalities (3.47), (3.48). In these inequalities, the following series occurs:

$$s_n = \frac{1}{\pi^2}\left[\frac{3\beta(1+\nu)}{2}\sigma^{(4)}_{n_2+1}\right]^{1/2}, \tag{3.143}$$

where $\sigma^{(4)}_m$ is the series defined by the relation (3.127) (the relation (3.143) can be obtained in a manner analogous to that used for (3.128)). In order to calculate the quantities $\left\|L^{-1/2} R_{H_1^n} q_1\right\|^2_{H_1}$, $\left\|K^{-1/2} R_{H_2^n} q_2\right\|_{H_2}$, we again use the relations (3.129), (3.130), in which the following relations appear:

$$\left\|(\nabla^2)^{-1} R_{H_1^n} q_1\right\|^2_{L_2(\Omega_1)} = \frac{4}{\pi^4}\left[\sum_{k,l=1}^{\infty} - \sum_{k,l=1}^{n_1} \left[(2k-1)^2 + (2l-1)^2\right]^{-2}\right.$$
$$\left. \times \left(\int_{\Omega_1} q_1(x, y) \cos(2k-1)\pi x \cos(2l-1)\pi y \, dx\, dy\right)^2\right], \tag{3.144}$$

$$\left\|\left(-\lambda^2\frac{\partial^2}{\partial x^2} - \lambda^2\frac{\partial^2}{\partial y^2} - \frac{\partial^2}{\partial z^2}\right)^{-1/2} R_{H_2^n} q_2\right\|_{L_2(\Omega_2)}$$
$$= \frac{2\sqrt{2}}{\pi}\left[\left(\sum_{k,l=1}^{\infty}\sum_{m=1}^{\infty} - \sum_{k,l=1}^{n_1}\sum_{m=1}^{n_2}\right)(\lambda^2(2k-1)^2 + \lambda^2(2l-1)^2 + 4m^2)^{-1}\right.$$
$$\left. \times \left(\int_{\Omega_2} q_2(x, y, z) \cos(2k-1)\pi x \cos(2l-1)\pi y \sin 2\pi m z \, dx\, dy\, dz\right)^2\right]^{1/2}. \tag{3.145}$$

The relations (3.3), (3.145) can be obtained using an expression of the function $(\nabla^2)^{-1} R_{H_1^n} q_1$ into a Fourier series in the basis (3.141) and of the functions $\left(-\lambda^2 \frac{\partial^2}{\partial x^2} - \lambda^2 \frac{\partial^2}{\partial y^2} - \frac{\partial^2}{\partial z^2}\right)^{-1/2} R_{H_2^n} q_2$, $R_{H_2^n} q_2$ into Fourier series in the basis (3.142).

Using (3.121)–(3.124), (3.129), (3.130), (3.140), (3.143), we can obtain estimates of the errors (3.47), (3.48) introduced by the Bubnov–Galerkin method for the problem (3.135)–(3.139). The following is one of the estimates obtained from (3.48):

$$\max_{0 \le t \le T} \left\{ \int\limits_{\Omega_1} \left[\frac{c}{\lambda^4} \left(\frac{\partial}{\partial t} \Delta w_n \right)^2 + \frac{1}{24(1-\nu^2)} (\nabla^2 \Delta w_n)^2 \right] d\Omega_1 + \frac{1}{\beta(1-\nu)} \int\limits_{\Omega_2} (\Delta \theta_n)^2 \, d\Omega_2 \right\}$$

$$\le \int\limits_{\Omega_1} \left[\frac{c}{\lambda^4} (R_{H_1^n} w_1)^2 + \frac{1}{12(1-\nu^2)} (R_{H_1^n} \nabla^2 w_0)^2 - 2 R_{H_1^n} q_1 R_{H_1^n} w_0 \right] d\Omega_1$$

$$+ \frac{1}{\beta(1-\nu)} \int\limits_{\Omega_2} (R_{H_1^n} \theta_0)^2 d\Omega_2 + 24(1-\nu^2) \left\| (\nabla^2)^{-1} R_{H_1^n} q_1 \right\|^2_{L_2(\Omega_1)}$$

$$+ \frac{\beta}{2(1-\nu)} \left\{ \frac{1}{\beta} \left\| \left(-\lambda^2 \frac{\partial^2}{\partial x^2} - \lambda^2 \frac{\partial^2}{\partial y^2} - \frac{\partial^2}{\partial z^2} \right)^{-1/2} R_{H_2^n} q_2 \right\|_{L_2(\Omega_2)} + \frac{c_n}{\pi^2} \left[\frac{3(1-\nu)}{2} \sigma^{(4)}_{\left[\frac{n_2}{2}\right]+1} \right]^{1/2} \right\}^2 T, \tag{3.146}$$

where $\Delta w_n = w_n - w$; $\Delta \theta_n = \theta_n - \theta$; $\{w, \theta\}$ is the exact solution of the problem; $\{w_n, \theta_n\}$ is the approximate solution obtained using the Bubnov–Galerkin method; c_n is the series defined by the relation (3.134), where by H_1^n, H_2^n we mean the linear neighbourhoods of the systems of basis functions $\{\tilde{w}^{(1)}_{kl} | 1 \le k \le l \le n_1\}$, $\{\tilde{w}^{(2)}_{klm} | 1 \le k \le l \le n_1, \quad 1 \le m \le n_2\}$;

$$\left\| (\nabla^2)^{-1} R_{H_1^n} q_1 \right\|^2_{L_2(\Omega_1)}, \quad \left\| \left(-\lambda^2 \frac{\partial^2}{\partial x^2} - \lambda^2 \frac{\partial^2}{\partial y^2} - \frac{\partial^2}{\partial z^2} \right)^{-1/2} R_{H_2^n} q_2 \right\|_{L_2(\Omega_2)},$$

are the quantities defined by (3.3), (3.145), respectively; and $\sigma^{(4)}_m$ is the series defined by the relation (3.127).

Assuming that the symmetry conditions (3.135)–(3.139) are satisfied, we compare the estimate (3.146) with the estimate (3.133). For the same values of n_1 and n_2, the right-hand side of the inequality (3.146) is smaller than the right-hand side of the inequality (3.133) (the operators $R_{H_1^n}$, $R_{H_2^n}$ in the right-hand side of (3.146) correspond to the functions $L_2(\Omega_1) \cap S_1$, $L_2(\Omega_2) \cap S_2$, with shorter remainders of the Fourier series than for the operators in the right hand-side of (3.133)). The number of basis functions used for the symmetric data, with the estimate (3.146) $\left(\frac{1}{2}(n_1+1\right) n_1$ basis functions in the series for the first component of the approximate solution $w_n(x, y, t)$ and $\frac{1}{2}(n_1+1)n_1 n_2$ basis functions in the series for the second component of the approximate solution $\theta_n(x, y, z, t)$), is also smaller than in the case without symmetry, with the estimate (3.133) (n_1^2 basis functions in the series for the first component of the approximate solution $w_n(x, y, z)$ and $n_1^2 n_2$ basis functions in the series for the second component of the approximate solution $\theta_n(x, y, z, t)$). Therefore, the estimate (3.146) of the error introduced by the Bubnov–Galerkin method is better in the case of symmetry.

The occurrence in the left-hand side of each of the inequalities (3.133), (3.146) of a term containing a harmonic operator gives us a possibility to estimate the error of the deflection function $w(x, y, t)$ in a unique norm because of the variables x, y, t, i.e. in a norm of the space $C(\overline{\Omega}_1 \times [0, T])$. Let us establish some following estimates. We denote by $\{\tilde{w}_n, \tilde{\theta}_n\}$ the exact solution of the intermediate initial–boundary value problem of the form (3.116)–(3.119) with free terms $P_{H_1^n} q_1$, $P_{H_1^n} q_2$ and initial data $P_{H_1^n} w_0$, $P_{H_1^n} w_1$, $P_{H_1^n} \theta_0$, where H_1^n denotes either a linear neighbourhood of the system of functions $\{\tilde{w}_{kl}^{(1)} | 1 \le k \le l \le n_1\}$ (if the input data of the problem (3.116)–(3.119) satisfy the symmetry conditions (3.135)–(3.139)) or a linear neighbourhood of the system of functions $\{\tilde{w}_{kl}^{(1,3)} | 1 \le k, l \le n_1\}$ (if the symmetry conditions (3.135)–(3.139) are not satisfied). We start with the obvious inequality

$$\|\Delta w_n\|_{C(\overline{\Omega}_1 \times [0,T])} \le \|\tilde{w}_n - w\|_{C(\overline{\Omega}_1 \times [0,T])} + \|w_n - \tilde{w}_n\|_{C(\overline{\Omega}_1 \times [0,T])}, \tag{3.147}$$

where w is the first component of the exact solution $\{w, \theta\}$ to the basic problem (3.116)–(3.119), w_n is the first component of the approximate solution to the initial problem constructed with the use of the Bubnov–Galerkin method, and $\Delta w_n = w_n - w$. Considering $\{\tilde{w}_n, \tilde{\theta}_n\}$ as the approximate solution of the problem (3.116)–(3.119) and $\{w_n, \theta_n\}$ as the approximate solution to the intermediate problem, we obtain the corresponding estimates of the error introduced by the Bubnov–Galerkin method either of type (3.133) (if the data of the initial problem (3.116)–(3.119) are nonsymmetric) or of type (3.146) (if the data of the initial problem satisfy the symmetry conditions (3.135)–(3.139)). In the left hand sides of the inequalities obtained we omit all terms except those containing a harmonic operator, and the following inequalities are obtained:

$$\max_{0 \le t \le T} \left\| \nabla^2 [\tilde{w}_n(t) - w(t)] \right\|_{L_2(\Omega_1)} \le \alpha_n, \tag{3.148}$$

$$\max_{0\le t\le T} \left\| \nabla^2[w_n(t) - \tilde{w}_n(t)] \right\|_{L_2(\Omega_1)} \le \beta_n. \tag{3.149}$$

Let us introduce the following notation:

$$\sigma_0^* = \sum_{k,l=1}^{\infty} (k^2 + l^2)^{-2}, \quad \sigma_0^{**} = \sum_{k,l=1}^{\infty} [(2k-1)^2 + (2l-1)^2]^{-2}, \tag{3.150}$$

$$\sigma_m^* = \left(\sum_{k,l=1}^{\infty} - \sum_{k,l=1}^{m} \right) (k^2 + l^2)^{-2}, \tag{3.151}$$

$$\sigma_m^{**} = \left(\sum_{k,l=1}^{\infty} - \sum_{k,l=1}^{m} \right) [(2k-1)^2 + (2l-1)^2]^{-2}. \tag{3.152}$$

Let $w(x, y)$ be an arbitrary function of the class $\overset{\circ}{W}{}_2^2(\Omega_1)$; let $w_n(x, y)$ be an arbitrary function of the class $\overset{\circ}{W}{}_2^2(\Omega_1)$ satisfying the condition

$$P_{H_1^n} w_n = 0, \tag{3.153}$$

where H_1^n is a linear neighbourhood of the system of functions $\{w_{kl}^{(1,3)} | 1 \le k,\ l \le n_1\}$; let $w^{(s)}(x, y)$ be an arbitrary function of $\overset{\circ}{W}{}_2^2(\Omega_1)$ satisfying the symmetry conditions (3.137); and let $w_{(n)}^{(s)}(x, y)$ be an arbitrary function of $\overset{\circ}{W}{}_2^2(\Omega_1)$ satisfying both the condition (3.153), where H_1^n is the linear neighbourhood of the system of functions $\{\tilde{w}_{kl}^{(1)} | 1 \le k \le l \le n_1\}$, and the symmetry conditions (3.137). The following inequalities hold (these inequalities are similar to those for $W_2^2(\Omega_1) \to C(\overline{\Omega_1})$):

$$\|w\|_{C(\overline{\Omega}_1)} \le \frac{2}{\pi^2} \sqrt{\sigma_0^*} \left\| \nabla^2 w \right\|_{L_2(\Omega_1)}, \tag{3.154}$$

$$\left\| w_{(n)} \right\|_{C(\overline{\Omega}_1)} \le \frac{2}{\pi^2} \sqrt{\sigma_{n_1}^*} \left\| \nabla^2 w_{(n)} \right\|_{L_2(\Omega_1)}, \tag{3.155}$$

$$\left\| w^{(s)} \right\|_{C(\overline{\Omega}_1)} \le \frac{2}{\pi^2} \sqrt{\sigma_0^{**}} \left\| \nabla^2 w^{(s)} \right\|_{L_2(\Omega_1)}, \tag{3.156}$$

$$\left\| w_{(n)}^{(s)} \right\|_{C(\overline{\Omega}_1)} \le \frac{2}{\pi^2} \sqrt{\sigma_{n_1}^{**}} \left\| \nabla^2 w_{(n)}^{(s)} \right\|_{L_2(\Omega_1)}. \tag{3.157}$$

The inequalities (3.154)–(3.157) are obtained by the series expansion of the functions w, $\nabla^2 w$, $w_{(n)}$, $\nabla^2 w_{(n)}$ into Fourier series in the basis $\{w_{kl}^{(1,3)}\}$ and by the series expansion of the functions $w^{(s)}$, $\nabla^2 w^{(s)}$, $w_{(n)}^{(s)}$, $\nabla^2 w_{(n)}^{(s)}$ into Fourier series in the basis $\{\tilde{w}_{kl}^{(1)}\}$, with successive applications of the Cauchy–Buniakowski inequality. From (3.154)–(3.157), we obtain

$$\|w\|_{C(\overline{\Omega}_1\times[0,T])} \leq \frac{2}{\pi^2}\sqrt{\sigma_0^*}\max_{0\leq t\leq T}\left\|\nabla^2 w(t)\right\|_{L_2(\Omega_1)}, \tag{3.158}$$

$$\left\|w_{[n]}\right\|_{C(\overline{\Omega}_1\times[0,T])} \leq \frac{2}{\pi^2}\sqrt{\sigma_{n_1}^*}\max_{0\leq t\leq T}\left\|\nabla^2 w_{[n]}(t)\right\|_{L_2(\Omega_1)}, \tag{3.159}$$

$$\left\|w^{[s]}\right\|_{C(\overline{\Omega}_1\times[0,T])} \leq \frac{2}{\pi^2}\sqrt{\sigma_0^{**}}\max_{0\leq t\leq T}\left\|\nabla^2 w^{[s]}(t)\right\|_{L_2(\Omega_1)}, \tag{3.160}$$

$$\left\|w_{[n]}^{[s]}\right\|_{C(\overline{\Omega}_1\times[0,T])} \leq \frac{2}{\pi^2}\sqrt{\sigma_{n_1}^{**}}\max_{0\leq t\leq T}\left\|\nabla^2 w_{[n]}^{[s]}(t)\right\|_{L_2(\Omega_1)}, \tag{3.161}$$

where $w(x, y, t)$ is an arbitrary function of the class $C([0, T]; \overset{\circ}{W}{}_2^2(\Omega_1))$, $w_{[n]}(x, y, t)$ is an arbitrary function of the class $C([0, T]; \overset{\circ}{W}{}_2^2(\Omega_1))$ satisfying the condition (3.153), in which H_1^n is the linear neighbourhood of the system of functions $\{w_{kl}^{(1,3)} | 1 \leq k,\ l \leq n_1\}$; $w^{[s]}(x, y)$ is an arbitrary function of the class $C([0, T]; \overset{\circ}{W}{}_2^2(\Omega_1))$ satisfying symmetry conditions of the type (3.137); and $w_{[n]}^{[s]}(x, y)$ is an arbitrary function of the class $C([0, T]; \overset{\circ}{W}{}_2^2(\Omega_1))$ satisfying simultaneously the condition (3.153), in which H_1^n is the linear neighbourhood of the system of functions $\{\tilde{w}_{kl}^{(1)} | \leq k \leq l \leq n_1\}$, and the symmetry condition (3.137).

It is easy to check that the components $\tilde{w}_n, \tilde{\theta}_n$ of the exact solution to the intermediate problem are projections of the corresponding components w, θ of the exact solution to the initial problem (3.116)–(3.119) into H_1^n, and, in particular, that $P_{H_1^n}(\tilde{w}_n - w) = 0$. We now use the inequalities (3.158), (3.159) with $w = w_n - \tilde{w}_n$, $w_{[n]} = \tilde{w}_n - w$ when the symmetry conditions (3.135), (3.139) are not satisfied, and the inequalities (3.160), (3.161) with $w^{[s]} = w_n - \tilde{w}_n$, $w_{[n]}^{[s]} = \tilde{w}_n - w$ when the symmetry conditions (3.135)–(3.139) are satisfied, and we estimate from above the right-hand side of the inequality (3.147). Combining the estimates of the norm $\|w\|_{C(\overline{\Omega}_1\times[0,T])}$ obtained and the inequalities (3.148), (3.149), we obtain

$$\|\Delta w_n\|_{C(\overline{\Omega}_1\times[0,T])} \leq \frac{2}{\pi^2}\left(\sqrt{\sigma_{n_1}^*}\alpha_n + \sqrt{\sigma_0^*}\beta_n\right) \tag{3.162}$$

when the symmetry conditions (3.135)–(3.139) are not satisfied, and

$$\|\Delta w_n\|_{C(\overline{\Omega}_1\times[0,T])} \leq \frac{2}{\pi^2}\left(\sqrt{\sigma_{n_1}^{**}}\alpha_n + \sqrt{\sigma_0^{**}}\beta_n\right) \tag{3.163}$$

when the symmetry conditions (3.135)–(3.139) are satisfied.

The inequalities (3.162), (3.163) are used to estimate the errors occurring during calculation of a deflection function in an equally measured norm. We now present the estimates in a fully worked-out form. When the symmetry conditions (3.135)–(3.139) are not satisfied, we have

$$\|\Delta w_n\|_{C(\overline{\Omega}_1\times[0,T])} \leq \frac{2}{\pi^2}\sqrt{\sigma^*_{n_1}}\left\{\int\limits_{\Omega_1}\left[\frac{24(1-\nu^2)c}{\lambda^4}(R_{H_1^n}w_1)^2\right.\right.$$

$$\left.+2(R_{H_1^n}\nabla^2 w_0)^2-48(1-\nu)R_{H_1^n}q_1R_{H_1^n}w_0\right]d\Omega_1$$

$$+\frac{24(1+\nu)}{\beta}\int\limits_{\Omega_2}(R_{H_1^n}\theta_0)^2 d\Omega_2+576(1-\nu^2)^2\left\|(\nabla^2)^{-1}R_{H_1^n}q_1\right\|^2_{L_2(\Omega_1)}$$

$$\left.+\frac{12(1+\nu)}{\beta}\left\|\left(-\lambda^2\frac{\partial^2}{\partial x^2}-\lambda^2\frac{\partial^2}{\partial y^2}-\frac{\partial^2}{\partial z^2}\right)^{-1/2}R_{H_1^n}q_2\right\|^2_{L_2(\Omega_1)}T\right\}^{1/2}$$

$$+\frac{2}{\pi^2}\sqrt{\sigma_0^*}\left\{\frac{24(1-\nu)}{\beta}\int\limits_{\Omega_1}(R_{H_1^n}P_{H_1^n}\theta_0)^2+12\beta(1+\nu)\right.$$

$$\times\left[\frac{1}{\beta}\left\|\left(-\lambda^2\frac{\partial^2}{\partial x^2}-\lambda^2\frac{\partial^2}{\partial y^2}-\frac{\partial^2}{\partial z^2}\right)^{-1/2}R_{H_1^n}q_2\right\|_{L_2(\Omega_1)}\right.$$

$$\left.\left.+\frac{c_n}{\pi^2}\left(\frac{3(1-\nu^2)}{2}\sigma^{(4)}_{\left[\frac{n_2}{2}\right]+1}\right)^{1/2}\right]^2 T\right\}^{1/2}, \tag{3.164}$$

where H_1^n is the linear neighbourhood of the system of functions $\{w_{kl}^{(1,3)}|1\leq k,l\leq n_1\}$; H_n^2 is the linear neighbourhood of the system of functions $\{w_{klm}^{(2)}|1\leq k,\ l\leq n_1, 1\leq m\leq n_2\}$; $\sigma_m^4, c_n, \sigma_m^*$ are the series defined by (3.127), (3.134), (3.151); $\left\|(\nabla^2)^{-1}R_{H_1^n}q_1\right\|^2_{L_2(\Omega_1)}$ is the quantity defined by the relation (3.131); and σ_0^* is the constant defined by the first relation of (3.150). We then have

$$\left\|\left(-\lambda^2\frac{\partial^2}{\partial x^2}-\lambda^2\frac{\partial^2}{\partial y^2}-\frac{\partial^2}{\partial z^2}\right)^{-1/2}R_{H_1^n}q_2\right\|^2_{L_2(\Omega_1)}$$

$$=\frac{8}{\pi^2}\left(\sum_{k,l=1}^{\infty}-\sum_{k,l=1}^{n_1}\right)\sum_{m=1}^{\infty}(\lambda^2k^2+\lambda^2l^2+m^2)^{-1}\left(\int\limits_{\Omega_2}q_2w_{klm}^{(2)}\,d\Omega_2\right)^2,$$

$$\left\| \left(-\lambda^2 \frac{\partial^2}{\partial x^2} - \lambda^2 \frac{\partial^2}{\partial y^2} - \frac{\partial^2}{\partial z^2} \right)^{-1/2} R_{H_1^n} P_{H_1^n} q_2 \right\|_{L_2(\Omega_1)}$$

$$= \frac{2\sqrt{2}}{\pi} \left[\sum_{k,l=1}^{n_1} \sum_{m=1}^{\infty} (\lambda^2 k^2 + \lambda^2 l^2 + m^2)^{-1} \left(\int_{\Omega_2} q_2 w_{klm}^{(2)} d\Omega_2 \right)^2 \right]^{1/2}.$$

When the symmetry conditions (3.135)–(3.139) are satisfied, we have

$$\|\Delta w_n\|_{C(\overline{\Omega}_1 \times [0,T])} \le \frac{2}{\pi^2} \sqrt{\sigma_{n_1}^{**}} \left\{ \int_{\Omega_1} \left[\frac{24(1-\nu^2)c}{\lambda^4} (R_{H_1^n} w_1)^2 \right. \right.$$

$$\left. + 2(R_{H_1^n} \nabla^2 w_0)^2 - 48(1-\nu) R_{H_1^n} q_1 R_{H_1^n} w_0 \right] d\Omega_1$$

$$+ \frac{24(1+\nu)}{\beta} \int_{\Omega_2} (R_{H_1^n} \theta_0)^2 d\Omega_2 + 576(1-\nu^2)^2 \left\| (\nabla^2)^{-1} R_{H_1^n} q_1 \right\|_{L_2(\Omega_1)}^2$$

$$\left. + \frac{12(1+\nu)}{\beta} \left\| \left(-\lambda^2 \frac{\partial^2}{\partial x^2} - \lambda^2 \frac{\partial^2}{\partial y^2} - \frac{\partial^2}{\partial z^2} \right)^{-1/2} R_{H_1^n} q_2 \right\|_{L_2(\Omega_1)}^2 T \right\}^{1/2}$$

$$+ \frac{2}{\pi^2} \sqrt{\sigma_0^*} \left\{ \frac{24(1+\nu)}{\beta} \int_{\Omega_1} (R_{H_1^n} P_{H_1^n} \theta_0)^2 + 12\beta(1+\nu) \right.$$

$$\times \left[\frac{1}{\beta} \left\| \left(-\lambda^2 \frac{\partial^2}{\partial x^2} - \lambda^2 \frac{\partial^2}{\partial y^2} - \frac{\partial^2}{\partial z^2} \right)^{-1/2} R_{H_1^n} P_{H_1^n} q_2 \right\|_{L_2(\Omega_2)} \right.$$

$$\left. \left. + \frac{c_n}{\pi^2} \left(\frac{3(1-\nu^2)}{2} \sigma_{n_2+1}^{(4)} \right)^{1/2} \right]^2 T \right\}^{1/2}, \tag{3.165}$$

where H_1^n is the linear neighbourhood of the system of functions $\{w_{kl}^{(1)} | 1 \le k \le l \le n_1\}$; H_2^n is the linear neighbourhood of the system of functions $\{\tilde{w}_{klm}^{(2)} | 1 \le k \le l \le n_1, 1 \le m \le n_2$; $\sigma_m^{(4)}$, c_n, σ_m^{**} are the series defined by the relations (3.127),

(3.134), (3.152); $\|(\nabla^2)^{-1} R_{H_1^n} q_1\|^2_{L_2(\Omega_1)}$ is the quantity defined by the relation (3.3); and σ_0^{**} is the constant defined by the second relation of (3.150). We then have

$$\left\| \left(-\lambda^2 \frac{\partial^2}{\partial x^2} - \lambda^2 \frac{\partial^2}{\partial y^2} - \frac{\partial^2}{\partial z^2} \right)^{-1/2} R_{H_1^n} q_2 \right\|^2_{L_2(\Omega_2)}$$

$$= \frac{8}{\pi^2} \left(\sum_{k,l=1}^{\infty} - \sum_{k,l=1}^{n_1} \right) \sum_{m=1}^{\infty} \left[\lambda^2 (2k-1)^2 + \lambda^2 (2l-1)^2 + 4m^2 \right]^{-1}$$

$$\times \left(\int_{\Omega_2} q_2(x,y,z) \cos(2k-1)\pi x \cos(2l-1)\pi y \sin 2\pi m z \, dx\, dy\, dz \right)^2,$$

$$\left\| \left(-\lambda^2 \frac{\partial^2}{\partial x^2} - \lambda^2 \frac{\partial^2}{\partial y^2} - \frac{\partial^2}{\partial z^2} \right)^{-1/2} R_{H_1^n} P_{H_1^n} q_2 \right\|_{L_2(\Omega_2)}$$

$$= \frac{2\sqrt{2}}{\pi} \left[\sum_{k,l=1}^{n_1} \sum_{m=1}^{\infty} \left(\lambda^2 (2k-1)^2 + \lambda^2 (2l-1)^2 + 4m^2 \right)^{-1} \right.$$

$$\left. \times \left(\int_{\Omega_2} q_2(x,y,z) \cos(2k-1)\pi x \cos(2l-1)\pi y \sin 2\pi m z \, dx\, dy\, dz \right)^2 \right]^{1/2}.$$

Remark 3.7 The method presented above for obtaining an estimate of the error of a calculation of the deflection function (with use of the intermediate function $\tilde{w}_n$) in a uniformly distributed norm leads to a better estimate than does a direct application of the inclusion theorem $W_2^2(\Omega_1) \to C(\overline{\Omega}_1)$ to the inequalities (3.133), (3.146).

Remark 3.8 The method of obtaining the estimates (3.47), (3.48) can be extended without major difficulties to cases of coupled thermoelastic problems for shallow shells with a constant thickness in the framework of the Kirchhoff–Love model with boundary conditions of free support on ribs that cannot be stretched or compressed in the tangential plane.

The computational examples showing the efficiency of the estimates described in this section are given in Chap. 4.

3.4 Coupled Problem of Thermoelasticity Within a Timoshenko-Type Model

In this section, a scheme leading to prior estimates given by (3.3), (3.19), (3.23), (3.39) of the errors introduced by the Bubnov–Galerkin method for a coupled problem of thermoelasticity of shallow shells, rectangular in plan and with variable thickness, in the framework of a kinematic model of Timoshenko type, is described.

Assuming that the conditions (2.64)–(2.68), (2.83) are satisfied, we consider the system of differential equations (2.20)–(2.23) with the homogeneous initial conditions (2.33)–(2.37), where $u, v, w, \Psi_x, \Psi_y, \theta$ are the sought functions, defined for $(x, y) \in \Omega_1, |z| \le \frac{1}{2}h(x, y)$, $t \in [0, T]$. We assume also that $\Omega_1 = (-\frac{a}{2}, \frac{a}{2}) \times (-\frac{b}{2}, \frac{b}{2})$, where a, b are positive constants. Any one of the sets of boundary conditions (2.48), (2.50), (2.52) is applied to the shell $\partial\Omega_2$, and on the surface of $\partial\Omega_2$ we apply a homogeneous condition of type (2.38).

We use the notation $H_1 = (L_2(\Omega_1))^5$, $H_2 = L_2(\Omega_2)$ and assume that the scalar product (3.98) is determined for the space $L_2(\Omega_2)$. The differential operator L in this case is defined on $D(L) = (\overset{\circ}{W}{}_2^2(\Omega_1))^5$ by the relation (2.72). The auxiliary operator L_0 with a known spectrum is sought in the form

$$L_0\overline{\omega} = \tilde{L}\overline{\omega} + P_L\overline{\omega}, \tag{3.166}$$

where $\overline{\omega} = \{u, v, w, \Psi_x, \Psi_y, \theta\}$, $\tilde{L}$ is the self-adjoint extension of the differential operator initially defined on $(\overset{\circ}{W}{}_2^2(\Omega_1))^5 \cap (C^2(\overline{\Omega}_1))^5$ in the sense of Friedrichs by the relation

$$\tilde{L}\overline{\omega} = \left\{ -\frac{\partial^2 u}{\partial x^2} - \frac{1-v}{2}\frac{\partial^2 u}{\partial y^2}, -\frac{\partial^2 v}{\partial y^2} - \frac{1-v}{2}\frac{\partial^2 v}{\partial x^2}, \right.$$

$$\left. -\nabla^2 w, -\frac{\partial^2 \Psi_x}{\partial x^2} - \frac{1-v}{2}\frac{\partial^2 \Psi_x}{\partial y^2}, -\frac{\partial^2 \Psi_y}{\partial y^2} - \frac{1-v}{2}\frac{\partial^2 \Psi_y}{\partial x^2} \right\},$$

and P_L is a positive constant to be found. The eigenvector functions $\overline{v}_{kl}^{(j)}$ $(j = 1, \ldots, 5)$ of the operator L_0 have the following form:

$$\overline{v}_{kl}^{(j)}(x, y) = \left\{\sigma_i^j \omega_{kl}^{(1,1)}\left(\frac{x}{a}, \frac{y}{b}\right)\right\}_{i=1}^5, j = 1, 4, k = 0, 1, \ldots, l = 1, 2, \ldots, \tag{3.167}$$

$$\overline{v}_{kl}^{(j)}(x, y) = \left\{\sigma_i^j \omega_{kl}^{(1,2)}\left(\frac{x}{a}, \frac{y}{b}\right)\right\}_{i=1}^5, j = 2, 5, k = 1, 2, \ldots, l = 0, 1, \ldots, \tag{3.168}$$

$$\overline{v}_{kl}^{(3)}(x, y) = \left\{\sigma_i^3 \omega_{kl}^{(1,3)}\left(\frac{x}{a}, \frac{y}{b}\right)\right\}_{i=1}^5, k, l = 1, 2, \ldots, \tag{3.169}$$

for the boundary conditions (2.48),

$$\overline{v}_{kl}^{(j)}(x, y) = \left\{\sigma_i^j w_{kl}^{(1,3)}\left(\frac{x}{a}, \frac{y}{b}\right)\right\}_{i=1}^{5}, j = 1, 2, 3, k, l = 1, 2, \ldots, \tag{3.170}$$

$$\overline{v}_{kl}^{(4)}(x, y) = \left\{\sigma_i^4 \omega_{kl}^{(1,1)}\left(\frac{x}{a}, \frac{y}{b}\right)\right\}_{i=1}^{5}, k = 0, 1, \ldots, l = 1, 2, \ldots, \tag{3.171}$$

$$\overline{v}_{kl}^{(5)}(x, y) = \left\{\sigma_i^5 \omega_{kl}^{(1,2)}\left(\frac{x}{a}, \frac{y}{b}\right)\right\}_{i=1}^{5}, k = 1, 2, \ldots, l = 1, 2, \ldots, \tag{3.172}$$

for the boundary conditions (2.50), and

$$\overline{v}_{kl}^{(j)}(x, y) = \left\{\sigma_i^j \omega_{kl}^{(1,3)}\left(\frac{x}{a}, \frac{y}{b}\right)\right\}_{i=1}^{5}, j = 1, \ldots, 5, k, l = 1, 2, \ldots, \tag{3.173}$$

for the boundary conditions (2.52).

In (3.167)–(3.173), $\omega_{kl}^{(1,j)}(j = 1, 2, 3)$ are the functions defined by the relations (3.67), (3.68); σ_i^j is the Kronecker delta. The eigenvalues $\lambda_{kl}^{(j)}$ of the operator L_0 (for all three cases (2.48), (2.50), (2.52)) are defined by the relations

$$\lambda_{kl}^{(1)} = \lambda_{kl}^{(4)} = \pi^2\left[\frac{k^2}{a^2} + \frac{(1-\nu)l^2}{2b^2}\right] + P_L, \tag{3.174}$$

$$\lambda_{kl}^{(2)} = \lambda_{kl}^{(5)} = \pi^2\left[\frac{(1-\nu)k^2}{2a^2} + \frac{l^2}{b^2}\right] + P_L, \tag{3.175}$$

$$\lambda_{kl}^{(3)} = \pi^2\left[\frac{k^2}{a^2} + \frac{l^2}{b^2}\right] + P_L. \tag{3.176}$$

Using a method similar to that described in [360] and the technique described in Sect. 3.2, it can be seen that for an appropriate choice of the constant P_L the operator (3.166) is convergent and forms an acute angle to the operator (2.72), and that estimates of types (3.93), (3.95), (3.96) hold.

The operators K and K_0 are defined by the relations (3.103), (3.100), and with an appropriate choice of p_k both of those operators are convergent and form an acute angle, and the estimates (3.104), (3.105) hold. The eigenfunctions and eigenvalues of the operator K_0 are defined by the relations (3.101), (3.102).

The operators I_1, M for the problem under consideration are defined by (2.70), (2.76). In accordance with (3.103), the operators I_1, N are defined by a differential equation obtained by multiplying both sides of (2.23) by $T_0^{-1}h(x, y)$, i.e. we take

$$I_2\theta = \frac{c_\varepsilon(1+\varepsilon)h}{T_0}\theta,$$

$$N\overline{\omega} = \frac{E\alpha_T h}{1-\nu}\left[\frac{\partial u}{\partial x} + \frac{\partial v}{\partial y} - (k_x + k_y)w + z\left(\frac{\partial \Psi_x}{\partial x} + \frac{\partial \Psi_y}{\partial y}\right)\right].$$

The free terms of (3.1), (3.2) can be obtained by the relations $g_1 = \{p_1, p_2, q_1, 0, 0\}$, $g_2 = T_0^{-1} h q_2$.

In an analogous way to that presented in Sect. 3.2, we obtain

$$\left\| I_1^{1/2} \right\|_{H_1 \to H_1} = \rho^{1/2} \tilde{H}_0^{1/2}, \tag{3.177}$$

where

$$\tilde{H}_0 = \max_{(x,y)\in\overline{\Omega}_1} \left\{ h(x,y), \frac{1}{12} h^3(x,y) \right\}.$$

We also obtain the second relation of (3.108), the inequalities (3.110), (3.111) and the inequality

$$\left\| M K^{-1/2} \right\|_{H_2 \to H_1} \le \tilde{B}_1, \tag{3.178}$$

where $\tilde{B}_1$ is a positive constant. In accordance with (3.178), for the problem under consideration (unlike the coupled thermoelastic problem of a shallow shell within the Kirchhoff–Love kinematic model), the estimate (3.19), (3.39) hold, together with the estimates of the error introduced by the Bubnov–Galerkin method (3.3), (3.23).

Using the relation (3.177), the second relation of (3.108), inequalities of the form (3.96), (3.110), (3.111), and the inequalities (3.105), (3.178), we estimate from above the constants c_i $(i = 1, \ldots, 5)$, $c_2^*, c_3^*, d_j (j = 1, 2, 3)$ and d_2^* defined by (3.5) and (3.40), as well as the values of e_1, e_2, e_1^* defined by (3.37), (3.38), (3.41). Then, substituting into the right-hand sides of (3.3), (3.19), (3.23), (3.39) the estimates of these constants and variables, we obtain prior estimates of the errors of the Bubnov–Galerkin method for the differential equations (2.20)–(2.23) with the homogeneous initial conditions (2.33)–(2.37) and homogeneous boundary conditions. The following is one of the estimates obtained, and is a specific case of (3.19):

$$\max_{0\le t\le T} \left\{ \int_{\Omega_1} \left[\rho h \left(\left(\frac{\partial}{\partial t} \Delta u_n \right)^2 + \left(\frac{\partial}{\partial t} \Delta v_n \right)^2 + \left(\frac{\partial}{\partial t} \Delta w_n \right)^2 \right) \right. \right.$$

$$+ \frac{1}{12} \rho h^3 \left(\left(\frac{\partial}{\partial t} \Delta \Psi_{xn} \right)^2 + \left(\frac{\partial}{\partial t} \Delta \Psi_{yn} \right)^2 \right) + \frac{Eh}{1+\nu} \left(\left(\frac{\partial}{\partial x} \Delta u_n \right. \right.$$

$$\left. \left. - k_x \Delta w_n \right)^2 + \left(\frac{\partial}{\partial y} \Delta v_n - k_y \Delta w_n \right)^2 + \frac{1}{2} \left(\frac{\partial}{\partial y} \Delta u_n + \frac{\partial}{\partial x} \Delta v_n \right)^2 \right)$$

$$+\frac{Eh\nu}{1-\nu^2}\left(\frac{\partial}{\partial x}\Delta u_n+\frac{\partial}{\partial y}\Delta v_n-(k_x+k_y)\Delta w_n\right)^2$$

$$+\frac{k^2Eh}{2(1+\nu)}\left(\left(\frac{\partial}{\partial x}\Delta w_n+\Delta\Psi_{xn}\right)^2+\left(\frac{\partial}{\partial y}\Delta w_n+\Delta\Psi_{yn}\right)^2\right)$$

$$+D(1-\nu)\left(\left(\frac{\partial}{\partial x}\Delta\Psi_{xn}\right)^2+\left(\frac{\partial}{\partial y}\Delta\Psi_{yn}\right)^2+\frac{1}{2}\left(\frac{\partial}{\partial y}\Delta\Psi_{xn}\right.\right.$$

$$\left.\left.+\frac{\partial}{\partial x}\Delta\Psi_{yn}\right)^2\right)+D\nu\left(\frac{\partial}{\partial x}\Delta\Psi_{xn}+\frac{\partial}{\partial y}\Delta\Psi_{yn}\right)^2\Bigg]d\Omega_1$$

$$\left.+\frac{c_\varepsilon(1+\varepsilon)}{T_0}\int\limits_{\Omega_2}(\Delta\theta_n)^2d\Omega_2+\frac{\lambda_q}{T_0}\int\limits_0^t\int\limits_{\Omega_2}|\nabla\Delta\theta_n|^2d\Omega_2d\tau\right\}$$

$$\leq 4A_LT^{1/2}(1+\alpha_L\beta_L^{-1})(M_g(T))^{1/2}\left\{\left[\int\limits_0^T\int\limits_{\Omega_1}\frac{1}{h}(p_1^2+p_2^2+q_1^2)\right.\right.$$

$$\left.\times d\Omega_1d\tau\right]^{1/2}+\frac{\sqrt{2}}{2}e^{1/2}T^{1/2}\tilde{B}_1\left[\frac{1}{\rho}\int\limits_0^T\int\limits_{\Omega_1}\frac{1}{h}(p_1^2+p_2^2+q_1^2)\right.$$

$$\left.\times d\Omega_1d\tau+\frac{T_0}{c_\varepsilon(1+\varepsilon)}\int\limits_0^T\int\limits_{\Omega_2}q_2^2d\Omega_2d\tau\right]^{1/2}$$

$$\left.+\rho^{1/2}\tilde{H}_0^{1/2}T^{1/2}(M_g(T))^{1/2}\right\}\tilde{\lambda}_{i(n)+1}^{-1/2}+A_k^2(1+\alpha_k\beta_k^{-1})^2$$

$$\times\left\{\left(\int\limits_0^T\int\limits_{\Omega_2}hq_2^2\,d\Omega_2\,d\tau\right)^{1/2}+[c_\varepsilon(1+\varepsilon)T_0^{-1}H_0+B_2^2]^{1/2}\right.$$

$$\left.\times T^{1/2}(M_g(T))^{1/2}\right\}^2\tilde{æ}_{j(n)+1}^{-1}\equiv 0(\max(\tilde{\lambda}_{i(n)+1}^{-1/2},\tilde{æ}_{j(n)+1}^{-1})),$$

where $\Delta u_n=u_n-u$, $\Delta v_n=v_n-v$, $\Delta w_n=w_n-w$, $\Delta\Psi_{xn}=\Psi_{xn}-\Psi_x$, $\Delta\Psi_{yn}=\Psi_{yn}-\Psi_y$, $\Delta\theta_n=\theta_n-\theta$, $\{u,v,w,\Psi_x,\Psi_y,\theta\}$ is the exact ("classical") solution to the problem, and $\{u_n, v_n, w_n, \Psi_{xn}, \Psi_{yn}, \theta_n\}$ is the approximate solution obtained using the Bubnov–Galerkin approach. This approximate solution is given by

$$\overline{\omega}_n(x, y, t) = \{u_n(x, y, t), v_n(x, y, t), w_n(x, y, t), \Psi_{xn}(x, y, t), \Psi_{yn}(x, y, t)\}$$
$$= \sum_{\overline{v}_{kl}^{(j)} \in H_1^n} \alpha_{kl}^{(j)}(t)\overline{v}_{kl}^{(j)}(x, y),$$

$$\theta_n(x, y, z, t) = \sum_{\eta_{klm} \in H_2^n} \beta_{klm}(t)\eta_{klm}(x, y, z),$$

$$\tilde{\lambda}_{i(n)+1} = \min\{\lambda_{kl}^{(j)} : \overline{v}_{kl}^{(j)} \overline{\in} H_1^n\},$$

$$\tilde{æ}_{i(n)+1} = \min\{æ_{kl}^{(j)} : \eta_{klm} \overline{\in} H_2^n\},$$

where H_1^n, H_2^n are the subspaces of the spaces $(L_2(\Omega_1))^5$ and $L_2(\Omega_2)$, spanned by $i(n)$ basis vector functions $\overline{v}_{kl}^{(j)}$ and defined in relation to the mechanical boundary conditions of the problem under consideration, either by the relations (3.167)–(3.169), by the relations (3.170)–(3.172), by the relation (3.173), or on $j(n)$ basis functions η_{klm} defined by the relations (3.101). The $\lambda_{kl}^{(j)}$, $æ_{klm}$ are the eigenvalues of the auxiliary operators L_0, K_0 defined by the relations (3.174)–(3.176), (3.102). $M_g(T)$ is the constant defined by the relation (3.113), in which $M[\cdot, \cdot, \cdot]$ is the value defined by the relation (3.114).

Remark 3.9 If the thickness of the shell is constant and the shell edges are supported freely on elastic ribs that cannot be stretched or compressed, and if both the mechanical and the thermal loads are constant in time, then for the problem considered here the specific scheme for estimation of the errors (3.47), (3.48) of the Bubnov–Galerkin method is similar to that presented in Sect. 3.3.

Chapter 4
Numerical Investigations of the Errors of the Bubnov–Galerkin Method

In Chap. 3, many prior error estimates for the Bubnov–Galerkin method applied to coupled thermoelastic problems of shallow shells and plates were derived. The estimates related to the general case of a shell with a variable thickness subjected to time-varying mechanical and thermal loads possess an important theoretical meaning (they guarantee, for a wide class of problems, strong convergence of successive approximations to the exact solution with a velocity larger than the estimated velocity). However, in applications, when specific real problems have to be considered the estimates possess a more generalized meaning. This question is addressed, for instance, in the case of the estimates obtained in Sect. 3.3 for vibrations of a simply supported plate with constant thickness, subjected to mechanical and thermal loads that are constant in time. In this chapter, numerical results are given and the efficiency of the estimate used is verified.

4.1 Vibration of a Transversely Loaded Plate

We consider the initial–boundary value problem (3.116)–(3.119) (the bars over nondimensional quantities are omitted). The behaviour of a plate without a thermal load will be considered, and the following three groups of data will be used:

$$\begin{aligned} &q_1(x, y) \not\equiv 0, \quad w_0(x, y) = w_1(x, y) \equiv 0, \\ &q_2(x, y, z) = \theta_0(x, y, z) \equiv 0, \end{aligned} \tag{4.1}$$

$$\begin{aligned} &q_1(x, y) = w_1(x, y) \equiv 0, \quad w_0(x, y) \not\equiv 0, \\ &q_2(x, y, z) = \theta_0(x, y, z) \equiv 0, \end{aligned} \tag{4.2}$$

J. Awrejcewicz, V. A. Krysko, *Elastic and Thermoelastic Problems in Nonlinear Dynamics of Structural Members*, Scientific Computation,
https://doi.org/10.1007/978-3-030-37663-5_4

$$q_1(x, y) = w_0(x, y) \equiv 0, \quad w_1(x, y) \not\equiv 0,$$
$$q_2(x, y, z) = \theta_0(x, y, z) \equiv 0. \tag{4.3}$$

It should be noted that the solution of an arbitrary initial–boundary value problem of the form (3.116)–(3.119) with nonhomogeneous mechanical and homogeneous thermal data (no thermal load) can be represented in the form of a sum of solutions of the three problems related to the data (4.1)–(4.3).

In this chapter, the notation used Sect. 3.3 is used here also.

In this section, the problem (3.116)–(3.119) with the data (4.1) is considered (the other variants of the data, i.e. (4.2) and (4.3), will be considered later). The problem relates to the vibrations of a transversely loaded simply supported plate. We assume that $q_1 \in L_2(\Omega_1)$, and the case of a symmetric load (3.135) will be considered separately.

We are going to solve the problem stated above using the Bubnov–Galerkin method by taking the basis functions either in the form (3.68), (3.99) (when the symmetry conditions (3.135) are not satisfied) or in the form (3.141), (3.142) (when the symmetry conditions are satisfied). Thus, the estimates (3.133), (3.146), (3.164) and (3.165) hold. Taking into account (4.1), the estimates (3.164), (3.165) read

$$\|w_n - w\|_{C(\overline{\Omega}_1 \times [0,T])} \leq \frac{96(1-\nu^2)}{\pi^4}\sqrt{\sigma_{n_1}^*}\left[\left(\sum_{k,l=1}^{\infty} - \sum_{k,l=1}^{n_1}\right)(k^2+l^2)^{-2} \times \left(\int_{\Omega_1} q_1 w_{kl}^{(1,3)}\, d\Omega_1\right)^2\right]^{1/2} + \frac{6\lambda^2}{\pi^4}\left[\frac{2\beta(1+\nu)(1-\nu^2)}{c} \times \sigma_0^* \sigma^{(4)}_{\left[\frac{n_2}{2}\right]+1} T\right]^{1/2} \|q_1\|_{L_2(\Omega_1)}, \tag{4.4}$$

$$\|w_n - w\|_{C(\overline{\Omega}_1 \times [0,T])} \leq \frac{96(1-\nu^2)}{\pi^4}\sqrt{\sigma_{n_1}^{**}}\left[\left(\sum_{k,l=1}^{\infty} - \sum_{k,l=1}^{n_1}\right)((2k-l)^2 + (2l-1)^2)^{-2} \left(\int_{\Omega_1} q_1(x,y)\cos(2k-1)\pi x \cos(2l-1)\pi y\, dx\, dy\right)^2\right]^{1/2} + \frac{6\lambda^2}{\pi^4}\left[\frac{2\beta(1+\nu)(1-\nu^2)\sigma_0^{**}\sigma^{(4)}_{0n_2+1} T}{c}\right]^{1/2} \|q_1\|_{L_2(\Omega_1)}. \tag{4.5}$$

In the left-hand side of each of the inequalities (3.133), (3.146), we omit the terms containing Δw_n, and the newly obtained inequalities, taking account of (4.1), have the form

$$\max_{0\le t\le T} \|\theta_n(t)-\theta(t)\|_{L_2(\Omega_2)} \le \frac{96(1-\nu)(1-\nu^2)}{\pi^4}$$

$$\times\left[\left(\sum_{k,l=1}^{\infty}-\sum_{k,l=1}^{n_1}\right)(k^2+l^2)^{-2}\left(\int_{\Omega_1} q_1 w_{kl}^{(1,3)}\,d\Omega_1\right)^2\right.$$

$$\left.+\frac{3\beta^2(1-\nu^2)\lambda^4}{4\pi^4 c}T\sigma^{(4)}_{\left[\frac{n_2}{2}\right]+1}\|q_1\|^2_{L_2(\Omega_1)}\right]^{1/2}, \tag{4.6}$$

$$\max_{0\le t\le T} \|\theta_n(t)-\theta(t)\|_{L_2(\Omega_2)} \le \frac{96(1-\nu)(1-\nu^2)}{\pi^4}\left[\left(\sum_{k,l=1}^{\infty}-\sum_{k,l=1}^{n_1}\right)\right.$$

$$\times\left((2k-1)^2(2l-1)^2\right)^{-2}\left(\int_{\Omega_1} q_1(x,y)\cos(2k-1)\pi x\right.$$

$$\left.\left.\times\cos(2l-1)\pi y\,dx\,dy\right)^2+\frac{3\beta^2(1-\nu^2)\lambda^4}{4\pi^4 c}T\sigma^{(4)}_{n_2+1}\|q_1\|^2_{L_2(\Omega_1)}\right]. \tag{4.7}$$

In (4.4)–(4.7), $\{w,\theta\}$ is the exact solution to the problem under consideration, and $\{w_n,\theta_n\}$ is its approximate solution constructed using the Bubnov–Galerkin method. The estimates (4.4), (4.6) correspond to a case of a nonsymmetric distribution of the transverse load, and the estimates (4.5), (4.7) correspond to a case of a symmetric distribution of the transverse load (3.135).

Let us remember that the plate under consideration is a square with side a and thickness h. For the purpose of computation we take $a = 7\times 10^{-2}$ m and $h = 10^{-3}$ m, and therefore $\lambda = \dfrac{h}{a} = 1/70$. Two materials for a plate will be considered: aluminium (Al) and copper (Cu). The material properties, taken from [193], are

$$E=\begin{cases}6.85\times 10^{10}\,\text{Pa} & \text{(Al)},\\ 11.2\times 10^{10}\,\text{Pa} & \text{(Cu)},\end{cases}$$

$$\nu=\begin{cases}0.36 & \text{(Al)},\\ 0.37 & \text{(Cu)},\end{cases} \tag{4.8}$$

$$\rho = \begin{cases} 2.70 \times 10^3 \text{ kg/m}^3 & \text{(Al)}, \\ 8.93 \times 10^3 \text{ kg/m}^3 & \text{(Cu)}, \end{cases} \quad \alpha_t = \begin{cases} 2.28 \times 10^{-5} \text{ K}^{-1} & \text{(Al)}, \\ 1.66 \times 10^{-5} \text{ K}^{-1} & \text{(Cu)}, \end{cases}$$

$$\lambda_q = \begin{cases} 207 \text{ W/(m K)} & \text{(Al)}, \\ 398 \text{ W/(m K)} & \text{(Cu)}, \end{cases} \quad c_p = \begin{cases} 902 \text{ J/(kg K)} & \text{(Al)}, \\ 386 \text{ J/(kg K)} & \text{(Cu)}. \end{cases}$$

The initial temperature of the plate is $T_0 = 293$ K. Using the known formula [131]

$$c_\varepsilon = c_\sigma - 3\alpha_T^2 E T_0/(1 - 2\nu),$$

where $c_\sigma = \rho c_p$ is the specific heat capacity for a constant stress tensor, we can calculate the specific heat for the two materials considered, for a constant strain tensor, as follows:

$$c_\varepsilon = \begin{cases} 2.324 \times 10^6 \text{ J/(m}^3 \text{ K)} & \text{(Al)}, \\ 3.343 \times 10^6 \text{ J/(m}^3 \text{ K)} & \text{(Cu)}. \end{cases}$$

The coupling coefficient is calculated using the second expression of (2.24):

$$\varepsilon = \begin{cases} 3.407 \times 10^{-2} & \text{(Al)}, \\ 2.262 \times 10^{-2} & \text{(Cu)}. \end{cases}$$

The values of the nondimensional parameters c, β for the materials are calculated using (3.115):

$$c = \begin{cases} 2.92 \times 10^{-10} & \text{(Al)}, \\ 1.08 \times 10^{-9} & \text{(Cu)}, \end{cases} \quad \beta = \begin{cases} 6.78 \times 10^{-3} & \text{(Al)}, \\ 4.20 \times 10^{-3} & \text{(Cu)}. \end{cases} \tag{4.9}$$

We consider time intervals related to the maximum period T^* of free shell vibrations $T^* = \frac{2\pi}{\omega^*}$, where ω^* is the minimum shell eigenfrequency obtained for uncoupled deformation and temperature fields. Finding ω^* from the equation governing elastic vibrations of a plate (without a thermal term), we obtain $T^* = \frac{2\sqrt{3c(1-\nu)^2}}{\pi\lambda^2}$; we obtain from this

$$T^* = \begin{cases} 0.08614 & \text{(Al)}, \\ 0.16496 & \text{(Cu)}. \end{cases} \tag{4.10}$$

As the transverse load we take $g_1(x, y) \equiv 1$, which is uniformly distributed over the plate surface. Therefore the symmetry conditions (3.135) are satisfied and the estimates for the Bubnov–Galerkin method (4.5), (4.7) hold. Substituting $\lambda = 1/70$, the values of ν, c, β given by (4.8), (4.9), and $g_1(x, y) \equiv 1$, we obtain

$$\|w_n - w\|_{C(\overline{\Omega_{11}}\times[0,T])} \le \Phi_1(n_1, n_2, T), \tag{4.11}$$

$$\max_{0\le t\le T} \|\theta_n(t) - \theta(t)\|_{L_2(\Omega_2)} \le \Psi_1(n_1, n_2, T), \tag{4.12}$$

where

$$\Phi_1(n_1, n_2, T) = \begin{cases} 0.34766\sqrt{\sigma_{n_1}^{**}\sigma_{n_1}} + 4.9537 \times 10^{-2}\sqrt{\sigma_{n_2+1}^{(4)}T} & \text{(Al)},\\ 0.34475\sqrt{\sigma_{n_1}^{**}\sigma_{n_1}} + 2.0262 \times 10^{-2}\sqrt{\sigma_{n_2+1}^{(4)}T} & \text{(Cu)},\end{cases}$$

$$\Psi_1(n_1, n_2, T) = \begin{cases} (6.1140 \times 10^{-4}\sigma_{n_1} + 4.3939 \times 10^{-5}\sigma_{n_2+1}^{(4)}T)^{1/2} & \text{(Al)},\\ (3.6970 \times 10^{-4}\sigma_{n_1} + 4.5207 \times 10^{-6}\sigma_{n_2+1}^{(4)}T)^{1/2} & \text{(Cu)},\end{cases}$$

$$\sigma_m = \left(\sum_{k,l=1}^{\infty} - \sum_{k,l=1}^{m}\right)(2k-1)^2(2l-1)^{-2}\left[(2k-1)^2 + (2l-1)^2\right]^{-2}.$$

Here $\sigma_m^{(4)}, \sigma_m^{**}$ are numerical series defined by the relations (3.172), (3.152); and $n = \{n_1, n_2\}$, where n_1, n_2 are natural numbers characterizing the number of terms taken in the components of the approximate solution w_n, θ_n (let us remember that w_n includes $\frac{1}{2}(n_1 + 1)n_1$ terms, whereas θ_n includes $\frac{1}{2}(n_1 + 1)n_1 n_2$ terms).

Let us now introduce the relative errors of the Bubnov–Galerkin method $\delta_{n,T}w, \delta_{n,T}\theta$, which characterize the accuracy of computation of the deflection function in terms of a norm that is uniform with respect to the variables x, y, t, and the accuracy of computation of the thermal function in relation to the average squared norm with respect to the spatial coordinates:

$$\delta_{n,T}w = \frac{\|w_n - w\|_{C(\overline{\Omega}_1\times[0,T])}}{\|w_n\|_{C(\overline{\Omega}_1\times[0,T])}}, \tag{4.13}$$

$$\delta_{n,T}\theta = \frac{\|\theta_n(T) - \theta(T)\|_{L_2(\Omega_2)}}{\|\theta_n(T)\|_{L_2(\Omega_2)}}. \tag{4.14}$$

Let us introduce the following notation:

$$\Delta_{n,T}w = \frac{\Phi_1(n_1, n_2, T)}{\|w_n\|_{C(\Omega_1\times[0,T])}}, \tag{4.15}$$

$$\Delta_{n,T}\theta = \frac{\Psi_1(n_1, n_2, T)}{\|\theta_n(T)\|_{L_2(\Omega_2)}}. \tag{4.16}$$

According to (4.11), (4.12) the inequalities $\delta_{n,T}w \le \Delta_{n,T}w$, $\delta_{n,T}\theta \le \Delta_{n,T}\theta$ hold, i.e. the quantities defined by the relations (4.15), (4.16) must be treated as theoretically guaranteed estimates from above of the relative errors $\delta_{n,T}w$, $\delta_{n,T}\theta$ of the Bubnov–Galerkin method, defined by the relations (4.13), (4.14).

The problem considered here has been solved numerically using the Bubnov–Galerkin method (the set of basis functions used changes in the intervals $1 \le n_1, n_2 \le 6$), with successive application of the Runge–Kutta method for solving the system of differential equations. The following computational scheme has been followed:

1. Using the maximal set of basis functions ($n_1 = n_2 = 6$), we calculate $w_n(0, 0, t)$ and $\theta_n(0, 0, 0.25, t)$ for $t = (i/160)T^*, i = 1, 2, \ldots, 160$, where T^* is the maximum period of the eigenvibrations of the plate (the value of T^* was calculated from (4.10)).
2. We find the time value $T_1 = (i_1/160)T^*$ for which the maximum of the function $i \to |w_n(0, 0, (i/160)T^*)|$ is achieved, and then we calculate the values $w_n(x, y, T_1)$ for $x = j/40$, $y = k/40$, $j, k = 0, 1, \ldots, 19$ (owing to the evenness of the function w_n in relation to x and y, we calculate only its values for $x, y \ge 0$, i.e. on a quadrant of the plate).
3. We find the value $T_2 = (i_2/160)T^*$ for which the first explicitly expressed extremum of the function $i \to (0, 0, 0.25, (i/160)T^*)$ appears, and calculate the values $\theta_n(x, y, z, T_2)$ for $x = j/40,\ y = k/40,\ z = l/40,\ j, k, l = 0, 1, \ldots, 19$ (because the function θ is even with respect to x and y and odd with respect to z, only the calculations for $x, y, z \ge 0$ are carried out, i.e. 1/8 of the plate is treated as a three-dimensional body). By "the first explicitly expressed extremum" of the function $i \to \theta(i)$ calculated on a set I of integers i, we mean that one of the two extrema of the functions $\max\limits_{i\in I} \theta(i)$, $\min\limits_{i\in I} \theta(i)$ which is achieved for the smaller value of the argument i.
4. Now we vary a number of basis functions in the intervals corresponding to $1 \le n_1, n_2 \le 6$ for each pair of n_1, n_2 and calculate the norm $\|w_n\|_{C(\overline{\Omega}_1[0,T_1])}$ according to the approximate formula
$$\|w_n\|_{C(\overline{\Omega}_1[0,T_1])} \approx \max_{\substack{1\le i\le i_1\\ 0\le j,\ k\le 19}} \left| w_n\left(\frac{j}{40}, \frac{k}{40}, \frac{i}{160}T^*\right)\right| .$$
We then calculate the guaranteed estimate $\Delta_{n,T_1}w$ using the formula (4.15).
5. The number of basis functions is varied in the intervals corresponding to $1 \le n_1, n_2 \le 6$ for each pair n_1, n_2, and for each pair of values n_1, n_2 we calculate the norm $\|\theta_n\|_{L_2(\Omega_2)}$ from the approximate formula
$$\|\theta\|_{L_2(\Omega_2)} \approx \left[\frac{1}{8000} \sum_{j,k,l=0}^{19} \left|\theta_n\left(\frac{j}{40}, \frac{k}{40}, \frac{i}{160}T_2\right)\right|^2\right]^{1/2} .$$

We then calculate the guaranteed estimate $\Delta_{n,T_2}\theta$ of the relative error of the calculation of the thermal function $\delta_{n,T_2}\theta$, using the formula (4.16).

The following values of T_1, T_2 have been obtained:

$$T_1 = 0.5T^* = \begin{cases} 4.307 \times 10^{-2} & \text{(Al)}, \\ 8.248 \times 10^{-2} & \text{(Cu)}, \end{cases} \tag{4.17}$$

$$T_2 = 0.4T^* = \begin{cases} 3.446 \times 10^{-2} & \text{(Al)}, \\ 6.598 \times 10^{-2} & \text{(Cu)}. \end{cases} \tag{4.18}$$

In Tables 4.1 and 4.2, the theoretically guaranteed estimates $\Delta_{n,T_1}w$ of the relative error of a computation of the deflection function, in terms of a uniformly distributed norm, for all pairs n_1, n_2 $(1 \le n_1, n_2 \le 6)$ for aluminium and copper plates are given as percentages. It can be seen that the estimates of the errors introduced by the Bubnov–Galerkin method guarantee values of the relative error of a computation of the deflection of 1–2% (or even smaller). For instance, in the case of an aluminium plate (see Table 4.1), the relative error of a computation of the deflection function is less than 1.838% for $n_1 = 3, n_2 = 2$, less than 1.538% for $n_1 = 2, n_2 = 3$, and less than 0.856% for $n_1 = 3, n_2 = 4$. In the case of a copper plate (see Table 4.2), the relative error of a computation of the deflection function is less than 1.464% for $n_1 = n_2 = 2$, 0.73% for $n_1 = n_2 = 3$. When the maximum

Table 4.1 Estimates of relative error (as a percentage) of a computation of the deflection function in terms of a uniformly distributed norm, $\Delta_{n,T_1}w$; problem (3.116)–(3.119); data set (4.1); $\lambda = 1/70$; $q_1(x, y) \equiv 1$; Al ($\nu = 0.36$, $c = 2.92 \times 10^{-10}$, $\beta = 6.78 \times 10^{-3}$)

$n_2 \backslash n_1$	1	2	3	4	5	6
1	6.935	3.971	3.607	3.528	3.500	3.489
2	5.206	2.197	1.838	1.757	1.730	1.719
3	4.564	1.538	1.180	1.099	1.072	1.061
4	4.248	1.214	0.856	0.775	0.748	0.737
5	4.066	1.028	0.670	0.589	0.562	0.551
6	3.951	0.909	0.522	0.471	0.444	0.433

Table 4.2 Estimates of relative error (as a percentage) of a computation of the deflection function in terms of a uniformly distributed norm, $\Delta_{n,T_1}w$; problem (3.116)–(3.119); data set (4.1); $\lambda = 1/70$; $q_1(x, y) \equiv 1$; Cu ($\nu = 0.37$, $c = 1.08 \times 10^{-9}$, $\beta = 4.20 \times 10^{-3}$)

$n_2 \backslash n_1$	1	2	3	4	5	6
1	5.478	2.476	2.116	2.036	2.008	1.997
2	4.491	1.464	1.106	1.025	0.998	0.986
3	4.125	1.088	0.730	0.649	0.622	0.611
4	3.994	0.903	0.546	0.465	0.438	0.426
5	3.841	0.796	0.439	0.358	0.331	0.320
6	3.775	0.729	0.372	0.291	0.264	0.252

Table 4.3 Estimates of relative error (as a percentage) of a computation of the temperature function in terms of a mean square norm, $\Delta_{n,T_2}\theta$; problem (3.116)–(3.119); data set (4.1); $\lambda = 1/70$; $q_1(x, y) \equiv 1$; Al ($\nu = 0.36, c = 2.92 \times 10^{-10}, \beta = 6.78 \times 10^{-3}$)

$n_2 \backslash n_1$	1	2	3	4	5	6
1	180.50	66.35	53.20	51.06	50.56	50.40
2	172.56	49.08	29.72	25.82	24.84	24.53
3	171.31	45.12	22.68	17.26	15.77	15.27
4	170.94	43.81	19.95	13.48	11.51	10.82
5	170.79	43.26	18.72	11.59	9.22	8.33
6	170.72	42.99	18.09	10.55	7.88	6.82

Table 4.4 Estimates of relative error (as a percentage) of a computation of the temperature function in terms of a mean square norm, $\Delta_{n,T_2}\theta$; problem (3.116)–(3.119); date set (4.1); $\lambda = 1/70$; $q_1(x, y) \equiv 1$; Cu ($\nu = 0.37, c = 1.08 \times 10^{-9}, \beta = 4.20 \times 10^{-3}$)

$n_2 \backslash n_1$	1	2	3	4	5	6
1	330.95	97.54	63.17	56.61	55.01	54.49
2	325.03	84.67	41.67	31.05	28.08	27.07
3	324.18	82.03	36.09	23.05	18.86	17.32
4	323.94	81.19	34.14	19.86	14.80	12.78
5	323.84	80.84	33.31	18.40	12.77	10.37
6	323.80	80.67	32.90	17.65	11.67	8.97

number of basis functions is used ($n_1 = n_2 = 6$), the guaranteed estimate of the relative error of a computation of the deflection function does not exceed 0.433% for a plate made from aluminium and does not exceed 0.242% for a plate made from copper.

In Tables 4.3 and 4.4, the theoretically guaranteed estimates $\Delta_{n,T_2}\theta$ of the relative error of a computation of the temperature function, in terms of a mean square norm with respect to the spatial variables, calculated for all pairs $n_1, n_2 (1 \leq n_1, n_2 \leq 6)$ are given as percentages.

It can be seen that the estimates of the error of the Bubnov–Galerkin method guarantee that the relative error of a computation of the temperature function is of the order of 10–20% or less. For example, in the case of an aluminium plate (see Table 4.3), the relative error of a computation of the temperature function is less than 19.95% for $n_1 = 3, n_2 = 4$, less than 17.26% for $n_1 = 4, n_2 = 3$, and less than 9.22% for $n_1 = n_2 = 5$. According to Table 4.4, for the copper plate, the relative error of a calculation of the temperature function is less than 19.86% for $n_1 = n_2 = 4$, and less than 8.97% for $n_1 = n_2 = 6$.

To conclude, the estimates of the accuracy of the Bubnov–Galerkin method allow numerical study (in the framework of the model used here) of one form of interesting behaviour caused by coupling of the temperature and deformation fields. In spite of the lack of a thermal load, a temperature field has been created inside a vibrating plate, as a result of only the deformation process.

4.2 Vibration of a Plate with an Imperfection in the Form of a Deflection

Let us consider the problem (3.116)–(3.119) with the data set (4.2). This problem concerns the vibration of a simply supported plate with a given initial deflection distribution (imperfection), with neither a mechanical nor a thermal load. The coupling of the thermal and deformation fields, as well as the nonlinearity of the temperature field distribution through the plate thickness, are taken into account.

Let us remember that the plate considered here is square, with side a and thickness h. For the computations, we take $a = 7 \times 10^{-2}$ m, $h = 10^{-3}$ m and $\lambda = \frac{h}{a} = 1/70$. We consider two materials from which the plate may be made: aluminium (Al) and copper (Cu). The values of the parameters ν, c, β for these materials have been taken from the relations (4.8), (4.9). Let us assume an initial deflection of the plate of the following form:

$$w_0(x, y) = \left(x^2 - \frac{1}{4}\right)\left(x^2 - \frac{5}{4}\right)\left(y^2 - \frac{1}{4}\right)\left(y^2 - \frac{5}{4}\right).$$

That is, we take the function w_0 to be the minimum-order polynomial which satisfies the boundary conditions of simple support (thus the problem is included in the class $\overset{\circ}{W}{}_2^4(\Omega_1)$).

The problem has been solved using the Bubnov–Galerkin method, with the basis functions $\tilde{w}_{kl}^{(1)}(x, y)(1 \le k \le l \le n_1)$, $\tilde{w}_{klm}^{(2)}(x, y, z)(1 \le k \le l \le n_1, 1 \le m \le n_2)$ in the form (3.141), (3.142). Because the symmetry conditions (3.137) are satisfied in the present case, the estimates (3.146), (3.165) hold. Taking into account the actual values of the parameters λ, ν, c, β and of the function $w_0(x, y)$, the estimates (3.146), (3.165) have the following form (similar to problem presented in Sect. 4.1 with a data set of type (4.1)):

$$\|w_n\|_{C(\overline{\Omega}_1 \times [0,T_1])} < \Phi_2(n_1, n_2, T), \tag{4.19}$$

$$\max_{0 \le t \le T} \|\theta_n(t) - \theta(t)\|_{L_2(\Omega_2)} < \Psi_2(n_1, n_2, T), \tag{4.20}$$

where $\{w, \theta\}$ is the exact solution of the problem, $\{w_n, \theta_n\}$ is the approximate solution obtained by the Bubnov–Galerkin method, and

$$\Phi_2(n_1, n_2, T) = 0.28658 \left[\sigma_{n_1}^{**} \left(0.47170\tilde{\sigma}_{n_1+1}^{(6)} + 0.94353\tilde{\sigma}_{n_1+1}^{(8)}\right.\right.$$

$$\left.\left. +0.47238\tilde{\sigma}_{n_1+1}^{(10)}\right)\right]^{1/2} + \begin{cases} 9.2628 \times 10^{-2}\sqrt{\sigma_{n_2+1}^{(4)} T} & \text{(Al)}, \\ 3.8210 \times 10^{-2}\sqrt{\sigma_{n_2+1}^{(4)} T} & \text{(Cu)}, \end{cases}$$

$$\Psi_2(n_1, n_2, T) = \begin{cases} [1.9597 \times 10^{-4}\tilde{\sigma}^{(6)}_{n_1+1} + 3.9198 \times 10^{-4}\tilde{\sigma}^{(8)}_{n_1+1} \\ \quad +1.9624 \times 10^{-4}\tilde{\sigma}^{(10)}_{n_1+1} + 1.5363 \times 10^{-4}\tilde{\sigma}^{(4)}_{n_2+1} T]^{1/2} \quad (\mathrm{Al}), \\ [1.2051 \times 10^{-4}\tilde{\sigma}^{(6)}_{n_1+1} + 2.4105 \times 10^{-4}\tilde{\sigma}^{(8)}_{n_1+1} \\ \quad +1.2068 \times 10^{-4}\tilde{\sigma}^{(10)}_{n_1+1} + 1.5241 \times 10^{-5}\tilde{\sigma}^{(4)}_{n_2+1} T]^{1/2} \quad (\mathrm{Cu}), \end{cases}$$

$$\tilde{\sigma}^{(\alpha)}_m = \sum_{k=m}^{\infty} (2k-1)^{-\alpha}, \quad \alpha = 6, 8, 10. \tag{4.21}$$

Here $\sigma^{(4)}_m$ and σ^{**}_m are the numerical series defined by the relations (3.127) and (3.152), and $n = \{n_1, n_2\}$.

Let us consider the relative errors $\delta_{n,T}w$, $\delta_{n,T}\theta$ of the Bubnov–Galerkin method defined by the relations (4.13), (4.14), which characterize both the accuracy of the deflection function calculation, in terms of a norm uniformly distributed with respect to x, y, t, and the accuracy of the temperature function calculation, in terms of a mean square norm with respect to the spatial coordinates. Let us denote

$$\Delta_{n,T}w = \frac{\Phi_2(n_1, n_2, T)}{\|w_n\|_{C(\overline{\Omega}_1\times[0,T])}}, \tag{4.22}$$

$$\Delta_{n,T}\theta = \frac{\Psi_2(n_1, n_2, T)}{\|\theta_n(T)\|_{L_2(\Omega_2)}}. \tag{4.23}$$

In accordance with (4.19), (4.20), the following inequalities hold: $\delta_{n,T}w < \Delta_{n,T}w$, $\delta_{n,T}\theta < \Delta_{n,T}\theta$, i.e. the quantities $\Delta_{n,T}w$, $\Delta_{n,T}\theta$ defined by (4.22), (4.23) can be considered as theoretically guaranteed estimates from above of the relative errors $\delta_{n,T}w$, $\delta_{n,T}\theta$ of the Bubnov–Galerkin method, defined by (4.13), (4.14).

The problem has been solved using the Bubnov–Galerkin method (the number of basis functions used was taken from the intervals $1 \le n_1, n_2 \le 6$), with successive application of the Runge–Kutta method for solving the ordinary differential equations. The numerical scheme presented in Sect. 4.1 was applied. Here we shall mention only that during performance of steps 1–3 of the numerical scheme, the following values of T_1, T_2 were obtained:

$$T_1 = 0.5T^* = \begin{cases} 4.307 \times 10^{-2} \quad (\mathrm{Al}), \\ 8.248 \times 10^{-2} \quad (\mathrm{Cu}), \end{cases} \tag{4.24}$$

$$T_2 = \begin{cases} 0.4T^* = 3.446 \times 10^{-2}\ (\mathrm{Al}), \\ 0.375T^* = 6.186 \times 10^{-2}\ (\mathrm{Cu}), \end{cases} \tag{4.25}$$

where T^* is the maximum period of free oscillations of the plate defined by (4.10).

Table 4.5 Estimates of relative error (as percentages) of deflection function calculations using the uniform norm $\Delta_{n,T_1}w$; problem (3.116)–(3.119); data set (4.2); $\lambda = 1/70$; $w_0(x, y) = \left(x^2 - \frac{1}{4}\right)\left(x^2 - \frac{5}{4}\right)\left(y^2 - \frac{1}{4}\right)\left(y^2 - \frac{5}{4}\right)$; Al ($\nu = 0.36, c = 2.92 \times 10^{-10}, \beta = 6.78 \times 10^{-3}$)

$n_2 \backslash n_1$	1	2	3	4	5	6
1	7.123	5.827	5.693	5.665	5.656	5.652
2	4.269	2.949	2.816	2.789	2.779	2.775
3	3.208	1.879	1.747	1.720	1.710	1.707
4	2.686	1.353	1.222	1.194	1.184	1.181
5	2.386	1.050	0.919	0.891	0.882	0.878
6	2.195	0.858	0.727	0.699	0.690	0.686

Table 4.6 Estimates of relative error (as percentages) of deflection function calculations using the uniform norm $\Delta_{n,T_1}w$; problem (3.116)–(3.119); data set (4.2); $\lambda = 1/70$; $w_0(x, y) = \left(x^2 - \frac{1}{4}\right)\left(x^2 - \frac{5}{4}\right)\left(y^2 - \frac{1}{4}\right)\left(y^2 - \frac{5}{4}\right)$; Cu($\nu = 0.37, c = 1.08 \times 10^{-9}, \beta = 4.20 \times 10^{-3}$)

$n_2 \backslash n_1$	1	2	3	4	5	6
1	4.718	3.402	3.269	3.241	3.232	3.228
2	3.088	1.759	1.627	1.599	1.590	1.586
3	2.483	1.148	1.017	0.989	0.980	0.976
4	2.185	0.848	0.717	0.689	0.679	0.676
5	2.014	0.675	0.544	0.516	0.507	0.503
6	1.905	0.565	0.435	0.406	0.397	0.393

In Tables 4.5 and 4.6, the theoretically guaranteed estimates $\Delta_{n,T_1}w$ from above of the relative error of a deflection function calculation using the uniform norm, for all pairs n_1, n_2 ($1 \le n_1, n_2 \le 6$) are given in percentages for aluminium and copper plates. Analysing the tables, one can conclude that the estimates given here guarantee a relative error of a computation of the deflection function of 1–2% or less. For instance, according to Table 4.5 (aluminium), the relative error has the following values: less than 1.879% for $n_1 = 2, n_2 = 3$, and less than 0.919% for $n_1 = 3, n_2 = 5$. Also, according to the results shown in Table 4.6 (copper), the relative error of a computation of the deflection function is less than 1.759% for $n_1 = n_2 = 4$, and less than 0.717% for $n_1 = 3, n_2 = 4$.

In Tables 4.7 and 4.8 (for aluminium and copper, respectively), the theoretically guaranteed estimates from above, $\Delta_{n,T_2}\theta$, of the relative error of a computation of the temperature function in terms of the mean square norm with respect to the spatial coordinates are given as percentages for all pairs n_1, n_2 in the intervals $1 \le n_1, n_2 \le 6$. An analysis of Tables 4.7 and 4.8 leads to the conclusion that the estimates given here of the error introduced by the Bubnov–Galerkin method guarantee a relative error of a computation of the temperature function of less than 10–15%. For instance, according to Table 4.7 (aluminium) the relative error of a temperature function computation is less than 14.61% for $n_1 = 2, n_2 = 3$ and less than 9.03% for $n_1 = 3, n_2 = 4$. Also according to Table 4.8 (copper), the relative

Table 4.7 Estimates of relative error (as percentages) of temperature function calculations using the mean square norm $\Delta_{n,T_2}\theta$; problem (3.116)–(3.119); data set (4.2); $\lambda = 1/70$; $w_0(x, y) = \left(x^2 - \frac{1}{4}\right)\left(x^2 - \frac{5}{4}\right)\left(y^2 - \frac{1}{4}\right)\left(y^2 - \frac{5}{4}\right)$; Al ($\nu = 0.36$, $c = 2.92 \times 10^{-10}$, $\beta = 6.78 \times 10^{-3}$)

$n_2\backslash n_1$	1	2	3	4	5	6
1	55.80	42.36	41.72	41.63	41.62	41.61
2	41.80	21.59	20.34	20.17	20.14	20.13
3	38.63	14.61	12.70	12.43	12.38	12.36
4	37.58	11.57	9.03	8.66	8.57	8.55
5	37.14	10.06	7.00	6.51	6.39	6.36
6	36.93	9.25	5.77	5.16	5.02	4.97

Table 4.8 Estimates of relative error (as percentages) of temperature function calculations using the mean square norm $\Delta_{n,T_2}\theta$; problem (3.116)–(3.119); data set (4.2); $\lambda = 1/70$; $w_0(x, y) = \left(x^2 - \frac{1}{4}\right)\left(x^2 - \frac{5}{4}\right)\left(y^2 - \frac{1}{4}\right)\left(y^2 - \frac{5}{4}\right)$; Cu ($\nu = 0.37$, $c = 1.08 \times 10^{-9}$, $\beta = 4.20 \times 10^{-3}$)

$n_2\backslash n_1$	1	2	3	4	5	6
1	79.87	43.71	41.57	41.29	41.23	41.21
2	70.69	24.71	20.77	20.22	20.09	20.05
3	68.85	18.98	13.47	12.59	12.40	12.34
4	68.27	16.77	10.12	8.92	8.64	8.55
5	68.03	15.78	8.38	6.88	6.51	6.39
6	67.91	15.28	7.39	5.63	5.17	5.03

error of a temperature function computation is less than 13.47% for $n_1 = n_2 = 3$ and less than 8.92% for $n_1 = n_2 = 4$.

To conclude, the estimates of the errors introduced by the Bubnov–Galerkin method given here confirm numerically that there is coupling between the temperature and deformation fields when, in the absence of any thermal interaction, a temperature field occurs owing to a deformation process.

4.3 Vibration of a Plate with a Given Variable Deflection Change

Let us consider the problem (3.116)–(3.119) with the data set (4.3). This problem concerns the vibration of a simply supported plate with an initial distribution of the velocity of change of the deflection, with neither a thermal nor a mechanical load, but with inclusion of coupling between the temperature and deformation fields and with a nonlinear distribution of the temperature field through the plate thickness.

Following Sects. 4.1, 4.2, we assume that the relative plate thickness is $\lambda = 1/70$ and the values of the parameters ν, c, β are defined by (4.8), (4.9). We assume also that the initial distribution of the velocity of deflection change over the plate surface has the form

$$w_1(x, y) = \left(x^2 - \frac{1}{4}\right)\left(y^2 - \frac{1}{4}\right),$$

i.e. we take the function w_1 to be equal to the polynomial of the lowest power which takes zero values on $\partial\Omega_1$ (it belongs to the class $\mathring{W}_2^2(\Omega_1)$).

We have solved the problem using the Bubnov–Galerkin method, with the basis functions $\tilde{w}_{kl}^{(1)}(x, y)(1 \leq k \leq l \leq n_1)$, $\tilde{w}_{klm}^{(2)}(x, y, z)(1 \leq k \leq l \leq n_1, 1 \leq m \leq n_2)$ in the form (3.141), (3.142). The symmetry conditions (3.137) are satisfied, and therefore the estimates (3.146), (3.165) can be applied. Taking into account the actual values of the parameters λ, ν, c, β and of the function $w_1(x, y)$, the estimates (3.146), (3.165), obtained analogously to the method used in Sect. 4.1 for the data set (4.1), read as follows:

$$\|w_n - w\|_{C(\overline{\Omega}_1 \times [0, T_1])} < \Phi_3(n_1, n_2, T), \tag{4.26}$$

$$\max_{0 \leq t \leq T} \|\theta_n(t) - \theta(t)\|_{L_2(\Omega_2)} < \Psi_3(n_1, n_2, T), \tag{4.27}$$

where $\{w, \theta\}$ is the exact solution of the problem, $\{w_n, \theta_n\}$ is its approximate solution obtained by the Bubnov–Galerkin method, and

$$\Phi_3(n_1, n_2, T) = \begin{cases} 3.6532 \times 10^{-3}\sqrt{\sigma_{n_1}^{**}\tilde{\sigma}_{n_1+1}^{(6)}} + 9.0151 \times 10^{-4}\sqrt{\sigma_{n_2+1}^{(4)}T} & \text{(Al)}, \\ 6.9961 \times 10^{-3}\sqrt{\sigma_{n_1}^{**}\tilde{\sigma}_{n_1+1}^{(6)}} + 7.1094 \times 10^{-4}\sqrt{\sigma_{n_2+1}^{(4)}T} & \text{(Cu)}, \end{cases}$$

$$\Psi_3(n_1, n_2, T) = \begin{cases} [6.7507 \times 10^{-8}\tilde{\sigma}_{n_1+1}^{(6)} + 1.4552 \times 10^{-8}\sigma_{n_2+1}^{(4)}T]^{1/2} & \text{(Al)}, \\ [1.5226 \times 10^{-7}\tilde{\sigma}_{n_1+1}^{(6)} + 5.5652 \times 10^{-9}\sigma_{n_2+1}^{(4)}T]^{1/2} & \text{(Cu)}. \end{cases}$$

Here $\sigma_m^{(4)}, \sigma_m^{**}, \tilde{\sigma}_m^{(6)}$ are the series of numbers defined by the relations (3.127), (3.152), (4.21), and $n = \{n_1, n_2\}$.

Let us consider the relative errors of the Bubnov–Galerkin method $\delta_{n,T}w$, $\delta_{n,T}\theta$, defined by the relations (4.13), (4.14), characterizing the accuracy of computation of the deflection function in terms of a uniformly distributed norm with respect to x, y, t, and of the temperature function in terms of a mean square norm with respect to the space variables. We use the notations

$$\Delta_{n,T}w = \frac{\Phi_3(n_1, n_2, T)}{\|w_n\|_{C(\overline{\Omega}_1) \times [0, T])}}, \tag{4.28}$$

$$\Delta_{n,T}\theta = \frac{\Psi_3(n_1, n_2, T)}{\|\theta_n(T)\|_{L_2(\Omega_2)}}. \tag{4.29}$$

In accordance with (4.26), (4.27), the inequalities $\delta_{n,T}w < \Delta_{n,T}w$, $\delta_{n,T}\theta < \Delta_{n,T}\theta$ hold, i.e. the quantities $\Delta_{n,T}w$, $\Delta_{n,T}\theta$ defined by (4.28), (4.29) can be considered as theoretically guaranteed estimates from above of the relative errors of the Bubnov–Galerkin method $\delta_{n,T}w$, $\delta_{n,T}\theta$, defined by (4.13), (4.14).

The problem has been solved numerically using the Bubnov–Galerkin method (the number of basis functions used was taken from intervals $1 \le n_1, n_2 \le 6$), and the Runge–Kutta method was applied to solve the ordinary differential equations. A computational experiment, similar to that presented in Sect. 4.1, was performed. During the computations, we found

$$T_1 = 0.25T^* = \begin{cases} 2.153 \times 10^{-2} & \text{(Al)}, \\ 4.124 \times 10^{-2} & \text{(Cu)}, \end{cases} \tag{4.30}$$

$$T_2 = \begin{cases} 0.175T^* = 1.507 \times 10^{-2} & \text{(Al)}, \\ 0.15T^* = 2.474 \times 10^{-2} & \text{(Cu)}, \end{cases} \tag{4.31}$$

where T^* is the maximum period of free vibration of the plate, given by (4.10).

In Tables 4.9 and 4.10, the theoretically guaranteed estimates from above of the relative error of a computation of the deflection function, for all pairs n_1, n_2 in the intervals $1 \le n_1, n_2 \le 6$, are presented. It can be seen from the tables that these estimates guarantee a relative error of the deflection function values of the order of 1–2% or less. Conclusions similar to those presented in the previous sections can be easily derived.

To conclude, in this case also the estimates of the error introduced by the Bubnov–Galerkin method given here allow us to detect and follow the distribution of a temperature field inside a vibrating plate that is due only to a deformation process, without heat sources.

Table 4.9 Estimates of relative error (as percentages) of deflection function calculations using the uniform norm $\Delta_{n,T_1}w$; problem (3.116)–(3.119); data set (4.3); $\lambda = 1/70$; $w_1(x, y) = \left(x^2 - \frac{1}{4}\right)\left(y^2 - \frac{1}{4}\right)$; Al ($\nu = 0.36$, $c = 2.92 \times 10^{-10}$, $\beta = 6.78 \times 10^{-3}$)

$n_2 \backslash n_1$	1	2	3	4	5	6
1	6.904	4.563	4.305	4.251	4.232	4.225
2	4.786	2.413	2.158	2.103	2.085	2.077
3	3.999	1.614	1.360	1.305	1.287	1.279
4	3.611	1.221	0.968	0.912	0.894	0.887
5	3.389	0.995	0.742	0.687	0.688	0.661
6	3.247	0.851	0.599	0.543	0.525	0.517

Table 4.10 Estimates of relative error (as percentages) of deflection function calculations using the uniform norm $\Delta_{n,T_1} w$; problem (3.116)–(3.119); data set (4.3); $\lambda = 1/70$; $w_1(x, y) = \left(x^2 - \frac{1}{4}\right)\left(y^2 - \frac{1}{4}\right)$; Cu ($\nu = 0.37, c = 1.08 \times 10^{-9}, \beta = 4.20 \times 10^{-3}$)

$n_2 \backslash n_1$	1	2	3	4	5	6
1	5.115	2.747	2.492	2.437	2.418	2.411
2	3.908	1.522	1.268	1.213	1.195	1.187
3	3.459	1.067	0.814	0.758	0.740	0.732
4	3.239	0.843	0.590	0.535	0.516	0.509
5	3.112	0.714	0.461	0.406	0.388	0.380
6	3.031	0.632	0.380	0.324	0.306	0.298

Chapter 5
Coupled Nonlinear Thermoelastic Problems

In this chapter, we formulate fundamental assumptions and relations similar to those presented in Chap. 2 for coupled linear thermoelasticity problems of shallow shells. A Timoshenko-type model including the inertial effect of rotation of shell elements is used. Both the generalized heat transfer equation and the equations governing vibration of a shell are formulated in Sect. 5.2, and then some special cases of these equations are analysed. In the next section, boundary and initial conditions are attached to the differential equations. In Sect. 5.4, the existence and uniqueness of a solution as well as the convergence of the Bubnov–Galerkin method, are rigorously discussed.

The question of solvability of the initial–boundary value problems of coupled thermoelasticity defined in Sects. 5.2 and 5.3, and the possibility of the approximate solution of these problems using the Bubnov–Galerkin method are addressed. The existence of a solution to the problem governed by the differential equations and initial and boundary conditions formulated earlier is proved. A few additional theorems are formulated.

5.1 Fundamental Relations and Assumptions

We consider again, as in Chap. 2, a shallow shell with a constant thickness h occupying an area Ω in plan, with a boundary $\partial\Omega_1$. We assume the shell material to be isotropic, homogeneous and elastic. We introduce an orthogonal system of coordinates x, y, z and, in the usual way [338], the coordinates x, y are tangent to the middle surface, and the coordinate z is normal to the middle surface, with its positive direction oriented toward the centre of curvature.

J. Awrejcewicz, V. A. Krysko, *Elastic and Thermoelastic Problems in Nonlinear Dynamics of Structural Members*, Scientific Computation,
https://doi.org/10.1007/978-3-030-37663-5_5

Let us denote the displacements of the middle surface along the axes x, y, z by $u(x, y, z)$, $v(x, y, z)$, $w(x, y, z)$, respectively. We denote the initial shell curvatures related to the coordinate axes x, y by k_x, k_y.

As the fundamental model, a Timoshenko-type model including the inertial effect of rotation of shell elements will be used [338]. It is assumed that shell fibres normal to the middle surface do not curve during a deformation process, but may cease to be normal to the middle surface. The deformation of fibres normal to the middle surface is defined by the condition of a plane stress state ($\sigma_{zz} = 0$). Within this model the following relations between deformations and displacements [338] hold:

$$\varepsilon_{11}^z = \varepsilon_{11} + z æ_{11}, \tag{5.1}$$

$$\varepsilon_{22}^z = \varepsilon_{22} + z æ_{22}, \tag{5.2}$$

$$\varepsilon_{12}^z = \varepsilon_{12} + z æ_{12} \qquad \left(-\frac{h}{2} \leq z \leq \frac{h}{2}\right), \tag{5.3}$$

$$\varepsilon_{13}^z = \varepsilon_{13}, \tag{5.4}$$

$$\varepsilon_{23}^z = \varepsilon_{23}, \tag{5.5}$$

where ε_{ij} $(i, j = 1, 2)$ are the tangential deformations of the middle surface, $\varepsilon_{13}, \varepsilon_{23}$ are the shear deformations, $æ_{ij}$ $(i, j = 1, 2)$ are the bending deformations, and Ψ_x, Ψ_y are the rotation angles in the xz and yz planes, respectively. We have

$$\varepsilon_{11} = \frac{\partial u}{\partial x} - k_x w + \frac{1}{2}\left(\frac{\partial w}{\partial x}\right)^2 \quad (x \leftrightarrow y), \tag{5.6}$$

$$\varepsilon_{12} = \frac{\partial u}{\partial y} + \frac{\partial v}{\partial x} + \frac{\partial w}{\partial x}\frac{\partial w}{\partial y}, \tag{5.7}$$

$$æ_{11} = \frac{\partial \Psi_x}{\partial x}, \quad æ_{22} = \frac{\partial \Psi_y}{\partial y}, \quad æ_{12} = \frac{\partial \Psi_x}{\partial y} + \frac{\partial \Psi_y}{\partial x}, \tag{5.8}$$

$$\varepsilon_{13} = \Psi_x + \frac{\partial w}{\partial x}, \quad \varepsilon_{23} = \Psi_y + \frac{\partial w}{\partial y}. \tag{5.9}$$

Note that the deformations $\varepsilon_{11}, \varepsilon_{22}, \varepsilon_{12}, \varepsilon_{23}, \varepsilon_{13}$ are assumed to be small in comparison with one, i.e. the derivatives of deflections are neglected.

Experiment shows that the deformation of a body is related to its internal heat capacity and vice versa. Therefore, one can believe that an appropriate modelling of both the deformation and temperature field will result in a better understanding of real processes inside bodies. This effect is particularly important in polymers [170], where interesting elastic waves occur.

Let us assume that the shell has a temperature T_0 in its unstrained, undeformed state. Its temperature then starts to change because of the action of surface and mass forces, internal heat sources and heat exchange with the surrounding medium. We denote by $\theta(x, y, z, t) = T_1(x, y, z, t) - T_0$ the temperature increase at the point of the shell point x, y, z at the time moment t; $T_1(x, y, z)$ is the absolute temperature at this point. We assume $\left|\frac{\theta}{T_0}\right| \ll 1$, i.e. that the change of temperature θ is so small that it does not affect the elastic and thermal constants of the shell material. The following notation will be used in what follows: E, Young modulus; ν, Poisson's ratio; α_T, coefficient of thermal expansion; ρ, density; λ_q, heat transfer coefficient; c_ε, specific heat transfer coefficient for a constant deformation. Assuming that the shell is in a local quasi-equilibrium condition [139] and taking into account the relations of irreversible thermodynamic processes for small values of deformation and of temperature increase (see [266]), the following Duhamel–Neumann relations can be obtained:

$$\varepsilon_{ij}^z = \frac{1+\nu}{E}\sigma_{ij} - \nu\frac{(\sigma_{11}+\sigma_{22})}{E}\delta_{ij} + \alpha_T Q\delta_{ij}, \quad i, j = 1, 2, 3,$$

$$\delta_{ij} = \begin{cases} 0, & i \neq j \\ 1, & i = j \end{cases}. \tag{5.10}$$

These relations establish the links between the stresses σ_{ij} and deformations ε_{ij}^z in the condition of a plane stress state ($\sigma_{33} = 0$). We take

$$\sigma_{xx} = \sigma_{11},\ \sigma_{yy} = \sigma_{22},\ \sigma_{zz} = \sigma_{33},\ \sigma_{xy} = \sigma_{12},\ \sigma_{xz} = \sigma_{13},\ \sigma_{yz} = \sigma_{23}.$$

Solving (5.1) for σ_{ij}, we obtain

$$\sigma_{11} = \frac{E}{1-\nu^2}(\varepsilon_{11}^z + \nu\varepsilon_{22}^z) - \frac{E}{1-\nu}\alpha_T Q, \tag{5.11}$$

$$\sigma_{22} = \frac{E}{1-\nu^2}(\varepsilon_{22}^z + \nu\varepsilon_{11}^z) - \frac{E}{1-\nu}\alpha_T Q, \tag{5.12}$$

$$\sigma_{12} = \frac{E}{2(1+\nu)}\varepsilon_{12}^z, \tag{5.13}$$

$$\sigma_{13} = \frac{E}{2(1+\nu)}\varepsilon_{13}^z, \tag{5.14}$$

$$\sigma_{23} = \frac{E}{2(1+\nu)}\varepsilon_{23}^z. \tag{5.15}$$

Integrating the stresses (5.11)–(5.15) with respect to z and taking into account (5.1), the following longitudinal and transverse forces in the middle surface are found:

$$N_x = \int_{-\frac{h}{2}}^{\frac{h}{2}} \sigma_{11}\, dz = \frac{Eh}{1-\nu^2}\left\{\frac{\partial u}{\partial x} - k_x w + \frac{1}{2}\left(\frac{\partial w}{\partial x}\right)^2 + \nu\left[\frac{\partial v}{\partial y} - k_y w + \frac{1}{2}\left(\frac{\partial w}{\partial y}\right)^2\right]\right\} - \frac{E\alpha_T}{1-\nu}\int_{-\frac{h}{2}}^{\frac{h}{2}} Q\, dz, \tag{5.16}$$

$$N_y = \int_{-\frac{h}{2}}^{\frac{h}{2}} \sigma_{22}\, dz = \frac{Eh}{1-\nu^2}\left\{\frac{\partial v}{\partial y} - k_y w + \frac{1}{2}\left(\frac{\partial w}{\partial y}\right)^2 + \nu\left[\frac{\partial u}{\partial x} - k_x w + \frac{1}{2}\left(\frac{\partial w}{\partial x}\right)^2\right]\right\} - \frac{E\alpha_T}{1-\nu}\int_{-\frac{h}{2}}^{\frac{h}{2}} Q\, dz, \tag{5.17}$$

$$S = \int_{-\frac{h}{2}}^{\frac{h}{2}} \sigma_{12}\, dz = \frac{Eh}{2(1+\nu)}\left(\frac{\partial u}{\partial y} + \frac{\partial v}{\partial x} + \frac{\partial w}{\partial x}\frac{\partial w}{\partial y}\right), \tag{5.18}$$

$$Q_x = \int_{-\frac{h}{2}}^{\frac{h}{2}} \sigma_{12}\, dz = k^2 \frac{Eh}{2(1+\nu)}\left(\Psi_x + \frac{\partial w}{\partial x}\right), \tag{5.19}$$

$$Q_y = \int_{-\frac{h}{2}}^{\frac{h}{2}} \sigma_{23}\, dz = k^2 \frac{Eh}{2(1+\nu)}\left(\Psi_y + \frac{\partial w}{\partial y}\right), \tag{5.20}$$

where

$$\frac{1}{k^2} = \frac{1}{h}\int_{-\frac{h}{2}}^{\frac{h}{2}} f^2(z)\, dz.$$

The function $f(z)$ characterizes the distribution of tangential stresses through the shell thickness. In order to obtain the moments, the stresses are integrated with weight z over the thickness:

$$M_x = \int_{-\frac{h}{2}}^{\frac{h}{2}} \sigma_{11} z\, dz = \frac{Eh^3}{12(1-\nu^2)} \left(\frac{\partial \Psi_x}{\partial x} + \nu \frac{\partial \Psi_y}{\partial y} \right) - \frac{E\alpha_T}{1-\nu} \int_{-\frac{h}{2}}^{\frac{h}{2}} zQ\, dz, \tag{5.21}$$

$$M_y = \int_{-\frac{h}{2}}^{\frac{h}{2}} \sigma_{22} z\, dz = \frac{Eh^3}{12(1-\nu^2)} \left(\frac{\partial \Psi_y}{\partial y} + \nu \frac{\partial \Psi_x}{\partial x} \right) - \frac{E\alpha_T}{1-\nu} \int_{-\frac{h}{2}}^{\frac{h}{2}} zQ\, dz, \tag{5.22}$$

$$H = \int_{-\frac{h}{2}}^{\frac{h}{2}} \sigma_{12} z\, dz = \frac{Eh^3}{12(1-\nu^2)} \frac{(1-\nu)}{2} \left(\frac{\partial \Psi_x}{\partial y} + \frac{\partial \Psi_y}{\partial x} \right). \tag{5.23}$$

5.2 Differential Equations

From the equation governing the entropy balance and taking into account the coupling between the components of heat flow and the thermodynamic forces associated with the Fourier law, we obtain the generalized heat transfer equation (see [266]):

$$c_\varepsilon \frac{\partial \theta}{\partial t} - \lambda_q \Delta \theta = -\underline{\frac{E\alpha_T T_0}{(1-2\nu)} \frac{\partial}{\partial t} (\varepsilon_{11}^z + \varepsilon_{22}^z + \varepsilon_{33}^z)} + g_2. \tag{5.24}$$

The underlined term links the temperature increase with the velocity of the volume change, and $g_2(x, y, z, t)$ is the quantity of heat transferred in a unit volume within a unit time interval.

In order to obtain the equations governing the vibration of a shell with respect to the displacements, we use the following differential equation, which governs the motion (including inertial forces) of the shell element $h\, dx\, dy$ in the x, y, z directions and the rotational inertia of the element with respect to x, y [338]:

$$\frac{\partial N_x}{\partial x} + \frac{\partial S}{\partial y} + p_1 - \rho h \frac{\partial^2 u}{\partial t^2} = 0, \tag{5.25}$$

$$\frac{\partial S}{\partial x} + \frac{\partial N_y}{\partial y} + p_2 - \rho h \frac{\partial^2 v}{\partial t^2} = 0, \tag{5.26}$$

$$\begin{aligned}&\frac{\partial Q_x}{\partial x}+\frac{\partial Q_y}{\partial y}+k_xN_x+k_yN_y+\frac{\partial}{\partial x}\left(N_x\frac{\partial w}{\partial x}+S\frac{\partial w}{\partial y}\right)\\&\quad+\frac{\partial}{\partial y}\left(S\frac{\partial w}{\partial x}+N_y\frac{\partial w}{\partial x}\right)+g_1-\rho h\frac{\partial^2 w}{\partial t^2}=0,\end{aligned} \tag{5.27}$$

$$\frac{\partial M_x}{\partial x}+\frac{\partial H}{\partial y}+Q_x-\rho\frac{h^3}{12}\frac{\partial^2\Psi_x}{\partial t^2}=0, \tag{5.28}$$

$$\frac{\partial H}{\partial x}+\frac{\partial M_y}{\partial y}+Q_y-\rho\frac{h^3}{12}\frac{\partial^2\Psi_y}{\partial t^2}=0, \tag{5.29}$$

where p_1, p_2, g_1 are the intensities of the external loads along the x, y, z axes; $\rho h\frac{\partial^2 u}{\partial t^2}dx\,dy$, $\rho h\frac{\partial^2 v}{\partial t^2}dx\,dy$, $\rho h\frac{\partial^2 w}{\partial t^2}dx\,dy$ are the components of the inertial force along the x, y, z axes; and $\rho\frac{h^3}{12}\frac{\partial^2\Psi_x}{\partial t^2}$, $\rho\frac{h^3}{12}\frac{\partial^2\Psi_y}{\partial t^2}$ are the moments of inertia around the x and y axes.

Substituting (5.16)–(5.23) into (5.25)–(5.29) and attaching the heat transfer equation (5.24), we obtain the full system of differential equations of thermoelasticity with respect to displacements, in the following form:

$$\begin{aligned}&\frac{Eh}{1-\nu^2}\frac{\partial}{\partial x}\left[\left(\frac{\partial u}{\partial x}-k_xw+\frac{1}{2}\left(\frac{\partial w}{\partial x}\right)^2\right)+\nu\left(\frac{\partial v}{\partial y}-k_yw+\frac{1}{2}\left(\frac{\partial w}{\partial y}\right)^2\right)\right]\\&\quad-\frac{E\alpha_T}{1-\nu}\frac{\partial}{\partial x}\left\{\int_{-\frac{h}{2}}^{\frac{h}{2}}\theta\,dz\right\}+\frac{Eh}{2(1+\nu)}\frac{\partial}{\partial y}\left[\frac{\partial u}{\partial y}+\frac{\partial v}{\partial x}+\frac{\partial w}{\partial x}\frac{\partial w}{\partial y}\right]\\&\quad+p_1-\rho h\frac{\partial^2 u}{\partial t^2}=0,\end{aligned} \tag{5.30}$$

$$\begin{aligned}&\frac{Eh}{1-\nu^2}\frac{\partial}{\partial y}\left[\left(\frac{\partial v}{\partial y}-k_yw+\frac{1}{2}\left(\frac{\partial w}{\partial y}\right)^2\right)+\nu\left(\frac{\partial u}{\partial x}-k_xw+\frac{1}{2}\left(\frac{\partial w}{\partial x}\right)^2\right)\right]\\&\quad-\frac{E\alpha_T}{1-\nu}\frac{\partial}{\partial y}\left\{\int_{-\frac{h}{2}}^{\frac{h}{2}}\theta\,dz\right\}+\frac{Eh}{2(1+\nu)}\frac{\partial}{\partial x}\left[\frac{\partial u}{\partial y}+\frac{\partial v}{\partial x}+\frac{\partial w}{\partial x}\frac{\partial w}{\partial y}\right]\\&\quad+p_2-\rho h\frac{\partial^2 v}{\partial t^2}=0,\end{aligned} \tag{5.31}$$

$$k^2\frac{Eh}{2(1+\nu)}\frac{\partial}{\partial x}\left[\Psi_x+\frac{\partial w}{\partial x}\right]+k^2\frac{Eh}{2(1+\nu)}\frac{\partial}{\partial y}\left[\Psi_y+\frac{\partial w}{\partial y}\right]+k_x\left[\frac{Eh}{1-\nu}\left\{\left(\frac{\partial u}{\partial x}\right.\right.\right.$$

$$-k_x w + \frac{1}{2}\left(\frac{\partial w}{\partial x}\right)^2\right) + \nu\left(\frac{\partial v}{\partial y} - k_y w + \frac{1}{2}\left(\frac{\partial w}{\partial y}\right)^2\right)\Bigg\} - \frac{E\alpha_t}{1-\nu}\int_{-\frac{h}{2}}^{\frac{h}{2}} \theta\, dz\Bigg]$$

$$+k_y\left[\frac{Eh}{1-\nu^2}\left\{\left(\frac{\partial v}{\partial y} - k_y w + \frac{1}{2}\left(\frac{\partial w}{\partial y}\right)^2\right) + \nu\left(\frac{\partial u}{\partial x} - k_x w + \frac{1}{2}\left(\frac{\partial w}{\partial x}\right)^2\right)\right\}\right.$$

$$\left. - \frac{E\alpha_T}{1-\nu}\int_{-\frac{h}{2}}^{\frac{h}{2}} \theta\, dz\right] + \frac{\partial}{\partial x}\left[\frac{\partial w}{\partial x}\left\{\frac{Eh}{1-\nu^2}\left(\left[\frac{\partial u}{\partial x} - k_x w + \frac{1}{2}\left(\frac{\partial w}{\partial x}\right)^2\right]\right.\right.\right.$$

$$\left.\left. + \nu\left[\frac{\partial v}{\partial y} - k_y w + \frac{1}{2}\left(\frac{\partial w}{\partial y}\right)^2\right]\right) - \frac{E\alpha_T}{1-\nu}\int_{-\frac{h}{2}}^{\frac{h}{2}} \theta\, dz\right\}$$

$$\left. + \frac{\partial w}{\partial y}\left\{\frac{Eh}{2(1+\nu)}\left[\frac{\partial u}{\partial y} + \frac{\partial v}{\partial x} + \frac{\partial w}{\partial x}\frac{\partial w}{\partial y}\right]\right\}\right]$$

$$+\frac{\partial}{\partial y}\left[\frac{\partial w}{\partial y}\left\{\frac{Eh}{1-\nu^2}\left(\left[\frac{\partial v}{\partial y} - k_y w + \frac{1}{2}\left(\frac{\partial w}{\partial y}\right)^2\right] + \nu\left[\frac{\partial u}{\partial x} - k_x w + \frac{1}{2}\left(\frac{\partial w}{\partial x}\right)^2\right]\right)\right.\right.$$

$$\left.\left. - \frac{E\alpha_T}{1-\nu}\int_{-\frac{h}{2}}^{\frac{h}{2}} \theta\, dz\right\} + \frac{\partial w}{\partial x}\left\{\frac{Eh}{2(1+\nu)}\left[\frac{\partial u}{\partial y} + \frac{\partial v}{\partial x} + \frac{\partial w}{\partial x}\frac{\partial w}{\partial y}\right]\right\}\right]$$

$$+g_1 - \rho h\frac{\partial^2 w}{\partial t^2} - 0, \tag{5.32}$$

$$D\frac{\partial}{\partial x}\left[\frac{\partial \Psi_x}{\partial x} + \nu\frac{\partial \Psi_y}{\partial y}\right] - \frac{E\alpha_T}{1-\nu}\frac{\partial}{\partial x}\left\{\int_{-\frac{h}{2}}^{\frac{h}{2}} \theta\, dz\right\} + D\frac{1-\nu}{2}\frac{\partial}{\partial y}\left[\frac{\partial \Psi_x}{\partial y}\right.$$

$$\left. + \frac{\partial \Psi_y}{\partial x}\right] - k^2\frac{Eh}{2(1+\nu)}\left(\Psi_x + \frac{\partial w}{\partial x}\right) - \rho\frac{h^3}{12}\frac{\partial^2 \Psi_x}{\partial t^2} = 0, \tag{5.33}$$

$$D\frac{\partial}{\partial y}\left[\frac{\partial \Psi_y}{\partial y} + \nu\frac{\partial \Psi_x}{\partial x}\right] - \frac{E\alpha_T}{1-\nu}\frac{\partial}{\partial y}\left\{\int_{-\frac{h}{2}}^{\frac{h}{2}} \theta\, dz\right\} + D\frac{1-\nu}{2}\frac{\partial}{\partial x}\left[\frac{\partial \Psi_x}{\partial y}\right.$$

$$\left. + \frac{\partial \Psi_y}{\partial x}\right] - k^2\frac{Eh}{2(1+\nu)}\left(\Psi_y + \frac{\partial w}{\partial y}\right) - \rho\frac{h^3}{12}\frac{\partial^2 \Psi_y}{\partial t^2} = 0, \tag{5.34}$$

$$C_0 \frac{\partial \theta}{\partial t} - \lambda_q \Delta \theta = -\frac{E \alpha_T T_0}{1-\nu} \frac{\partial}{\partial t} \left[\frac{\partial u}{\partial x} - k_x w + \frac{1}{2} \left(\frac{\partial w}{\partial x} \right)^2 \right.$$

$$\left. + z \frac{\partial \Psi_x}{\partial x} + \frac{\partial v}{\partial y} - k_y w + \frac{1}{2} \left(\frac{\partial w}{\partial y} \right)^2 + z \frac{\partial \Psi_y}{\partial y} \right] + g_2, \tag{5.35}$$

$$C_0 = C_\varepsilon + \frac{E \alpha_T^2 T_0 (1+\nu)}{(1-\nu)(1-2\nu)}, \quad D = \frac{E h^3}{12(1-\nu^2)}.$$

The system of differential equations (5.30)–(5.35) governs the coupled dynamical problem of thermoelasticity taking into account the geometric nonlinearity [1]. Let us emphasize that during the derivation of the equations, no limitations on the temperature distribution through the shell thickness have been assumed, and therefore a system of equations with different dimensions has been obtained. The variable θ occurring in the heat transfer equation (a parabolic-type equation) depends on the three spatial variables x, y, z and the time t. The variables Ψ_x, Ψ_y, w, u, v occurring in the system of equations governing the motion of a shell element (a hyperbolic-type equation) depend on the two spatial variables x, y and the time t.

Let us now consider some special cases of the equations above. First, we do not take transverse shear effects into account, i.e. we consider the coupled dynamical problem of thermoelasticity within the Kirchhoff–Love model. In this case the rotation angles Ψ_x, Ψ_y of the normal to the middle surface have the following form [338]:

$$\Psi_x = -\frac{\partial w}{\partial x}, \quad \Psi_y = -\frac{\partial w}{\partial y}. \tag{5.36}$$

The transverse forces and moments now have the form

$$Q_x = \frac{\partial M_x}{\partial x} + \frac{\partial H}{\partial y} - \rho \frac{h^3}{12} \frac{\partial^2}{\partial t^2} \left(\frac{\partial w}{\partial x} \right), \tag{5.37}$$

$$Q_y = \frac{\partial M_y}{\partial y} + \frac{\partial H}{\partial x} - \rho \frac{h^3}{12} \frac{\partial^2}{\partial t^2} \left(\frac{\partial w}{\partial y} \right), \tag{5.38}$$

$$M_x = -D \left[\frac{\partial^2 w}{\partial x^2} + \nu \frac{\partial^2 w}{\partial y^2} \right] - \frac{E \alpha_T}{1-\nu} \int_{-\frac{h}{2}}^{\frac{h}{2}} \theta z \, dz, \tag{5.39}$$

$$M_y = -D \left[\frac{\partial^2 w}{\partial y^2} + \nu \frac{\partial^2 w}{\partial x^2} \right] - \frac{E \alpha_T}{1-\nu} \int_{-\frac{h}{2}}^{\frac{h}{2}} \theta z \, dz, \tag{5.40}$$

$$H = -D(1-\nu) \frac{\partial^2 w}{\partial x \partial y}. \tag{5.41}$$

According to (5.36)–(5.41), the equations governing the motion of a shell element have now the form

$$\frac{\partial N_x}{\partial x}+\frac{\partial S}{\partial y}+p_1-\rho\frac{h^3}{12}\frac{\partial^2 u}{\partial t^2}=0,$$

$$\frac{\partial S}{\partial x}+\frac{\partial N_y}{\partial y}+p_2-\rho\frac{h^3}{12}\frac{\partial^2 v}{\partial t^2}=0,$$

$$\frac{\partial^2 M_x}{\partial x^2}+2\frac{\partial^2 H}{\partial x\,\partial y}+\frac{\partial^2 M_y}{\partial y^2}+k_x N_x+k_y N_y$$

$$+\frac{\partial}{\partial x}\left(N_x\frac{\partial w}{\partial x}+S\frac{\partial w}{\partial y}\right)+\frac{\partial}{\partial y}\left(S\frac{\partial w}{\partial x}+N_y\frac{\partial w}{\partial y}\right)$$

$$+g_1+\rho\frac{h^3}{12}\frac{\partial^2}{\partial t^2}(\Delta w)-\rho h\frac{\partial^2 w}{\partial t^2}=0. \qquad (5.42)$$

The volume deformation is governed by the following equation:

$$\varepsilon_{11}^z+\varepsilon_{22}^z+\varepsilon_{33}^z=\left[\frac{\partial u}{\partial x}-k_x w+\frac{1}{2}\left(\frac{\partial w}{\partial x}\right)^2+\frac{\partial v}{\partial y}-k_y w\right.$$

$$\left.+\frac{1}{2}\left(\frac{\partial w}{\partial y}\right)^2-z\left(\frac{\partial^2 w}{\partial x^2}+\frac{\partial^2 w}{\partial y^2}\right)\right]\frac{1-2\nu}{1-\nu}+\frac{1+\nu}{1-\nu}\alpha_T\theta. \qquad (5.43)$$

Substituting (5.16)–(5.18) and (5.39)–(5.41) into (5.42), attaching the generalized equation of heat transfer (5.24) and taking into account (5.43), the following system of equations is obtained:

$$\frac{Eh}{1-\nu^2}\frac{\partial}{\partial x}\left[\left(\frac{\partial u}{\partial x}-k_x w+\frac{1}{2}\left(\frac{\partial w}{\partial x}\right)^2\right)+\nu\left(\frac{\partial v}{\partial y}-k_y w+\frac{1}{2}\left(\frac{\partial w}{\partial y}\right)^2\right)\right]$$

$$-\frac{E\alpha_T}{1-\nu}\frac{\partial}{\partial x}\left\{\int_{-\frac{h}{2}}^{\frac{h}{2}}\theta\,dz\right\}+\frac{Eh}{2(1+\nu)}\frac{\partial}{\partial y}\left[\frac{\partial u}{\partial y}+\frac{\partial v}{\partial x}+\frac{\partial w}{\partial x}\frac{\partial w}{\partial y}\right]$$

$$+p_1-\rho h\frac{\partial^2 u}{\partial t^2}=0, \qquad (5.44)$$

$$\frac{Eh}{1-\nu^2}\frac{\partial}{\partial y}\left[\left(\frac{\partial v}{\partial y}-k_y w+\frac{1}{2}\left(\frac{\partial w}{\partial y}\right)^2\right)+\nu\left(\frac{\partial u}{\partial x}-k_x w+\frac{1}{2}\left(\frac{\partial w}{\partial x}\right)^2\right)\right]$$

$$-\frac{E\alpha_T}{1-\nu}\frac{\partial}{\partial y}\left\{\int\limits_{-\frac{h}{2}}^{\frac{h}{2}}\theta\,dz\right\}+\frac{Eh}{2(1+\nu)}\frac{\partial}{\partial x}\left[\frac{\partial u}{\partial y}+\frac{\partial v}{\partial x}+\frac{\partial w}{\partial x}\frac{\partial w}{\partial y}\right]$$

$$+p_2-\rho h\frac{\partial^2 v}{\partial t^2}=0, \tag{5.45}$$

$$-D\frac{\partial^2}{\partial x^2}\left[\frac{\partial^2 w}{\partial x^2}+\nu\frac{\partial^2 w}{\partial y^2}\right]-2D(1-\nu)\frac{\partial^2}{\partial x\partial y}\left[\frac{\partial^2 w}{\partial x\partial y}\right]-D\frac{\partial^2}{\partial y^2}\left[\frac{\partial^2 w}{\partial y^2}+\nu\frac{\partial^2 w}{\partial x^2}\right]$$

$$-\frac{E\alpha_T}{1-\nu}\Delta\left\{\int\limits_{-\frac{h}{2}}^{\frac{h}{2}}\theta z\,dz\right\}+\rho\frac{h^3}{12}\frac{\partial^2}{\partial t^2}(\Delta w)$$

$$+k_x\left[\frac{Eh}{1-\nu}\left\{\left(\frac{\partial u}{\partial x}-k_x w+\frac{1}{2}\left(\frac{\partial w}{\partial x}\right)^2\right)+\nu\left(\frac{\partial v}{\partial y}-k_y w+\frac{1}{2}\left(\frac{\partial w}{\partial y}\right)^2\right)\right\}\right.$$

$$\left.-\frac{E\alpha_t}{1-\nu}\int\limits_{-\frac{h}{2}}^{\frac{h}{2}}\theta\,dz\right]+k_y\left[\frac{Eh}{1-\nu^2}\left\{\left(\frac{\partial v}{\partial y}-k_y w+\frac{1}{2}\left(\frac{\partial w}{\partial y}\right)^2\right)\right.\right.$$

$$\left.\left.+\nu\left(\frac{\partial u}{\partial x}-k_x w+\frac{1}{2}\left(\frac{\partial w}{\partial x}\right)^2\right)\right\}-\frac{E\alpha_T}{1-\nu}\int\limits_{-\frac{h}{2}}^{\frac{h}{2}}\theta dz\right]$$

$$+\frac{\partial}{\partial x}\left[\frac{\partial w}{\partial x}\left\{\frac{Eh}{1-\nu^2}\left(\left[\frac{\partial u}{\partial x}-k_x w+\frac{1}{2}\left(\frac{\partial w}{\partial x}\right)^2\right]+\nu\left[\frac{\partial v}{\partial y}-k_y w+\frac{1}{2}\left(\frac{\partial w}{\partial y}\right)^2\right]\right)\right.\right.$$

$$\left.\left.-\frac{E\alpha_T}{1-\nu}\int\limits_{-\frac{h}{2}}^{\frac{h}{2}}\theta\,dz\right\}+\frac{\partial w}{\partial y}\left\{\frac{Eh}{2(1+\nu)}\left[\frac{\partial u}{\partial y}+\frac{\partial v}{\partial x}+\frac{\partial w}{\partial x}\frac{\partial w}{\partial y}\right]\right\}\right]$$

$$+\frac{\partial}{\partial y}\left[\frac{\partial w}{\partial y}\left\{\frac{Eh}{1-\nu^2}\left(\left[\frac{\partial v}{\partial y}-k_y w+\frac{1}{2}\left(\frac{\partial w}{\partial y}\right)^2\right]+\nu\left[\frac{\partial u}{\partial x}-k_x w+\frac{1}{2}\left(\frac{\partial w}{\partial x}\right)^2\right]\right)\right.\right.$$

$$\left.\left.-\frac{E\alpha_T}{1-\nu}\int\limits_{-\frac{h}{2}}^{\frac{h}{2}}\theta\,dz\right\}+\frac{\partial w}{\partial x}\left\{\frac{Eh}{2(1+\nu)}\left[\frac{\partial u}{\partial y}+\frac{\partial v}{\partial x}+\frac{\partial w}{\partial x}\frac{\partial w}{\partial y}\right]\right\}\right]$$

$$+g_1-\rho h\frac{\partial^2 w}{\partial t^2}=0, \tag{5.46}$$

$$C_0\frac{\partial\theta}{\partial t}-\lambda_q\Delta\theta=-\frac{E\alpha_T T_0}{1-\nu}\frac{\partial}{\partial t}\left[\frac{\partial u}{\partial x}-k_x w+\frac{1}{2}\left(\frac{\partial w}{\partial x}\right)^2\right.$$
$$\left.-z\left(\frac{\partial^2 w}{\partial x^2}+\frac{\partial^2 w}{\partial y^2}\right)+\frac{\partial v}{\partial y}-k_y w+\frac{1}{2}\left(\frac{\partial w}{\partial y}\right)^2\right]+g_2. \tag{5.47}$$

Therefore, the system of differential equations (5.44)–(5.47) governs the coupled problem of thermoelasticity in the framework of the Kirchhoff–Love model, taking into account the geometrical nonlinearity and rotational inertia of the shell elements. It is clear that in the system (5.44)–(5.47), different types of equations (parabolic and hyperbolic) with different dimensions appear again.

The equations for linear problems discussed in Chap. 2 can be derived from the above system as a particular case.

5.3 Boundary and Initial Conditions

The differential equations given above must be completed by initial and boundary conditions. We need to include the initial state of the shell and to formulate the conditions describing its influence on the surrounding medium. For the mathematical model of thermoelastic processes in shells used here, we take into account only mechanical and thermal phenomena between the shell and the neighbouring medium.

As the initial conditions, we take the distribution (for $t=0$) of the displacements w, u, v, the angles of rotation of the normal Ψ_x, Ψ_y, the velocities of those displacements and angles $\frac{\partial w}{\partial t}, \frac{\partial u}{\partial t}, \frac{\partial v}{\partial t}, \frac{\partial \Psi_x}{\partial t}, \frac{\partial \Psi_y}{\partial t}$ and the temperature increase θ (the last condition is equivalent to introducing the temperature T_1) inside the shell.

$$w|_{t=0}=\varphi_1(x,y),\quad u|_{t=0}=\varphi_2(x,y),$$
$$v|_{t=0}=\varphi_3(x,y),\quad \theta|_{t=0}=\varphi_4(x,y,z), \tag{5.48}$$

$$\Psi_x|_{t=0}=\varphi_5(x,y),\quad \Psi_y|_{t=0}=\varphi_6(x,y), \tag{5.49}$$

$$\left.\frac{\partial w}{\partial t}\right|_{t=0}=\Psi_1(x,y),\quad \left.\frac{\partial u}{\partial t}\right|_{t=0}=\Psi_2(x,y),\quad \left.\frac{\partial v}{\partial t}\right|_{t=0}=\Psi_3(x,y), \tag{5.50}$$

$$\left.\frac{\partial \Psi_x}{\partial t}\right|_{t=0}=\Psi_4(x,y),\quad \left.\frac{\partial \Psi_y}{\partial t}\right|_{t=0}=\Psi_5(x,y). \tag{5.51}$$

As an example of the mechanical conditions (boundary conditions) on the shell boundary, we present the following set, which will be used later (the formulation for other boundary conditions is known and can be found in the series of monographs [198, 338, 339]):

$$w|_{\partial\Omega_1} = f_1(x, y, t), \qquad \left.\frac{\partial w}{\partial n}\right|_{\partial\Omega_1} = f_2(x, y, t), \tag{5.52}$$

$$u|_{\partial\Omega_1} = f_3(x, y, t), \quad v|_{\partial\Omega_1} = f_4(x, y, t), \tag{5.53}$$

$$\Psi_x|_{\partial\Omega_1} = f_5(x, y, t), \quad \Psi_y|_{\partial\Omega_1} = f_6(x, y, t), \tag{5.54}$$

$$(x, y) \in \partial\Omega_1, \ t \in [0, T],$$

where $[0, T]$ denotes the time interval during which the shell vibrations are investigated; $f_i(x, y, t)$, $i = 1, \ldots, 6$, are known functions of the coordinates and time; and $\frac{\partial}{\partial n}$ is the differential operator acting along an external normal to $\partial\Omega_1$. For $f_i = 0$, $i = 1, \ldots, 6$, these conditions correspond to a shell clamped on the contour $\partial\Omega_1$.

As the heat transfer conditions, one of the following [188] can be used:

1. A temperature distribution is assumed on the shell surface $\partial\Omega_2$:

$$\theta(x, y, z, t) = f_7(x, y, z, t), \quad t \in [0, T], \quad (x, y, z) \in \partial\Omega_2. \tag{5.55}$$

2. The density of the heat flow through the surface $\partial\Omega_2$ is given:

$$\frac{\partial\theta(x, y, z, t)}{\partial n} = f_8(x, y, z, t), \quad t \in [0, T], \quad (x, y, z) \in \partial\Omega_2,$$

   where $\frac{\partial}{\partial n}$ denotes the differential operator along an external normal to $\partial\Omega_2$.
3. A condition of free heat exchange with the surrounding medium on the surface $\partial\Omega_2$ is specified:

$$\left(\frac{\partial}{\partial n} + \alpha\right)\theta(x, y, z, t) = f_9(x, y, z, t), \quad t \in [0, T], \quad (x, y, z) \in \partial\Omega_2,$$

   where $\alpha = const$, $f_i(x, y, z, t)$, $i = 7, 8, 9$, are known functions of the coordinates and time.

5.4 On the Existence and Uniqueness of a Solution and on the Convergence of the Bubnov–Galerkin Method

We introduce the following notations and definitions.

Ω_1 is a bounded volume of the Euclidean space E_2 ($\overline{x} = (x, y)$ denotes a point of E_2), i.e. Ω_1 is a coupled open set; $\overline{\Omega_1} = \Omega_1 \cup \partial\Omega_1$, and $\partial\Omega_1$ is the boundary of Ω_1. $Q_1 = \Omega_1 \times (0, T)$ is a cylinder of height $T > 0$ in the space $R_3 = E_2 \times (-\infty < t < +\infty)$. $G = \partial\Omega_1 \times [0, T]$ is the surface of the cylinder Q_1. $\Omega_2 = \Omega_1 \times \left(-\frac{h}{2}, \frac{h}{2}\right)$ is a cylinder of height h in the space E_3 ($\overline{x} = (x, y, z)$ is a point of E_3). $\overline{\Omega_2} = \Omega_2 \cup \partial\Omega_2$, where $\partial\Omega_2$ is the boundary surface of Ω_2. $Q_2 = \Omega_2 \times (0, T)$ is a cylinder of height $T > 0$ in the space $R_4 = E_3 \times \{-\infty < t < +\infty\}$. $S = \partial\Omega_2 \times [0, T]$ is the surface of the cylinder Q_2.

In our investigations of nonstationary problems, we use the following functional space: $L^p(0, T; X)$ is the space of functions $f(t)$ with values in the space X, measured with respect to the Lebesgue measure dt of the interval $[0, T]$ and possessing the finite norm

$$\|f\| = \left[\int_0^T \|f\|_x^p\right]^{1/p},$$

where $\|f\|_x$ is the norm of elements in X. In this section, we use as X the spaces $\overset{\circ}{W}{}_n^m(\Omega_2)$, $\overset{\circ}{W}{}_n^m(\Omega_1)$, $W_n^m(\Omega_1)$, $L_n(\Omega_1)$, $L_n(\Omega_2)$.

The aim of this section is to investigate the solvability of the initial–boundary value problems of coupled thermoelasticity which were defined in the Sects. 5.2 and 5.3 and the possibility of their approximate solution using the Bubnov–Galerkin method. Firstly, we prove the existence of a solution to the problem defined by the equations (5.44)–(5.47), the initial conditions (5.48)–(5.51) and the boundary conditions (5.52)–(5.54). Using the notation defined above, the initial and boundary conditions have the following form (here we consider only homogeneous boundary conditions, i.e. $f_i = 0$, $i = 1, 2, 3, 4, 7$):

$$w|_{t=0} = \varphi_1(x, y), \quad u|_{t=0} = \varphi_2(x, y), \quad v|_{t=0} = \varphi_3(x, y),$$

$$(x, y) \in \partial\Omega_1, \quad t \in [0, T],$$

$$\left.\frac{\partial w}{\partial t}\right|_{t=0} = \Psi_1(x, y), \quad \left.\frac{\partial u}{\partial t}\right|_{t=0} = \Psi_2(x, y), \quad \left.\frac{\partial v}{\partial t}\right|_{t=0} = \Psi_3(x, y), \tag{5.56}$$

$$w|_G = \left.\frac{\partial w}{\partial n}\right|_G = u|_G = v|_G = 0, \tag{5.57}$$

$$\theta|_{t=0} = \varphi_4(x, y, z), \quad (x, y, z) \in \Omega_2, \tag{5.58}$$

$$\theta|_S = 0. \tag{5.59}$$

Theorem 5.1 *Let $\partial\Omega_1$ be a sufficiently smooth boundary and let $p_1, p_2, g_1 \in L^2(0, T; (L_2(\Omega_1)),\ g_2 \in L^2(0, T; L_2(\Omega_2)),\ \varphi_1 \in \mathring{W}_2^2(\Omega_1),\ \varphi_2, \varphi_3 \in \mathring{W}_2^1(\Omega_1),\ \varphi_4 \in \mathring{W}_2^1(\Omega_2),\ \Psi_1 \in \mathring{W}_2^1(\Omega_1),\ \Psi_2, \Psi_3 \in L_2(\Omega_1)$. Then:*

1. *There exists at least one set of four functions $\{w, u, v, \theta\}$ that are a solution to the problem (5.44)–(5.47), (5.56)–(5.59), and the following relations hold:*

$$w \in L^2(0, T; \mathring{W}_2^2(\Omega_1)), \quad \frac{\partial w}{\partial t} \in L^2(0, T; \mathring{W}_2^1(\Omega_1)), \tag{5.60}$$

$$u, v \in L^2(0, T; \mathring{W}_2^1(\Omega_1)), \quad \frac{\partial u}{\partial t}, \frac{\partial v}{\partial t} \in L^2(0, T; L_2(\Omega_1)), \tag{5.61}$$

$$\theta \in L^2(0, T; \mathring{W}_2^1(\Omega_2)). \tag{5.62}$$

2. *An approximate solution to the problem stated above can be found with the help of the Bubnov–Galerkin method in the form*

$$w_n = \sum_{i=1}^{n} g_{1i}(t)\chi_{1i}(x, y), \quad u_n = \sum_{i=1}^{n} g_{2i}(t)\chi_{2i}(x, y),$$
$$v_n = \sum_{i=1}^{n} g_{3i}(t)\chi_{2i}(x, y), \quad \theta_n = \sum_{i=1}^{n} g_{4i}(t)\chi_{3i}(x, y). \tag{5.63}$$

Here $\{\chi_{1i}\}$ is a basis system in $\mathring{W}_2^2(\Omega_1)$, orthonormalized with the norm $\mathring{W}_2^1(\Omega_1)$, where the norm (in this case) has the form

$$\|a\|_{\mathring{W}_2^1(\Omega_1)} = \left[\int_{\Omega_1} \left\{\rho h a^2 + \rho\frac{h^3}{12}\left[(\frac{\partial a}{\partial x})^2 + (\frac{\partial a}{\partial y})^2\right]\right\} d\Omega_1\right]^{1/2},$$

for an arbitrary element $a \in \mathring{W}_2^1(\Omega_1)$; $\{\chi_{2i}\}$ is a basis system in $\mathring{W}_2^1(\Omega_1)$, orthonormalized in $L_2(\Omega_1)$; $\{\chi_{3i}\}$ is a basis system in $\mathring{W}_2^1(\Omega_2)$, orthonormalized in $L_2(\Omega_2)$; and for arbitrary i, $\chi_{3i} = a_i(x, y)b_i(z)$ holds. In order to define the unknown $g_{ij}(t)$, $j = 1, \ldots, 4$, we obtain a system of ODEs solvable for arbitrary n on the whole interval $[0, T]$.

3. *The whole set of approximate solutions $\{w_n, u_n, v_n, \theta_n\}$ is weakly compact in the spaces corresponding to (5.60)–(5.62).*

Note that sufficient smoothness of $\partial\Omega_1$ is required by the theorems on inclusion and on representation of the basis functions in $\mathring{W}_2^1(\Omega_2)$ which are used in the proof.

Proof The proof of the above theorem will be constructed here using the method of compactness (in the sense given in [226]). Fundamental investigations of this method and its application to problems of shells were performed by Vorovich and Morozov [252, 253, 343].

Step 1 Construction of the approximate solution using the Bubnov–Galerkin method.

Substituting the functions (5.63) into the system (5.44)–(5.47), we define the unknown $g_{ij}(t)$, $1, \dots, 4$, by the following ordinary differential equations:

$$
\begin{aligned}
-\frac{Eh}{1-\nu^2}&\left(\left[\left(\frac{\partial u_n}{\partial x} - k_x w_n + \frac{1}{2}\left(\frac{\partial w_n}{\partial x}\right)^2\right) + \nu\left(\frac{\partial v_n}{\partial y} - k_y w_n + \frac{1}{2}\left(\frac{\partial w_n}{\partial y}\right)^2\right)\right],\right. \\
&\left.\frac{\partial \chi_{2l}}{\partial x}\right)_{L_2(\Omega_1)} + \frac{E\alpha_T}{1-\nu}\left(\int_{-\frac{h}{2}}^{\frac{h}{2}} \theta_n\, dz, \frac{\partial \chi_{2l}}{\partial x}\right)_{L_2(\Omega_1)} \\
&-\frac{Eh}{2(1+\nu)}\left(\left[\frac{\partial u_n}{\partial y} + \frac{\partial v_n}{\partial x} + \frac{\partial w_n}{\partial x}\frac{\partial w_n}{\partial y}\right], \frac{\partial \chi_{2l}}{\partial x}\right)_{L_2(\Omega_1)} \\
&+(p_1, \chi_{2l})_{L_2(\Omega_1)} = \rho h\left(\frac{\partial^2 u_n}{\partial t^2}, \chi_{2l}\right)_{L_2(\Omega_1)},
\end{aligned}
\tag{5.64}
$$

$$
\begin{aligned}
-\frac{Eh}{1-\nu^2}&\left(\left[\left(\frac{\partial v_n}{\partial y} - k_y w_n + \frac{1}{2}\left(\frac{\partial w_n}{\partial y}\right)^2\right) + \nu\left(\frac{\partial u_n}{\partial x} - k_x w_n + \frac{1}{2}\left(\frac{\partial w_n}{\partial x}\right)^2\right)\right],\right. \\
&\left.\frac{\partial \chi_{2l}}{\partial y}\right)_{L_2(\Omega_1)} + \frac{E\alpha_T}{1-\nu}\left(\int_{-\frac{h}{2}}^{\frac{h}{2}} \theta_n\, dz, \frac{\partial \chi_{2l}}{\partial y}\right)_{L_2(\Omega_1)} \\
&-\frac{Eh}{2(1+\nu)}\left(\left[\frac{\partial u_n}{\partial y} + \frac{\partial v_n}{\partial x} + \frac{\partial w_n}{\partial x}\frac{\partial w_n}{\partial y}\right], \frac{\partial \chi_{2l}}{\partial x}\right)_{L_2(\Omega_1)} \\
&+(p_1, \chi_{2l})_{L_2(\Omega_1)} = \rho h\left(\frac{\partial^2 v_n}{\partial t^2}, \chi_{2l}\right)_{L_2(\Omega_1)},
\end{aligned}
\tag{5.65}
$$

$$
\begin{aligned}
-D&\left(\left[\frac{\partial^2 w_n}{\partial x^2} + \nu\frac{\partial^2 w_n}{\partial y^2}\right], \frac{\partial^2 \chi_{1l}}{\partial x^2}\right)_{L_2(\Omega_1)} - 2D(1-\nu)\left(\frac{\partial^2 w_n}{\partial x \partial y}, \frac{\partial^2 \chi_{1l}}{\partial x \partial y}\right)_{L_2(\Omega_1)} \\
&-D\left(\left[\frac{\partial^2 w_n}{\partial y^2} + \nu\frac{\partial^2 w_n}{\partial x^2}\right], \frac{\partial^2 \chi_{1l}}{\partial x^2}\right)_{L_2(\Omega_1)} \\
&-\frac{E\alpha_T}{1-\nu}\left(\int_{-\frac{h}{2}}^{\frac{h}{2}} \theta_n z\, dz, \left[\frac{\partial^2 \chi_{1l}}{\partial x^2} + \frac{\partial^2 \chi_{1l}}{\partial y^2}\right]\right)_{L_2(\Omega_1)} \\
&+\frac{Eh}{1-\nu^2}\left(\left[\left(\frac{\partial u_n}{\partial x} - k_x w_n + \frac{1}{2}\left(\frac{\partial w_n}{\partial x}\right)^2\right) + \nu\left(\frac{\partial v_n}{\partial y} - k_y w_n\right.\right.\right.
\end{aligned}
$$

$$+\frac{1}{2}\left(\frac{\partial w_n}{\partial y}\right)^2\Bigg)\Bigg], k_x\chi_{1l}\Bigg)_{L_2(\Omega_1)} - \frac{E\alpha_T}{1-\nu}\left(\int\limits_{-\frac{h}{2}}^{\frac{h}{2}}\theta_n\,dz, [k_x+k_y]\chi_{1l}\right)_{L_2(\Omega_1)}$$

$$+\frac{Eh}{1-\nu^2}\Bigg(\Bigg[\left(\frac{\partial v_n}{\partial y}-k_y w_n+\frac{1}{2}\left(\frac{\partial w_n}{\partial y}\right)^2\right)+\nu\Bigg(\frac{\partial u_n}{\partial x}-k_x w_n$$

$$+\frac{1}{2}\left(\frac{\partial w_n}{\partial x}\right)^2\Bigg)\Bigg], k_y\chi_{1l}\Bigg)_{L_2(\Omega_1)} - \frac{Eh}{1-\nu^2}\Bigg(\frac{\partial w_n}{\partial x}\Bigg[\frac{\partial u_n}{\partial x}-k_x w_n+\frac{1}{2}\left(\frac{\partial w_n}{\partial x}\right)^2$$

$$+\nu\left(\frac{\partial v_n}{\partial y}-k_y w_n+\frac{1}{2}\left(\frac{\partial w_n}{\partial y}\right)^2\right)\Bigg], \frac{\partial \chi_{1l}}{\partial x}\Bigg)_{L_2(\Omega_1)}$$

$$-\frac{Eh}{2(1+\nu)}\left(\frac{\partial w_n}{\partial y}\left[\frac{\partial u_n}{\partial y}+\frac{\partial v_n}{\partial x}+\frac{\partial w_n}{\partial x}\frac{\partial w_n}{\partial y}\right], \frac{\partial \chi_{1l}}{\partial x}\right)_{L_2(\Omega_1)}$$

$$-\frac{Eh}{1-\nu^2}\Bigg(\frac{\partial w_n}{\partial y}\Bigg[\frac{\partial v_n}{\partial y}-k_y w_n+\frac{1}{2}\left(\frac{\partial w_n}{\partial y}\right)^2+\nu\Bigg(\frac{\partial u_n}{\partial x}-k_x w_n$$

$$+\frac{1}{2}\left(\frac{\partial w_n}{\partial x}\right)^2\Bigg)\Bigg], \frac{\partial \chi_{1l}}{\partial y}\Bigg)_{L_2(\Omega_1)}$$

$$-\frac{Eh}{2(1+\nu)}\left(\left[\frac{\partial u_n}{\partial y}+\frac{\partial v_n}{\partial x}+\frac{\partial w_n}{\partial x}\frac{\partial w_n}{\partial y}\right], \frac{\partial \chi_{1l}}{\partial y}\right)_{L_2(\Omega_1)}$$

$$+\frac{E\alpha_T}{1-\nu}\left(\frac{\partial w_n}{\partial x}\int\limits_{-\frac{h}{2}}^{\frac{h}{2}}\theta_n\,dz, \frac{\partial \chi_{1l}}{\partial x}\right)_{L_2(\Omega_1)}$$

$$+\frac{E\alpha_T}{1-\nu}\left(\frac{\partial w_n}{\partial y}\int\limits_{-\frac{h}{2}}^{\frac{h}{2}}\theta_n\,dz, \frac{\partial \chi_{1l}}{\partial y}\right)_{L_2(\Omega_1)} + (g_1,\ \chi_{1l})_{L_2(\Omega_1)}$$

$$=\rho h\left(\frac{\partial^2 w_n}{\partial t^2}, \chi_{1l}\right)_{L_2(\Omega_1)} - \frac{\rho h^3}{12}\left(\frac{\partial^2}{\partial t^2}\left[\frac{\partial^2 w_n}{\partial x^2}+\frac{\partial^2 w_n}{\partial y^2}\right], \chi_{1l}\right)_{L_2(\Omega_1)}, \tag{5.66}$$

$$-\lambda_q\left[\left(\frac{\partial \theta_n}{\partial x}, \frac{\partial \chi_{3l}}{\partial x}\right)_{L_2(\Omega_2)}+\left(\frac{\partial \theta_n}{\partial y}, \frac{\partial \chi_{3l}}{\partial y}\right)_{L_2(\Omega_2)}+\left(\frac{\partial \theta_n}{\partial z}, \frac{\partial \chi_{3l}}{\partial z}\right)_{L_2(\Omega_2)}\right]$$

$$-\frac{E\alpha_T T_0}{1-\nu}\left(\frac{\partial}{\partial t}\left[\frac{\partial u_n}{\partial x}-k_x w_n+\frac{1}{2}\left(\frac{\partial w_n}{\partial x}\right)^2+\nu\left(\frac{\partial v}{\partial y}-k_y w_n+\frac{1}{2}\left(\frac{\partial w_n}{\partial y}\right)^2\right)\right.\right.$$

$$-z\left\{\frac{\partial^2 w_n}{\partial x^2}+\frac{\partial^2 w_n}{\partial y^2}\right\}\Bigg],\ \chi_{3l}\Bigg)_{L_2(\Omega_2)}+(g_2,\ \chi_{3l})_{L_2(\Omega_2)}=c_0\left(\frac{\partial\theta_n}{\partial t},\ \chi_{3l}\right)_{L_2(\Omega_2)}, \tag{5.67}$$

$$l=1,\ldots,n.$$

We complete the system of equations (5.64)–(5.67) in the usual way [226, 343] with the following initial conditions:

$$g_{1l}(0)=(\varphi_1,\ \chi_{1l})_{L_2(\Omega_1)},\quad g_{2l}(0)=(\varphi_2,\ \chi_{2l})_{L_2(\Omega_1)},$$
$$g_{3l}(0)=(\varphi_3,\ \chi_{2l})_{L_2(\Omega_1)},$$
$$\frac{dg_{1l}(0)}{dt}=(\Psi_1,\ \chi_{1l})_{L_2(\Omega_1)},\quad \frac{dg_{2l}(0)}{dt}=(\Psi_2,\ \chi_{2l})_{L_2(\Omega_1)},$$
$$\frac{dg_{3l}(0)}{dt}=(\Psi_3,\ \chi_{2l})_{L_2(\Omega_1)},\quad g_{4l}(0)=(\varphi_4,\ \chi_{3l})_{L_2(\Omega_2)},\quad l=1,\ldots,n. \tag{5.68}$$

We now prove[1] the solvability of (5.64)–(5.68) using Schauder's principle of a fixed point, which reads:

> If a fully continuous operator A maps a bounded closed convex manifold S of the Banach space E onto a part of the space, then a fixed point of this mapping exists, i.e. $x\in S$ such that $Ax=x$.

First, taking into account the orthogonality property of the basis functions $\{\chi_{ji}\}$, $j=1,2,3$, in the spaces $L_2(\Omega_1)$, $L_2(\Omega_2)$, $\overset{\circ}{W}{}_2^1(\Omega_1)$, we transform (5.64)–(5.68) to the form

$$\frac{d^2g_{1l}(0)}{dt^2}=f_{1l}\left(\overline{g_1},\overline{g_2},\overline{g_3},\overline{g_4}\right)+(g_1,\ \chi_{1l})_{L_2(\Omega_1)},$$
$$\frac{d^2g_{2l}(0)}{dt^2}=\frac{1}{\rho h}f_{2l}\left(\overline{g_1},\overline{g_2},\overline{g_3},\overline{g_4}\right)+\frac{1}{\rho h}(p_1,\ \chi_{2l})_{L_2(\Omega_1)},$$
$$\frac{d^2g_{3l}(0)}{dt^2}=\frac{1}{\rho h}f_{3l}\left(\overline{g_1},\overline{g_2},\overline{g_3},\overline{g_4}\right)+\frac{1}{\rho h}(g_1,\ \chi_{2l})_{L_2(\Omega_1)},$$
$$\frac{d^2g_{4l}(0)}{dt^2}=\frac{1}{c_0}f_{4l}\left(\overline{g_1},\overline{g_1},\overline{g_2},\overline{g_3},\overline{g_4}\right)+\frac{1}{c_0}(g_2,\ \chi_{3l})_{L_2(\Omega_2)}, \tag{5.69}$$
$$l=1,2,\ldots,n,$$

[1] In general, the problem stated here is related to the proof of the Peano theorem about the existence of solutions to the Cauchy problem for a system of ordinary differential equations. However, for problems related to the theory of shells, a detailed proof of this theorem is not known to the authors.

where $\overline{g}_i = (g_{i1}, g_{i2}, \ldots, g_{in})$, $\dot{\overline{g}}_i = \frac{d\overline{g}_i}{dt}$, $i = 1, \ldots, 4$, and f_{il} denotes the left-hand sides of (5.69). It is not difficult to see that all f_{il} are continuous with respect to $g_{ij}(t)$, $\dot{g}_{ij}(t)$, $i = 1, \ldots, 4$, $j = 1, 2, 3, 4, \ldots, n$, for all $-\infty < g_{ij}(t) < +\infty$, $-\infty < \dot{g}_{ij}(t) < +\infty$ and for arbitrary $t \in [0, T]$, and therefore it is sufficient to consider only the terms in which g_{ij} and $\dot{g}_{ij}$ appear in a nonlinear way. Let us consider the following

$$R = \int\limits_{\Omega_1} \frac{\partial w_n}{\partial y} \frac{\partial v_n}{\partial y} \frac{\partial \chi_{1l}}{\partial y} d\Omega_1 = \int\limits_{\Omega_1} \left(\sum_{i=1}^{n} g_{1i} \frac{\partial \chi_{1i}}{\partial y} \sum_{j=1}^{n} g_{3j} \frac{\partial \chi_{2i}}{\partial y} \right) \frac{\partial \chi_{1l}}{\partial y} d\Omega_1$$

$$= \sum_{i,j=1}^{n} g_{1i} g_{3j} \int\limits_{\Omega_1} \frac{\partial \chi_{1i}}{\partial y} \frac{\partial \chi_{1l}}{\partial y} \frac{\partial \chi_{2j}}{\partial y} d\Omega_1 = \sum_{i,j=1}^{n} g_{1i} g_{3j} a_{ilj},$$

where a_{ilj} are arbitrary constants. It is obvious that the function R is continuous on the set of variables $g_{1i} g_{3j}$. The rest of the nonlinear terms can be analysed similarly. Let us now move from the system (5.69) to an equivalent form using integral equations:

$$g_{1l}(t) = g_{1l}(0) + \int\limits_0^t \dot{g}_{1l}(\tau)\, d\tau,$$

$$\dot{g}_{1l}(t) = \dot{g}_{1l}(0) + \int\limits_0^t f_{1l}(\overline{g}_1, \overline{g}_2, \overline{g}_3, \overline{g}_4)\, d\tau + \int\limits_0^t \int\limits_{\Omega_1} g_1 \chi_{1l}\, d\Omega_1\, d\tau,$$

$$g_{jl}(t) = g_{jl}(0) + \int\limits_0^t \dot{g}_{jl}(\tau)\, d\tau, \quad j = 2, 3,$$

$$\dot{g}_{jl}(t) = \dot{g}_{jl}(0) + \frac{1}{\rho h} \int\limits_0^t f_{jl}(\overline{g}_1, \overline{g}_2, \overline{g}_3, \overline{g}_4)\, d\tau + \frac{1}{\rho h} \int\limits_0^t \int\limits_{\Omega_1} p_k \chi_{2l}\, d\Omega_1\, d\tau,$$

$$g_{4l}(t) = g_{4l}(0) + \frac{1}{c_0} \int\limits_0^t f_{4l}(\overline{g}_1, \dot{\overline{g}}_1, \overline{g}_2, \overline{g}_3, \overline{g}_4)\, d\tau + \frac{1}{c_0} \int\limits_0^t \int\limits_{\Omega_2} g_2 \chi_{3l}\, d\Omega_2\, d\tau,$$

$$l = 1, \ldots, n, \quad k = 1, 2. \tag{5.70}$$

We take the right-hand sides of the system (5.70) as the operator A used in the Schauder principle. This operator is defined in the cube

$$D_0 = \Big\{\overline{g}_i, \overline{g}_i, g_4 |\, |g_{ij}(t) - g_{ij}(0)| \le b, \quad |g_{4j}(t) - g_{4j}(0)| \le b,$$
$$|\dot{g}_{ij}(t) - \dot{g}_{ij}(0)| \le b, \quad i = 1, 2, 3, \quad j = 1, \dots, n\Big\}$$

of the space $C[0, T]$ of the continuous functions $g_{ij}, \dot{g}_{ij}, g_{4j}$, with the norm

$$\big\| g_{ij}, \dot{g}_{ij} \big\|_C = \max_{t \in [0,T]} \{|g_{ij}|, |\dot{g}_{ij}|\},$$
$$\big\| g_{4j}, 0 \big\|_C = \max_{t \in [0,T]} \{|g_{4j}|, 0\}.$$

In order to show that the operator A defined in this way is fully continuous in D_0, we need to show that:

(a) A is continuous on D_0,
(b) A is compact on D_0, i.e. A maps an arbitrary bounded set from D_0 to a compact one.

It is clear that the operator A will be continuous if all its components are continuous. We consider only the component f_{1l}, because the proof for other components works similarly. Continuity of f_{1l} implies that $\forall \varepsilon$, $\exists \delta$, and if $|\overline{g}_i - \overline{g}_i'| < \delta$, then

$$|f_{1l}(\overline{g}_1, \overline{g}_2, \overline{g}_3, \overline{g}_4) - f_{1l}(\overline{g}_1', \overline{g}_2', \overline{g}_3', \overline{g}_4')| < \varepsilon. \tag{5.71}$$

Let us consider an arbitrary uniformly convergent series $\{\overline{g}^m\} \to \overline{g}^0$, $m \to \infty$, $\overline{g}^m = \{\overline{g}_1^m, \overline{g}_2^m, \overline{g}_3^m, \overline{g}_4^m\}$, i.e. $\forall \varepsilon_1$, $\exists N(\varepsilon_1)$ such that for $m \ge N(\varepsilon_1)$ the inequality

$$\left|\overline{g}^m(t) - \overline{g}^0(t)\right| < \varepsilon_1, \quad \forall t \in [0, T],$$

holds. Choosing N large enough for ε_1 to be smaller than δ, we obtain from (5.71):

$$\left| f_{1l}(\overline{g}^m(t)) - f_{1l}(\overline{g}^0(t)) \right| < \varepsilon, \ \forall t \in [0, T]. \tag{5.72}$$

Using (5.72), we then obtain

$$\left\| \dot{g}_{1l}^m(t) - \dot{g}_{1l}^0(t) \right\|_c = \max_{t \in [0,T]} \left| \int_0^t \left[f_{1l}(\overline{g}^m(t)) - f_{1l}(\overline{g}^0(t)) \right] d\tau \right|$$
$$\le \max_{t \in [0,T]} \int_0^t \left| f_{1l}(\overline{g}^m(t)) - f_{1l}(\overline{g}^0(t)) \right| d\tau \le \varepsilon T. \tag{5.73}$$

The inequality (5.73) proves the continuity of one of the components of the operator A. Thus, the condition (a) has been proved. We shall now give a proof of the condition (b) using this component. We use the Arcel theorem [345]. First, let us observe that because f_{il}, $i = 1, \dots, 4$, are continuous functions on D_0, they also achieve their maximal values there. Defining

$$M = \max_{l=1,\dots,n}\left[\max_{\overline{g},\dot{\overline{g}}\in D_0}\{|f_{1l}|, |f_{2l}|, |f_{3l}|, |f_{4l}|\}\right],$$

we obtain inequalities corresponding to the conditions of the Arcel theorem.

For an arbitrary $\overline{g}(t) \in D_0$, we have

$$\left|\dot{g}_{1l}(0) + \int_0^t f_{1l}(\overline{g})\,d\tau + \int_0^t\int_{\Omega_1} g_1\chi_{1l}\,d\Omega_1\,d\tau\right| \le |\dot{g}_{1l}(0)|$$
$$+ \int_0^t |f_{1l}(\overline{g})|\,d\tau + \frac{1}{2}\int_0^t \|g_1\|^2_{L_2(\Omega_1)}\,d\tau + \frac{1}{2}\int_0^t \|\chi_{1l}\|^2_{L_2(\Omega_1)}\,d\tau$$
$$\le |\dot{g}_{1l}(0)| + MT + \frac{1}{2}\int_0^T \|g_1\|^2_{L_2(\Omega_1)}\,d\tau + \frac{T}{2}\|\chi_{1l}\|^2_{L_2(\Omega_1)}. \tag{5.74}$$

For arbitrary $t_1, t_2 \in [0, T]$ $(t_1 > t_2)$, we have

$$\left|\dot{g}_{1l}(0) + \int_0^{t_2} f_{1l}(\overline{g})\,d\tau + \int_0^{t_2}\int_{\Omega_1} g_1\chi_{1l}d\Omega_1 d\tau - \dot{g}_{1l}(0) - \int_0^{t_1} f_{1l}(\overline{g})\,d\tau\right.$$
$$\left. - \int_0^{t_1}\int_{\Omega_1} g_1\chi_{1l}\,d\Omega_1\,d\tau\right| = \left|\int_{t_1}^{t_2} f_{1l}(\overline{g})\,d\tau + \int_{t_1}^{t_2}\int_{\Omega_1} g_1\chi_{1l}\,d\Omega_1\,d\tau\right|$$
$$\le M(t_2 - t_1) + \left[\int_{t_1}^{t_2} \|g_1\|^2_{L_2(\Omega_1)}\,d\tau\right]^{1/2}\left[\int_{t_1}^{t_2} \|\chi_{1l}\|^2_{L_2(\Omega_1)}\,d\tau\right]^{1/2}$$
$$\le M(t_2 - t_1) + \left[\int_0^T \|g_1\|^2_{L_2(\Omega_1)}\,d\tau\right]^{1/2}\sqrt{t_2 - t_1}\,\|\chi_{1l}\|^2_{L_2(\Omega_1)}. \tag{5.75}$$

The inequalities (5.74), (5.75) show that the set of patterns of the component of the operator A under consideration is equally bounded and equally continuous (the proof is analogous for the other components). Taking the largest constant from those which equally bound each of the operator components (i.e. we use estimates in the sense of (5.74)), we obtain in accordance with the Arcel theorem, the result that the operator A is compact. Thus, the operator is fully continuous.

According to the definition given above, the set D_0 is convex and closed. In order to finish our treatment of the solvability of (5.64)–(5.68), we use the Schauder principle and show that the operator maps the set D_0 into itself. The last observation is true if each of the components of the operator A maps D_0 into itself. A proof of the latter requirement is given here using f_{1l} as an example:

$$|\dot{g}_{1l}(t) - \dot{g}_{1l}(0)| = \left| \int_0^t f_{1l}(\overline{g})\, d\tau + \int_0^t \int_{\Omega_1} g_1 \chi_{1l}\, d\Omega_1\, d\tau \right|$$

(we use the Cauchy inequality)

$$\leq Mt + \frac{\varepsilon}{2} \int_0^t \|g_1\|^2_{L_2(\Omega_1)}\, d\tau + \frac{1}{2\varepsilon} \int_0^t \|\chi_{1l}\|^2_{L_2(\Omega_1)}\, d\tau$$

$$\leq \left(M + \frac{1}{2\varepsilon} \|\chi_{1l}\|^2_{L_2(\Omega_1)} \right) t + \frac{\varepsilon}{2} \int_0^T \|g_1\|^2_{L_2(\Omega_1)}\, d\tau, \quad 0 \leq t \leq T, \tag{5.76}$$

where ε is an arbitrary positive number. From (5.76), we conclude that the component maps D_0 into itself when

$$\left(M + \frac{1}{2\varepsilon} \|\chi_{1l}\|^2_{L_2(\Omega_1)} \right) t + \frac{\varepsilon}{2} \int_0^T \|g_1\|^2_{L_2(\Omega_1)}\, d\tau = b. \tag{5.77}$$

Now, letting ε be small enough that the inequality

$$b - \frac{\varepsilon}{2} \int_0^T \|g_1\|^2_{L_2(\Omega_1)}\, d\tau > 0$$

is satisfied, we find the upper boundary of the time interval $[0, t_0]$ for which the condition of mapping D_0 into itself is satisfied:

$$t_0 = \frac{b - \frac{\varepsilon}{2} \int_0^T \|g_1\|^2_{L_2(\Omega_1)}\, d\tau}{M + \frac{1}{2\varepsilon} \|\chi_{1l}\|^2_{L_2(\Omega_1)}}. \tag{5.78}$$

In particular, if we take

$$\varepsilon = \frac{b}{\int\limits_0^T \|g_1\|^2_{L_2(\Omega_1)}\,d\tau}$$

in (5.78), then we obtain

$$t_0 = \frac{\frac{b}{2}}{M + \frac{1}{2b}\left(\int\limits_0^T \|g_1\|^2_{L_2(\Omega_1)}\,d\tau\right)\|\chi_{1l}\|^2_{L_2(\Omega_1)}} = \frac{b}{2M + \frac{1}{b}\|\chi_{1l}\|^2_{L_2(\Omega_1)}\int\limits_0^T \|g_1\|^2_{L_2(\Omega_1)}\,d\tau}. \tag{5.79}$$

Analysis of (5.76) and (5.79) leads to the conclusion that during the investigation of the other components only the $\|\chi_{1l}\|^2_{L_2(\Omega_1)}\int\limits_0^T \|g_1\|^2_{L_2(\Omega_1)}\,d\tau$ term can be changed, and therefore we take

$$c_1 = \max_{\substack{l=1,\ldots,n\\k=1,2}}\left\{\|\chi_{1l}\|^2_{L_2(\Omega_1)}\int\limits_0^T \|g_1\|^2_{L_2(\Omega_1)}\,d\tau;\ \int\limits_0^T \|p_k\|^2_{L_2(\Omega_1)}\,d\tau;\ \int\limits_0^T \|g_2\|^2_{L_2(\Omega_2)}\,d\tau\right\},$$

while the systems of functions $\{\chi_{2l}\}$, $\{\chi_{3l}\}$ are orthonormalized in $L_2(\Omega_1)$, $L_2(\Omega_2)$. Therefore

$$t_0 = \frac{b^2}{2Mb + c_1}. \tag{5.80}$$

It should be emphasized that owing to its definition, the quantity M is a function of $\overline{g}(0)$, $\overline{g}(0)$, b. Therefore, the following lemma has been proved.

Lemma 5.1 *If $t \in [0, T_0]$, where $T_0 = \min\{T, t_0\}$, then there exists in a cube of the space $C[0, T]$ at least one solution of the problem (5.64)–(5.68).*

Later in a typical way [341], we shall extend the solution to the whole interval $[0, T]$ using prior estimates.

Step 2. Prior Estimates In order to obtain prior estimates we multiply (5.64) by $\dot{g}_{1l}(t)$, (5.65) by $\dot{g}_{2l}(t)$, (5.66) by $\dot{g}_{3l}(t)$ and (5.67) by $\dot{g}_{4l}(t)$. We then sum each of the resulting expressions with respect to l, and add the first three equations. We obtain

$$\frac{1}{2}\frac{d}{dt}\Bigg\{\rho h\left\|\frac{\partial w_n(t)}{\partial t}\right\|^2_{L_2(\Omega_1)}+\frac{\rho h^3}{12}\left\|\frac{\partial w_n(t)}{\partial t}\right\|^2_{\overset{\circ}{W}{}^1_2(\Omega_1)}+\rho h\left\|\frac{\partial u_n(t)}{\partial t}\right\|^2_{L_2(\Omega_1)}$$
$$+\left\|\frac{\partial v_n(t)}{\partial t}\right\|^2_{L_2(\Omega_1)}+D\left\|w_n\right\|^2_{\overset{\circ}{W}{}^2_2(\Omega_1)}+\int\limits_{\Omega_1}\Bigg(\frac{Eh}{1-\nu^2}\left[\frac{\partial u_n}{\partial x}-k_x w_n\right.$$
$$\left.+\frac{1}{2}\left(\frac{\partial w_n}{\partial x}\right)^2\right]^2+\frac{Eh}{1-\nu^2}\left[\frac{\partial v_n}{\partial y}-k_y w_n+\frac{1}{2}\left(\frac{\partial w_n}{\partial y}\right)^2\right]^2$$
$$+\frac{2\nu Eh}{1-\nu^2}\left[\frac{\partial u_n}{\partial x}-k_x w_n+\frac{1}{2}\left(\frac{\partial w_n}{\partial x}\right)^2\right]\left[\frac{\partial v_n}{\partial y}-k_y w_n\right.$$
$$\left.+\frac{1}{2}\left(\frac{\partial w_n}{\partial y}\right)^2\right]+\frac{Eh}{2(1+\nu)}\left[\frac{\partial u_n}{\partial y}-\frac{\partial v_n}{\partial x}+\frac{\partial w_n}{\partial x}\frac{\partial w_n}{\partial y}\right]^2\Bigg)\,d\Omega_1\Bigg\}$$
$$-\frac{E\alpha_t}{1-\nu}\int\limits_{\Omega_1}\int\limits_{-\frac{h}{2}}^{\frac{h}{2}}\theta_n\frac{\partial}{\partial t}\left[\frac{\partial u_n}{\partial x}-k_x w_n+\frac{1}{2}\left(\frac{\partial \Omega_n}{\partial x}\right)^2+\frac{\partial v_n}{\partial y}-k_y w_n\right.$$
$$\left.+\frac{1}{2}\left(\frac{\partial \Omega_n}{\partial y}\right)^2-z\left(\frac{\partial^2 w_n}{\partial x^2}+\frac{\partial^2 w_n}{\partial y^2}\right)\right]dz\,d\Omega_1$$
$$=\int\limits_{\Omega_1}\left(g_1\frac{\partial w_n}{\partial t}+p_1\frac{\partial u_n}{\partial t}+p_2\frac{\partial v_n}{\partial t}\right)d\Omega_1, \tag{5.81}$$

$$\frac{1}{2}\frac{d}{dt}\left\{\frac{c_0}{T_0}\left\|\theta_n\right\|^2_{L_2(\Omega_2)}\right\}+\frac{\lambda_q}{T_0}\left\|\theta_n\right\|^2_{\overset{\circ}{W}{}^1_2(\Omega_2)}$$
$$+\frac{E\alpha_t}{1-\nu}\int\limits_{\Omega_2}\frac{\partial}{\partial t}\left[\frac{\partial u_n}{\partial x}-k_x w_n+\frac{1}{2}\left(\frac{\partial w_n}{\partial x}\right)^2+\frac{\partial v_n}{\partial y}\right.$$
$$\left.-k_y w_n+\frac{1}{2}\left(\frac{\partial w_n}{\partial y}\right)^2-z\left(\frac{\partial^2 w_n}{\partial x^2}+\frac{\partial^2 w_n}{\partial y^2}\right)\right]d\Omega_2$$
$$=\int\limits_{\Omega_2} g_2\theta_n\,d\Omega_2. \tag{5.82}$$

Summing (5.81) and (5.82), we obtain the following (note that in the left-hand sides, the terms containing temperature cancel):

$$\frac{1}{2}\frac{d}{dt}\left\{\rho h\left(\left\|\frac{\partial w_n(t)}{\partial t}\right\|^2_{L_2(\Omega_1)}+\left\|\frac{\partial u_n(t)}{\partial t}\right\|^2_{L_2(\Omega_1)}+\left\|\frac{\partial v_n(t)}{\partial t}\right\|^2_{L_2(\Omega_1)}\right)\right.$$

$$+\frac{\rho h^3}{12}\left\|\frac{\partial w_n(t)}{\partial t}\right\|^2_{\mathring{W}_2^1(\Omega_1)}+D\left\|w_n\right\|^2_{\mathring{W}_2^1(\Omega_1)}+\frac{c_0}{T_0}\left\|\theta_n\right\|^2_{L_2(\Omega_2)}$$

$$+\int\limits_{\Omega_1}\left(\frac{Eh}{1-\nu^2}\left[\frac{\partial u_n}{\partial x}-k_x w_n+\frac{1}{2}\left(\frac{\partial w_n}{\partial x}\right)^2\right]^2+\left[\frac{\partial v_n}{\partial y}-k_y w_n\right.\right.$$

$$\left.+\frac{1}{2}\left(\frac{\partial w_n}{\partial y}\right)^2\right]^2+2\nu\left[\frac{\partial u_n}{\partial x}-k_x w_n+\frac{1}{2}\left(\frac{\partial w_n}{\partial x}\right)^2\right]^2\left[\frac{\partial v_n}{\partial y}\right.$$

$$\left.-k_y w_n+\frac{1}{2}\left(\frac{\partial w_n}{\partial y}\right)^2\right]+\frac{Eh}{2(1+\nu)}\left[\frac{\partial u_n}{\partial x}+\frac{\partial v_n}{\partial x}+\frac{\partial w_n}{\partial x}\right.$$

$$\left.\left.\left.\times\frac{\partial w_n}{\partial y}\right]^2\right)d\Omega_1\right\}+\frac{\lambda_q}{T_0}\left\|\theta_n\right\|^2_{\mathring{W}_2^1(\Omega_2)}=\int\limits_{\Omega_1}\left(g_1\frac{\partial w_n}{\partial t}+p_1\frac{\partial u_n}{\partial t}\right.$$

$$\left.+p_2\frac{\partial v_n}{\partial t}\right)d\Omega_1+\int\limits_{\Omega_2}g_2\theta_n\,d\Omega_2. \tag{5.83}$$

Integrating (5.83) with respect to t in the interval from 0 to τ, $\tau\le T$, and then transforming the right-hand side of (5.83) by use of the Cauchy inequality, we obtain

$$\frac{1}{2}\rho h\left(\left\|\frac{\partial u_n(\tau)}{\partial t}\right\|^2_{L_2(\Omega_1)}+\left\|\frac{\partial v_n(\tau)}{\partial t}\right\|^2_{L_2(\Omega_1)}+\left\|\frac{\partial w_n(\tau)}{\partial t}\right\|^2_{L_2(\Omega_1)}\right)$$

$$+\frac{1}{2}\frac{\rho h^3}{12}\left\|\frac{\partial w_n(\tau)}{\partial t}\right\|^2_{\mathring{W}_2^1(\Omega_1)}+\frac{1}{2}D\left\|w_n(\tau)\right\|^2_{\mathring{W}_2^2(\Omega_1)}+\frac{\lambda_q}{T_0}\int\limits_0^{\tau}\left\|\theta_n\right\|^2_{\mathring{W}_2^1(\Omega_2)}dt$$

$$+\frac{1}{2}\frac{c_0}{T_0}\left\|\theta_n\right\|^2_{L_2(\Omega_2)}$$

$$\le\frac{1}{2}\rho h\left(\left\|\frac{\partial u_n(0)}{\partial t}\right\|^2_{L_2(\Omega_1)}+\left\|\frac{\partial v_n(0)}{\partial t}\right\|^2_{L_2(\Omega_1)}+\left\|\frac{\partial w_n(0)}{\partial t}\right\|^2_{L_2(\Omega_1)}\right)$$

$$+\frac{1}{2}\frac{\rho h^3}{12}\left\|\frac{\partial w_n(0)}{\partial t}\right\|^2_{\mathring{W}_2^1(\Omega_1)}+\frac{1}{2}D\left\|w_n(0)\right\|^2_{\mathring{W}_2^2(\Omega_1)}+\frac{1}{2}\frac{c_0}{T_0}\left\|\theta_n(0)\right\|^2_{L_2(\Omega_2)}$$

$$
+\frac{1}{2}\frac{1}{\rho h}\int_0^{\tau}\int_{\Omega_1}\left[(g_1)^2+(p_1)^2+(p_2)^2\right]d\Omega_1\,dt+\frac{1}{2}\int_0^{\tau}\left\{\rho h\left(\left\|\frac{\partial u_n}{\partial t}\right\|^2_{L_2(\Omega_1)}\right.\right.
$$

$$
\left.+\left\|\frac{\partial v_n}{\partial t}\right\|^2_{L_2(\Omega_1)}+\left\|\frac{\partial w_n}{\partial t}\right\|^2_{L_2(\Omega_1)}\right)+\frac{\rho h^3}{12}\left\|\frac{\partial w_n}{\partial t}\right\|^2_{\overset{\circ}{W}{}_2^1(\Omega_1)}
$$

$$
\left.+D\left\|w_n\right\|^2_{\overset{\circ}{W}{}_2^2(\Omega_1)}\right\}dt+\frac{1}{2}\frac{T_0}{c_0}\int_0^{\tau}\int_{\Omega_2}(g_2)^2\,d\Omega_2+\frac{1}{2}\frac{c_0}{T_0}\int_0^{\tau}\left\|\theta_n\right\|^2_{L_2(\Omega_2)}dt
$$

$$
+\int_{\Omega_1}\left(\frac{Eh}{1-\nu^2}\left\{\left[\frac{\partial u_n(0)}{\partial x}-k_x w_n(0)+\frac{1}{2}\left(\frac{\partial w_n(0)}{\partial x}\right)^2\right]^2\right.\right.
$$

$$
+\left[\frac{\partial v_n(0)}{\partial y}-k_y w_n(0)+\frac{1}{2}\left(\frac{\partial w_n(0)}{\partial y}\right)^2\right]^2+2\nu\left[\frac{\partial u_n(0)}{\partial x}\right.
$$

$$
\left.\left.-k_x w_n(0)+\frac{1}{2}\left(\frac{\partial w_n(0)}{\partial x}\right)^2\right]\left[\frac{\partial v_n(0)}{\partial y}-k_y w_n(0)+\frac{1}{2}\left(\frac{\partial w_n(0)}{\partial y}\right)^2\right]\right\}
$$

$$
\left.+\frac{Eh}{2(1+\nu)}\left[\frac{\partial u_n(0)}{\partial y}+\frac{\partial v_n(0)}{\partial x}+\frac{\partial w_n(0)}{\partial x}\frac{\partial w_n(0)}{\partial y}\right]^2\right)d\Omega_1
$$

$$
\le\frac{1}{2}\rho h\left(\left\|\frac{\partial u_n(0)}{\partial t}\right\|^2_{L_2(\Omega_1)}+\left\|\frac{\partial v_n(0)}{\partial t}\right\|^2_{L_2(\Omega_1)}+\left\|\frac{\partial w_n(0)}{\partial t}\right\|^2_{L_2(\Omega_1)}\right)
$$

$$
+\frac{1}{2}\frac{\rho h^3}{12}\left\|\frac{\partial w_n(0)}{\partial t}\right\|^2_{\overset{\circ}{W}{}_2^1(\Omega_1)}+\frac{1}{2}D\left\|w_n(0)\right\|^2_{\overset{\circ}{W}{}_2^2(\Omega_1)}+\frac{1}{2}\frac{c_0}{T_0}\left\|\theta_n(0)\right\|^2_{L_2(\Omega_2)}
$$

$$
+\frac{1}{2}\int_0^{T}\left[\left(\|g_1\|^2_{L_2(\Omega_1)}+\|p_1\|^2_{L_2(\Omega_1)}+\|p_2\|^2_{L_2(\Omega_1)}\right)+\frac{T_0}{c_0}\|g_2\|^2_{L_2(\Omega_2)}\right]dt
$$

$$
+\frac{Eh}{1-\nu^2}c_2\left(\left\|\frac{\partial u_n(0)}{\partial x}\right\|^2_{L_2(\Omega_1)}+\left\|\frac{\partial v_n(0)}{\partial y}\right\|^2_{L_2(\Omega_1)}\right.
$$

$$
\left.+\frac{1-\nu}{2}\left\|\frac{\partial u_n(0)}{\partial y}\right\|^2_{L_2(\Omega_1)}+\frac{1-\nu}{2}\left\|\frac{\partial v_n(0)}{\partial x}\right\|^2_{L_2(\Omega_1)}+1+\left\|w_n(0)\right\|^4_{\overset{\circ}{W}{}_2^2(\Omega_1)}\right)
$$

$$
+\frac{1}{2}\int_0^{\tau}\left\{\frac{1}{2}\rho h\left(\left\|\frac{\partial u_n(t)}{\partial t}\right\|^2_{L_2(\Omega_1)}+\left\|\frac{\partial v_n(t)}{\partial t}\right\|^2_{L_2(\Omega_1)}+\left\|\frac{\partial w_n(t)}{\partial t}\right\|^2_{L_2(\Omega_1)}\right)\right.
$$

$$+\frac{\rho h^3}{12}\left\|\frac{\partial w_n(t)}{\partial t}\right\|^2_{\overset{\circ}{W}{}^1_2(\Omega_1)} + \frac{c_0}{T_0}\left\|\theta_n(t)\right\|^2_{L_2(\Omega_2)} + D\left\|w_n(t)\right\|^2_{\overset{\circ}{W}{}^2_2(\Omega_1)}\Bigg\} dt, \qquad (5.84)$$

$$c_2 = const > 0.$$

In order to obtain the chain of inequalities (5.84), the following inequalities and inclusion theorems have been used (in addition to the Cauchy inequality):

(a)

$$\int\limits_{\Omega_1}\left(\left[\frac{\partial u}{\partial x} - k_x w + \frac{1}{2}\left(\frac{\partial w}{\partial x}\right)^2\right]^2 + \left[\frac{\partial v}{\partial y} - k_y w + \frac{1}{2}\left(\frac{\partial w}{\partial y}\right)^2\right]^2\right.$$

$$\left.+2\nu\left[\frac{\partial u}{\partial x} - k_x w + \frac{1}{2}\left(\frac{\partial w}{\partial x}\right)^2\right]\left[\frac{\partial v}{\partial y} - k_y w + \frac{1}{2}\left(\frac{\partial w}{\partial y}\right)^2\right]^2\right) d\Omega_1$$

$$\geq \int\limits_{\Omega_1}(1-\nu)\left(\left[\frac{\partial u}{\partial x} - k_x w + \frac{1}{2}\left(\frac{\partial w}{\partial x}\right)^2\right]^2 + \left[\frac{\partial v}{\partial y} - k_y w + \frac{1}{2}\left(\frac{\partial w}{\partial y}\right)^2\right]^2\right) d\Omega_1 \geq 0,$$

$$0 < \nu < \frac{1}{2};$$

(b) $\forall u \in \overset{\circ}{W}{}^1_2(\Omega_1)$, $\Omega_1 \subset E_2$, the following applies

$$\int\limits_{\Omega_1} u^4\, d\Omega_1 \leq c\left[\int\limits_{\Omega_1}\left\{\left(\frac{\partial u}{\partial x}\right)^2 + \left(\frac{\partial u}{\partial y}\right)^2\right\} d\Omega_1\right]^2.$$

Using one of the integrals as an example, we illustrate below how the estimates in the chain (5.84) have been obtained:

$$\int\limits_{\Omega_1}\left[\frac{\partial v_n(0)}{\partial y} - k_y w_n(0) + \frac{1}{2}\left(\frac{\partial w_n(0)}{\partial y}\right)^2\right]^2 d\Omega_1$$

(using the Cauchy inequality)

$$\leq \int\limits_{\Omega_1}\left[\left(\frac{\partial v_n(0)}{\partial y} - k_y w_n(0)\right)^2 + \left(\frac{\partial w_n(0)}{\partial y}\right)^4\right] d\Omega_1$$

$$\leq \int\limits_{\Omega_1}\left[4\left(\frac{\partial w_n(0)}{\partial y}\right)^2 + 4k_y^2 w_n^2(0) + \left(\frac{\partial w_n(0)}{\partial y}\right)^4\right] d\Omega_1$$

(completing the second term in the integral to the full norm in the space $\mathring{W}_2^2(\Omega_1)$, and using the inequality (b) for the third term)

$$\leq c \left(\left\| \frac{\partial v_n(0)}{\partial y} \right\|^2_{L_2(\Omega_1)} + \|w_n(0)\|^2_{\mathring{W}_2^2(\Omega_1)} + \|w_n(0)\|^4_{\mathring{W}_2^2(\Omega_1)} \right)$$

$$\leq c \left(\left\| \frac{\partial v_n(0)}{\partial y} \right\|^2_{L_2(\Omega_1)} + \|w_n(0)\|^4_{\mathring{W}_2^2(\Omega_1)} + \frac{1}{2} \right).$$

The remaining terms of (5.84) are estimated in an analogous way, and c_2 is expected to be the largest constant from all c.

Furthermore, let us observe that the arbitrary elements $u_n(0)$, $v_n(0)$, $w_n(0)$, $\theta_n(0)$ occurring in (5.84) can be considered as elements of convergent series with a corresponding norm, i.e. they approximate the initial conditions and, as the elements of convergent series, they are bounded by a constant independent of the number n. Finally, taking into account the conditions applied to g_1, g_2, p_1, p_2 in (5.84), we obtain:

$$\frac{1}{2}\rho h \left(\left\| \frac{\partial u_n(\tau)}{\partial t} \right\|^2_{L_2(\Omega_1)} + \left\| \frac{\partial v_n(\tau)}{\partial t} \right\|^2_{L_2(\Omega_1)} + \left\| \frac{\partial w_n(\tau)}{\partial t} \right\|^2_{L_2(\Omega_1)} \right)$$

$$+\frac{1}{2}\rho h^3 \left\| \frac{\partial w_n(\tau)}{\partial t} \right\|^2_{\mathring{W}_2^1(\Omega_1)} + \frac{1}{2} D \|w_n(\tau)\|^2_{\mathring{W}_2^2(\Omega_1)}$$

$$+\frac{\lambda_q}{T_0} \int_0^T \|\theta_n\|^2_{\mathring{W}_2^1(\Omega_2)} dt + \frac{1}{2}\frac{c_0}{T_0} \|\theta_n(\tau)\|^2_{L_2(\Omega_2)}$$

$$\leq c + \frac{1}{2} \int_0^\tau \left\{ \frac{1}{2}\rho h \left(\left\| \frac{\partial u_n(t)}{\partial t} \right\|^2_{L_2(\Omega_1)} + \left\| \frac{\partial v_n(t)}{\partial t} \right\|^2_{L_2(\Omega_1)} \right.\right.$$

$$\left. + \left\| \frac{\partial w_n(t)}{\partial t} \right\|^2_{L_2(\Omega_1)} \right) + \frac{\rho h^3}{12} \left\| \frac{\partial w_n(t)}{\partial t} \right\|^2_{\mathring{W}_2^1(\Omega_1)}$$

$$\left. +\frac{c_0}{T_0} \|\theta_n(t)\|^2_{L_2(\Omega_2)} + D \|w_n(t)\|^2_{\mathring{W}_2^2(\Omega_1)} \right\} dt. \tag{5.85}$$

Rejecting the term $\frac{\lambda_q}{T_0}\int\limits_0^T \|\theta_n\|^2_{\mathring{W}_2^1(\Omega_2)}\,dt$ and using Gronwill's lemma[2] [213], we obtain

$$\rho h\left(\left\|\frac{\partial u_n(t)}{\partial t}\right\|^2_{L_2(\Omega_1)}+\left\|\frac{\partial v_n(t)}{\partial t}\right\|^2_{L_2(\Omega_1)}+\left\|\frac{\partial w_n(t)}{\partial t}\right\|^2_{L_2(\Omega_1)}\right)$$

$$+\frac{\rho h^3}{12}\left\|\frac{\partial w_n(t)}{\partial t}\right\|^2_{\mathring{W}_2^1(\Omega_1)}+D\,\|w_n(t)\|^2_{\mathring{W}_2^2(\Omega_1)}+\frac{c_0}{T_0}\,\|\theta_n(t)\|^2_{L_2(\Omega_2)}\le c, \tag{5.86}$$

$$\forall t\in[0,T].$$

Returning to (5.85) and (5.83), with the help of (5.86) we obtain

$$\frac{\lambda_q}{T_0}\int\limits_0^T \|\theta_n\|^2_{\mathring{W}_2^1(\Omega_2)}\,dt\le c,\qquad c=const>0. \tag{5.87}$$

(The symbol c is used here mean various constants). From (5.83), using (5.86), (5.87), we obtain

$$\int\limits_{\Omega_1}\left[\left(\frac{\partial u_n(t)}{\partial x}-k_x w_n(t)+\frac{1}{2}\left(\frac{\partial w_n(t)}{\partial x}\right)^2\right)^2+\left(\frac{\partial v_n(t)}{\partial y}\right.\right.$$

$$\left.-k_y w_n(t)+\frac{1}{2}\left(\frac{\partial w_n(t)}{\partial y}\right)^2\right)^2+2\nu\left(\frac{\partial u_n(t)}{\partial x}-k_x w_n(t)\right.$$

$$\left.+\frac{1}{2}\left(\frac{\partial w_n(t)}{\partial x}\right)^2\right)\left(\frac{\partial v_n(t)}{\partial y}-k_y w_n(t)+\frac{1}{2}\left(\frac{\partial w_n(t)}{\partial y}\right)^2\right)$$

$$\left.+\frac{1-\nu}{2}\left(\frac{\partial u_n(t)}{\partial y}+\frac{\partial v_n(t)}{\partial x}+\frac{\partial w_n(t)}{\partial x}\frac{\partial w_n(t)}{\partial y}\right)^2\right]d\Omega_1\le c. \tag{5.88}$$

---

[2]Gronwill's lemma may be stated as follows. Let $v(t)$ be a nonnegative function continuous on $[a,b]$, let $c\ge 0$ be a constant, and in addition let $v(t)\le c+\int\limits_a^t v(\sigma)\,d\sigma$, $a\le t\le b$. Then $v(t)\le c_1$, where $c_1=\exp(\int\limits_a^t d\sigma)$, and if $c=0$ then $v=0$.

From (5.88), using the Cauchy inequality, we obtain

$$\int\limits_{\Omega_1} \left(\frac{\partial u_n(t)}{\partial x} \right)^2 d\Omega_1 \le c, \quad \int\limits_{\Omega_1} \left(\frac{\partial v_n(t)}{\partial y} \right)^2 d\Omega_1 \le c,$$

$$\int\limits_{\Omega_1} \left(\frac{\partial u_n(t)}{\partial y} + \frac{\partial v_n(t)}{\partial x} \right)^2 d\Omega_1 \le c. \tag{5.89}$$

Furthermore, using (5.89) and Korn's inequality [245, 248], we obtain

$$\int\limits_{\Omega_1} \left[\left(\frac{\partial u_n}{\partial x} \right)^2 + \left(\frac{\partial u_n}{\partial y} \right)^2 + \left(\frac{\partial v_n}{\partial x} \right)^2 + \left(\frac{\partial v_n}{\partial y} \right)^2 \right] d\Omega_1$$

$$\le c \int\limits_{\Omega_1} \left[\left(\frac{\partial u_n}{\partial x} \right)^2 + \left(\frac{\partial v_n}{\partial y} \right)^2 + \left(\frac{\partial u_n}{\partial y} + \frac{\partial v_n}{\partial x} \right)^2 \right] d\Omega_1 \le c, \tag{5.90}$$

$$c = const > 0.$$

Remark 5.1 With the help of the prior estimates (5.86), (5.90) it is possible to extend the solution of the system (5.70) to the whole interval $[0, T]$. Owing to the orthonormalization of the basis functions in the spaces $L_2(\Omega_1)$, $\overset{\circ}{W}{}_2^1(\Omega_1)$, $L_2(\Omega_2)$, we obtain

$$c \ge \left\| \frac{\partial u_n(t)}{\partial t} \right\|^2_{L_2(\Omega_1)} = \sum_{i=1}^{n} (\dot{g}_{2i}(t))^2, \quad \forall t \in [0, T]. \tag{5.91}$$

Because $\|u_n\|^2_{L_2(\Omega_1)} \le c$, $\|u_n\|^2_{\overset{\circ}{W}{}_2^1(\Omega_1)} \le c$, we have

$$c \ge \|u_n(t)\|^2_{L_2(\Omega_1)} = \sum_{i=1}^{n} (g_{2i}(t))^2, \quad \forall t \in [0, T], \tag{5.92}$$

and estimates for other functions can be obtained analogously. The estimates (5.91), (5.92) show that $g_{2i}(t)$ and $\dot{g}_{2i}(t)$ lie in a certain sphere R of the space $C[0, T]$ and, in what follows, a maximal value of M with respect to all $g_{2i}(t)$, $\dot{g}_{2i}(t) \in R$ can be obtained. Substituting the value of $\max\{M\}$ into (5.80) we obtain the time interval $[0, t_0]$ on which the lemma on the solvability of (5.70) guarantees a solution for arbitrary initial conditions $\overline{g}(T_0)$, $\overline{g}(T_0) \in R$. Therefore, the whole interval $[0, T]$ is covered by a finite number of intervals with length t_0, which implies that the system (5.70) has a solution in the whole interval $[0, T]$.

Step 3. Limiting Transition It is known that an arbitrary bounded set in a Hilbert space is weakly compact. Thus, using the prior estimates (5.86), (5.87), (5.90) obtained in the step 2, we can separate out the following series with the following properties:

$$\frac{\partial u_{n_k}}{\partial t} \to \frac{\partial u}{\partial t} \quad \text{weak in} \quad L_2(0,T;L_2(\Omega_1)),$$

$$\frac{\partial v_{n_k}}{\partial t} \to \frac{\partial v}{\partial t} \quad \text{weak in} \quad L_2(0,T;L_2(\Omega_1)),$$

$$\frac{\partial w_{n_k}}{\partial t} \to \frac{\partial w}{\partial t} \quad \text{weak in} \quad L_2(0,T;\overset{\circ}{W}{}^1_2(\Omega_1)),$$

$$u_{n_k} \to u \quad \text{weak in} \quad L_2(0,T;\overset{\circ}{W}{}^1_2(\Omega_1)),$$

$$v_{n_k} \to v \quad \text{weak in} \quad L_2(0,T;\overset{\circ}{W}{}^1_2(\Omega_1)),$$

$$w_{n_k} \to w \quad \text{weak in} \quad L_2(0,T;\overset{\circ}{W}{}^2_2(\Omega_1)),$$

$$\theta_{n_k} \to \theta \quad \text{weak in} \quad L_2(0,T;\overset{\circ}{W}{}^1_2(\Omega_1)).$$

Let us construct, for the arbitrary functions $\alpha_e(t) \in C^1([0,T])$, $\alpha_e(T) = 0$, the following operations:

$$\Phi_1 = \sum_{i=1}^{n} \alpha_i(t) \otimes \chi_{1i}(x,y), \;\; \Phi_2 = \sum_{i=1}^{n} \alpha_i(t) \otimes \chi_{2i}(x,y),$$

$$\Phi_3 = \sum_{i=1}^{n} \alpha_i(t) \otimes \chi_{3i}(x,y).$$

We multiply each of the Eqs. (5.64)–(5.67) by $\alpha_l(t)$, sum them over l from 1 to n, and then integrate with respect to t. We obtain

$$-\rho h \int_0^T \left(\frac{\partial u_n}{\partial t}, \frac{\partial \Phi_2}{\partial t}\right)_{L_2(\Omega_1)} dt + \frac{Eh}{1-\nu^2} \int_0^T \left(\varepsilon_{11}^n + \nu \varepsilon_{22}^n, \frac{\partial \Phi_2}{\partial x}\right)_{L_2(\Omega_1)} dt$$

$$+\frac{Eh}{2(1+\nu)} \int_0^T \left(\varepsilon_{12}^n, \frac{\partial \Phi_2}{\partial y}\right)_{L_2(\Omega_1)} dt - \frac{E\alpha_T}{1-\nu} \int_0^T \left(\int_{-\frac{h}{2}}^{\frac{h}{2}} \theta_n \, dz, \frac{\partial \Phi_2}{\partial x}\right)_{L_2(\Omega_1)} dt$$

$$= \int_0^T (p_1, \Phi_2)_{L_2(\Omega_1)} dt + \left(\frac{\partial u_n}{\partial t}, \Phi_2\right)_{L_2(\Omega_1)}\Bigg|_{t=0},$$

$$-\rho h\int\limits_0^T\left(\frac{\partial v_n}{\partial t},\frac{\partial \Phi_2}{\partial t}\right)_{L_2(\Omega_1)}dt+\frac{Eh}{1-\nu^2}\int\limits_0^T\left(\varepsilon_{22}^n+\nu\varepsilon_{11}^n,\frac{\partial \Phi_2}{\partial y}\right)_{L_2(\Omega_1)}dt$$

$$+\frac{Eh}{2(1+\nu)}\int\limits_0^T\left(\varepsilon_{12}^n,\frac{\partial \Phi_2}{\partial x}\right)_{L_2(\Omega_1)}dt-\frac{E\alpha_T}{1-\nu}\int\limits_0^T\left(\int\limits_{-\frac{h}{2}}^{\frac{h}{2}}\theta_n\,dz,\frac{\partial \Phi_2}{\partial y}\right)_{L_2(\Omega_1)}dt$$

$$=\int\limits_0^T(p_2,\Phi_2)_{L_2(\Omega_1)}\,dt+\left.\left(\frac{\partial v_n}{\partial t},\Phi_2\right)_{L_2(\Omega_1)}\right|_{t=0},$$

$$-\rho h\int\limits_0^T\left(\frac{\partial w_n}{\partial t},\frac{\partial \Phi_1}{\partial t}\right)_{L_2(\Omega_1)}dt-\frac{\rho h^3}{12}\int\limits_0^T\left(\frac{\partial^2 w_n}{\partial t\,\partial x},\frac{\partial^2 \Phi_1}{\partial x\,\partial t}\right)_{L_2(\Omega_1)}dt$$

$$-\frac{\rho h^3}{12}\int\limits_0^T\left(\frac{\partial^2 w_n}{\partial t\,\partial y},\frac{\partial^2 \Phi_1}{\partial y\,\partial t}\right)_{L_2(\Omega_1)}dt+D\int\limits_0^T(w_n,\Phi_1)_{\mathring{W}_2^2(\Omega_1)}\,dt$$

$$+\frac{E\alpha_T}{1-\nu}\int\limits_0^T\left(\int\limits_{-\frac{h}{2}}^{\frac{h}{2}}\theta_n z\,dz,\frac{\partial^2 \Phi_1}{\partial x^2}+\frac{\partial^2 \Phi_1}{\partial y^2}\right)_{L_2(\Omega_1)}dt$$

$$-\frac{Eh}{1-\nu^2}\int\limits_0^T(\varepsilon_{11}^n+\nu\varepsilon_{22}^n,k_x\Phi_1)_{L_2(\Omega_1)}\,dt-\frac{Eh}{1-\nu^2}\int\limits_0^T(\varepsilon_{22}^n+\nu\varepsilon_{11}^n,k_y\Phi_1)_{L_2(\Omega_1)}\,dt$$

$$+\frac{E\alpha_T}{1-\nu2}\int\limits_0^T\left(\int\limits_{-\frac{h}{2}}^{\frac{h}{2}}\theta_n\,dz,[k_x+k_y]\Phi_1\right)_{L_2(\Omega_1)}dt+\frac{Eh}{1-\nu^2}\int\limits_0^T\left(\frac{\partial w_n}{\partial x}\left(\varepsilon_{11}^n\right.\right.$$

$$\left.\left.+\nu\varepsilon_{22}^n,\frac{\partial \Phi_1}{\partial x}\right)\right)_{L_2(\Omega_1)}dt+\frac{Eh}{1-\nu^2}\int\limits_0^T\left(\frac{\partial w_n}{\partial y}\left(\varepsilon_{22}^n+\nu\varepsilon_{11}^n,\frac{\partial \Phi_1}{\partial y}\right)\right)_{L_2(\Omega_1)}dt$$

$$+\frac{Eh}{2(1+\nu)}\int\limits_0^T\left(\frac{\partial w_n}{\partial y}\varepsilon_{12}^n,\frac{\partial \Phi_1}{\partial x}\right)_{L_2(\Omega_1)}dt+\frac{Eh}{2(1+\nu)}\int\limits_0^T\left(\frac{\partial w_n}{\partial x}\varepsilon_{12}^n,\frac{\partial \Phi_1}{\partial y}\right)_{L_2(\Omega_1)}dt$$

$$-\frac{E\alpha_T}{1-\nu}\int\limits_0^T\left(\frac{\partial w_n}{\partial x}\theta_n,\frac{\partial \Phi_1}{\partial x}\right)_{L_2(\Omega_1)}dt-\frac{E\alpha_T}{1-\nu}\int\limits_0^T\left(\frac{\partial w_n}{\partial y}\theta_n,\frac{\partial \Phi_1}{\partial y}\right)_{L_2(\Omega_1)}dt$$

$$= \int_0^T (g_1, \Phi_1)_{L_2(\Omega_1)}\,dt + \left(\frac{\partial w_n}{\partial x}, \Phi_1\right)_{L_2(\Omega_1)}\Bigg|_{t=0} + \left(\frac{\partial^2 w_n}{\partial t\,\partial x}, \frac{\partial \Phi_1}{\partial x}\right)_{L_2(\Omega_1)} dt\Bigg|_{t=0}$$

$$+ \left(\frac{\partial^2 w_n}{\partial t\,\partial y}, \frac{\partial \Phi_1}{\partial y}\right)_{L_2(\Omega_1)} dt\Bigg|_{t=0}, \tag{5.93}$$

$$-c_0 \int_0^T \left(\theta_n, \frac{\partial \Phi_3}{\partial t}\right)_{L_2(\Omega_1)} dt - \lambda_q \int_0^T (\theta_n, \Phi_3)_{\mathring{W}_2^1(\Omega_2)}\,dt$$

$$= -\frac{E\alpha_T T_0}{1-\nu} \int_0^T \Bigg\{ -\left(\frac{\partial u_n}{\partial t}, \frac{\partial \Phi_3}{\partial x}\right)_{L_2(\Omega_2)} - \left([k_x - k_y]\frac{\partial w_n}{\partial t}, \Phi_3\right)_{L_2(\Omega_2)}$$

$$+ \left(\frac{\partial^2 w_n}{\partial t\,\partial x}\frac{\partial w_n}{\partial x} + \frac{\partial^2 w_n}{\partial t\,\partial y}\frac{\partial w_n}{\partial y}, \Phi_3\right)_{L_2(\Omega_2)} - \left(\frac{\partial v_n}{\partial t}, \frac{\partial \Phi_3}{\partial y}\right)_{L_2(\Omega_2)}$$

$$+ \left(\frac{\partial^2 w_n}{\partial t\,\partial x}, z\frac{\partial \Phi_3}{\partial x}\right)_{L_2(\Omega_2)} + \left(\frac{\partial^2 w_n}{\partial t\,\partial y}, z\frac{\partial \Phi_3}{\partial y}\right)_{L_2(\Omega_2)} \Bigg\} dt$$

$$+ \int_0^T (g_2, \Phi_3)_{L_2(\Omega_2)}\,dt + (\theta_n, \Phi_3)_{L_2(\Omega_2)}\Big|_{t=0},$$

$$\varepsilon_{11}^n = \frac{\partial u_n}{\partial x} - k_x w_n + \frac{1}{2}\left(\frac{\partial w_n}{\partial x}\right)^2, \quad \varepsilon_{22}^n = \frac{\partial v_n}{\partial y} - k_y w_n + \frac{1}{2}\left(\frac{\partial w_n}{\partial y}\right)^2,$$

$$\varepsilon_{12}^n = \frac{\partial u_n}{\partial y} + \frac{\partial v_n}{\partial y} + \frac{\partial w_n}{\partial x}\frac{\partial w_n}{\partial y}.$$

We take the limit of (5.93) in accordance with (5.92). It is clear that attention needs to be focused only on the nonlinear terms, and therefore we consider some of the more typical nonlinear terms in detail. Before proceeding let us note that the transition to the limit for the nonlinear terms is carried out using the following lemmas:

Lemma 5.2 *An arbitrary series that is weakly convergent in* $\mathring{W}_2^{1,1}(Q_T)$, $Q_T = \Omega \times (0, T)$, *is strongly compact in the space* $L_4(Q_T)$ *when* Ω *is a bounded area of* E_2.

Lemma 5.3 *Let* $a(x, y, t) \in \mathring{W}_2^{2,1}(Q_T)$, $Q_T = \Omega \times (0, T)$, $\Omega \subset E_2$. *In this case* $\frac{\partial a}{\partial x}, \frac{\partial a}{\partial y} \in L_4(Q_T)$, *and* $\|a\|_{L_2(\Omega)}$ *is a continuous function of time.*

Lemma 5.4 *If the series* $\{a_n(x, y, t)\}$ *is weakly convergent in* $\overset{\circ}{W}{}_2^{2,1}(Q_T)$, $Q_T = \Omega \times (0, T)$, $\Omega \subset E_2$, *to a certain function* $a_0 \in \overset{\circ}{W}{}_2^{2,1}(Q_T)$, *then the series* $\left\{\frac{\partial a_n}{\partial x}\right\}$, $\left\{\frac{\partial a_n}{\partial y}\right\}$ *are strongly convergent in the space* $L_4(Q_T)$ *to the functions* $\frac{\partial a_0}{\partial x}$, $\frac{\partial a_n}{\partial y}$, *respectively.*

In these lemmas, the space $\overset{\circ}{W}{}_m^{l_1,l_2}(Q_T)$ has been used, which is the subspace of $W_m^{l_1,l_2}(Q_T)$ consisting of those elements which are equal to zero on $S_T = \partial\Omega \times (0, T)$, $\overline{\Omega} = \partial\Omega \cup \Omega$. $W_m^{l_1,l_2}(Q_T)$ is the Banach space of the functions $u(\overline{x}, t)$, $\overline{x} = (x, y)$, belonging to $L_m(Q_T)$, together with their derivatives with respect to $\overline{x}$ of order l_1, and their derivatives with respect to t of order l_2. In this space, the norm can be defined in the following way:

$$\|u\|_{W_m^{l_1,l_2}} = \left\{ \int\limits_{Q_T} \left(\sum_{k=0}^{l_1} \sum_{(k)} |D_{\overline{x}u}^k|^m + \sum_{k=1}^{l_2} |D_t^k u|^m \right) d\overline{x}\, dt \right\}^{1/m},$$

where $D_{\overline{x}}^k$ and D_t^k denote the derivatives with respect to the corresponding variables.

Let us consider the limiting transition in the following terms (by w_0 and θ_0 we mean the weak limits of the series $\{w_n\}$ and $\{\theta_n\}$ in the sense of (5.92)):

1.

$$\int\limits_0^T \int\limits_{\Omega_1} \left(\frac{\partial w_n}{\partial x} \frac{1}{2} \left(\frac{\partial w_n}{\partial x} \right)^2 \frac{\partial \Phi_1}{\partial x} - \frac{\partial w_0}{\partial x} \frac{1}{2} \left(\frac{\partial w_0}{\partial x} \right)^2 \frac{\partial \Phi_1}{\partial x} \right) d\Omega_1\, dt$$

$$= \frac{1}{2} \int\limits_0^T \int\limits_{\Omega_1} \left[\left(\frac{\partial w_n}{\partial x} - \frac{\partial w_0}{\partial x} \right) \left(\frac{\partial w_n}{\partial x} \right)^2 \frac{\partial \Phi_1}{\partial x} \right.$$

$$\left. + \left(\frac{\partial w_n}{\partial x} - \frac{\partial w_0}{\partial x} \right) \left(\frac{\partial w_n}{\partial x} + \frac{\partial w_0}{\partial x} \right) \frac{\partial w_0}{\partial x} \frac{\partial \Phi_1}{\partial x} \right] d\Omega_1\, dt$$

(using the Holder inequality)

$$\leq \frac{1}{2} \int\limits_0^T \left\{ \left[\int\limits_{\Omega_1} \left(\frac{\partial w_n}{\partial x} - \frac{\partial w_0}{\partial x} \right)^2 \left(\frac{\partial \Phi_1}{\partial x} \right)^2 d\Omega_1 \right]^{1/2} \left[\int\limits_{\Omega_1} \left(\frac{\partial w_n}{\partial x} \right)^4 d\Omega_1 \right]^{1/2} \right.$$

$$+ \left[\int\limits_{\Omega_1} \left(\frac{\partial w_n}{\partial x} - \frac{\partial w_0}{\partial x} \right)^2 \left(\frac{\partial w_n}{\partial x} - \frac{\partial w_0}{\partial x} \right)^2 d\Omega_1 \right]^{1/2}$$

$$\times \left[\int_{\Omega_1} \left(\frac{\partial w_{0n}}{\partial x} \right)^2 \left(\frac{\partial \Phi_1}{\partial x} \right)^2 d\Omega_1 \right]^{1/2} \Bigg\} dt$$

$$\leq \frac{1}{2} \left\| \frac{\partial w_n}{\partial x} - \frac{\partial w_0}{\partial x} \right\|_{L_4(Q_1)} \left\| \frac{\partial w_n}{\partial x} \right\|_{L_4(Q_1)} \left\| \frac{\partial \Phi_1}{\partial x} \right\|_{L_4(Q_1)}$$

$$+ \frac{1}{2} \left\| \frac{\partial w_n}{\partial x} - \frac{\partial w_0}{\partial x} \right\|_{L_4(Q_1)} \left[8 \int_0^T \int_{\Omega_1} \left(\left(\frac{\partial w_n}{\partial x} \right)^4 \right.\right.$$

$$\left.\left. + \left(\frac{\partial w_0}{\partial x} \right)^4 \right) d\Omega_1 \, dt \right]^{1/4} \left\| \frac{\partial w_0}{\partial x} \right\|_{L_4(Q_1)} \left\| \frac{\partial \Phi_1}{\partial x} \right\|_{L_4(Q_1)}.$$

It is clear that, owing to Lemma 5.3, all of the integrals in the expression above exist and are bounded, whereas, owing to Lemma 5.4, the whole expression approaches zero.

2.

$$\int_0^T \int_{\Omega_1} \left\{ \frac{\partial w_n}{\partial x} \left[\int_{-\frac{h}{2}}^{\frac{h}{2}} \theta_n \, dz \right] \frac{\partial \Phi_1}{\partial x} - \frac{\partial w_0}{\partial x} \left[\int_{-\frac{h}{2}}^{\frac{h}{2}} \theta \, dz \right] \frac{\partial \Phi_1}{\partial x} \right\} d\Omega_1 \, dt$$

$$= \int_0^T \int_{\Omega_1} \left(\frac{\partial w_n}{\partial x} - \frac{\partial w_0}{\partial x} \right) \frac{\partial \Phi_1}{\partial x} \left[\int_{-\frac{h}{2}}^{\frac{h}{2}} \theta_n \, dz \right] d\Omega_1 \, dt$$

$$+ \int_0^T \int_{\Omega_1} \frac{\partial w_0}{\partial x} \frac{\partial \Phi_1}{\partial x} \left[\int_{-\frac{h}{2}}^{\frac{h}{2}} (\theta_n - \theta_0) \, dz \right] d\Omega_1 \, dt$$

(using again the Holder inequality)

$$\leq \left\{ \int_0^T \int_{\Omega_1} \left(\frac{\partial w_n}{\partial x} - \frac{\partial w_0}{\partial x} \right)^2 \left(\frac{\partial \Phi_1}{\partial x} \right)^2 d\Omega_1 \, dt \right\}^{1/2} \left\{ \int_0^T \int_{\Omega_1} \left[\int_{-\frac{h}{2}}^{\frac{h}{2}} \theta_n \, dz \right]^2 d\Omega_1 \, dt \right\}^{1/2}$$

$$+ \int_0^T \int_{\Omega_1} \int_{-\frac{h}{2}}^{\frac{h}{2}} \frac{\partial w_0}{\partial x} \frac{\partial \Phi_1}{\partial x} (\theta_n - \theta_0) dz \, d\Omega_1 \, dt$$

$$\leq \left\| \frac{\partial w_n}{\partial x} - \frac{\partial w_0}{\partial x} \right\|_{L_4(Q_1)} \left\| \frac{\partial \Phi_1}{\partial x} \right\|_{L_4(Q_1)} (h)^{1/2} \|\theta_n\|_{L_2(Q_1)}$$

$$+ \int_0^T \int_{\Omega_2} (\theta_n - \theta_0) \frac{\partial w_0}{\partial x} \frac{\partial \Phi_1}{\partial x} d\Omega_2 \, dt.$$

The first term in the last expression approaches zero according to Lemmas 5.3, 5.4, whereas the second approaches zero owing to the convergence of the series $\{\theta_n\}$ to θ_0 in $L^2(0, T; L_2(\Omega_2))$, or $\frac{\partial w_1}{\partial x} \frac{\partial \Phi_1}{\partial x} \in L_2(\Omega_2)$.

3.

$$\int_0^T \int_{\Omega_1} \int_{-\frac{h}{2}}^{\frac{h}{2}} \left(\frac{\partial^2 w_n}{\partial t \, \partial x} \frac{\partial w_n}{\partial x} \Phi_3 - \frac{\partial^2 w_0}{\partial t \, \partial x} \frac{\partial w_0}{\partial x} \Phi_3 \right) dz \, d\Omega_1 \, dt$$

$$= \int_0^T \int_{\Omega_1} \int_{-\frac{h}{2}}^{\frac{h}{2}} \left\{ \left(\frac{\partial w_n}{\partial x} - \frac{\partial w_{0n}}{\partial x} \right) \frac{\partial^2 w_n}{\partial t \, \partial x} \Phi_3 + \left(\frac{\partial^2 w_n}{\partial t \, \partial x} - \frac{\partial^2 w_0}{\partial t \, \partial x} \right) \frac{\partial w_0}{\partial x} \Phi_3 \right\} dz \, d\Omega_1 \, dt$$

Note that owing to the special form of the function Φ_3 in the first and the second term, we initially integrate the components of Φ_3 along z. As a result, Φ_3 will contain only components depending on the two spatial variables x, y and on the time t. Therefore

$$\Phi_3^1 \in \mathring{W}_2^{1,1}(Q_1), \quad \Phi_3^1 = \sum_{i=1}^{n} \alpha_i(t) \otimes a_i(x, y) c_i, \quad c_i = \int_{-\frac{h}{2}}^{\frac{h}{2}} b_i(z) \, dz,$$

and Lemma 5.2 is applicable to Φ_3.

Taking into account the above considerations, we extend the inequality as follows:

$$\leq \left\| \frac{\partial^2 w_n}{\partial t \, \partial x} \right\|_{L_2(Q_1)} \left\| \frac{\partial w_n}{\partial x} - \frac{\partial w_0}{\partial x} \right\|_{L_4(Q_1)} \|\Phi_3\|_{L_4(Q_1)}$$

$$+ \int_0^T \int_{\Omega_2} \left(\frac{\partial^2 w_n}{\partial t \partial x} - \frac{\partial^2 w_0}{\partial t \partial x} \right) \frac{\partial w_0}{\partial x} \Phi_3^1 \, d\Omega_2 \, dt.$$

The first term approaches zero because of Lemma 5.4, and the second term approaches zero either because of a weak convergence of $\frac{\partial^2 w_n}{\partial t\,\partial x}$ in $L_2(Q_1)$ or because $\frac{\partial w}{\partial x}\Phi_3^1 \in L_2(Q_1)$.

For the remaining nonlinear terms, the limiting transition is performed in a similar way.

Let us focus on the initial conditions. The expressions

$$\left(\frac{\partial w_n}{\partial t} - \Psi_1, \Phi_1\right)_{L_2(Q_1)}\Bigg|_{t=0}, \qquad (\theta_n - \theta_0, \Phi_3)_{L_2(Q_2)}\big|_{t=0},$$

and others approach zero when the approximate solution to the problem is constructed. The rest of the initial conditions must be understood in the usual way [213], i.e.

$$\lim_{t\to 0}\|w_0 - \varphi_1\|_{L_2(Q_1)} = 0,\ \lim_{t\to 0}\|u_0 - \varphi_2\|_{L_2(Q_1)} = 0,\ \lim_{t\to 0}\|v_0 - \varphi_3\|_{L_2(Q_1)} = 0,$$

where $w_0,\ v_0,\ u_0$ is the solution to the problem under consideration, and the existence of the limits results from a theorem about traces of the space $W_2^{1,1}(Q_1)$ [213] and a theorem on inclusion that states $W_2^{2,1} \subset W_2^{1,1}(Q_1)$.

This step ends the proof of the theorem about the existence of a generalized solution (in the sense of (5.60)–(5.62)) to the problem.

Let us now formulate a new theorem. For a class of sufficiently smooth initial data, the solution to the problem (5.42)–(5.47) is unique. This holds also for nonlinear problem if, in (5.42)–(5.47), the nonlinear terms (5.56)–(5.59) are neglected (and in addition, the term related to the rotational inertia of the shell element $\rho\frac{h^3}{12}\frac{\partial^2}{\partial t^2}(\Delta w)$ is ignored).

Theorem 5.2 *Let $\partial\Omega_1$ satisfy the conditions of Theorem 5.1 and, in addition*

$$p_1, p_2, g_1, \frac{\partial p_1}{\partial t}, \frac{\partial p_2}{\partial t}, \frac{\partial g_1}{\partial t} \in L^2(0, T; L_2(\Omega_1)),$$

$$g_2, \frac{\partial g_2}{\partial t} \in L^2(0, T; L_2(\Omega_2)),$$

$$\varphi_1 \in \mathring{W}_2^2(\Omega_1) \cap W_2^4(Q_1), \quad \Psi_1 \in \mathring{W}_2^2(\Omega_1),$$

$$\varphi_2, \varphi_3 \in \mathring{W}_2^1(\Omega_1) \cap W_2^2(Q_1), \quad \varphi_4 \in \mathring{W}_2^1(\Omega_2) \cap W_2^2(Q_2),$$

$$\Psi_2, \Psi_3 \in \mathring{W}_2^1(\Omega_1).$$

Then, a unique choice of the four functions $\{w, u, v, \theta\}$ that is the solution to the problem (5.56)–(5.59) exists, and these functions satisfy the following conditions:

$$w, \frac{\partial w}{\partial t} \in L^2(0, T; \overset{\circ}{W}{}_2^2(\Omega_1)), \quad \frac{\partial^2 w}{\partial t^2} \in L^2(0, T; L_2(\Omega_1)),$$

$$u, \frac{\partial u}{\partial t} \in L^2(0, T; \overset{\circ}{W}{}_2^1(\Omega_1)), \quad \frac{\partial^2 u}{\partial t^2} \in L^2(0, T; L_2(\Omega_1)),$$

$$v, \frac{\partial v}{\partial t} \in L^2(0, T; \overset{\circ}{W}{}_2^1(\Omega_1)), \quad \frac{\partial^2 v}{\partial t^2} \in L^2(0, T; L_2(\Omega_1)),$$

$$\theta \in L^2(0, T; \overset{\circ}{W}{}_2^1(\Omega_2)), \quad \frac{\partial \theta}{\partial t} \in L^2(0, T; \overset{\circ}{W}{}_2^1(\Omega_2)).$$

Proof The proof of this theorem again relies on the Bubnov–Galerkin method, and therefore verifies its convergence. Because of the uniqueness of the solution, all series of approximate solutions (of the type given in Theorem 5.1) converge to the exact solution of the problem. Since a method for investigating the smoothness of the generalized solution is known [214, 226] and the principal elements of the proof have already been investigated carefully in the treatment of Theorem 5.1, only the fundamental steps of the proof will be outlined.

Step 1 We construct the system of ordinary differential equations related to the problem (5.56)–(5.59) according to the Faedo–Galerkin method. This system coincides with the system (5.64)–(5.67) if both the term responsible for rotational inertia and the nonlinear terms are ignored. We look for the following approximate solution:

$$w_n = \sum_{i=1}^{n} g_{5i}(t)\chi_{4i}(x, y), \quad u_n = \sum_{i=1}^{n} g_{6i}(t)\chi_{5i}(x, y),$$

$$v_n = \sum_{i=1}^{n} g_{7i}(t)\chi_{5i}(x, y), \quad \theta_n = \sum_{i=1}^{n} g_{8i}(t)\chi_{6i}(x, y, z),$$

where $\{\chi_{4i}\}$ is a full system in $\overset{\circ}{W}{}_2^2(\Omega_1) \cap W_2^4(Q_1)$, orthonormalized in $L_2(\Omega_1)$; $\{\chi_{5i}\}$ is a full system in $\overset{\circ}{W}{}_2^1(\Omega_1) \cap W_2^2(Q_1)$, orthonormalized in $L_2(\Omega_1)$; and $\{\chi_{6i}\}$ is a full system in $\overset{\circ}{W}{}_2^1(\Omega_2) \cap W_2^2(Q_2)$, orthonormalized in $L_2(\Omega_2)$ (it should be noted that a special representation for the function χ_{6i} is not needed now). Therefore, the solvability of the system is evident and, moreover, owing to the linear set of ordinary differential equations, this solution is unique for arbitrary n.

Step 2. First Prior Estimate In the same way as described in the treatment of Theorem 5.1, we obtain the prior estimates (5.86), (5.87), (5.90) (of course, without the term related to the rotational inertia).

Step 3. Second Prior Estimate We differentiate each of the equations of the linearized system (5.64)–(5.67) with respect to t (this is possible because $(\frac{\partial p_1}{\partial t}, \frac{\partial p_2}{\partial t}, \frac{\partial g_1}{\partial t}) \in L^2(Q_1)$, $\frac{\partial g_2}{\partial t} \in L^2(Q_2)$ owing to the conditions of the theorem). We multiply each of the resulting equations by $\frac{\partial^2 g_{il}}{\partial t^2}$, $j = 5, \ldots, 8$, and sum them over l from 1 to n. Because the problem is a linear one, we can repeat the procedure outlined above for the first prior estimate to construct a second prior estimate of the form (5.86), (5.87), (5.90). The only difference is that now, in each expression within the norm symbols, we have time derivatives:

$$\rho h \left(\left\| \frac{\partial^2 u_n(t)}{\partial t^2} \right\|^2_{L_2(\Omega_1)} + \left\| \frac{\partial^2 v_n(t)}{\partial t^2} \right\|^2_{L_2(\Omega_1)} + \left\| \frac{\partial^2 w_n(t)}{\partial t^2} \right\|^2_{L_2(\Omega_1)} \right)$$

$$+ D \left\| \frac{\partial^2 w_n(t)}{\partial t^2} \right\|^2_{\mathring{W}_2^2(\Omega_1)} + \frac{C_0}{T_0} \left\| \frac{\partial \theta_n(t)}{\partial t} \right\|^2_{L_2(\Omega_2)} \leq C, \quad t \in [0, T],$$

$$\frac{\lambda_q}{T_0} \int_0^T \left\| \frac{\partial \theta_n(t)}{\partial t} \right\|^2_{\mathring{W}_2^1(\Omega_2)} dt \leq C,$$

$$\left\| \frac{\partial u_n(t)}{\partial t} \right\|^2_{\mathring{W}_2^1(\Omega_1)} + \left\| \frac{\partial v_n(t)}{\partial t} \right\|^2_{\mathring{W}_2^1(\Omega_1)} \leq C. \tag{5.94}$$

Step 4. Limiting Transition On the basis of the first and the second (5.94) prior estimates, we pass to the limit as in the treatment of Theorem 5.1. In this case, the initial conditions are completed by the following relations:

$$\|w(x, y, t) - \Psi_1\|_{L_2(\Omega_1)} \to 0, \quad t \to 0,$$

$$\|u(x, y, t) - \Psi_2\|_{L_2(\Omega_1)} \to 0, \quad t \to 0,$$

$$\|v(x, y, t) - \Psi_3\|_{L_2(\Omega_1)} \to 0, \quad t \to 0,$$

and the proof is analogous to that in the case of Theorem 5.1.

Step 5. Uniqueness It is not difficult to check that in the framework of the theorem under consideration the following "energetical" equality holds:

$$\frac{1}{2} \left\{ \rho h \left(\left\| \frac{\partial u}{\partial t} \right\|^2_{L_2(\Omega_1)} + \left\| \frac{\partial v}{\partial t} \right\|^2_{L_2(\Omega_1)} + \left\| \frac{\partial w}{\partial t} \right\|^2_{L_2(\Omega_1)} \right) + D \|w\|^2_{\mathring{W}_2^2(\Omega_1)} \right.$$

$$+ \int_{\Omega_1} \frac{Eh}{1 - \nu^2} \left(\left[\frac{\partial u}{\partial x} - k_x w \right]^2 + \left[\frac{\partial v}{\partial y} - k_y w \right]^2 + 2\nu \left[\frac{\partial u}{\partial x} - k_x w \right] \left[\frac{\partial v}{\partial y} - k_y w \right] \right.$$

$$+\frac{1-\nu}{2}\left[\frac{\partial u}{\partial y}+\frac{\partial v}{\partial x}\right]^2\Bigg) d\Omega_1 + \frac{C_0}{T_0}\|\theta\|^2_{L_2(\Omega_2)}\Bigg\}\Bigg|_{t=0}^{t=t_1} + \frac{\lambda_q}{T_0}\int_0^{t_1}\|\theta\|^2_{\overset{\circ}{W}{}_2^1(\Omega_2)}\,dt$$

$$=\int_0^{t_1}\left[\int_{\Omega_1}\left(p_1\frac{\partial u}{\partial t}+p_2\frac{\partial v}{\partial t}+g_1\frac{\partial w}{\partial t}\right)d\Omega_1+\int_{\Omega_1}p_2\theta\,d\Omega_2\right]dt.$$

In order to obtain this equality, we proceed in the following stages:

1. The nonlinear terms in the system (5.44)–(5.47) are neglected.
2. Equation (5.44) is multiplied by a function equal to zero for $t \in [t_1, T]$ and to $\frac{\partial u}{\partial t}$ for $t \in [0, t_1]$.
3. Equation (5.45) is multiplied by a function equal to zero for $t \in [t_1, T]$ and to $\frac{\partial v}{\partial t}$ for $t \in [0, t_1]$.
4. Equation (5.46) (without the rotational-inertia term) is multiplied by a function equal to zero for $t \in [t_1, T]$ and to $\frac{\partial w}{\partial t}$ for $t \in [0, t_1]$.
5. Equation (5.47) is multiplied by a function equal to zero for $t \in [t, T]$ and to θ for $t \in [0, t_1]$;
6. Each of the equations obtained is integrated with respect to time and transformed by integration by parts.
7. The resulting equations are summed.

As a result, we obtain the "energetical" equation. Applying the Cauchy inequality and the Gronwill lemma to the "energetical" equation, we obtain

$$\Bigg\{\rho h\left(\left\|\frac{\partial u}{\partial t}\right\|^2_{L_2(\Omega_1)}+\left\|\frac{\partial v}{\partial t}\right\|^2_{L_2(\Omega_1)}+\left\|\frac{\partial w}{\partial t}\right\|^2_{L_2(\Omega_1)}\right)+D\,\|w\|^2_{\overset{\circ}{W}{}_2^2(\Omega_1)}$$
$$+\frac{C_0}{T_0}\|\theta\|^2_{L_2(\Omega_2)}\Bigg\}\Bigg|_{t=t_1}\le e^{t_1}c_3,$$

$$\frac{Eh}{1-\nu^2}\int_{\Omega_1}\Bigg[\left(\frac{\partial u}{\partial x}-k_x w\right)^2+\left(\frac{\partial v}{\partial y}-k_y w\right)^2+2\nu\left(\frac{\partial u}{\partial x}-k_x w\right)\left(\frac{\partial v}{\partial y}-k_y w\right)$$

$$+\frac{1-\nu}{2}\left(\frac{\partial u}{\partial y}+\frac{\partial v}{\partial x}\right)^2\Bigg]d\Omega_1\Bigg|_{t=t_1}\le C_3+\frac{1}{2}\int_0^{t_1}\Bigg\{\rho h\left(\left\|\frac{\partial u}{\partial t}\right\|^2_{L_2(\Omega_1)}+\left\|\frac{\partial v}{\partial t}\right\|^2_{L_2(\Omega_1)}\right.$$
$$\left.+\left\|\frac{\partial w}{\partial t}\right\|^2_{L_2(\Omega_1)}\right)+\frac{C_0}{T_0}\|\theta\|^2_{L_2(\Omega_2)}\Bigg\}dt,$$

$$\frac{\lambda_q}{T_0}\int_0^{t_1}\|\theta\|^2_{\mathring{W}_2^1(\Omega_2)}\,dt \le C_3+\frac{1}{2}\int_0^{t_1}\left\{\rho h\left(\left\|\frac{\partial u}{\partial t}\right\|^2_{L_2(\Omega_1)}+\left\|\frac{\partial v}{\partial t}\right\|^2_{L_2(\Omega_1)}\right.\right.$$
$$\left.\left.+\left\|\frac{\partial w}{\partial t}\right\|^2_{L_2(\Omega_1)}\right)+\frac{C_0}{T_0}\|\theta\|^2_{L_2(\Omega_2)}\right\}dt, \tag{5.95}$$

$$C_3=\rho h\left(\left\|\frac{\partial u}{\partial t}\right\|^2_{L_2(\Omega_1)}+\left\|\frac{\partial v}{\partial t}\right\|^2_{L_2(\Omega_1)}+\left\|\frac{\partial w}{\partial t}\right\|^2_{L_2(\Omega_1)}\right)+\frac{C_0}{T_0}\|\theta\|^2_{L_2(\Omega_2)}$$
$$+\frac{Eh}{1-\nu^2}\int_{\Omega_1}\left[\left(\frac{\partial u}{\partial x}-k_xw\right)^2+\left(\frac{\partial v}{\partial y}-k_yw\right)^2+2\nu\left(\frac{\partial u}{\partial x}-k_xw\right)\left(\frac{\partial v}{\partial y}-k_yw\right)\right.$$
$$\left.+\frac{1-\nu}{2}\left(\frac{\partial u}{\partial y}+\frac{\partial v}{\partial x}\right)^2\right]d\Omega_1\bigg|_{t=0}+\int_0^{t_1}\left\{\frac{1}{\rho h}\left(\|p_1\|^2_{L_2(\Omega_1)}+\|p_2\|^2_{L_2(\Omega_1)}\right.\right.$$
$$\left.\left.+\|g_1\|^2_{L_2(\Omega_1)}\right)+\frac{T_0}{C_0}\|g_2\|^2_{L_2(\Omega_2)}\right\}dt. \tag{5.96}$$

Let us consider now two solutions $\{w_1, u_1, v_1, \theta_1\}$ and $\{w_2, u_2, v_2, \theta_2\}$. It is clear that $\{w_1-w_2, u_1-u_2, v_1-v_2, \theta_1-\theta_2\}$ is also a solution to the problem, where we must take

$$p_1=p_2=g_1=g_2=\varphi_1=\varphi_2=\varphi_3=\varphi_4=\Psi_1=\Psi_2=\Psi_3=0,$$

which means that $c_3=0$. Substituting these values into (5.95), we obtain

$$w_1=w_2,\quad u_1=u_2,\quad v_1=v_2,\quad \theta_1=\theta_2,$$

which proves the uniqueness of the solution. This ends the proof.

Two new theorems related to the solvability of the problem (5.48)–(5.54) (resulting from linearization of (5.44)–(5.47)) will now be formulated for homogeneous boundary conditions. The proofs are omitted as they repeat fully the proofs of Theorems 5.1 and 5.2.

Theorem 5.3 *Let $\partial\Omega_1$ satisfy the conditions of Theorem 5.1, and let*

$$p_1, p_2, g_1\in L^2(0,T;L_2(\Omega_1)),\quad g_2\in L^2(0,T;L_2(\Omega_2)),\quad \varphi_i\in\mathring{W}_2^1(\Omega_1),$$
$$i=1,2,3,5,6,\quad \varphi_4\in\mathring{W}_2^1(\Omega_2),\quad \Psi_j\in L_2(\Omega_1),\quad j=1,\dots,5.$$

Then:

1. *There exists at least one set of six functions* $\{w, u, v, \Psi_x, \Psi_y, \theta\}$ *that is the solution to (5.48)–(5.54) (the linearized form of (5.44)–(5.47)), and the following conditions are satisfied:*

$$w, u, v, \Psi_x, \Psi_y \in L^2(0, T; \mathring{W}_2^1(\Omega_1)), \quad \theta \in L^2(0, T; \mathring{W}_2^1(\Omega_2)),$$
$$\frac{\partial w}{\partial t}, \frac{\partial u}{\partial t}, \frac{\partial v}{\partial t}, \frac{\partial \Psi_x}{\partial t}, \frac{\partial \Psi_y}{\partial t} \in L^2(0, T; L_2(\Omega_1)). \tag{5.97}$$

2. *An approximate solution to the specified problem can be found with the help of the Bubnov–Galerkin method. A set of approximate solutions is weakly compact in the spaces corresponding to (5.97), and each of the weak limits is a solution to the problem.*

Theorem 5.4 *Let the conditions of Theorem 5.3 be satisfied, and let*

$$\frac{\partial p_1}{\partial t}, \frac{\partial p_2}{\partial t}, \frac{\partial g_1}{\partial t} \in L^2(0, T; L_2(\Omega_1)), \quad \frac{\partial g_2}{\partial t} \in L^2(0, T; L_2(\Omega_2)),$$
$$\varphi_1, \varphi_2, \varphi_3, \varphi_5, \varphi_6 \in \mathring{W}_2^1(\Omega_1) \cap W_2^2(Q_1), \quad \varphi_4 \in \mathring{W}_2^1(\Omega_2) \cap W_2^2(Q_2),$$
$$\Psi_1, \Psi_2, \Psi_3, \Psi_4, \Psi_5 \in \mathring{W}_2^1(\Omega_1).$$

Then there exists a unique set of six functions $\{w, u, v, \Psi_x, \Psi_y, \theta\}$ *that is the solution to the problem* (5.48)–(5.54). *These functions satisfy the following conditions:*

$$w, u, v, \Psi_x, \Psi_y, \frac{\partial w}{\partial t}, \frac{\partial u}{\partial t}, \frac{\partial v}{\partial t}, \frac{\partial \Psi_x}{\partial t}, \frac{\partial \Psi_y}{\partial t} \in L^2(0, T; \mathring{W}_2^1(\Omega_1)),$$
$$\frac{\partial^2 w}{\partial t^2}, \frac{\partial^2 u}{\partial t^2}, \frac{\partial^2 v}{\partial t^2}, \frac{\partial^2 \Psi_x}{\partial t^2}, \frac{\partial^2 \Psi_y}{\partial t^2} \in L^2(0, T; L_2(\Omega_1)),$$
$$\theta \in L^2(0, T; \mathring{W}_2^1(\Omega_2)), \quad \frac{\partial \theta}{\partial t} \in L^2(0, T; \mathring{W}_2^1(\Omega_2)).$$

Remark 5.2 The contour bounding a shell that is rectangular or elliptic in plan can serve as a boundary contour satisfying the conditions of the above theorems.

Note that the method presented above for proving Theorems 5.1, 5.2, 5.3 and 5.4 can be applied also in other cases of a simply supported shell with displacements equal to zero on its contour.

To conclude, in this chapter we have stated and proved some mathematical theorems governing the process of thermoelastic deformation of a shallow shell with an arbitrary distribution of the temperature field through the shell thickness. This means that the initial–boundary value problems corresponding to the models used here are solvable. However, in order to judge how well those models correspond to the real processes occurring in shells, one needs to compare the numerical results with real experimental data.

Finally, it should be emphasized that the method of construction of an approximate solution used in Theorem 5.1 generalizes a known (in the theory of thermoelasticity) rule for approximation of the temperature function by an arbitrary polynomial with respect to the variable z [187].

Chapter 6
Theory with Physical Nonlinearities and Coupling

In this chapter, a theory of shells with physical nonlinearities and coupling is outlined. In Sect. 6.1, the fundamental assumptions and relations are introduced.

In Sect. 6.2, the variational equations of physically nonlinear coupled problems are derived. The equations are then reformulated into equations in terms of displacements (Sect. 6.3), which are suitable for a series of other thermoelastic problems.

6.1 Fundamental Assumptions and Relations

Similarly to previous chapters, a shallow shell with a constant thickness h having a shape Ω in plane and a boundary $d\Omega_1$ is considered. The shell material is assumed to be isotropic and physically nonlinear. An orthogonal system of coordinates x_1, x_2, x_3 is used here instead of the orthogonal system of coordinates x, y, z used in Chap. 5. As earlier, the x_1, x_2 coordinate axes coincide with the directions of the principal curvatures of the middle surface of the shell, and the x_3 coordinate is oriented towards the centre of curvature of the middle surface and is normal to the middle surface. The displacements of the points of the middle surface along x_1, x_2 and x_3 are denoted by $u_1(x_1, x_2, t)$, $u_2(x_1, x_2, t)$ and $w(x_1, x_2, t)$, respectively. The initial curvatures with respect to x_1, x_2 are denoted by k_1, k_2, respectively.

We shall now to derive the governing equations on the basis of the kinematic Kirchhoff–Love model, for which the following relations between the deformation of the middle surface and an arbitrary surface hold [336]:

$$\varepsilon_{ij}^{z} = \varepsilon_{ij} + x_3 æ_{ii} \quad (i, j = 1, 2) \tag{6.1}$$
$$\left(-\frac{h}{2} \leq x_3 \leq \frac{h}{2}\right).$$

J. Awrejcewicz, V. A. Krysko, *Elastic and Thermoelastic Problems in Nonlinear Dynamics of Structural Members*, Scientific Computation,
https://doi.org/10.1007/978-3-030-37663-5_6

Here ε_{ij} $(i, j = 1, 2)$ are the tangential deformations of the middle surface, and $æ_{ij}$ are its bending deformations, and the following relations hold:

$$\varepsilon_{ii} = \frac{\partial u_i}{\partial x_i} - k_1 w + \frac{1}{2}\left(\frac{\partial w}{\partial x_i}\right)^2 \quad (i = 1, 2), \tag{6.2}$$

$$\varepsilon_{12} = \frac{\partial u_1}{\partial x_2} + \frac{\partial u_2}{\partial x_1} + \frac{\partial w}{\partial x_1}\frac{\partial w}{\partial x_2}, \tag{6.3}$$

$$æ_{ii} = -\frac{\partial^2 w}{\partial {x_i}^2} \quad (i = 1, 2), \quad æ_{12} = -2\frac{\partial^2 w}{\partial x_1 \partial x_2}. \tag{6.4}$$

It should be noted that the deformations ε_{ij} are considered to be small in comparison with one. Thus, the derivatives of deformations can be neglected in comparison with the deformations.

The physical relations are assumed to be of the following form [172]:

$$\sigma_{ii} = 3Kæ(e_0)e_0 + 2G\gamma(e_i)(e_{ii} - e_0) \quad (i = 1, 2, 3), \tag{6.5}$$

$$\sigma_{ii} = G\gamma(e_1)e_{12}, \tag{6.6}$$

where K, G are the moduli of volume compression and shear, $e_0 = \frac{1}{3}(e_{11}^z + e_{22}^z + e_{33}^z)$ is the average deformation, $\gamma(e_1)$ is the shear function, $æ(e_0)$ is the elongation function, and

$$e_i = \frac{\sqrt{2}}{3}\sqrt{(e_{11}^z - e_{22}^z)^2 + (e_{22}^z - e_{33}^z)^2 + (e_{33}^z - e_{11}^z)^2 + \frac{3}{2}(e_{12}^z)^2}$$

is the deformation intensity (in accordance with [172], we take $æ(e_0) \equiv 1$).

Using the hypothesis of a plane stress–strain state ($\sigma_{33} = 0$) and taking into account the influence of temperature via the Duhamel–Neumann hypothesis (similarly to previous chapters), we obtain

$$\sigma_{11} = \frac{E\gamma(e_i)}{(1+\nu)(1-\tilde{\mu})}\left[\varepsilon_{11}^z + \tilde{\mu}\varepsilon_{22}^z - (1+\tilde{\mu})\alpha_t\theta\right], \tag{6.7}$$

$$\sigma_{22} = \frac{E\gamma(e_i)}{(1+\nu)(1-\tilde{\mu})}\left[\varepsilon_{22}^z + \tilde{\mu}\varepsilon_{11}^z - (1+\tilde{\mu})\alpha_t\theta\right], \tag{6.8}$$

$$\sigma_{12} = \frac{E\gamma(e_i)}{2(1+\nu)}\varepsilon_{12}^z, \tag{6.9}$$

where E is Young's modulus, ν is Poisson's ratio, $\theta(x_1, x_2, x_3, t) = T_1(x_1, x_2, x_3, t) - T_0$ is the temperature increase at the point x_1, x_2, x_3 at the time moment t, and $T_1(x_1, x_2, x_3, t)$ is the absolute temperature at this point at time t. We assume that $\left|\frac{\theta}{T_0}\right| \ll 1$, i.e. the temperature increase is so small that all of the thermoelastic parameters can be treated as constant (independent of temperature). The quantity α_t

is the coefficient of linear thermal expansion, and

$$\tilde{\mu} = \frac{3 - 2g\gamma(e_i)}{2(3 + g\gamma(e_i))}, \tag{6.10}$$

where $g = \frac{G}{K}$.

We introduce (similarly to previous chapters) the following integral stress characteristics:

$$T_{ij} = \int_{-\frac{h}{2}}^{\frac{h}{2}} \sigma_{ij}\, dx_3, \quad M_{ij} = \int_{-\frac{h}{2}}^{\frac{h}{2}} \sigma_{ij}\, x_3 dx_3. \tag{6.11}$$

Substituting (6.1) into (6.7)–(6.9) and then into (6.11), we obtain after a series of transformations

$$T_{ii} = \frac{Eh}{1 - \nu^2}(\varepsilon_{ii} + \nu\varepsilon_{jj}) + \Delta T_{ii} - N_T - \Delta_T T, \tag{6.12}$$

$$M_{ii} = \frac{Eh^3}{12(1 - \nu^2)}(æ_{ii} + \nu æ_{jj}) + \Delta M_{ii} - M_T - \Delta_T M \tag{6.13}$$

$$(i, j = 1, 2, \quad i \neq j),$$

$$T_{12} = \frac{Eh}{2(1 + \nu)}\varepsilon_{12} + \Delta T_{12}, \tag{6.14}$$

$$M_{12} = \frac{Eh^3}{24(1 + \nu)}æ_{12} + \Delta M_{12}, \tag{6.15}$$

$$N_T = \frac{E\alpha_t}{1 - \nu}\int_{-\frac{h}{2}}^{\frac{h}{2}} \theta\, dx_3, \quad M_T = \frac{E\alpha_t}{1 - \nu}\int_{-\frac{h}{2}}^{\frac{h}{2}} \theta x_3\, dx_3, \tag{6.16}$$

where

$$\Delta T_{ii} = \frac{Eh}{1 + \nu}\left[\varepsilon_{ii} p_1^1 + \varepsilon_{jj} p_1^2 + h(æ_{ii} p_2^1 + æ_{jj} p_2^2)\right] \tag{6.17}$$

$$\Delta M_{ii} = \frac{Eh^2}{1 + \nu}\left[\varepsilon_{ii} p_2^1 + \varepsilon_{jj} p_2^2 + h(æ_{ii} p_3^1 + æ_{jj} p_3^2)\right] \tag{6.18}$$

$$(i, j = 1, 2, \quad i \neq j),$$

$$\Delta T_{12} = \frac{Eh}{2(1 + \nu)}(\varepsilon_{12} p_1^3 + hæ_{12} p_2^3), \tag{6.19}$$

$$\Delta M_{12} = \frac{Eh^2}{2(1+\nu)}(\varepsilon_{12}p_2^3 + h æ_{12}p_2^3), \tag{6.20}$$

$$\Delta_T T = E\alpha_T W_1, \quad \Delta_T M = E\alpha_T W_2, \tag{6.21}$$

$$P_k^n = \frac{1}{h^k}\int_{-\frac{h}{2}}^{\frac{h}{2}} F_n x_3^{k-1}\, dx_3 \tag{6.22}$$

$$(n, k = 1, 2, 3),$$

$$W_i = \int_{-\frac{h}{2}}^{\frac{h}{2}} F_i \theta x_3^{i-1}\, dx_3 \quad (i = 1, 2), \tag{6.23}$$

$$F_1 = \frac{(1-\nu)\gamma(e_i) - (1-\tilde{\mu})}{(1-\nu)(1-\tilde{\mu})}, \tag{6.24}$$

$$F_2 = \frac{\tilde{\mu}(1-\nu)\gamma(e_i) - \nu(1-\tilde{\mu})}{(1-\nu)(1-\tilde{\mu})}, \tag{6.25}$$

$$F_3 = \gamma(e_i) - 1, \tag{6.26}$$

$$F_4 = \frac{(1+\tilde{\mu})(1-\nu)\gamma(e_i) - (1+\nu)(1-\tilde{\mu})}{(1-\nu^2)(1-\tilde{\mu})}. \tag{6.27}$$

Solving (6.12) and (6.14) for the deformations of the middle surface $\varepsilon_{11}, \varepsilon_{22}, \varepsilon_{12}$, and (6.13), (6.15) for $æ_{ii}, æ_{12}$, we obtain

$$\varepsilon_{ii} = \frac{1}{Eh}\Big[(T_{ii} - \nu T_{jj}) - (\Delta T_{ii} - \nu \Delta T_{jj}) + (1-\nu)(N_T + \Delta_T T)\Big], \tag{6.28}$$

$$\varepsilon_{12} = \frac{2(1+\nu)}{Eh}(T_{12} - \Delta T_{12}), \tag{6.29}$$

$$æ_{ii} = \frac{12}{Eh^3}\Big[(M_{ii} - \nu M_{jj}) - (\Delta M_{ii} - \nu \Delta M_{jj}) + (1-\nu)(M_T + \Delta_T M)\Big] \tag{6.30}$$

$$æ_{12} = \frac{24(1+\nu)}{Eh^3}(M_{12} - \Delta M_{12}) \tag{6.31}$$

$$(i, j = 1, 2, \quad i \neq j).$$

The theory of small elasto-plastic deformations yields

$$G\gamma(e_i) = \frac{1}{3}\frac{\sigma_i(e_i)}{e_i}, \tag{6.32}$$

$$\gamma(e_i) = \frac{\sigma_i(e_i)}{3Ge_i}, \tag{6.33}$$

where $\sigma_i(e_i)$ characterizes the relation between the stress intensity and the deformation intensity.

6.2 Variational Equations of Physically Nonlinear Coupled Problems

We apply the Biot variational principle [68]

$$\delta V_1 + \delta D + \delta K = \iint\limits_{\Omega_2} (\mathbf{F}_0\,\delta\mathbf{u}_0 - \theta\mathbf{n}\,\delta\mathbf{S})\,d\Omega_2, \tag{6.34}$$

where $\delta V_1, \delta D, \delta K$ are the variations of the generalized free vibrations of the dissipative function and of the kinetic energy, respectively; $\mathbf{F}_0$, $\mathbf{u}_0$ are the surface force and displacement; $\mathbf{n}$ is the external normal; and Ω_2 is the surface of a bounded body. The change of the entropy vector $\mathbf{S}$ is related to the increase of the entropy φ in a volume unit [68] by

$$\varphi = -div\,\mathbf{S} - -\sum_i \frac{\partial S_i}{\partial x_i}. \tag{6.35}$$

Using the results presented in [68], we obtain

$$\delta V_1 = \iiint\limits_V \left(\sum_{ij} \delta_{ij}\delta e_{ij} + \theta\,\delta\varphi\right) dV, \tag{6.36}$$

$$\delta D = \iiint\limits_V \frac{T_0}{R_T}\mathbf{S}\,\delta\mathbf{S}\,dV, \tag{6.37}$$

where R_T is the thermal conductivity, and V denotes the volume occupied by the shell.

Substituting (6.35) into (6.36) and integrating by parts, we obtain

$$\iiint\limits_V \theta\, \delta\varphi\, dV \equiv \iiint\limits_V grad\, \theta\, \delta\mathbf{S}\, dV - \iint\limits_{\Omega_2} \theta \mathbf{n}\, \delta\mathbf{S}\, d\Omega_2. \tag{6.38}$$

The surface integrals in (6.38) and (6.34) cancel each other, and from (6.38) we obtain

$$\iiint\limits_V \theta\, \delta\varphi\, dV + \delta D \equiv \iiint\limits_V \left(grad\, \theta + \frac{T_0}{R_T}\mathbf{S} \right) \delta\mathbf{S}\, dV. \tag{6.39}$$

From (6.38), we can deduce the following vector equation:

$$grad\, \theta + \frac{T_0}{R_T}\mathbf{S} = 0. \tag{6.40}$$

Taking the divergence of (6.40) and taking into account a result given in [68]

$$\varphi = \frac{c\theta}{T_0} + \beta e, \tag{6.41}$$

we obtain

$$\nabla^2 \theta = \frac{T_0}{R_T} \frac{\partial}{\partial t} \left(\frac{c\theta}{T_0} + \beta e \right), \tag{6.42}$$

where c is the specific heat capacity, $e = \varepsilon^z_{11} + \varepsilon^z_{22} + \varepsilon^z_{33}$, ∇^2 is the three-dimensional Laplace operator, $\beta = k\alpha_t$, and t denotes time.

Therefore, a heat transfer equation which includes coupling between the temperature and the deformation has been obtained.

In order to reduce the order of the integrals occurring in (6.34), we integrate the terms that are not used with respect to $-\frac{h}{2} \le x_3 \le \frac{h}{2}$. Neglecting the terms $\ddot{u}_1$, $\ddot{u}_2$, and taking into account (6.1), (6.11) and the condition $\sigma_{33} = 0$, we obtain

$$\delta K = -\rho h \iint\limits_{\Omega} \ddot{w}\, \delta w\, dx_1 dx_2, \tag{6.43}$$

$$\begin{aligned} \iiint\limits_V \sum_{i,j} \sigma_{ij}\, \delta e_{ij}\, dV = &\iint\limits_{\Omega_2} (T_{11}\, \delta\varepsilon_{11} + T_{22}\, \delta\varepsilon_{22} + T_{12}\, \delta\varepsilon_{12})\, dx_1 dx_2 \\ &+ \iint\limits_{\Omega_2} (M_{11}\, \delta æ_{11} + M_{22}\, \delta æ_{22} + M_{12}\, \delta æ_{12})\, dx_1 dx_2, \end{aligned} \tag{6.44}$$

$$\iint\limits_{\Omega} (\mathbf{F}_0\, \delta \mathbf{u}_0)\, d\Omega = \iint\limits_{\Omega} (q - \rho h \varepsilon \dot{w})\, \delta w\, dx_1 dx_2 + \int T_{11}^0\, \delta u_1\, dx_2$$
$$+ \int T_{22}^0\, \delta u_2\, dx_1 + \int T_{12}^0\, \delta u_1\, dx_1 + \int T_{12}^0\, \delta u_2\, dx_2, \tag{6.45}$$

where ε is a damping coefficient; q denotes the intensity of the transverse load; T_{11}^0, T_{22}^0, are given longitudinal loads along x_1, x_2, respectively; and T_{12}^0 is a given shear load.

In order to obtain the initial differential equation in a hybrid form, we transform the first term in the right-hand side of the relations (6.44) to the form

$$\iint\limits_{\Omega_2} (T_{11}\, \delta\varepsilon_{11} + T_{22}\, \delta\varepsilon_{22} + T_{12}\, \delta\varepsilon_{12})\, dx_1 dx_2$$
$$= \delta \iint\limits_{\Omega_2} (T_{11}\varepsilon_{11} + T_{22}\varepsilon_{22} + T_{12}\varepsilon_{12})\, dx_1 dx_2 \tag{6.46}$$
$$- \iint\limits_{\Omega_2} (\varepsilon_{11}\, \delta T_{11} + \varepsilon_{22}\, \delta T_{22} + \varepsilon_{12}\, \delta T_{12})\, dx_1 dx_2.$$

We introduce also the stress function F (Airy's stress function) in the form

$$T_{ii} = \frac{\partial F}{\partial x_{jj}^2}, \quad T_{12} = \frac{\partial^2 F}{\partial x_1 \partial x_2} \quad (i, j = 1, 2, \quad i \neq j), \tag{6.47}$$

as well as the known differential operators

$$\nabla_k^2(\cdot) = K_2 \frac{\partial^2(\cdot)}{\partial x_1{}^2} + K_1 \frac{\partial^2(\cdot)}{\partial x_2{}^2}, \tag{6.48}$$

$$L((\cdot), (\cdot)) = \frac{\partial^2(\cdot)}{\partial x_1{}^2} \frac{\partial^2(\cdot)}{\partial x_2{}^2} - 2 \frac{\partial^2(\cdot)}{\partial x_1 \partial x_2} \frac{\partial^2(\cdot)}{\partial x_1 \partial x_2} + \frac{\partial^2(\cdot)}{\partial x_2{}^2} \frac{\partial^2(\cdot)}{\partial x_1{}^2}. \tag{6.49}$$

We substitute the values of the deformations (6.2), (6.3) into the first integral on the right-hand side of (6.46), and substitute the values of $æ_{ij}$ from (6.4) and ε_{ij} from (6.28), (6.29) into the second integral on the right-hand side of each of (6.44), (6.46). In the transformed expression (6.44), we integrate by parts and take the variations, taking into account both (6.47) and the rule that the contour integrals in (6.45) cancel with the contour integrals obtained via transformation of (6.46). We move from the variational principle (6.34) to the following variational equation:

$$
\begin{aligned}
&\iiint\limits_V (grad\,\theta + \frac{T_0}{R_T}\mathbf{S})\,\delta\mathbf{S}\,dV - \iint\limits_\Omega \Big\{\Big\langle \nabla_k^2 w + \frac{1}{2}L(w,w) \\
&+\frac{1}{Eh}\Big[\nabla^2\,\nabla^2\,F - (\Delta T_{11} - \nu\Delta T_{22})_{x_2x_2} - (\Delta T_{22} - \nu\Delta T_{11})_{x_1x_1} \\
&+2(1+\nu)(\Delta T_{12})_{x_1x_2} + (1-\nu)\,\nabla^2\,(N_T + \Delta_T T)\Big]\Big\rangle\delta F + \Big[\nabla_k^2\,F \\
&+\frac{1}{2}L(w,F) - D\,\nabla^2\,\nabla^2 w - \nabla^2(M_T + \Delta_T M) + (\Delta M_{11})_{x_1x_1} \\
&+ (\Delta M_{22})_{x_2x_2} + 2(\Delta M_{12})_{x_1x_2} - \rho h(\ddot{w} + \varepsilon\dot{w}) + q\Big]\delta w\Big\}\,dx_1dx_2 \\
&-\int\left[k_1 w - \frac{1}{2}\left(\frac{\partial w}{\partial x_1}\right)^2 + \varepsilon_{11}\right](\delta F)_{x_2}\Bigg|_{x_2=0}^{x_2=b} dx_1 \\
&-\int\left[k_2 w - \frac{1}{2}\left(\frac{\partial w}{\partial x_2}\right)^2 + \varepsilon_{22}\right](\delta F)_{x_1}\Bigg|_{x_1=0}^{x_1=a} dx_2 + \int\Big[k_2 w\frac{\partial w}{\partial x_2} \\
&+\frac{\partial^2 w}{\partial {x_1}^2}\frac{\partial w}{\partial x_2} - \frac{\partial\varepsilon_{12}}{\partial x_1}\frac{\partial\varepsilon_{11}}{\partial x_2}\Big]\delta F\Bigg|_{x_2=0}^{x_2=b} dx_1 + \int\Big[k_2 w\frac{\partial w}{\partial x_1} \\
&+\frac{\partial^2 w}{\partial {x_2}^2}\frac{\partial w}{\partial x_1} - \frac{\partial\varepsilon_{12}}{\partial x_2}\frac{\partial\varepsilon_{22}}{\partial x_1}\Big]\delta F\Bigg|_{x_1=0}^{x_1=a} dx_2 + \left(\varepsilon_{12} - \frac{\partial w}{\partial x_1}\frac{\partial w}{\partial x_2}\right) \\
&\times\delta F\Bigg|_{x_1=0}^{x_1=a}\Bigg|_{x_2=0}^{x_2=b} - 2M_{12}\,\delta w\Bigg|_{\substack{x_1=0\\x_2=0}}^{\substack{x_1=a\\x_2=b}} - \int M_{11}(\delta w)_{x_1}\Bigg|_{x_1=0}^{x_1=a} dx_2 \\
&-\int M_{22}(\delta w)_{x_2}\Bigg|_{x_2=0}^{x_2=b} dx_1 + \int\left(\frac{\partial M_{11}}{\partial x_1} + 2\frac{\partial M_{12}}{\partial x_2} + T_{11}\frac{\partial w}{\partial x_1}\right. \\
&\left.+T_{12}\frac{\partial w}{\partial x_2}\right)\delta w\Bigg|_{x_1=0}^{x_1=a} dx_2 + \int\left(\frac{\partial M_{22}}{\partial x_2} + 2\frac{\partial M_{12}}{\partial x_1} + T_{22}\frac{\partial w}{\partial x_2}\right. \\
&\left.+T_{12}\frac{\partial w}{\partial x_2}\right)\delta w\Bigg|_{x_2=0}^{x_2=b} dx_1 = 0.
\end{aligned}
\tag{6.50}
$$

The subscripts x_i in (6.50) denote differentiation with respect to the coordinate x_i. The functional (6.50) has been written for a rectangular space $\Omega \in \{x_1x_2|0 \le x_1 \le a,\ 0 \le x_2 \le b\}$.

Equating to zero the expressions accompanying $\delta \mathbf{S}, \delta w$ and δF, we obtain from the variational equation (6.50) a system of three differential equations with different dimensions: a three-dimensional heat transfer equation with coupling of deformation and temperature, and the two-dimensional equations for the shell motion.

The system of equations obtained reads

$$\begin{aligned}
&\frac{\partial^2\theta}{\partial {x_1}^2} + \frac{\partial^2\theta}{\partial {x_2}^2} + \frac{\partial^2\theta}{\partial {x_3}^2} = \frac{1}{2}\dot{\theta} + \frac{T_0 E \alpha_T}{3(1-2\nu)R_T}\dot{e},\\
&D\,\nabla^2\,\nabla^2 w - L(w,F) - \nabla_k^2 F + \nabla^2(M_T + \Delta_T M) - (\Delta M_{11})_{x_1x_1}\\
&-(\Delta M_{22})_{x_2x_2} - 2(\Delta M_{12})_{x_1x_2} - q + \rho h(\ddot{w} + \varepsilon \dot{w}) = 0,\\
&\frac{1}{Eh}\nabla^2\,\nabla^2 F = -\,\nabla_k^2\, w - \frac{1}{2}L(w,w) + \frac{1}{Eh}[(\Delta T_{11} - \nu \Delta T_{22})_{x_2x_2}\\
&+(\Delta T_{22} - \nu\Delta T_{11})_{x_1x_1} - 2(1+\nu)(\Delta T_{12})_{x_1x_2}\\
&-(1-\nu)\,\nabla^2\,(N_T + \Delta_T T)],
\end{aligned} \tag{6.51}$$

where $\alpha = \frac{R_T}{\rho}$ is the thermal conductivity coefficient.

6.3 Equations in Terms of Displacements

For some series of problems (for example, in order to investigate thermoelastic stress wave transitions), it is much more convenient to use the governing equations expressed in terms of the displacements u_1, u_2 and w. However, omission of $\ddot{u}_1$, $\ddot{u}_2$ is not permitted in this case.

The general form of the heat transfer equation is not changed, but we need to substitute (6.1), (6.2) and (6.4) in the expression $e = \varepsilon_{11}^z + \varepsilon_{22}^z$.

In the variational principle (6.34), some terms are transformed to another form; in particular,

$$\delta K = -\rho h \iint\limits_{\Omega_2} \left(\sum_{i=1}^{2} \ddot{u}_i\,\delta u_i + \ddot{w}\,\delta w \right) dx_1 dx_2. \tag{6.52}$$

The expression (6.43) does not change, but one needs to substitute ε_{ij} as represented by (6.2) and (6.3) into the transformed equation (6.46). Then, constructing the standard variational equation including the influence of damping, we obtain

$$\iiint\limits_V (grad\,\theta + \frac{T_0}{R_T}\mathbf{S})\,\delta\mathbf{S}\,dV - \iint\limits_\Omega \left\{ \sum_{\substack{i,j=1\\ i\neq j}}^{2} \left(\frac{\partial T_{ii}}{\partial x_i} + \frac{\partial T_{ij}}{\partial x_j} \right.\right.$$
$$\left. -\rho h \ddot{u}_i \right)\delta u_i + \left(\sum_{\substack{i,j=1\\ i\neq j}}^{2} \left[\frac{\partial}{\partial x_i}\left(T_{ii}\frac{\partial w}{\partial x_i} + T_{ij}\frac{\partial w}{\partial x_j} \right) k_i T_{ii} \right.\right.$$
$$\left.\left.\left. + \frac{\partial^2 M_{ii}}{\partial {x_i}^2} + \frac{\partial^2 M_{ii}}{\partial x_i \partial x_j} \right] + q - \rho h \ddot{w} \right)\delta w \right\} dx_1 dx_2$$
$$+ \int \sum_{\substack{i,j=1\\ i\neq j}}^{2} \left\{ \left[T_{ii}\frac{\partial w}{\partial x_i} + T_{ij}\left(\frac{\partial w}{\partial x_i} + \frac{\partial w}{\partial x_j} \right) + T_{jj}\frac{\partial w}{\partial x_j} \right] \delta w \right. \tag{6.53}$$
$$\left. + M_{ii}(\delta w)_{x_i} + M_{ij}(\delta w)_{x_j} \right\} dx_j = 0.$$

From the variational equation (6.53), we obtain the system of differential equations governing the shell dynamics,

$$\begin{aligned} &\frac{\partial T_{ii}}{\partial x_i} + \frac{\partial T_{ij}}{\partial x_j} = \rho h \ddot{u}_i \quad (1 \leftrightarrow 2) \\ &\frac{\partial^2 M_{11}}{\partial {x_1}^2} + 2\frac{\partial^2 M_{12}}{\partial x_i \partial x_j} + \frac{\partial^2 M_{22}}{\partial {x_2}^2} + K_1 T_{11} \\ &+ K_2 T_{22} + \frac{\partial}{\partial x_1}\left(T_{11}\frac{\partial w}{\partial x_1} + T_{12}\frac{\partial w}{\partial x_2} \right) \\ &+ \frac{\partial}{\partial x_2}\left(T_{12}\frac{\partial w}{\partial x_1} + T_{22}\frac{\partial w}{\partial x_2} \right) + q = \rho h \ddot{w}, \end{aligned} \tag{6.54}$$

and the heat transfer equation,

$$\begin{aligned} &\frac{\partial^2 \theta}{\partial {x_1}^2} + 2\frac{\partial^2 \theta}{\partial {x_2}^2} + \frac{\partial^2 \theta}{\partial {x_3}^2} + \frac{1}{2}\dot{\theta} + \frac{T_0 E \alpha_T}{3(1-2\nu)R_T}\left[\frac{\partial \dot{u}_1}{\partial x_1} + \frac{\partial \dot{u}_2}{\partial x_2} \right. \\ &\left. -(K_1 + K_2)\dot{w} - x_3\ \nabla^2\ \dot{w} + \frac{1}{2}\left(\left(\frac{\partial \dot{w}}{\partial x_1} \right)^2 + \left(\frac{\partial \dot{w}}{\partial x_2} \right)^2 \right) \right]. \end{aligned} \tag{6.55}$$

Using (6.12)–(6.17), (6.2) and (6.13), the following equations in terms of displacements are obtained:

$$\frac{\partial}{\partial x_1}\left[\left(\frac{1}{1-\nu}+p_1^1\right)\left(\frac{\partial u_1}{\partial x_1}-k_1 w+\frac{1}{2}(\frac{\partial w}{\partial x_1})^2\right)\right.$$
$$+\left(\frac{\nu}{1-\nu}+p_1^2\right)\left(\frac{\partial u_2}{\partial x_2}-k_2 w+\frac{1}{2}\left(\frac{\partial w}{\partial x_2}\right)^2\right)$$
$$\left.-h\left(\frac{\partial^2 w}{\partial x_1{}^2}p_2^1+\frac{\partial^2 w}{\partial x_2{}^2}p_2^2\right)\right]+\frac{\partial}{\partial x_2}\left[\frac{p_1^3+1}{2}\right.$$
$$\left.\times\left(\frac{\partial u_1}{\partial x_1}+\frac{\partial u_2}{\partial x_1}+\frac{\partial w}{\partial x_1}\frac{\partial w}{\partial x_1}\right)-h\frac{\partial^2 w}{\partial x_1\partial x_2}p_2^3\right]$$
$$=\frac{1+\nu}{E}\left[\rho\ddot{u}_i+\frac{1}{h}\frac{\partial}{\partial x_1}(N_T+\Delta_T T)\right]\quad(1\leftrightarrow 2),$$

$$\sum_{\substack{i,j=1\\ i\neq j}}^{2}\left\langle\left(\frac{\partial u_i}{\partial x_i}-k_i w+\frac{1}{2}\left(\frac{\partial w}{\partial x_i}\right)^2\right)k_i\left(\frac{1}{1-\nu}+p_1^1\right)\right.$$
$$+\left(\frac{\partial u_j}{\partial x_j}-k_j w+\frac{1}{2}\left(\frac{\partial w}{\partial x_j}\right)^2\right)k_i\left(\frac{\nu}{1-\nu}+p_1^2\right)$$
$$-h\left(\frac{\partial^2 w}{\partial x_i{}^2}p_2^1+\frac{\partial^2 w}{\partial x_i{}^2}p_2^2\right)+\frac{\partial^2}{\partial x_i{}^2}\left[h\left(\frac{\partial u_i}{\partial x_i}-k_i w\right.\right.$$
$$\left.+\frac{1}{2}\left(\frac{\partial w}{\partial x_i}\right)^2\right)p_2^1+\left(\frac{\partial u_j}{\partial x_j}-k_j w+\frac{1}{2}\left(\frac{\partial w}{\partial x_j}\right)^2\right)p_2^2$$
$$\left.-h^2\frac{\partial^2 w}{\partial x_i{}^2}\left(\frac{1}{12(1-\nu)}+p_3^1\right)-h^2\frac{\partial^2 w}{\partial x_j{}^2}\left(\frac{1}{12(1-\nu)}+p_3^2\right)\right]$$
$$+\frac{\partial^2}{\partial x_i\,\partial x_j}\left[\frac{h}{2}\left(\frac{\partial u_i}{\partial x_j}+\frac{\partial u_j}{\partial x_i}+\frac{\partial w}{\partial x_i}\frac{\partial w}{\partial x_j}\right)p_2^3\right.$$
$$\left.-h^2\frac{\partial^2 w}{\partial x_i\partial x_j}\left(\frac{1}{12(1-\nu)}+p_3^3\right)\right]+\frac{\partial}{\partial x_i}\left\{\frac{\partial w}{\partial x_i}\right.$$
$$\times\left[\left(\frac{\partial u_i}{\partial x_i}-k_i w+\frac{1}{2}\left(\frac{\partial w}{\partial x_i}\right)^2\right)\left(p_1^1+\frac{1}{1-\nu}\right)+\left(\frac{\partial u_j}{\partial x_j}\right.\right.$$
$$\left.-k_j w+\frac{1}{2}\left(\frac{\partial w}{\partial x_j}\right)^2\right)\left(p_1^2+\frac{\nu}{1-\nu}\right)-h\left(\frac{\partial^2 w}{\partial x_i{}^2}p_2^1\right.$$

$$+\frac{\partial^2 w}{\partial {x_j}^2} p_2^2\bigg)\bigg] + \frac{\partial w}{\partial x_j}\bigg[\bigg(\frac{\partial u_i}{\partial x_j} + \frac{\partial u_j}{\partial x_i} + \frac{\partial w}{\partial x_i}\frac{\partial w}{\partial x_j}\bigg)$$

$$\times \frac{1+p_1^3}{2} - \frac{h}{2}\frac{\partial^2 w}{\partial x_i \partial x_j} p_2^3\bigg]\bigg]\bigg\}\bigg\rangle = \frac{1+\nu}{Eh}\bigg\{\rho h \ddot{w} + q$$

$$+\sum_{i=1}^{2}\bigg[k_i(N_T + \Delta_T T) + \frac{\partial}{\partial x_i}\bigg(\frac{\partial w}{\partial x_i}(N_T + \Delta_T T)\bigg)$$

$$+\frac{\partial^2}{\partial {x_i}^2}(M_T + \Delta_T M)\bigg]\bigg\}, \tag{6.56}$$

where $\Delta_T N$, $\Delta_T M$, p_n^k $(n = 1, 2, 3,\ k = 1, 2, 3)$ are nonlinear terms defined by the relations (6.16)–(6.27).

The boundary and initial conditions have been formulated in Sects. 2.1.3 and 5.3. The mechanical boundary conditions corresponding to certain types of support of the shell edge are cited below (because they will be used later):

$$w = \frac{\partial w}{\partial x_1} = \varepsilon_{22} = \frac{\partial \varepsilon_{22}}{\partial x_1} - \frac{\partial \varepsilon_{12}}{\partial x_2} = 0, \quad x_1 = const, \tag{6.57}$$

$$\begin{aligned} M_{11} = F = \frac{\partial F}{\partial x_1} = 0, \quad x_1 = const, \\ \frac{\partial M_{11}}{\partial x_1} + 2\frac{\partial M_{12}}{\partial x_1} = 0, \quad x_1 = const. \end{aligned} \tag{6.58}$$

The last condition is equivalent to the conditions

$$T_{11}\frac{\partial w}{\partial x_1} + T_{12}\frac{\partial w}{\partial x_2} = 0, \quad x_1 = const, \tag{6.59}$$

$$w = M_{11} = \varepsilon_{22} = \frac{\partial \varepsilon_{22}}{\partial x_1} - \frac{\partial \varepsilon_{12}}{\partial x_2} = 0, \quad x_1 = const, \tag{6.60}$$

$$w = \frac{\partial w}{\partial x_1} = F = \frac{\partial F}{\partial x_1} = 0, \quad x_1 = const, \tag{6.61}$$

$$w = M_{11} = F = \frac{\partial F}{\partial x_1} = 0, \quad x_1 = const. \tag{6.62}$$

In addition to the boundary conditions given above and those given in Sect. 2.3, there exist many more possibilities.

Chapter 7
Nonlinear Problems of Hybrid-Form Equations

This chapter is devoted to the analysis of some nonlinear problems governed by the hybrid form of the differential equations obtained earlier.

In Sect. 7.1, a method of solution for nonlinear coupled problems is addressed. The finite-difference method is used to discretize the derivatives in order to obtain a system of ordinary differential and algebraic equations. A procedure for solution of the initial–boundary value problem is described. In Sect. 7.2, a relaxation method which reduces the problem to a Cauchy problem of ordinary differential equations is presented. Some advantages and disadvantages of the method are discussed. In addition, the high efficiency of the method is illustrated using examples of problems of plates and shells with physical and physical–geometrical nonlinearities. In Sect. 7.3, numerical investigations are described, and the reliability of the solution of the three-dimensional nonstationary heat transfer equation obtained using the relaxation method (later referred to as the "set-up" method) is shown. In addition, a solution to a coupled thermoelastic problem of a flexible shallow shell without any prior assumptions about the temperature distribution through the thickness is given.

The use of the three-dimensional heat transfer equation is addressed. The efficiency of the method used here when applied to the solution of integral–differential equations with different dimensions (three-dimensional equations related to the Kirchhoff–Love model) and of different type (heat transfer equations and the hyperbolic equations of shell theory) is demonstrated.

In Sect. 7.4, the vibration of an isolated shell subjected to an impulse load that is constant in time and uniformly distributed is addressed.

Many computational results are reported and discussed. In Sect. 7.5, the dynamic stability of shells under thermal shock is analysed. Many computational examples are included and many conclusions are derived. In Sect. 7.6, the influence of coupling and rotational inertia on stability is analysed. Theorems related to the existence of a solution of a nonlinear coupled thermoelastic problem within the Kirchhoff–Love model, including the rotational inertia of shell elements, are formulated and proved. Also, a numerical algorithm is proposed for solving the

J. Awrejcewicz, V. A. Krysko, *Elastic and Thermoelastic Problems in Nonlinear Dynamics of Structural Members*, Scientific Computation,
https://doi.org/10.1007/978-3-030-37663-5_7

problem, and many numerical examples are given. In the next section, results of numerical tests are provided. In Sect. 7.8, the influence of two control parameters, i.e. damping and excitation amplitude, is investigated. Also, some tools to analyse the regular and chaotic behaviour of a plate are introduced. Section 7.9 is devoted to the analysis of spatial–temporal symmetric chaos. Nonsymmetric dissipative oscillations of a square isotropic plate subjected to a longitudinal, one-sided periodic load are reported in Sect. 7.10. The scenarios leading to chaos are illustrated. The last section, Sect. 7.11, deals with solitary waves in thin flexible plates within the kinematic Kirchhoff–Love model.

7.1 Method of Solution for Nonlinear Coupled Problems

We transform the system of differential equations (6.51) to the following nondimensional form:

$$\begin{aligned}
&\lambda_1^2\frac{\partial^2\theta}{\partial {x_1}^2}+\lambda_2^2\frac{\partial^2\theta}{\partial {x_2}^2}+\frac{\partial^2\theta}{\partial {x_3}^2}=\dot{\theta}+\dot{\beta}e,\\
&\frac{1}{12(1-\nu^2)}\left(\lambda^{-2}\frac{\partial^4 w}{\partial {x_1}^4}+2\frac{\partial^4 w}{\partial x_1^2\partial x_2^2}+\lambda^2\frac{\partial^4 w}{\partial {x_2}^4}\right)-L(w,F)\\
&-\nabla_k^2F+\lambda^{-1}(M_T+\Delta_T M)_{x_1x_1}+\lambda(M_T+\Delta_T M)_{x_2x_2}\\
&-\lambda^{-1}(\Delta M_{11})_{x_1x_1}-\lambda(\Delta M_{22})_{x_2x_2}-2(\Delta M_{12})_{x_1x_1}\\
&-q+\text{æ}(\ddot{w}+\varepsilon\dot{w})=0, \qquad\qquad (7.1)\\
&\lambda^{-2}\frac{\partial^4 F}{\partial {x_1}^4}+2\frac{\partial^4 F}{\partial x_1^2\,\partial x_2^2}+\lambda^2\frac{\partial^4 F}{\partial {x_2}^4}+(1-\nu)\Big[\lambda^{-1}(N_T\\
&+\Delta_T T)_{x_1x_1}+\lambda(N_T+\Delta_T T)_{x_2x_2}\Big]+\nabla_k^2 w+\frac{1}{2}L(w,w)\\
&-\lambda^{-1}(\Delta T_{22}-\nu\Delta T_{11})_{x_1x_1}-\lambda(\Delta T_{11}-\nu\Delta T_{22})_{x_2x_2}\\
&+2(1+\nu)(\Delta T_{12})_{x_1x_2}=0.
\end{aligned}$$

The notation used in (7.1) is described in Chap. 6.

As the initial conditions we take

$$\begin{aligned}
&w|_{t=0}=\varphi_1, \quad \theta|_{t=0}=\varphi_2,\\
&w|_{t=0}=\Psi_1,
\end{aligned} \qquad (7.2)$$

and as the boundary conditions we take[1]

$$
\begin{aligned}
w = M_{11} = \varepsilon_{22} = \frac{\partial \varepsilon_{22}}{\partial x_1} - \frac{\partial \varepsilon_{12}}{\partial x_2} = 0, \quad x_1 = const, \\
w = M_{22} = \varepsilon_{11} = \frac{\partial \varepsilon_{11}}{\partial x_2} - \frac{\partial \varepsilon_{12}}{\partial x_1} = 0, \quad x_2 = const,
\end{aligned} \tag{7.3}
$$

$$
\begin{aligned}
\frac{\partial \theta}{\partial x_3}|_{x_3=0.5} = q_T, \quad \frac{\partial \theta}{\partial x_3}|_{x_3=-0.5} = 0, \\
\frac{\dot{\partial} \theta}{\partial h}|_{\Gamma} = 0.
\end{aligned} \tag{7.4}
$$

The boundary conditions (7.4) characterize a thermal shock with an intensity q_T on the surface $x_3 = 0.5$ (the other surfaces of the shell are thermally isolated). We take $t \in [0, t_1]$, where t_1 is the observation time of the behaviour of the shell; $\Omega_1 = (0, 1) \times (0, 1)$ is the space in which the independent variables x_1, x_2 vary, which is referred to as the middle surface; $\overline{\Omega}_1 = \Omega_1 \cup \partial\Omega_1$; $\partial\Omega_1$ is the edge of the middle surface; $Q_1 = \Omega_1 \times (0, t_1)$; $\Gamma = \partial\Omega_1 \times [0, t_1]$; $\Omega_2 = \Omega_1 \times \left(-\frac{h}{2}, \frac{h}{2}\right)$; h is the constant shell thickness; $\overline{\Omega}_2 = \Omega_2 \cup \partial\Omega_2$; $\partial\Omega_2$ is the surface surrounding the shell volume in three-dimensional space; $Q_2 = \Omega_2 \times (0, t_1)$; and $S = \partial\Omega_2 \times (0, t_1)$. The following nondimensional parameters have been used:

$$
\begin{aligned}
&\overline{x}_1 = \frac{x_1}{a}, \ \overline{x}_2 = \frac{x_2}{b}, \ \overline{x}_3 = \frac{x_3}{h}, \ \overline{w} = \frac{w}{h}, \ \lambda_1 = \frac{h}{a}, \ \lambda_2 = \frac{h}{b}, \ \lambda = \frac{a}{b}, \\
&\overline{k}_1 = k_1 \frac{a^2}{h}, \ \overline{k}_2 = k_2 \frac{b^2}{h}, \ \overline{F} = \frac{F}{Eh^3}, \ \overline{\varepsilon} = \varepsilon \frac{h^2}{\alpha}, \ \overline{t} = t \frac{\alpha}{h^2}, \\
&æ = \frac{a^2 b^2 \rho \alpha^2}{Eh^6}, \quad \overline{q}_T = q_T \frac{ab\alpha_T}{hk_t}, \quad \overline{\theta} = \theta \frac{\alpha_T ab}{h^2}, \\
&\Delta \overline{T}_{ij} = \Delta T_{ij} \frac{ab}{Eh^3}, \quad \Delta \overline{M}_{ij} = \Delta M_{ij} \frac{ab}{Eh^4}, \\
&\beta = \frac{T_0 E \alpha_T}{3(1 - 2\nu)\rho h^2}.
\end{aligned} \tag{7.5}
$$

In (7.1)–(7.4), the bars over nondimensional expressions have been omitted.

The first equation of (7.1) is three-dimensional and of parabolic type, whereas the second and third ones are two-dimensional. In addition, they are also integral-type equations. In order to reduce the partial differential equations to ordinary differential equations with respect to time, the finite-difference method is used to discretize the derivatives along the spatial coordinates x_1, x_2, x_3 with an $O(h^2)$ approximation.

[1] It is possible to apply any boundary conditions with the algorithm described here.

This method results in ODEs for w and θ, and a system of algebraic equations for the Airy function F.

The following difference operators have been used in relation to the spatial coordinates x_1, x_2, x_3, with a mesh of uniform spacings h_1, h_2, h_3, respectively:

$$
\begin{aligned}
\Lambda_i y &= \frac{1}{h_i^2}\left[y(x_i - h_i) - 2y(x_i) + y(x_i + h)\right]_{i=1,2,3},\\
\Lambda_{ij} y &= \frac{1}{4h_i h_j}\left[y(x_i + h_i, x_j + h_j) + y(x_i - h_i, x_j - h_j)\right.\\
&\quad \left. - y(x_i + h_i, x_j - h_j) - y(x_i - h_i, x_j + h_j)\right],\\
\Lambda_i^2 y &= \frac{1}{h_i^4}\left[y(x_i - 2h_i) - 4y(x_i - h_i) + 6y_i(x_i)\right.\\
&\quad \left. - 4y(x_i + h_i) + y(x_i + 2h_i)\right],\\
\Lambda_{ij}^2 y &= \frac{1}{h_i^2 h_j^2}\left[y(x_i - h_i, x_j - h_j) - 2y(x_i - h_i, x_j)\right.\\
&\quad + y(x_i - h_i, x_j + h_j) - 2y(x_i, x_j - h_j) + 4y(x_i, x_j)\\
&\quad - 2y(x_i, x_j + h_j) + y(x_i + h_i, x_j - h_j) - 2y(x_i + h_i, x_j)\\
&\quad \left. + y(x_i + h_i, x_j + h_j)\right].
\end{aligned}
\tag{7.6}
$$

Using the difference operators (7.6) and introducing a new variable $\frac{dw}{dt} = \dot{w}$, the system of ODEs is reduced to first-order ODEs with respect to time and to algebraic equations:

$$
\begin{aligned}
\left[1 + \frac{1+\nu}{1-\nu}\beta\right]\frac{d\theta_{ijk}}{dt} &= (1-2\nu)\beta\frac{d}{dt}\left[\lambda^{-1}\Lambda_1 F_{ij}\right.\\
&\quad + \lambda\Lambda_2 F_{ij} - (\Delta T_{11} + \Delta T_{22} - 2N_T - 2\Delta_T T)_{ij}\\
&\quad \left. - \frac{x_{3k}}{1-\nu}(\lambda^{-1}\Lambda_1 w_{ij} + \lambda\Lambda_2 w_{ij})\right]
\end{aligned}
\tag{7.7}
$$

$$
(i = 0, \ldots, n; \quad j = 0, \ldots, m; \quad k = 0, \ldots, p,)
$$

$$
\frac{dw_{ij}}{dt} = \dot{w}_{ij},
$$

$$\frac{d\dot{w}_{ij}}{dt} + \varepsilon \dot{w}_{ij} = (-1/æ)[A(w) + B(w, F) + C(M_T, \Delta_T M, \Delta M_{11}, \Delta M_{22}, \Delta M_{12})] + \lambda_1^2 \Lambda_1 \theta_{ijk} + \lambda_2^2 \Lambda_2 \theta_{ijk} + \Lambda_3 \theta_{ijk}, \tag{7.8}$$

$$D(F) = E(w) + G(N_T, \Delta_T T, \Delta T_{11}, \Delta T_{22}, \Delta T_{12}). \tag{7.9}$$

In (7.8) and (7.9), the following notation has been used:

$$A(w) = \frac{1}{12(1-\nu^2)}(\lambda^{-2}\Lambda_1^2 w_{ij} + 2\Lambda_{12}^2 w_{ij} + \lambda^2 \Lambda_2^2 w_{ij}),$$
$$B(w, F) = K_1 \Lambda_2 F_{ij} + K_2 \Lambda_1 F_{ij} + \Lambda_1 w_{ij} \Lambda_2 F_{ij} + \Lambda_2 w_{ij} \Lambda_1 F_{ij} - \Lambda_{12} w_{ij} \Lambda_{12} F_{ij},$$

$$\begin{aligned} C(M_T, \Delta_T M, \Delta M_{11}, \Delta M_{22}, \Delta M_{12}) &= \lambda^{-1}\Lambda_1(M_T \\ +\Delta_T M - \Delta M_{11})_{ij} + \lambda\Lambda_2(M_T + \Delta_T M - \Delta M_{22})_{ij} &- \Lambda_{12}(\Delta M_{12})_{ij}, \\ D(F) &= 12(1-\nu^2)A(F), \\ E(w) &= -K_1\Lambda_2 w_{ij} - K_2 \Lambda_1 w_{ij} - \Lambda_1 w_{ij}\Lambda_2 w_{ij} + (\Lambda_{12} w_{ij})^2, \end{aligned} \tag{7.10}$$

$$G(N_T, \Delta_T T, \Delta T_{11}, \Delta T_{22}, \Delta T_{12}) = -(1-\nu)\lambda^{-1}\Lambda_1(N_T + \Delta_T T)_{ij} + \lambda(1-\nu)\Lambda_2(N_T + \Delta_T T)_{ij} + \lambda^{-1}\Lambda_1(\Delta T_{22} - \nu \Delta T_{11})_{ij} + \lambda\Lambda_2(\Delta T_{11} - \nu\Delta T_{22})_{ij} - \frac{1+\nu}{2}\Lambda_{12}(\Delta T_{12})_{ij}.$$

The following relation has been used in the first equation of (7.1) in order to describe the volume extension e:

$$e = (1-2\nu)[\lambda^{-2} F_{x_1x_1} + \lambda F_{x_2x_2} - \Delta T_{11} - \Delta T_{22} + 2(N_T + \Delta_T T)] - \frac{1-2\nu}{1-\nu} x_3(\lambda^{-1} w_{x_1x_1} + \lambda w_{x_2x_2}) + \frac{1+\nu}{1-\nu}\theta. \tag{7.11}$$

This leads to (7.7).

In the procedure for solution of the coupled problem the time derivative in the right-hand side of (7.7) is approximated by a one-sided finite-difference relation:

$$\frac{dy_{ijk}}{dt} = \frac{1}{12h_t}(3y_{ijk}^{e-4} - 16y_{ijk}^{e-3} + 36y_{ijk}^{e-2} + 25y_{ijk}^{e}) + O(h_t^4), \tag{7.12}$$

where h_t is a time interval.

We need to attach to the difference equations (7.7)–(7.10) a group of boundary conditions, which must be formulated using a central finite-difference relation (it is necessary to find the values of the fundamental functions on nodes at the edge of the shell).

A procedure for solution of the initial–boundary value problem has been constructed in the following way.

Using the results found at the previous time step, w_{ij}, N_{Tij}, ΔT_{11ij}, $\Delta_T T_{ij}$, ΔT_{22ij}, ΔT_{12ij}, the system of linear algebraic equations (7.9) is solved for Airy's function F_{ij}. The values of F_{ij} obtained and values of N_{Tij}, $\Delta_T T_{ij}$, M_{Tij}, $\Delta_T M_{ij}$, ΔM_{11ij}, ΔM_{22ij}, ΔM_{12ij} found at the previous time step are substituted into the right-hand sides of (7.7)–(7.8). The ODEs are then integrated with respect to time using a method which will be explained later.

Using the temperature distribution in the shell volume found above, and integrating over the thickness with the help of Simpson's rule,

$$\int_{-\frac{1}{2}}^{\frac{1}{2}} \theta(x_1, x_2, x_3)\, dx_3 = \frac{1}{6\Delta h}\Big[(\theta_0 + \theta_{2n} + 4(\theta_1 + \cdots + \theta_{2n-1}) + 2(\theta_2 + \cdots + \theta_{2n-2})\Big], \tag{7.13}$$

the temperature terms N_T and M_T defined by (6.16) are found. They are used in later calculations.

Then, for each point of the shell volume, the deformation intensity is calculated:

$$e_i = \frac{\sqrt{2}}{3}\left[(e_{11} - e_{22})^2 + (e_{22} - e_{33})^2 + (e_{22} - e_{11})^2 + \frac{3}{2}e_{12}^2\right]^{1/2}, \tag{7.14}$$

where the e_{ij} are calculated using the formulas given in the previous chapter. A dependence $\sigma_i(e_i)$ is assumed (this depends on the shell material used), and σ_1 is defined at each point of the shell volume. The shear function $\gamma(e_i)$ calculated from (6.32) is used for calculation of the nonlinear terms (6.16)–(6.26), which are used during the second step of the algorithm.

In the first time step, owing to the initial conditions (7.2) $N_T = M_T = 0$, $\gamma \equiv 1$, all nonlinear terms are equal to zero. Use of N_T and M_T obtained in the previous step allows us to solve the integral–differential equations (7.1) as a system of differential equations.

We emphasize that we need to solve the system of algebraic equations (7.9) for each time step, but the matrix of this system remains constant. The Gauss method can be used to solve (7.9), and a transformation to a triangular form of the matrix is performed only once during the computations, in the first step of the calculation. The choice of the Gauss method possesses many advantages with respect to both high accuracy and a relatively short computational time. For instance, the use of a relaxation method requires computations five time longers to solve the problem under discussion.

The use of the three-dimensional heat transfer equation (7.7) is essential, because it makes any additional hypothesis about the temperature distribution through the thickness unnecessary.

During the integration of the ODEs (7.7), (7.8), a very careful verification of each computational step is needed. In addition, an economical choice of the integration method used for the ODEs (7.7), (7.8) is important. Consideration of the approaches proposed in this book has led the authors to apply a combination of the explicit and implicit Adams methods, known in the literature as the Adams–Bashford or prediction–correction method.

The methods used can be represented using the following formula:

$$P_k(EC_{k+1})^n\, E, \tag{7.15}$$

where P_k means prediction, i.e. finding a solution using the explicit Adams method; k denotes the order of accuracy; E means to computation of the right-hand sides of the ODEs; C_{k+1} means correction, i.e. improvement of the accuracy using the implicit Adams method, to an accuracy of order $(k + 1)$ (here the solution obtained after the operation P_k is used as an initial approximation); and n denotes the required number of iterations.

When the operations $P_k(EC_{k+1})$ have been carried out, it is recommended to compute once more the right-hand sides of the ODEs in order to use them during the next steps of computation.

The right-hand sides of the ODEs are computed $(n + 1)$ times, where n is independent of k; the Runge–Kutta method of the kth order needs computation of the right-hand sides to be performed k times.

In order to compare the accuracy and computation times of the methods, a few first-order ODEs have been solved using the Runge–Kutta method of the fourth order, and the Adams method in accordance with the schemes P_3EC_4E and P_5EC_6E. Results have been obtained for some equations with known solution, including $y = e^x \sin 10x$, and $y = \exp\left(2\frac{x^{11}}{11} - \frac{x^{12}}{12}\right)$. For $x \in [0; 3]$, the following results were obtained: for $h_t = 0.1$, the Runge–Kutta method gives more accurate results; for $h_t = 0.01$, the Runge–Kutta method gives a solution similar to that obtained by the scheme P_5EC_6E; for $h_t = 0.002$, the Adams methods give solutions with better accuracy in comparison with the Runge–Kutta method. Note that a solution obtained by the scheme $P_kEC_{k+1}E$ with $k = 3$ does not differ (practically) from that obtained with $k = 5$, but a test of the scheme with $k = 3$ in comparison with the fourth order Runge–Kutta method under conditions of a shortage of computational resources achieved about 45% better accuracy. This was observed for all model solutions used.

On the basis of these obtained results, the scheme (7.15) with $k = 3$ was used to integrate the system (7.7)–(7.9). This scheme, with one iteration, achieved 40% of the computational time of the fourth-order Runge–Kutta method, when applied to the nonlinear coupled problem.

7.2 Relaxation Method

Current methods applied to the solution of nonlinear problems of shells (with geometrical and physical nonlinearities) generally use a projection onto a system of nonlinear differential or nonlinear algebraic equations, with successive linearization using either Newton–Raphson or other iterative methods (for instance, the relaxation method). Nowadays there exists a wide literature devoted to such methods and their algorithms.

In 1963 Fedosev [110] proposed a dynamical approach to solve a problem related to the stability of shells. From a mathematical point of view, this method is called the "set-up" method [127]. The main idea of this method is that the solution of the nonlinear partial differential equations is reduced to a Cauchy problem of ODEs which is linear in time. This means that this method linearizes the nonlinear equations and decreases their dimension.

We discuss briefly an advantage of this method here. From a mathematical point of view, the "set-up" method can be treated as an iterative method to solve nonlinear algebraic equations, where each time step provides a new approximation to the exact solution. Like all iterative methods, this one is characterized by a high accuracy of computation. In addition, it does not have the common disadvantage of iterative methods of a high sensitivity to the choice of the initial approximation. Additionally, the "set-up" method not only gives a very simple rule for obtaining nonunique solutions of static problems, but also allows one to find the stable and unstable branches of the equilibrium position of the system under consideration and to capture all process of the jumping behaviour of a shell.

In the process of solution of homogeneous equations via traditional methods, in order to obtain a nontrivial solution one needs to introduce an artificial excitation (in the theory of shells this corresponds, for instance, to a small transverse load, a small curvature or some other initial imperfection). However, this influences (sometimes significantly) the results obtained. In the case of the "set-up" method, the initial conditions play the role of the initial excitations, and small changes of these conditions do not influence the static solution obtained. Another advantage of the method is related to its simple realization, because nowadays there are many effective algorithms and programs devoted to solution of the Cauchy problem.

Let us clarify the method for obtaining unstable solutions using an example of an arbitrary nonlinear algebraic equation,

$$f(x) = 0. \tag{7.16}$$

We construct two differential equations from for (7.16),

$$c\left(\frac{d^2x}{dt^2} + \varepsilon\frac{d\dot{x}}{dt}\right) = \pm f(x). \tag{7.17}$$

Table 7.1 Comparison of SU and Newton methods

Exact values	$x_1 = 0$	$x_2 = 3$	$x_3 = 4$	$x_4 = 5$
Initial approximation (x_{in})	$x_1 = 2.5$	$x_2 = 3.5$	$x_3 = 4.5$	$x_4 = 10$
SU	$x_1 = 0$	$x_2 = 3$	$x_3 = 4$	$x_4 = 5$
N	$x_1 = 5$	$x_2 = 5$	$x_3 = 3$	$x_4 = 5$
Iteration number I(SU)	68	54	87	60
Iteration number I(N)	7	12	9	10

In order to obtain the complete set of roots of (7.16), we need to solve the two differential equations (7.17). However, in practice we can do this in the following way.

We take an arbitrary initial condition (the initial approximation) $x_{in} = x_0$ for (7.17) with the positive right-hand side, and solving the equation we obtain x_1. Then we take the initial condition $x_0 = x_1 + \delta$ and solve (7.17) with the negative value of the right-hand side, and so on. As an example, we consider the solution of the equation

$$f(x) = x^4 - 12x^3 + 47x^2 - 60x = 0. \tag{7.18}$$

This equation has four real roots $x_1 = 0$, $x_2 = 3$, $x_3 = 4$, $x_4 = 5$. We have solved this equation by the "set-up" (SU) and Newton (N) methods. In Table 7.1 the roots found by both methods, the number of iterations (I) needed to obtain the solution and the initial approximations are reported.

It can be seen that the SU method, in contrast to the Newton method, finds more exactly a stable root of Eq. (7.18) close to the initial approximation.

Let us consider one more (transcendental) equation:

$$f(x) = \arctan x = 0. \tag{7.19}$$

This equation has a root $x = 0$. It is known that for (7.19) the Newton method approaches this root only for initial values such that $|x| < 1.39$, but the "set-up" method approaches this root for $\forall x_{in}$. For instance, taking $x_{in} = 2$, the root $x = 0$ is found after 33 iterations.

We now illustrate the high efficiency of the "set-up" method using a series of nonlinear problems of the theory of plates and shells, which are characterized by a wide range of properties with respect to different nonlinearity types (geometrical, physical and geometrical–physical) as well as different models (Kirchhoff–Love or Timoshenko).

We formulate now a nonlinear equation governing the dynamics of a shallow shell including transverse shear effects, in hybrid form:

$$
\begin{aligned}
&\nabla_k^2 F + L(w, F) + \frac{2}{3}\left[\lambda_1\left(\frac{\partial \gamma_x}{\partial x_1} + \frac{\partial^2 w}{\partial x_1{}^2}\right) + \lambda_2\left(\frac{\partial \gamma_y}{\partial x_2} + \frac{\partial^2 w}{\partial x_2{}^2}\right)\right] \\
&\qquad\qquad + q - \frac{\partial^2 w}{\partial t^2} - \varepsilon\frac{\partial w}{\partial t} = 0, \\
&\frac{1}{5}\left[\lambda^{-2}\frac{\partial^2 \gamma_x}{\partial x_1{}^2} + A_{1212}\frac{\partial^2 \gamma_y}{\partial y_2{}^2} + (A_{1122} + A_{1212})\frac{\partial^2 \gamma_y}{\partial x_1 \partial x_2}\right] \\
&\qquad - \frac{1}{20}\left[\lambda^{-1}\frac{\partial^2 w}{\partial x_1{}^2} + (A_{1122} + A_{1212})\frac{\partial^2 w}{\partial x_1 \partial x_2}\right] \\
&\qquad - 2\lambda_1\left(\gamma_x + \frac{\partial w}{\partial x_1}\right) = 0 \quad (x_1 \leftrightarrow x_2), \\
&\nabla^4 F + \nabla_x^2 w + \frac{1}{2}L(w, w) = 0.
\end{aligned}
\tag{7.20}
$$

The symbol $(x_1 \leftrightarrow x_2)$ denotes that a third equilibrium equation can be obtained from the second equation by a cyclic change of indices. The system (7.20) has been transformed to a nondimensional form as earlier. Equations (7.20) include several geometrical and physical–geometrical quantities; in particular, $k_{x_1} = \frac{a^2}{R_{y_1}(2h)}$ and $k_{x_2} = \frac{a^2}{R_{y_2}(2h)}$ (geometrical parameters characterizing the curvatures $\left(\frac{1}{R_{x_1}} \text{ and } \frac{1}{R_{x_2}}\right)$ of the shell and its size $(a, b, 2h)$), and $\lambda_1 = \frac{G_{13}}{A_{1111}}\left(\frac{b}{2h}\right)^2$ and $\lambda_2 = \frac{G_{23}}{A_{1111}}\left(\frac{a}{2h}\right)^2$ are the physical–geometrical parameters characterizing the effect of transverse shear on the solution.

The following boundary conditions are attached to (7.20):

$$w = M_n = \gamma_{\overline{n}} = T_n = T_{\overline{n}}|_\Gamma = 0,$$

where $n = x$, $\overline{n} = y$ define the shell boundary; $t \in [0, T]$, where T is the observation time; $\Omega_1 = (0; 1) \times (0; 1)$ is the space in which the independent parameters x_1 and x_2 vary, defining the middle surface; $\overline{\Omega}_1 = \Omega_1 \cup \partial\Omega_1$, where $\partial\Omega_1$ is the contour (edge) of the middle surface; $Q_1 = \Omega_1 \times (0; T)$; and $\Gamma = \partial\Omega_1 \times [0; 1]$.

The following initial conditions are applied:

$$w|_{t=0} = \frac{\partial w}{\partial t}|_{t=0} = 0. \tag{7.21}$$

The problem defined by (7.20)–(7.21) is reduced to a Cauchy problem using higher approximations of the Bubnov-Galerkin method:

$$
\begin{aligned}
w &= \sum_{i,j=1}^{N} A_{ij}(t) \sin i\pi x_1 \sin j\pi x_2, \\
F &= \sum_{i,j=1}^{N} B_{ij}(t) \sin i\pi x_1 \sin j\pi x_2, \\
\gamma_{x_1} &= \sum_{i,j=1}^{N} C_{ij}(t) \cos i\pi x_1 \sin j\pi x_2, \\
\gamma_{x_2} &= \sum_{i,j=1}^{N} D_{ij}(t) \sin i\pi x_1 \cos j\pi x_2.
\end{aligned}
\tag{7.22}
$$

As a result we obtain a system of N ODEs for $A_{ij}(t)$ and a system of algebraic equations for $B_{ij}(t)$, $C_{ij}(t)$ and $D_{ij}(t)$, which fortunately can be solved in closed form. The integrals over the middle surface $\Omega_1 = (0; 1) \times (0; 1)$ were computed analytically for each of the problems considered. The system of ODEs for $A_{ij}(t)$ with the initial conditions (7.21) was solved using the Runge–Kutta method with an automatic choice of the integration step.

Let us analyse the influence of the damping coefficient ε appearing in the first equation of the system (7.20) on the value of the dynamic critical load, when an impulse load of infinitesimal duration is applied (Fig. 7.1). We definite $\varepsilon = \varepsilon_{cr}$ as $\min \varepsilon$, when the dynamic critical load $q = q_{cr}^d$ is equal to the static critical load $q = q_{cr}^s$ (i.e. $q_{cr}^s = q_{cr}^d$). The criterion for a dynamic load given in [22] is used here. Plots of $q(w)$ and $t(w)$ are shown in Fig. 7.2 for a cylindrical shell ($k_x = 0$, $k_{x_2} = 48$, $\lambda_1 = \lambda_2 = 1000$). Curve 2 corresponds to $\varepsilon = 3$, whereas curve 1 corresponds to a static solution for $q_{cr}^d < q_{cr}^s$. Both parts of the figure need to be analysed simultaneously. For $\varepsilon = \varepsilon_n$ and for a given load q, the deflection approaches a constant value with increase of time. Having a set of values

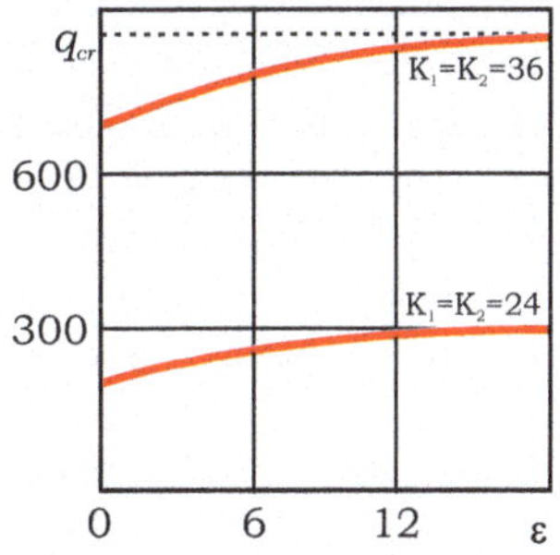

Fig. 7.1 Influence of damping coefficient ε on the dynamic critical load

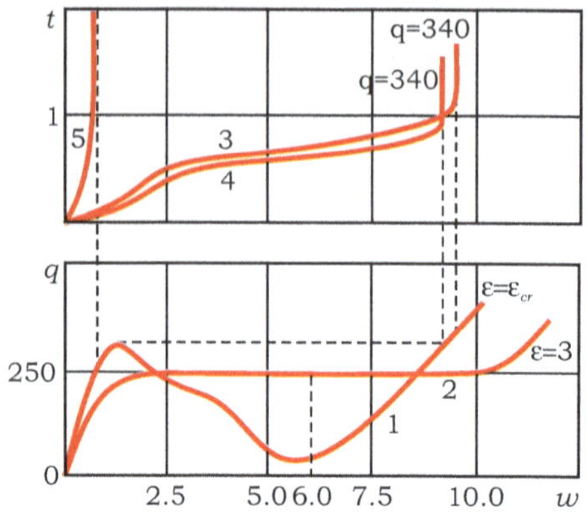

Fig. 7.2 Dependences $q(w)$, $t(w)$ for a cylindrical shell ($N = 5$)

$\{t(w_i), q_i)\}$ one can construct a dependence $q(w)$. Curves 3, 4, 5 for $t(w)$, were obtained for $q = 255, 320$ and 350, with $\varepsilon = \varepsilon_{cr}$. From Fig. 7.1 it can be seen that for $\varepsilon \rightarrow \varepsilon_{cr}$, $q_{cr}^d \rightarrow q_{cr}^s$, for an impulse load that is uniformly distributed over the shell surface and constant in time (the results were obtained from (7.22) with $N = 5$). A further increase of the number of terms in (7.22) does not change the fundamental functions or their derivatives up to second order. The efficiency and high accuracy of the Bubnov–Galerkin method when applied to both static and dynamic problems were demonstrated long ago (see, for instance [198]).

In solving nonlinear problems of composite shells exhibiting weak shear stiffness within the Timoshenko-type theory, an increase in the number of modes of equilibrium for a given load occurs, and local stability loss is observable. The stable equilibrium configuration was chosen here from the set of various possible configurations using the "set-up" method. Plots of $q(w)$ for $k_{x_1} = k_{x_2} = 24$, $\lambda = 1$, $\lambda_1 = \lambda_2 = 40, 20, 10, 5$ (corresponding to curves 1, 2, 3, 4, respectively) are presented in Fig. 7.3 (obtained from (7.22) with $N = 7$). The $q(w)$ curves were obtained using a static method, i.e. a large set of nonlinear algebraic equations for A_{ij}, B_{ij}, C_{ij} and D_{ij} was solved using the Newton–Raphson method for a fixed value of the deflection in the centre of the shell. The arrows indicate external static loads which have been detected through the "set-up" method (the other equilibrium configurations are not physically realized).

In practice, one can find shallow shell structures with values of k_{x_1} and k_{x_2} even greater than 500. However, in the literature, only problems related to $k_{x_1}, k_{x_2} < 50$ have been addressed. The reason is clear. When the geometric parameters are increased, the number of possible equilibrium configurations increases greatly. This requires one to increase the number of terms in the approximating functions (and nodes) used when either variational or finite-difference methods are applied to static problems, in order to properly describe the complex surface shape. The convergence of the solution obtained for the nonlinear algebraic equations dramatically decreases, because their order increases.

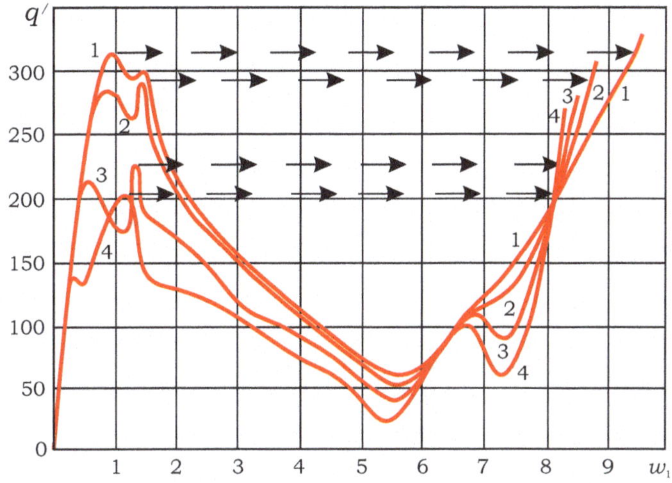

Fig. 7.3 Dependences $q(w)$, $t(w)$ for a cylindrical shell ($N = 7$)

The "set-up" method avoids the above disadvantages. We do not need to solve any system of equations at all, and in particular, not a nonlinear one. There is no problem in analysing shells with $24 \leq k_{x_1} = k_{x_2} \leq 500$ (we have taken $N = 15$ in (7.21)). The numerical results agree fully with experiment for $k_{x_1} = k_{x_2} = 409.1$. The difference in q^s_{cr} is 15%, and qualitatively the picture of stability loss is the same [318].

In Fig. 7.4, contours of equal relative deflection $w(x_1, x_2)$, $0 \leq x_1, x_2 \leq 0.5$, for $k_{x_1} = k_{x_2} = 409.1$ are drawn. Cases a–e are related to the pre-critical state, whereas f corresponds to a post-critical stability loss. With an increase of k_{x_1} and k_{x_2}, the shell loses its stability first in the corner zones, and during "jump" behaviour, holes appear on the axial curves, which is also indicated by experimental data.

We now illustrate the high efficiency of the "set-up" method using an example of a physically and geometrically nonlinear problem of a plate. We recall (7.1) without the temperature terms:

$$\begin{aligned}
&\frac{1}{12(1-\nu^2)} \nabla^4 w - L(w, F) - \nabla_k^2 F - \lambda^{-1}(\Delta M_{11})_{x_1 x_1} \\
&- 2(\Delta M_{12})_{x_1 x_2} - \lambda(\Delta M_{22})_{x_2 x_2} = q - æ(\ddot{w} + \varepsilon \dot{w}), \\
&\nabla^4 F + \nabla_k^2 w + \frac{1}{2} L(w, w) - \lambda^{-1}(\Delta T_{22} - \nu \Delta T_{11})_{x_1 x_1} \\
&+ 2(1+\nu)(\Delta T_{12})_{x_1 x_2} - \lambda(\Delta T_{11} - \nu \Delta T_{22})_{x_2 x_2} = 0.
\end{aligned} \tag{7.23}$$

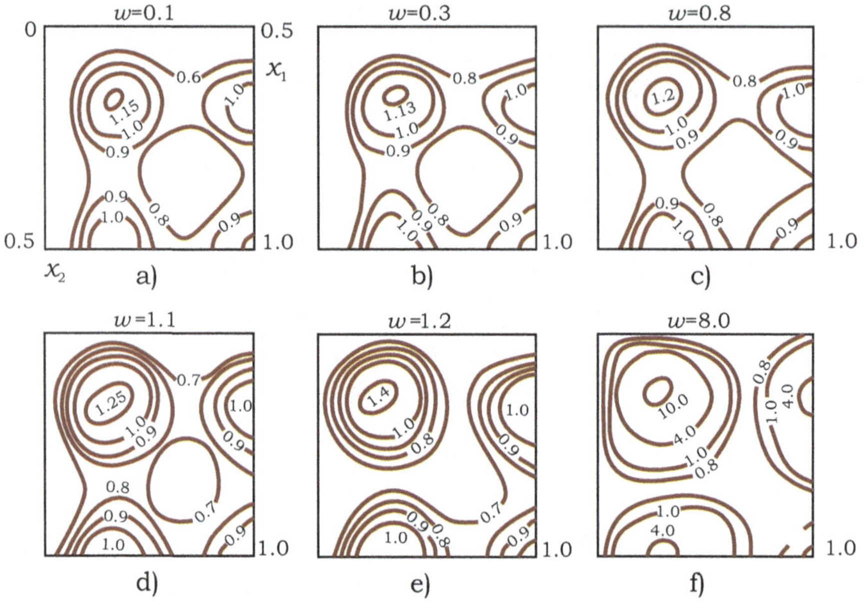

Fig. 7.4 **(a–f)** Diagrams for equal relative deflections $w(x_1, x_2)$

The construction of a solution to (7.23) using the "set-up" method has been described in Sect. 7.1. We consider the geometrically and physically nonlinear problem of bending of a square plate ($\lambda = 1$). We assume that a uniformly distributed load with intensity q on the surface $x_3 = -0.5$. The physical parameters of the plate material (AMC alloy) have been taken to be as follows: $E = 69$ GPa, $\mu = 0.3$, $\rho = 2800\,\text{kg/m}^3$, $\sigma_i(e_i)$, and

$$\begin{aligned} \sigma_i &= E_1 e_i, \quad e_i \le e_s, \\ \sigma_i &= E_1 e_s + E_2(e_i - e_s), \quad e_i > e_s. \end{aligned} \tag{7.24}$$

For the shear function, we obtain

$$\begin{aligned} \gamma(e_i) &= \frac{\sigma_i(e_i)}{3Ge_i}, \\ \gamma &\equiv 1, \quad e_i \le e_s, \\ \gamma &= \frac{E_1}{E} + \left(1 - \frac{E_1}{E}\right)\left(\frac{e_s}{e_i}\right), \quad e_i > e_s, \\ \frac{E_1}{E} &= 0.57735, \quad e_s = 0.98 \cdot 10^{-3}. \end{aligned} \tag{7.25}$$

Table 7.2 Comparison of computational methods

q	$w(0.5; 0.5)$		
	1	2	3
23.4	0.96	0.95	1.04%
60.0	2.03	2.00	1.96%
70.3	2.16	2.23	3.24%

The geometrical parameters of the plate were taken to be as follows $a = b = 0.1$ m, $\frac{a}{h} = 50$. The boundary conditions were defined by (6.58), and the initial conditions were $w|_{t=0} = 0$, $\dot{w}|_{t=0} = 0$.

The results were compared with results obtained using the method of variational iterations and variable elastic parameters [349]. In the "set-up" method, the physical nonlinearity was taken into account through the theory of plasticity using Iliushin's method of elastic solutions. Here we remark that, from a mathematical point of view, the latter method corresponds to some variants of the simple iteration method. A numerical realization of Iliushin's method of elastic solutions possesses many advantages in comparison with the method of variable elastic parameters. In Table 7.2, the values of the deflection at the centre of the plate obtained by the method described in [22] (column "1") and by the "set-up" method (column "2") are reported, for three values of the transverse-load parameter. Column 3 reports the difference as a percentage.

First of all, it can be concluded from Table 7.2 that the results obtained by these two methods are similar. Some of the difference is caused by the specific structures of two algorithms. The static approach relies on a reduction from partial differential equations (PDEs) to ODEs as a first approximation and on application of the variable-elasticity method. The dynamical approach is based on the "set-up" method and the method of elastic solutions.

We consider now the stability of a geometrically nonlinear elastic shell. The following parameters were chosen $k_{x_1} = k_{x_2} = 24$, $E = 69$ GPa, $\mu = 0.3$, $a = b = 0.1$ m, $h = 0.83 \times 10^{-3}$ m, $\sigma_i(e_i)$ a defined by (7.24), $E_1/E = 0.4478$, $e_s = 1.35 \times 10^{-3}$. The boundary conditions were governed by (6.58) and zero-value initial conditions were applied.

The dependence $q(w(0.5; 0.5))$ obtained is shown in Fig. 7.5.

7.3 Numerical Investigations and Reliability of the Results Obtained

In the previous section we have shown that reliable solutions of problems of shallow-shell theory with both geometrical and geometrical–physical nonlinearities can be obtained when the method described in Sect. 7.1 is applied. In this section, we show the reliability of the solution to the three-dimensional nonstationary heat transfer equation obtained when the "set-up" method is used. In addition, we also

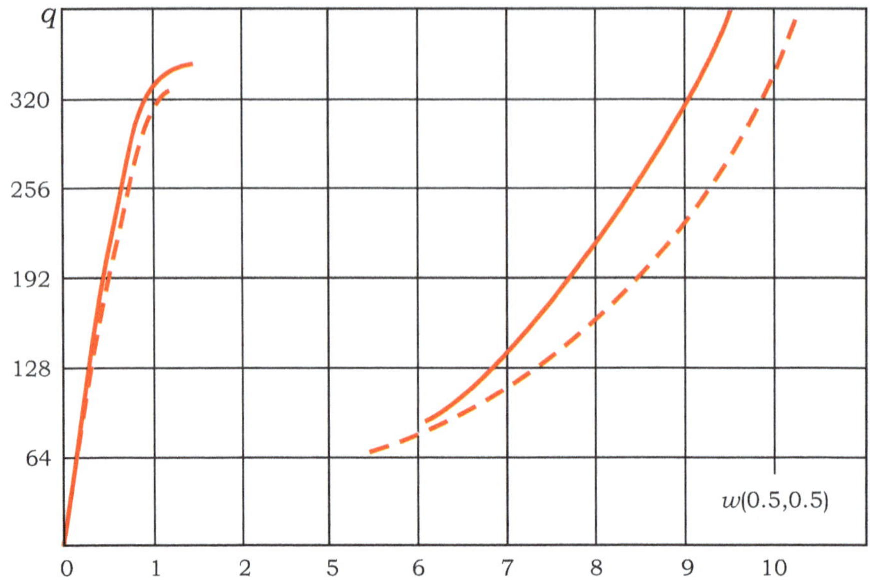

Fig. 7.5 Dependence $q(w(0.5; 0.5))$

discuss and illustrate (using the technique described here) a solution to a coupled thermoelastic problem of a flexible shallow shell without any prior assumptions about the temperature distribution through the thickness, i.e. using the three-dimensional heat transfer equation. As a matter of fact, we need to demonstrate the efficiency of the method when it is applied to the solution of integral–differential equations with different dimensions (three-dimensional heat transfer equations and two-dimensional equations related to the Kirchhoff–Love model) and of different types (heat transfer equations and hyperbolic equations of shell the theory).

In the equations governing the shell motion, a temperature does not appear directly, but its integral characteristics (thermal forces and moments) are used; these appear because of the reduction of the three-dimensional heat transfer equation to a two-dimensional one.

The solution to the three-dimensional heat transfer equation has been sought using the finite-difference method. A number of researchers have emphasized that neglecting the nonlinear form of the temperature variation through the shell thickness greatly influences the solution to a nonstationary thermoelastic problem of a shell or a plate. Therefore, a solution to the three-dimensional nonstationary heat transfer equation without any introduction of hypotheses about the temperature distribution through the thickness possesses important practical meaning.

Let us recall the first equation of the system (7.1):

$$\frac{\partial \theta}{\partial t} = \lambda_1^2 \frac{\partial^2 \theta}{\partial {x_1}^2} + \lambda_2^2 \frac{\partial^2 \theta}{\partial {x_2}^2} + \frac{\partial^2 \theta}{\partial {x_3}^2} + w_0. \tag{7.26}$$

In this given equation, there is no term representing the shell geometry; however, the validity of this approach has been proved [187–189]. Using the finite-difference method along the spatial coordinates with an $O(h^2)$ approximation, we obtain the following ODEs with respect to time:

$$\frac{d}{dt}\frac{\partial \theta_{ijk}}{\partial t} = \lambda_1 \Lambda_1 \theta_{ijk} + \lambda_2 \Lambda_2 \theta_{ijk} + \Lambda_3 \theta_{ijk} + w_{0ijk}. \tag{7.27}$$

We need to attach to the system (7.27) the initial and boundary conditions formulated earlier, represented in a suitable (finite-difference) form.

We consider a numerical example (second type of boundary conditions) and convective heat transfer (third type of boundary conditions). The space was divided into $(6 \times 6 \times 6)$, $(8 \times 8 \times 8)$ and $(12 \times 12 \times 12)$ parts. The results obtained were compared with the analytical solution proposed by Kovalenko [189], showing very good agreement.

The results of a problem related to a thermal shock on a shell surface which is thermally isolated on its other sides are given in Table 7.3. The results where the surface was partitioned into $(12 \times 12 \times 12)$ and $(8 \times 8 \times 8)$ parts differ from the analytical results by no more than 0.2% and 0.5%, respectively.

We consider one of the methods of reduction of the three-dimensional problem to a two-dimensional one. In the case of linear problems of thermoelasticity, application of the operator method for thin-walled structures has been proposed [278], where thin-walled conditions from the point of view of heat transfer theory have also been proposed. Let us begin with the first invariant of the deformation tensor:

$$e = e_{11} + e_{22} + e_{33}. \tag{7.28}$$

Table 7.3 Comparison of results obtained for different mesh sizes

		Time		
Solution method	Coordinates	$t = 0.2$	$t = 0.4$	$t = 0.6$
Exact	$z = 0.5$	67.19	97.01	124.06
	$z = 0$	21.06	47.66	74.26
	$z = -0.5$	8.18	31.55	57.71
$(12 \times 12 \times 12)$	$z = 0.5$	67.04	96.92	123.97
	$z = 0$	20.98	47.58	74.17
	$z = -0.5$	8.16	31.49	57.62
$(8 \times 8 \times 8)$	$z = 0.5$	66.86	96.80	123.87
	$z = 0$	20.89	47.48	74.07
	$z = -0.5$	8.14	31.41	57.53
$(4 \times 4 \times 4)$	$z = 0.5$	65.87	96.17	123.33
	$z = 0$	20.38	46.96	73.56
	$z = -0.5$	8.08	32.00	57.03

The quantity e_{33} can be found from the condition of a plane strain state ($\sigma_{33} = 0$), and it reads

$$e_{33} = -\frac{\nu}{E}(\sigma_{11} + \sigma_{22}) + \alpha_T \theta. \tag{7.29}$$

We shall now formulate the heat transfer equations in the absence of physical nonlinearities. Substituting σ_{11} and σ_{22} into (7.29), for a physically linear body, we obtain

$$e_{33} = -\frac{\nu}{1-\nu}(e_{11} + e_{22}) + \frac{1+\nu}{1-\nu}\alpha_T \theta. \tag{7.30}$$

Substituting (7.29) into (7.28), we obtain the first invariant of the deformation tensor:

$$\begin{aligned} e &= \frac{1-2\nu}{1-\nu}(e_{11} + e_{22}) + \frac{1+\nu}{1-\nu}\alpha_T \theta \\ &= \frac{1-2\nu}{1-\nu}\left[\frac{1-\nu}{Eh}(T_{11} + T_{22}) + 2\alpha\tau N\tau \right. \\ &\left. + z(æ_{11} + æ_{22})\right] + \frac{1+\nu}{1-\nu}\alpha_T \theta. \end{aligned} \tag{7.31}$$

Substituting (7.31) into the heat transfer equation, we obtain

$$\begin{aligned} &K_T \frac{\partial \theta}{\partial t} - K_T \left(\frac{\partial^2 \theta}{\partial {x_1}^2} + \frac{\partial^2 \theta}{\partial {x_2}^2} + \frac{\partial^2 \theta}{\partial {x_3}^2} \right) \\ &= \frac{E \alpha_T T_0}{1-2\nu} \frac{\partial}{\partial t}(e_{11} + e_{22} + e_{33}) + W_0, \end{aligned} \tag{7.32}$$

and we introduce the following notation:

$$\alpha_1 = \frac{K_T}{C}, \quad \gamma_0 = \frac{\alpha_T E T_0}{C(1-2\nu)}, \quad \beta = \frac{\alpha_1}{1 + \alpha_T \gamma_0 \frac{1+\nu}{1-\nu}},$$

$$\gamma = \frac{\gamma_0 (1-2\nu)}{(1-\nu)\left(1 + \gamma_0 \alpha_T \frac{1+\nu}{1-\nu}\right)}, \quad D^2 = \nabla^2 - \frac{1}{\beta}\frac{\partial}{\partial t},$$

$$D_1^2 = \frac{1}{\beta}\frac{\partial}{\partial t}, \quad M = æ_{11} + æ_{22}, \quad N = \frac{1-\nu}{Eh}(T_{11} + T_{22}) + 2\alpha_T N_T,$$

$$N_T = \frac{1}{h}\int_{-\frac{h}{2}}^{\frac{h}{2}} \theta \, dx_3. \tag{7.33}$$

In operator form, the heat transfer equation reads

$$\frac{\partial^2\theta}{\partial {x_3}^2} + D^2\theta - \gamma D_1^2(N + x_3 M) = 0. \tag{7.34}$$

The solution to (7.34) obtained using the symbolic method reads

$$\theta = \frac{\sin x_3 D}{D}\left\{\left.\frac{\partial\theta}{\partial x_3}\right|_{x_3=0} - \gamma\frac{D_1^2}{D^2}M\right\} + \cos x^3 D\left\{\theta\Big|_{x_3=0} - \gamma\frac{D_1^2}{D^2}N\right\} + \gamma\frac{D_1^2}{D^2}(N + zM). \tag{7.35}$$

Integrating with respect to x_3, we obtain

$$N_T = \frac{1}{h}\int_{-\frac{h}{2}}^{\frac{h}{2}} \theta\, dx_3 = \frac{\sin\frac{h}{2}D}{\frac{h}{2}D}\left\{\theta\Big|_{x_3=0} - \gamma\frac{D_1^2}{D^2}N\right\} + \frac{1}{\gamma}\frac{D_1^2}{D^2}N, \tag{7.36}$$

$$M_T = \frac{12}{h^3}\int_{-\frac{h}{2}}^{\frac{h}{2}} \theta x_3\, dx_3 = \frac{12}{h^2D^2}\frac{\sin\frac{h}{2}D}{\frac{h}{2}D}\left\{\left.\frac{\partial\theta}{\partial x_3}\right|_{x_3=0} - \gamma\frac{D_1^2}{D^2}M\right\} - \frac{12}{h^2D^2}\cos\frac{h}{2}D\left\{\left.\frac{\partial\theta}{\partial x_3}\right|_{x_3=0} - \gamma\frac{D_1^2}{D^2}M\right\} - \gamma\frac{D_1^2}{D^2}M. \tag{7.37}$$

Removing $\theta|_{x_3=0}$ and $\left.\frac{\partial\theta}{\partial x_3}\right|_{x_3=0}$ from (7.37) and (7.36) and substituting into (7.35), we obtain the temperature θ expressed in terms of the integral characteristics:

$$\theta(x_1, x_2, x_3, t) = \frac{\frac{h}{2}D}{\sin\frac{h}{2}D}\cos x_3 D N_T + \frac{h^3D^2}{24}\frac{\sin x_3 D}{\sin\frac{h}{2}D - \frac{h}{2}D\cot\frac{h}{2}D}M_T + \gamma\frac{D_1^2}{D^2}\left\{\frac{\frac{h}{2}D\cos zD}{\sin\frac{h}{2}D} - 1\right\}N - \frac{\gamma}{3}\frac{D_1^2}{D^2}\left\{\frac{\frac{h^3}{8}D^3\cot\frac{h}{2}D}{1 - \frac{h}{2}\cot\frac{h}{2}D} - 3\right\}M. \tag{7.38}$$

After the boundary conditions on the surfaces

$$x_3 = \pm\frac{h}{2} : \left.\frac{\partial\theta}{\partial x_3}\right|_{x_3=-0.5} = 0 \quad \text{and} \quad \left.\frac{\partial\theta}{\partial x_3}\right|_{x_3=+0.5} = 0$$

are satisfied, the following equations are obtained:

$$
\begin{aligned}
&-\frac{h}{2}D^2N_T+\frac{h^3D^3}{24}\frac{\cot\frac{h}{2}D}{1-\frac{h}{2}D\cot\frac{h}{2}D}M_T+\gamma\frac{h}{2}D_1^2N\\
&-\frac{\gamma}{3}\frac{D_1^2}{D^2}\left\{\frac{\frac{h^3}{8}D^3\cot\frac{h}{2}D}{1-\frac{h}{2}\cot\frac{h}{2}D}-3\right\}M=0,
\end{aligned}
\tag{7.39}
$$

$$
\begin{aligned}
&\frac{h}{2}D^2N_T+\frac{h^3D^3}{24}\frac{\cot\frac{h}{2}D}{1-\frac{h}{2}D\cot\frac{h}{2}D}M_T+\gamma\frac{h}{2}D_1^2N\\
&-\frac{\gamma}{3}\frac{D_1^2}{D^2}\left\{\frac{\frac{h^3}{8}D^3\cot\frac{h}{2}D}{1-\frac{h}{2}\cot\frac{h}{2}D}-3\right\}M=0.
\end{aligned}
\tag{7.40}
$$

These equations yield an equation for the thermal stresses N_T,

$$
D^2N_T-\gamma D_1^2N=0,
\tag{7.41}
$$

and an equation for the thermal moments M_T,

$$
h^2D^2M_T-12M_T-\gamma h^2D_1^2N=0.
\tag{7.42}
$$

Equation (7.42) has been obtained by keeping only two terms of the series of $\cot\frac{h}{2}D$, which corresponds to a cubic form of the temperature distribution (7.38) through the thickness.

In terms of the nondimensional parameters introduced earlier, (7.41) and (7.42) read

$$
\begin{aligned}
&\lambda_1\lambda_2\left(\lambda^{-1}\frac{\partial^2N_T}{\partial {x_1}^2}+\lambda\frac{\partial^2N_T}{\partial {x_2}^2}\right)-\frac{\alpha_T^2ET_0}{C}\left(\frac{\partial}{\partial t}\left(\lambda^{-1}\frac{\partial^2F}{\partial {x_1}^2}\right.\right.\\
&\left.\left.+\lambda\frac{\partial^2F}{\partial {x_2}^2}\right)\right)=\left(1+\gamma_0\alpha_T\frac{1+\nu}{1-\nu}\right)\left(1+\frac{2\gamma_0\alpha_T}{1+\gamma_0\alpha_t\frac{1+\nu}{1-\nu}}\right)\frac{\partial N_T}{\partial t},
\end{aligned}
\tag{7.43}
$$

$$
\begin{aligned}
&\lambda_1\lambda_2\left(\lambda^{-1}\frac{\partial^2M_T}{\partial {x_1}^2}+\lambda\frac{\partial^2M_T}{\partial {x_2}^2}\right)-12M_T\\
&+\frac{\alpha_T^2ET_0}{C(1-\nu)}\frac{\partial}{\partial t}\left(\lambda^{-1}\frac{\partial^2w}{\partial {x_1}^2}+\lambda\frac{\partial^2w}{\partial {x_2}^2}\right)\\
&=\left(1+\gamma_0\alpha_T\frac{1+\nu}{1-\nu}\right)\frac{\partial M_T}{\partial t}.
\end{aligned}
\tag{7.44}
$$

We need to attach to these equations the equations governing the motion of the shell and the continuity equation:

$$\frac{1}{12(1-\nu^2)}\nabla^2\,\nabla^2 w - \nabla_k^2 F - L(w,F)$$
$$+\frac{a^2b^2\rho\alpha_1^2}{Eh^3}\frac{\partial^2 w}{\partial t^2}+\frac{1}{12(1-\nu^2)}\left(\lambda^{-1}\frac{\partial^2 M_T}{\partial x_1{}^2}+\lambda\frac{\partial^2 M_T}{\partial x_2{}^2}\right)=q, \tag{7.45}$$
$$\nabla^2\,\nabla^2 F=-\lambda^{-1}\frac{\partial^2 N_T}{\partial x_1{}^2}-\lambda\frac{\partial^2 N_T}{\partial x_2{}^2}-\nabla_k^2 w-\frac{1}{2}L(w,w).$$

As a result, we have obtained a system of four PDEs (7.43)–(7.45) governing the shell motion with coupling of both the deformation and the temperature fields. Unlike the integral–differential set of equations obtained in the previous chapter, the system above is only differential and has one dimension. It s approximate, because in the series of cot $\frac{h}{2}D$ only the first two terms are taken, which in practice leads to a cubic distribution of temperature through the shell thickness. However, the system remains a hybrid one, i.e. parabolic and hyperbolic.

We need to attach to the system (7.43)–(7.45) the boundary and initial conditions. In order to solve the system (similarly to the integral–differential system (7.1)), we have applied a finite-difference method with respect to the spatial coordinates with an $O(h^2)$ approximation. This finally results in ODEs for w_{ij}, N_{Tij}, M_{Tij} and a system of algebraic equations for F_{ij}. The system of algebraic equations was (again) solved using the Gauss method for every time step, and for the system of differential equations, the Runge–Kutta method with an automatically chosen step of integration was chosen. Unlike the situation for the "set-up" method, where the values of the thermal loads occurring in the three-dimensional heat transfer equation must be taken from the previous step of the integration in time, we do not need to take these values from the previous step now.

In order to estimate the influence of the assumption that we have made, the stress–strain state of a shell made from AMC alloy with the boundary conditions (6.59) has been analysed.

It has been assumed also that on the surfaces $x_1 = 0, x_2 = 0, x_1 = 1, x_2 = 1$, the following boundary condition of the first type for temperature has been used: $\theta = 0$, i.e. $N_T = M_T = 0$. At the initial time instant $t = 0$ a heat load, extending over an infinite time, uniformly distributed over the shell surface, was applied, with $q = 81$ and $k_{x_1} = k_{x_2} = 24$. The results obtained are presented in Table 7.4.

Method 1 in Table 7.4 corresponds to the solution of the system (7.43)–(7.45). Method 2 corresponds to the solution of the integral–differential equations. The results obtained indicate the very high efficiency of the approaches presented here for the solution of coupled thermoelastic problems of flexible shells subjected to transverse impulse-type loads.

Table 7.4 Comparison of results obtained for different methods

	Time							
	$t = 0.1$		$t = 0.2$		$t = 0.3$		$t = 0.4$	
	Method							
Function	1	2	1	2	1	2	1	2
$w(0.5; 0.5)$	0.99	0.99	1.78	1.78	0.85	0.85	0.02	0.02
$w(0.25; 0.25)$	0.68	0.68	1.07	1.07	0.68	0.68	0.03	0.03
$F(0.5; 0.5)$	0.38	0.38	0.63	0.63	0.33	0.33	0.00	0.00
$F(0.25; 0.25)$	0.15	0.15	0.23	0.23	0.14	0.14	0.004	0.004
$w_{x_1x_2}(0.5; 0.5)$	−2.34	−2.34	−12.86	−12.86	0.90	0.94	−0.75	−0.74
$w_{x_1x_2}(0.25; 0.25)$	−8.94	−8.94	−10.52	−10.52	−9.78	−9.77	−0.36	−0.37
$F_{x_1x_2}(0.5; 0.5)$	−4.63	−4.63	−8.46	−8.46	−3.44	−3.44	−0.03	−0.02
$F_{x_1x_2}(0.25; 0.25)$	−1.83	−1.83	−2.38	−2.38	−1.93	−1.93	−0.12	−0.12
$N_T(0.5; 0.5)$	0.124	0.126	0.228	0.330	0.092	0.093	0.001	0.001
$M_T(0.5; 0.5)$	−0.101	−0.101	0.298	0.310	0.353	0.344	0.080	0.084

7.4 Vibration of Isolated Shell Subjected to Impulse

We consider now a problem related to the dynamical response to an impulse load with intensivity q, constant in time and uniformly distributed, suddenly applied to the surface $x_3 = -0.5$. The problem is formulated in Sect. 7.1. The boundary conditions of the heat transfer equation (7.1) now correspond to thermal isolation of all shell surfaces (7.4). The boundary conditions associated with (7.3) correspond to free support on flexible ribs that are nonstretchable in their tangential plane. The initial conditions are $w|_{t=0} = \dot{w}|_{t=0} = \theta|_{t=0} = 0$.

As in earlier cases, all results presented from now on correspond to nondimensional quantities (except when it is specifically mentioned otherwise). We use the following notation: coupled geometrically nonlinear problem, GN(c); uncoupled geometrically nonlinear problem, GN; coupled geometrically and physically nonlinear problem, GPN(c); uncoupled geometrically and physically nonlinear problem, GPN. These problems are also referred to as problems 1, 2, 3 and 4, respectively; this latter notation is used in the figures.

In Fig. 7.6, the deflection in the shell centre versus time under the precritical load $q = 220$ is shown for these four problems. For both physically linear and nonlinear problems, the inclusion of coupling results in a decrease of the oscillation amplitude, which agrees with results obtained a long time ago (see, for instance, [214, 267]). In the first period of oscillation, the decrease of the amplitude for problems 1 and 2 (GN(c) and GN) is 3.8%, whereas this decrease if 6% in the case of problems 3 and 4 (GPN(c) and GPN). In the second period of oscillation, the damping (caused by energy dissipation) is more pronounced and is 13% for problems 1 and 2, and 21% for problems 3 and 4.

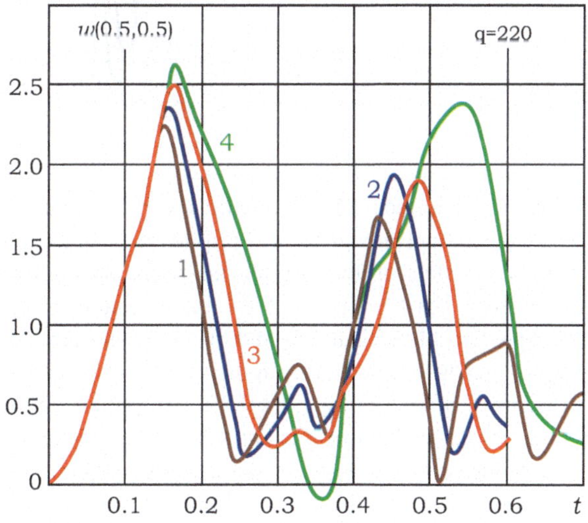

Fig. 7.6 Deflection in the shell centre for $q = 220$

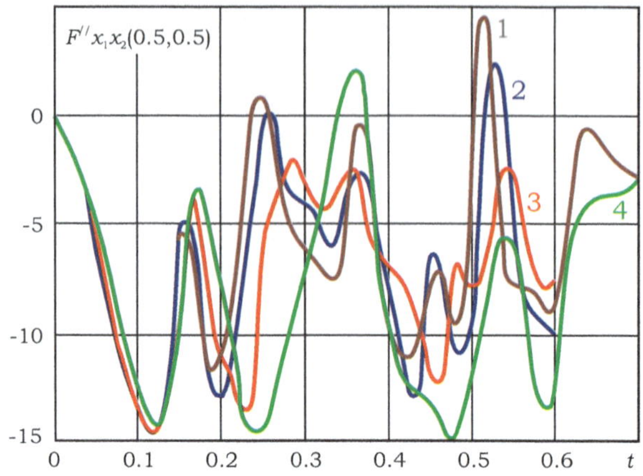

Fig. 7.7 Dependence of $F''_{x_1x_1}$ in the shell centre for $q = 220$

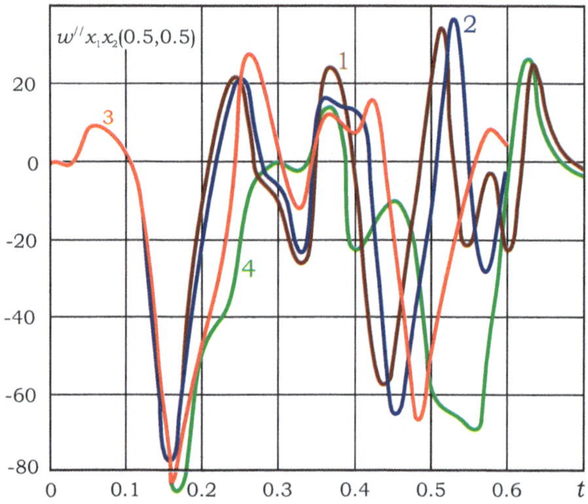

Fig. 7.8 Dependence of $W''_{x_1x_1}$ in the shell centre for $q = 220$

In Figs. 7.7 and 7.8, the dependences of $F''_{x_1x_1}$ and $w''_{x_1x_1}$ on time in the shell centre are reported. These functions vary with time in a way that depends on the mathematical model used. In spite of the fact that the shell is very thin, the stress–strain state of the shell has an important influence on the effects of physical nonlinearity. For $t = 0.19 - 0.24$, the extremal stresses increase by 9.7% (for the uncoupled problems 2 and 4) and by 13% (for coupled problems). The inclusion of coupling leads to a decrease (in the time interval considered) in the maximum load (in terms of modulus) $F''_{x_1x_2}$ of 9.2% (for physically linear problems) and of 6.9% for the physically nonlinear problems. At the time moments $t = 0.25, 0.36, 0.48$ the $F''_{x_1x_1}$ values corresponding to the extreme deflection values differ by 2–2.5 times. Therefore, when the physical nonlinearity and the coupling are taken into account, the stress distribution changes in time significantly, and both their maximal and minimal values of the stresses are changed.

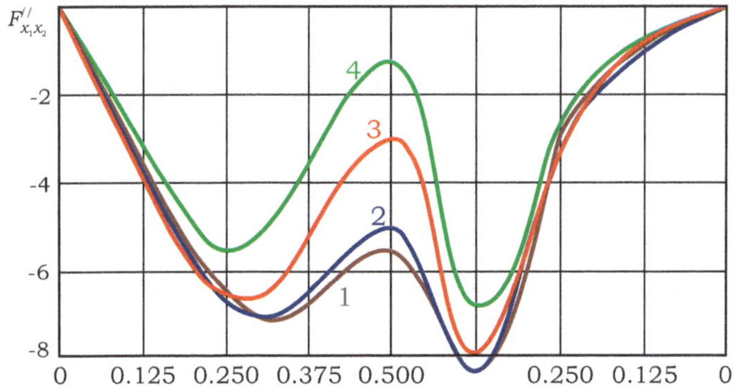

Fig. 7.9 Distribution of $F''_{x_1x_1}$ along axis of symmetry for $t = 0.16$

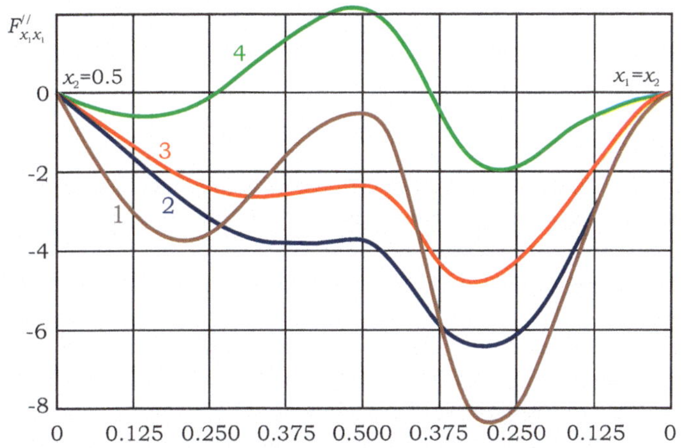

Fig. 7.10 Distribution of $F''_{x_1x_1}$ along axis of symmetry for $t = 0.36$

The distribution of $F''_{x_1x_1}$ along the axis of symmetry for the time moments $t = 0.16$ and $t = 0.36$ is shown in Figs. 7.9 and 7.10. It is apparent that the results obtained with different mathematical models have significant differences for $t = 0.16$, and these differences are more evident for $t = 0.36$. In the latter case, the solution corresponding to problem 4 differs from those for problems 1–3 not only quantitatively but also in the sign of the stresses.

In spite of the thinness of the plate, the influence of physical nonlinearity is also important. The maximal deflection in the shell centre obtained taking account of physical nonlinearity increases during the first period of oscillation by 9.9% (for the coupled problem) and 11% (for the uncoupled problems).

Consequently, in order to obtain a reliable estimate of the stress–strain state of a thin shallow shell with the geometrical and physical parameters considered above during oscillations caused by a load close to the critical value, one needs to consider

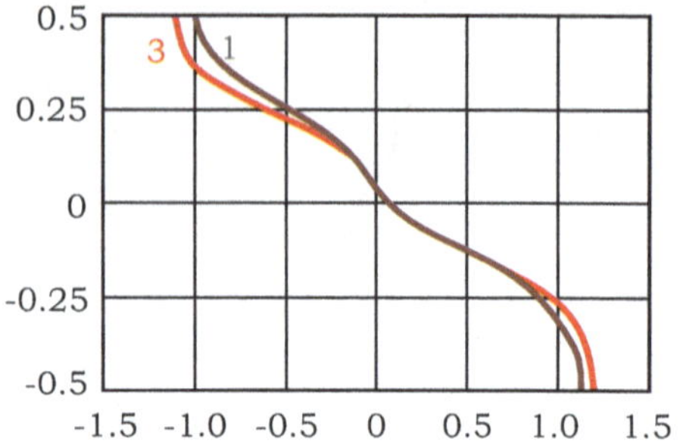

Fig. 7.11 Distribution of temperature in a shell centre along thickness

the coupling of deformation and temperature. Although the deflections change only slightly, the stresses occurring in the shell can be arbitrary large owing to coupling effects.

In Fig. 7.11, the temperature distribution at the shell centre through the thickness of the shell is presented for problems 1 and 3. The maximum temperature rise of the shell due to the mechanical work is 4.6° in dimensional units for the physically nonlinear problem, and of 4.3° for the physically linear one. This heating is observed when the shell reaches its maximum deflection at $t = 0.16$ on the surface $x_3 = -0.5$ (Fig. 7.9 corresponds to this case). The coupling between deformation and temperature leads to a change of the temperature distribution in the shell volume. In Fig. 7.11, a negative increase of temperature can also be seen in certain parts of the shell, which is in agreement with the theoretical [278] and experimental [65] data obtained earlier.

The temperature distribution parallel to the middle surface of the shell at the same time moment $t = 0.16$ is shown in Figs. 7.12 and 7.13 for problems 1 and 3, respectively, for three different surfaces. Qualitatively, figures are similar. Inclusion of physical nonlinearity increases the self-heating of the shell only slightly. Here also, a negative temperature increase, as well as a sudden temperature decrease in passing through the shell thickness, is observed. The maximum decrease is observed at $t = 0.16$ in the shell centre, as can be seen from the Figs. 7.12 and 7.13. For problem 1 this decrease is equal to 7.4°, whereas for problem 3 it is equal to 8.3°. The time moment $t = 0.16$ corresponds to the maximum shell deflections.

In Fig. 7.14, the temperature distribution parallel to the middle surface for problem 3 at $t = 0.26$ is reported (at this time $w(0.5, 0.5) = 0.54$). Figure 7.14 are again qualitatively similar to each other (the temperature distribution in the surfaces $x_3 = \pm 0.5$ at $t = 0.26$ is shown). The sudden temperature decrease at the shell centre reaches 4.9° at $t = 0.26$. Note that the parts of the shell that heat up and cool down here changed their positions in comparison with those for $t = 0.16$. This observation is in agreement with the results given in [65].

We shall now consider the effect of coupling between the deformation and temperature fields and physical nonlinearity on the critical values of the stress–strain state for a flexible shell with geometric parameters $k_{x_1} = k_{x_2} = 24$.

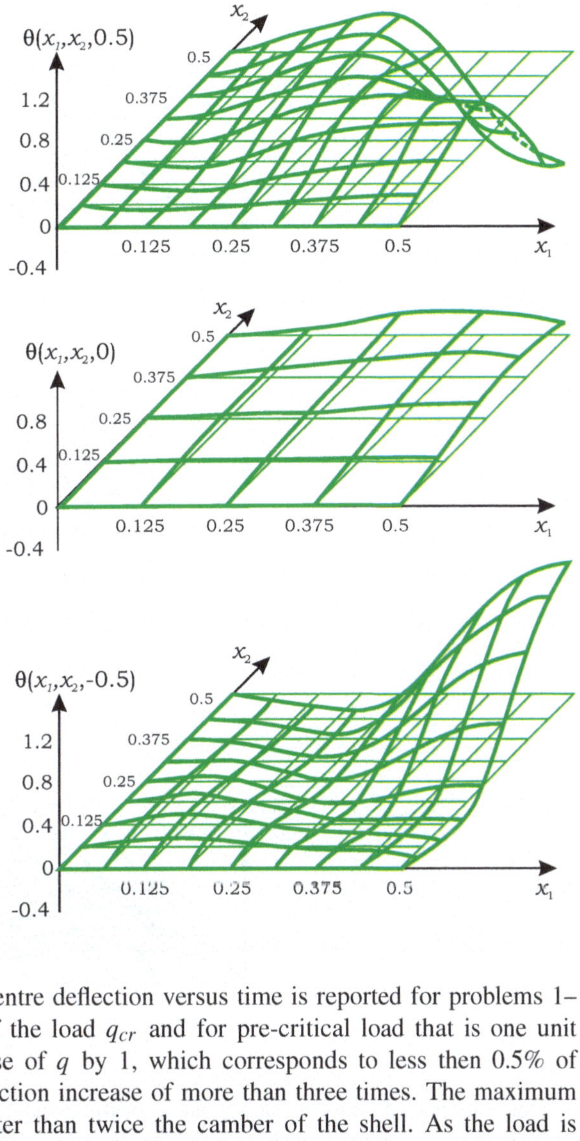

Fig. 7.12 Distribution of temperature for coupled geometrically nonlinear problem ($t = 0.16$)

In Fig. 7.15, the shell centre deflection versus time is reported for problems 1–4, for the critical value of the load q_{cr} and for pre-critical load that is one unit less. Therefore, an increase of q by 1, which corresponds to less then 0.5% of q_{cr}, corresponds to a deflection increase of more than three times. The maximum deflection values are greater than twice the camber of the shell. As the load is increased above the critical load, the time needed to achieve the maximum deflection decreases. The q values obtained are in agreement with the criteria of Vol'mir and Kantor and are referred to as the critical ones. The following critical values have been found: 230 for problems 1 and 2, 225 for problem 3 and 222 for problem 4. The Sho–Sung–Roth criterion has been applied in the case presented in Fig. 7.16. In this figure, the curve labelled 1 shows the deflection versus time at the shell centre for problem 4, with $q = 222$. Curve 2 shows the corresponding result for $q = 233$, curve 3 for $q = 224$.

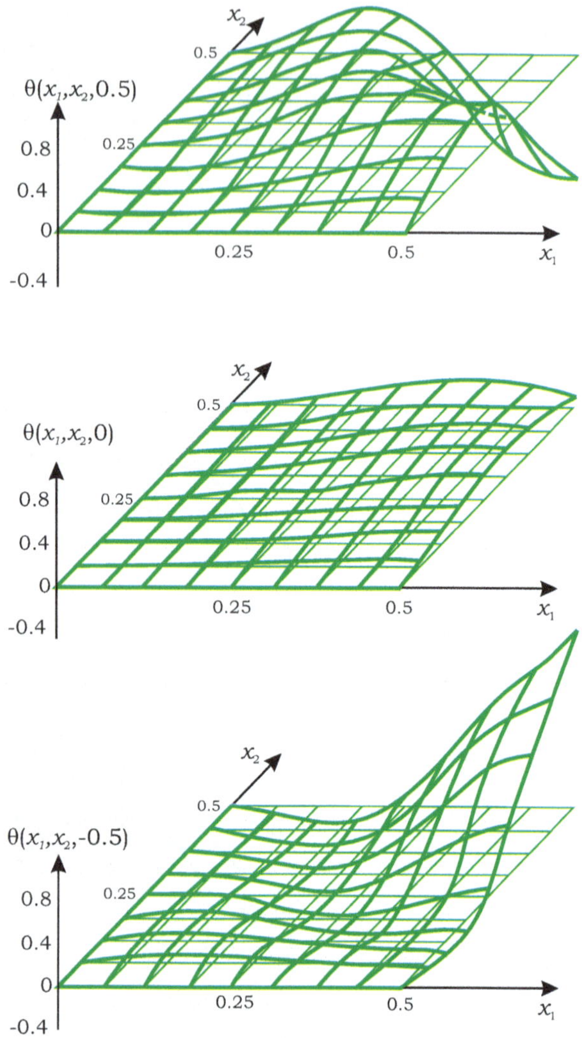

Fig. 7.13 Distribution of temperature for coupled geometrically and physically nonlinear problem ($t = 0.16$)

The influence of coupling on the value of the critical load is not of significance. For the physically nonlinear, problems with the given geometrical and physical shell parameters, coupling increases q_{cr} by only 1.4%. The effect cannot be detected at all for problems 1 and 2. The increase of q_{cr} in the coupled problems can be explained by the part played by energy dissipation in the shell.

When we take into account the physical nonlinearity of the material, we find that q_{cr} for problem 4 is 3% smaller than for problem 2, and q_{cr} for problem 2 is 2.2% smaller than for problem 1. To conclude, the influence of physical nonlinearities on q_{cr} for a shell with the given parameters is also of small significance.

Fig. 7.14 Distribution of temperature for coupled geometrically and physically nonlinear problem ($t = 0.26$)

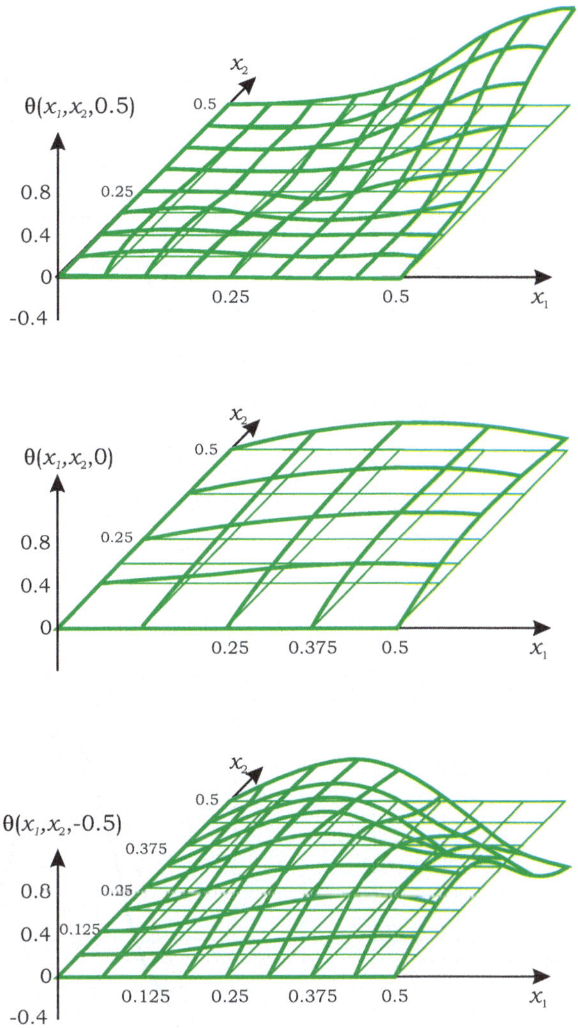

On the other hand, the choice of mathematical model can lead to qualitatively different results for a load close to the critical value. Let us consider, for instance, the stress–strain state obtained for problems 3 and 4 with $q = 223$ at the time moments $t = 0.16$ and $t = 0.36$. Although at $t = 0.16$ the shell deflection (Fig. 7.17) and the stress state differ only slightly, at $t = 0.36$ (Figs. 7.18, 7.19 and 7.20) they differ qualitatively.

The shell centre deflection obtained from the solution to problem 4 (GPN) is 7% less than in the coupled doubly nonlinear problem at $t = 0.16$. The difference in $w''_{x_1x_1}$ at this chosen time moment for problem 3 and 4 (GN(c) and GPN) at the shell centre is almost negligible. However, a large difference has been observed in $F''_{x_1x_1}$.

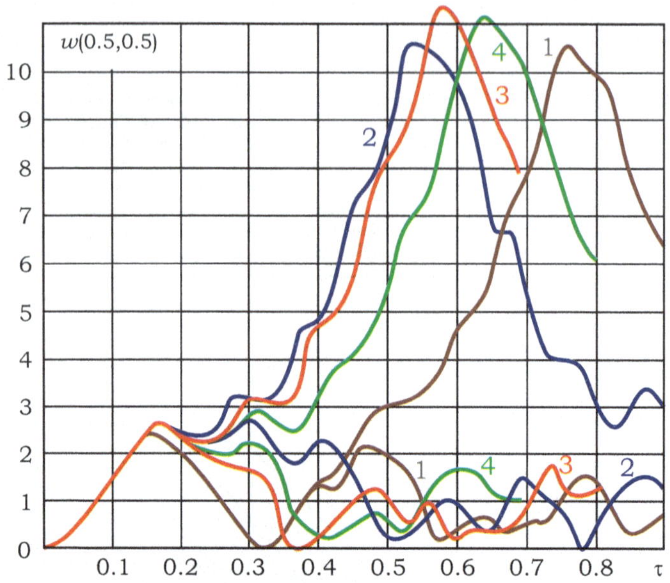

Fig. 7.15 Shell centre deflection for q_{cr} and $q_{cr} - 1$

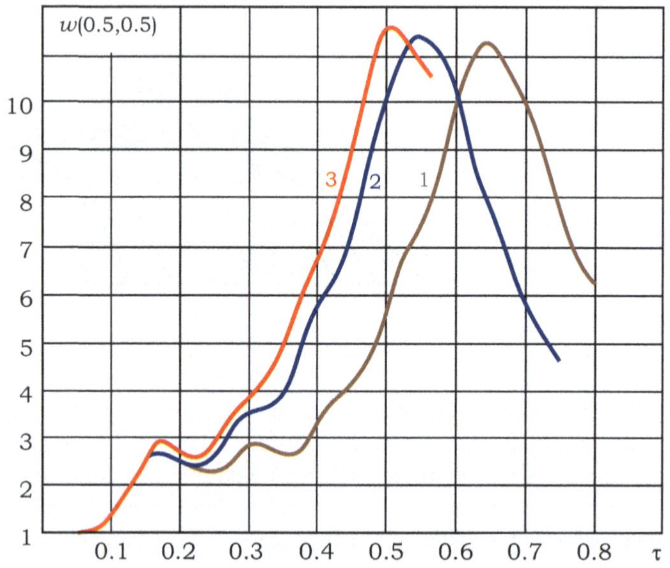

Fig. 7.16 Deflection at the shell centre for the uncoupled geometrically and physically nonlinear problem (curve 1 for $q = 222$, 2 for $q = 223$, 3 for $q = 224$)

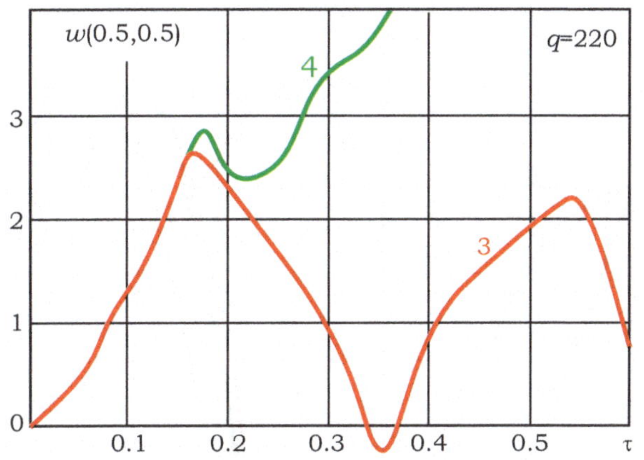

Fig. 7.17 Deflection at the shell centre for $q = 223$ and $t = 0.16$

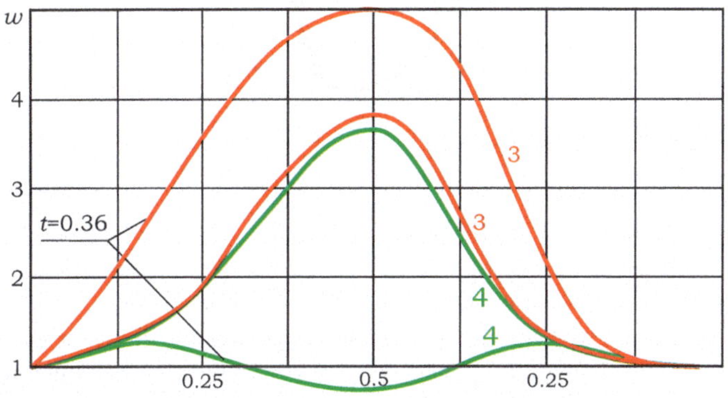

Fig. 7.18 Distribution of w for $t = 0.36$

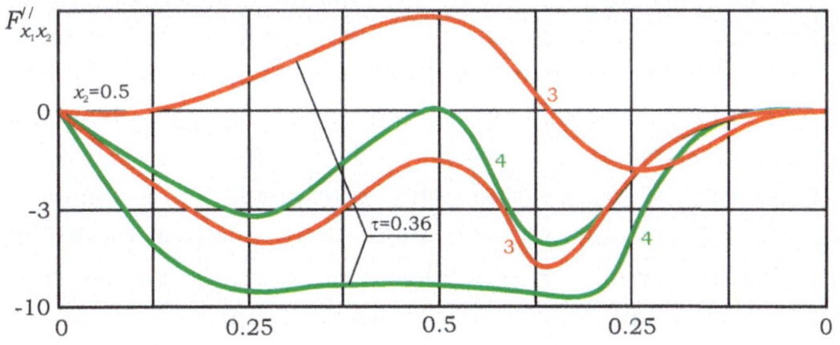

Fig. 7.19 Distribution of $F''_{x_1x_1}$ for $t = 0.36$

For problem 4, $F''_{x_1x_1}$ (0, 0) is close to zero, whereas for problem 3 $F''_{x_1x_1}(0, 0) = -2.5$. It should be noted that in this case also, the coupling between deformation and temperature is mainly exhibited in the stresses.

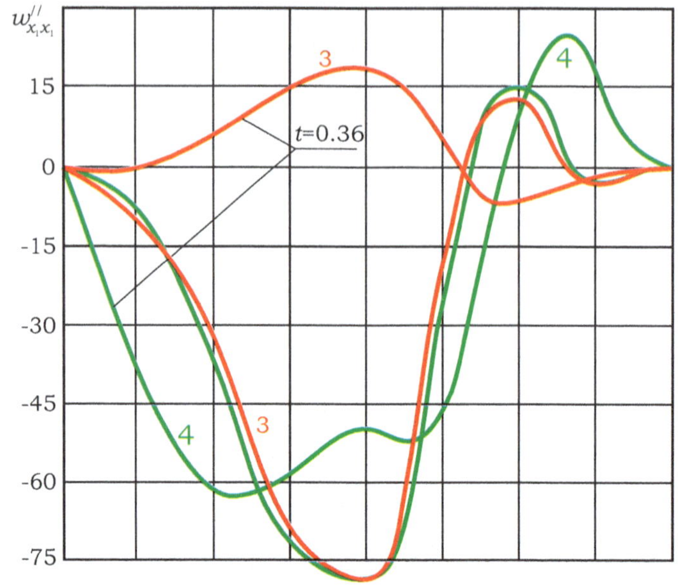

Fig. 7.20 Distribution of $W''_{x_1x_1}$

The time instant $t = 0.36$, for problem 3 with $q = 223$, corresponds to a case where the shell achieves negative deflection values during pre-critical vibrations. Note that at the same time moment, jumping behaviour occurs in the case of problem 4, and the maximum deflection here is of the order of four times the thickness. In what follows, the values of $F''_{x_1x_1}$ and $w''_{x_1x_1}$ not only have quantitative differences but also have different signs in a large part of the shell.

The jumping behaviour of the shell at the critical load has a time variation that depends on the model used. In Fig. 7.15, the maximum deflections are achieved at the critical load at $t = 0.76$ (problem 1), $t = 0.53$ (problem 2), $t = 0.56$ (problem 3) and $t = 0.64$ (problem 4). Therefore, if we compare the stress–strain states of the shell at the critical load at the same time moment, we can again obtain qualitatively different results. This statement is proved by the graphs shown in Figs. 7.21, 7.22, 7.23 and 7.24, where the stress–strain states corresponding to problems 1–4 at $t = 0.53$ are reported. The maximum deflections of the point on the middle surface (0.5,0.5) are 10.5 (problem 2), 9.2 (problem 3), 6.9 (problem 4) and 3.2 (problem 1). The different values of deflections correspond to different stress–strain states (the functions shown in Figs. 7.21, 7.22, 7.23, and 7.24 often differ from each other by a factor of several times at the same characteristic points).

The values of $w''_{x_1x_1}$, $F''_{x_1x_1}$ and σ_{11} for problems 1 and 2 differ not only in magnitude but also in sign over a large part of the shell. The values of $w''_{x_1x_1}$ for problems 3 and 4 differ at the shell centre by a factor of almost two, and at the point (0.125, 0.125) they differ by almost of three in terms of absolute values; and in addition, they have different signs. Although the values of $F''_{x_1x_1}$ for those problems coincide at the point (0.5, 0.5), at other points the stresses very often have different

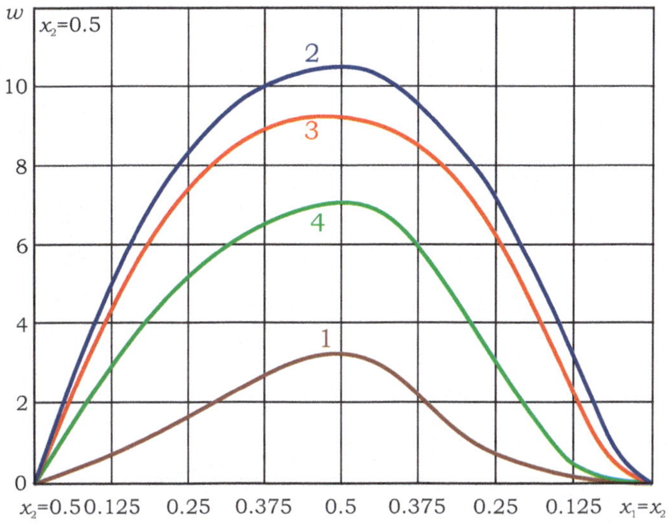

Fig. 7.21 Distribution of w for $t = 0.53$

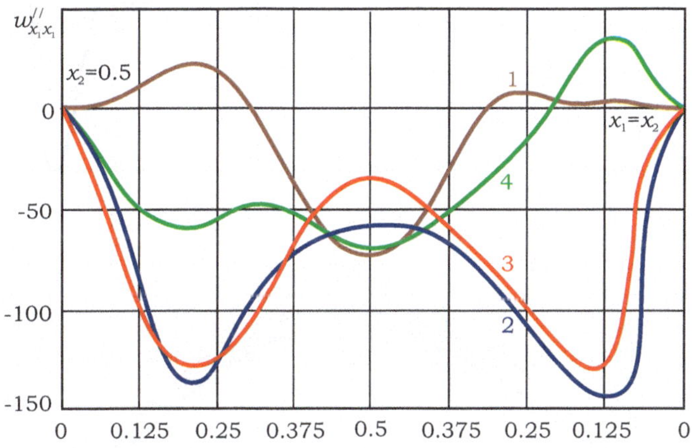

Fig. 7.22 Distribution of $w''_{x_1x_2}$ for $t = 0.53$

signs and their values differ significantly. The stresses σ_{11} obtained at q_{cr} are very close to each other along the $x_1 = 0.5$ axis for problems 3 and 4, but they often differ in sign.

To conclude, it has been shown that the stress–strain states obtained with different models can differ significantly, either quantitatively or qualitatively.

The discussion up to this point has illustrated the fact that taking account of coupling and of physical nonlinearities has a significant influence on the stress–strain state of a shell. The comparison has been carried out for problems 1–4 under the same pre-critical load. Now we are going to show how coupling and physical nonlinearities affect the stress–strain state of a shell during dynamic stability loss.

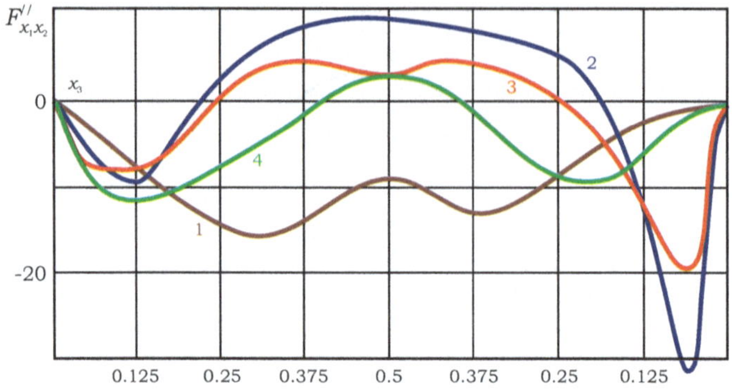

Fig. 7.23 Distribution of $F''_{x_1x_2}$ for $t = 0.53$

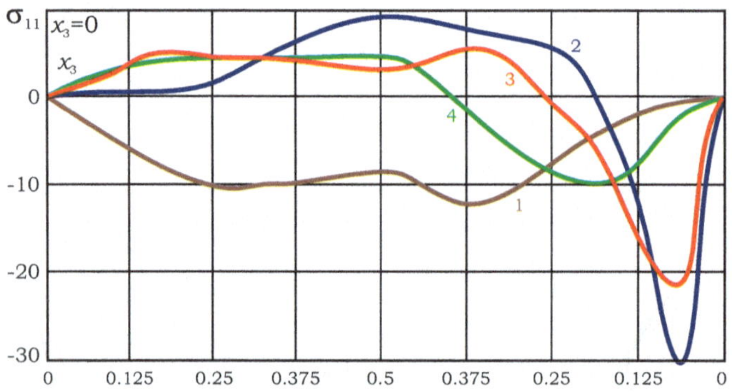

Fig. 7.24 Distribution of σ_{11} for $t = 0.53$

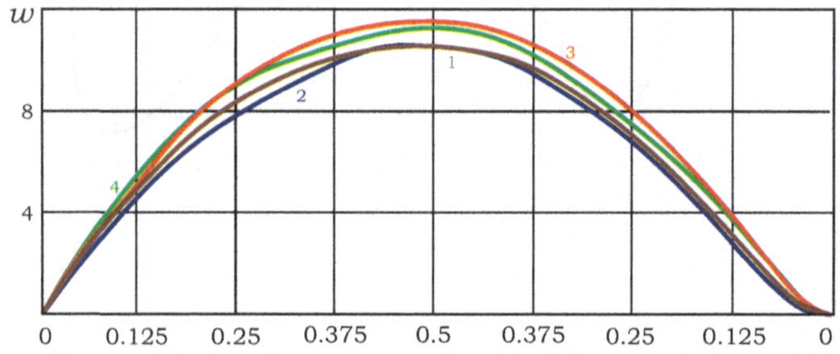

Fig. 7.25 Distribution of w at the time moment of maximum deflection

The stress–strain state will be analysed at the time moment characteristic of the four types of problems discussed, i.e. when the maximum deflections are achieved. The various functions corresponding to this time value are reported in Figs. 7.25, 7.26, 7.27, 7.28, 7.29, 7.30 and 7.31. As earlier, on the left in the figures the functions are plotted along one of the symmetry axes of the shell crossing its centre ($x_2 = 0.5$, for

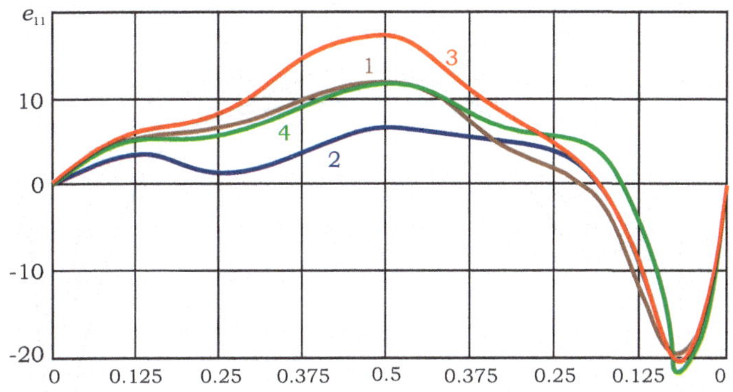

Fig. 7.26 Distribution of ε_{11}

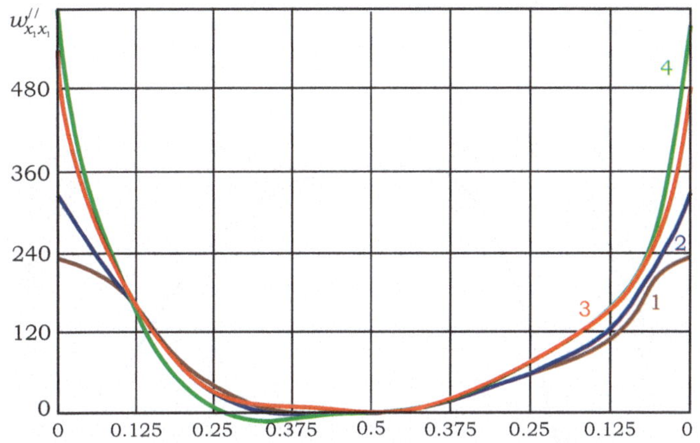

Fig. 7.27 Distribution of $w''_{x_1x_1}$

instance). On the right, the function is plotted along a diagonal through the origin at (0, 0). The data are reported for only one-quarter of the middle surface of the shell (because of symmetry).

The shell deflections shown in Fig. 7.25 at the time of maximum deflection differ only slightly (the values of $w(0.5, 0.5)$ for problems 1 and 2 are 6–7% less than for the physically nonlinear problems). However, the deformations of the middle surface differ significantly: $\varepsilon_{11}(0.5, 0.5)$ increases by 77% for the physically linear problem and by 57% for the physically nonlinear problem when coupling is assumed. In the neighbourhood of the corner, for given boundary conditions and at the critical load, a stress concentration and an increase of deformation is observed: a change of sign of ε_{11} sign and a notable increase of the deflection values occur. Performing the computations for the simple mathematical model (problem 2) leads to a slight increase of the results (by 4.8%). However, at the shell centre, ε_{11} decreases by approximately a factor of two in comparison with problem 3.

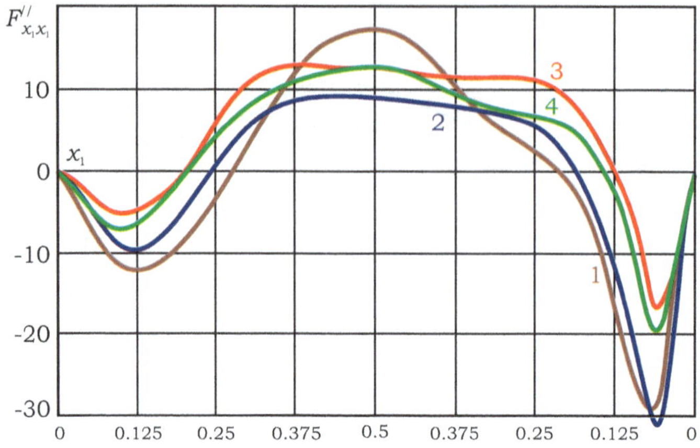

Fig. 7.28 Distribution of $F''_{x_1x_1}$

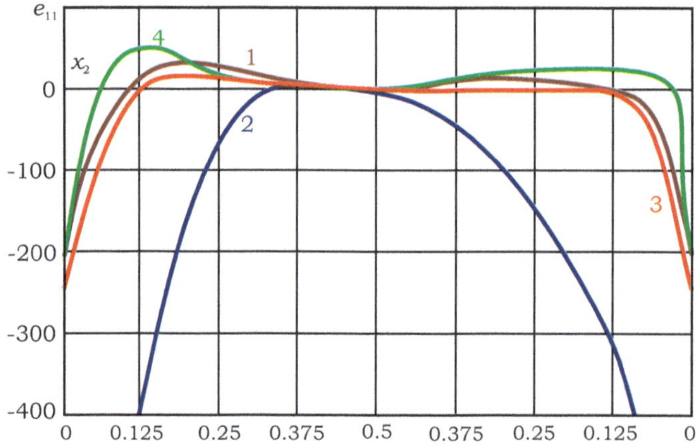

Fig. 7.29 Distribution of ε_{12}

Calculations of the shear deformations of the middle surface for problem 2 give results different from those for the other problems (Fig. 7.29). A similar behaviour occurs for $F''_{x_1x_2}$ (Fig. 7.31). The values of ε_{12} and $F''_{x_1x_2}$ in the corner for problems 2 and 1 differ by factors of 4 and 4.7, respectively. Coupling has an important influence on $F''_{x_1x_2}$ and ε_{12}. For instance, at the point (0.125, 0.125) $F''_{x_1x_2}$ values differ by more than 30% for problems 3 and 4, and the difference in the ε_{12} values is even larger close to a corner. The calculations of ε_{12} and $F''_{x_1x_2}$ without coupling give lower values of those functions.

A similar picture is seen in Fig. 7.27 for the distribution of $w''_{x_1x_2}$ over the middle surface. Close to (0, 0) the values of $w''_{x_1x_2}$ obtained for different models differ significantly (for example, for problems 3 and 4 they differ by 14.4%).

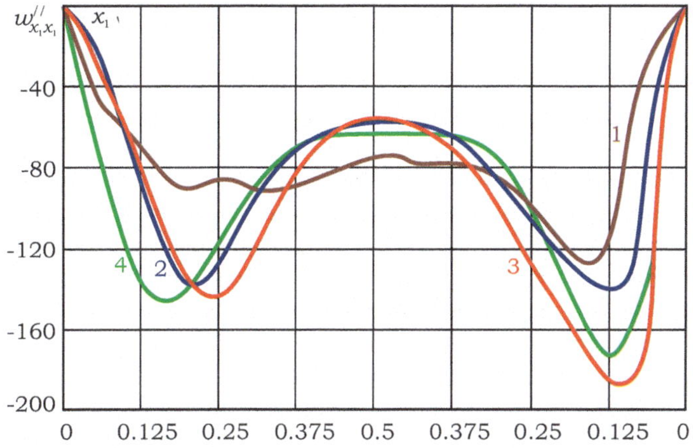

Fig. 7.30 Distribution of $w''_{x_1x_2}$

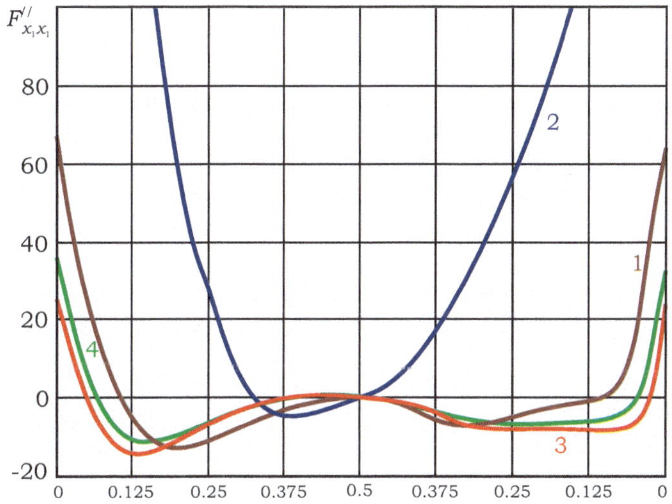

Fig. 7.31 Distribution of $F''_{x_1x_2}$

In Figs. 7.28 and 7.30, the distribution of the second derivative with respect to x_1 of the load function and the deflection along the symmetry axes of the shell is reported. The calculations show that the use of a physically nonlinear model leads to a notable improvement of the results obtained: for $F''_{x_1x_1}$ at the centre of the shell, problems 3 and 4 give the same results, but for problem 2 we obtain a value of $F''_{x_1x_1}$ (0, 0) that is less by 26%, and for problem 1, a value that is greater by 40%. At the point (0.0625, 0.0625), where stress concentration has been observed, the results for problem 2 are larger by 1.6 times than those for problem 4 (taking coupling into account decreases the stress by 15% at this point).

It turns out that the maximum values of $w''_{x_1x_1}$ are located close to the corner, for all problems. The value of $w''_{x_1x_1}$ (0.125, 0.125) is 7% larger for problem 3 than for problem 4. In contrast, in the physically linear problems, the coupling decreases the stresses concentration at this point. The difference between the results for problems 1 and 3 is 37%, whereas for problems 2 and 4 it is 18%.

Now we are able to formulate the following conclusions. Calculations for a load in the neighbourhood of the critical load can lead to qualitatively different results for different models. When coupling is taken into account, the value of the critical load increases owing to energy dissipation, which leads to heating of the shell. Inclusion of both coupling and physical nonlinearity of the material leads to changes of the times of jumping processes of the shell, and a comparison of the stress–strain states obtained with different mathematical models for the same time can also exhibit qualitative differences. The coupling and physical nonlinearity greatly influence the stress–strain state of the shell. When a simple mathematical model is used (without coupling or physical nonlinearity), it can lead to results improved theories (for instance, in the case of shear deformations).

The coupling between deformation and temperature helps us to analyse the influence of mechanical loads on the self-heating of the shell (remember that all shell surfaces are assumed to be thermally isolated).

In Figs. 7.32, 7.33 and 7.34, the temperature distribution through the shell thickness at the centre for problem 3 at $t = 0.3, 0.5, 0.56$ and for problem 1 at $t = 0.54, 0.72, 0.76$ is shown. The time moments for the two problems have been chosen in order to obtain similar shell deflections in problems 3 and 1. The physical nonlinearity changes the temperature distribution θ obtained earlier. The maximum shell heating is observed at the time of maximum deflection close to the corner, i.e. in the place of maximum stress (deformation) concentration. The distribution of θ over the middle surface of the shell at times $t = 0.3, 0.5, 0.56$ for problem 3

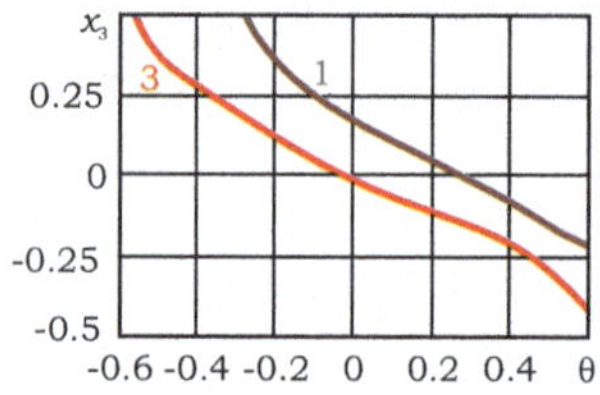

Fig. 7.32 Temperature distribution through the shell thickness at the centre of the shell: problem 1, $t = 0.54$; problem 3, $t = 0.3$

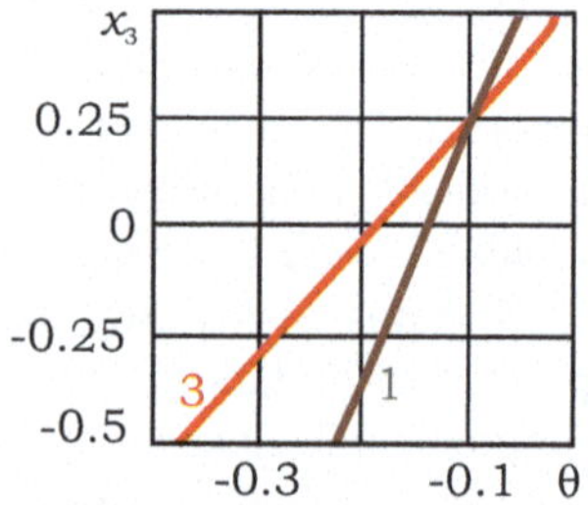

Fig. 7.33 Temperature distribution through the shell thickness at the centre of the shell: problem 1, $t = 0.72$; problem 3, $t = 0.5$

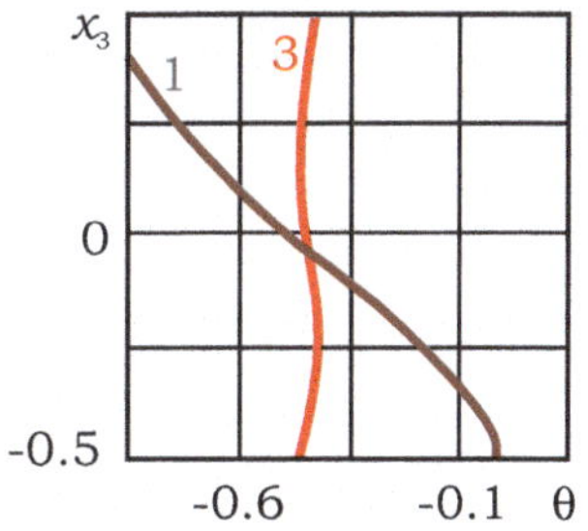

Fig. 7.34 Temperature distribution through the shell thickness at the centre of the shell problem 1, $t = 0.76$; problem 3, $t = 0.56$

and at $t = 0.76$ for problem 1 are shown in Figs. 7.35, 7.36, 7.37 and 7.38. The maximum self-heating is 12.1° for problem 3 and 8.3° for problem 1. Qualitatively, the temperature distributions over the middle surface for problem 3 (Fig. 7.37) and for the problem 1 (Fig. 7.38) are similar (the maximal self-heating is observed in the surface $x_3 = 0.5$ both cases).

It is easy to check that the areas where θ increases on one of the shell surfaces correspond to areas where θ decreases on the opposite surface. Therefore, the temperature change across the thickness at the point (0.0625, 0.0625) reaches 19.5° in problem 3, whereas at the point (0.125, 0.125) it reaches 13.2° in problem 1.

The shift of the area of maximum shell heating from the centre of the shell to the corner is caused by the increase in the load and deflection and by the type of jumping behaviour during stability loss. Incidentally, for a post-critical load ($q = 230$, problem 3), a value of $\theta = -0.78$ was reached at the shell centre on the surface $x_3 = 0.5$ (and $\theta = -0.46$ for $q = 225$), and the change of the solution for θ at the point (0.0625, 0.0625) reached 19.9°.

In Fig. 7.39, the temperature distribution θ through the shell thickness at the point (0.0625, 0.0625), at the time moments $t = 0.1, 0.3, 0.5, 0.56, 0.6, 0.625$ is shown for problem 3. Curves 1–6 correspond to these times in increasing order, and curve 2 (not shown) almost coincides with the axis $\theta = 0$. The distributions of θ presented in Fig. 7.39 are sometimes (curves 1–3) close to linear, and sometimes (curves 4–6) very different from linear. Therefore, the assumption of a linear distribution of θ through the shell thickness in an analysis of the self-heating of a shell subjected to mechanical loads will introduce errors.

7.5 Dynamic Stability of Shells Under Thermal Shock

We consider the dynamic stability of a thin shallow shell ($k_1 = k_2 = 24$) subjected to a suddenly applied load q, constant in time and uniformly distributed, in the surface $x_3 = 0.5$ (see Sect. 7.1). The boundary conditions attached to (7.1)–(7.3) are defined by (6.60). At $t = 0$, a thermal shock is applied on the internal surface $x_3 = 0.5$. The other shell surfaces are isolated (7.4). As the initial condition, zero values of $w, \dot{w}, \theta$ have been taken.

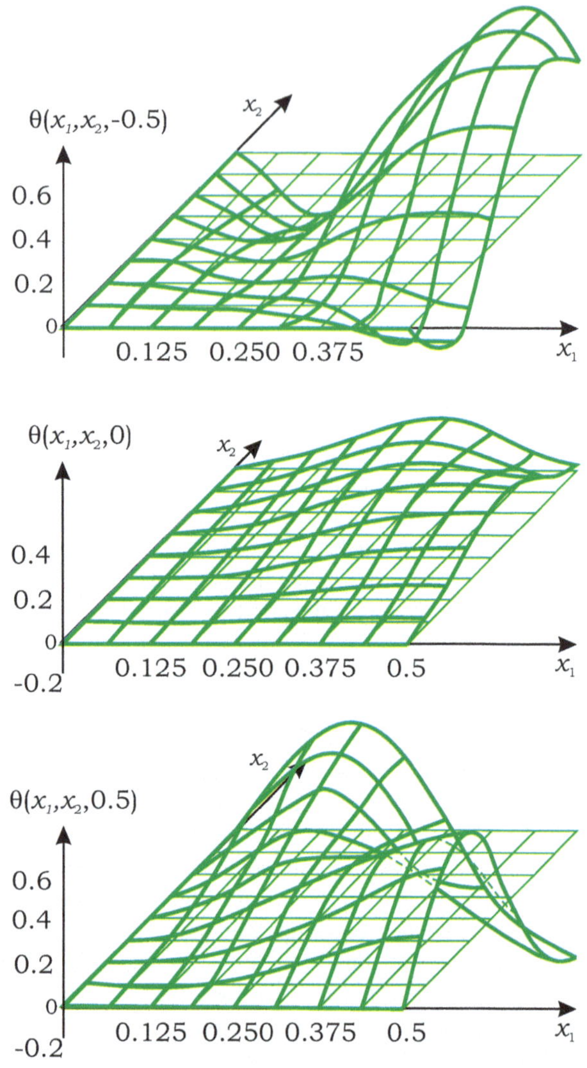

Fig. 7.35 Distribution of temperature over the middle surface of the shell, problem 3, $t = 0.3$

The values of the critical load obtained for problems 1–4 for full thermal isolation of the shell and for different q_T values, are presented in Table 7.5. The intensity q_T of the thermal shock has been chosen in such a way as to satisfy the heat transfer equation (7.1) obtained for the condition $\frac{T-T_0}{T_0} \ll 1$, i.e. for $T_0 = 320°$ K the increase θ is treated as small up to 100°.

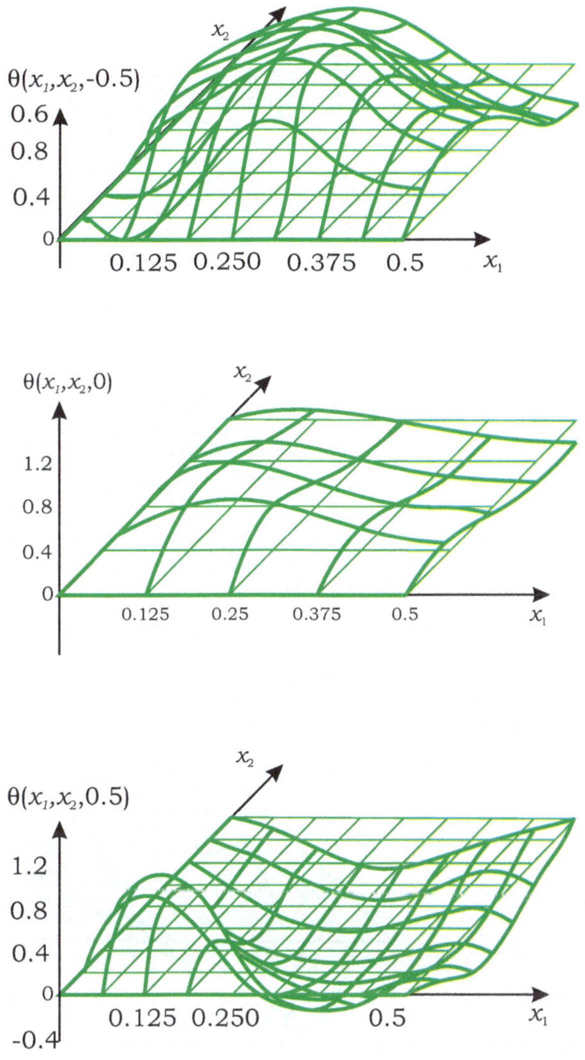

Fig. 7.36 Distribution of temperature over the middle surface of the shell, problem 3, $t = 0.5$

In Fig. 7.40, the values of the deflection at the shell centre versus time are shown for the values of q_{cr} listed in Table 7.5 and for certain pre-critical loads differing by one unit from q_{cr}. Curves 1–4 refer to problems 1–4 with a thermal shock $q_T = 25$. "Problem 5" refers to the uncoupled physically nonlinear problem with $q_T = 12$, and "problem 6" refers to the coupled physically nonlinear problem with $q_T = 12$.

It is observed that when q_T is increased, the influence of physical nonlinearities on the value of the critical loads increases. For $q_T = 0$, the difference between the values of q_T for problems 1 and 3 is 2.2% and the difference for problems 2 and 4 is 3.5%. For $q_T = 25$, the corresponding differences are 6.1% and 5.3%, respectively.

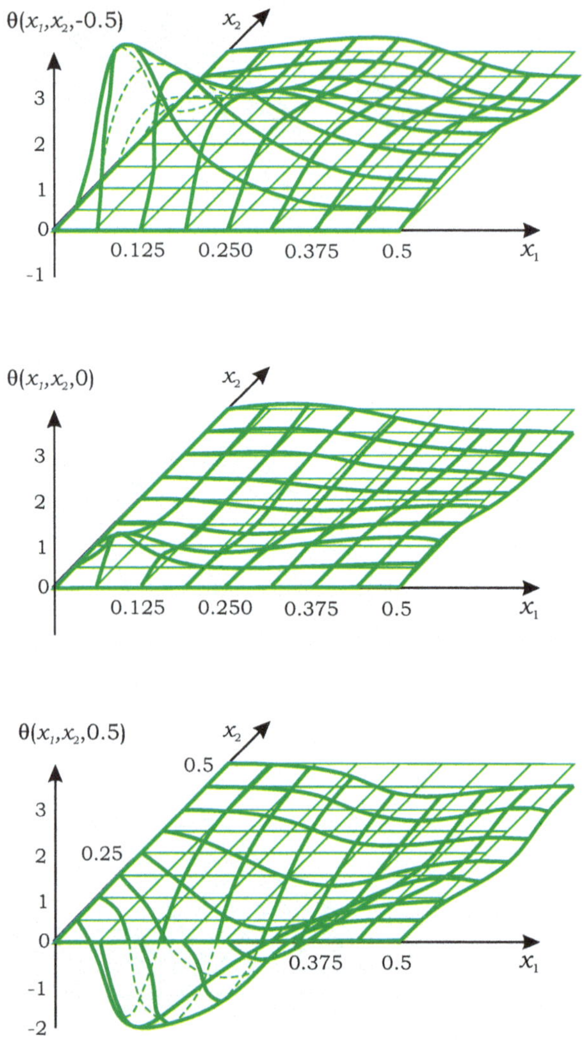

Fig. 7.37 Distribution of temperature over the middle of the shell, problem 3, $t = 0.56$

The influence of coupling onto "jumping" phenomenon greatly decreases with increase of q_T. Curves 1 and 2 in Fig. 7.40 almost overlap. For problems 5 and 6, the difference between the times required to achieve the maximum deflection is about 0.1. Increasing q_T by a factor of two (problems 3 and 4) causes this difference to decrease to 0.04.

We now compare the stress–strain states of the shell for problems 1–4 and for problems 5 and 6 at the times of maximum deflection: $t = 0.53$ for problems 1 and 2 (GN(c) and GN, respectively); $t = 0.54$ for problem 3 (GPN(c)); $t = 0.50$ for problems 4 and 5 (GPN problems for different q_T values); and $t = 0.60$ for problem 6 (GPN(c)).

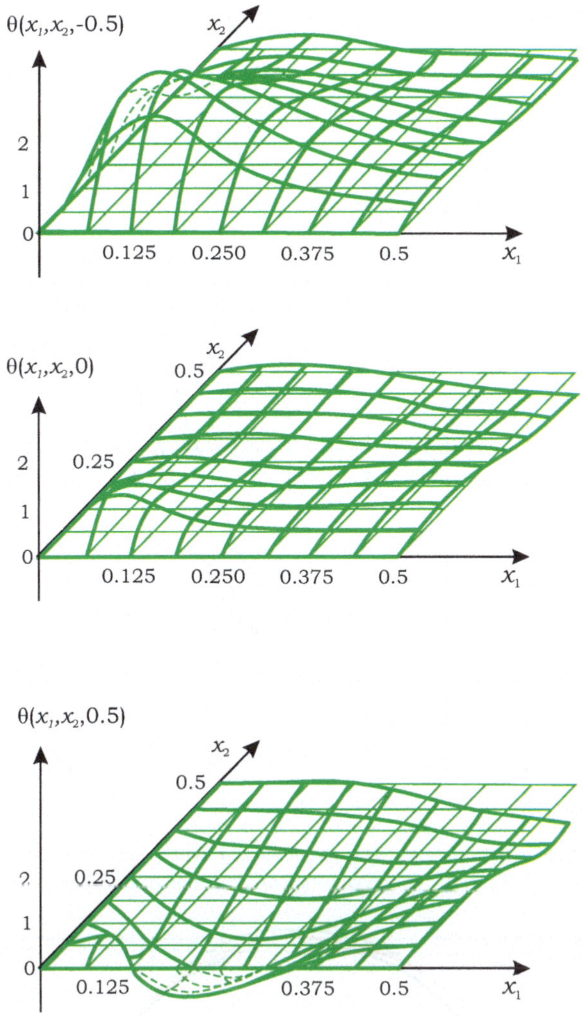

Fig. 7.38 Distribution of temperature over the middle shell, problem 1, $t = 0.76$

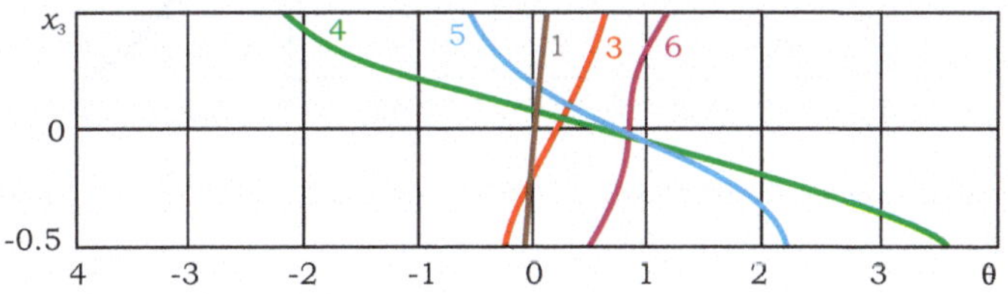

Fig. 7.39 Distribution of temperature θ through shell thickness at the point (0.0625, 0.0625), at the time moments $t = 0.1, 0.3, 0.5, 0.56, 0.6, 0.625$

Table 7.5 Critical load q_{cr} for different boundary conditions of heat transfer equation

Problem	$q_T = 0$	$q_T = 12$	$q_T = 25$
1	230	–	260
2	230	–	257
3	225	236	245
4	222	232	244

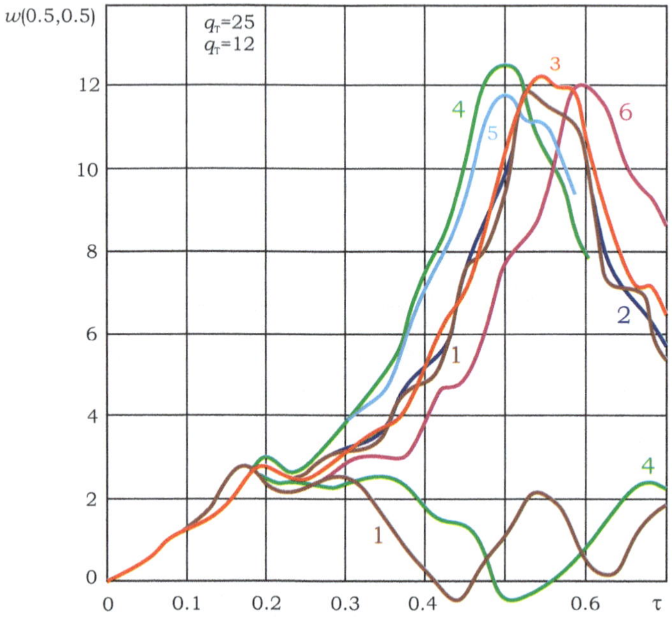

Fig. 7.40 Deflection of the shell centre for problems 1–4 with a thermal shock $q_T = 25$, and for uncoupled and coupled physically nonlinear problems with a thermal shock $q_T = 12$

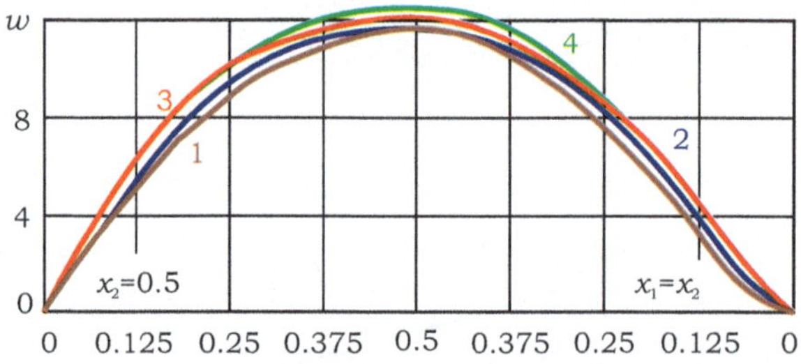

Fig. 7.41 Shell deflections for problem 1, $t = 0.53$; problem 2, $t = 0.53$; problem 3, $t = 0.54$; problem 4, $t = 0.50$

The deflections of the shell along its symmetry axes obtained in for problems 1–4 at these time moments are presented in Fig. 7.41. For all four problems, the values of w obtained are close to each other.

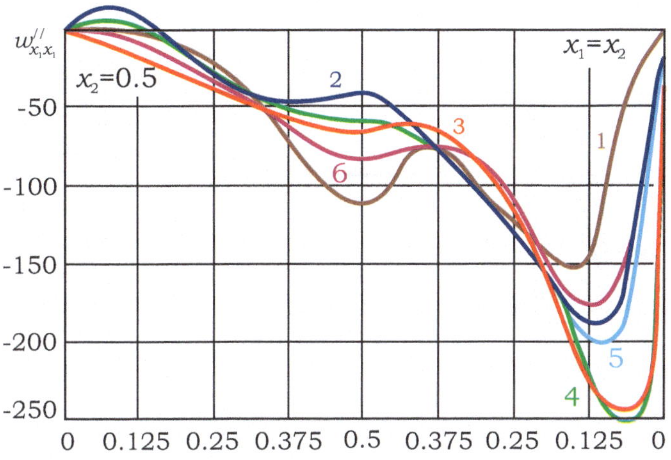

Fig. 7.42 Distribution of $w''_{x_1x_1}$ for problems 1–6

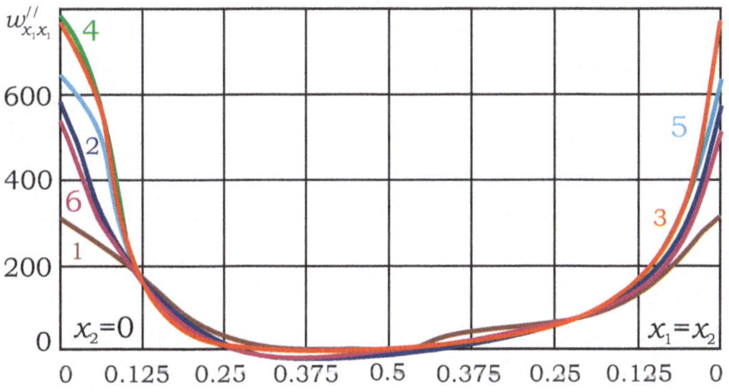

Fig. 7.43 Distribution of $w''_{x_1x_2}$ for problems 1–6

In Figs. 7.42, 7.43, 7.44, 7.45 and 7.46, some of the functions characterizing the stress-strain states for problems 1–6 are shown. It can be seen that with an increase of q_T, the influence of material physical nonlinearity increases and the effect of coupling decreases. In these figures, there are large differences between the curves corresponding to problems 1 and 2. The differences are much smaller for problems 5 and 6, and even more so for problems 3 and 4.

For $q_T = 0$, the values of ε_{12}, $w''_{x_1x_2}$ and $F''_{x_1x_2}$ have rather large quantitative and qualitative differences in the neighbourhood of the corners. However, computation with simple mathematical model (problem 2) does not give a qualitative difference. At the point (0, 0) a difference in ε_{12} is 25% (problems 1 and 2), 24% (problems 5 and 6) and 21% (problems 3 and 4). The values of $w''_{x_1x_2}$ differ by 46% (problems 1 and 2), 14.5% (problems 5 and 6), and 1.2% (problems 3 and 4). A similar tendency

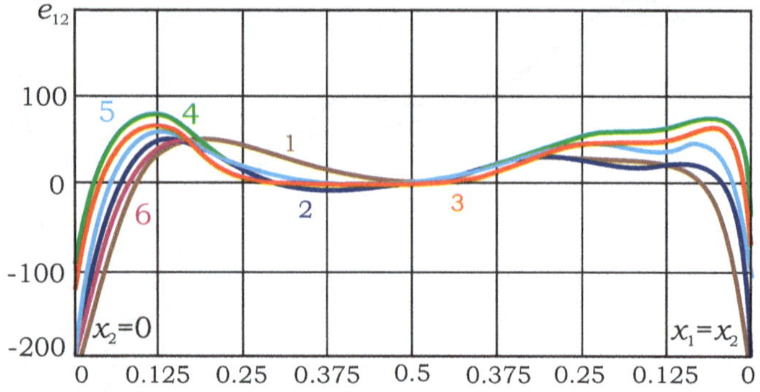

Fig. 7.44 Distribution of e''_{12} for problems 1–6

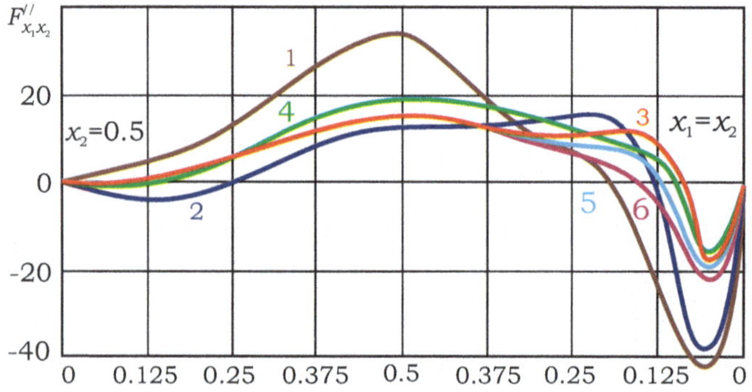

Fig. 7.45 Distribution of $F''_{x_1x_2}$ for problems 1–6

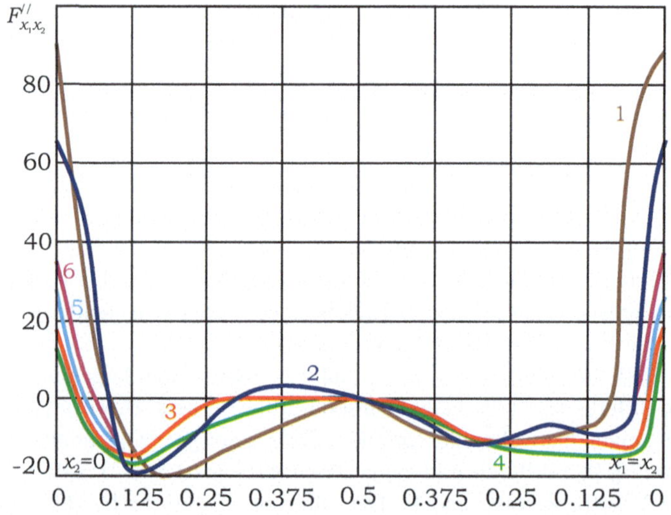

Fig. 7.46 Distribution of $F''_{x_1x_2}$ for problems 1–6

appears for $F''_{x_1x_2}$ (0, 0): the difference are 25%, 21.5% and 15.9% for problems 1 and 2, 5 and 6, and 3 and 4, respectively.

For both $w''_{x_1x_1}$ and $F''_{x_1x_1}$, the effects of coupling are large in the physically linear problems. At the shell centre, $F''_{x_1x_2}$ for problem 1 is 2.7 times the corresponding value for problem 2. At the point (0.0625, 0.0625), where a large stress concentration is observed, the difference is 11.9%. For the physically nonlinear problems, the influence of coupling is of less importance: for problems 5 and 6, the influence of coupling on $F''_{x_1x_2}$ is 29% and 10.5%; for problems 3 and 4, the corresponding numbers are 21.5% and 2.9%. For $w''_{x_1x_2}$, the corresponding increases are 2.7 times and 21.7% for problems 1 and 2; 30% and 8% for problems 5 and 6, and 11.8% and 1.9% for problems 3 and 4. The increase in of q_{cr} for $q_T \neq 0$ does not affect qualitatively the self-heating of the shell caused by the mechanical load. Comparing the thermal fields of the coupled and noncoupled problems, one can observe that the maximum difference occurs at the moment corresponding to maximum deflection, close to the corner of the surface $x_3 = 0.5$. For problems 3 and 4, the difference at the point (0.0625, 0.0625) reaches 13.06°, whereas for problems 1 and 2 the difference reaches 10.7°. The differences observed in the temperature gradient through the shell thickness due to warming up of part of the surface $x_3 = -0.5$ and cooling down of part of the surface $x_3 = 5$ reach 22.5° for problems 3 and 4 and 15° for problems 1 and 2.

Inclusion of coupling causes changes in the thermal field. It also decreases the strain and stress values. However, it brings about an important increase of the exactness of the results obtained. For problems 3 and 4 with $q_T = 25$, the differences in $F''_{x_1x_2}$ (0, 0), $F''_{x_1x_1}$ (0.5, 0.5), $w''_{x_1x_1}$ (0.5; 0.5) reach values of 15.9%, 21.5% and 11.8%, respectively.

The temperature distribution through the shell thickness is reported in Fig. 7.47. The dashed curve corresponds to θ for the uncoupled problem 4 at $t = 0.54$ the heating is caused by external sources, and this dependence holds for an arbitrary point (x_1, x_2). Curve 1 corresponds to the distribution of θ at the point (0.0625, 0.0625), and curve 2 corresponds to the distribution of θ at the point (0.5, 0.5) for problem 3. The dash–dot curves 3 and 4 correspond to the temperature distribution at $t = 0.625$ for problems 3 and 4 at the same point as for curve 1.

The coupling effect significantly influences distribution of θ along x_3, especially at the moment when the maximum deflection is reached. For a shell with given physical and geometrical parameters, the influence of coupling during thermal shock is most evident in the neighbourhood of a corner point.

Fig. 7.47 Distribution of temperature θ through the shell thickness

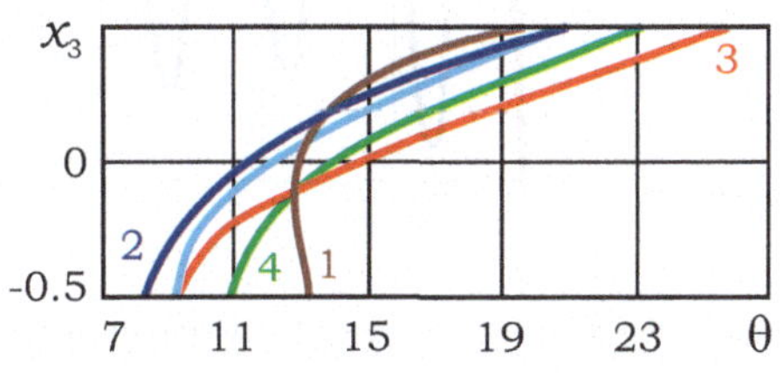

Now we consider a problem related to the dynamic reaction of a thin, shallow ($K_1 = K_2 = 48$) physically nonlinear shell to the load specified in Sect. 7.1, applied to the surface $x_3 = -0.5$. The boundary conditions (7.3) are used. The other geometrical and physical parameters are the same as earlier.

Also, as earlier we assume that the shell satisfies the thinness condition $f/a < 1/8$. According to the definition given in [333], a spherical shell is shallow if the angle subtended at the centre by a chord which forms a side of a shell, square in plan, does not exceed 60° (in our case, for $K_1 = K_2 = 48$, this angle is 26°).

The deflections at the shell centre versus time for $q = 900$, for problems 1–4, are shown in Fig. 7.48. It can be seen from the figure that the use of physically linear and physically nonlinear models gives qualitatively different results. The maximum deflections at the shell centre for problems 3 and 4 in the first period are more than twice the deflections for problems 1 and 2; such deflections have not been observed for a shell with $K_1 = K_2 = 24$ under pre-critical loads.

It is clear that for $K_1 = K_2 = 48$ and loads close to the critical load, the use of a physically nonlinear model improves the accuracy of the results of computations.

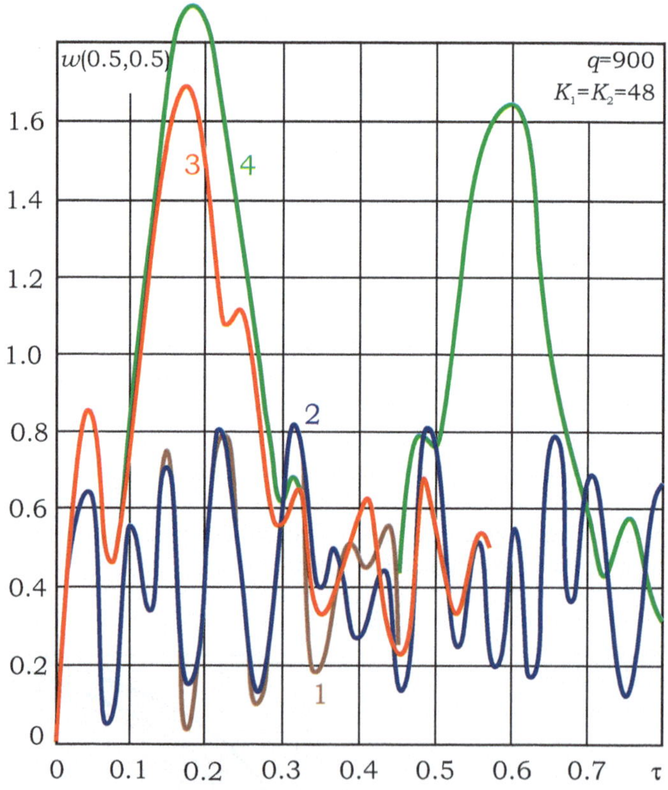

Fig. 7.48 Deflection of the shell centre for problems 1–4, $q = 900$, $K_1 = K_2 = 48$

For $K_1 = K_2 = 24$, the influence of physical nonlinearity on the deflections is not significant.

In the physically nonlinear problem, coupling decreases the deflection at the shell centre by 10.5% during the first period of oscillation. In problem 3 (GPN(c)), the dynamic reaction of the shell has been traced up to $t = 1.3$ (in the figure, results for a shorter time are reported), and after the first period the deflection of the shell at its centre does not exceed 0.8 times the shell thickness. This means that the deflections are comparable to those obtained in problems 1 and 2. In the problem 4 (GPN), damping of oscillations does not occur and the values of $w(0.5; 0.5)$ in the subsequent periods are similar.

It turns out that the load $q = 900$ cannot be treated as critical in problems 3 and 4. During a small increase of the load from this value a small deflection increase was observed. The time needed to achieve the maximum deflection increases and the deflection at the shell centre does not reach twice the value of the height of the shell over the plane.

Now we compare the stress–strain states of the shell in problems 3 and 4 at $t = 0.17$, when the deflections (and stresses) are maximum. The results are presented in Figs. 7.49, 7.50, 7.51, 7.52, 7.53 and 7.54.

In these figures (except for Figs. 7.53 and 7.54), the stress–strain state of the shell is reported for problems 1 and 2 at $t = 0.22$ and at $t = 0.17$ (a picture for $t = 0.22$ is omitted here). As Fig. 7.48 illustrates, at $t = 0.22$ the deflections in problems 1 and 2 are maximum, and the shell is working in extremely difficult circumstances; therefore this moment of time has been chosen in order to make comparisons with problems 3 and 4. The data for $t = 0.17$ illustrate the stress–strain state of the shell obtained from different models at the same moment of time.

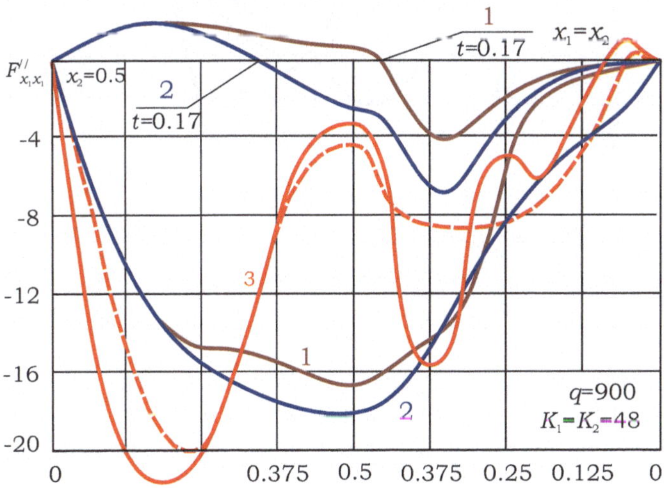

Fig. 7.49 Distribution of $F''_{x_1x_1}$, for $t = 0.17$, $q = 900$, $K_1 = K_2 = 48$

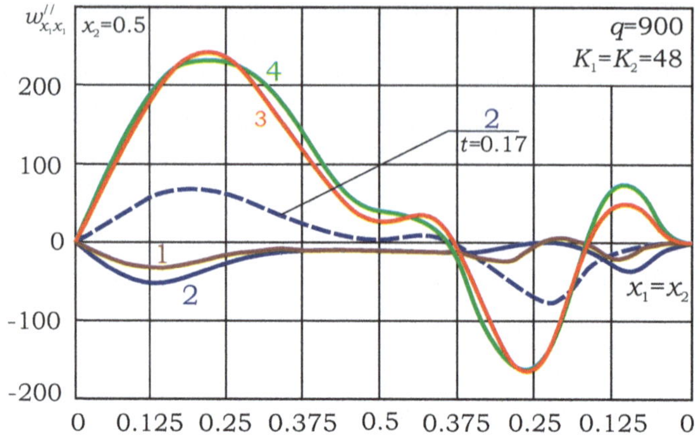

Fig. 7.50 Distribution of $W''_{x_1x_1}$, for $t = 0.17$, $q = 900$, $K_1 = K_2 = 48$

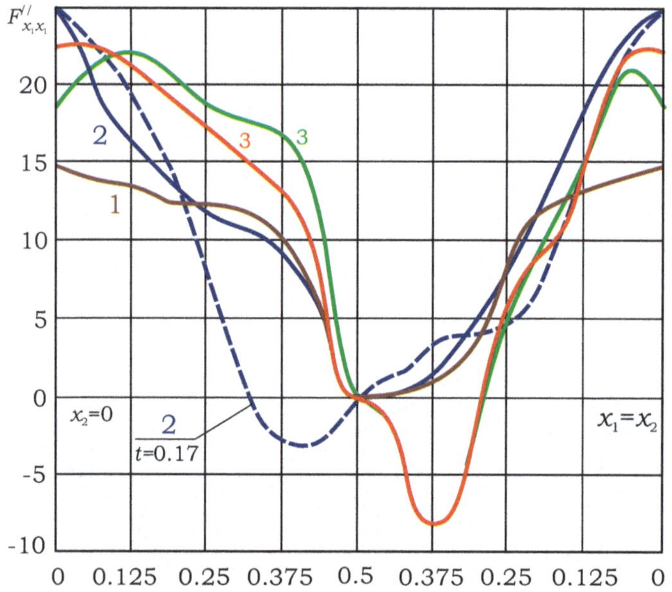

Fig. 7.51 Distribution of $F''_{x_1x_1}$, for $t = 0.17$

Additional information about the data is included in the figures. The results shown in Figs. 7.49, 7.50, 7.51 and 7.52 indicate a large difference between the solutions obtained for problems 3 and 4 and for problems 1 and 2. The distributions of the various functions over the middle surface of the shell possess complex character. The values of $F_{x_1x_1}$ in problems 3 and 4 at the shell centre are more than four times smaller than in problems 1 and 2, for $t = 0.22$. At the point (0.1875, 0.5),

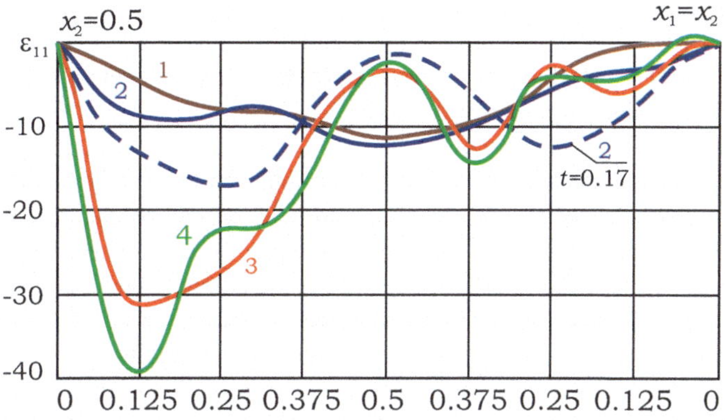

Fig. 7.52 Distribution of ε_{11}, for $t = 0.17$

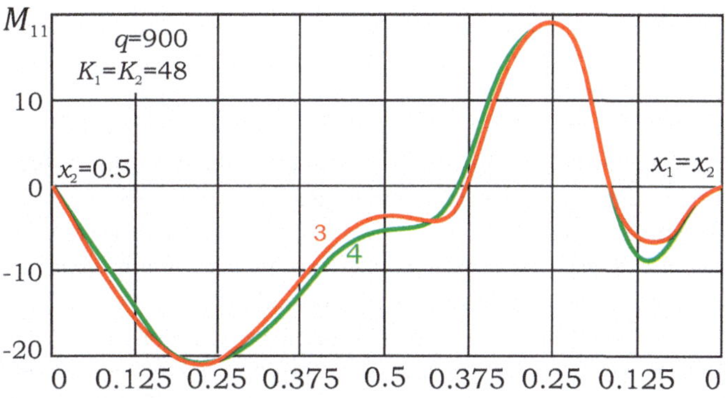

Fig. 7.53 Distribution of M_{11}, for $q = 900$, $K_1 = K_2 = 48$

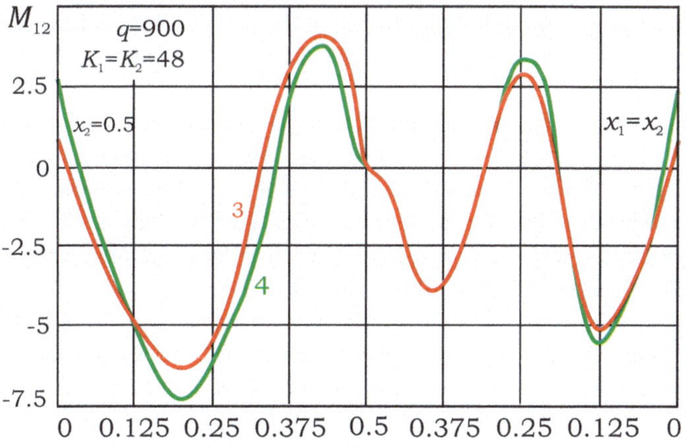

Fig. 7.54 Distribution of M_{12}, for $q = 900$, $K_1 = K_2 = 48$

$F''_{x_1x_1}$ in problem 4 (GPN) is almost 36% higher than in problems 1 and 2. The difference between the value for problem 3 and those for problems 1 and 2 is 26%. A comparison of $F''_{x_1x_1}$ along the x_2 axis in problems 3 and 4 and problems 1 and 2 for $t = 0.17$ gives an even greater difference.

The distribution of $w''_{x_1x_1}$ along the axis $x_2 = 0.5$ in problems 3 and 4 also differs qualitatively from that obtained in problems 1 and 2 at the time moments chosen. Because of the complex variation of $F''_{x_1x_1}$ and $w''_{x_1x_1}$ along the diagonal $x_1 = x_2$, it is difficult to compare these functions (the difference varies from a few to a hundred percent). Also the distributions of $F''_{x_1x_2}$ and ε_{11} are complex and a large difference between problems 3 and 4 and problems 1 and 2 is observed.

Therefore, we draw the following conclusion. The physically linear model when used in the computation of the stress–strain state of a shell with the given curvature parameters under loads close to the critical load, leads to unacceptable errors.

The influence of coupling on the stress–strain state of the shell in problems 3 and 4 is presented in Figs. 7.49, 7.50, 7.51, 7.52, 7.53 and 7.54. It appears that coupling does not change significantly the bending and rotational moments (Figs. 7.53 and 7.54), nor $w''_{x_1x_1}$ (Fig. 7.50). For instance, at the point (0.25, 0.5) the difference between the values of $w''_{x_1x_1}$ for problems 3 and 4 is 4.3% (a similar observation can be made for M_{11} and M_{12}). More obvious differences occur at the points (0.5, 0.5) and (0.125, 0.125). The differences between the values of $w''_{x_1x_1}$ for problems 3 and 4 at these two points are 30% and 36%, respectively. The corresponding differences for m_{11} are 23% and 20%, respectively. However, at those points the values of $w''_{x_1x_1}$ and M_{11} are far from the maximum values achieved at that time moment on the middle surface of the shell.

More evident coupling effects are exhibited by the functions $F''_{x_1x_1}$ and ε_{11} (Figs. 7.49, 7.51 and 7.52). The behaviour reported here has been reported earlier. For a shell with the given parameters, when the maximum deflection is achieved the coupling decreases the maximum absolute stress $F''_{x_1x_1}$ by 14% (for instance, at the point 0.375, 0.375, the value of $F''_{x_1x_1}$ is reduced by a factor of two). The maximum absolute value of ε_{11} in the middle surface at the point (0.125, 0.125) is decreased by 21.5%. A significant change in $F''_{x_1x_2}$ of the two shell sides is also induced by coupling (Fig. 7.52).

The complicated behaviour of the functions shown in Figs. 7.49, 7.50, 7.51, 7.52, 7.53 and 7.54 is caused by the complicated form of the middle surface during vibration. In the problems considered earlier, when the shell reaches its maximum (in a given period) deflection at the point (0.5, 0.5), it does not have any other points of the middle surface with higher values of w. Here we observe a much more complicated behaviour, which is illustrated in Fig. 7.55.

Curves 1 and 2 in Fig. 7.55 correspond to the distribution of w along the axes in problems 1 and 2 for $t = 0.22$. The maximum value of w is achieved in the shell centre. Curves 3 and 4 correspond to the distribution of w in problems 3 and 4 when, at the shell centre, the maximum deflection has been achieved within the vibration period considered. As has been mentioned, taking account of coupling decreases the amplitude of vibration greatly not only in the centre but also over all the middle surface. Curves 5 and 6 correspond to the distributions for problems 3 and 4 at

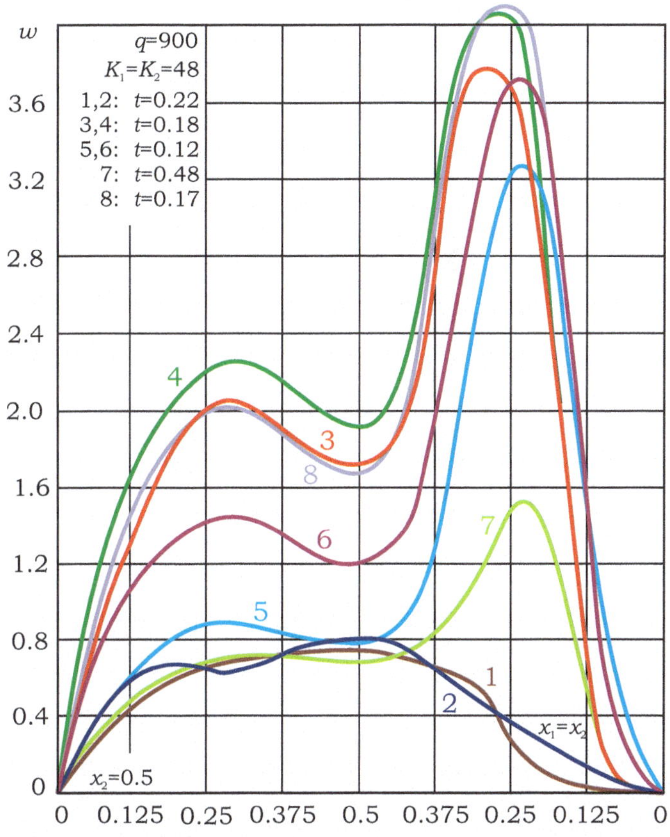

Fig. 7.55 Distribution of w for different problems and time instants, $q = 900$, $K_1 = K_2 = 48$

the time moment $t = 0.12$, and curves 7 and 8 correspond to the distributions at $t = 0.48$ and $t = 0.6$, respectively.

The deflections indicated by curves corresponding to uncoupled physically nonlinear problem practically overlap at the point (0.25, 0.25), but in the centre (Fig. 7.49) they show a significant difference. In the coupled problem, the values of w in the first period of vibration (curve 3) and in the next period (curve 7) differ greatly. For $t = 0.48$, which corresponds to the highest deflection value in one of the periods, $w(0.5, 0.5)$ differs from the value related to the first period by 2.5 times, and $w(0.25, 0.25)$ differs by 2.16 times.

A significant increase of the deflection in the neighbourhood of the point (0.25, 0.25) is caused by zones of plastic deformation. In order to illustrate the complicated shape of the shell deflection, contour lines linking the same deflection values for problem 3 at $t = 0.17$ have been shown in Fig. 7.56 (a similar like behaviour has been observed for problem 4, but with slightly higher w values).

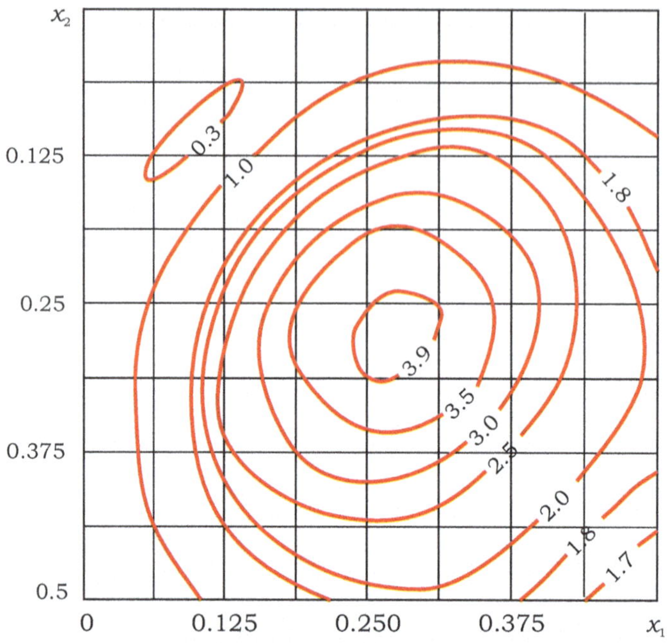

Fig. 7.56 Distribution of w for problem 3, $t = 0.17$

The temperature distribution is much more complicated than in the previously considered coupled problem. A few areas of heating and cooling occur on the external shell surfaces $x_3 = \pm 0.5$. As an example, the distribution of θ over the middle surface for problem 3, at the time instant of maximum deflection, is reported in Fig. 7.57. Now, the maximum heating of the shell does not occur at this time moment. In Fig. 7.58, the change of temperature θ with time at the points (0.5, 0.5) and (0.25, 0.25) (curves 1 and 2, respectively), on the surface $x_3 = 0.5$ is reported. The distribution of θ through the thickness at the time moments $t = 0.07, 0.017$ and 0.39 (curves 1, 2 and 3, respectively) at the points (0.25; 0.25) (solid curves) and (0.5, 0.5) (dashed curves) is shown in Fig. 7.59. The distribution of θ through thickness at the point (0.5, 0.25) at the same times is presented in Fig. 7.60. The self-heating of the shell changes slightly. For $t = 0.07$ the self-heating reaches 12.8°, for $t = 0.17$ it reaches about 6° and for $t = 0.39$ it reaches little more than 6.5°. The maximum heating of the shell in the surface $x_3 = -0.5$ is observed at this point at $t = 0.07$ (it is equal to 7.5°).

Now we consider the influence of coupling on the stress–strain state and the self-heating of the shell with an increase of the load q up to its critical value. Our numerical experiments show that for problems 3 and 4, the critical loads q_{cr} are 945 and 925, respectively. The change of the deflection at the shell centre versus time is shown in Fig. 7.61 for problems 3 and 4 with loads equal to q_{cr} and for the pre-critical loads $q = 940$ and $q = 920$.

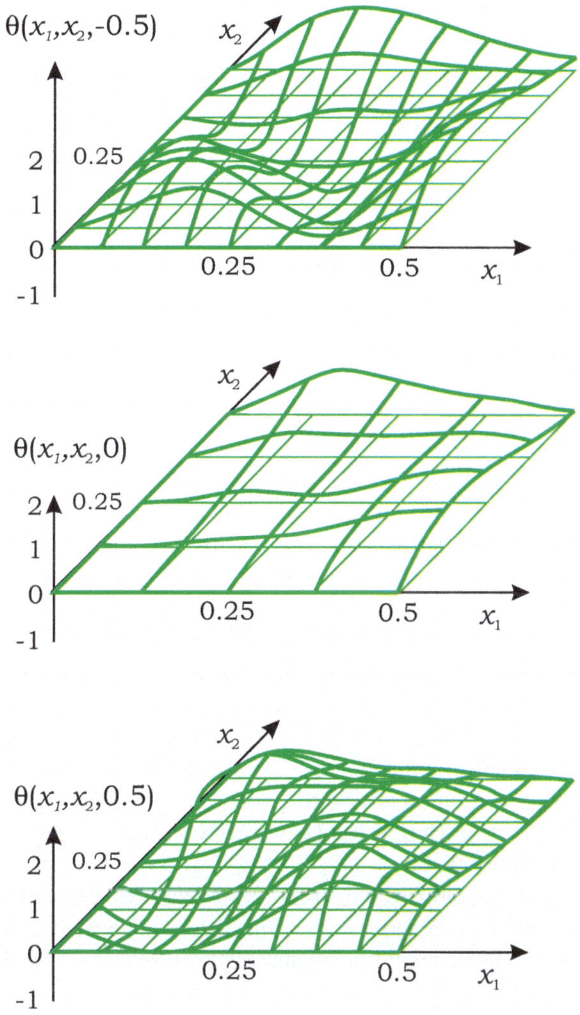

Fig. 7.57 Distribution of temperature over middle surface and through shell thickness

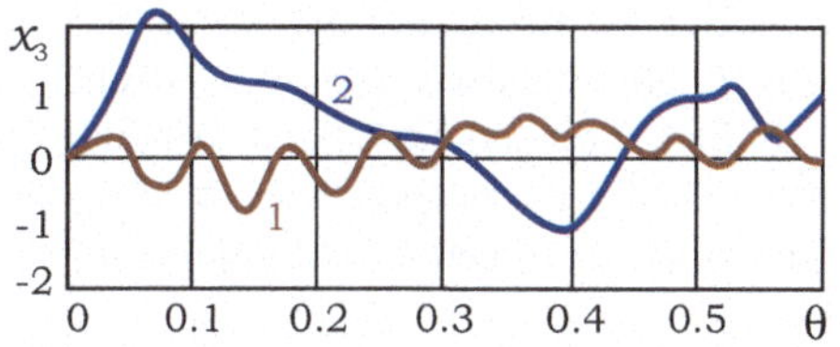

Fig. 7.58 Temperature θ on the surface $x_3 = 0.5$; curve 1, (0.5, 0.5); curve 2, (0.25, 0.25)

As earlier, coupling slightly increases the critical load for problems 3 and 4 by about 2.2% (an increase of 1.4% for problems 3 and 4 was reported in Sect. 7.3). Physically linear models (problems 1 and 2) give qualitatively different results and cannot be used to define the critical load. In Fig. 7.61, the shell centre deflection for

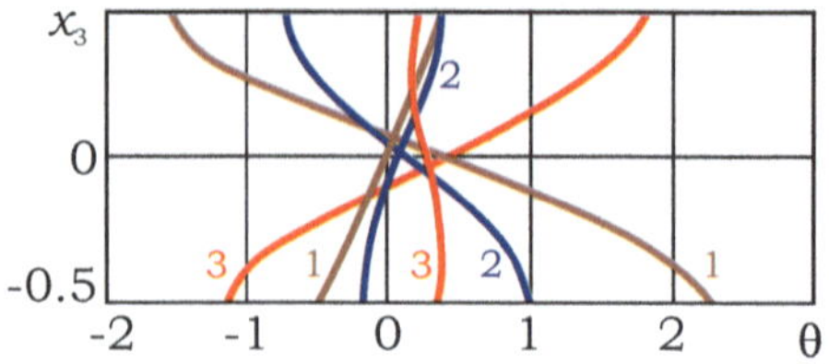

Fig. 7.59 Temperature distribution through the shell thickness for $t = 0.07, 0.17, 0.39$ (curves 1, 2, 3); solid curves (0.25, 0.25); dashed curves (0.5, 0.5)

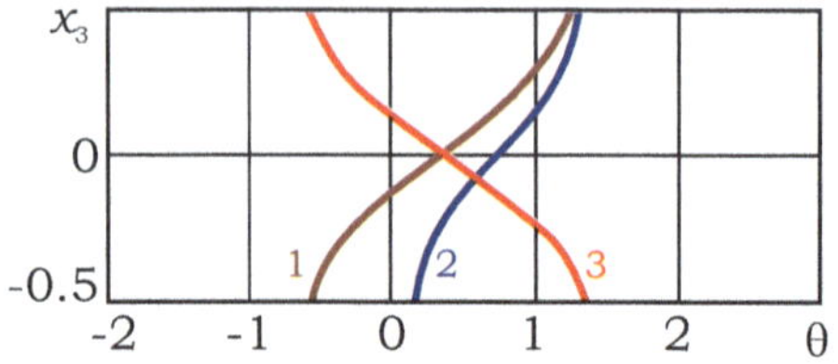

Fig. 7.60 Temperature distribution through the shell thickness for $t = 0.07, 0.17, 0.39$ (curves 1, 2, 3) at point (0.5, 0.25)

problems 1 and 2 with $q = 1080$ are also presented. This means that the load is greater than q_{cr} by 14% and 17% for problems 3 and 4. The loads in problems 1 and 2 are 14% and 17% larger than those in problems 3 and 4. The dynamic reaction of the shell in problem 1 (GN(c)) has been traced until $t = 0.07$, and in problem 2 (GN) it has been traced up to $t = 1.4$, and a deflection value equal to three shell thickness has never been observed. For $q = 900$ unacceptable divergences of the results for problems 1 and 2 and problems 3 and 4 have been observed.

To conclude, it has been shown that computations in the framework of different mathematical models for a load close to the critical load lead to qualitatively different results because the load can be, depending on the model, either pre- or post-critical. Here it has been clearly shown that the use of linear models must be strictly avoided.

Let us compare the stress–strain states of the shell obtained from calculations for problems 3 and 4 at the critical load. The shell deflections during the jumping process at different time moments are presented in Fig. 7.62. Curves 1, 3 and 5 show the deflections for problem 3 at the moments $t = 0.19, 0.3$ and 0.33, respectively. Curves 2, 4 and 6 correspond to problem 4 at $t = 0.312, 0.43, 0.47$. The coupling decreases the maximum deflection by 11.5% in spite of the fact that q_{cr} for the coupled problem exceeds q_{cr} for the uncoupled one.

In Figs. 7.63 and 7.64, the distributions of $F''_{x_1x_1}$, $w''_{x_1x_1}$ along the symmetry axes of the shell for problems 3 and 4 at the moments corresponding to the maximum deflections are shown. In addition in Fig. 7.63, the values of $F''_{x_1x_1}$ for problems 3 and 4 at $t = 0.14$ are reported. The latter results illustrate the fact that coupling, in the initial stages of jumping, practically no influence on the stress–strain state. However, when the shell reaches its maximum deflection, the coupling has an important effect.

Analysing, in addition, Figs. 7.65 and 7.66, where the distributions of $F''_{x_1x_2}$ and $w''_{x_1x_2}$ are reported, one can conclude that for the problem considered in this section, coupling greatly decreases the maximum values of $F''_{x_1x_1}$, $w''_{x_1x_1}$, $F''_{x_1x_2}$, $w''_{x_1x_2}$ at the time when the shell reaches its maximum deflection. The following decrease are

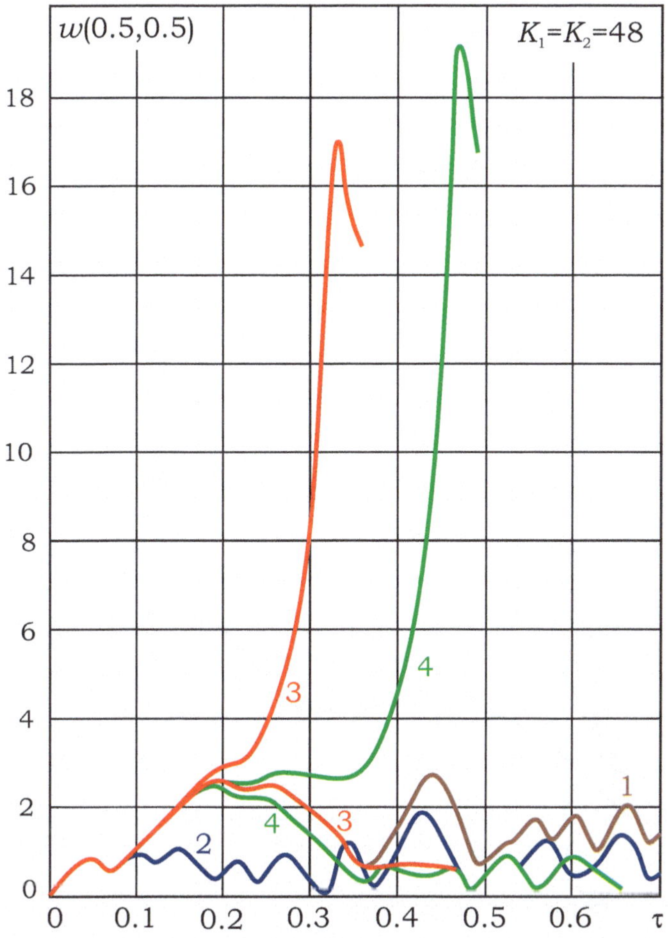

Fig. 7.61 Deflection at the shell centre for critical ($q_{cr} = 945$ and $q_{cr} = 925$) and pre-critical ($q = 940$ and $q = 920$) loads, for problems 3 and 4, and for $q = 1080$ for problems 1 and 2

obtained: for $F''_{x_1x_1}$ at the point (0.5, 0.5), 14%; for $w''_{x_1x_1}$ at the point (0.5, 0.5), 33%.

For $F''_{x_1x_2}$, at the point (0, 0.25) the decrease is 70%, whereas at the point (0.25, 0.25), it is 35%. For $w''_{x_1x_2}$, at the point (0.2, 0.2), the difference is 14%. Even larger differences in $w''_{x_1x_2}$ and $F''_{x_1x_2}$ have been observed at other points of the shell. The value of $F''_{x_1x_2}$ reported in Fig. 7.65 for problems 3 and 4 at $t = 0.14$ prove again that, at the initial moment of shell jumping, coupling induces negligible changes.

However, coupling greatly changes the temperature during jumping phenomena. In Fig. 7.67, the distribution of the temperature θ over the middle surface at the moment of maximum deflection is shown. In the centre of the shell, at this time,

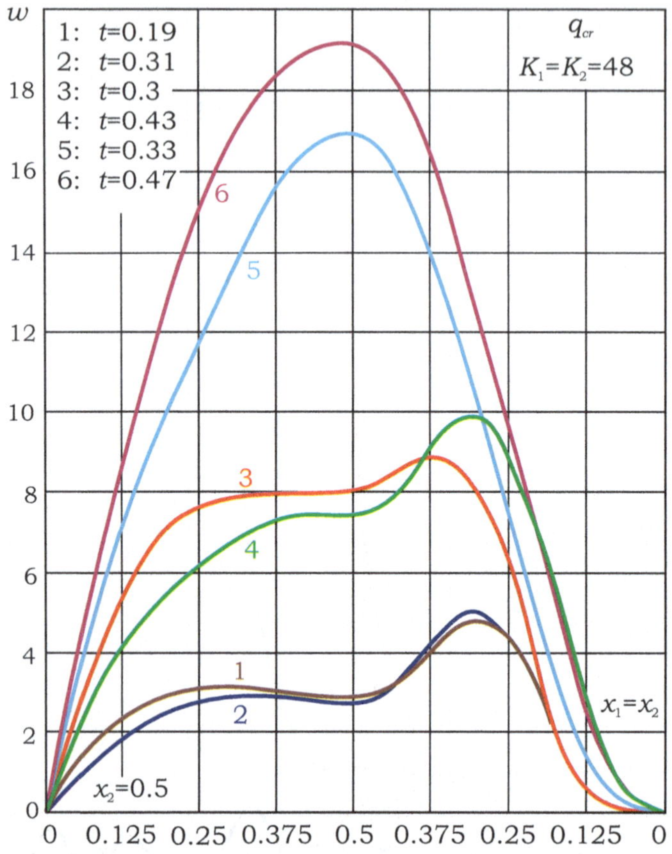

Fig. 7.62 Distribution of w during jumping

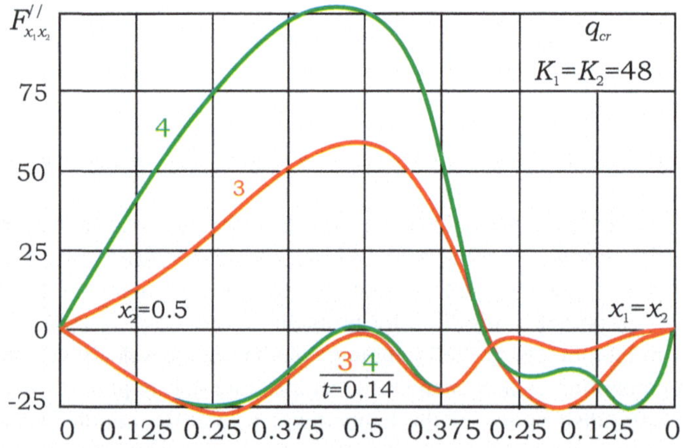

Fig. 7.63 Distribution of $F''_{x_1x_1}$ at the time instants corresponding to maximum deflection

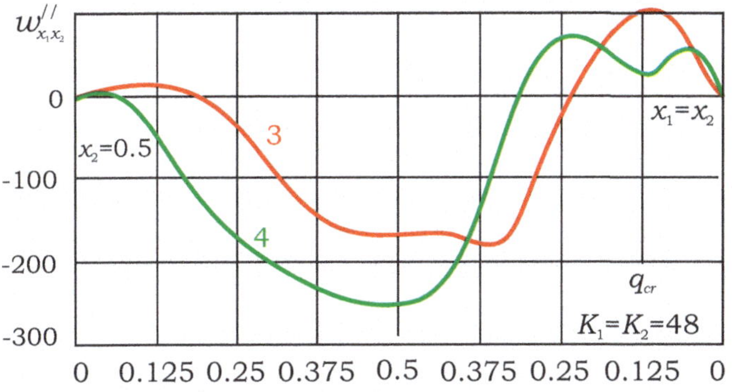

Fig. 7.64 Distribution of $w''_{x_1x_1}$ at the time instants corresponding to maximum deflection

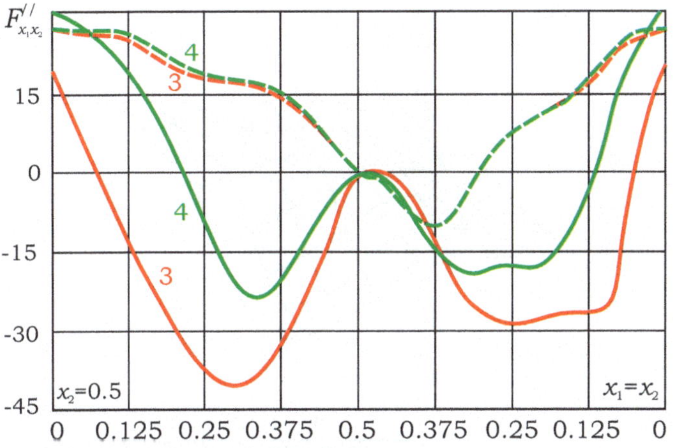

Fig. 7.65 Distribution of $F''_{x_1x_2}$ for time instants corresponding to maximal deflections

a relatively large area of cooling occurs owing to a sudden increase of the volume expansion e. The fall of θ at this time in the centre of the shell is almost 26°.

In Figs. 7.68, 7.69 and 7.70, the dependence of θ on time on the surface $x_3 = -0.5$ (solid curves) and $x_3 = 0.5$ (dashed curves) at the points (0.5, 0.5), (0.5, 0.2) and (0.1875, 0.0625) are reported.

The figures show the appearance of large temperature gradients at the moment of the sudden increase of shell deflection. For instance, at the point (0.5, 0.25) at $t = 0.28$, the fall in θ is 28.6°, and the fall in θ shown in Fig. 7.70 at $t = 0.35$ is even more larger and is over 35°. Therefore, in the problem under consideration during shell jumping a zone of cooling appears in the central part, which extends from the surface $x_3 = 0.5$ to the surface $x_3 = -0.125$. Relatively large gradients of θ occur in the neighbourhood of the sides of the shell, as a result of self-heating on the surface $x_3 = -0.5$ and cooling on the surface $x_3 = 0.5$.

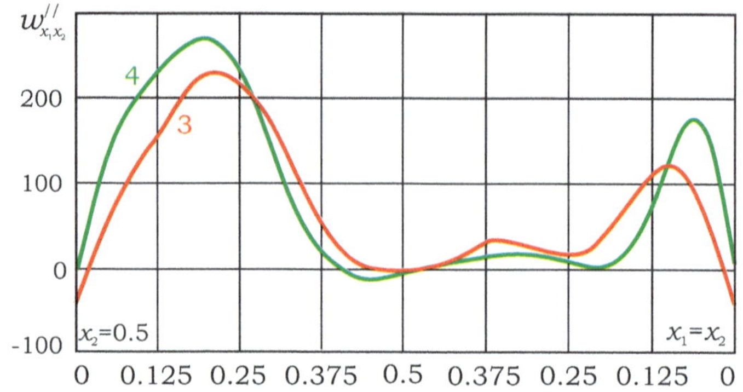

Fig. 7.66 Distribution of $w''_{x_1x_2}$ for time instants corresponding to maximal deflections

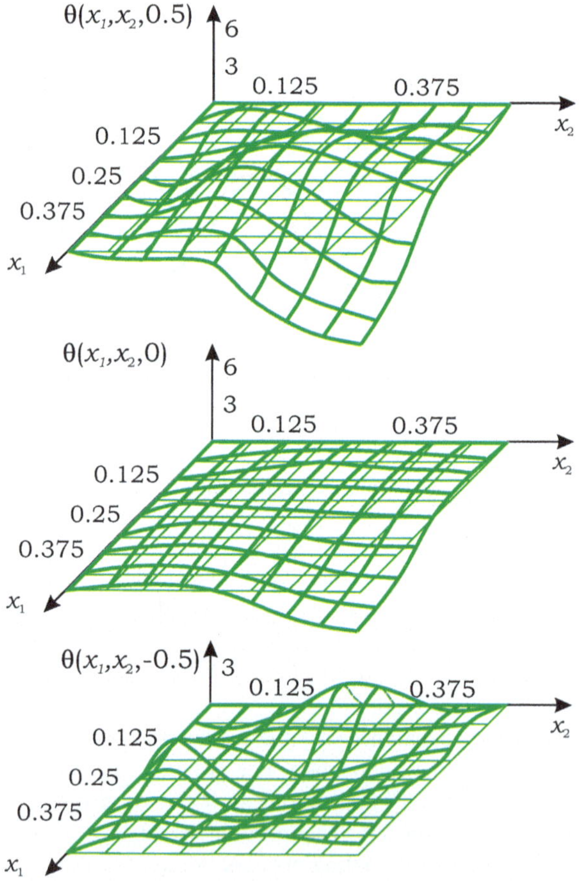

Fig. 7.67 Distribution of temperature θ over the shell for time instant corresponding to maximum deflection

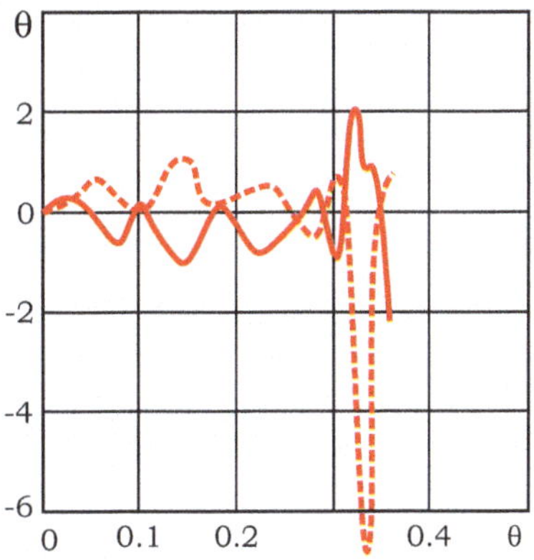

Fig. 7.68 Temperature θ versus time for the point (0.5, 0.5): *solid curve*, $x_3 = -0.5$; *dashed curve*, $x_3 = 0.5$

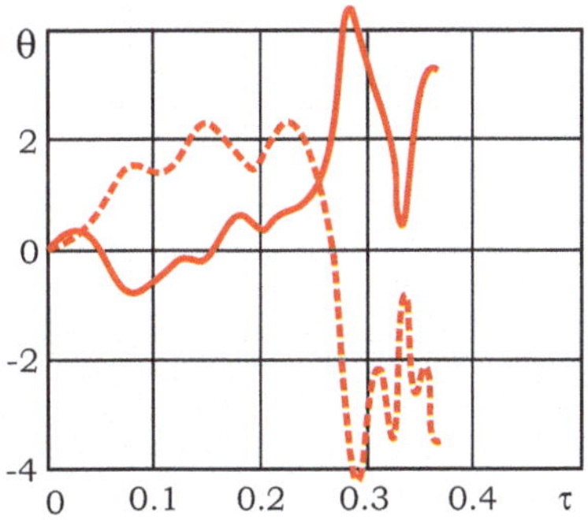

Fig. 7.69 Temperature θ versus time for the point (0.5, 0.2): *solid curve*, $x_3 = -0.5$; *dashed curve*, $x_3 = 0.5$

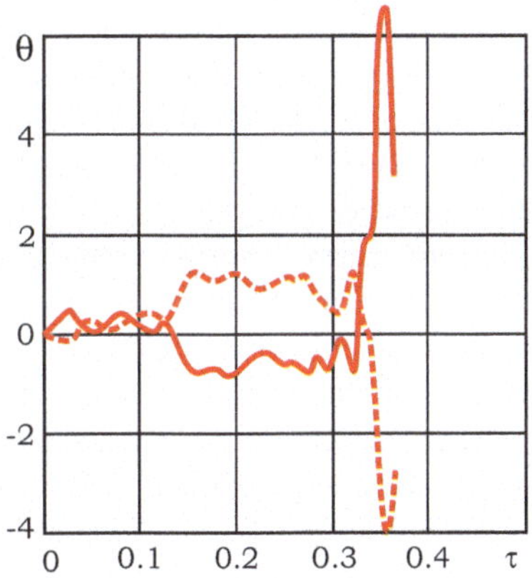

Fig. 7.70 Temperature θ versus time for the point (0.1875, 0.0625): *solid curve*, $x_3 = -0.5$; *dashed curve*, $x_3 = 0.5$

The distribution of θ through the shell thickness at some points are reported for time moments $t = 0.33$ and $t = 0.28$ in Figs. 7.71 and 7.72, respectively. Curves 1, 2, 3 and 4 show the distributions of θ at the points (0.5, 0.5), (0.5, 0.2), (0.0625, 0.25) and (0.2, 0.2), respectively. This distribution qualitatively repeats the results obtained earlier.

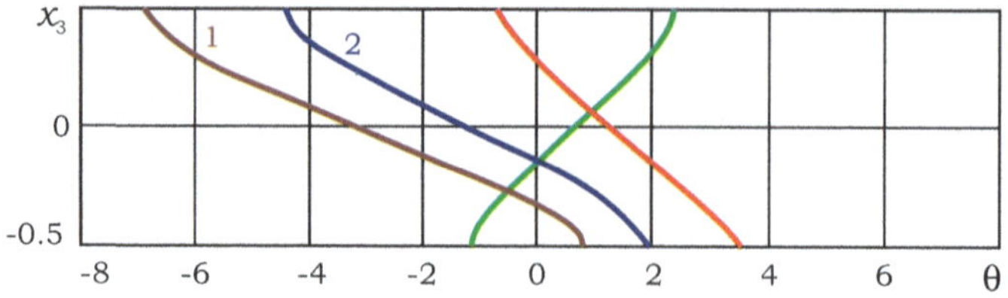

Fig. 7.71 Temperature distribution through shell thickness for $t = 0.33$

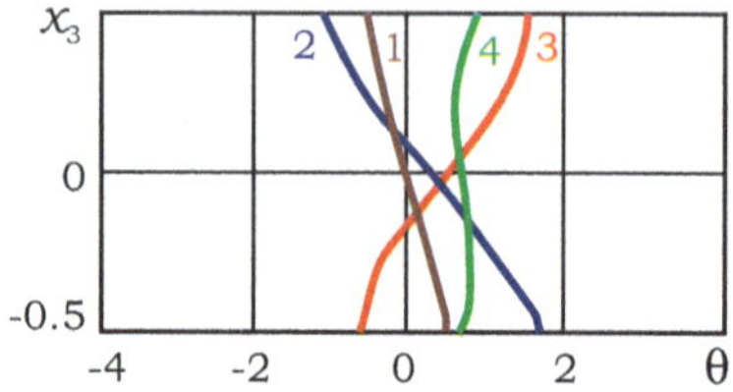

Fig. 7.72 Temperature distribution through shell thickness for $t = 0.28$

7.6 Influence of Coupling and Rotational Inertia on Stability

In previous chapters of this book, the problem of the stress–strain state of a shallow shell in a nonuniformly distributed temperature field has been stated. The problem is governed by a system of nonlinear, coupled equations of parabolic and hyperbolic type with different dimensions. It is caused essentially by the nonlinear distribution of the thermal field through the shell thickness. We have proposed and discussed methods of solution to this problem and their efficiency. In addition the existence of a solution to the nonlinear coupled thermoelastic problem within a Timoshenko-type model has been proved.

In this section, we prove the existence of a solution to the nonlinear coupled thermoelastic problem in the framework of the Kirchhoff–Love model including the rotational inertia of the shell elements. In addition, we present a numerical algorithm to solve problems of this class, which will be illustrated by numerical examples.

The physically linear equations of shell theory with coupling between the temperature and deformation fields read as follows (the equations are given in the general form for a physically nonlinear material in Sect. 6.3):

$$A(w) \equiv \rho h \frac{\partial^2 w}{\partial t^2} + D\Delta^2 w + \frac{E\alpha_T}{1-\nu}\Delta\left\{\int_{-\frac{h}{2}}^{\frac{h}{2}} x_3\theta\, dx_3\right\}$$

$$-\rho\frac{h^3}{12}\Delta\left(\frac{\partial^2 w}{\partial t^2}\right) - k_1\frac{Eh}{1-\nu^2}[\varepsilon_{11} + \nu\varepsilon_{22}]$$

$$-k_2\frac{Eh}{1-\nu^2}[\varepsilon_{22}+\nu\varepsilon_{11}]+(k_1+k_2)\frac{E\alpha_T}{1-\nu}\int_{-\frac{h}{2}}^{\frac{h}{2}}\theta\,dx_3$$

$$-\frac{\partial}{\partial x_1}\left(\frac{Eh}{1-\nu^2}[\varepsilon_{11}+\nu\varepsilon_{22}]\frac{\partial w}{\partial x_1}+\frac{Eh}{2(1+\nu)}\varepsilon_{12}\frac{\partial w}{\partial x_2}\right)$$

$$-\frac{\partial}{\partial x_2}\left(\frac{Eh}{1-\nu^2}[\varepsilon_{22}+\nu\varepsilon_{11}]\frac{\partial w}{\partial x_2}+\frac{Eh}{2(1+\nu)}\varepsilon_{12}\frac{\partial w}{\partial x_1}\right)$$

$$+\frac{E\alpha_T}{1-\nu}\left(\frac{\partial}{\partial x_1}\left\{\frac{\partial w}{\partial x_1}\int_{-\frac{h}{2}}^{\frac{h}{2}}\theta\,dx_3\right\}+\frac{\partial}{\partial x_2}\left\{\frac{\partial w}{\partial x_2}\int_{-\frac{h}{2}}^{\frac{h}{2}}\theta\,dx_3\right\}\right)=q_1,$$

$$B(u)\equiv\rho h\frac{\partial^2 u}{\partial t^2}-\frac{Eh}{1-\nu^2}\frac{\partial}{\partial x_1}[\varepsilon_{11}+\nu\varepsilon_{22}]$$

$$+\frac{E\alpha_T}{1-\nu}\frac{\partial}{\partial x_1}\left\{\int_{-\frac{h}{2}}^{\frac{h}{2}}\theta\,dx_3\right\}-\frac{Eh}{2(1+\nu)}\frac{\partial w}{\partial x_2}\varepsilon_{12}=p_1, \tag{7.46}$$

$$C(v)\equiv\rho h\frac{\partial^2 v}{\partial t^2}-\frac{Eh}{1-\nu^2}\frac{\partial}{\partial x_2}[\varepsilon_{22}+\nu\varepsilon_{11}]$$

$$+\frac{E\alpha_T}{1-\nu}\frac{\partial}{\partial x_2}\left\{\int_{-\frac{h}{2}}^{\frac{h}{2}}\theta\,dx_3\right\}-\frac{Eh}{2(1+\nu)}\frac{\partial w}{\partial x_1}\varepsilon_{12}=p_2,$$

$$D_1(\theta)\equiv C\frac{\partial\theta}{\partial t}-k_T\Delta\theta+\frac{E\alpha_T T_0}{1-\nu}[\varepsilon_{11}+\varepsilon_{22}-x_3\Delta w]=q_2,$$

where $\varepsilon_{11}, \varepsilon_{22}, \varepsilon_{12}$ have the form (6.2) and (6.3), $k_1, k_2 = const \geq 0$.

The following initial and boundary conditions are applied:

$$w|_{t=0}=\varphi_1(x_1,x_2),\ u|_{t=0}=\varphi_2(x_1,x_2),\ v|_{t=0}=\varphi_3(x_1,x_2),$$

$$\theta|_{t=0}=\varphi_4(x_1,x_2,x_3),\ \left.\frac{\partial w}{\partial t}\right|_{t=0}=\Psi_1(x_1,x_2),$$

$$\left.\frac{\partial u}{\partial t}\right|_{t=0}=\Psi_2(x_1,x_2),\ \left.\frac{\partial v}{\partial t}\right|_{t=0}=\Psi_3(x_1,x_2), \tag{7.47}$$

$$w|_\Gamma=\left.\frac{\partial w}{\partial n}\right|_\Gamma=u|_\Gamma=v|_\Gamma=0,\ \theta_S|_S=0.$$

Here and in what follows $t \in [0; T]$; Ω_1 is the space in which x_1 and x_2 vary; $\Omega_1 \subset R^2$; $\overline{\Omega}_1 = \Omega_1 \cup \partial\Omega_1$; $\Omega_1 = \Omega_1 \times (0 \times T)$; $\Gamma = \partial\Omega_1 \times [0, T]$; $\Omega_2 = \Omega_1 \times \left(-\frac{h}{2}, \frac{h}{2}\right)$; $\overline{\Omega}_2 = \Omega_2 \cup \partial\Omega_2$; $Q_2 = Q_2 \times (0 \times T)$; $S = \partial\Omega_2 \times [0, T]$; $E, h, \rho, c, k_T, \alpha_T, T_0, \nu, D$ are positive constants (all much greater than 0); $0 < \nu \ll 1/2$; $w(x_1, x_2, t)$, $u(x_1, x_2, t)$, $v(x_1, x_2, t)$, $\theta(x_1, x_2, x_3, t)$ are the sought functions; and $g_1(x_1, x_2, t)$, $p_1(x_1, x_2, t)$, $p_2(x_1, x_2, t)$, $g_2(x_1, x_2, t)$ are given functions; $\Delta^2 = \Delta\Delta$ is the biharmonic operator; and Δ is the Laplace operator (two- or three-dimensional).

We introduce the space $L^P(0, T; X)$, which is the space of functions $f(t)$ with values in the space X measured through the Lebesgue measure dt in the interval $[0, T]$ and having a finite norm $\|f\| = [\int\limits_0^T \|f\|_x^p dt]^{1/p}$, where $\|f\|_x$ is the norm of the elements in the space X (various spaces will be used as X) [254].

For the problem (7.46), (7.47) formulated above, the following theorem holds:

Theorem 7.1 *Let $\partial\Omega_1$ be a sufficiently smooth boundary of Ω_1; let $g_1, p_1, p_2 \in L^2(0, T; L_2(\Omega_1))$, $g_2 \in L^2(0, T; L_2(\Omega_1))$, $\varphi_1 \in \mathring{W}_2^2(\Omega_1)4$, $\varphi_2, \varphi_3 \in \mathring{W}_2^1(\Omega_1)$, $\varphi_4 \in \mathring{W}_2^1(\Omega_2)$, $\Psi_1 \in \mathring{W}_2^1(\Omega_2)$, and $\Psi_2, \Psi_3 \in L_2(\Omega_1)$. Then:*

1. There exists at least one choice from the four functions $\{w, u, v, \theta\}$ that is a solution to the problem (7.46), (7.47), and the following conditions are satisfied:

$$w \in L^2(0, T; \mathring{W}_2^2(\Omega_1)), \quad \frac{\partial w}{\partial t} \in L^2(0, T; W_2^1(\Omega_1)), \tag{7.48}$$

$$u, v \in L^2(0, T; \mathring{W}_2^1(\Omega_1)), \quad \frac{\partial u}{\partial t}, \frac{\partial v}{\partial t} \in L^2(0, T; L_2(\Omega_1)), \tag{7.49}$$

$$\theta \in L^2(0, T; \mathring{W}_2^1(\Omega_2)). \tag{7.50}$$

2. An approximate solution to the problem defined above can be found with the help of the Bubnov–Galerkin (Bubnov–Galerkin) method in the form

$$\begin{aligned} w_n &= \sum_{i=1}^n g_{1i}(t) X_{1i}(x_1, x_2), \quad & u_n &= \sum_{i=1}^n g_{2i}(t) X_{2i}(x_1, x_2), \\ v_n &= \sum_{i=1}^n g_{3i}(t) X_{2i}(x_1, x_2), \quad & \theta_n &= \sum_{i=1}^n g_{4i}(t) X_{3i}(x_1, x_2), \end{aligned} \tag{7.51}$$

where $\{X_{1i}\}$ is a basis system in $\mathring{W}_2^2(\Omega_1)$, orthonormalized through the norm $W_2^1(\Omega_1)$; $\{X_{2i}\}$ is a basis function in $\mathring{W}_2^1(\Omega_1)$, orthonormalized in $L_2(\Omega_1)$; and $\{X_{3i}\}$ is a basis function in $\mathring{W}_2^1(\Omega_2)$, orthonormalized in $L_2(\Omega_2)$. In addition, we assume that $\forall i$, $X_{3i} = a_i(x_1, x_2) b_i(x_2)$ (note that the existence of such a basis system implies result given in [255]). We can find the unknown $g_{ji}(t)$, $j = 1, 2, 3, 4$, from the following system of ODEs given below, which is solvable for $\forall n$ in the whole interval $[0, T]$.

3. *The whole set of approximate solutions* $\{w_n, u_n, v_n, \theta_n\}$ *is weakly compact in the spaces corresponding to* (7.48)–(7.50).

We present here only the fundamental steps of the proof, based on the method of compactness [226].

Step 1. Construction of the Approximating Solution Using the Bubnov–Galerkin Method We substitute the functions (7.51) into the system (7.46), (7.47) and find the unknown $g_{ji}(t)$, $j = 1, 2, 3, 4$, from the following system of ODEs:

$$\begin{aligned}(A(w_n) - g_1, X_{1k})_{L_2(\Omega_1)} &= 0,\\ (B(u_n) - p_1, X_{2k})_{L_2(\Omega_1)} &= 0,\\ (C(v_n) - p_2, X_{2k})_{L_2(\Omega_1)} &= 0,\\ (D_1(\theta_n) - g_2, X_{3k})_{L_2(\Omega_2)} &= 0\\ (k = 1, 2, \ldots, n),&\end{aligned} \tag{7.52}$$

with the boundary conditions

$$\begin{aligned}&g_{1k}(0) = (\varphi_1, X_{1k})_{L_2(\Omega_1)}, \quad g_{2k}(0) = (\varphi_2, X_{2k})_{L_2(\Omega_1)},\\ &g_{3k}(0) = (\varphi_3, X_{3k})_{L_2(\Omega_1)}, \quad g_{4k}(0) = (\varphi_4, X_{3k})_{L_2(\Omega_2)},\\ &\frac{dg_{1k}(0)}{dt} = (\Psi_1, X_{1k})_{L_2(\Omega_1)}, \quad \frac{dg_{2k}(0)}{dt} = (\Psi_2, X_{2k})_{L_2(\Omega_1)},\\ &\frac{dg_{3k}(0)}{dt} = (\Psi_3, X_{2k})_{L_2(\Omega_1)}, \quad k = 1, 2, \ldots, n.\end{aligned} \tag{7.53}$$

Similarly to the Peano theorem [144] on the existence of a solution to the Cauchy problem, the existence of a solution to the problem (7.52), (7.53) on $[0, T_l]$, $T_l \le T$, in the space $W_2^2(0, T_l)$, can be proved.

Step 2. Prior Estimates In order to obtain prior estimates, we multiply each of the equations in (7.52) by $\frac{dg_{1k}(0)}{dt}$, $\frac{d2_{1k}(0)}{dt}$, $\frac{dg_{3k}(0)}{dt}$ and $g_{4k}(0)$, respectively, sum the resulting over k and add the four equations obtained. We then obtain

$$\begin{aligned}\frac{1}{2}\frac{d}{dt}\Bigg\{&\rho h\left(\left|\frac{\partial w_n}{\partial t}\right|^2_{L_2(\Omega_1)} + \left|\frac{\partial u_n}{\partial t}\right|^2_{L_2(\Omega_1)} + \left|\frac{\partial v_n}{\partial t}\right|^2_{L_2(\Omega_1)}\right)\\ &+\rho\frac{h^3}{12}\left|\frac{\partial w_n}{\partial t}\right|^2_{\mathring{W}_2^1(\Omega_1)} + D\left|w_n\right|^2_{\mathring{W}_2^2(\Omega_1)} + \frac{C}{T_0}\left|\theta_n\right|^2_{L_2(\Omega_2)}\\ &+\int_{\Omega_1}\left(\frac{Eh}{1-\nu^2}[\varepsilon_{11}^2 + \varepsilon_{22}^2 + 2\nu\varepsilon_{11}\varepsilon_{22}] + \frac{Eh}{2(1+\nu)}\varepsilon_{12}^2\right)d\Omega_1\Bigg\}\end{aligned} \tag{7.54}$$

$$+\frac{k_T}{T_0}\left|\theta_n\right|^2_{\mathring{W}_2^1(\Omega_1)}=\int\limits_{\Omega_1}\left(g_1\frac{\partial w_n}{\partial t}+p_1\frac{\partial u_n}{\partial t}+p_2\frac{\partial v_n}{\partial t}\right)d\Omega_1$$

$$+\int\limits_{\Omega_2} g_2\theta_n\,d\Omega_2.$$

Denoting the right-hand side of (7.54) by R, we estimate it using the Cauchy inequality:

$$\begin{aligned}R\le{}&\frac{1}{2\rho h}\int\limits_{\Omega_1}|(g_1)^2+(p_1)^2+(p_2)^2|\,d\Omega_1+\frac{T_0}{2C}\int\limits_{\Omega_1}(g_2)^2\,d\Omega_2\\&+\frac{1}{2}\rho h\left(\left|\frac{\partial w_n}{\partial t}\right|^2_{L_2(\Omega_1)}+\left|\frac{\partial u_n}{\partial t}\right|^2_{L_2(\Omega_1)}+\left|\frac{\partial v_n}{\partial t}\right|^2_{L_2(\Omega_1)}\right)\\&+\frac{1}{2}\frac{C}{T}\left|\theta_n\right|^2_{L_2(\Omega_2)}+\frac{1}{2}\rho\frac{h^3}{12}\left|\frac{\partial w_n}{\partial t}\right|^2_{\mathring{W}_2^1(\Omega_1)}+\frac{1}{2}D\left|w_n\right|^2_{\mathring{W}_2^2(\Omega_1)}\\&+\int\limits_{\Omega_1}\left(\frac{Eh}{1-\nu^2}[\varepsilon_{11}^2+\varepsilon_{22}^2+2\nu\varepsilon_{11}\varepsilon_{22}]+\frac{Eh}{2(1+\nu)}\varepsilon_{12}^2\right)d\Omega_1.\end{aligned}\tag{7.55}$$

The last three terms in (7.55) have been added for convenience of later considerations. Integrating (7.54) with respect the variable t over an interval $(0,\tau)$, $\tau\le T_1$, taking (7.55) into account and using Gronwill's lemma [144] the following inequality is obtained:

$$\begin{aligned}&\rho h\left(\left|\frac{\partial w_n}{\partial t}\right|^2_{L_2(\Omega_1)}+\left|\frac{\partial u_n}{\partial t}\right|^2_{L_2(\Omega_1)}+\left|\frac{\partial v_n}{\partial t}\right|^2_{L_2(\Omega_1)}\right)\\&+\rho\frac{h^3}{12}\left|\frac{\partial w_n}{\partial t}\right|^2_{\mathring{W}_2^1(\Omega_1)}+D\left|w_n\right|^2_{\mathring{W}_2^2(\Omega_1)}+\frac{C}{T}\left|\theta_n\right|^2_{L_2(\Omega_2)}\\&\int\limits_{\Omega_1}\left(\frac{Eh}{1-\nu^2}[\varepsilon_{11}^2+\varepsilon_{22}^2+2\nu\varepsilon_{11}\varepsilon_{22}]+\frac{Eh}{2(1+\nu)}\varepsilon_{12}^2\right)d\Omega_1<\beta,\end{aligned}\tag{7.56}$$

$$\beta=const>0.$$

To obtain the constant β, Lemma 1 from [254] is used. Returning to (7.54) and taking into account (7.56), obtain

$$\frac{k_T}{T_0}\int_0^{\tau} |\theta_n|^2_{\overset{\circ}{W}{}^1_2(\Omega_1)}\,dt \le \beta. \tag{7.57}$$

From the last term of the left-hand side of the inequality (7.56), the following estimates are obtained using the Cauchy inequality:

$$\int_{\Omega_1}\left(\frac{\partial u_n}{\partial x_1}\right)^2 d\Omega_1 \le \beta, \quad \int_{\Omega_1}\left(\frac{\partial v_n}{\partial x_2}\right)^2 d\Omega_1 \le \beta,$$

$$\int_{\Omega_1}\left(\frac{\partial u_n}{\partial x_2}+\frac{\partial v_n}{\partial x_1}\right) d\Omega_1 \le \beta.$$

After applying the Korn inequality, these estimates yield

$$|u_n|^2_{\overset{\circ}{W}{}^1_2(\Omega_1)} \le \beta, \quad |v_n|^2_{\overset{\circ}{W}{}^1_2(\Omega_1)} \le \beta. \tag{7.58}$$

The inequalities (7.56)–(7.58) are prior estimates of the solution to the problem (7.47), (7.48). Using these estimates, the solution to the system (7.52), (7.53) can be extended to the whole interval $[0, T]$.

Step 3. Limiting Transition As is known, an arbitrary bounded set in a Hilbert space is weakly compact. Consequently, using the prior estimates (7.56)–(7.58), one can extract the following subseries:

$$\frac{\partial u_{n_k}}{\partial t} \to \frac{\partial u}{\partial t}, \ \frac{\partial v_{n_k}}{\partial t} \to \frac{\partial v}{\partial t} \quad \text{weak in } L^2(0, T; L_2(\Omega_1)),$$

$$\frac{\partial w_{n_k}}{\partial t} \to \frac{\partial w}{\partial t} \quad \text{weak in } L^2(0, T; \overset{\circ}{W}{}^1_2(\Omega_1)),$$

$$\begin{aligned} &w_{n_k} \to w \quad \text{weak in } L^2(0, T; \overset{\circ}{W}{}^2_2(\Omega_1)),\\ &\theta_{n_k} \to \theta \quad \text{weak in } L^2(0, T; \overset{\circ}{W}{}^1_2(\Omega_1)),\\ &u_{n_k} \to u, \ v_{n_k} \to v \quad \text{weak in } L^2(0, T; \overset{\circ}{W}{}^1_2(\Omega_1)). \end{aligned} \tag{7.59}$$

A limiting transition in the spaces related to (7.48)–(7.50) can be performed using (7.59) for the hyperbolic-type equations (7.46) similarly to the liming transition presented in [226]. Therefore, we shall examine only one term in the parabolic-type system. For an arbitrary function $\Phi = \sum_{i=1}^{n} \alpha_i(t) \otimes X_{3i}(x_1, x_2, x_3) \forall i\ \alpha_i(t) \in C'([0, T]), \alpha_i(t) = 0$, we have

$$\int_0^T \int_{\Omega_1} \int_{-\frac{h}{2}}^{\frac{h}{2}} \left(\frac{\partial^2 w_n}{\partial t \partial x_1} - \frac{\partial w_n}{\partial x_1}\Phi - \frac{\partial^2 w_n}{\partial t \partial x_1} - \frac{\partial w_n}{\partial x_1}\Phi \right) dx_3\, d\Omega_1\, dt$$

$$= \int_0^T \int_{\Omega_1} \int_{-\frac{h}{2}}^{\frac{h}{2}} \left\{ \left(\frac{\partial w_n}{\partial x_1} - \frac{\partial w}{\partial x_1} \right) \frac{\partial^2 w_n}{\partial t \partial x_1} + \left(\frac{\partial^2 w_n}{\partial t \partial x_1} \right.\right.$$

$$\left.\left. - \frac{\partial^2 w}{\partial t \partial x_1} \right) \frac{\partial w}{\partial x_1} \right\} \Phi\, dx_3\, d\Omega_1\, dt \le \left| \frac{\partial^2 w}{\partial t v \partial x_1} \right|_{L_2(Q_1)} \left| \frac{\partial w_n}{\partial x_1} \right.$$

$$\left. - \frac{\partial w}{\partial x_1} \right|_{L_2(Q_1)} \left| \Phi_1 \right|_{L_2(Q_1)} + \int_0^T \int_{\Omega_1} \left(\frac{\partial^2 w_n}{\partial t \partial x_1} - \frac{\partial^2 w}{\partial t \partial x_1} \right) \frac{\partial w}{\partial x_1} \Phi_1\, d\Omega_1\, dt. \tag{7.60}$$

Note that, owing to the special choice of the basis function $X_{3i}(x_1, x_2, x_3)$, we first integrate the components of Φ along x_3 (because there is no other function depending on x_3 in the expression under consideration). So, only the components that depend on two spatial coordinates x_1, x_2 and on time remain. Therefore $\Phi_1 \in W_2^{1,1}(Q_1)$, where $\Phi_1 = \sum_{i=1}^{n} \alpha_i(t) \otimes a_i(x_1, x_2)c_i$, $c_i = \int_{-\frac{h}{2}}^{\frac{h}{2}} b_i(x_3)\, dx_3$, $X_{3i} = a_i(x_1, x_2)b_i(x_3)$. In what follows, Lemma 10 from [253], on the compactness of the elements $W_2^{1,1}(Q_1)$ included in $L_4(Q_1)$, $\Omega_1 \subset R^2$, can be used.

The first term in (7.60) approaches zero, owing to Lemma 5.2 from [253]. The second term also approaches zero because of the weak convergence of $\frac{\partial w_n}{\partial t}$ in $L^2(0, T; \mathring{W}_2^1(\Omega_1)$, or of $\frac{\partial w}{\partial x_1} \times \Phi_1 \in L_2(Q_1)$. In the remaining nonlinear terms, the limiting transition can be carried out in a similar way.

The satisfaction of the initial conditions is understood in a generalized sense [188].

Remark 7.1 A weak convergence in (7.59) takes place for practical purposes in the spaces $L^\infty(0, T; H)$; however, it is not important to prove this.

Remark 7.2 Note that in the linearized system of equations (7.46), (7.47) the uniqueness of the solution is apparent for initial data that are smoother than those required by the theorem. However, if the energetic equality holds in the conditions of the theorem related to the linearized problem then the uniqueness of the solutions to the problem (7.46), (7.47) is not difficult to prove using the methodology described in [253].

In order to solve the ODEs (7.46) using the Bubnov–Galerkin method in its higher approximations, the system of equations is transformed to the following non-dimensional form:

$$
\begin{aligned}
&\frac{1}{1-\nu^2}\frac{\partial}{\partial x_1}[\lambda^{-1}\varepsilon_{11}+\nu\lambda\varepsilon_{22}]-\frac{1}{1-\nu}\frac{\partial}{\partial x_1}\left\{\int\limits_{-\frac{1}{2}}^{\frac{1}{2}}\theta\,dx_3\right\}\\
&+\frac{\lambda}{2(1+\nu)}\frac{\partial}{\partial x_2}\varepsilon_{12}-\frac{\overline{c}}{\lambda_1\lambda_2}\frac{\partial^2 u}{\partial t^2}+P_{x_1}=0 \qquad (x_1\leftrightarrow x_2),\ (1\leftrightarrow 2),\\
&\frac{1}{12(1-\nu^2)}\left\{-\frac{\partial^2}{\partial x_1^2}\left[\lambda^{-2}\frac{\partial^2 w}{\partial x_1^2}+\nu\frac{\partial^2 w}{\partial x_2^2}\right]-2(1-\nu)\right.\\
&\left.\times\frac{\partial^2}{\partial x_1\partial x_2}\left[\frac{\partial^2 w}{\partial x_1\partial x_2}\right]-\frac{\partial^2}{\partial x_2^2}\left[\lambda^2\frac{\partial^2 w}{\partial x_2^2}+\nu\frac{\partial^2 w}{\partial x_1^2}\right]\right\}\\
&-\frac{1}{1-\nu}\left(\lambda^{-1}\frac{\partial^2}{\partial x_1^2}\left\{\int\limits_{-\frac{1}{2}}^{\frac{1}{2}}\theta x_3\,dx_3\right\}+\lambda\frac{\partial^2}{\partial x_2^2}\left\{\int\limits_{-\frac{1}{2}}^{\frac{1}{2}}\theta x_3\,dx_3\right\}\right)\\
&+\frac{\overline{c}}{12}\frac{\partial^2}{\partial t^2}\left[\lambda_2^{-2}\frac{\partial^2 w}{\partial x_1^2}+\lambda_1^{-2}\frac{\partial^2 w}{\partial x_2^2}\right]+k_1\left[\frac{1}{1-\nu^2}\{\lambda^{-2}\varepsilon_{11}\right.\\
&\left.+\nu\varepsilon_{22}\}-\frac{1}{\lambda(1-\nu)}\int\limits_{-\frac{1}{2}}^{\frac{1}{2}}\theta\,dx_3\right]+k_2\left[\frac{1}{1-\nu^2}\{\lambda^{-2}\varepsilon_{22}+\nu\varepsilon_{11}\}\right.\\
&\left.+\frac{\lambda}{1-\nu}\int\limits_{-\frac{1}{2}}^{\frac{1}{2}}\theta\,dx_3\right]+\frac{\partial}{\partial x_1}\left[\frac{\partial w}{\partial x_1}\left\{\frac{1}{1-\nu^2}\{\lambda^{-2}\varepsilon_{11}+\nu\varepsilon_{22}\}\right.\right.\\
&\left.\left.-\frac{1}{\lambda(1-\nu)}\int\limits_{-\frac{1}{2}}^{\frac{1}{2}}\theta\,dx_3\right\}+\frac{\partial w}{\partial x_2}\left\{\frac{1}{2(1+\nu)}\varepsilon_{12}\right\}\right]+\frac{\partial}{\partial x_2}\left[\frac{\partial w}{\partial x_2}\right.\\
&\times\left\{\frac{1}{1-\nu^2}\{\lambda^{-2}\varepsilon_{22}+\nu\varepsilon_{11}\}-\frac{\lambda}{1-\nu}\int\limits_{-\frac{1}{2}}^{\frac{1}{2}}\theta\,dx_3\right\}+\frac{\partial w}{\partial x_1}\left\{\frac{1}{2(1+\nu)}\right.
\end{aligned}
$$

$$\times\,\varepsilon_{12}\Big\}\Big] - \frac{\overline{c}}{\lambda_1^2\lambda_2^2}\frac{\partial^2 w}{\partial t^2} + g_1 = 0, \tag{7.61}$$

$$\frac{\partial\theta}{\partial t} - \left(\lambda_1^2\frac{\partial^2\theta}{\partial {x_1}^2} + \lambda_2^2\frac{\partial^2\theta}{\partial {x_2}^2} + \frac{\partial^2\theta}{\partial {x_3}^2}\right)$$

$$= -\gamma\frac{\partial}{\partial t}\times\left[\lambda^{-1}\varepsilon_{11} + \lambda\varepsilon_{22} - x_3\left(\lambda^{-1}\frac{\partial^2 w}{\partial {x_1}^2} + \lambda\frac{\partial^2 w}{\partial {x_2}^2}\right)\right] + g_2.$$

Equation (7.61) are in nondimensional form (the overbars are omitted). In what follows, we add the following additional nondimensional parameters (which have not been used earlier):

$$u = \frac{a}{h^2}\overline{u},\ v = \frac{b}{h^2}\overline{v},\ t = \frac{K_T}{ch^2},\ \theta = \frac{\alpha_T ab}{h^2}\overline{\theta},$$

$$\gamma = \frac{E\alpha_T^2 T_0}{(1-\nu)c}\overline{\theta},\ \overline{c} = \frac{\rho K_T^2}{c^2Eh^2},\ g_1 = \frac{a^2b^2}{Eh}\overline{g_1},$$

$$g_2 = \frac{ab\alpha_t}{K_T}\overline{g_2},\ P_{x_1} = \frac{a^2b}{Eh^3}\overline{P_{x_1}},\ P_{x_1} = \frac{ab^2}{Eh^3}\overline{P_{x_2}}.$$

The mechanical boundary conditions for the system (7.61) are the same as those for (7.46)–(7.47). For the heat transfer equation, besides the condition (7.47), we solve the problem for the condition

$$\left.\frac{\partial\theta}{\partial n}\right|_S = 0. \tag{7.62}$$

We have solved the systems (7.61), (7.46) and (7.61), (7.62) using the Bubnov–Galerkin method in its higher approximations. The series of approximating functions $\{w_n, u_n, v_n, \theta_n\}$ were approximated in the following way:

$$\begin{aligned}
w_n &= \sum_i \cos i\pi x_1 \cos^2 \pi x_1 \sum_j \cos j\pi x_2 \cos^2 \pi x_2 \times g_{1ij}(t)\\
&\quad (i, j = 1, 3, 5),\\
u_n &= \sum_i \cos \pi x_1 \sin i\pi x_1 \sum_j \cos \pi x_2 \cos j\pi x_2 \times g_{2ij}(t)\\
&\quad (i, j = 1, 2, 3),\\
v_n &= \sum_{i=1} \cos \pi x_1 \cos i\pi x_1 \sum_j \cos \pi x_2 \cos j\pi x_2 \times g_{3ij}(t)\\
&\quad (i, j = 1, 2, 3).
\end{aligned} \tag{7.63}$$

For the first type of thermal part we have used

$$\theta_n = \sum_l \left(x_1^2 - \frac{1}{4}\right) x_1^l \sum_k \left(x_2^2 - \frac{1}{4}\right) x_2^k \sum_m \left(x_3^2 - \frac{1}{4}\right) x_3^m \times g_{4lkm}(t) \tag{7.64}$$

$$(l, k, m = 0, 1).$$

For the second type we have used

$$\theta_n = \left(1 + \sum_i \sin i\pi x_1\right)\left(1 + \sum_j \sin j\pi x_2\right)\left(1 + \sum_k \sin k\pi x_3\right) \times g_{4ijk}(t)$$

$$(i, j, k = 1, 2). \tag{7.65}$$

After application of the procedure of the Bubnov–Galerkin method to the systems (7.61), (7.46) and (7.61), (7.62), we obtained second-order ODEs for g_{1ij}, g_{2ij}, g_{3ij}, g_{4ijk}. The second-order equations were reduced by a change of variables to first-order ODEs, which were then solved using a Runge–Kutta method of the fourth order. The integrals related to S and $\Omega_1 = \left(-\frac{1}{2}, \frac{1}{2}\right) \times \left(-\frac{1}{2}, \frac{1}{2}\right)$ were computed analytically.

We have considered a spherical shell, rectangular in plan, clamped on its edge, made from an isotropic material, with boundary conditions for the thermal part of the problem of the first and the second type, $\Psi_1 = \Psi_2 = \Psi_3 = 0$, $\varphi_1 = \varphi_2 = \varphi_3 = \varphi_4 = 0$, $g_1 = const$, $g_2 = 0$. The following values of the dimensionless parameters were used: a_1 is a scaling multiplier, $\lambda = 1$ $\lambda_1 = 1/50$, $\lambda_2 = 1/50$, $\nu = 0.3$, $k_1 = k_2 = 9$ and 15, $a_1 = 300$, $\gamma = 0.03a_1$, $\overline{c} = 0.3 \times 10^{-9} \times a_1^2$.

The results of the calculations are presented in Figs. 7.73, 7.74, 7.75 and 7.76. In Fig. 7.73, the deflection function in the central point $w(0, 0, t)$ of the shell for $K_1 = K_2 = 9$ is shown. Curves 1–4 correspond to the uncoupled problem for $q = 50, 57, 70$ and 100, respectively. Curves 1–8 correspond to the first type of thermal boundary conditions, and curve 9 corresponds to the critical load $q_{cr} = 69$

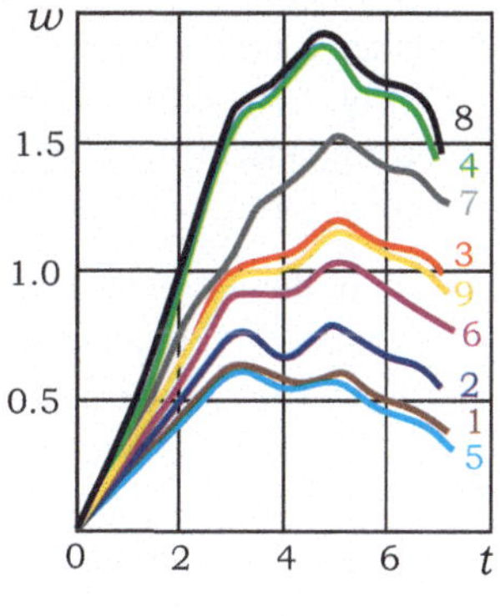

Fig. 7.73 Deflection of the shell centre $\theta(0, 0, 0, t)$ for different loads and boundary conditions ($K_1 = K_2 = 9$)

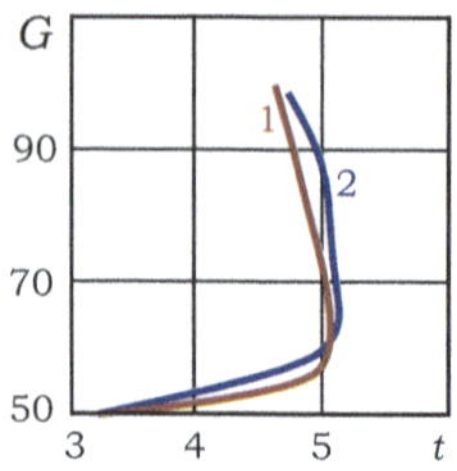

Fig. 7.74 Deflection of the shell centre $\theta(0, 0, 0, t)$ for $K_1 = K_2 = 15$: 1, uncoupled problem; 2, coupled problem

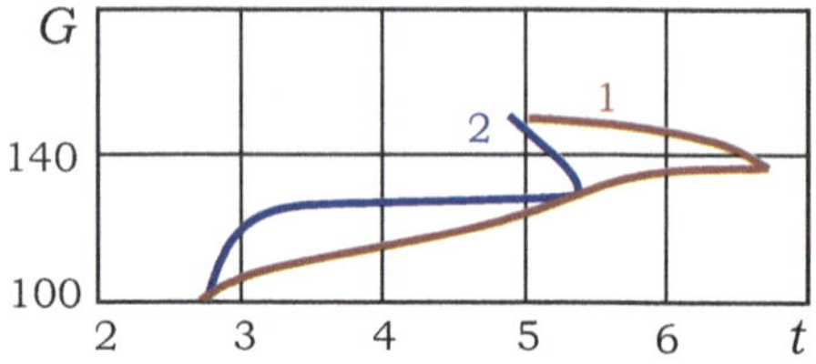

Fig. 7.75 Deflection of the shell centre $\theta(0, 0, 0, t)$ for $K_1 = K_2 = 9$: 1, uncoupled problem; 2, coupled problem

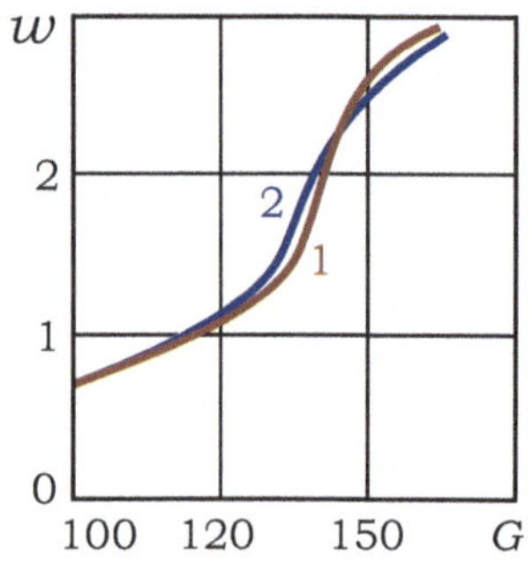

Fig. 7.76 Deflection of the shell centre $\theta(0, 0, 0, t)$ for $K_1 = K_2 = 15$: 1, uncoupled problem; 2, coupled problem; the Budiansky–Roth criterion has been applied

for boundary conditions of the second type. Curves 2 and 6 correspond to the critical loads for the uncoupled and the coupled problem, respectively. In an analogous way, computations have been carried out for both the coupled and the uncoupled problem for $k_1 = k_2 = 15$. Two approaches have been used to detect the q_{cr} value:

1. In Figs. 7.74 and 7.75, curve 1 corresponds to the uncoupled problem and curve 2 corresponds to the coupled problem. The Shian at al. [297] criterion has been applied for $k_1 = k_2 = 15$ and 9.
2. In Fig. 7.76, curve 1 corresponds to the uncoupled problem, and curve 2 corresponds to the coupled problem. The Budiansky–Roth criterion has been applied for $k_1 = k_2 = 15$ [77].

Both of these criteria give similar results. For the uncoupled problem with $k_1 = k_2 = 15$, $q_{cr} = 130$, and for $k_1 = k_2 = 9$, $q_{cr} = 57$. For the coupled problem with $k_1 = k_2 = 15$, $q_{cr} = 137$, and for $k_1 = k_2 = 9$, $q_{cr} = 65$. The graphs in Figs. 7.74, 7.75, 7.76 and 7.77 correspond to the first type of thermal boundary conditions.

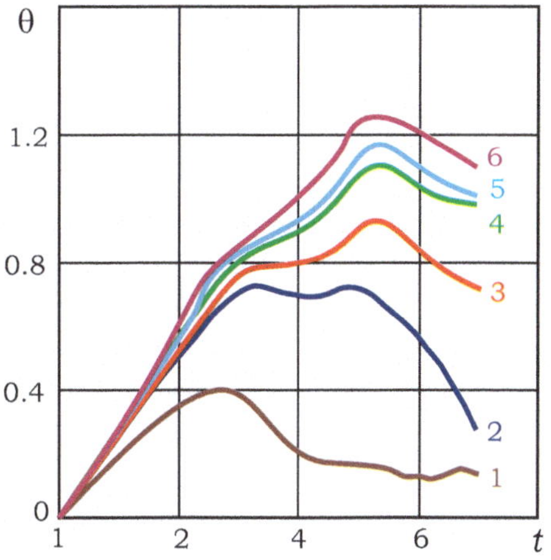

Fig. 7.77 Temperature field in the shell centre $\theta(0, 0, 0, t)$ for different loads and first type of the thermal boundary conditions ($K_1 = K_2 = 15$)

Table 7.6 Comparison of effects of rotational inertia

Uncoupled						Coupled					
Rotational inertia						Rotational inertia					
Yes			No			Yes			No		
g	w_{max}	t	g	w_{max}	t	g	w_{max}	t	g	w_{max}	t
100	0.7725	2.7	100	0.7728	2.7	100	0.7333	2.7	100	0.7337	2.7
126	1.3808	3.2	125	1.3462	3.2	130	1.5347	5.3	130	1.5367	5.3
129	1.666	5.3	130	1.7835	5.3	136	2.2369	5.4	136	2.2363	5.4
–	–	–	140	2.5324	5.2	137	2.3211	6.7	137	2.3236	6.7
150	2.8793	4.8	150	2.8778	4.3	150	2.9189	5.0	150	2.9210	5.0

In Fig. 7.77, the curves corresponding to the temperature field in the centre of the shell $\theta(0, 0, 0, t)$ versus time for $k_1 = k_2 = 15$ with the first type of thermal boundary conditions are shown. Curves 1–6 are related to the coupled problem without rotational inertia of the shell elements for $q = 100, 130, 133, 136, 137$ and 139, respectively. It should be noted, that in the computations, a dependence of the thermal field in the shell on the thermal boundary conditions is observed. In particular, for $k_1 = k_2 = 9$ and the first type of thermal boundary conditions with $q_{cr} = 65$, the maximum of the function $\theta(0, 0, 0, t)$ versus t is $\theta(0, 0, 0; 4.9) = 0.333$, at $t = 4.9$, whereas for the second type of thermal boundary conditions $q_{cr} = 69$, and the maximum of $\theta(0, 0, 0, t)$ is achieved at $t = 5$, and is equal to $\theta(0, 0, 0; 5) = 0.1618$.

Inclusion of the affect of rotational inertia has practically no influence on the value of the dynamic critical load. In Table 7.6, the maximum values of the deflection in the centre $w(0, 0, t)$ versus time, and the corresponding loads with and without the effect of rotational inertia, are reported for $k_1 = k_2 = 15$.

7.7 Numerical Tests

In order to provide numerical tests, we have reduced some high-dimensional problems of plate theory to one-dimensional problems (ODEs with respect to time), using the finite-differences method (with respect to the spatial coordinates and with the help of two approximations, $O(h^4)$ and $O(h^2)$). Hence, a question concerning the convergence of the methods with respect to the spatial and time coordinates, as well as about the optimal choice of a solution to algebraic equations arises. We have solved the first of this problem using the Runge principle: solutions obtained with different integration steps with respect to the spatial coordinates are compared.

For the numerical computations, we have used the following two types of boundary conditions:

1. Clamped edge–ball support on ribs that are unstretchable in the tangential plane:

$$
\begin{aligned}
w &= \frac{\partial^2 w}{\partial x^2} = F = \frac{\partial^2 F}{\partial x^2} = 0 \quad \text{for} \quad x = 0 \text{ and } x = 1, \\
w &= \frac{\partial^2 w}{\partial y^2} = F = \frac{\partial^2 F}{\partial y^2} = 0 \quad \text{for} \quad y = 0 \text{ and } y = 1.
\end{aligned} \tag{7.66}
$$

2. Loosely clamped edge ribs that are unstretchable in the tangential plane:

$$
\begin{aligned}
w &= \frac{\partial w}{\partial x} = F = \frac{\partial^2 F}{\partial x^2} = 0 \quad \text{for} \quad x = 0 \text{ and } x = 1, \\
w &= \frac{\partial w}{\partial y} = F = \frac{\partial^2 F}{\partial y^2} = 0 \quad \text{for} \quad y = 0 \text{ and } y = 1.
\end{aligned} \tag{7.67}
$$

The initial conditions must satisfy the boundary conditions (7.66), and (7.67) and they are as follows:

1. For the boundary conditions (7.66),

$$
w|_{t=0} = A \sin \pi x \sin \pi y, \; A = const, \qquad \left.\frac{\partial w}{\partial t}\right|_{t=0} = 0. \tag{7.68}
$$

2. For the boundary conditions (7.67),

$$
w|_{t=0} = A \sin^2 \pi x \sin^2 \pi y, \; A = const, \qquad \left.\frac{\partial w}{\partial t}\right|_{t=0} = 0. \tag{7.69}
$$

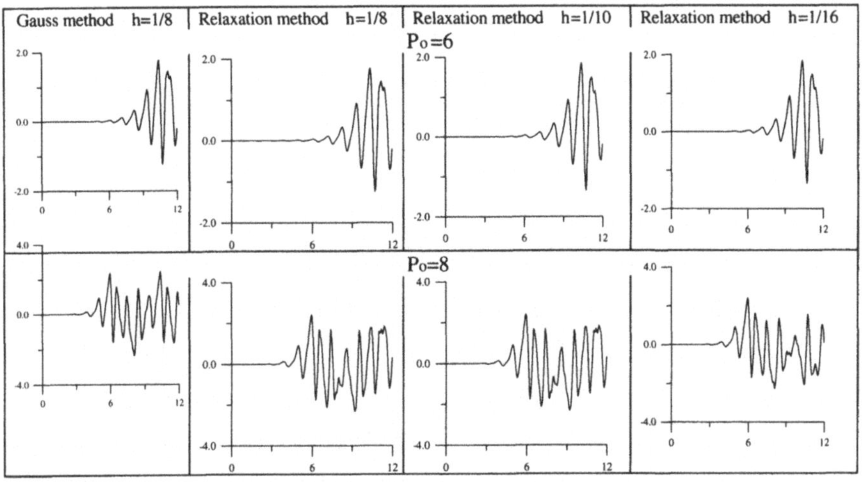

Fig. 7.78 Influence of different methods (Gauss or relaxation) and different spatial steps h on time history responses

A square plate ($\lambda = a/b = 1$) with the conditions (7.66) and (7.68) and with an initial excitation amplitude $A = 1 \times 10^{-4}$ was studied as an example. A longitudinal load $P_x = P_0 \sin \omega t$ (frequency $\omega = 5.72$, damping coefficient $\varepsilon = 1$) was applied. Only one quadrant of the plate was taken into account, owing to the two axes of symmetry along Ox and Oy, i.e. only symmetric solutions with respect to Ox and Oy were considered. The calculations were carried out using the algorithm, related an $O(h^4)$ approximation. The computational step was taken as $h = \{1/8, 1/10, 1/16\}$. The algebraic equations were solved using the Gauss method and the relaxation method. The load excitation amplitude P_0 was taken as 6 and 8. The fundamental characteristics $w(0.5, 0.5, t)$ and $w_t(w)$ for $x = y = 0.5$, fast Fourier transforms (FFTs) and power spectra are presented in Figs. 7.78 and 7.79.

An analysis of the results obtained leads to the conclusion that for further investigations it is sufficient to take $h = 1/8$ with respect to the spatial coordinates, not only for harmonic and quasi-periodic solutions, but also for solutions corresponding to crisis and chaotic states. The Gauss method, when applied to solve the system of linear algebraic equations, leads to a decrease of the computational time. The result of both qualitative and quantitative investigations of complex plate oscillations, excited harmonically and longitudinally, practically overlap for this step h. According to the Runge rule, $\Delta t = 2 \times 10^{-4}$ has been used in the Runge–Kutta method.

Now we consider the computational results obtained using the finite-difference method with $O(h^4)$ and $O(h^2)$ approximations. We shall consider the results for two areas, which we refer to as area II and area IV (crisis); these correspond to $P_0 = 4$ and $P_0 = 8.25$, respectively (an explanation of these areas will be given

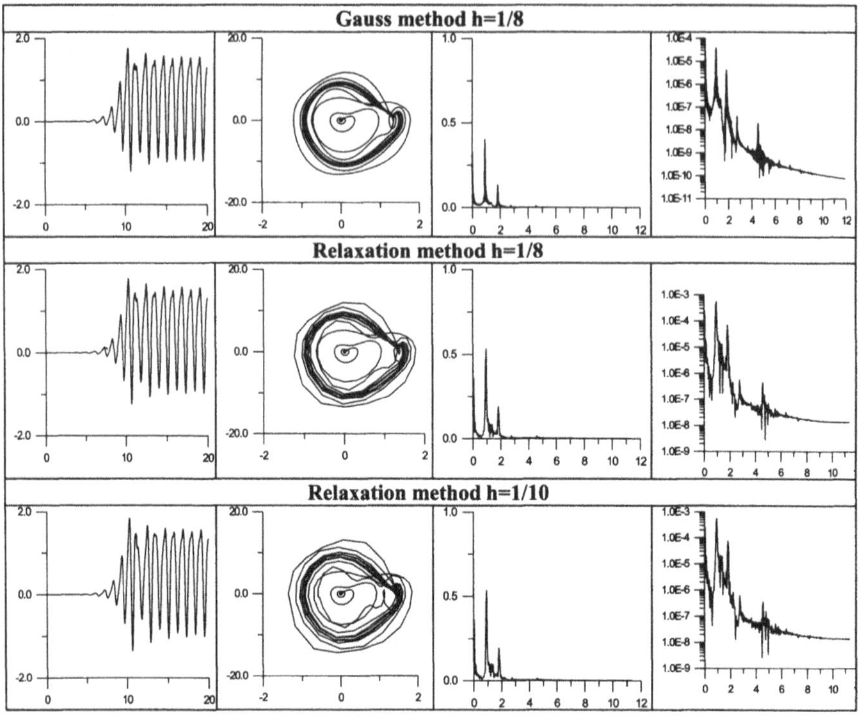

Fig. 7.79 Influence of different methods and different spatial step on phase portraits, FFTs and power spectra

later). In Fig. 7.80, the time histories $w(0.5, 0.5; t)$, phase portraits and FFTs for the point $x = y = 0.5$ and these two values of P_0 are reported. An analysis of the characteristics shown here indicates that, qualitatively and quantitatively, the results are in very good agreement for both the $O(h^4)$ and $O(h^2)$ approximations; this agreement is clearly visible in the phase portraits and the FFTs.

7.8 Influence of Damping ε and Excitation Amplitude A

The influence of the control parameters A and ε has been investigated for $\lambda = 1$, $\nu = 0.3$, the boundary conditions (7.66) and the initial conditions (7.68). The analysis was carried out for $A = 1 \times 10^{-4}$, 1.25 and 1.375 for the fixed values $\varepsilon = 1$ and $P_0 = 6$. The time history $w(0.5, 0.5; t)$, the phase portrait, the FFT and the power spectrum for $A = 1 \times 10^{-4}$ are shown in Fig. 7.79. For $A = 1.25$ and $A = 1.375$, these characteristics are shown in Fig. 7.81.

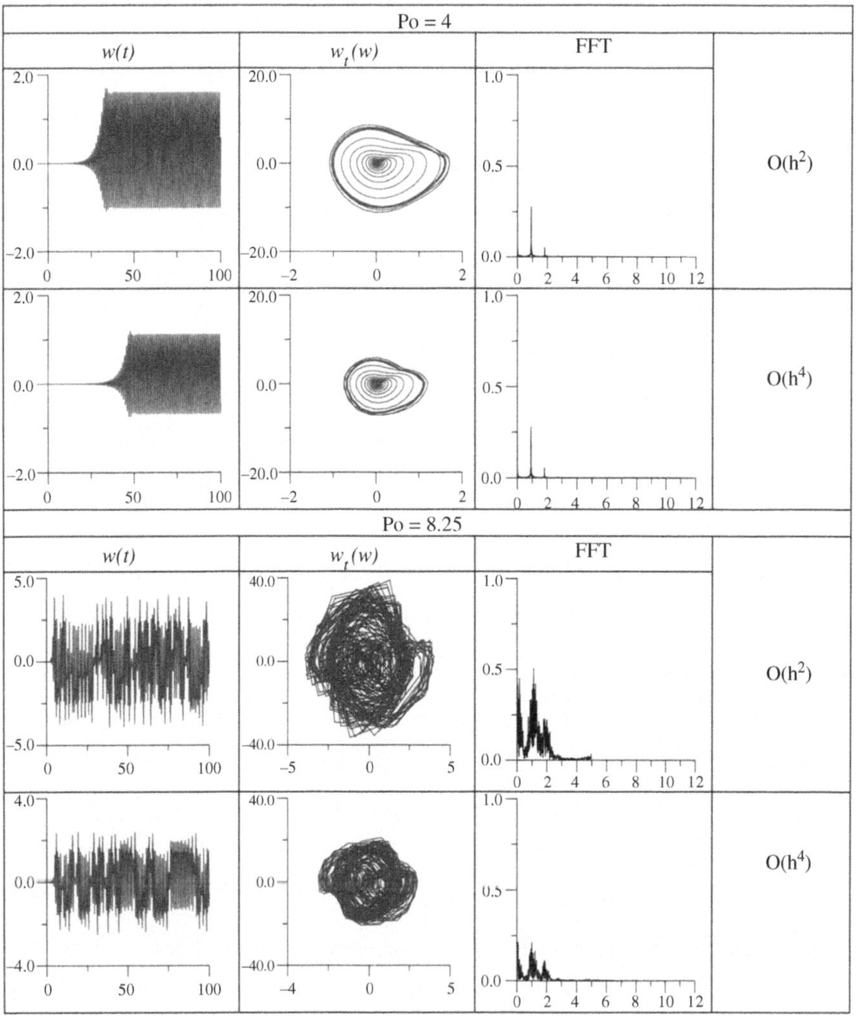

Fig. 7.80 Time histories, phase portraits and FFTs for different values of P_0 and two $O(h^k)$ approximations, $k = 2, 4$

In Fig. 7.81, the three-dimensional phase portraits (w_{tt}, w_t, w) and their projections $w_{tt}(w_t)$, $w_{tt}(w)$, $w_t(w)$ are reported, as well as the modal portraits $w_x(w)$ for the point $x = y = 0.25$, and the dependence of $w_{in} = \int_0^1 \int_0^1 w(x, y, t)\, dx\, dy$ on time.

In Fig. 7.82, the Poincaré pseudosections, the Poincaré sections, FFTs and the power spectra are shown for the three values of A considered. An analysis of the computational results shows that oscillations occur at two frequencies, and their

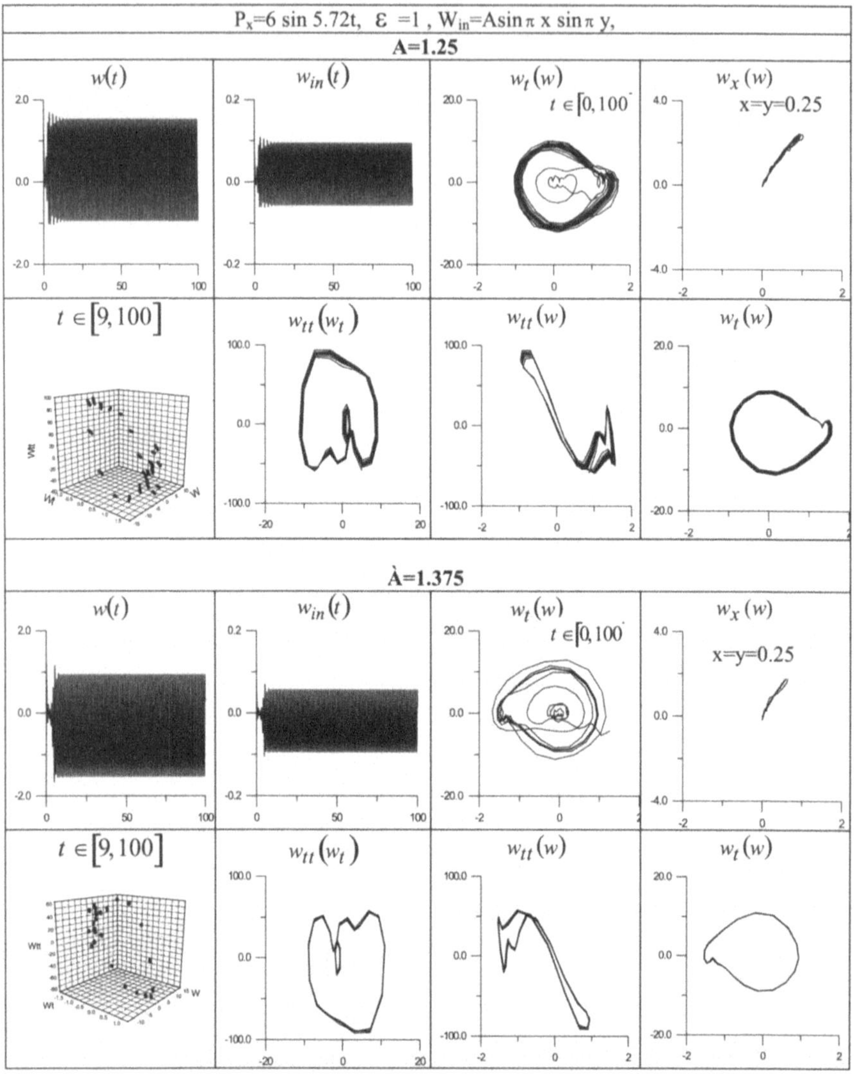

Fig. 7.81 Time histories, modal portraits, and three- and two-dimensional phase portraits, for different values of the amplitude A

values do not depend on the initial excitation amplitude. For each of the amplitudes considered, an internal synchronization at the frequency w_{17} (Table 7.7) occurs.

In comparing the computational results for $A = 1.25$ and $A = 1.375$, it should be emphasized that the phase portraits in a plane and in three-dimensional space are rotated relative to one another by about 180°. The indicator $w_{av} = \frac{w_{in}^{+}+w_{in}^{-}}{2}$, which depends on the initial excitation amplitude of the symmetric and nonsymmetric

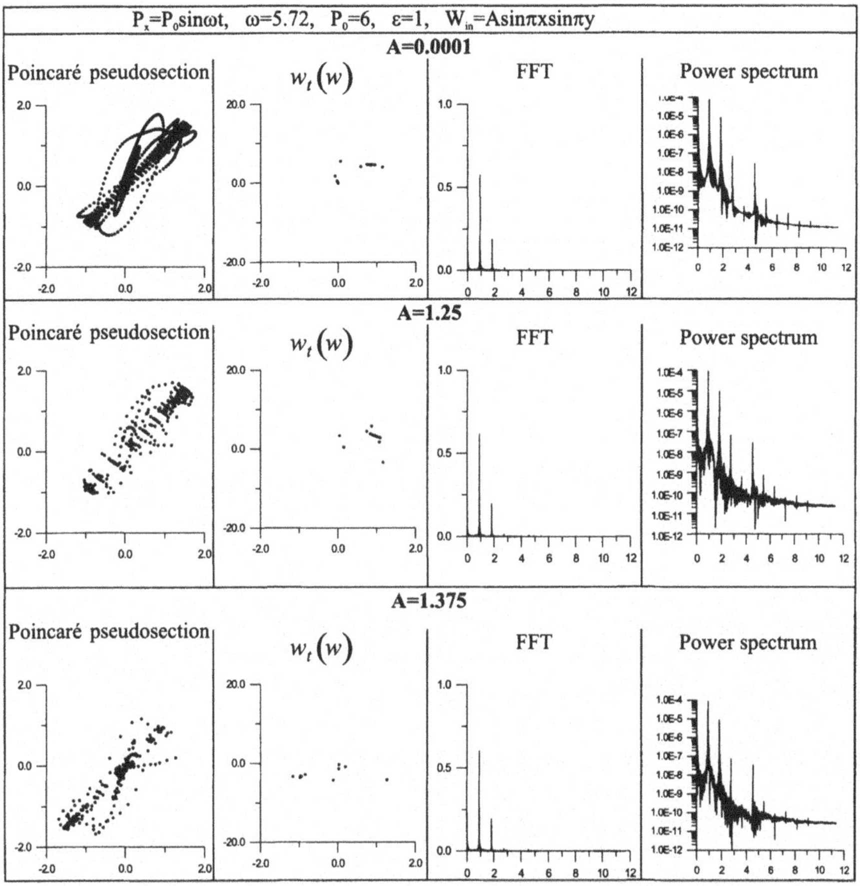

Fig. 7.82 Poincaré sections and pseudosections, $w_t(\omega)$, FFTs and power spectra for $A = 0.0001$, 1.25, and 1.375

Table 7.7 Synchronization frequencies for ω_{17} ($\otimes$—fundamental frequency, $\times$—independent frequency, $N\,\omega_k$—linear combinations)

No ω	ω	$A = 0.0001$	$A = 1.25$	$A = 1.375$
2	0.00555	$\times$	$\times$	$\times$
17	0.90998	$\otimes$	$\otimes$	$\otimes$
32	1.81996	$2\omega_{17}$	$2\omega_{17}$	$2\omega_{17}$

oscillation modes (Fig. 7.83) was constructed (nonsymmetric oscillations will be discussed later). Here the critical values of A, at which oscillations of the plate around a new equilibrium state occur, i.e. a slight change of A leads to a change of the state of the plate. For nonsymmetric oscillations (Fig. 7.83), there are two such values of A. This corresponds to a rotation by 180° of the phase portraits, Poincaré pseudosections and Poincaré sections $w_t(w)$ for every period of the load that is causing excitation. The Poincaré pseudosection is defined as $w(t)$ plotted against the deflection shifted in time, i.e. $[w(t), w(t+T)]$. The deflection $w(t+T)$

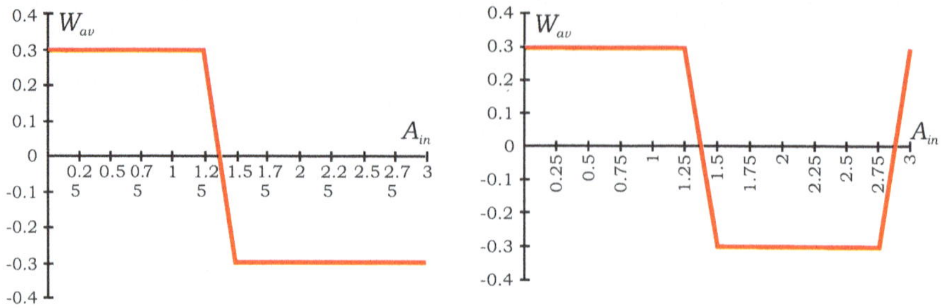

Fig. 7.83 Symmetric and nonsymmetric oscillations ($P_x = 6 \sin 5.72t$, $w_{in} = A \sin \pi x \sin \pi y$, $\varepsilon = 1$)

is related to the velocity $w_t(t)$ and, as a result, we obtain similar properties to those obtained when a full phase plane $[w(t), w_t(t)]$ is used.

We must emphasize that the above considerations are valid for fixed values of P_0 and the boundary and initial conditions applied here as well as for the period of the longitudinal load used here.

The nonlinear phase dynamics have been investigated for the fixed values $\varepsilon = 1$ and $\varepsilon = 5$, with the boundary and initial conditions given above, for $A = 1 \times 10^{-4}$ and $P_0 \in \{15, 20, 26\}$. In Fig. 7.84 the time histories $w(0.5, 0.5, t)$, phase portraits and FFTs for the plate centre ($x = y = 0.5$) are reported.

The numerical analysis leads to the conclusion that an increase of ε leads to an important change of the scenario leading to chaos using P_0 as the control parameter. For instance, for $\varepsilon = 1$ and $P_0 = 26$ we have a chaotic state, whereas for $\varepsilon = 5$ the plate undergoes post-crisis oscillations.

For $P_0 = 15$ and $\varepsilon = 5$ the plate is in a crisis state, and the number of quasi-periodic windows increases, but their durations in time greatly decrease, i.e. we observe intermittency with a higher number of oscillation modes in comparison with $\varepsilon = 1$.

7.9 Spatial–Temporal Symmetric Chaos

We have analysed the complex dynamics of symmetric, flexible plates with harmonic longitudinal excitation. Although scenarios leading to chaotic dissipative oscillations have been reported in many publications, the problems of plates investigated here belong to a different class of dynamical systems: we have to work with high-dimensional systems possessing two spatial coordinates and a dependence of time, i.e. oscillations with infinitely many degrees of freedom must be taken into account.

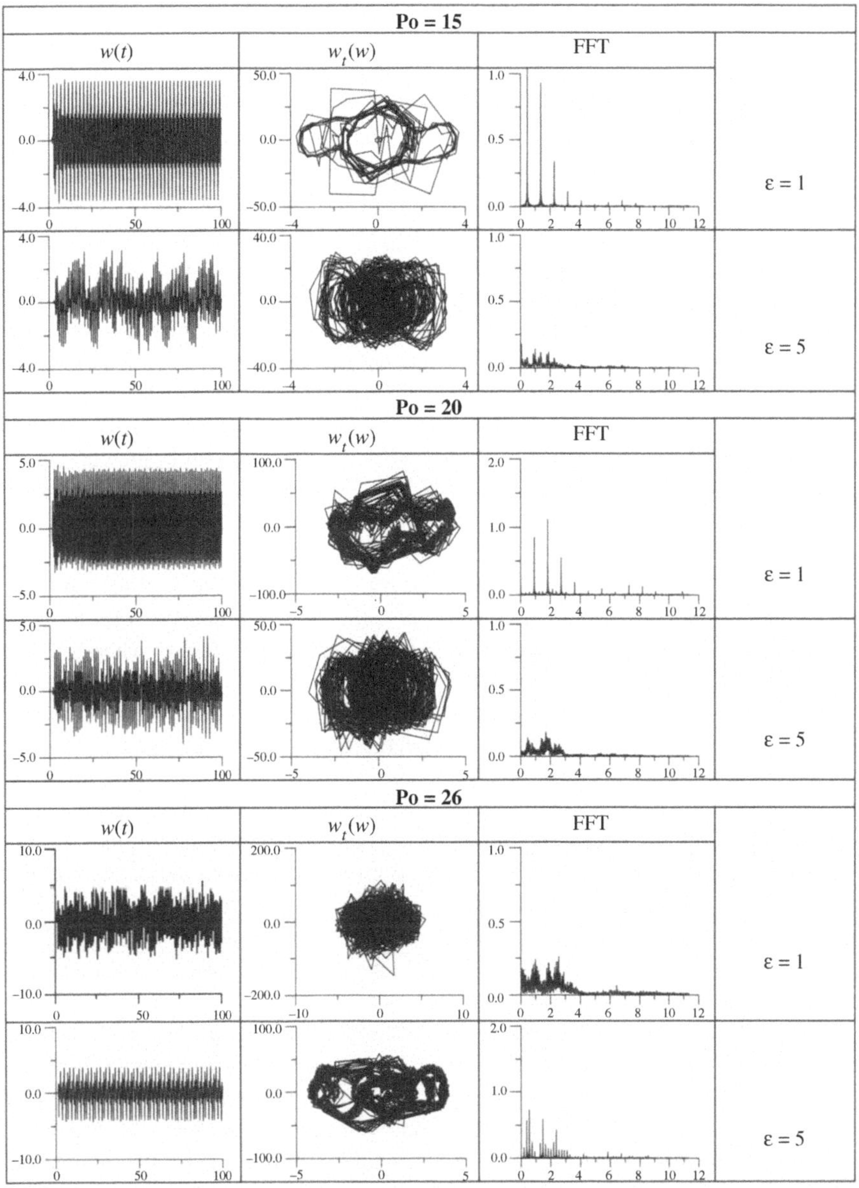

Fig. 7.84 Time histories $w(t)$, phase portraits $w_t(\omega)$ and FFTs for different values of P_0 and ε

In order to analyse multifrequency and chaotic oscillations, we need to use many dynamical characteristics, such as the oscillation processes at all points of the plate, phase portraits, FFTs, power spectra, Poincaré sections and pseudosections, and bifurcation diagrams.

In contrast to investigations in the fields of radiophysics, electronics, etc., where usually the transition to temporal chaos is studied, in problems related to plate theory a key point is the detection and analysis of spatial–temporal chaos.

A problem related to spatial chaos can be investigated relatively simply if the problem is homogeneous with respect to x and y. This situation occurs when x and y do not explicitly appear in the differential equations of the problem. If we consider one variable as time, then the system is autonomous. This means that from the point of view of the mathematics, the problem is analogous to that of temporal chaos. However, in the theory of plates with finite dimensions (in our case $\lambda = \frac{a}{b} = 1$), it is impossible to proceed in this way [6].

If we consider the deflection of the plate $w(x, y)$, then we have a physical interpretation up to the fourth derivative. The first derivatives $w_x(x, y)$ and $w_y(x, y)$ correspond to the tangents of the angles between the slopes of the deflection and the corresponding axes at the points where the derivatives are calculated. The second-order derivatives $w_{xx}(x, y)$, $w_{yy}(x, y)$ and $w_{xy}(x, y)$ correspond to the curvatures of the deflection functions at the points where the derivatives are calculated. If they are multiplied by the corresponding constants and summed, then the bending and torsion moments are obtained.

Therefore, by analysing those characteristics as a function of time, one can draw conclusions about the spatial variation of the plate surface. We refer to these characteristics later as the characteristics of the "modal portraits" when solving problems of plates using the method of finite differences. In a phase portrait the relation between a deflection and its velocity is reported for each time moment, whereas in a modal portrait we have a relation between the deflection and the tangent of its slope relative to the corresponding coordinate axis in a plane. If we consider the phase portraits in space, then for each time moment we have a relation between the deflection, velocity and acceleration. If we consider the modal portraits in space, then each for each point of the plate we have information about the deflection, the tangents of its slope and the curvatures. As a result, we can obtain all required information about the character of the deflection of the plate surface. For the analysis of temporal chaos, one needs to consider the intersection of phase portraits with the hyperplanes at the required time intervals.

The intersections of the trajectories of modal portraits allow extraction of information about multifrequency oscillations and spatial chaos. The chaotic motions include regular and stochastic parts. In the modal portraits there exist saddle equilibrium states, saddle periodic orbits and other saddle-type invariant structures.

Different spatial structures are expected (equilibrium, periodic or chaotic behaviour) depending on whether we have a saddle-type equilibrium, saddle-type periodic orbit or saddle chaotic motion. The occurrence of periodic or chaotic spatial structures is analogous to the occurrence of periodic or stochastic self-excitations. A difference arises in investigations of the stability of regular motion and investigations of the saddle character of a solution.

The tools described above can be used to trace temporal and spatial order, as well as temporal and spatial chaos.

We must emphasize that it is rather difficult to distinguish between regular and chaotic motion, especially in cases of intermittency. In addition, an arbitrary chaotic motion is characterized by a certain regularity, certain time rules and certain spatial structures. One problem of investigations of chaos is to detect and quantify temporal and spatial behaviour possessing both deterministic and stochastic aspects.

The observations discussed above have been proved by tracing a scenario of development of spatial–temporal chaos by changing the parameters of the longitudinal load $\{P_0, \omega\}$ along the Ox axis. We fixed the value of $\omega = 5.72$ and varied $P_0 \in [0, 30]$ for $t \in [0, 100]$ (the plate oscillations were analysed only for this time interval). We considered the symmetric form of the plate oscillations, i.e. one-quarter of the plate was considered. The following indicators are reported in Figs. 7.85, 7.86, 7.87, 7.88, 7.89, 7.90, and 7.91: the time history $w(t)$ for the centre of the plate; $w_{in}(t) = \int\int_S w(x, y, t)ds$; and Poincaré pseudosections for the whole time interval, as well as for some time intervals where periodic and quasi-periodic orbits were observed.

The corresponding data for the modal spaces (w_{xx}, w_x, w) and their projections on the planes $w_{xx}(w_x)$, $w_{xx}(w)$, $w_x(w)$, the mapping of the phase and modal portraits with respect to the period of the excitation, the FFT and power spectrum, the deflection of the plate surface at the characteristic time moments, and the values of the fundamental frequencies ω_i versus P_0 (i corresponds to the frequency number) are reported in Table 7.8. Graphs of $t_{in}(P_0)$ and $N\omega_i(P_0)$ are shown in Fig. 7.92, where t_{in} is the time interval for which the system is initially driven and $N\omega_i$ corresponds to the number of independent frequencies which govern the plate oscillations.

The one-parameter scenario leading to chaos, the corresponding values of $w_{av}(P_0)$, and various two-parameter portraits of the scenario leading to chaos, which will be discussed later, are shown in Figs. 7.93, 7.94 and 7.95.

The following parameters are used here: $w_{av}(P_0)$, $w_{av}(t)$, $w^*_{av} = \int_0^T w_{av}(t)\,dt$, $w_{in} = \int\int_S w(x, y, t)\,ds$ and $w_{av} = \frac{1}{2}(w^+_{in,n} + w^-_{in,n})$. The dependence $w_{av}(t)$ allows identification of the character of the oscillation after a bifurcation, i.e. whether the oscillation of the plate takes place in the neighbourhood of an initially well-defined equilibrium, or the oscillation takes place in a regime of jumps between two equilibria. The characteristics $w_{in}(t)$ and $w_{av}(t)$ are reported in Fig. 7.96 for a series of P_0 values. Analysis of the dependence $w_{av}(t)$ leads to the following conclusions: after a bifurcation and until a crisis, the oscillations take place around the initial equilibrium state, whereas in a state of crisis a series of jumps (the plate turns inside out) are observed, ending in the inside-out state, with oscillations around the positive or negative equilibrium state. After the crisis, up to the pre-chaotic state, the oscillations are accompanied by continuous turning inside out. In the pre-chaotic state, only oscillations around the initial equilibrium state take place.

The interval $P_0 \in [0, 30]$ has been divided into a series of subintervals, within each of which the oscillations have similar general properties, as listed below.

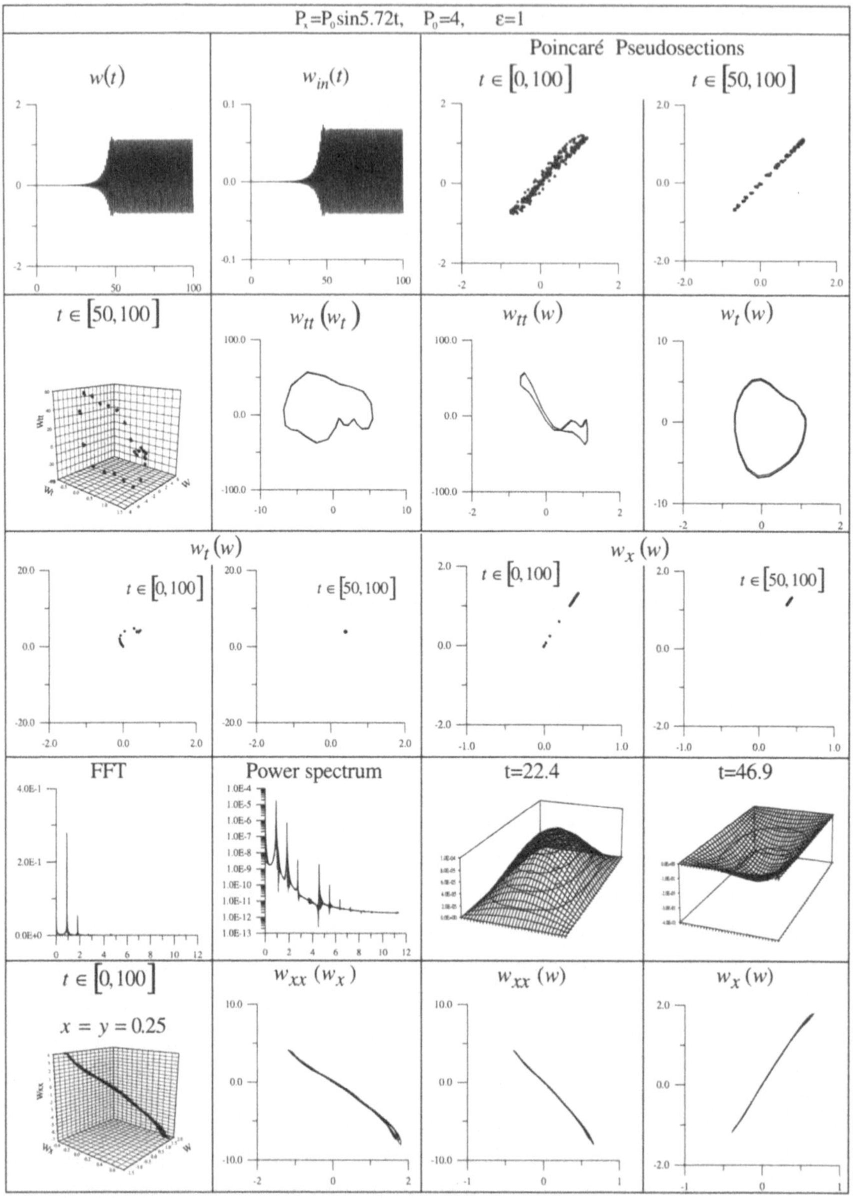

Fig. 7.85 Time histories, Poincaré sections and pseudosections, FFT, power spectrum, and modal portraits for the parameters and spatial configuration indicated ($P_0 = 4$)

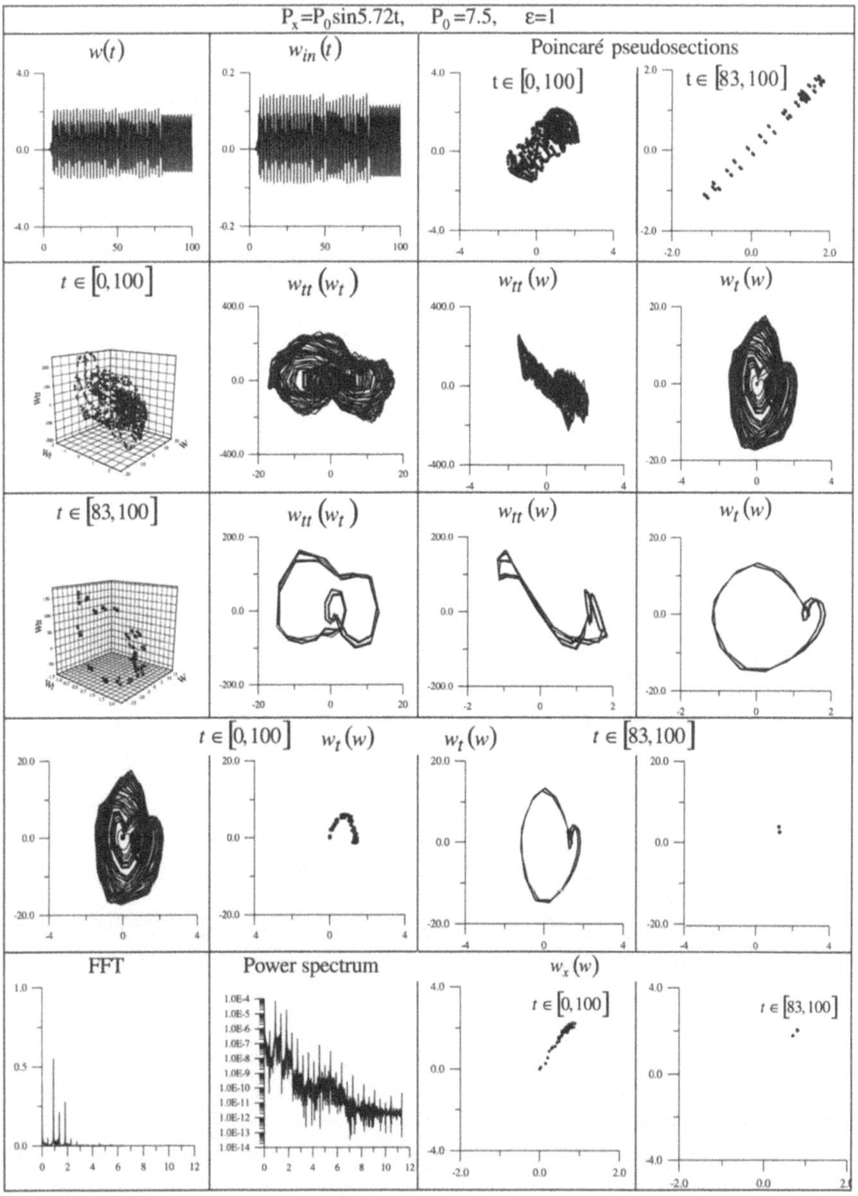

Fig. 7.86 Time histories, Poincaré sections and pseudosections, FFT, power spectrum, and modal portraits for the parameters indicated ($P_0 = 7.5$)

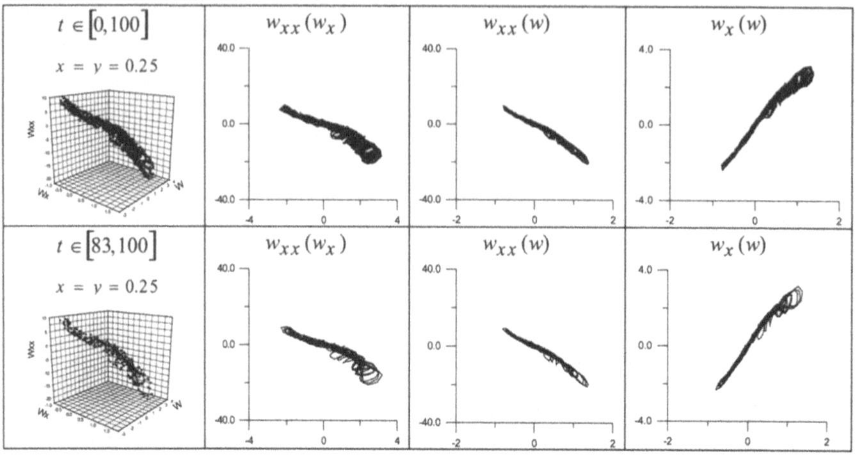

Fig. 7.86 (continued)

I. Stable Fixed Point for $P_0 \in [0, 35]$ In this subinterval, the solution of the equations does not undergo qualitative changes, and the origin of $w_t(w)$ remains a singular point (for $P_0 < 3.5$ it is an attractor).

II. Bifurcation, and Harmonic or Quasi-Periodic Oscillations, $P_0 \in [3.5, 7.25]$ *(see Fig. 7.85)* The main features of the oscillations in this interval are as follows:

1. For $P_0 \geq 3.5$, the first bifurcation occurs, and the plate oscillates around a new equilibrium state. The time needed for bifurcation to occur is largest for $P_0 = 3.5$, and this time is referred to as the critical time t_{cr}. With increase of P_0, i.e. $P_0 \in [3.5, 7.25]$, t_{cr} significantly decreases. When the static problem is solved using the relaxation method, we obtain $P_{0_{cr}} = 3.61$.
2. In this interval of P_0, oscillations take place in the time interval $t \in [0, 100]$ at two frequencies ω_{17}, and internal synchronization occurs, because the third frequency is $\omega_{32} = 2\omega_{17}$ (see Table 7.9).
3. In the interval $t \in [50, 100]$ the steady state is characterized by one frequency ω_{17}, because $\omega_{32} = 2\omega_{17}$, and the amplitude corresponding to ω_2 is much less than those for ω_{17} and $\omega_{32} = 2\omega_{17}$ (as shown by the FFT), and we obtain one point in the Poincaré map (see Fig. 7.85 for $P_0 = 4$).
4. The modal portrait $w_x(w)$ for the plate point $x = y = 0.25$, as well as for other points, forms an angle of 45° with the coordinates. Deflected plate surfaces are given for the two time moments $t = 22.4$ and $t = 46.9$. The projections of the space modal portrait (w_{xx}, w_x, w) are almost straight lines, which implies almost periodic oscillations in the steady state (see Fig. 7.85).
5. After bifurcation, the plate oscillates in an excited regime, the duration of which decreases with increase of P_0, and for $P_0 = 7$ this duration is at its minimum (see Fig. 7.92). The phase portrait trajectories and the sequence of points of the

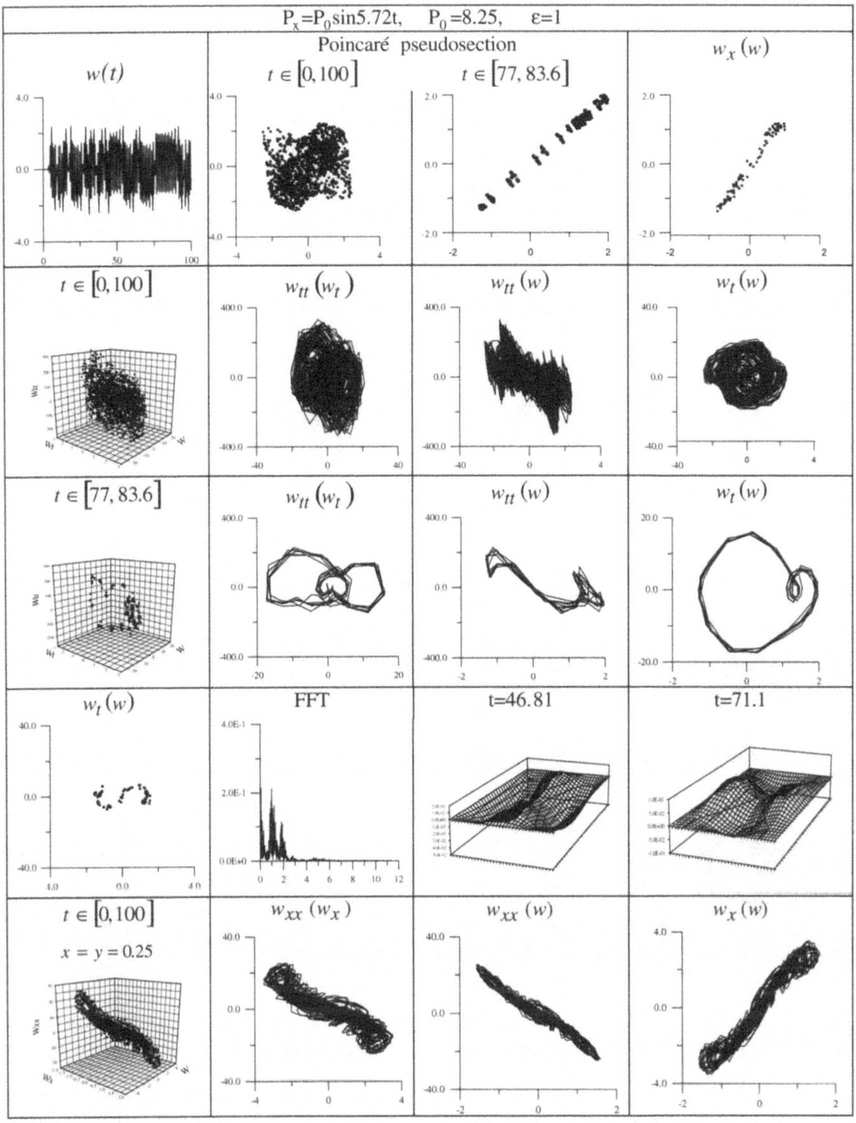

Fig. 7.87 Time histories, Poincaré sections and pseudosections, FFT, and modal portraits for the parameters and spatial configuration indicated ($P_0 = 8.25$)

Poincaré series create a sequence of curves and points, which are attached to an ellipse and one point.

6. The Poincaré pseudomaps $w^{t+T}(w^{(t)})$, which are defined by the deflection values at the times t and $t+T$, where T is the excitation period, also characterize the oscillations of the system.

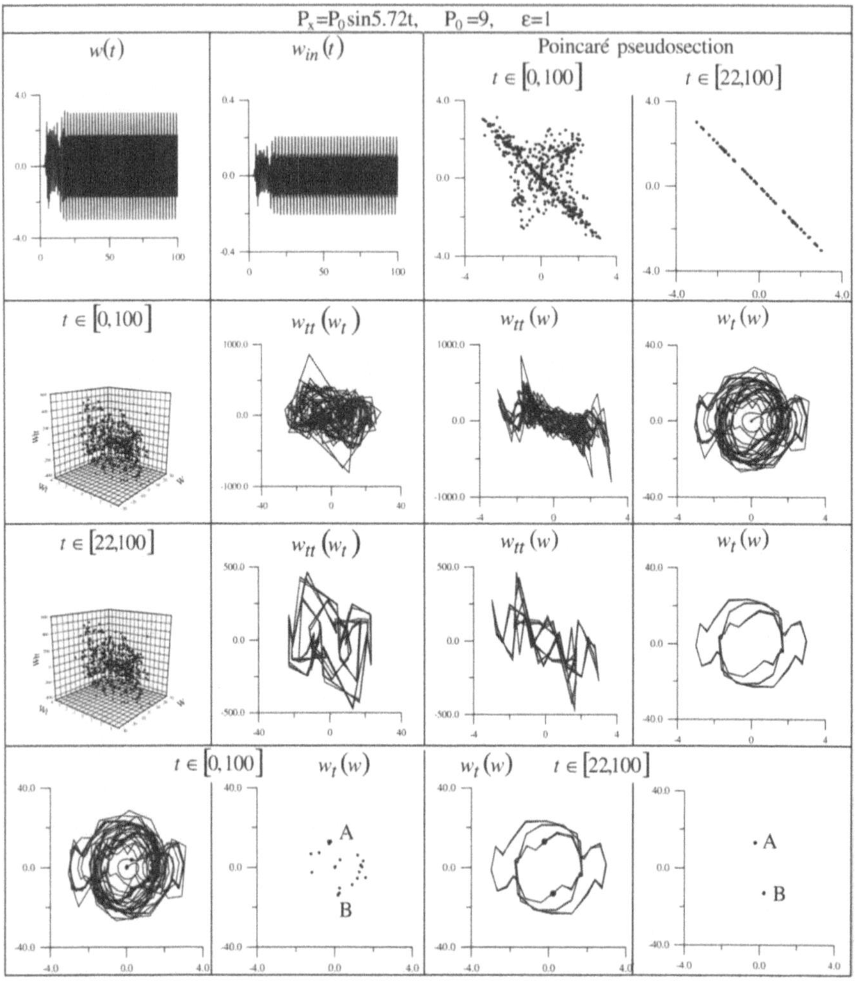

Fig. 7.88 Time histories, Poincaré sections and pseudosections, FFTs, power spectra, and modal portraits for the parameters indicated ($P_0 = 9$)

III. Pre-crisis—$P_0 \in [7.25; 7.95]$ This dynamical state is described by the example of $P_0 = 7.5$ (see Fig. 7.86). The dynamical behaviour for other values of $P_0 \in [7.25, 7.95]$ is qualitatively similar. The main features of the oscillations in this interval are as follows:

1. Here we observe oscillations at three frequencies, and a transition from two frequencies of oscillation to three frequencies occurs at $P_0 = 7.25$ via a jump, where the character of the oscillations changes greatly (see Table 7.9). It is known that at least three frequencies are needed for the occurrence of chaos. Here we have the following frequencies: ω_2, ω_{17} and ω_{28}; ω_2 and ω_{17} appeared

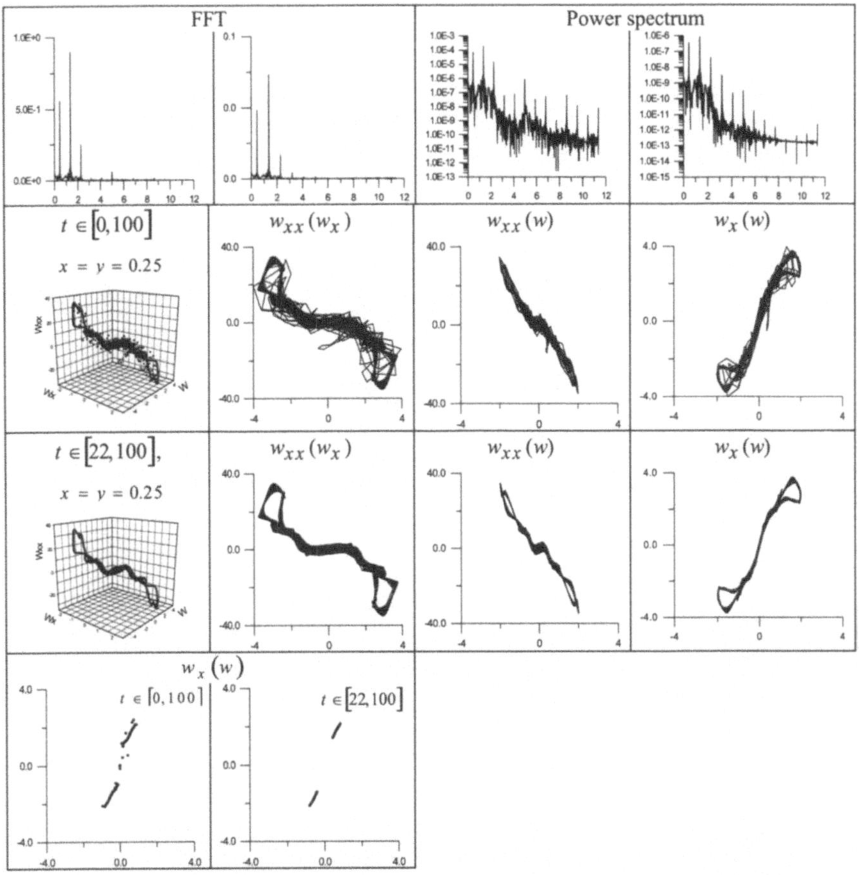

Fig. 7.88 (continued)

also in the previous interval of P_0, which implies an intermittent character of the oscillations. Similarly to the previous scenario, an interval synchronization of the form $\omega_{32} = 2\omega_{17}$ occurs here. The additional frequency ω_{28} also appears, which causes another occurrence of internal synchronization. In addition, two new frequencies, dependent on ω_{28} and ω_{17}, appear: $\omega_{11} = \omega_{28} - \omega_{17}$ and $\omega_{41} = \omega_{28} + \omega_{17}$. To conclude, the number of frequencies increases, but the three frequencies ω_2, ω_{17} and ω_{28} play the fundamental role.

2. The excited oscillations are transformed to chaotic oscillations, and their duration greatly increases. The occurrence of chaos is characterized by the volume and plane phase portraits, the power spectrum (which is broadband and continuous), and the structure of the Poincaré pseudomaps. Analysis of the modal portraits and bent plate surfaces shows the complex form of the surface of the bending plate. The modal portraits $w_x(w)$ for $x = y = 0.25$ are broadbanded, and at the boundary, steady state oscillating loops appear. The oscillations are unsymmetric

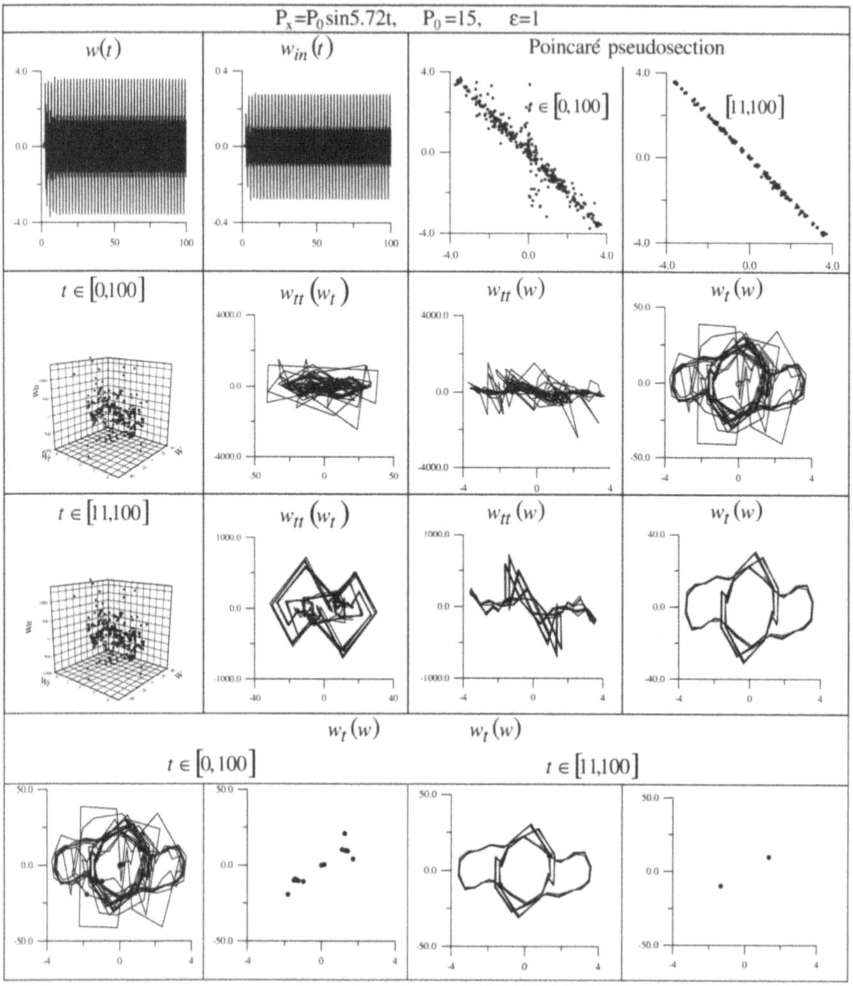

Fig. 7.89 Time histories, Poincaré sections and pseudosections, FFT, power spectrum, and modal portraits for the parameters indicated ($P_0 = 15$)

with respect to the equilibrium position, and they decrease with increase of P_0. Also, the angle of the tangent line to the deflection at the point $x = y = 0.25$ is changed.

3. In the centre of the interval [0, 83], a strange attractor is visible in the Poincaré map, which proves the existence of chaotic motion.
4. In the interval $t \in [83, 100]$, quasi-periodic orbits occur, which is testified by the space phase portrait and its projection onto the three planes. In the phase portraits there are loops, which correspond to a large increase of the second-order harmonics (for instance ω_{28}).

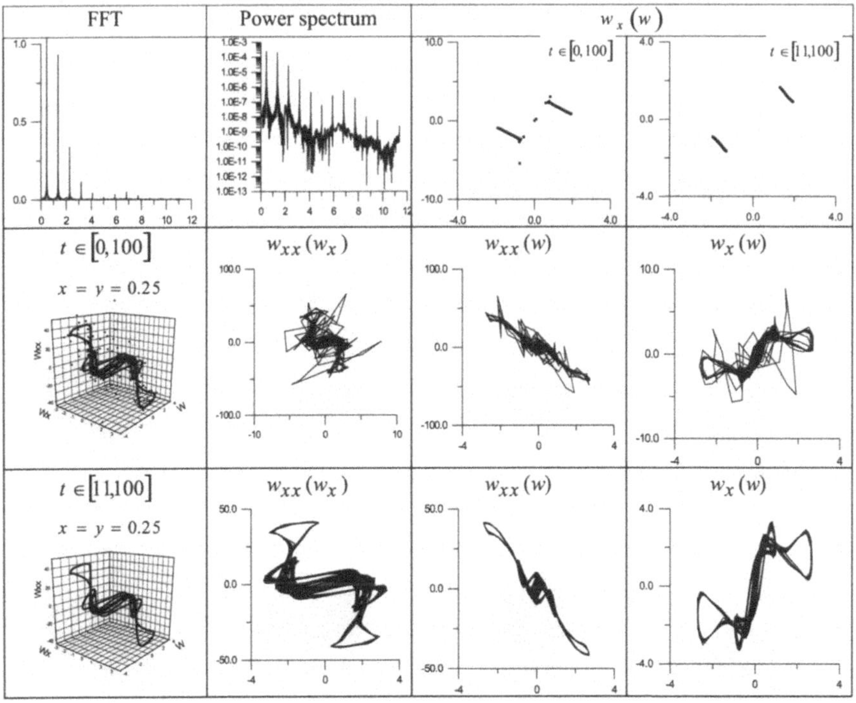

Fig. 7.89 (continued)

5. There are two points of the phase portrait in the cross-section of the phase portrait in the time interval $t \in [83, 100]$, which indicate that oscillations occur at the frequencies ω_{17} and ω_{28}.
6. The points of the Poincaré pseudosections are close to a line with a slope of 45°, as in the previously discussed case.
7. A sudden decrease is observed in the dependence $w_{av}^*(P_0)$, indicating that the system is approaching a new bifurcation point. The summed deflection $w_{av}^* = \int_1^{100}\int_0^1\int_0^1 w(x, y, t)\,dx\,dy\,dt$ over the interval $t \in [0, 100]$ changes its sign (Fig. 7.94).

IV. Crisis $P_0 \in [7.95; 8.5]$ (see Fig. 7.87) The main features of the oscillations in this interval are as follows:

1. Two quasi-periodic windows are observed within the ocean of chaos for $t \in [0, 100]$, and intermittency behaviour is also observed (see Fig. 7.92).
2. The phase portraits of the quasi-periodic windows are similar to those of the pre-crisis state (see Figs. 7.86 and 7.87).

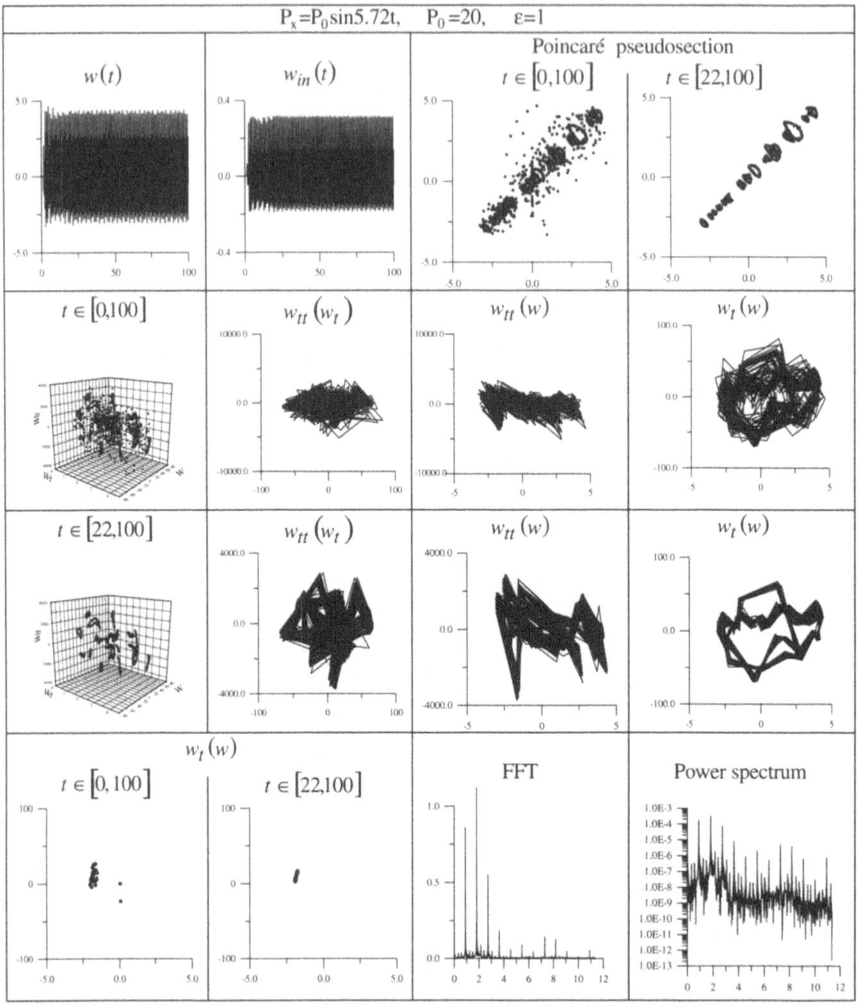

Fig. 7.90 Time histories, Poincaré sections and pseudosections, FFT, power spectrum, and modal portraits for the parameters indicated ($P_0 = 20$)

3. The oscillations in the crisis state occur at three or four independent frequencies (see Fig. 7.92).
4. In the cross-sections of the phase portrait $w_t(w)$, a structure corresponding to a strange attractor is present.
5. A strange attractor is visible in the intersections of the modal portraits (see Fig. 7.87).
6. The modal portraits $w_x(w)$ for $x = y = 0.25$, and for other points of the plate, are broadband, and various loops occur, located in the vicinity of the diagonal

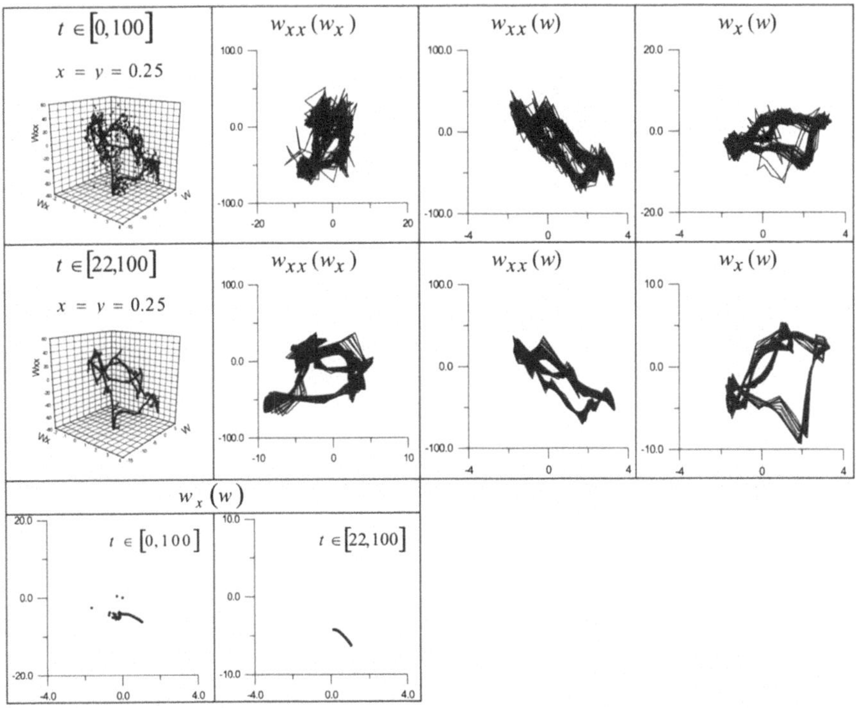

Fig. 7.90 (continued)

of the graph of $w_x(w)$. The complexity of the surface is shown in Fig. 7.87 for $t = 46.81$ and $t = 71.1$.

7. The slope of the Poincaré pseudosections with respect to the w_t axis is 45° in the case of quasi-periodicity. Otherwise, there is a set of points with a slope of 135° with respect to the w_t axis for $t \in [0, 100]$ (see Fig. 7.93).
8. Crisis is a necessary behaviour if a chaotic state is to be achieved, and it depends on many parameters.
9. In Fig. 7.94 ($w_{av}(t)$ against P_0), a series of bifurcations can be observed i.e. a continuous change of the sign of the deflection occurs in the time interval considered. Only in the case of quasi-periodic oscillations does w_{av} become constant, with either a positive or a negative sign.

V. Post-crisis, $P_0 \in [8.5, 10.25]$ *(see Figs. 7.88, 7.92, 7.93, 7.94 and 7.95)* The "post-crisis state" describes a narrow zone next to the crisis state of the mechanical system. The characterisitc properties of the state of the plate state are given in Fig. 7.88 (for $P_0 = 9$). The main features of the oscillations in this post-crisis interval are as follows:

1. The number of independent frequencies at which oscillation occurs decreases (see Table 7.8). Here we have three frequencies, internal synchronization occurs

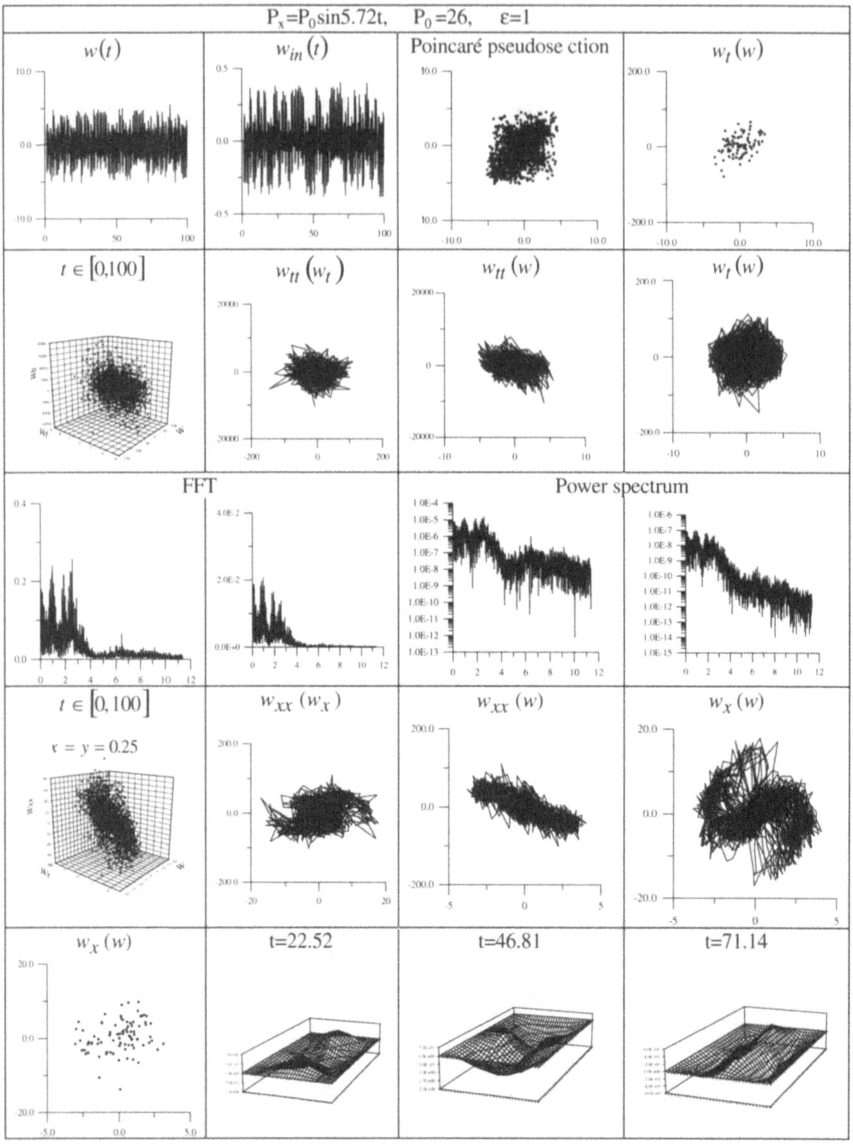

Fig. 7.91 Time histories, Poincaré sections and pseudosections, FFTs, power spectra, and modal portraits for the parameters and spatial configurations ($P_0 = 26$)

Table 7.8 Frequencies of oscillations ($P_x = P_0 \sin 5.72t$) (⊗—fundamental frequency, ×—independent frequency, $N\,\omega_k$—linear combinations)

No		P_0									
ω	ω	4	7	7.5	8	8.25	8.5	9	10	15	20
1	9.45×10^{-14}										×
2	0.00555	×	×	×	×		×				
3	0.0111							×			
4	0.02219										
11	0.45499			$\omega_{28} - \omega_{17}$				⊗	⊗	⊗	
14	0.8323					×					
15	0.88778										
17	0.90998	⊗	⊗	⊗					$2\omega_{11}$		⊗
18	0.91553					×					
19	0.99321				×						
20	0.99876										
21	1.04869			×							
26	1.31503					×					
28	1.36497			⊗				$3\omega_{11}$	$3\omega_{11}$	$3\omega_{11}$	
32	1.81996	$2\omega_{17}$	$2\omega_{17}$	$2\omega_{17}$					$4\omega_{11}$		$2\omega_{17}$
33	1.82551					×					
34	1.84215				×						
36	1.95867			×							
41	2.27495			$\omega_{28} + \omega_{17}$				$5\omega_{11}$	$5\omega_{11}$	$5\omega_{11}$	
43	2.72495										
44	2.72994			$2\omega_{28}$					$6\omega_{11}$		$3\omega_{11}$
45	3.17938									⊗	
46	3.63437										×
47	4.08936									$\omega_{45} + 2\omega_{11}$	
48	4.54434										×
51	4.99933							×			
52	5.00488								×		
54	5.45432										×
55	6.81929									×	
56	7.27428										×
57	7.72927									×	
58	8.18426										×

on ω_{11}: $\omega_{28} = 3\omega_{11}$, $\omega_{41} = 5\omega_{11}$. In a steady state, two frequencies of oscillation are observed (see the cross-section $w_t(w)$ for $t \in [22, 100]$).

2. The time needed to achieve quasi-periodic solutions decreases and becomes equal to that observed in the pre-crisis state.
3. The phase portrait is changed qualitatively. Higher-order harmonics occur, as demonstrated by the FFT and power spectrum.

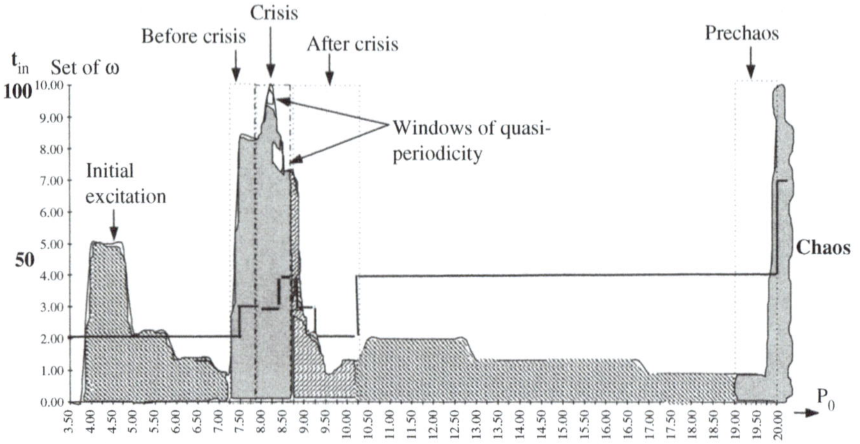

Fig. 7.92 Scenario to chaos versus P_0 (thick line characterizes the dependence $N\omega_i(P_0)$)

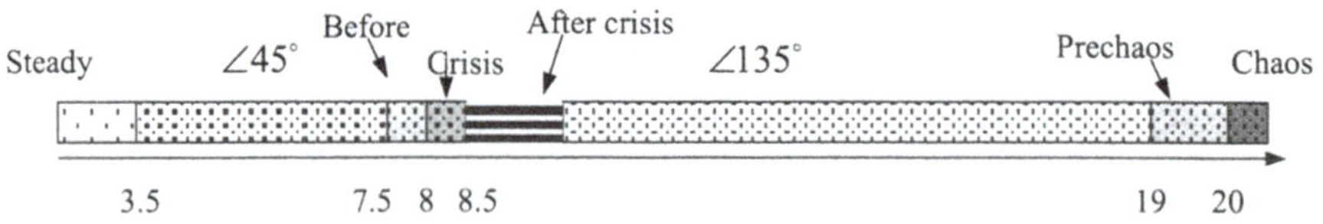

Fig. 7.93 Dynamics versus P_0 ($\omega = 5.72$)

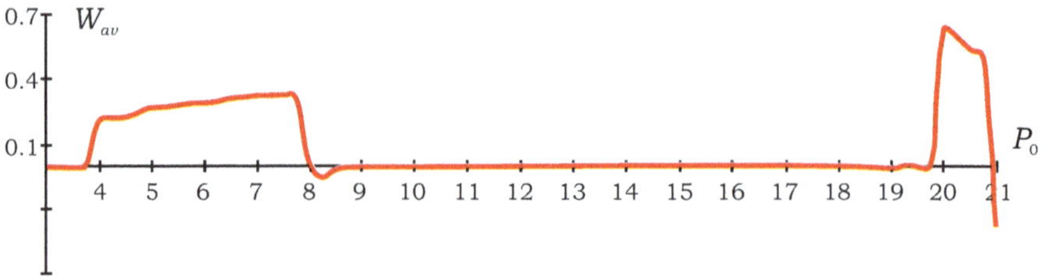

Fig. 7.94 w_{av} versus P_0 (see text)

4. The oscillations at $x = y = 0.5$ are symmetric with respect to the zero equilibrium position, i.e. jumps appear (Fig. 7.96). At the maximum displacement, only one loop occurs in the modal portrait (Fig. 7.88). Also, symmetric loops appear in the $w_{xx}(w)$ dependence.
5. The Poincaré pseudosection has a slope of 135° with respect to the w axis, but in the excited state there is a set of points that have a slope with respect to w_t of 45° (see Fig. 7.93).

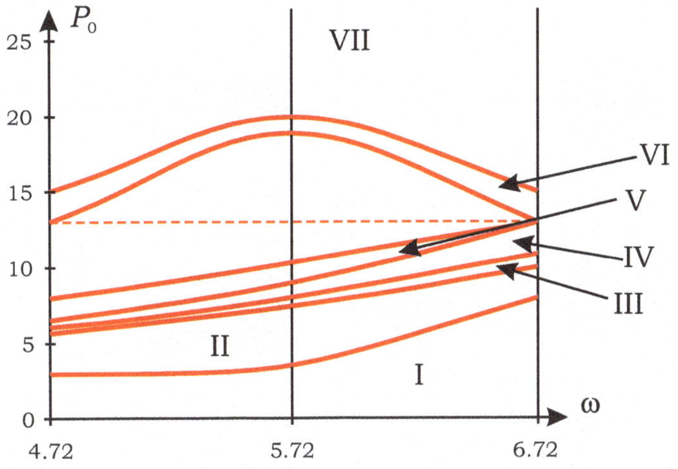

Fig. 7.95 Various dynamical states within the two-parameter space (see text for a detailed description)

6. The pre-crisis and post-crisis states can be treated as intervals of crisis, where the plate oscillations are transformed to the crisis state and then into a steady state with two or three frequencies, respectively.

VI. Steady Post-crisis State, $P_0 \in [10.25; , 18.9]$ In this case the oscillations are qualitatively similar to those of the post-crisis state (see Fig. 7.89). The phase and modal portraits are similar to that of the previous state, but the loops in their limiting states in the modal portraits are larger.

VII. Pre-chaotic State, $P_0 \in [18.9, 20]$ *(see Fig. 7.90)* The main features of the oscillations in this interval are as follows:

1. A number of independent frequencies occurs (see Table 7.8, and the FFT and power spectrum in Fig. 7.90). An internal synchronization on ω_{17}, i.e. $\omega_{32} = 2\omega_{17}$, $\omega_{44} = 3\omega_{17}$, occurs.
2. The points of the Poincaré pseudo-section are close to a line with a slope of 45° with respect to the w_t axis. The phase and modal portraits have closed subspaces.
3. A series of bifurcations occurs and the equilibrium states change (Fig. 7.96).
4. The deflection surfaces have a chaotic character.

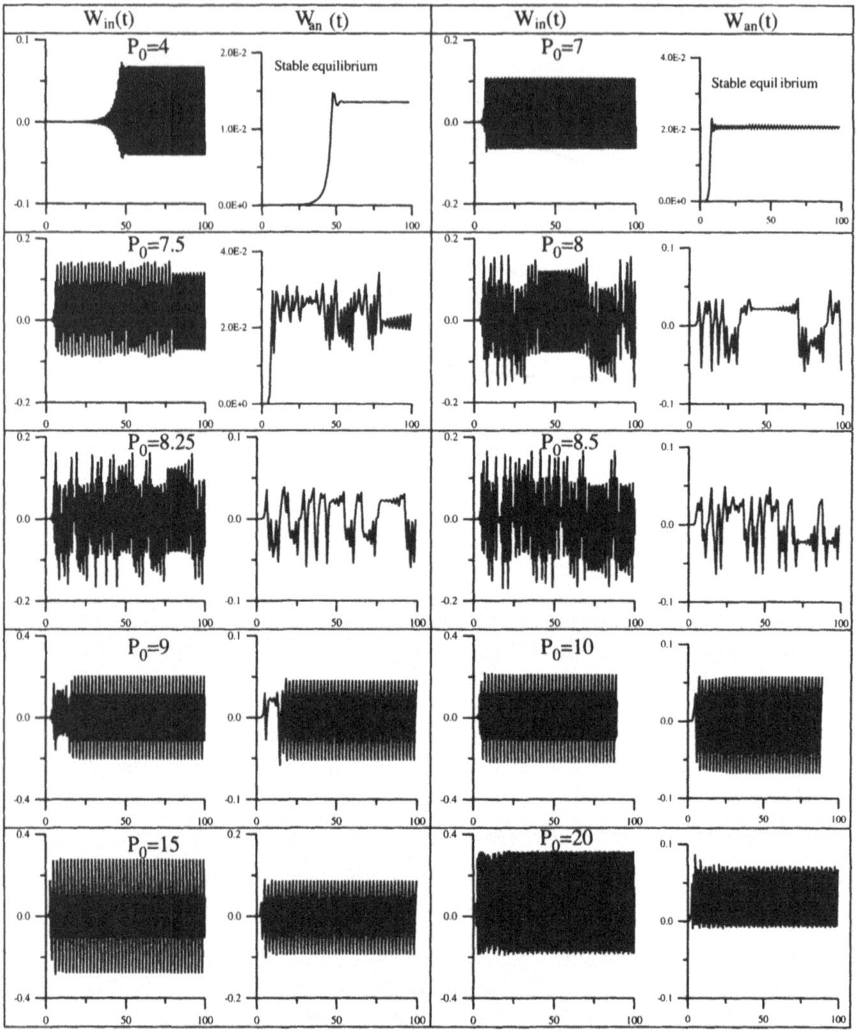

Fig. 7.96 w_{in} and w_{av} (see text) for different values of P_0

VIII. Chaos, $P_0 > 20$ (see Fig. 7.91) The term "chaos" is applied in deterministic problems, where there are no unpredictable or stochastic forces or parameters. For $P_0 > 20$, a slight change of P_0 leads to large changes of all characteristics of the system analysed.

Table 7.9 Frequencies of oscillations for various values of P_0 (see Table 7.8 for more details)

No		P_0									
ω	ω	4	7	7.5	8	8.25	8.5	9	10	10.25	10.5
2	0.00555	×	×	×	⊗	⊗	⊗		×	⊗	
3	0.0111										
5	0.02774										×
6	0.03329										
7	0.04439							×			
8	0.05549										
9	0.32737					×					
10	0.4217										
11	0.45499		$\omega_{28} - \omega_{17}$				⊗	⊗	⊗		
12	0.46609									⊗	
13	0.47164										×
15	0.88778							×			
16	0.89333										
17	0.90998	⊗	⊗	⊗	⊗						
18	0.91553					⊗					
22	1.13747		×								
23	1.21515				×						
24	1.23735					⊗					
25	1.24845			×							
27	1.33168										
28	1.36497		⊗				$3\omega_{11}$	$3\omega_{11}$	$3\omega_{11}$		
29	1.37607									⊗	

(continued)

Table 7.9 (continued)

No		P_0									
ω	ω	4	7	7.5	8	8.25	8.5	9	10	10.25	10.5
30	1.38161										×
31	1.44265									×	
32	1.81996	$2\omega_{17}$	$2\omega_{17}$	$2\omega_{17}$	$2\omega_{17}$			$4\omega_{11}$			
33	1.82551					$2\omega_{18} - \omega_2$					
35	1.85325										
37	2.15228					$\omega_{24} - \omega_{18}$					
38	2.24165										
39	2.2583									$\omega_{29} - \omega_{12} - \omega_2$	
40	2.26694								×		
41	2.27495		$\omega_{28} + \omega_{17}$				$5\omega_{11}$	$5\omega_{11}$			
42	2.28604										×
43	2.72439				$3\omega_{17} - \omega_2$						
44	2.72994		$2\omega_{28}$	$3\omega_{17}$			$7\omega_{11} - \omega_2$				
45	3.17938										
48	4.54434		×								
49	4.82178					×					
50	4.85507				×						
51	4.99933								×		
52	5.00488						×	×			
53	5.02708										
54	5.45432		×								

7.10 Dissipative Nonsymmetric Oscillations

We have considered nonsymmetric oscillations of a square, isotropic plate ($\lambda = 1$) subjected to the action of a longitudinal one-sided periodic load. The scenarios leading to chaotic state with a change of $\{P_0, \omega\}$ is investigated. The problem was solved using the algorithm described earlier. The fourth-order $0(h^4)$ approximation was used, and a condition of axial symmetry in the mesh space $\bar{G}_h$ was not been applied. The whole mesh space $\bar{G}_h$ was used, and the number of nodes was increased by a factor of two in comparison with the symmetric case. Therefore, we were able to find all possible symmetric and nonsymmetric solutions to the differential equations considered. The frequencies of the exciting force $P_x = P_0 \sin \omega t$ were taken from the interval $\{4.72, 672\}$. Figures 7.97, 7.98, 7.99, 7.100, 7.101, 7.102, 7.103, 7.104, 7.105, 7.106 and 7.107 correspond to the symmetric oscillations considered earlier.

To illustrate the scenarios leading to chaos in a flexible plate we have taken $\omega = 5.72$, $\varepsilon = 1$, $A = 1 \times 10^{-4}$. In Table 7.9, the frequencies of oscillation of the plate for a series of values of P_0 are given. The dependences $t_*(P_0)$ and $N\omega_i(P_0)$ are shown in Fig. 7.103. Figure 7.106 shows a classification of the oscillations, which is similar to that in the case of symmetric oscillations (Fig. 7.95). We shall describe only some of the differences. States I and II are qualitatively similar in the symmetric and nonsymmetric cases, but in state II the plate oscillates about a symmetric form. The corresponding intervals of P_0 in the nonsymmetric case are reported in Figs. 7.103 and 7.105. In state II we have a pre-crisis, and the oscillations exhibit 3–6 frequencies. The fundamental characteristics of this state are presented in Fig. 7.97.

In the phase portrait, three loops and three points of the Poincaré map correspond to quasi-periodic orbits. A strange attractor is visible for $t \in [0, 100]$. The mode portrait (Fig. 7.97) includes a series of closed orbits with complicated shapes.

One of the peculiarities of the crisis (see Figs. 7.98 and 7.99) is an exchange of symmetric and nonsymmetric forms. At the initial time moment, the oscillations begin symmetrically and then approach a nonsymmetric state. This is clearly manifested when the dependence $w(0.5, 0.5, t)$ and the deflection surface are analysed. In the crisis state, the deflection surfaces have extremely complicated shapes for both the symmetric and the nonsymmetric oscillations (Fig. 7.98).

In the quasi-periodic windows, the oscillations of the plate occur in the neighbourhood of various equilibrium states, and a series of space bifurcations appears (Fig. 7.107). This means that very often the magnitude and sign of the equilibrium deflection change, and the oscillations take place in the vicinity of this equilibrium deflection.

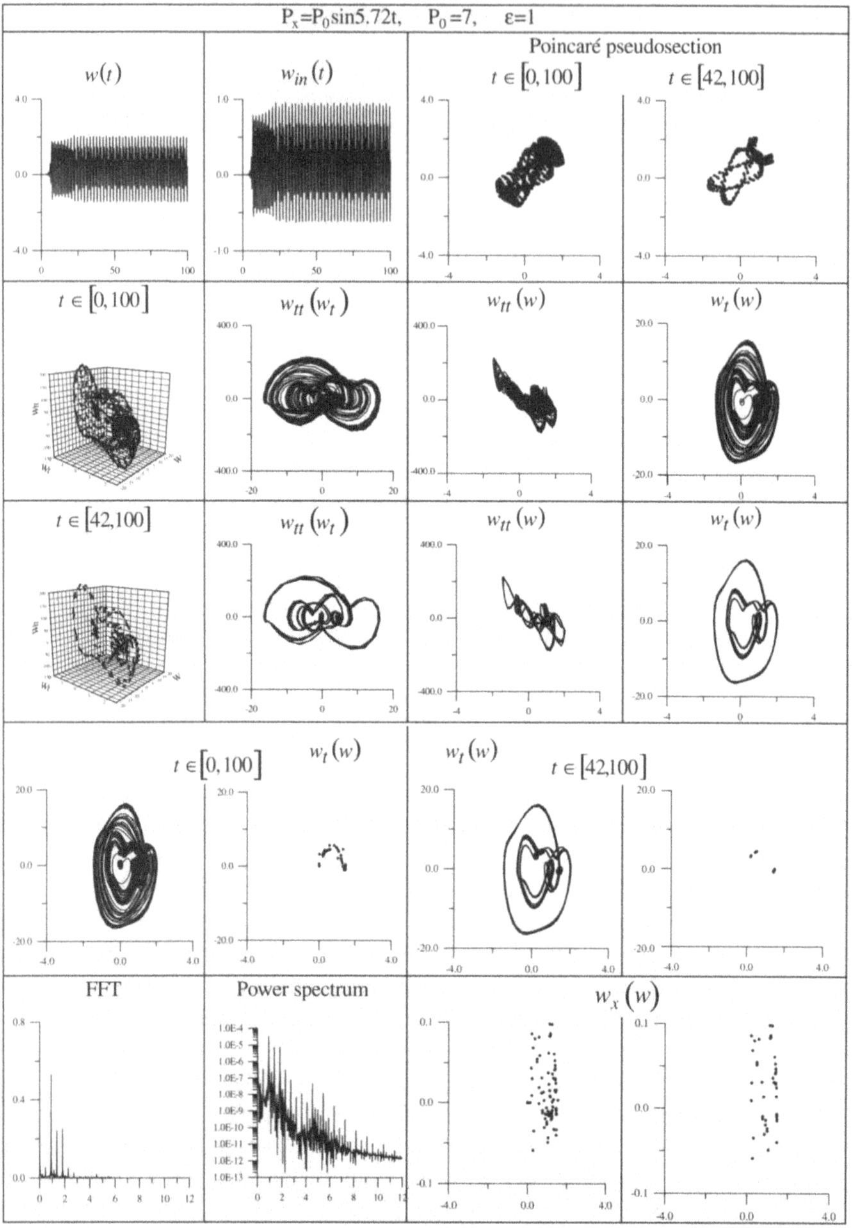

Fig. 7.97 Time histories, Poincaré sections and pseudosections, modal portraits, FFT and power spectrum ($P_0 = 7$)

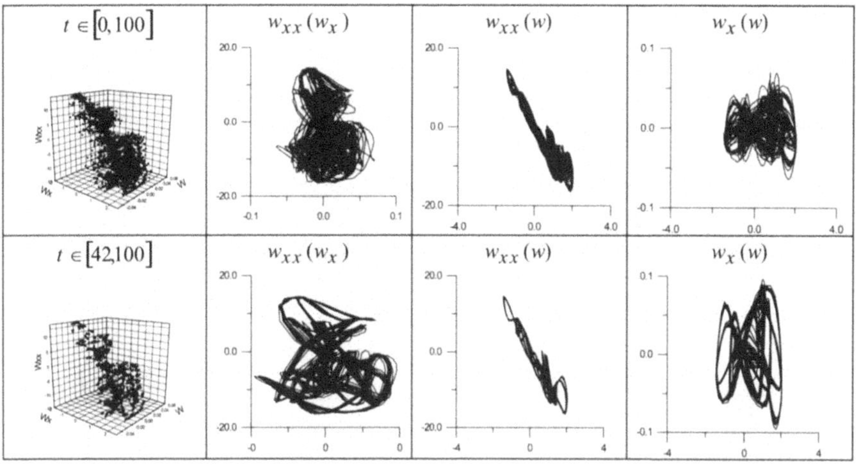

Fig. 7.97 (continued)

In state V, a post-crisis state is observed (Figs. 7.100 and 7.101), and oscillations occur in a regime of concavity and convexity. The strange attractors collapse, and there are two or four points in the intersections of the phase and modal portraits.

The corresponding characteristics for the chaotic state are reported in Fig. 7.102. The modal portrait is composed of a set of arbitrarily distributed curves, which create cross, which is interpreted as the creation of two mutually perpendicular waves at different times (see Fig. 7.102, surfaces for $t = 22.44$ and $t = 46.772$). The oscillations occur in the longitudinal direction with the creation of two or four waves, and in the transverse direction with the creation of one, three or five waves.

7.11 Solitary Waves

In this section we describe some results of applying a method described earlier to reduce the dimension of the PDEs by projection to ODEs, i.e. the method of finite differences with a spatial approximation of order $O(h^4)$. This approach has been used to study standing and travelling waves in a space occupied by a thin plate subjected to the action of a one-sided periodic longitudinal load.

Infinite gradients can appear in a wave profile in various physical situations. For instance, a wave on a fluid surface, may collapse by spurting.

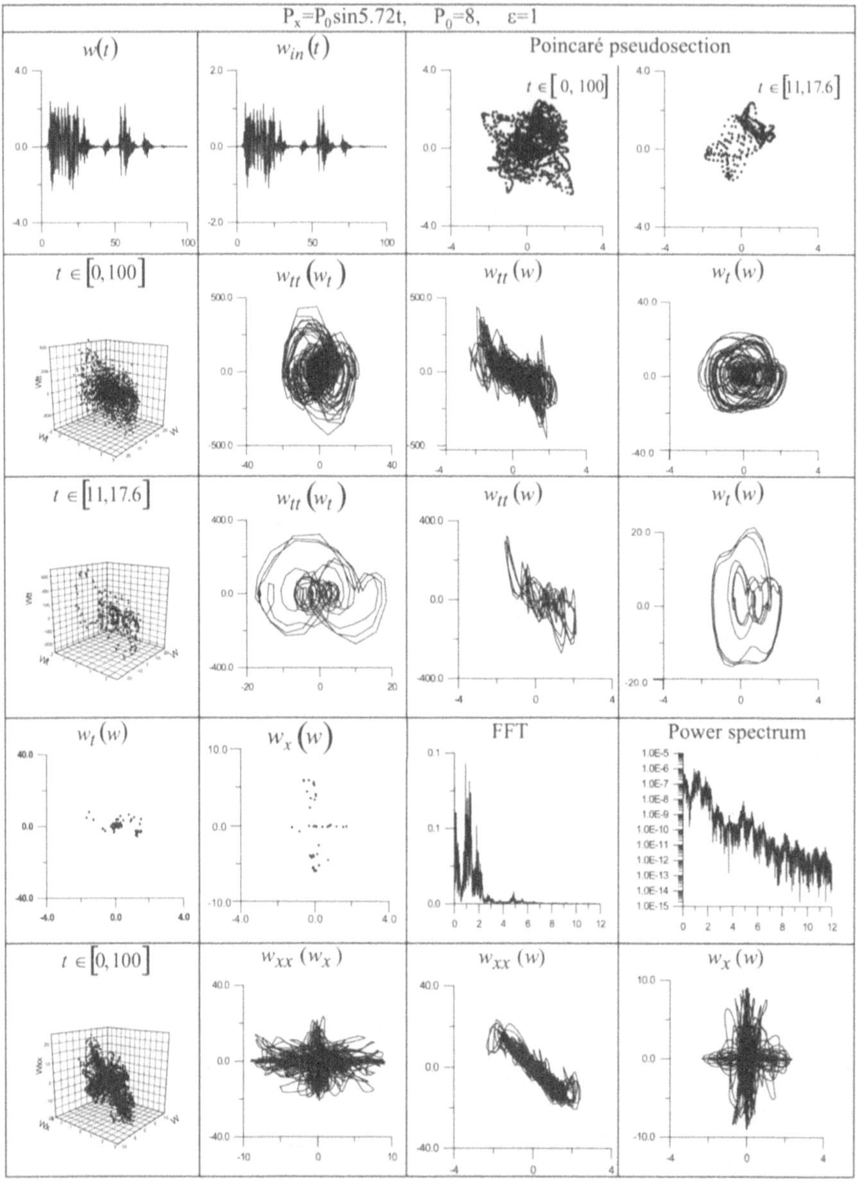

Fig. 7.98 Time histories, Poincaré sections and pseudosections, modal portraits, FFT and power spectrum, and spatial configurations for specified time moments ($P_0 = 8$)

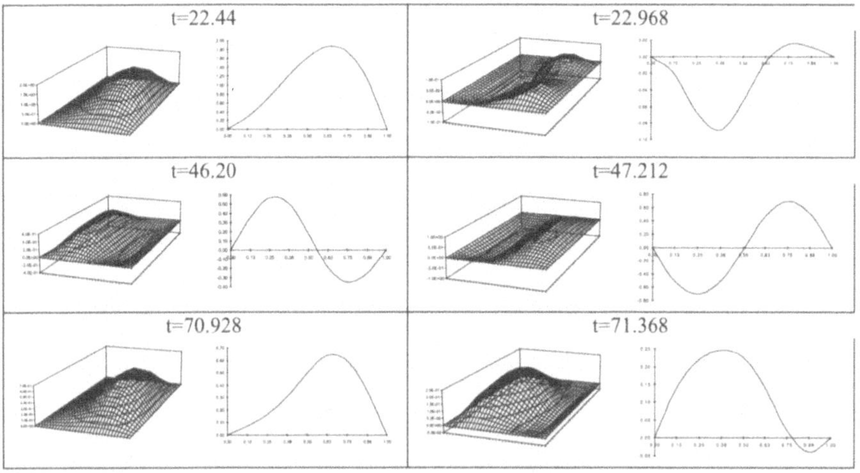

Fig. 7.98 (continued)

If the flow is composed of independent particles, then nonuniqueness appears in the wave profile. After the occurrence of a discontinuity in the fundamental flow, several different flows appear, moving with vary different velocities (multiflow). For a sound or electromagnetic field, where nonuniqueness is not allowed, the further development of the nonlinear wave depends on whether dissipative or dispersive effects will dominate in the neighbourhood of a rapid field change. The following one-wave PDE has been analysed:

$$\frac{\partial u}{\partial t} + v(u)\frac{\partial u}{\partial x} + \beta\frac{\partial^3 u}{\partial x^3} - \alpha\frac{\partial^2 u}{\partial x^2} = 0,$$

where u denotes the average velocity.

This equation can be reduced either to the Kortweg–de Vries equation ($\alpha = 0$) or to Burger's equation ($\beta = 0$). Owing to high-frequency dissipation, a discontinuity will not lead to stability loss. In addition, the dispersion bounds the width of the discontinuity, and as a result a "stationary" travelling wave with a constant profile can appear. The stationary waves of the Kortweg–de Vries equation correspond to a conservative nonlinear oscillator.

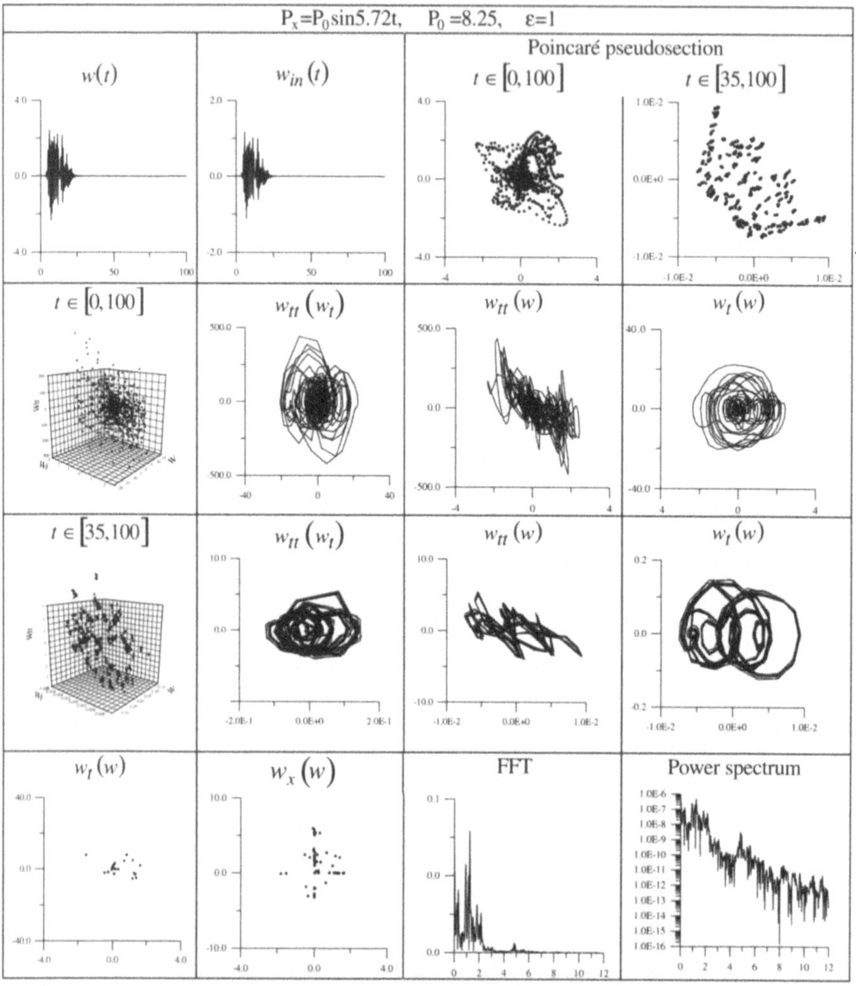

Fig. 7.99 Time histories, Poincaré sections and pseudosections, modal portraits, FFT and power spectrum ($P_0 = 8.25$)

Periodic motions in the neighbourhood of separatrices are waves. A spatially solution that can be described by one soliton is associated with a separatrix.

The one-dimensional stationary waves that appear in one-dimensional continuous systems (e.g. transmission lines) and in the case of plane waves (for instance, solitons on water governed by the Kortweg–de Vries equation) are well studied. On the other hand, it is clear that solitons in a falling water layer and ionic-sound

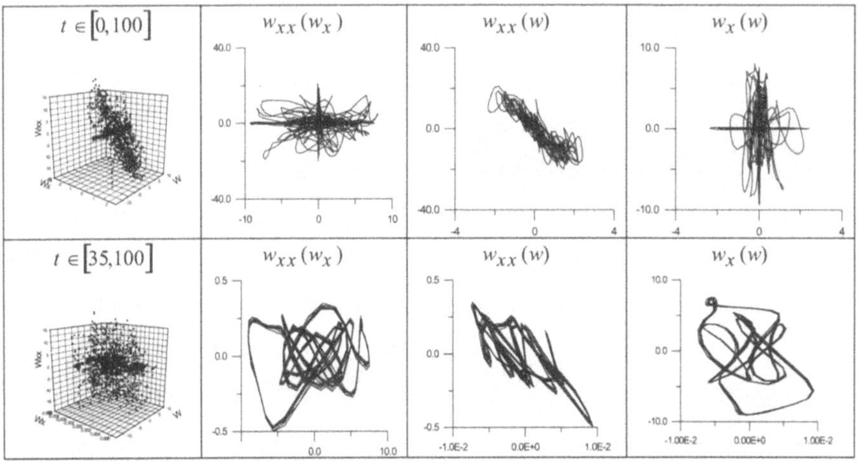

Fig. 7.99 (continued)

solitons in a plasma must depend on two spatial coordinates. A simple model, which is a generalization of the Kortweg–de Vries equation of the form

$$\frac{\partial}{\partial x} + (u_t + uu_x - \delta u_{xxx}) = \gamma u_{yy}.$$

is given in [274].

As has been mentioned in [261], a particular feature of the atmosphere of Jupiter is the Great Red Spot, which is a two-dimensional Rossby soliton. Rossby waves, in a linear approximation, correspond to waves in a rotating atmosphere, and the main reason for their existence is related to the change with a latitude of the horizontal projection of the Coriolis force.

For a medium with dissipation, the Chochleva–Zabolotskij equation plays a representative role in describing the waves that occur [281]. However, in the case of thin flexible plates, solitary waves have not previously been detected within the kinematic Kirchhoff–Love model. Here, we illustrate and analyse this behaviour.

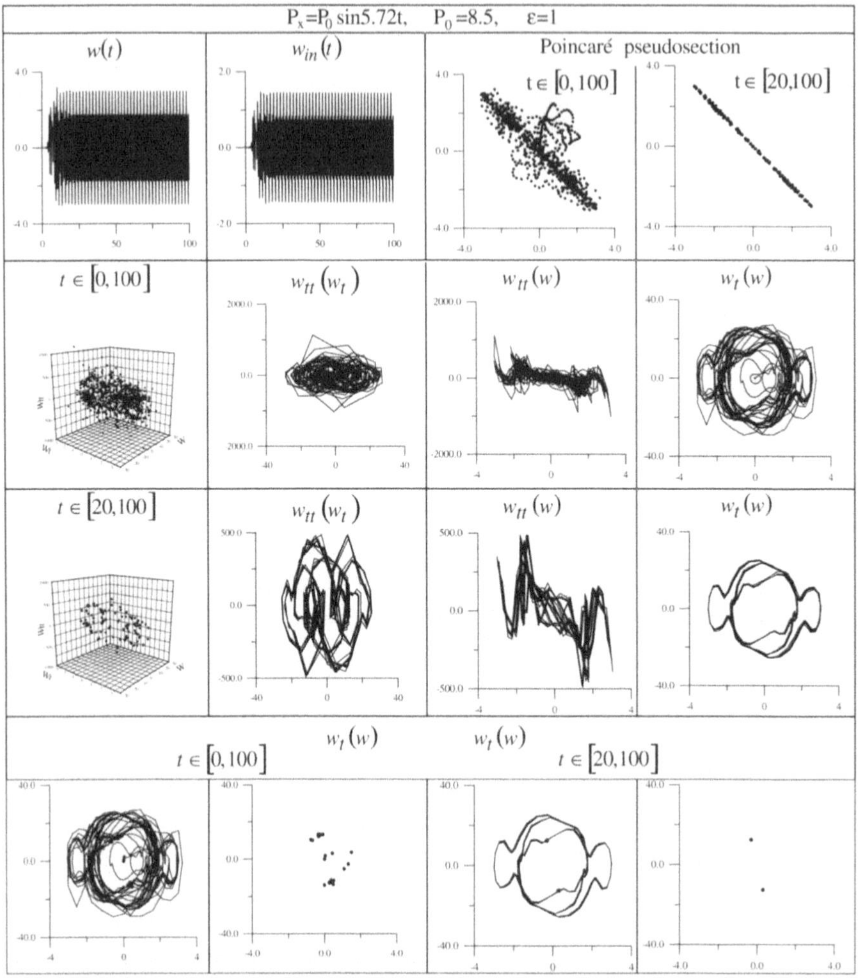

Fig. 7.100 Time histories, Poincaré sections and pseudosections, modal portraits, FFT and power spectrum ($P_0 = 8.5$)

The boundary and initial conditions (7.67) and (7.69) were applied, and the full mesh space $\bar{G}_h$ was used without symmetry conditions. A spatial step $h = 1/16$ and a time step $\Delta t = 2 \times 10^{-4}$ were used.

To analyse the very sensitive structure of multifrequency and stochastic oscillations, and to study transitions between different oscillation regimes, we have used the same characteristics as used earlier.

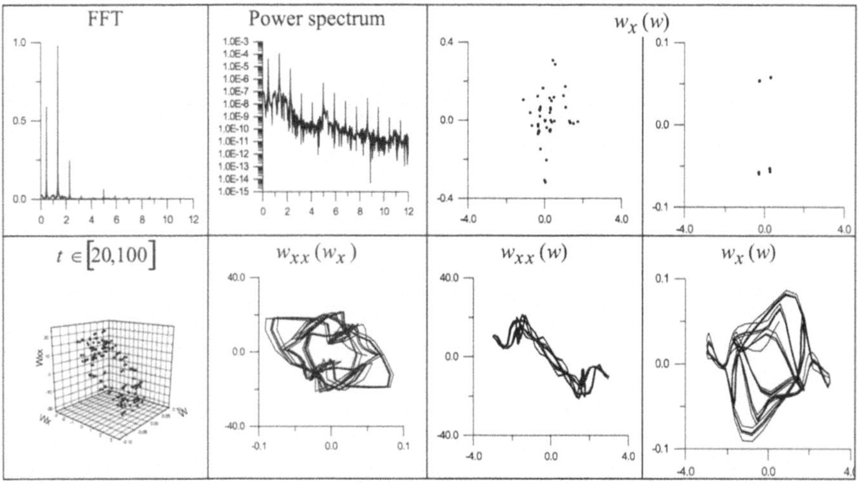

Fig. 7.100 (continued)

We took P_0 as the control parameter, and the other parameters were fixed: $\omega = 10.47$, $\lambda = \varepsilon = 1$, $\nu = 0.3$. In Figs. 7.107 and 7.108, the deflections $w(0.5, 0.5, t)$, phase portraits, Poincaré maps and FFTs are reported for various values of P_0. Beginning with $P_0 \geq 16$, the mechanical system analysed here is in a chaotic state, and a series of transitions between symmetric and nonsymmetric oscillation forms is observed.

For $P_0 > 19.25$, neither solitons nor jump phenomena appear. Instead, travelling waves and regular oscillations are observed for $t \in [48, 100]$.

We shall now focus our attention on the interval $P_0 \in [18.95; 19.25]$. In this interval, an oscillation jump is observed, which results in a change of the spatial–temporal configuration of the dynamical state, and in the occurrence of standing and travelling waves. For $P_0 = 18.95$ and $P_0 = 19$, the fundamental characteristics are reported in Figs. 7.109 and 7.110. The time before a jump occurs is different in the two cases. The oscillations before the jump are chaotic, as is indicated by

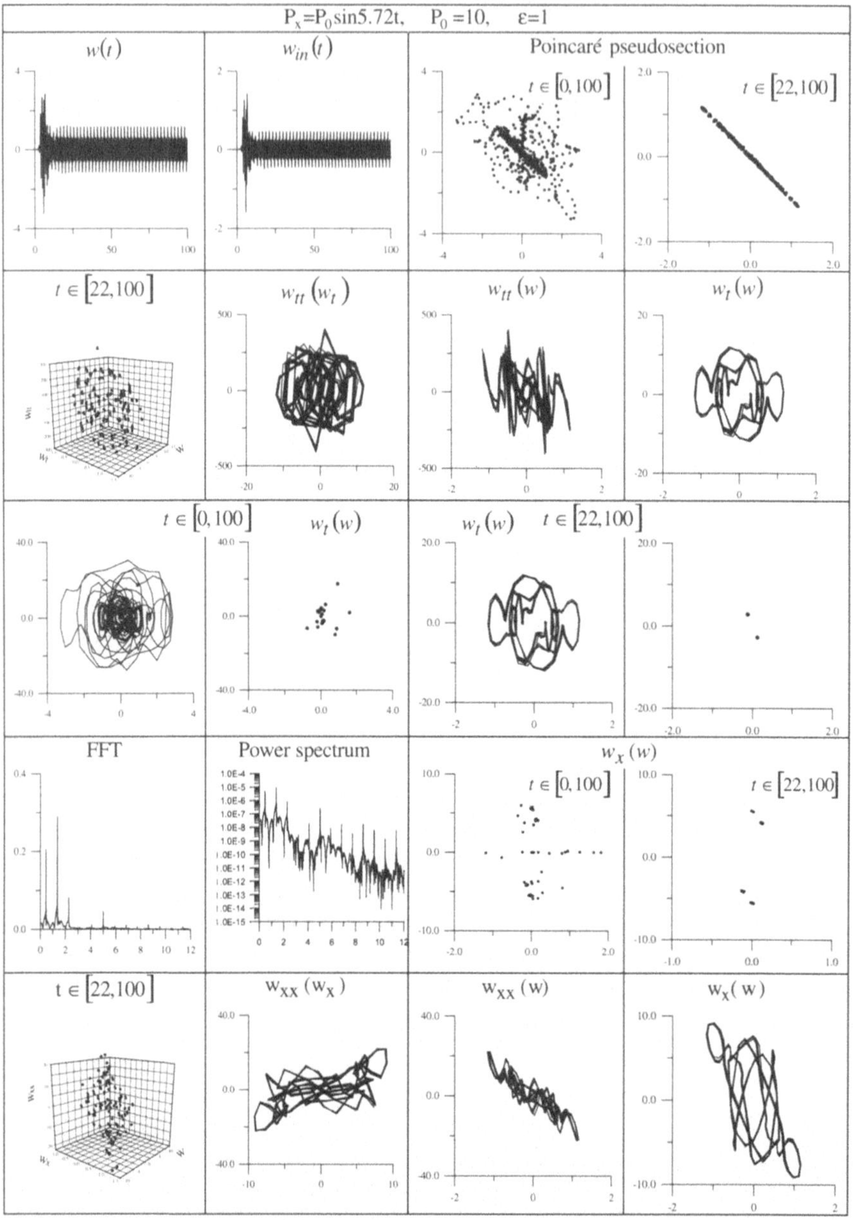

Fig. 7.101 Time histories, Poincaré sections and pseudosections, modal portraits, FFT and power spectrum ($P_0 = 10$)

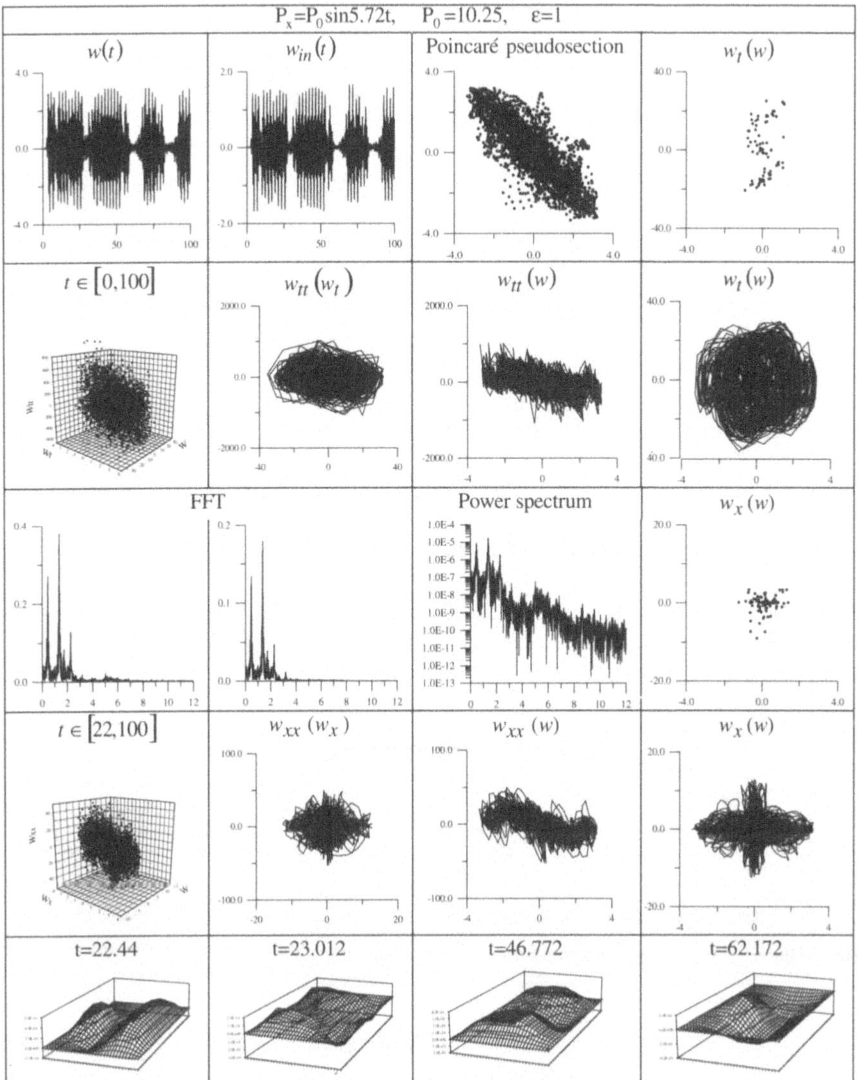

Fig. 7.102 Time histories, phase portraits, Poincaré sections and pseudosections, modal portraits, FFTs, power spectrum and spatial configurations ($P_0 = 10.25$)

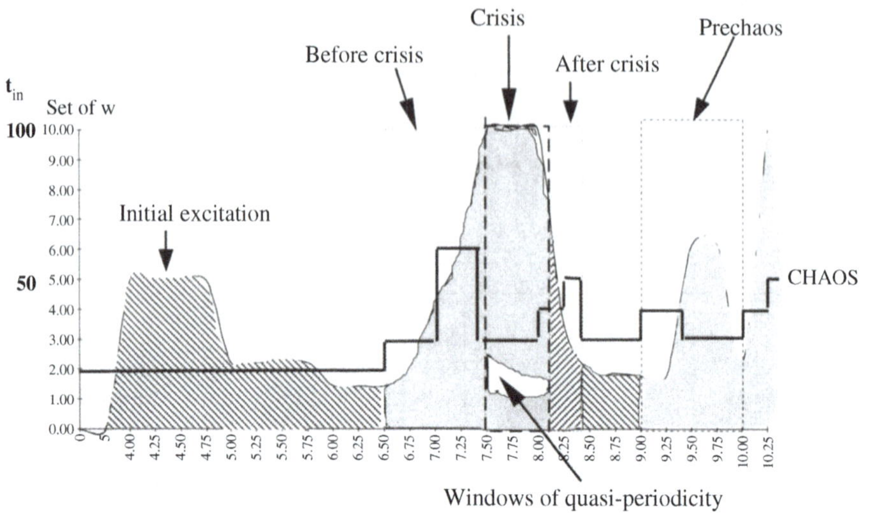

Fig. 7.103 Scenario leading to chaos versus P_0 (nonsymmetric case)

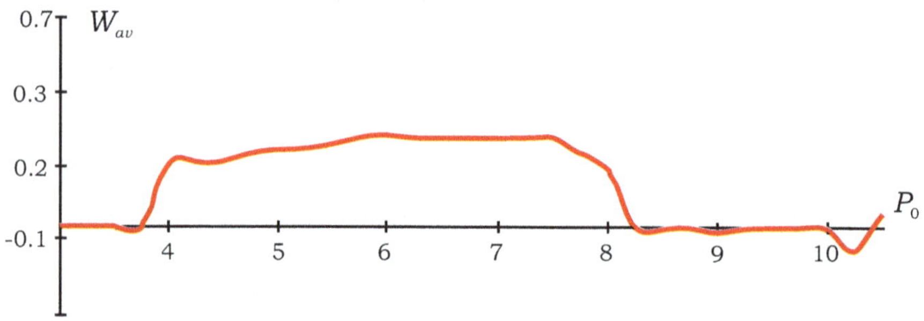

Fig. 7.104 w_{av} versus P_0 (nonsymmetric case); see text

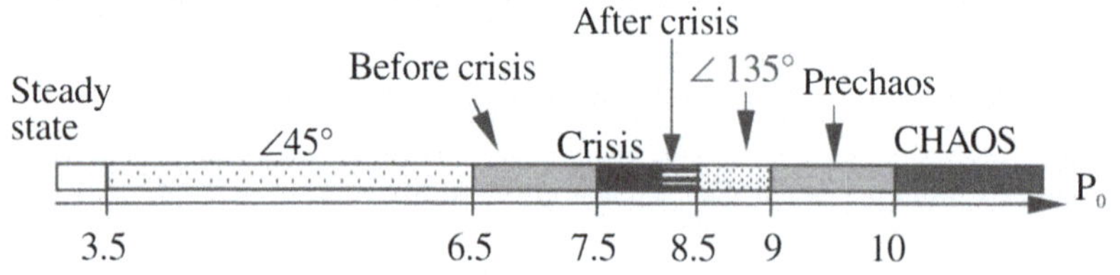

Fig. 7.105 Dynamics versus P_0 ($\omega = 5.72$) (nonsymmetric case)

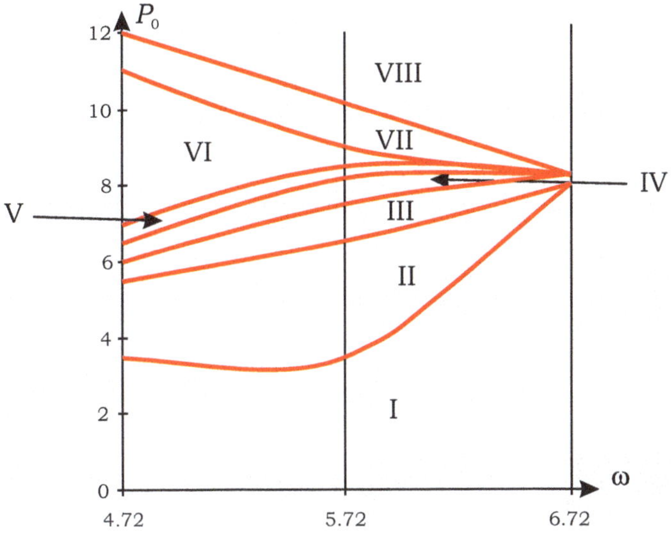

Fig. 7.106 Various dynamical states within the two-parameter space (nonsymmetric case); see text for a detailed description

the characteristics presented. Travelling bending waves are observed, shown in Figs. 7.109 and 7.110 for selected time moments, and the maximum deflections occur in the direction of two mutually perpendicular symmetry axes. The travelling waves become a standing wave (Fig. 7.111, $t = 10.04$, $P_0 = 19$), and then a jump to another deflection level occurs, as indicated by the characteristics shown in Fig. 7.110. At the centre of the plate, the effect of a discontinuity mentioned above is observed in the phase and modal portraits. The space phase and modal portraits and their projections into the planes show that chaos occurs (as indicated by the broadband character of the power spectrum). But a collapse of the standing waves starting from $t = 10.04$, $P_0 = 9$ does not appear. A new type of wave self-organization occurs in the standing wave shown in Fig. 7.110 at $t = 20$. This type of wave does not change with time and is practically stable over $t \in [10.04; 100]$. This observation leads to the conclusion that, within a chaotic state, a new self-organized behaviour, which we call two-dimensional standing solitary waves, can appear.

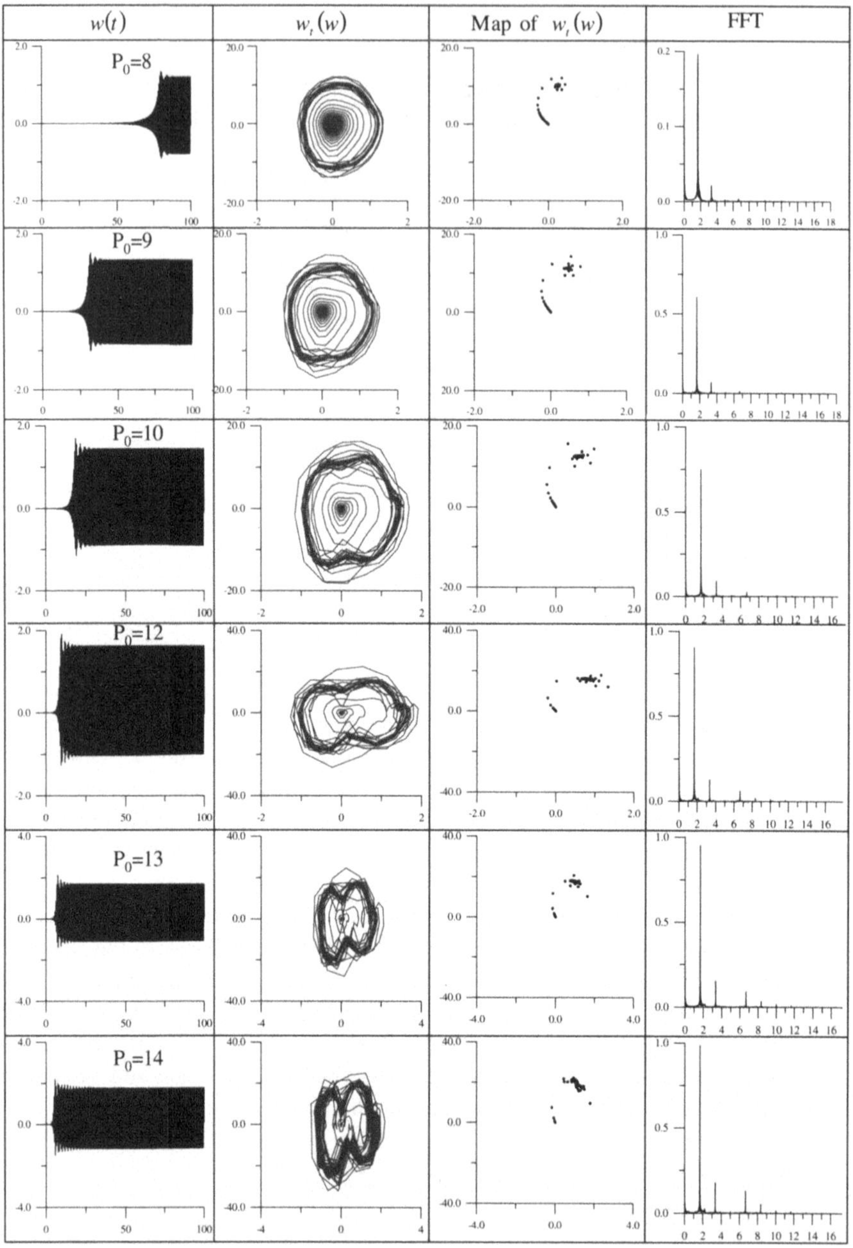

Fig. 7.107 Time histories of w_{in} and w_{av} for different values of P_0 (compare with Fig. 7.96)

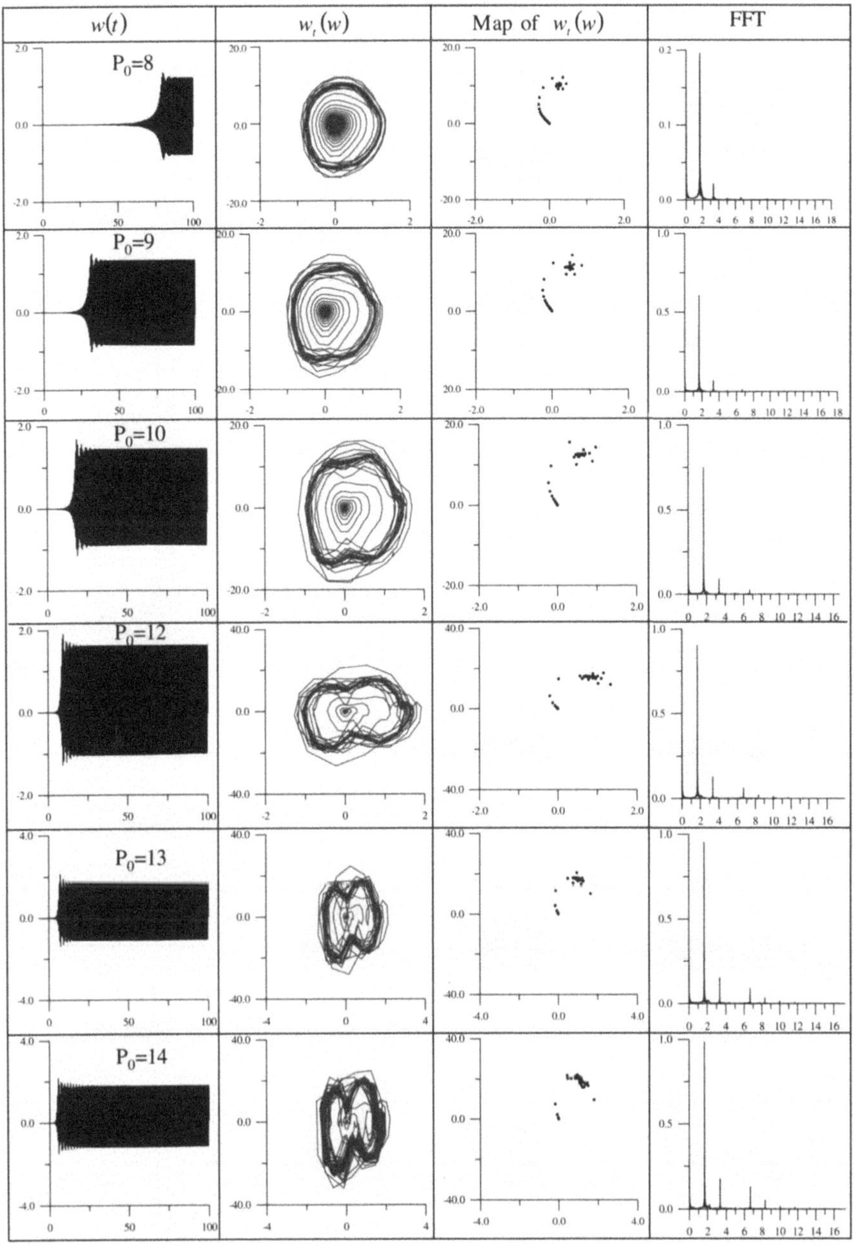

Fig. 7.108 Time histories of $w(t)$, phase portraits and Poincaré maps of $w_t(w)$, and FFTs for different P_0 values

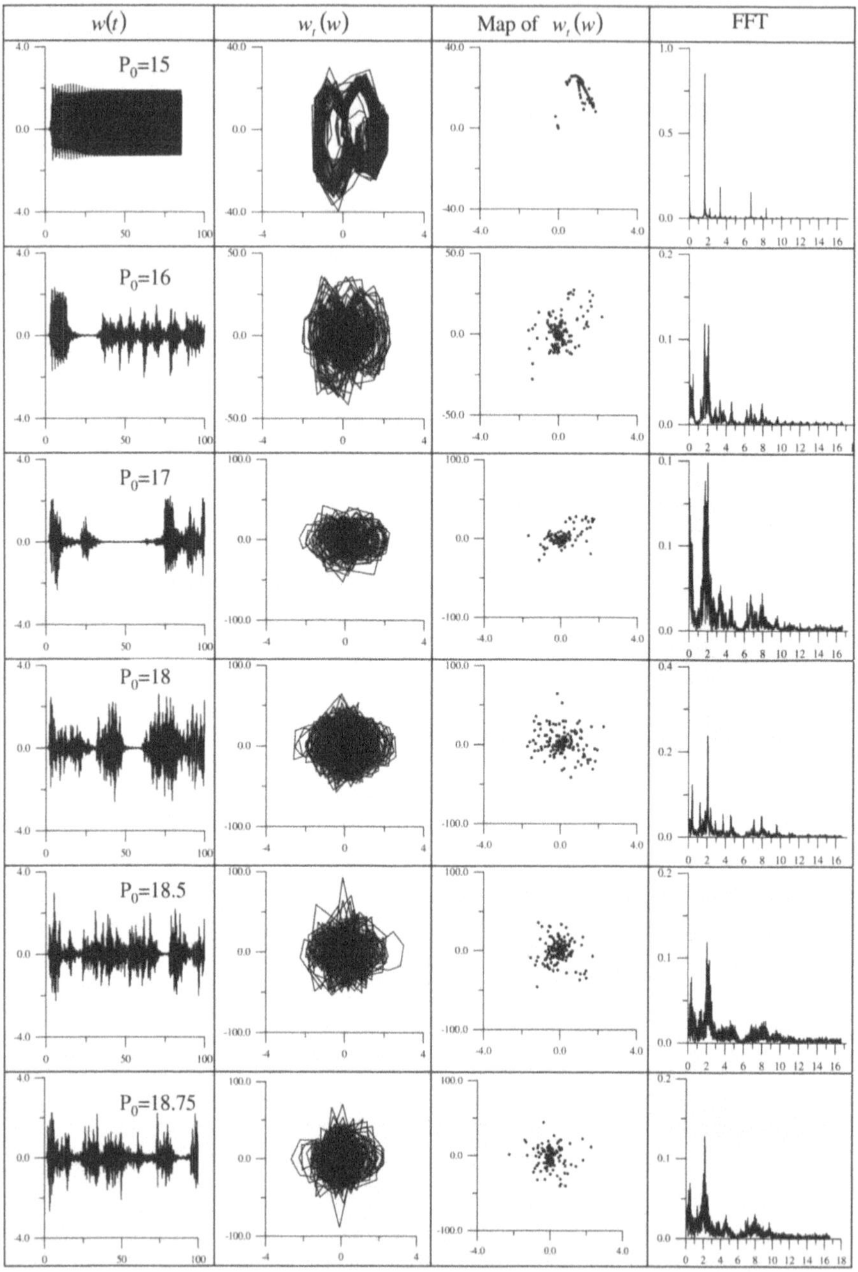

Fig. 7.108 (continued)

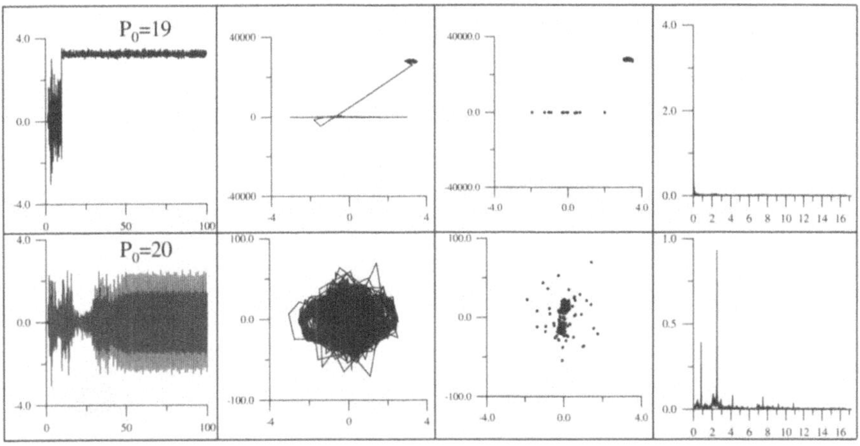

Fig. 7.108 (continued)

A periodically excited flexible plate is a complex dynamical system which can exhibit various dynamical behaviours. The vibrational process can be characterized by complex resonance structures, a collapse of a vibrational regime leading to a change of the spatial–temporal state, the occurrence of standing or travelling waves, stability loss with respect to either symmetric or nonsymmetric modes, and other features. This various features have been discussed and illustrated in this section.

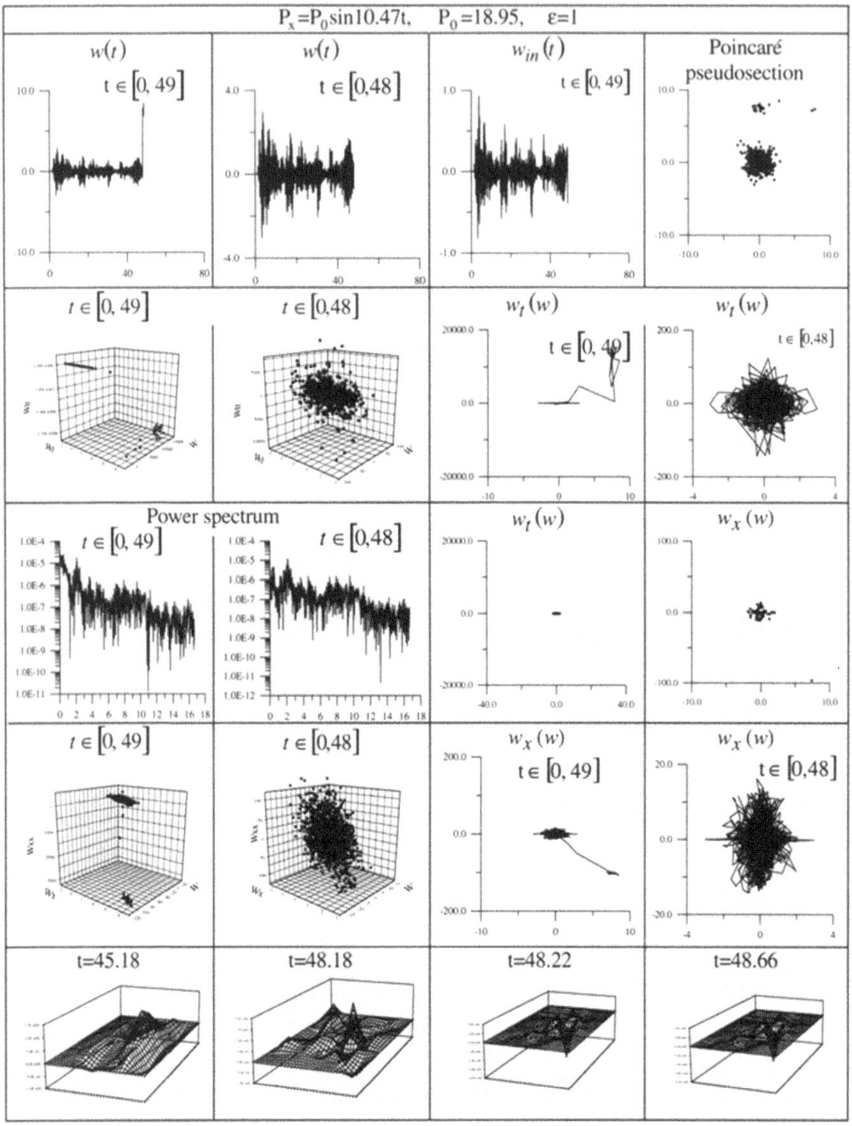

Fig. 7.109 Time histories of $w(t)$ and $w_{in}(t)$, phase portraits and Poincaré maps of $w_t(w)$, power spectra, modal portraits, and spatial configurations for the time moments and intervals indicated

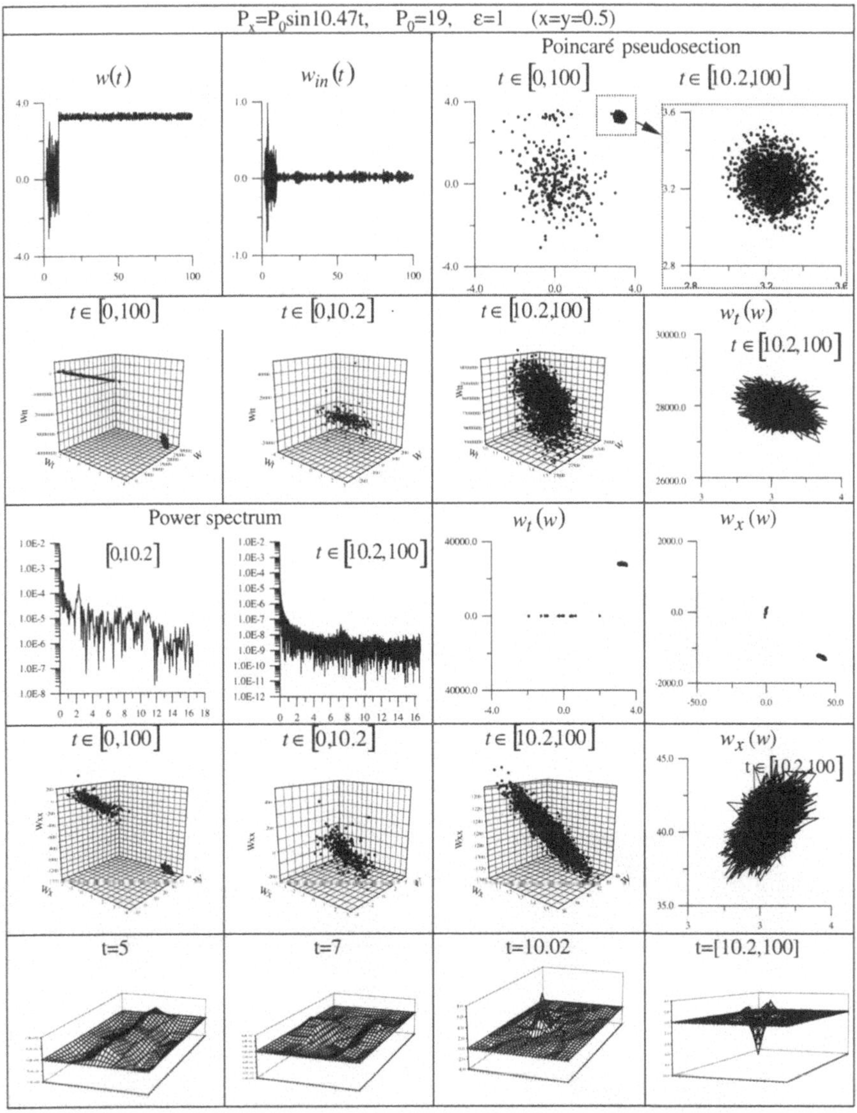

Fig. 7.110 Time histories of $w(t)$ and $w_{in}(t)$, Poincaré pseudosections, modal portraits, power spectra, and spatial configurations for the time moments and intervals indicated

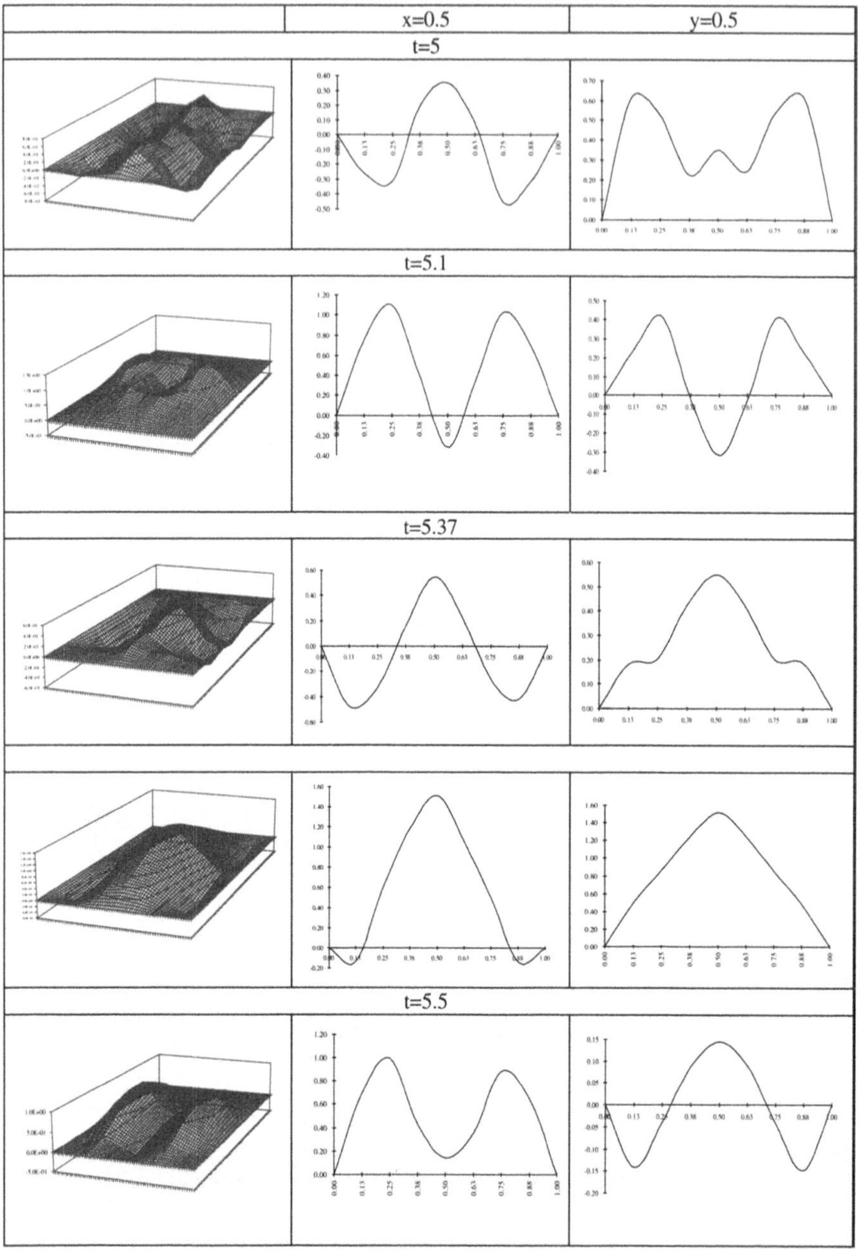

Fig. 7.111 Spatial configurations and cross-sections for the times and values of x and y indicated (see text)

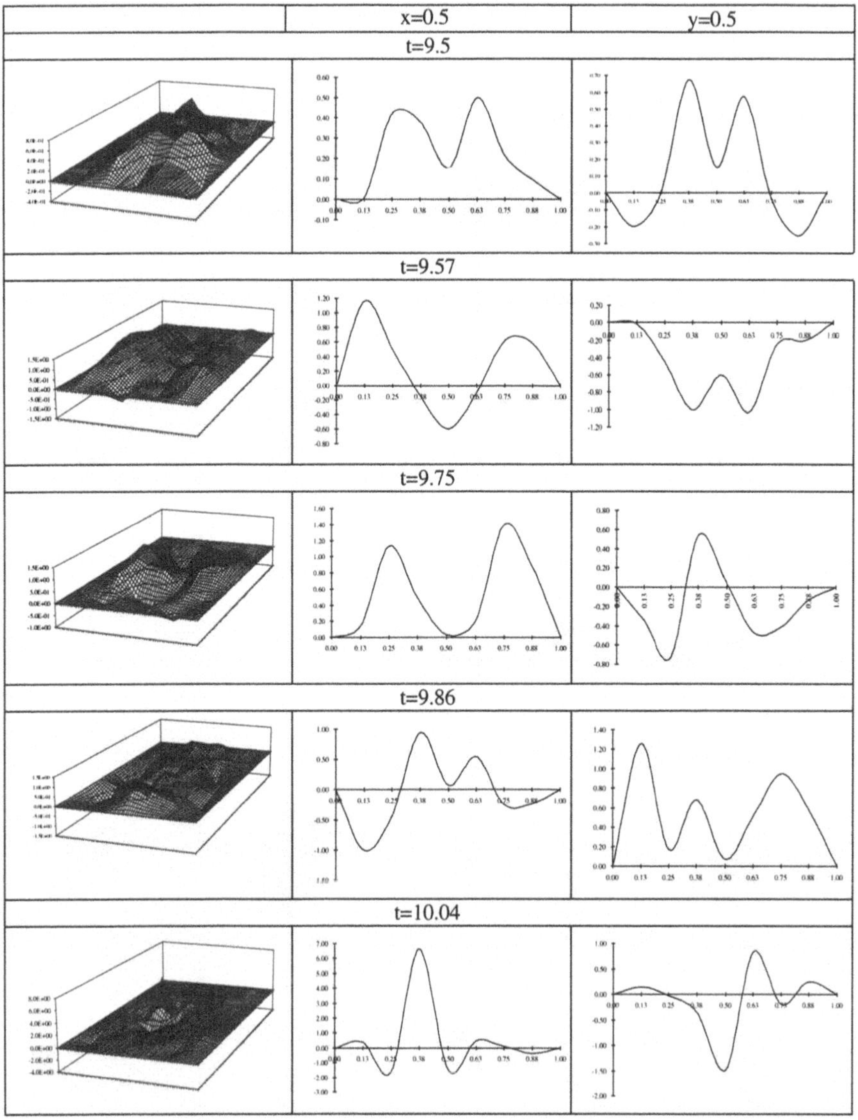

Fig. 7.111 (continued)

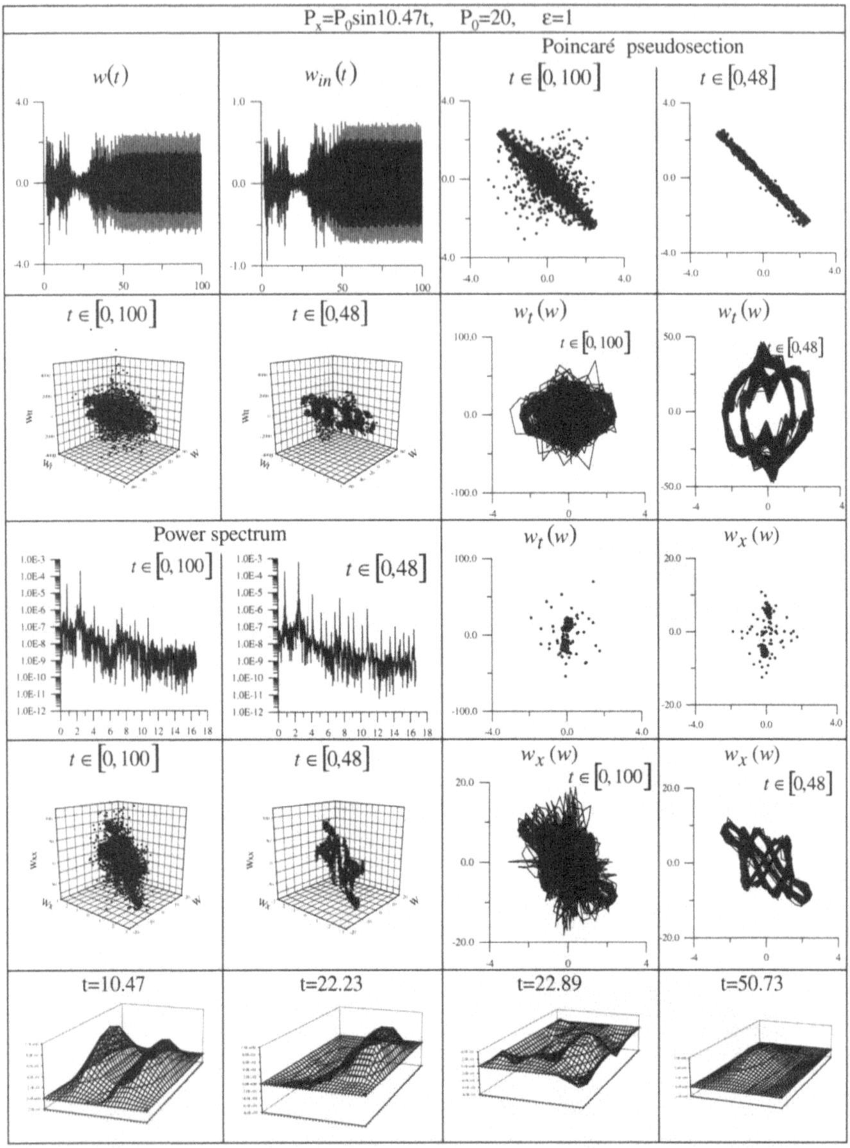

Time histories of $w(t)$ and $w_{in}(t)$, Poincaré pseudosections, phase portraits and Poincaré sections of $w_t(w)$, power spectra, modal portraits and maps, and spatial configurations for the time moments indicated

Chapter 8
Dynamics of Thin Elasto-Plastic Shells

In Chap. 6, a mathematical model governing the oscillations of a flexible shell with physical nonlinearity and coupling between the thermal and deformation fields has been presented. However, the loading and unloading processes overlap in the $\sigma_i(e_i)$ diagram, and the remaining elasto-plastic deformation has not been taken into account. In a theoretical treatment of complicated shell oscillations, in order to analyse the stress–strain state properly, the elasto-plastic deformation should be considered, as well as the fatigue behaviour of the material. Only in this case can the mathematical model be close to the real behaviour of a structure. The aim of this chapter is to describe the development of complex mathematical models of structures, including elasto-plastic deformation and cyclic loading.

The fundamental relations of the theory of small elastic–plastic deformations with a cyclic load are formulated in Sect. 8.1. In Sect. 8.2, a method of solution of the equations governing the elasto–plastic deformation of flexible shells is described. The oscillations and stability of elasto-plastic shells are considered in Sect. 8.3, where many computational examples are reported.

8.1 Fundamental Relations

We present below the fundamental relations of the theory of small elastic–plastic deformations when a cyclic load is applied, using the Maugin principle as generalized by Moskvitin [254, 255]:

$$\begin{aligned} \sigma_{11} &= 3Ke_0 + \frac{2}{3}\frac{\sigma_i}{e_i}(e_i - e_0), \quad (\overleftrightarrow{1, 2, 3}) \\ \sigma_{12} &= \frac{2}{3}\frac{\sigma_i}{e_i}e_{12}. \end{aligned} \tag{8.1}$$

J. Awrejcewicz, V. A. Krysko, *Elastic and Thermoelastic Problems in Nonlinear Dynamics of Structural Members*, Scientific Computation,
https://doi.org/10.1007/978-3-030-37663-5_8

Here (as earlier), K, e_0 and ν correspond to the modulus of volume compression, the average deformation and Poisson's ratio;

$$e_i = \left\{\frac{2}{3}\left[(e_{11} - e_0)^2 + (e_{22} - e_0)^2 + (e_{33} - e_0)^2 + e_{12}^2 + e_{23}^2 + e_{31}^2\right]\right\}^{1/2}$$

characterizes the intensity of the deformation. We use also the relation $\sigma_i = f(e_i) = 3G(1-\overline{w})e_i$, introduced earlier, where $\overline{w}$ is the Iliushin function. The relation above can be generalized in the case of a cycle load to

$$\overline{\sigma}_i^{(n)} = \alpha_n \sigma_T \Phi' \left(\frac{\overline{e}^{(n)}}{\alpha_n e_T}\right), \tag{8.2}$$

for the nth cycle of deformation. The elasto-plastic properties for an arbitrary number of load cycles are defined by the initial strain diagram. In (8.2),

$$\alpha_n = \alpha_2 (n-1)^{\gamma}, \tag{8.3}$$

where $\gamma < 1$ and α_n is a constant associated with the cycle. If we take $\alpha_2 = 2$ and $\gamma = 0$, then (8.3) describes the properties of a material that is cyclically ideal with respect to the Maugin principle. In this case the formula (8.2) has the following form in the first loading cycle:

$$\sigma_i^{(n)} = \sigma_T \Phi' \left(\frac{e^{(1)}}{e_T}\right). \tag{8.4}$$

For variable loads, we obtain

$$\overline{\sigma}_i^{(n)} = 2\sigma_T \Phi' \left(\frac{\overline{e}^{(n)}}{2e_T}\right). \tag{8.5}$$

Comparing these two relations, one can conclude that the stress intensity in the system of coordinates for the nth cycle is equal to twice the value of the stress intensity during the first loading. The latter corresponds to half the value of the deformation intensity in the initial coordinate system. Using Moskvitin's generalized model related to elastic–plastic unloading, for the nth loading cycle Iliushin's relation can be rewritten in the following form [155]:

$$\begin{aligned} \overline{\sigma}_{11}^{(n)} &= 3K\overline{e}_0^{(n)} + \frac{2}{3}\left(\frac{\overline{\sigma}^{(n)}}{\overline{e}_i^{(n)}}\right)(\overline{e}_{11}^{(n)} - \overline{e}_0^{(n)}), \\ \overline{\sigma}_{12}^{(n)} &= \frac{2}{3}\frac{\overline{\sigma}_i^{(n)}}{\overline{e}_i^{(n)}} \quad (\overleftrightarrow{1, 2, 3}), \end{aligned} \tag{8.6}$$

where

$$\overline{e}_{11}^{(n)} = e_{11}^{(n-1)} - e_{11}^{(n)}, \quad \overline{\sigma}_{11}^{(n)} = \sigma_{11}^{(n-1)} - \sigma_{11}^{(n)} \quad (\overleftrightarrow{1,2,3}),$$
$$\overline{e}_{12}^{(n)} = e_{12}^{(n-1)} - e_{12}^{(n)}, \quad \overline{\sigma}_{12}^{(n)} = \sigma_{12}^{(n-1)} - \sigma_{12}^{(n)}. \tag{8.7}$$

Now $\sigma_{11}^{(n)}, e_{11}^{(n)}, \sigma_{12}^{(n)}, e_{12}^{(n)}$ $(\overleftrightarrow{1,2,1})$ are components of the stress and deformation tensors in the nth loading cycle in the initial coordinate system; $\sigma_{11}^{(n-1)}, e_{11}^{(n-1)}, \sigma_{12}^{(n-1)}, e_{12}^{(n-1)}$ $(\overleftrightarrow{1,2,3})$ are the stress tensor components at the end of the previous, $(n-1)$th loading cycle in the initial coordinate system; and $\overline{\sigma}_{11}^{(n)}, \overline{e}_{11}^{(n)}, \overline{\sigma}_{12}^{(n)}, \overline{e}_{12}^{(n)}$ $(\overleftrightarrow{1,2,3})$ are the components of the stress and deformation tensors in the nth cycle in the moving coordinate system where the origin coincides with the origin at the beginning of the unloading process and where the axes are parallel to those of the initial system of coordinates.

The expressions for the deformation intensity $\overline{e}_i^{(n)}$, the average stress values $\overline{\sigma}_i^{(n)}$ and the average deformation $\overline{e}_0^{(n)}$ have the following forms for a cyclic load:

$$\overline{e}_i^{(n)} = \left\{\frac{2}{3}\left[(\overline{e}_{11}^{(n)} - \overline{e}_0^{(n)})^2 + (\overline{e}_{22}^{(n)} - \overline{e}_0^{(n)})^2 + (\overline{e}_{33}^{(n)} - \overline{e}_0^{(n)})^2 \right.\right.$$
$$\left.\left. + (\overline{e}_{12}^{(n)})^2 + (\overline{e}_{23}^{(n)})^2 + (\overline{e}_{31}^{(n)})^2\right]\right\}^{1/2}, \tag{8.8}$$

$$\overline{\sigma}_0^{(n)} = \frac{1}{3}(\overline{\sigma}_{11}^{(n)} + \overline{\sigma}_{22}^{(n)} + \overline{\sigma}_{33}^{(n)}), \tag{8.9}$$

$$\overline{e}_0^{(n)} = \frac{1}{3}(\overline{e}_{11}^{(n)} + \overline{e}_{22}^{(n)} + \overline{e}_{33}^{(n)}). \tag{8.10}$$

The coupling between the spherical stress and deformation tensors for the nth loading cycle has the form

$$\overline{\sigma}_0^{(n)} = 3K\overline{e}_0^{(n)}. \tag{8.11}$$

When the theory of small elasto-plastic deformations is used, the relation between the stress and the deformation intensity reads:

$$\overline{\sigma}_0^{(n)} = 3G(1 - \overline{w}^{(n)})\overline{e}_i^{(n)}, \tag{8.12}$$

where $\overline{w}^{(n)} = f(\overline{e}_i^{(n)})$ is the Iliushin function [155] for the nth loading cycle, which does not change on moving from one cycle to another. This function depends on the material and on the type of approximation used. The various $\sigma_i(e_i)$ relations are described and discussed in the monograph [198]. We present only the form for $\overline{w}^{(n)}(\overline{e}_i^{(n)})$ related to the bilinear approximation of $\sigma_i(e_i)$ in the moving coordinate system, which will be used later:

$$\overline{w}^{(n)} = \begin{cases} 0 \quad \text{for} \quad \overline{e}_i^{(n)} \leq \alpha_n e_T, \\ \left(1 - \frac{G_2}{G_1}\right)\left(1 - \frac{\alpha_n e_T}{\overline{e}_i^{(n)}}\right) \quad \text{for} \quad \overline{e}_i^{(n)} > \alpha_n e_T, \end{cases} \tag{8.13}$$

where G_1 and G_2 are the shear and tangent moduli of elasticity.

The Kirchhoff–Love hypothesis will be used to describe the shell structure in what follows. This hypothesis assumes that $\sigma_{33} = 0$ and $e_{23} = e_{31} = 0$ (it is also possible to obtain equations using the Timoshenko and Reussner kinematic hypotheses, and others).

For the Kirchhoff–Love kinematic model, the relations (8.6) and (8.7) for the nth loading cycle read

$$\begin{aligned} e_{33}^{(n)} &= e_{33}^{(n-1)} + \frac{1 + \nu - (1 - 2\nu)(1 - \overline{w}^{(n)})}{1 + \nu + (1 - 2\nu)(1 - \overline{w}^{(n)})}(e_{11}^{(n-1)} \\ &+ e_{22}^{(n-1)} + e_{11}^{(n)} + e_{22}^{(n)}), \end{aligned} \tag{8.14}$$

where $e_{33}^{(n)}$ is obtained by assuming a plane stress–strain state ($\sigma_{33} = 0$). This quantity appears in the expression for the deformation intensity, together with other components of the deformation tensor:

$$\begin{aligned} \sigma_{11}^{(n)} &= \frac{E(1 - \overline{w}^{(n)})}{(1 + \nu)(1 - \overline{\beta}^{(n)})}(e_{11}^{(n)} + \overline{\beta}^{(n)} e_{22}^{(n)}) + \sigma_{11}^{(n-1)} \\ &- \frac{E(1 - \overline{w}^{(n)})}{(1 + \nu)(1 - \overline{\beta}^{(n)})}(e_{11}^{(n-1)} + \overline{\beta}^{(n)} e_{22}^{(n-1)}) \quad (1 \leftrightarrow 2), \\ \sigma_{12} &= \frac{E(1 - \overline{w}^{(n)})}{2(1 + \nu)} e_{12}^{(n)} + \overline{\sigma}_{12}^{(n-1)} - \frac{E(1 - \overline{w}^{(n)})}{2(1 + \nu)} e_{11}^{(n-1)}, \end{aligned} \tag{8.15}$$

where

$$\overline{\beta}^{(n)} = \frac{1}{2}\left(\frac{3 - 2\alpha(1 - \overline{w}^{(n)})}{3 + 2\alpha(1 + \overline{w}^{(n)})}\right), \quad \alpha = \frac{3}{2}\frac{1 - 2\nu}{1 + \nu}. \tag{8.16}$$

By integrating the stresses (8.15) with respect to $z\left(-\frac{h}{2} \le z \le \frac{h}{2}\right)$, we obtain the following stresses in the middle surface:

$$
\begin{aligned}
T_{11}^{(n)} &= \frac{Eh}{1-\nu^2}(\varepsilon_{11}^{(n)} + \nu\varepsilon_{22}^{(n)}) + \Delta T_{11}^{(n)} \quad (1 \leftrightarrow 2),\\
T_{12}^{(n)} &= \frac{Eh}{2(1+\nu)}\varepsilon_{12}^{(n)} + \Delta T_{12}^{(n)},
\end{aligned}
\tag{8.17}
$$

where

$$
\begin{aligned}
\Delta T_{11}^{(n)} &= \frac{Eh}{1+\nu}(\overline{A}_1^{(n)}\,\varepsilon_{11}^{(n)} + \overline{B}_1^{(n)}\,\varepsilon_{22}^{(n)})\\
&+ \frac{Eh^2}{1+\nu}(\overline{A}_2^{(n)}\,æ_{11} + \overline{B}_2^{(n)}\,æ_{22}) + \int\limits_{-\frac{h}{2}}^{\frac{h}{2}} \sigma_{11}^{(n-1)}\,dx_3\\
&- \frac{E}{1+\nu}\int\limits_{-\frac{h}{2}}^{\frac{h}{2}} \frac{1-\overline{w}^{(n)}}{1-\overline{\beta}^{(n)}}(e_{11}^{(n-1)} + \overline{\beta}^{(n)}\,e_{22}^{(n-1)})\,dx_3 \quad (1 \leftrightarrow 2),\\
\Delta T_{12}^{(n)} &= \frac{Eh}{2(1+\nu)}\overline{C}_1^{(n)}\,\varepsilon_{12}^{(n)} + \frac{Eh^3}{1+\nu}\overline{C}_2^{(n)}\,æ_{12}^{(n)}\\
&+ \int\limits_{-\frac{h}{2}}^{\frac{h}{2}} \sigma_{12}^{(n-1)}\,dx_3 - \frac{E}{2(1+\nu)}\int\limits_{-\frac{h}{2}}^{\frac{h}{2}} (1-\overline{w}^{(n)})e_{12}^{(n-1)}\,dx_3.
\end{aligned}
\tag{8.18}
$$

In (8.18), the following notation has been used:

$$
\begin{aligned}
&\overline{F}_1^{(n)} = \frac{1-\overline{w}^{(n)}}{1-\overline{\beta}^{(n)}} - \frac{1}{1-\nu}, \quad \overline{F}_2^{(n)} = \frac{1-\overline{w}^{(n)}}{1-\overline{\beta}^{(n)}}\overline{\beta}^{(n)} - \frac{1}{1-\nu},\\
&\overline{A}_m^{(n)} = \frac{1}{n^j}\int\limits_{-\frac{h}{2}}^{\frac{h}{2}} \overline{F}_1^{(n)}\,x_3^{(j-1)}\,dx_3, \quad \overline{B}_m^{(n)} = \frac{1}{n^j}\int\limits_{-\frac{h}{2}}^{\frac{h}{2}} \overline{F}_2^{(n)}\,x_3^{(j-1)}\,dx_3,\\
&\overline{C}_m^{(n)} = \frac{1}{n^j}\int\limits_{-\frac{h}{2}}^{\frac{h}{2}} (-\overline{w}^{(n)})x_3^{(j-1)}\,dx_3 \quad (m = 1, 2, 3).
\end{aligned}
\tag{8.19}
$$

Multiplying $\sigma_{11}^{(n)}$, $\sigma_{22}^{(n)}$, $\sigma_{12}^{(n)}$ in (8.15) by x_3 and integrating over the thickness, we obtain the following moments:

$$
\begin{aligned}
M_{11}^{(n)} &= D(æ_{11}^{(n)} + \nu æ_{22}^{(n)}) + \Delta M_{11}^{(n)} \quad (1 \leftrightarrow 2),\\
M_{12}^{(n)} &= Dæ_{12}^{(n)} + \Delta M_{12}^{(n)},
\end{aligned}
\tag{8.20}
$$

where

$$D = \frac{Eh^3}{12(1-\nu^2)}$$

is the stiffness of the plate,

$$
\begin{aligned}
\Delta M_{11}^{(n)} &= \frac{Eh^2}{1+\nu}(\overline{A}_2^{(n)}\,\varepsilon_{11}^{(n)} + \overline{B}_2^{(n)}\,\varepsilon_{22}^{(n)})\\
&\quad + 12(1-\nu)D(\overline{A}_3^{(n)}\,æ_{11}^{(n)} + \overline{B}_3^{(n)}\,æ_{22}^{(n)}) + \int\limits_{-\frac{h}{2}}^{\frac{h}{2}} \sigma_{11}^{(n-1)}\,x_3\,dx_3\\
&\quad - \frac{E}{1+\nu}\int\limits_{-\frac{h}{2}}^{\frac{h}{2}} \frac{1-\overline{w}^{(n)}}{1-\overline{\beta}^{(n)}}(e_{11}^{(n-1)} + \overline{\beta}^{(n)}\,e_{22}^{(n-1)})x_3\,dx_3 \quad (1 \leftrightarrow 2),
\end{aligned}
\tag{8.21}
$$

$$
\begin{aligned}
\Delta M_{12}^{(n)} &= \frac{Eh^2}{2(1+\nu)}\overline{C}_2^{(n)}\,\varepsilon_{12}^{(n)} + 12(1-\nu)D\overline{C}_3^{(n)}\,æ_{12}^{(n)}\\
&\quad + \frac{E}{2(1+\nu)}\int\limits_{-\frac{h}{2}}^{\frac{h}{2}} (1-\overline{w}^{(n)})e_{12}^{(n-1)}\,x_3\,dx_3.
\end{aligned}
$$

The terms ΔN_{ij} and ΔM_{ij} $(i, j = 1, 2)$ include the final plastic deformations.

Similarly to the approach presented in Sect. 6.1, we solve the Eqs. (8.17) for the deformations $\varepsilon_{ij}^{(n)}$ $(i, j = 1, 2)$ of the middle surface, and we obtain

$$
\begin{aligned}
\varepsilon_{11}^{(n)} &= \frac{1}{Eh}\left[(T_{11}^{(n)} - \nu T_{22}^{(n)}) - (\Delta T_{11}^{(n)} - \nu\Delta T_{22}^{(n)})\right] \quad (1 \leftrightarrow 2),\\
\varepsilon_{12}^{(n)} &= \frac{2(1+\nu)}{Eh}(T_{12}^{(n)} - \Delta T_{12}^{(n)}).
\end{aligned}
\tag{8.22}
$$

Using the variational equations introduced in Sect. 6.2 and the earlier described methodology of derivation of equations, we obtain (analogously to the differential equations (6.51)) the following equations governing the dynamics of an elasto-plastic flexible shell:

$$\begin{aligned}
&\frac{1}{Eh}\nabla^2\nabla^2 F = -\nabla_k^2 w - \frac{1}{2}L(w,w) + \frac{1}{Eh}[(\Delta T_{11}^{(n)} - \nu\Delta T_{22}^{(n)})_{x_1x_2} \\
&\qquad + (\Delta T_{22}^{(n)} - \nu\Delta T_{11}^{(n)})_{x_1x_1} - 2(1+\nu)(\Delta T_{12}^{(n)})_{x_1x_1}], \\
&D\nabla^2\nabla^2 w - L(w,F) - \nabla_k^2 F - (\Delta M_{11}^{(n)})_{x_1x_1} - (\Delta M_{22}^{(n)})_{x_2x_2} \\
&\qquad - 2(\Delta M_{12}^{(n)})_{x_1x_2} - q + \rho h(\ddot{w} + \varepsilon\dot{w}) = 0.
\end{aligned} \tag{8.23}$$

The Airy function is introduced in the following way:

$$N_{11}^{(n)} = \frac{\partial^2 F}{\partial x_2{}^2}, \quad N_{22}^{(n)} = \frac{\partial^2 F}{\partial x_1{}^2}, \quad N_{12}^{(n)} = \frac{\partial^2 F}{\partial x_1 \partial x_2}. \tag{8.24}$$

Analogously to Sect. 6.3, we obtain the following differential equations for the displacements when we take account of elastic–plastic deformation:

$$\begin{aligned}
&\rho_0\frac{h}{g}\ddot{u}_1 = \Omega_1 + \Delta_1^{(n)}, \\
&\rho_0\frac{h}{g}\ddot{u}_2 = \Omega_2 + \Delta_2^{(n)}, \\
&\rho\frac{h}{g}(\ddot{w} + \varepsilon\dot{w}) = q - D\nabla^2\nabla^2 w + \Omega_3 + \Delta_3^{(n)},
\end{aligned} \tag{8.25}$$

where the following notation has been used:

$$\begin{aligned}
\Omega_1 = &\frac{Eh}{1-\nu^2}\left[\frac{\partial^2 u_1}{\partial x_1{}^2} + \frac{1-\nu}{2}\frac{\partial^2 u_1}{\partial x_2{}^2} + \frac{1+\nu}{2}\left(\frac{\partial^2 u_2}{\partial x_1 \partial x_2}\right.\right. \\
&\left. + \frac{\partial w}{\partial x_2}\frac{\partial^2 w}{\partial x_1 \partial x_2}\right) + \frac{\partial w}{\partial x_1}\left(\frac{\partial^2 w}{\partial x_1{}^2} + \frac{1-\nu}{2}\frac{\partial^2 w}{\partial x_2{}^2}\right) \\
&\left. -(k_1 + \nu k_2)\frac{\partial^2 w}{\partial x_1{}^2}\right] \quad (1 \leftrightarrow 2),
\end{aligned} \tag{8.26}$$

$$\Omega_3 = \frac{Eh}{1-\nu^2}\left\{\left[\frac{\partial u_1}{\partial x_1} + \nu\frac{\partial u_1}{\partial x_2} + \frac{1}{2}\left[\left(\frac{\partial w}{\partial x_1}\right)^2\right.\right.\right.$$

$$\left.\left.+\nu\left(\frac{\partial w}{\partial x_2}\right)^2\right] - w(k_1+\nu k_2)\right]\left(k_1 + \frac{\partial^2 w}{\partial x^2}\right)$$

$$+\left[\frac{\partial u_2}{\partial x_2} + \nu\frac{\partial u_1}{\partial x_1} + \frac{1}{2}\left[\left(\frac{\partial w}{\partial x_2}\right)^2 + \nu\left(\frac{\partial w}{\partial x_1}\right)^2\right]\right.$$

$$\left.-w(k_2+\nu k_1)\right]\left(k_2 + \frac{\partial^2 w}{\partial {x_2}^2}\right) + (1-\nu)\left(\frac{\partial u_2}{\partial x_1}\right.$$

$$\left.\left.+\frac{\partial u_1}{\partial x_2} + \frac{\partial w}{\partial x_1}\frac{\partial u_2}{\partial x_2}\right)\frac{\partial^2 w}{\partial x_1 \partial x_2}\right\}, \tag{8.27}$$

$$\Delta_1^{(n)} = (\Delta T_{11}^{(n)})_{x_1} + (\Delta T_{12}^{(n)})_{x_2} \quad (1 \leftrightarrow 2), \tag{8.28}$$

$$\Delta_3^{(n)} = \left(k_1 + \frac{\partial^2 w}{\partial {x_1}^2}\right)\Delta T_{11}^{(n)} + \left(k_2 \frac{\partial^2 w}{\partial {x_2}^2}\right)\Delta T_{22}^{(n)}$$
$$+\frac{\partial^2 w}{\partial x_1 \partial x_2}\Delta T_{12}^{(n)} + (\Delta M_{11}^{(n)})_{x_1 x_2} + 2(\Delta M_{12}^{(n)})_{x_1 x_2} + (\Delta M_{22}^{(n)})_{x_2 x_2}. \tag{8.29}$$

The boundary conditions are formulated in a way similar to that to the described in Sect. 6.3. We need only to replace the deformations ε_{ij} by $\varepsilon_{ij}^{(n)}$, replace the moments M_{ij} by $M_{ij}^{(n)}$ and replace the loads on the middle surface T_{ij} by $T_{ij}^{(n)}$. The initial conditions used in Chap. 6 can now be applied.

8.2 Method of Solution

Similarly to the methodology used in Chap. 6, the hybrid partial differential equations obtained above are reduced to ODEs with respect to time, with the help of a finite-difference method with respect to the spatial coordinates x_1, x_2 that uses an $O(h^2)$ approximation.

The system of equations (8.23), after application of the difference operators (7.6) and the substitution of variables $\frac{dw}{dt} = \dot{w}$, is reduced to a system of first-order ODEs and a system of linear algebraic equations similar to (7.8) and (7.9):

$$\frac{dw_{ij}}{dt} = \dot{w}_{ij},$$

$$\frac{d\dot{w}_{ij}}{dt} + \varepsilon \dot{w}_{ij} = -\frac{1}{æ}\Big[A(w) + B(w, F) + C(\Delta M_{11}^{(n)}, \Delta M_{22}^{(n)}, \Delta M_{12}^{(n)})\Big], \tag{8.30}$$

$$D(F_{ij}) = E(w_{ij}) + G(\Delta T_{11}^{(n)}, \Delta T_{22}^{(n)}, \Delta T_{12}^{(n)}). \tag{8.31}$$

The finite-difference operators appearing in (8.30) and (8.31) are similar to those in (7.10). One must simply replace N_T, $\Delta_T T$, M_T, $\Delta_T M$ by zero and replace the operators $\Delta T_{11}, \Delta T_{22}, \Delta T_{12}, \Delta M_{11}, \Delta M_{22}, \Delta M_{12}$ by $\Delta T_{11}^{(n)}$, $\Delta T_{22}^{(n)}$, $\Delta T_{12}^{(n)}$, $\Delta M_{11}^{(n)}$, $\Delta M_{22}^{(n)}$, $\Delta M_{12}^{(n)}$, respectively. Equations (8.30) and (8.31) are reduced to nondimensional form as in Chap. 7.

The algorithm that we have used for calculation of the elastic–plastic deformations of a flexible shell is similar to that used for geometrically and physically nonlinear problems described in Chap. 6. But this algorithm has some peculiarities and therefore will be considered in some detail.

The stress–strain state of the shell structure is computed for a series of successive steps in time Δt, the magnitude of which is chosen to satisfy requirements on the stability and convergence of the numerical process. In the present case, the conditions of dynamic equilibrium are considered at separate time moments corresponding to the beginning and end of each step.

At the initial time moment ($t_0 = 0$), the initial conditions $w_{ij}|_{t_0=0} = f_1$, $\dot{w}_{ij}|_{t_0=0} = f_2$ are applied. The stress field F_{ij} is then defined by the solution to the linear system of algebraic equations (8.31). The plastic supplements to the stresses $\Delta T_{11}, \Delta T_{22}, \Delta T_{12}$ and to the moments $\Delta M_{11}, \Delta M_{22}, \Delta M_{12}$ are taken to be zero. At the points (i, j, k) of the volume mesh, the deformation values e_{11}, e_{22}, e_{12} and the deformation intensities are equal to zero. We then consider a time step: by solving the system of algebraic equations (8.31) for F_{ij} and solving the ODEs (8.30), we find the values of F_{ij} at the end of the step Δt. Using w_{ij} and F_{ij}, we define the deformations e_{11}, e_{22}, e_{12} and also $æ_{11}, æ_{22}, æ_{12}$ at the points (i, j) of the middle surface. Using the Kirchhoff–Love hypothesis, we obtain the deformations in the shell volume, and also the deformation intensity $e_i|_{t+\Delta t}$ corresponding to the end of the interval Δt. This deformation intensity is compared with the value of e_i^t corresponding to the beginning of the interval. If $e_i|_{t+\Delta t} \leq e_i^t$, i.e. unloading has taken place, then a comparison of $e_i|_{t+\Delta t}$ with $\overline{e}^{(n)}|_t = \alpha_n e_T$ is made. Here two cases are possible: (1) if $e_i|_{t+\Delta t} \leq e_T^{(n)}$, then the shell point (i, j, k) is on an elastic branch of the deflection diagram (this means that the Iliushin function is equal to zero at this point); (2) if $e_i|_{t+\Delta t} > e_T^{(n)}$, then the point (i, j, k) is on the cyclic deformation diagram and the Iliushin function $\overline{w}^{(n)}$ is defined via (8.13), taking into account the number of the deformation cycle.

If $e_i|_{t+\Delta t} > e_i|_t$, then we have also two different cases: (1) for $e_i|_{t+\Delta t} \le e_T^{(n)}$, the deformation proceeds on an elastic branch of the deformation diagram, i.e. the Iliushin function $\overline{w}^{(n)} = 0$; (2) for $e_i|_{t+\Delta t} > e_T^{(n)}$, active loading on the strengthening branch takes place and the Iliushin function $\overline{w}^{(n)}$ is defined by (8.13), taking into account the number of the deformation cycle.

When all $\overline{w}_{ijk}^{(n)}$ have been found, the values of the function $\overline{F}_1^{(n)}$ are computed for all points in the volume of the shell. The values of $\overline{F}_2^{(n)}$ at the points (i, j, k) of the shell volume and the functions $\overline{A}_m^{(n)}, \overline{B}_m^{(n)}, \overline{C}_m^{(n)}$ on the middle surface (i, j) are also computed. Then, through (8.18) and (8.21), the plastic supplements $\Delta T_{11}^{(n)}, \Delta T_{22}^{(n)}, \Delta T_{12}^{(n)}, \Delta M_{11}^{(n)}, \Delta M_{22}^{(n)}, \Delta M_{12}^{(n)}$ to the stresses and moments are found. The stresses and moments themselves are calculated from (8.24) and (8.20). The stresses in the volume (i, j, k) are defined by (8.15). Finally the next step in time is considered out and the procedure is repeated beginning with computation of the stress functions F_{ij} and the deflections w_{ij}.

The method and algorithm described above can be used to perform computations for shell structures with various mathematical models, including models of flexible elastic–plastic constructions (FEPC models), physically and geometrically nonlinear (PGN) models, physically nonlinear (PN) models, geometrically nonlinear (GN) models and linear (L) models. In addition, various $\sigma_i(e_i)$ relations and various boundary and initial conditions can be used. Static and dynamic problems can be solved, and also structures in either stationary or nonstationary fields can be analysed.

As an testing example, we have considered a square, uniformly loaded plate. The $\sigma_i(e_i)$ relation used has been defined earlier. We have taken $e_T = 0.98 \times 10^{-3}$, $\frac{E_2}{E_1} = 0.57735$, $a = b = 0.1$ m, $\frac{a}{h} = 50$. Free support was taken as the boundary condition. Three different mathematical models were analysed: elastic–plastic body (EPB), nonhomogeneous body (NB) and nonlinear elastic body (NEB). Computations for the EPB and NB models were carried out using the method and algorithm described earlier. The results for the NB model were obtained using the method of variable elastic parameters and of variational iterations, in the first approximation [198]. Some results of the computations are given in Table 8.1.

One can conclude, taking into account the results shown in Table 8.1, that all three mathematical models give reliable results.

Table 8.1 Results of computations of deflection in the centre of a square plate for different mathematical models

	Mathematical model			Differences	
	$w(0.5, 0.5)$				
q (uniformly distributed)	EPB	NEB [198]	NB	2 from 3 (%)	2 from 4 (%)
1	2	3	4	5	6
23.4	0.98	0.96	0.95	+2.0	+3.0
60.0	1.97	2.03	2.00	−3.0	−2.0
70.3	2.26	2.16	2.23	+4.0	+1.0

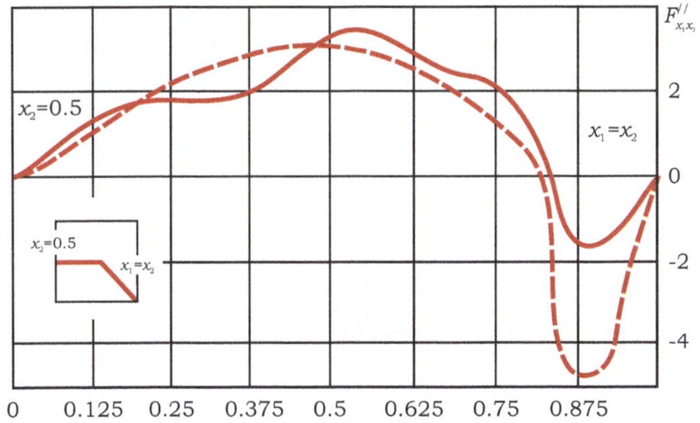

Fig. 8.1 Distribution of $F''_{x_1x_2}$ for $q = 60$, $t = 0.5$ taking into account the unloading cycle (*solid curve*) and without taking into account the unloading cycle (*dashed curve*)

Now we shall describe an investigation of the oscillations of this plate but with a suddenly applied load of infinite duration ($q \in \{23.4; 60.0; 70.3\}$), with the same geometrical and physical parameters. Damping ($\varepsilon = 30$) has been taken into account.

We have analysed the influence of choice of mathematical model on the forces $T_{22} = \frac{\partial^2 F}{\partial x_1{}^2}$ and the curvatures $w''_{x_1x_1}$ along the lines $x_2 = 0.5$, $x_1 \in [0, 0.5]$ and $x_1 = x_2 \in [0.5, 1.0]$. Graphs are shown in Figs. 8.1 and 8.2 for $q = 60.0$ and time $t = 0.5$, and the data listed above. The deformation fields in the corresponding cross-sections are given in Fig. 8.3 both with (Fig. 8.3a) or without (Fig. 8.3b) taking into account the unloading cycle. The various types of hatching used in Fig. 8.3 are explained in Fig. 8.4. An increase of the transverse load greatly changes the

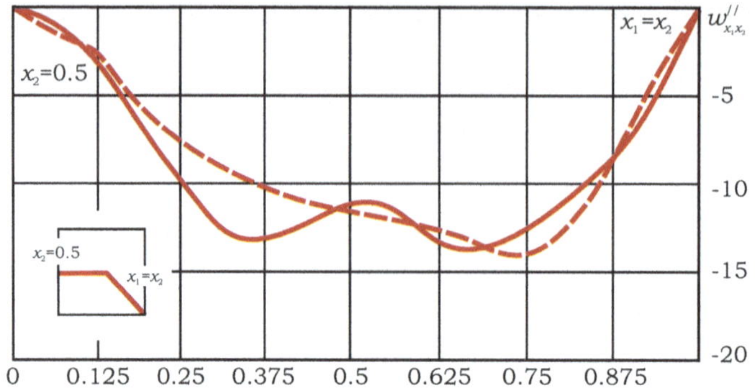

Fig. 8.2 Distribution of $w''_{x_1x_2}$ for $q = 60$, $t = 0.5$ taking into account the unloading cycle (*solid curve*) and without taking into account the unloading cycle (*dashed curve*)

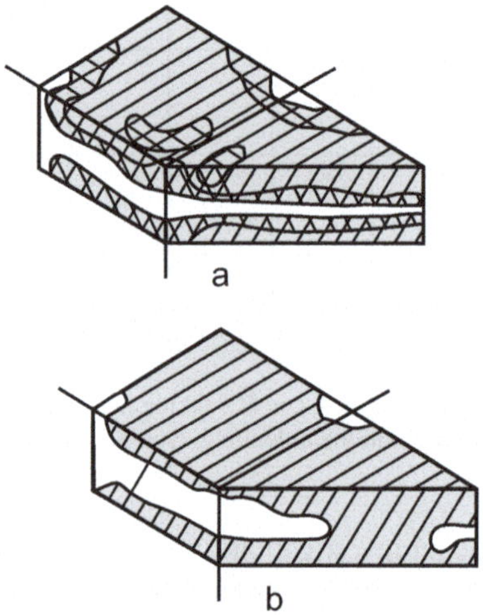

Fig. 8.3 Deformation fields in cross-sections corresponding to Figs. 8.1 and 8.2: (**a**) taking into account the unloading cycle; (**b**) without taking into account the unloading cycle

stress–strain state of the structure. In addition, the differences between the stresses at different times in the middle surface increase greatly, for separate time moments, which causes mixing of elastic and plastic zones.

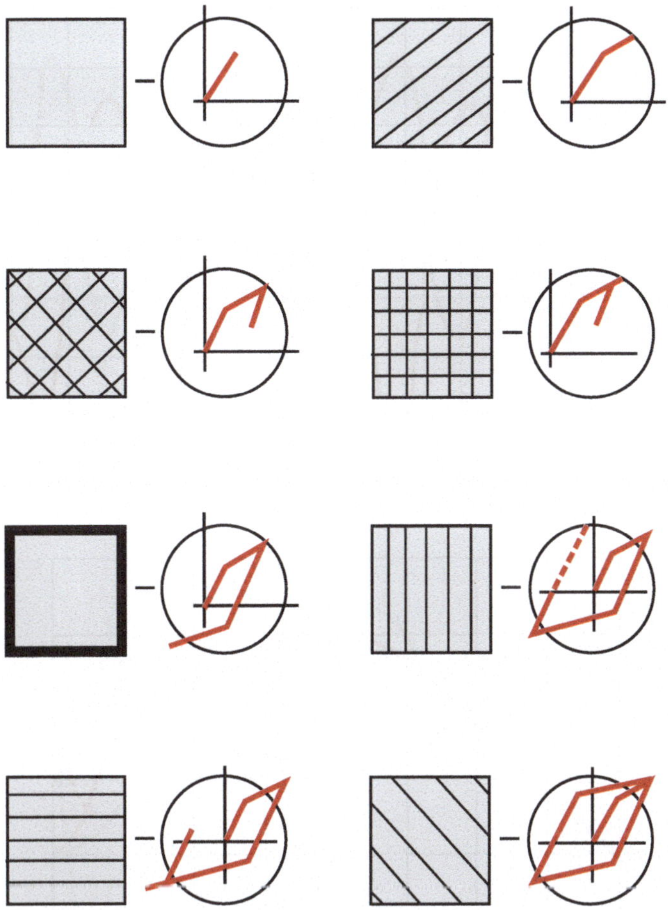

Fig. 8.4 Types of hatching used in Fig. 8.3

8.3 Oscillations and Stability of Elasto-Plastic Shells

We consider the oscillations of a simply supported plate (with the boundary condition (6.60)), with the initial conditions $w|_{t=0} = \dot{w}|_{t=0} = 0$. Two different types of load have been considered:

$$q = const, \tag{8.32}$$

$$q = \begin{cases} q = const, & t < t_1, \\ 2q = const, & t \geq t_1. \end{cases} \tag{8.33}$$

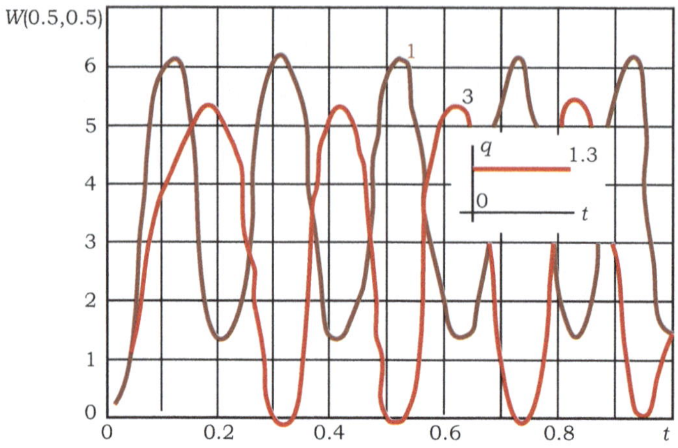

Fig. 8.5 Deflection of the plate centre $w(0.5, 0.5)$ as a function of time for different materials: 1, elastic–plastic material; 3, linear elastic material

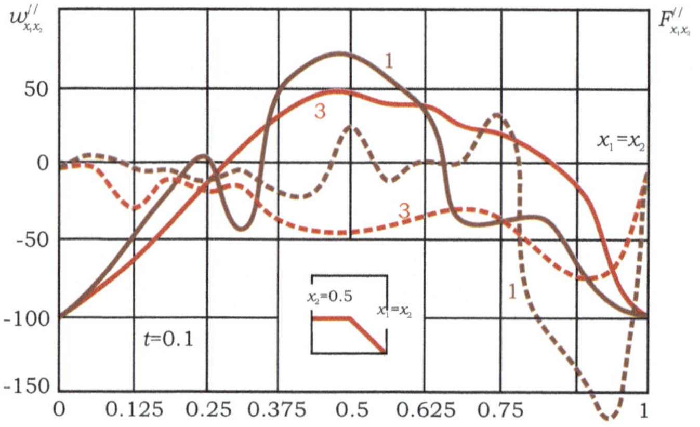

Fig. 8.6 Distribution of $F''_{x_1x_1}$ (*solid lines*) and $w''_{x_1x_1}$ (*dashed lines*) for $q = 503.7$, $t = 0.1$; 1, elastic–plastic material; 3, linear elastic material

The results obtained are presented in nondimensional form in Figs. 8.5, 8.6, 8.7, 8.8, 8.9, 8.10, 8.11, 8.12, 8.13, 8.14, 8.15, 8.16, 8.17, 8.18, 8.19, 8.20, 8.21, 8.22, 8.23, 8.24, 8.25, 8.26, 8.27, 8.28, 8.29, 8.30, 8.31, and 8.32, where the following notation has been used: curves labelled 1 and 2 correspond to an elastic–plastic material with $\frac{a}{h} = 111$ and 50, respectively; curves labelled 3 and 4 ($\frac{a}{h} = 111$) correspond to a linear elastic and an elastic–plastic material, for a step-like load (8.33); and curves labelled 5 correspond to an elastic–plastic material ($\frac{a}{h} = 111$, $\varepsilon = 35$). The numbers 1–5 are also used to refer to the corresponding

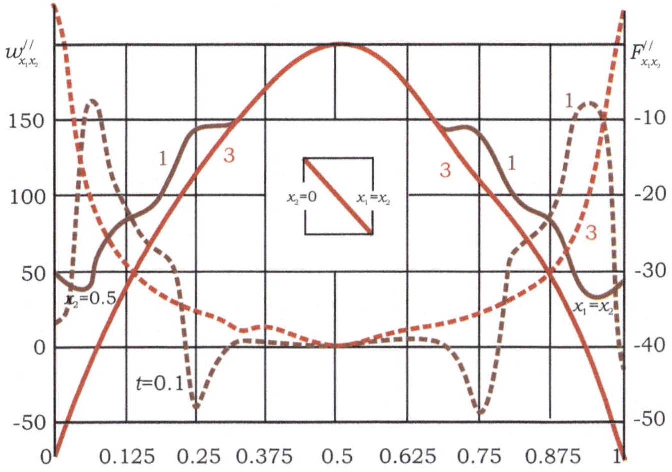

Fig. 8.7 Distribution of $F''_{x_1x_2}$ (*solid lines*) and $w''_{x_1x_2}$ (*dashed lines*) for $q = 503.7$, $t = 0.1$; 1, elastic–plastic material; 3, linear elastic material

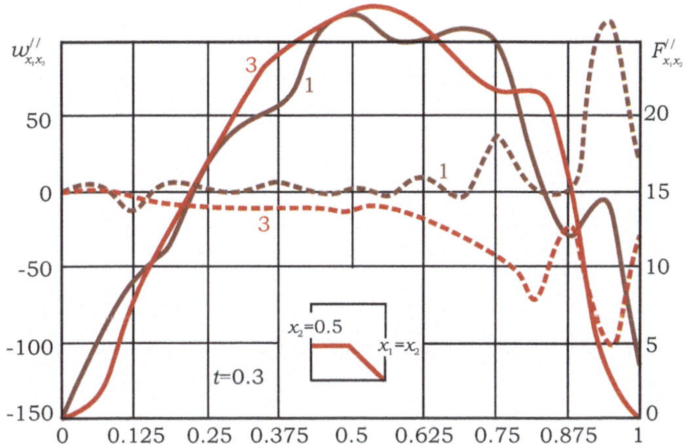

Fig. 8.8 Distribution of $F''_{x_1x_1}$ (*solid lines*) and $w''_{x_1x_1}$ (*dashed lines*) for $q = 503.7$, $t = 0.3$; 1, elastic–plastic material; 3, linear elastic material

models of the behaviour of the material. The following material parameters have been used:

$$
\begin{aligned}
E_1 &= 69\ \text{GPa}, \quad \nu = 0.3, \quad \rho_0 = 28\ \text{kN/m}^3, \\
e_T &= 1.35 \times 10^{-3}, \quad \frac{E_2}{E_1} = 0.4478.
\end{aligned}
\tag{8.34}
$$

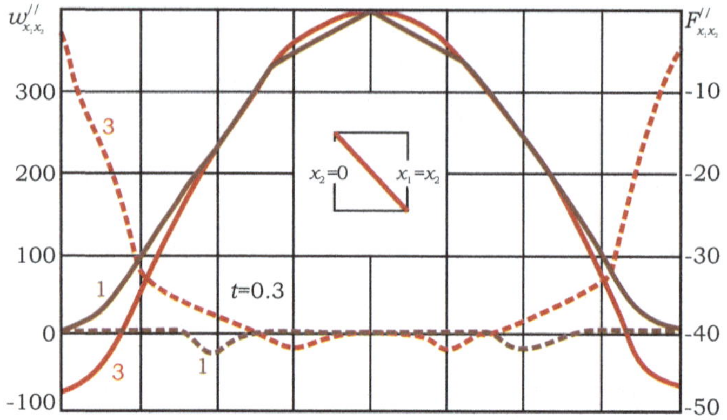

Fig. 8.9 Distribution of $F''_{x_1x_2}$ (*solid lines*) and $w''_{x_1x_2}$ (*dashed lines*) for $q = 503.7$, $t = 0.3$; 1, elastic–plastic material; 3, linear elastic material

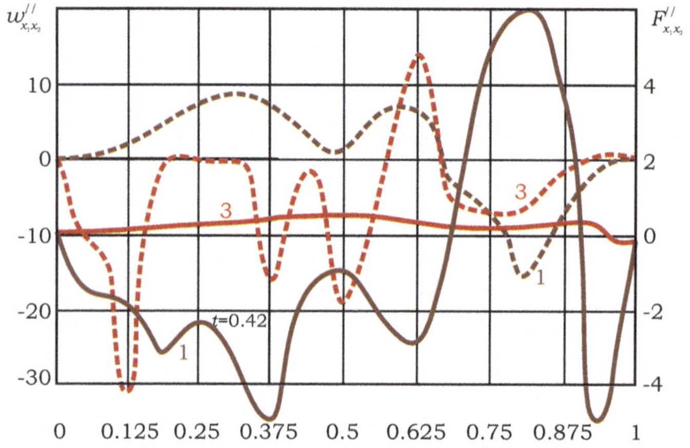

Fig. 8.10 Distribution of $F''_{x_1x_1}$ (*solid lines*) and $w''_{x_1x_1}$ (*dashed lines*) for $q = 503.7$, $t = 0.42$; 1, elastic–plastic material; 3, linear elastic material

The plate dimensions were the following: $a = b = 0.1$ m, $\frac{a}{h} = \{50, 111\}$. The shell volume was covered by a $16 \times 16 \times 8$ mesh. Analysis of the stress–strain state was carried out for various time moments. The integration step used in the Adams method was $\Delta t^k = 5 \times 10^{-4}$. The curvatures $w''_{x_1x_1}$, $w''_{x_1x_2}$ are denoted by dashed curves, whereas the stresses $T_{22} = \frac{\partial^2 F}{\partial {x_1}^2}$, $T_{12} = -\frac{\partial^2 F}{\partial x_1 \partial x_2}$ are denoted by solid curves. Graphs are presented for the cross-sections shown in Fig. 8.3.

The dependences $w(0.5, 0.5, t)$ for an elastic–plastic material (solid curve, labelled 1) and a linear elastic material (dashed curve, labelled 3), for $q = 503.7$ (load model (8.32)) are shown in Fig. 8.5. The use of model 1 causes a decrease of the static deflection by a factor of 1.4 and an increase of the amplitude by 10% (by the static deflection, we mean the deflection caused by a static load).

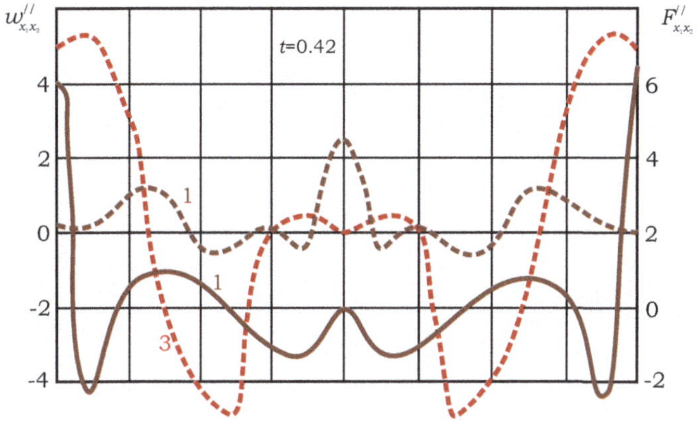

Fig. 8.11 Distribution of $F''_{x_1x_2}$ (*solid lines*) and $w''_{x_1x_2}$ (*dashed lines*) for $q = 503.7$, $t = 0.42$; 1, elastic–plastic material; 3, linear elastic material

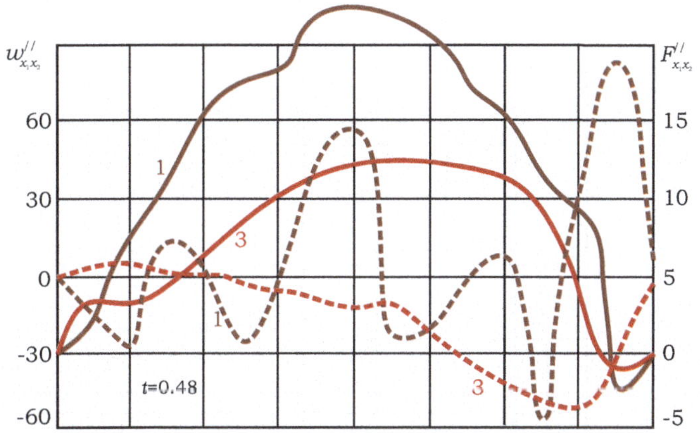

Fig. 8.12 Distribution of $F''_{x_1x_1}$ (*solid lines*) and $w''_{x_1x_1}$ (*dashed lines*) for $q = 503.7$, $t = 0.48$; 1, elastic–plastic material; 3, linear elastic material

For model 1, the plate oscillates around the static-equilibrium position ($w_{st} = 3.6$) with a period of 0.23 and an amplitude of 5.3. It should be noted that the period of free oscillations of the elastic plate is 16% higher than that of the elastic–plastic. The stress–strain states of the plate differ greatly between the mathematical models considered here. The maximum stress for model 1 is 12% larger than for model 3. In one quarter of the plate, the stresses obtained with the elastic–plastic model are a factor of two smaller than those obtained with model 3. The values of the curvature $w''_{x_1x_1}$ values obtained with different models differ by 50% in the corner zones. The difference in $F''_{x_1x_1}$ reaches 65% on the shell edge. The graphs of the curvatures $w''_{x_1x_2}$ obtained with different models differ not only quantitatively but also qualitatively (Figs. 8.6 and 8.7). We obtain a negative curvature using the

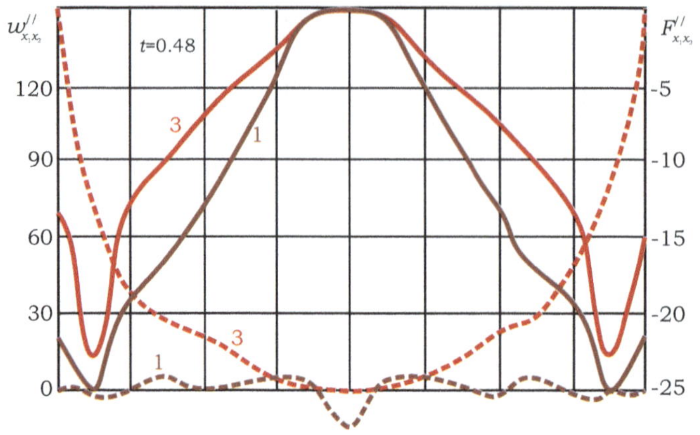

Fig. 8.13 Distribution of $F''_{x_1x_2}$ (*solid lines*) and $w''_{x_1x_2}$ (*dashed lines*) for $q = 503.7$, $t = 0.48$; 1, elastic–plastic material; 3, linear elastic material

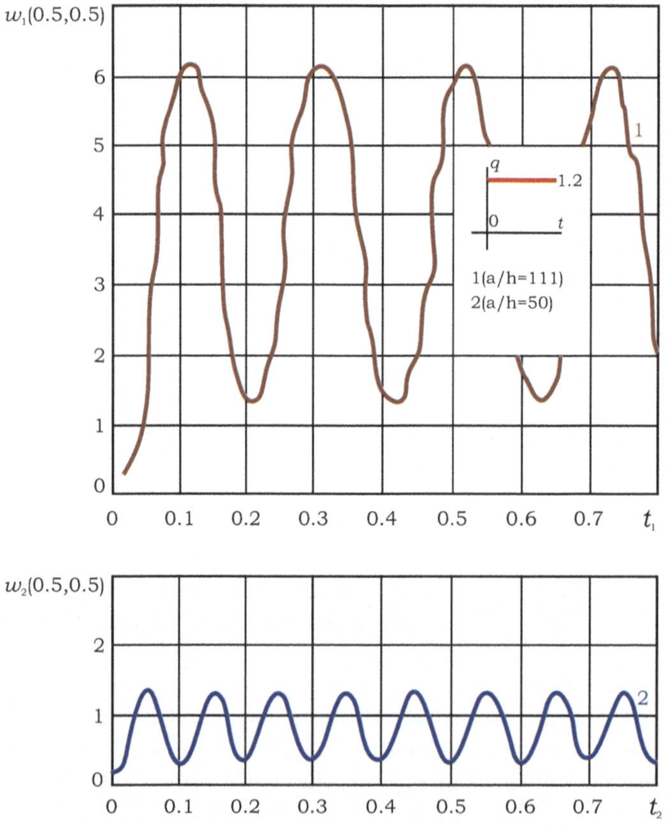

Fig. 8.14 Deflection of the plate centre $w(0.5, 0.5)$ as a function of time for $\frac{a}{h} = 111$ (*solid curve*) and for $\frac{a}{h} = 50$ (*dashed curve*)

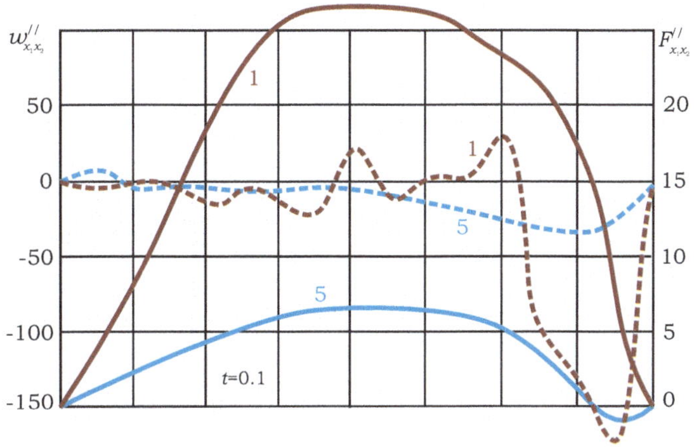

Fig. 8.15 Distribution of $F''_{x_1x_1}$ and $w''_{x_1x_1}$ for $\frac{a}{h} = 111$

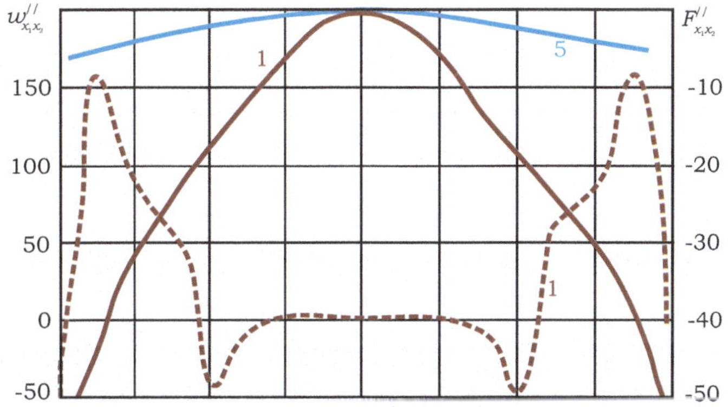

Fig. 8.16 Distribution of $F''_{x_1x_2}$ and $w''_{x_1x_2}$ for $\frac{a}{h} = 50$

elastic–plastic model, whereas for the elastic model the curvature is positive (on the shell edge). The difference can be explained by the fact that for $t = 0.1$, $w = 6.0$ for the elastic–plastic model, and $w = 3.8$ for the elastic model (Fig. 8.5).

Diagrams of $F''_{x_1x_1}$ and $w''_{x_1x_1}$ are shown in the Figs. 8.8 and 8.9 for the time instant $t = 0.3$. At this time, the two models give opposite phases for the position of the plate. There is no qualitative difference in the plots of $F''_{x_1x_1}$, but there are qualitative and quantitative differences between the plots of the curvatures in Fig. 8.8. The values of $w''_{x_1x_2}$ obtained with the different models differ in sign. The difference between the stresses $F''_{x_1x_2}$ reaches 17% at the edge.

Graphs for $t = 0.42$ are presented in Figs. 8.10 and 8.11. At this time, the two models again give opposite phases. These results show that graphs of $F''_{x_1x_1}$

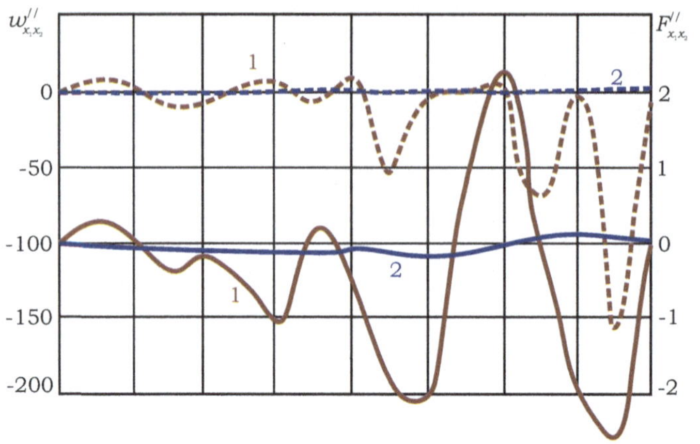

Fig. 8.17 Distribution of $F''_{x_1x_1}$ and $w''_{x_1x_1}$ for the smallest deflection

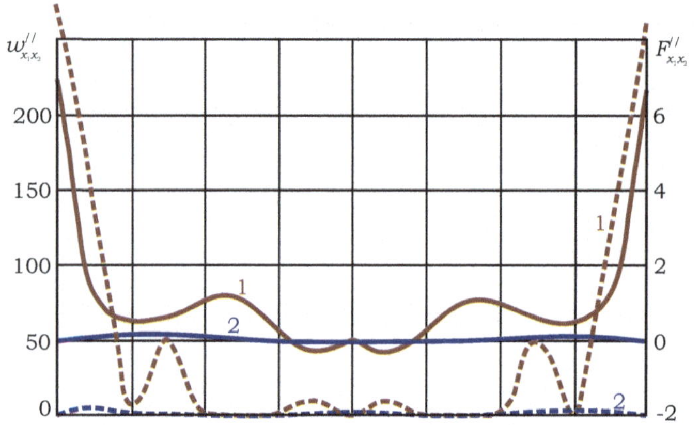

Fig. 8.18 Distribution of $F''_{x_1x_2}$ and $w''_{x_1x_2}$ for the smallest deflection

can differ even qualitatively for different models. Note that when the linear elastic model is used, the graphs of both stress and curvature are smoother than they are for model 1.

Graphs for $t = 0.48$ are presented in Figs. 8.12 and 8.13. A qualitative agreement of the diagrams is observable, but the stresses $F''_{x_1x_1}$ in the centre of the plate obtained with the elastic–plastic model are 39% larger than the stresses obtained with the elastic model. The difference in the stresses reaches 38% on the edge of the plate, although in one quarter it reaches only 8%.

From the results presented in Fig. 8.5, one can conclude that the plates oscillate around their equilibrium positions. The amplitude of oscillation of the plate is smaller with the elastic–plastic model, and the value of the deflection from the undeformed state w_{st} is larger. Taking account of elastic–plastic material properties

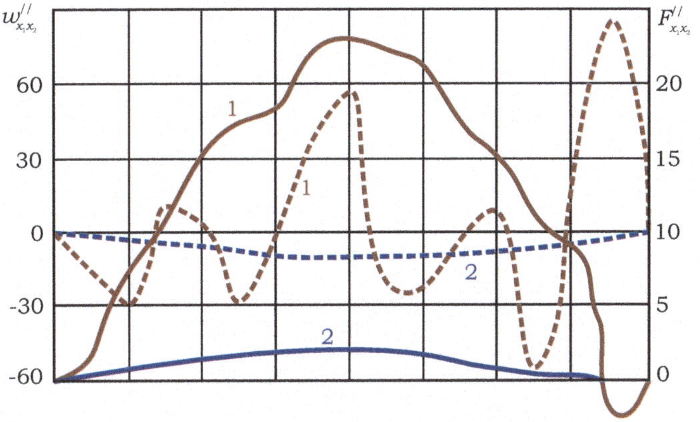

Fig. 8.19 Distribution of $F''_{x_1x_1}$ and $w''_{x_1x_1}$ for the greatest deflection

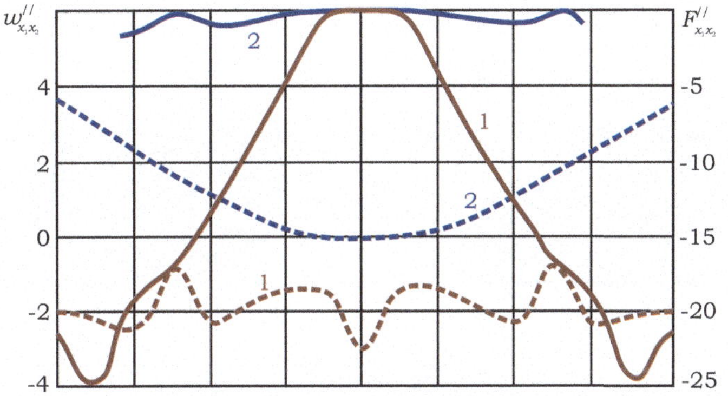

Fig. 8.20 Distribution of $F''_{x_1x_2}$ and $w''_{x_1x_2}$ for the greatest deflection

leads to remarkable mixing of different types of stresses with increase of time, caused by plastic deformation due to the impulse-type load.

The influence of the geometrical parameters on the oscillation of the plate and on the stress distribution is demonstrated in Figs. 8.14, 8.15, 8.16, 8.17, 8.18, 8.19, 8.20, 8.21, 8.22, and 8.23. In Fig. 8.14, the deflection at the centre of the shell is shown for the control parameter $\frac{a}{h} = 111$ (solid curve) and $\frac{a}{h} = 50$ (dashed curve). The static deflection in the first case is three times bigger than in the second case, and the amplitude of oscillation is 2.3 times larger.

In Figs. 8.15 and 8.16, graphs of the stress and curvature at the time, when both plates are in the neighbourhood of their maximum deflection are shown. The stress $F''_{x_1x_1}$ at the plate centre for $\frac{a}{h} = 111$ is more than 30 times the stress for $\frac{a}{h} = 50$. The difference in curvature in one quarter of the plate reaches 95%. The difference in the stress $F''_{x_1x_2}$ at the edge reaches 96%.

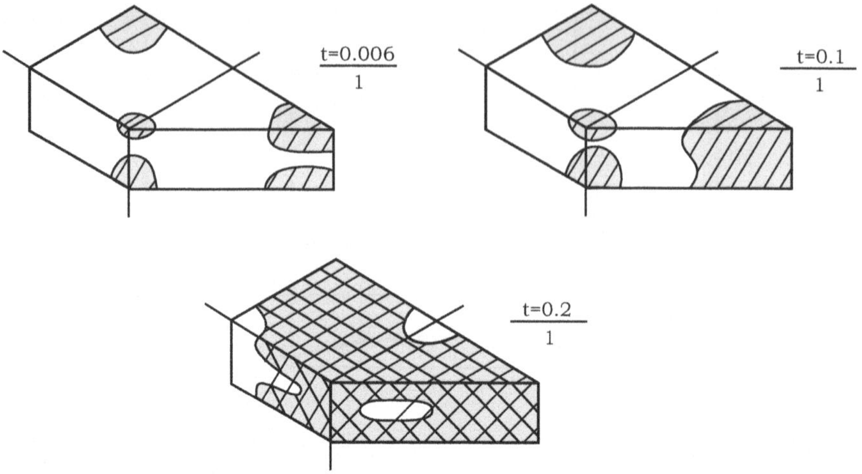

Fig. 8.21 Distribution of plastic zones at $t = 0.006, 0.1$ and 0.2 for a plate with $\frac{a}{h} = 111$

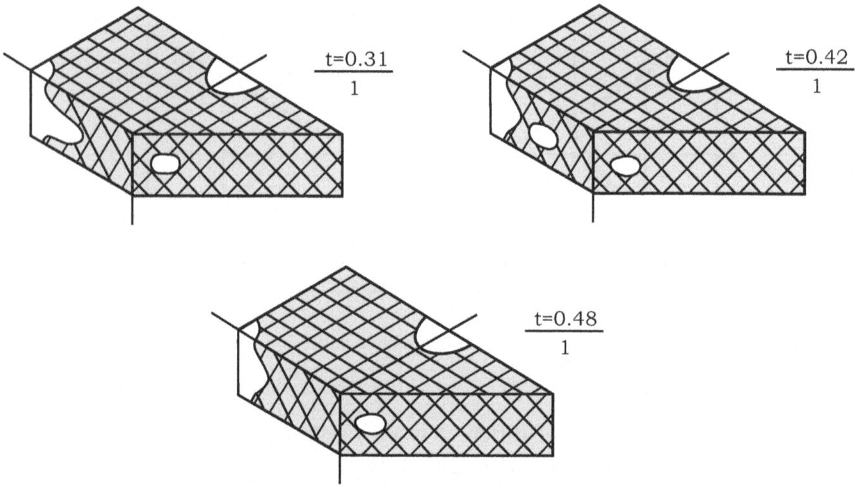

Fig. 8.22 Distribution of plastic zones at $t = 0.31, 0.42$ and 0.48 for a plate with $\frac{a}{h} = 111$

In Figs. 8.17 and 8.18, graphs are shown for the time when the deflection $w(0.5, 0.5)$ is at its minimum. Here we observe a difference in the characters of the graphs of stress and curvatures for plates with different thickness. However, for plates with the same thickness, the qualitative similarities of the graphs of $F''_{x_1x_1}$ and $w''_{x_1x_1}$, as well as those of $F''_{x_1x_2}$ and $w''_{x_1x_2}$ are remarkable. We note that larger changes of the stress–strain state have been observed for the more slender plate.

In Figs. 8.19 and 8.20, graphs of the stress and curvature at the time corresponding to the next maximum are shown. Here the difference in the stresses in the plate

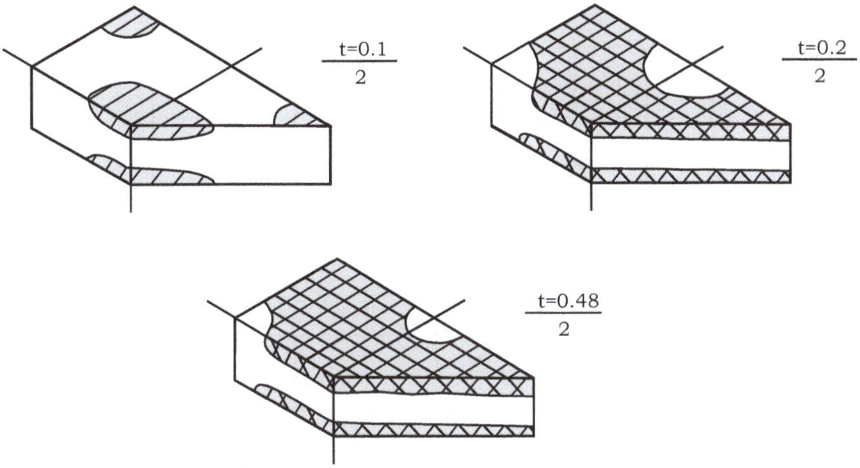

Fig. 8.23 Distribution of plastic zones at $t = 0.1, 0.2$ and 0.48 for a plate with $\frac{a}{h} = 50$

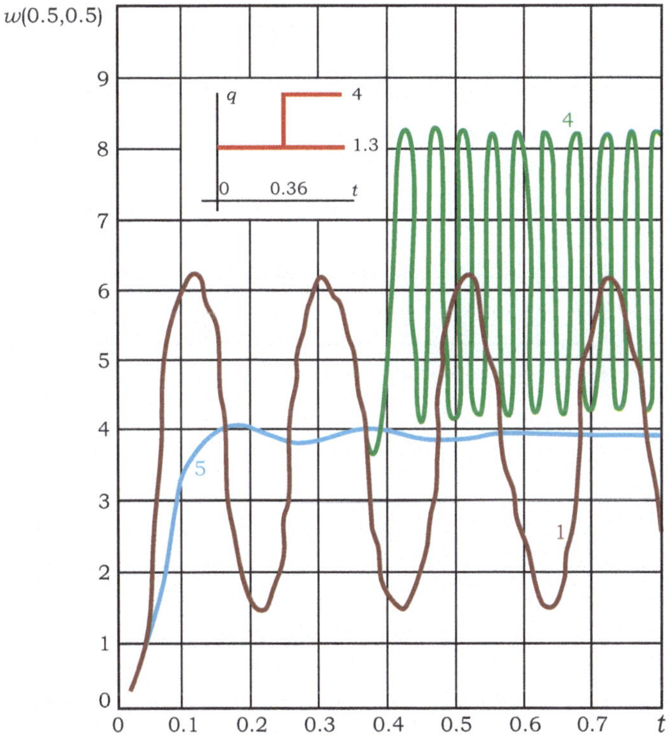

Fig. 8.24 Deflection of the plate centre $w(0.5, 0.5)$ for $\frac{a}{h} = 111$ under different loads: 1, impulse load $q = 503.7$ of type (8.32), 4, impulse load of type (8.33), 5, impulse load of type (8.32) with damping

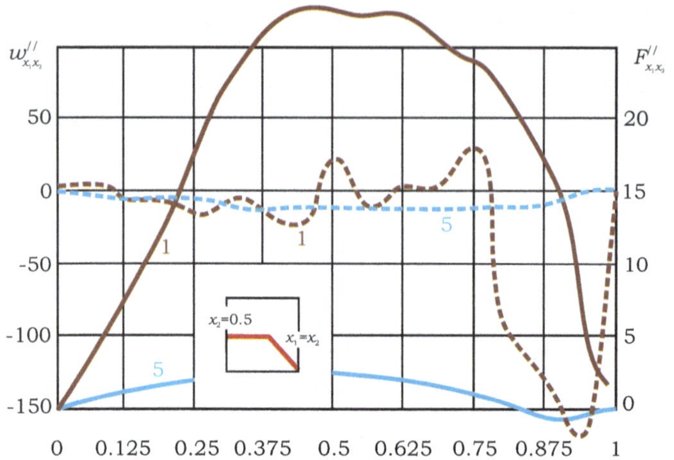

Fig. 8.25 Distribution of $F''_{x_1x_1}$ and $w''_{x_1x_1}$ at $t = 0.18$ for type (8.32) load: 1, without damping; 5, with damping

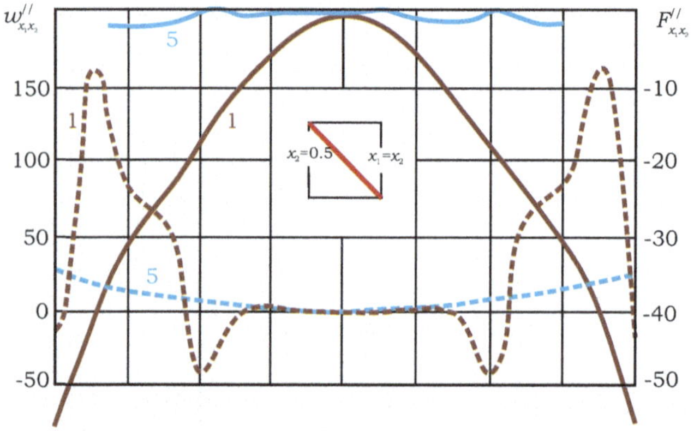

Fig. 8.26 Distribution of $F''_{x_1x_2}$ and $w''_{x_1x_2}$ at $t = 0.18$ for type (8.32) load: 1, without damping; 5, with damping

centre is 80%, and the difference in the curvature is 60%. It can be seen that a larger change in the stress–strain state occurs in the more slender plate at this time moment.

Figures 8.21, 8.22 and 8.23 show how the plastic zones develop in the shell volume with time (Figs. 8.21 and 8.22 correspond to $\frac{a}{h} = 111$, and Fig. 8.23 to $\frac{a}{h} = 50$). It can be seen how plastic zones develop in an initial period (up to $t = 0.12$), and then a stable distribution of plastic zones occurs, which for practical purposes does not change with time. Therefore, one can conclude that after achieving maximum deflection, the plate starts to oscillate like a linearly elastic one

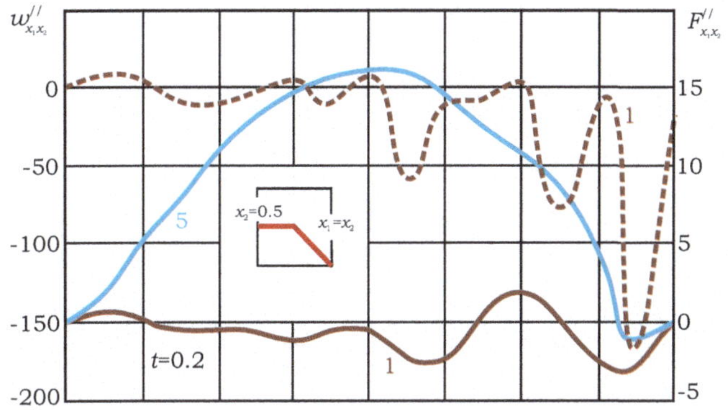

Fig. 8.27 Distribution of $F''_{x_1x_1}$ and $w''_{x_1x_1}$ at $t = 0.2$ for type (8.32) load: 1, without damping; 5, with damping

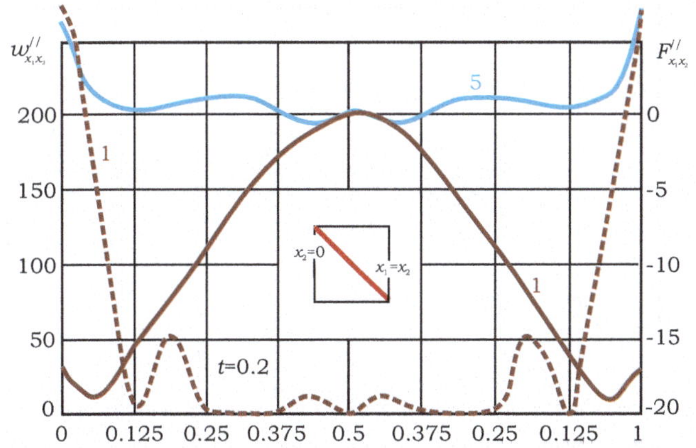

Fig. 8.28 Distribution of $F''_{x_1x_2}$ and $w''_{x_1x_2}$ at $t = 0.2$ for type (8.32) load: 1, without damping; 5, with damping

but with the plastic deformation that has occurred earlier. For the plate with $\frac{a}{h} = 50$ a stabilized distribution of plastic zones is observed from $t = 0.1$.

Comparing the character of the distribution of plastic zones in the volume of plates with different thickness, one can conclude that in relatively thin plates the plastic zones are distributed throughout the whole thickness, whereas for thicker plates the plastic zones are located close to the surfaces. In addition, in thinner plates nonbending stresses occur, which lead to a more uniform stress distribution through the thickness (Fig. 8.22). For thicker plates, bending stresses dominate and cause a nonuniform stress distribution through the thickness (Fig. 8.23), which leads to the occurrence of plastic zones but not through the whole thickness.

Now we consider the variation with time of the deflection of the shell centre for three cases: model 1, with an impulse load $q = 503.7$ of type (8.32) with

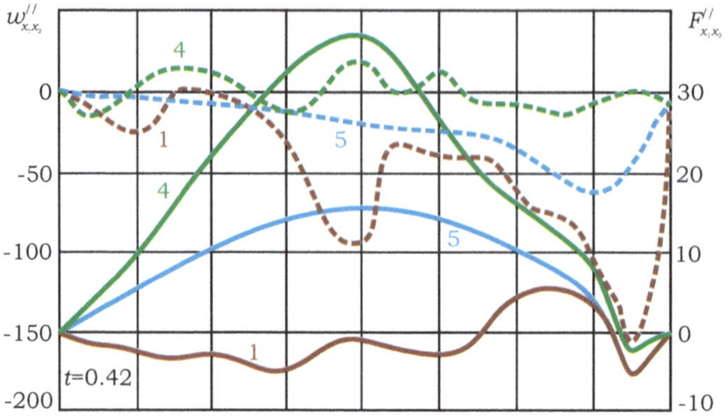

Fig. 8.29 Distribution of $F''_{x_1x_1}$ and $w''_{x_1x_1}$ at $t = 0.42$ for type (8.33) load

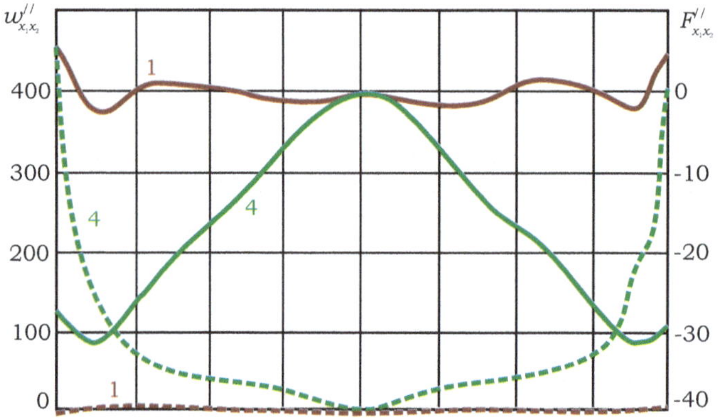

Fig. 8.30 Distribution of $F''_{x_1x_2}$ and $w''_{x_1x_2}$ at $t = 0.42$ for type (8.33) load

successive oscillations; model 4, with an impulse load of type (8.33); and model 5, with an impulse load of type (8.32) with damping. In all cases $\frac{a}{h} = 111$ and the material is elastic–plastic (8.34). The curves shown in Fig. 8.24 illustrate that the static deflection corresponding to the load $2q$ of type (8.33) is 1.6 times larger than the static deflection corresponding to the load of type (8.32). After the extra load has been applied both the amplitude and the frequency of the oscillations change greatly. The amplitude decreases by a factor of 1.3 whereas the frequency increased 5 times. Analysis of the curve for model 5 (with damping) shows that intense damping of the oscillations takes place up to $t = 0.15$, and that there are practically no oscillations after $t = 0.5$. The deflection reached after damping has occurred (curve 5) corresponds to the equilibrium state, around which the plate oscillates (curve 1) under the load of type (8.32).

In Figs. 8.25 and 8.26, graphs of the stress and curvature for the elastic–plastic model are shown, for the impulse-type load (8.32) without damping (model 1) and

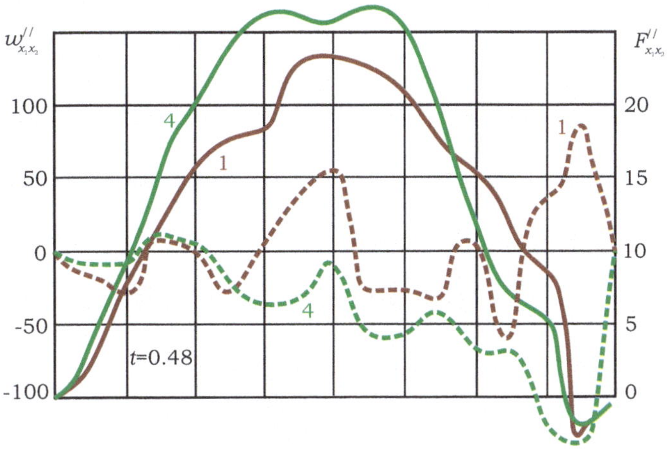

Fig. 8.31 Distribution of $F''_{x_1x_1}$ and $w''_{x_1x_1}$ at $t = 0.48$ for type (8.33) load

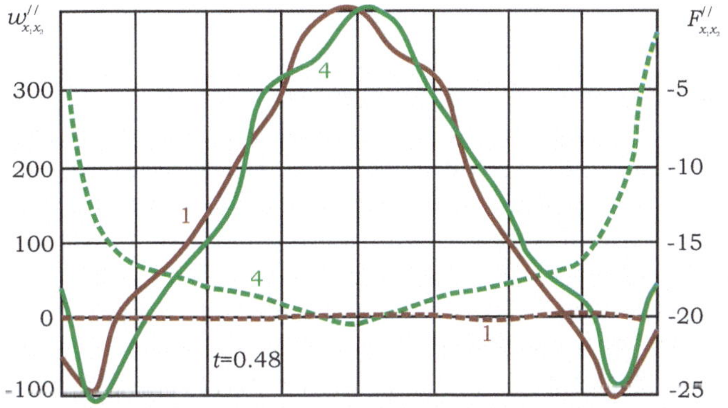

Fig. 8.32 Distribution of $F''_{x_1x_2}$ and $w''_{x_1x_2}$ at $t = 0.48$ for type (8.33) load

with damping (model 5). It can be seen that damping causes important changes in the stress–strain state (the deflection decreases by a factor of 1.8, and the stress $F''_{x_1x_1}$ in the plate centre decreases by a factor of 5). Damping smooths the graphs of the curvature and decreases the value of the curvature by a factor of about 4.8. Graphs of the stress and curvature at the time instant $t = 0.2$ for the same two cases are shown in Figs. 8.27 and 8.28. The divergence mentioned above has increased.

In Figs. 8.29 and 8.30, graphs of the stress and curvature at $t = 0.42$ for models 1 and 4, corresponding to a load of type (8.33), are presented. Results for model 5 are also shown in Fig. 8.29. A comparison of the graphs shows that the stress–strain state of the plate after the second step of the loading process differs significantly from the stress–strain state of the plate subjected only to the extended oscillations caused by an initial impulse load of type (8.32). A difference is visible not only in

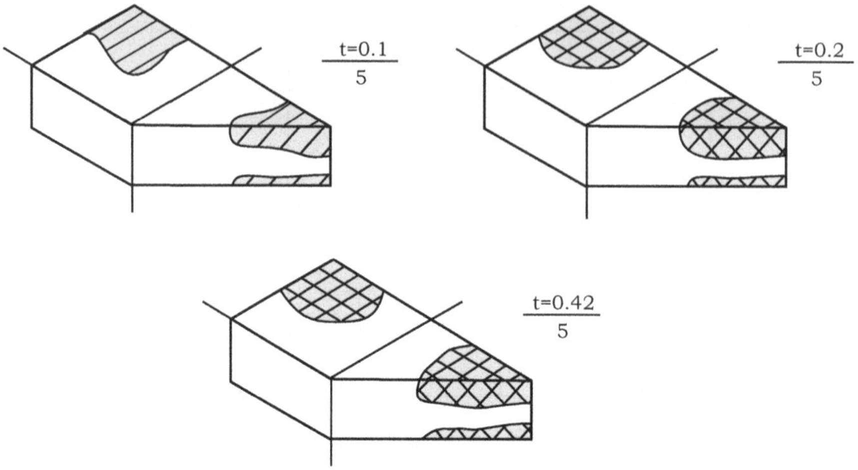

Fig. 8.33 Distribution of elastic and plastic zones at $t = 0.1, 0.2, 0.42$ (with damping)

the values but also of the stresses in their sign. For model 5, the graphs of stress and curvature are located between those for models 1 and 4.

At $t = 0.48$, the positions of the plates for models 1 and 4 approach each other more closely than they do at $t = 0.42$, and the difference in character of the graphs becomes smaller. For instance, in the centre of the loaded plate, the stress $F''_{x_1x_1}$ is 17% higher than the stress for the load reported in Fig. 8.31. The curvatures of the shell for the two models differ in sign, because at $t = 0.42$ the plates are on different sides of their equilibrium states (Fig. 8.24). The difference in the values of the shear stress can be neglected (Fig. 8.32).

In Fig. 8.33, the distribution of elastic and plastic zones in the plate, in the case with damping, is presented for three time moments. Because the intense oscillations have already been damped at $t = 0.1$, difference in the distribution of plastic zones is not observed. Comparison of the distribution of plastic zones at $t = 0.1$ for models 1 and 5 leads to the conclusion that plastic deformation occurs only near corners and in zones lying close to the plate surfaces. In the centre of the plate and also close to the middle surface, plastic deformation is not observed.

In Fig. 8.34, the distributions of plastic zones in the plate volume at the time instants $t = 0.42$, $t = 0.46$ and $t = 0.48$ for model 4 are presented. The latter two distributions differ from each other. This is evidence of the stabilization of the distribution of plastic zones, which has occurred after the maximum deflection has occurred under the extra load. Comparing the character of the distribution of plastic zones at $t = 0.48$ for models 1 and 4, we discovered that after the additional loading process the elastic zones remain only around the centres of the sides and vanish completely in the rest of the plate volume. Also, additional plastic deformation occurs in the corner zones and around the plate centre.

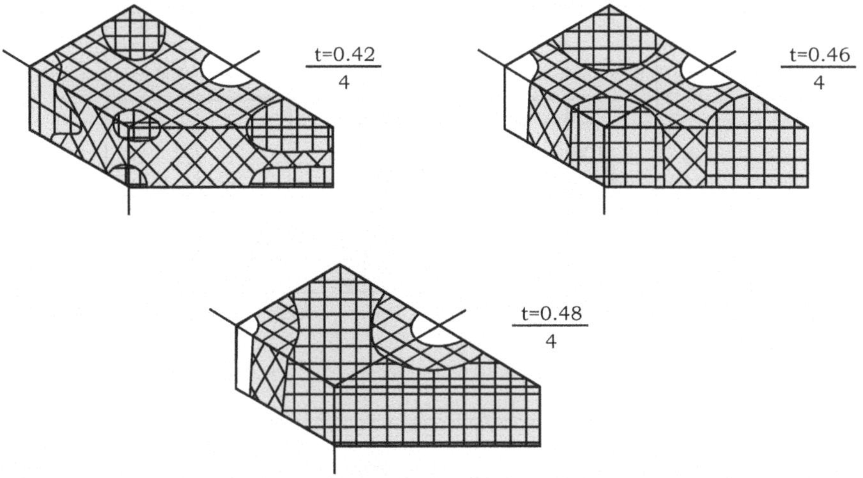

Fig. 8.34 Distribution of elastic and plastic zones at $t = 0.42, 0.46, 0.48$

We now investigate the dynamic stability and the stress–strain state of a flexible elastic–plastic shell under an impulse load of type (8.32). The initial equations were solved in the hybrid form (8.23). The computations were carried out for the boundary conditions (2.116) and the initial conditions (4.1). The program developed to do this, however, allows us to investigate other boundary conditions such as (7.33). Shells with the following geometric parameters have been considered: $a = 0.1$ m, $\lambda = 1$, $\frac{h}{a} = \frac{1}{111}$; the physical characteristics (8.34) were used. The nondimensional curvature parameters were $k_x = k_y = 24$ and 48. A mesh of size $16 \times 16 \times 8$ was used, and the integration step used in the Adams method was $\Delta t_k = 5 \times 10^{-4}$. The mesh size and integration step were determined by numerical experiments so as to satisfy the condition of stability of the solution. Dynamic stability criteria described earlier were applied. At the moment of a stability loss, an increase of 1% in the load above its critical value caused an increase in the deflection by more than three times. The maximum deflection exceeded twice the camber of the shell and the time required to reach the maximum deflection greatly decreased.

Analysis of the stress–strain state of a flexible rectangular plate showed that the results depend mainly on the material model used. In this section we investigate the influence of three models of material deformation (linear elastic, nonlinear elastic and elastic–plastic) on the stress–strain state, the distribution of plastic zones and the dynamic stability of a flexible, shallow, rectangular shell. In the figures, the data are presented in nondimensional form using the following notation: 1, results obtained using the linear elastic model; 2, results obtained using the nonlinear elastic model; 3, results obtained using the elastic–plastic model.

A shell with $K_{x_1} = K_{x_2} = 24$ in a pre-critical state was investigated. An analysis of plots of $w(t)$ for the central point of a shallow shell under a pre-critical load $q = 198$ (Fig. 8.35) showed that the point describes almost the same trajectory for

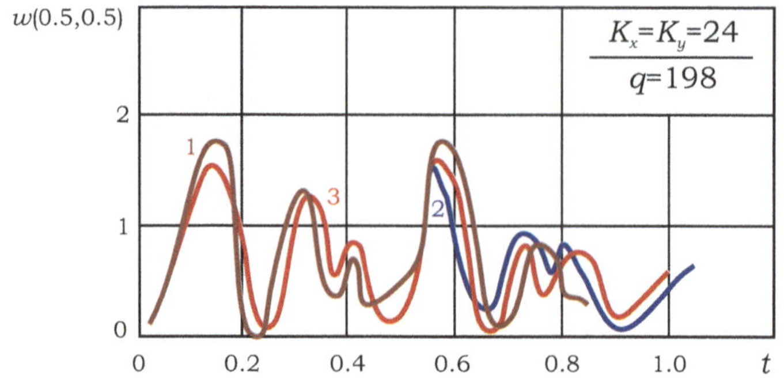

Fig. 8.35 Deflection of the plate centre $w(0.5, 0.5)$ for $K_1 = K_2 = 24$, $q = 198$

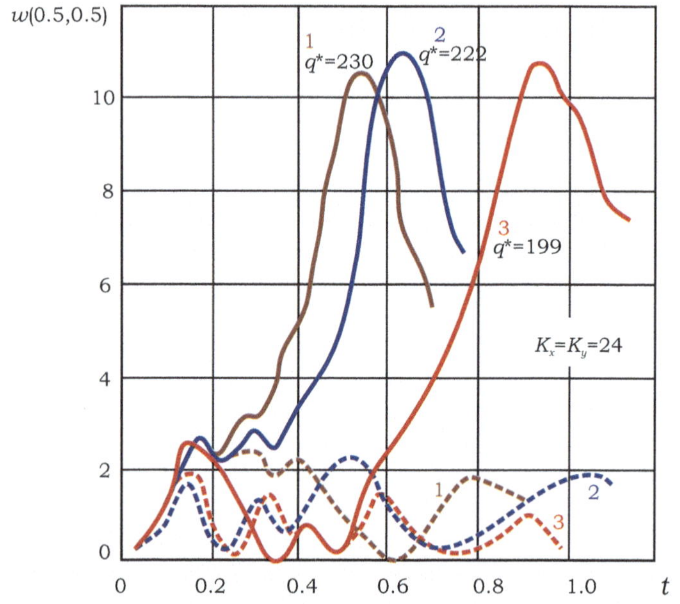

Fig. 8.36 Deflection of the plate centre $w(0.5, 0.5)$ for $K_1 = K_2 = 24$ and different loads

all three models. We examined the changes of the stresses $F''_{x_1x_1}$, $F''_{x_1x_2}$ and of the curvatures $w''_{x_1x_1}$, $w''_{x_1x_2}$ along the axis $x_2 = 0.5$ and along the diagonal using the "stop frame" method for the time moments $t = 0.14, 0.5, 0.58$ and 0.74.

In Fig. 8.36, the variation with time of the deflection at the shell centre is presented for all three models, for a pre-critical load (dashed line) and the critical load (solid line). It can be seen that buckling occurs at $t = 0.54$ for model 1, at $t = 0.64$ for model 2 and at $t = 0.92$ for model 3. Therefore, the use of a more simplified model of material deformation gives a higher value of the critical load and an earlier occurrence of buckling.

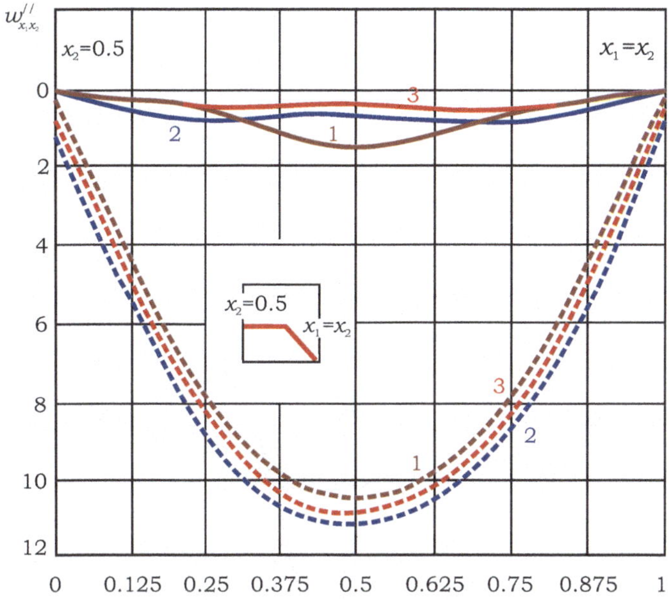

Fig. 8.37 Distribution of deflection w

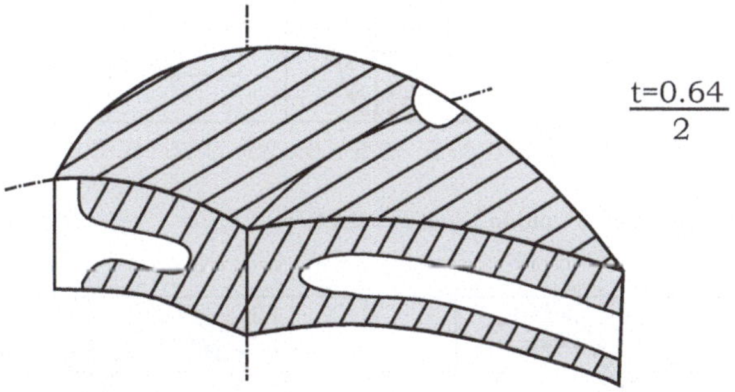

Fig. 8.38 Distribution of plastic zones at $t = 0.64$

Graphs of the deflection of the shell in the pre-critical state (solid curve) and the post-critical state (dashed curve) states are shown in Fig. 8.37 for all three models. We observed that during buckling the maximum deflection increases by factors of 6.4 times (model 1), 18 times (model 2) and 26 times (model 3).

At the time when the jump occurs, the largest stresses are observed for model 3; the stresses for models 2 and 1 are smaller by factors of 1.6 and 2, respectively. In zones adjacent to the corners, the sign of the curvature obtained using model 3 differs from that obtained using models 1 and 2.

The distribution of plastic zones in the shell volume at $t = 0.64$ for model 2 is reported in Fig. 8.38.

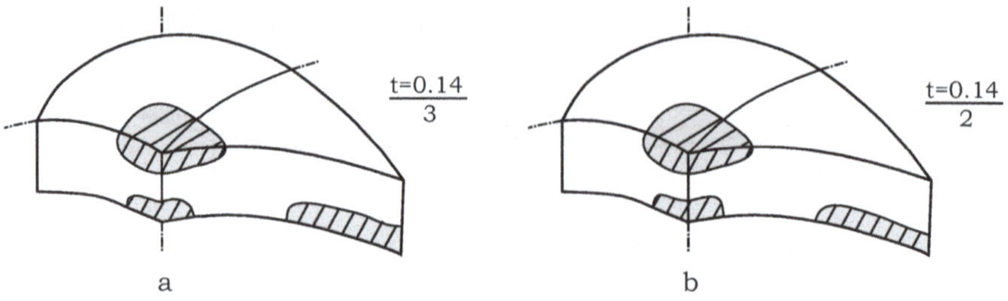

Fig. 8.39 (**a, b**) Distribution of plastic zones at $t = 0.14$

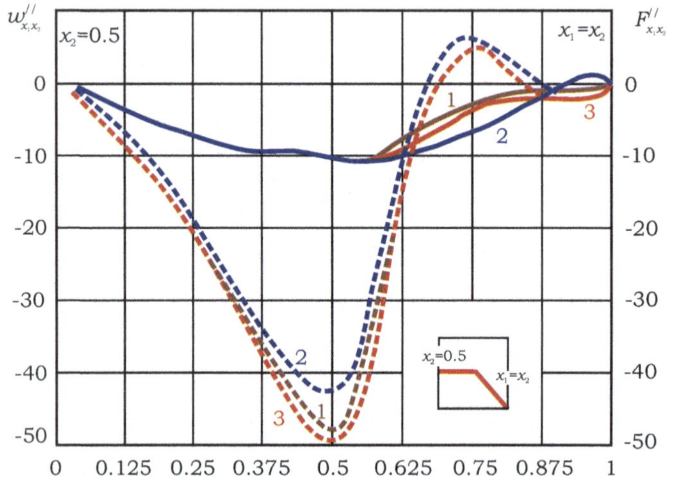

Fig. 8.40 Distribution of $F''_{x_1x_1}$ and $w''_{x_1x_1}$ at $t = 0.14$

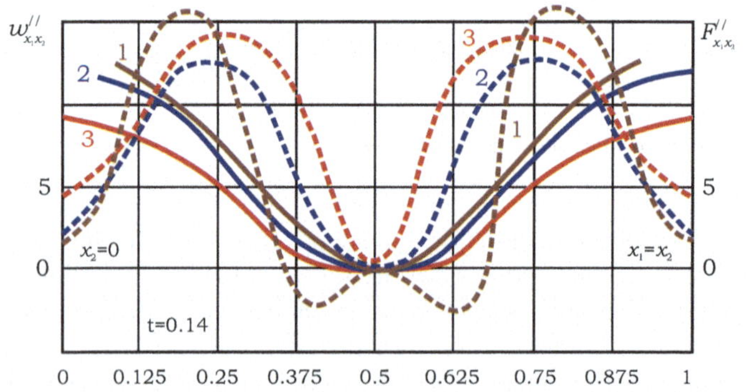

Fig. 8.41 Distribution of $F''_{x_1x_2}$ and $w''_{x_1x_2}$ at $t = 0.14$

The difference between the stress–strain states for the different models is slight; the results are shown in Figs. 8.40 and 8.41. In these figures, plots of $F''_{x_1x_1}$ (solid curves) and the curvatures $w''_{x_1x_2}$ (dashed curves) for $t = 0.14$ are shown.

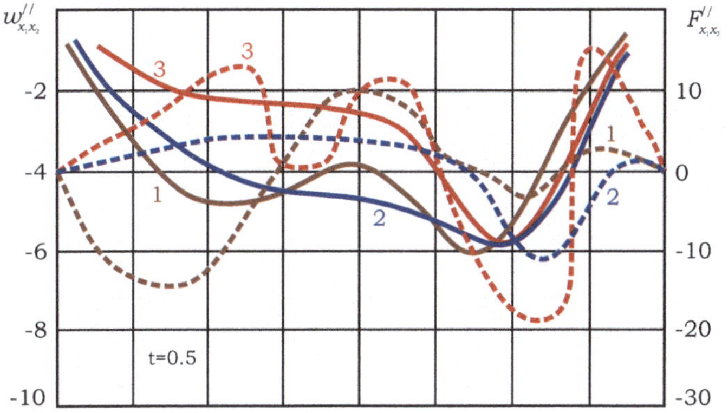

Fig. 8.42 Distribution of $F''_{x_1x_1}$ and $w''_{x_1x_1}$ at $t = 0.5$

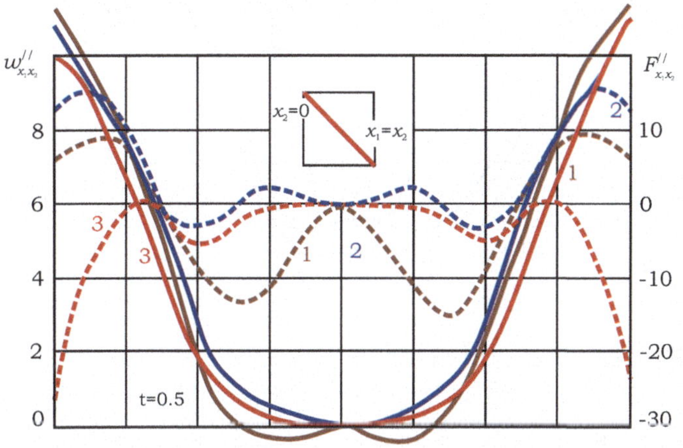

Fig. 8.43 Distribution of $F''_{x_1x_2}$ and $w''_{x_1x_2}$ at $t = 0.5$

The distribution of plastic zones obtained by use of the elastic–plastic model is shown in Fig. 8.39a, and the distribution obtained with the nonlinear elastic model (where the stress exceeds σ_T) is shown in Fig. 8.39b. The similarity in the distribution of plastic zones in Fig. 8.39 is due to overlapping of the maximum deflections for all three models at $t = 0.14$ (Fig. 8.35).

In Figs. 8.42 and 8.43, graphs of the stress and curvature at $t = 0.5$ are shown. Before this time instant, a large difference in the character of the plots is observed. For instance, at the point with coordinates $x_2 = 0.5$, $x_1 = 0.25$, the values of $F''_{x_1x_1}$ obtained with model 1 have a different sign from those calculated with model 3. The curvatures obtained with model 2 are 2–2.5 times smaller than those obtained with model 3 (Fig. 8.43).

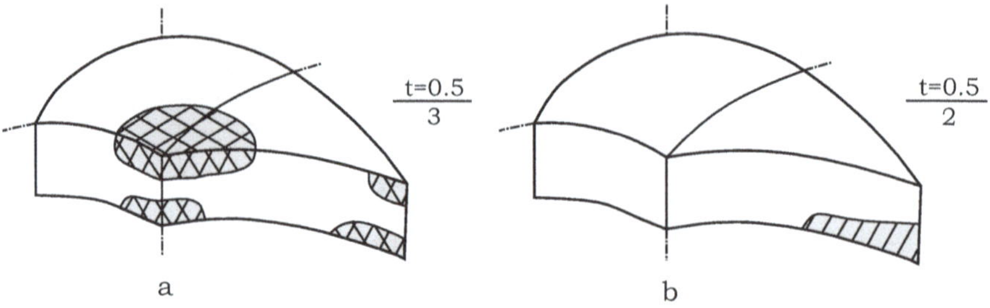

Fig. 8.44 (**a, b**) Distribution of plastic zones at $t = 0.5$ for models 3 and 2

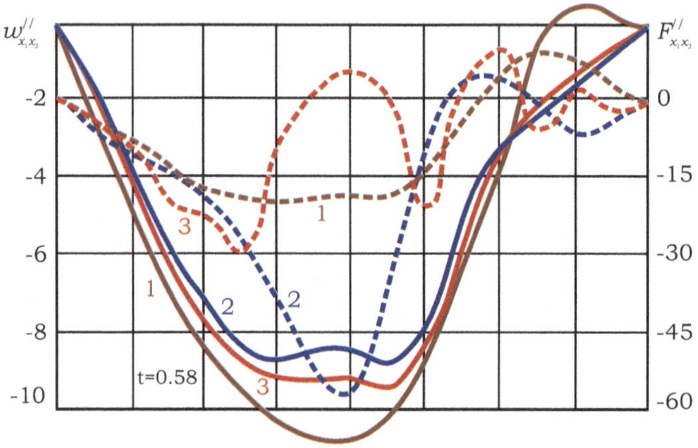

Fig. 8.45 Distribution of $F''_{x_1x_1}$ and $w''_{x_1x_1}$ at $t = 0.58$

Figure 8.44 illustrates the distribution of plastic zones according to model 3 (Fig. 8.44a) and the zone where $\sigma_i \geq \sigma_T$ according to model 2 (Fig. 8.44b). The differences are due to the fact, that in model 2, a loading release takes place along the same deformation curve as for the loading.

Graphs of stress and curvature for $t = 0.58$ are shown in Figs. 8.45 and 8.46. The results of computation using model 3 lie between those obtained using model 1 and model 2, and the differences are 12% and 9%, respectively. The difference between the deformations for models 3 and 2 is shown in Fig. 8.47 for the case of a loading release.

In Figs. 8.48 and 8.49, graphs of stress and curvature for $t = 0.74$ are shown. As Fig. 8.48 shows, the difference between the stresses computed with models 1 and 2 is insignificant, but these computed stresses differ strongly from the results obtained with model 3. In a large part of the shell, the stresses obtained with model 3 differ in sign from those obtained with models 1 and 2. The character of the distribution of plastic zones for models 3 and 2, shown in Fig. 8.50, supports this observation.

In general, when we analyse the results for the pre-critical state, although the trajectories of the shell centre are close to each other for all three models, a large

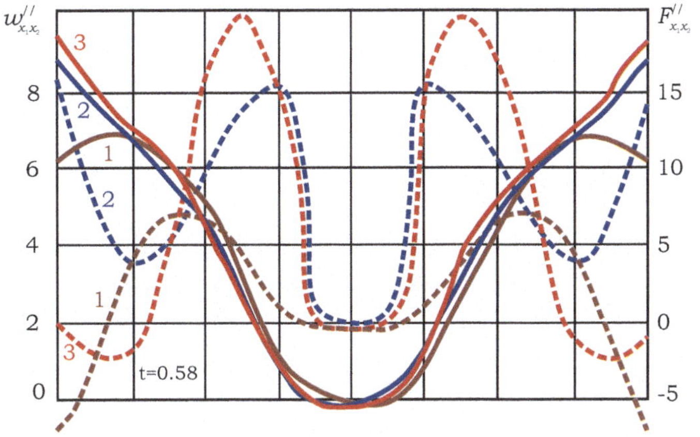

Fig. 8.46 Distribution of $F''_{x_1x_2}$ and $w''_{x_1x_2}$ at $t = 0.58$

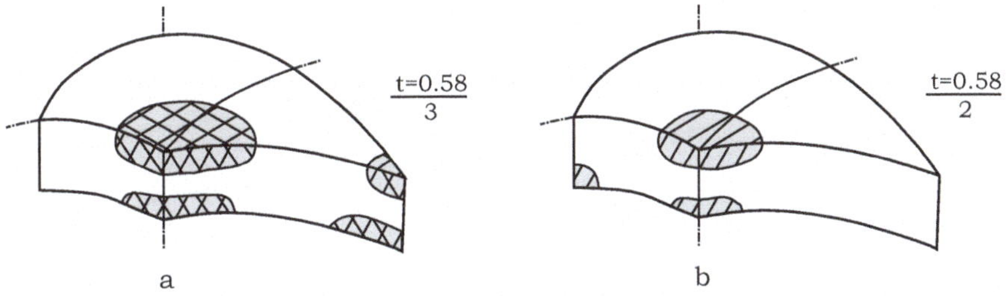

Fig. 8.47 Deformations for (**a**) the model 3 and (**b**) model 2

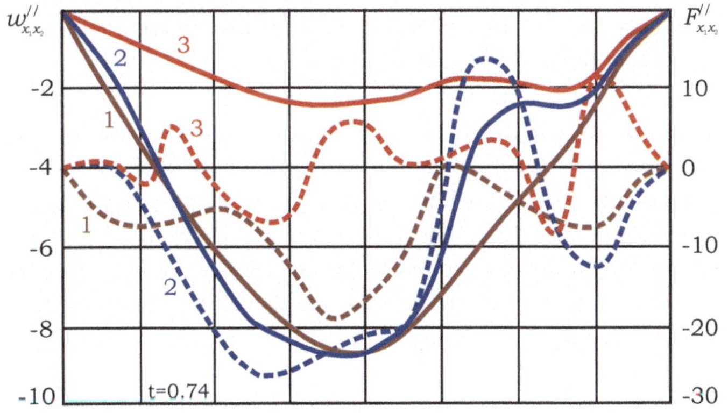

Fig. 8.48 Distribution of $F''_{x_1x_1}$ and $w''_{x_1x_1}$ at $t = 0.74$

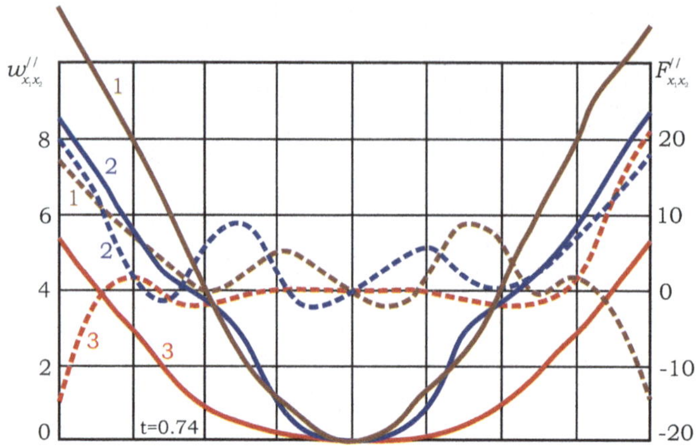

Fig. 8.49 Distribution of $F''_{x_1x_2}$ and $w''_{x_1x_2}$ at $t = 0.74$

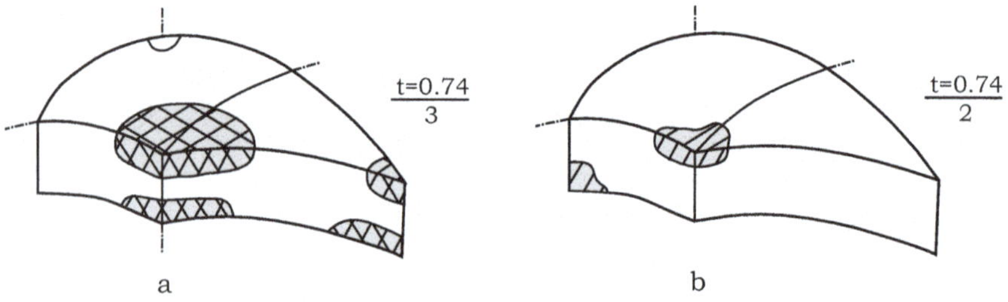

Fig. 8.50 (**a**, **b**) Distribution of plastic zones at $t = 0.58$ for models 3 and 2

difference in the stress–strain state can be observed at any given time moment. Evidently, however, the differences between the results obtained with models 2 and 3 are smaller than with models 1 and 3.

Now we shall describe the results of applying all three computational models to the problem of the dynamic stability of a shallow shell. According to models 1 and 2, the shell starts to exhibit buckling almost as soon as the critical load q^* is applied, in contrast to model 3, where the first "jump" is observed after oscillations around an equilibrium position. The values of the critical load obtained with the various models for a shell with geometric parameters $K_{x_1} = K_{x_2} = 24$ are given in Table 8.2. The mathematical models of elastic–plastic deformations plays the role of the reference model.

Table 8.2 Critical loads q^* for a shell with $K_{x_1} = K_{x_2} = 24$, calculated for different models

No	Model	Critical load q^*	%
1	Linear-elastic	230	15%
2	Non linear elastic	222	11%
3	Elastic–plastic	199	–

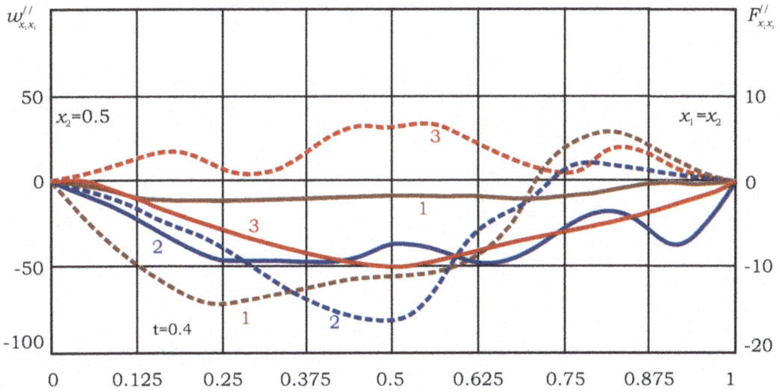

Fig. 8.51 Distribution of $F''_{x_1x_1}$ and $w''_{x_1x_1}$ for critical load at $t = 0.4$

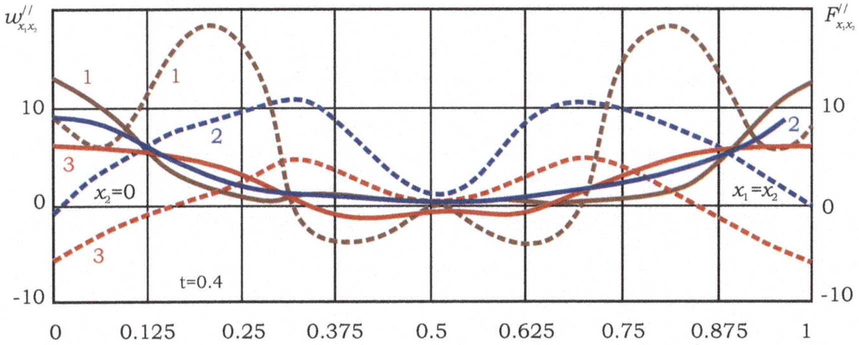

Fig. 8.52 Distribution of $F''_{x_1x_2}$ and $w''_{x_1x_2}$ for critical load at $t = 0.4$

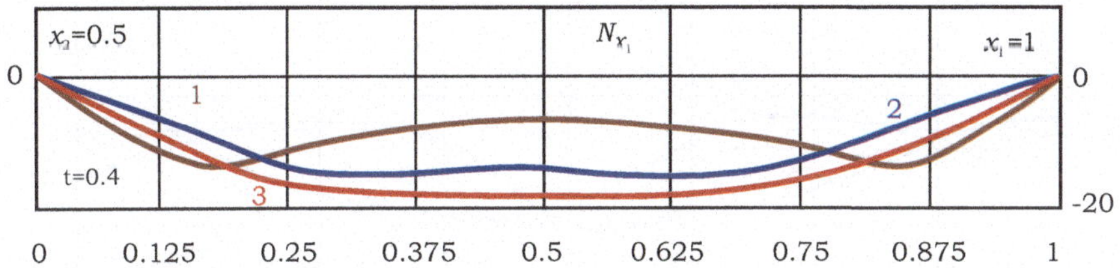

Fig. 8.53 Distribution of N_x for critical load at $t = 0.4$

Graphs of stress and curvature at $t = 0.4$ and for the critical load are shown in Figs. 8.51, 8.52 and 8.53. The results agree qualitatively for all models. However, $F''_{x_1x_1}$ is larger by a factor of 1.27 for model 2 and smaller by a factor of 5 times for model 1 than in the elastic–plastic model. The boundaries and locations of the

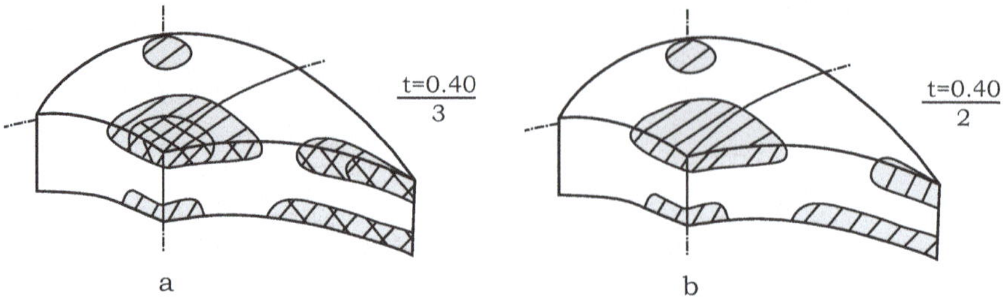

Fig. 8.54 (**a, b**) Distribution of plastic zones at $t = 0.4$ for models 3 and 2

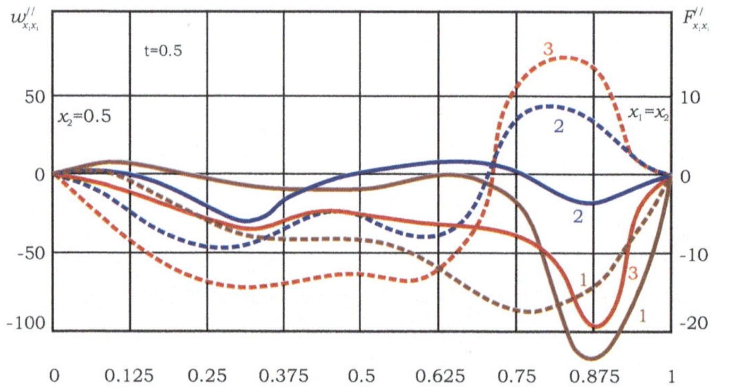

Fig. 8.55 Distribution of $F''_{x_1x_1}$ and $w''_{x_1x_1}$ for critical load at $t = 0.5$

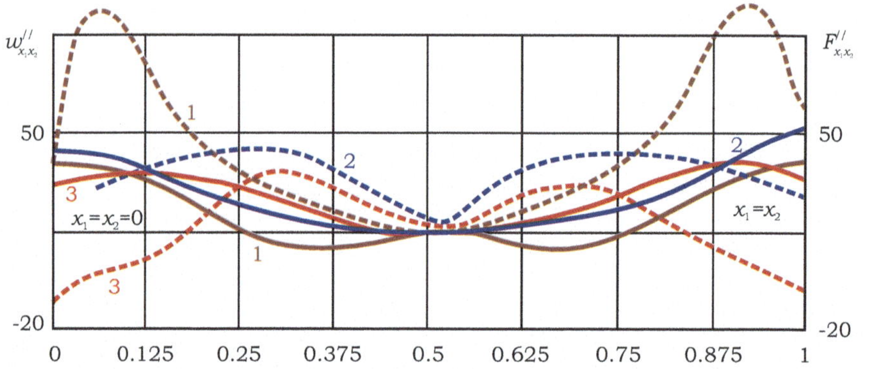

Fig. 8.56 Distribution of $F''_{x_1x_2}$ and $w''_{x_1x_2}$ for critical load at $t = 0.5$

plastic zones obtained with models 2 and 3 at $t = 0.4$ are shown in Fig. 8.54. The results are similar, although in the case of model 3 we observe that the material has already experienced load release.

The results of computation of the stresses and curvatures at $t = 0.5$ for the critical load are illustrated in Figs. 8.55, 8.56 and 8.57. As earlier, one can observe

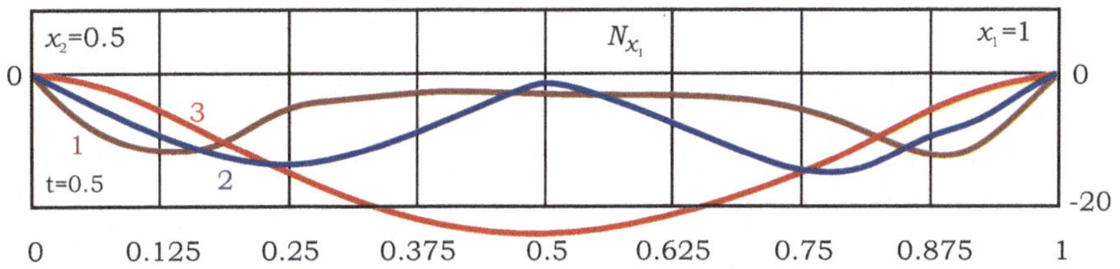

Fig. 8.57 Distribution of N_x for critical load at $t = 0.5$

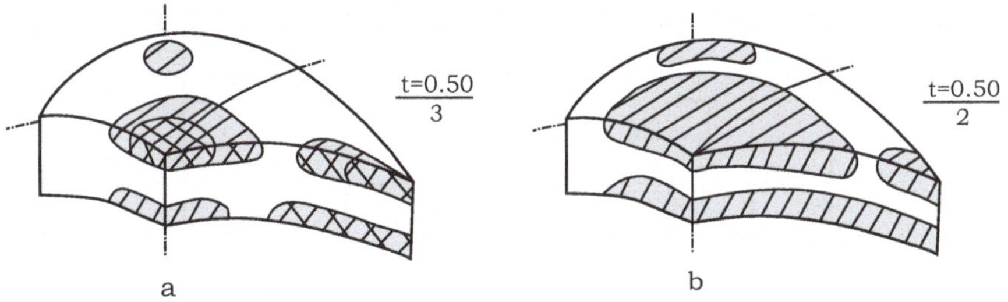

Fig. 8.58 (**a, b**) Distribution of plastic zones at $t = 0.5$ for models 3 and 2

a qualitative similarity here also. However, a qualitative difference can be observed in the graphs of $F''_{x_1x_1}$, where the stresses for model 3 lie between those for models 1 and 2. The stress in the corner is larger by a factor of 1.22 for model 1 and smaller by a factor of 6 times smaller for model 2 than that for model 3. In the case of the curvature $w''_{x_1x_1}$ in the corner zone, the curvature for model 2 is smaller by a factor of 1.6 than that for model 3. A computation with model 1 gives a curvature of different sign in this zone. An analogous effect can be detected in relation to the curvatures of rotation computed with models 1 and 3.

The distributions of plastic zones for models 3 and 2 are shown in Fig. 8.58. A qualitative similarity of the zones is observed, but the volume occupied by the plastic zone for model 2 is larger than that for model 3. This is because, at the time considered, the deflection for model 2 exceeds the deflection for model 3 by a factor of 14 (Fig. 8.36).

The graphs of the stress $F''_{x_1x_1}$ and N_{x_1} in Figs. 8.59 and 8.60 correspond to the time moments $t = 0.54$ (model 1), $t = 0.64$ (model 2) and $t = 0.92$ (model 3) and illustrate the differences described earlier (see Fig. 8.37).

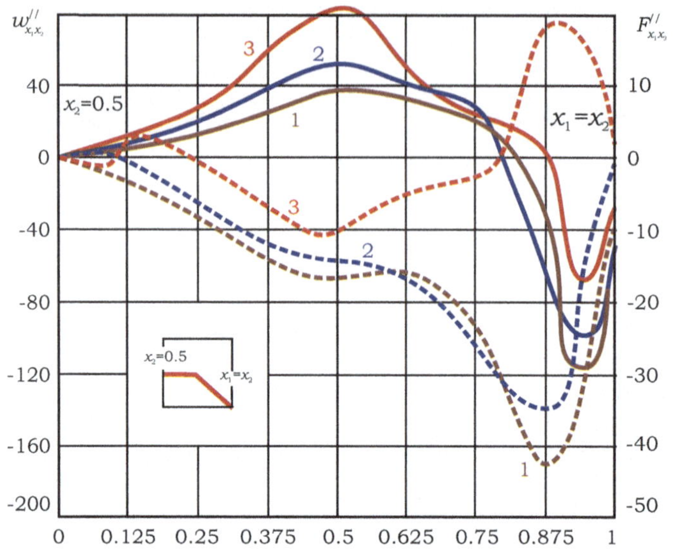

Fig. 8.59 Distribution of $F''_{x_1x_1}$ and $w''_{x_1x_1}$ at $t = 0.54, 0.64, 0.92$

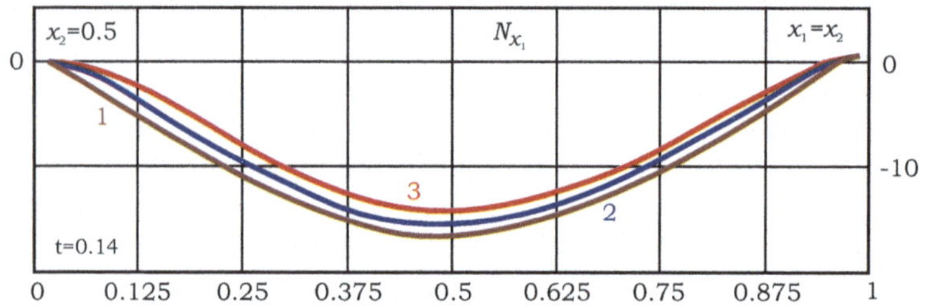

Fig. 8.60 Distribution of N_{x_1} for critical load at $t = 0.54, 0.64, 0.92$

A sequence of diagrams illustrating the development of plastic zones under the critical load versus time is presented in Figs. 8.61 and 8.62. Beginning at $t = 0.85$, the plastic zones develop through the thickness of the shell in its centre. The zones of elastic deformation are located at the centres of the sides of the shell before and after the jump.

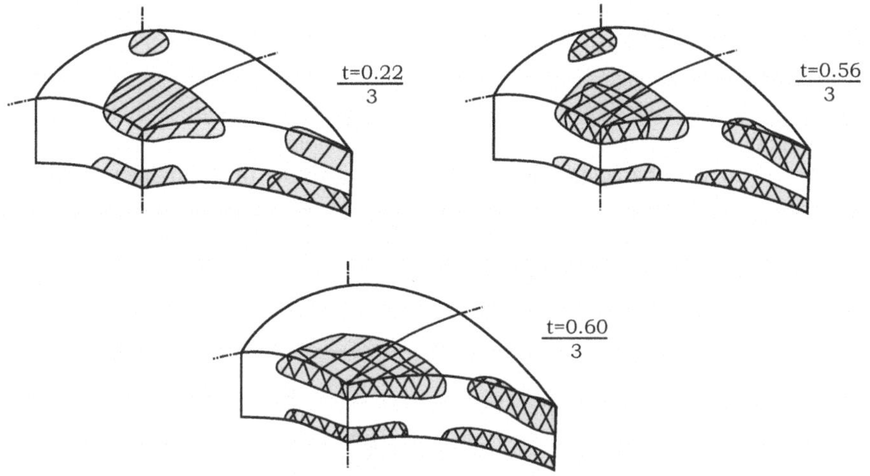

Fig. 8.61 Distribution of plastic zones for model 3 at $t = 0.22; 0.56; 0.60$

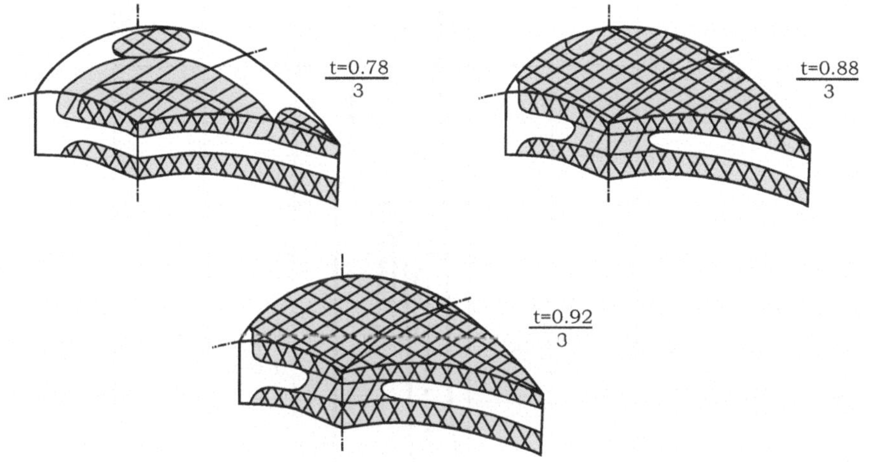

Fig. 8.62 Distribution of plastic zones for model 3 at $t = 0.78; 0.88; 0.92$

In Fig. 8.38, the distribution of plastic zones in the shell volume at $t = 0.64$ for model 2 is shown. Comparing this distribution with that shown in Fig. 8.62 for $t = 0.92$ (this time corresponds to the jump of the shell for model 3), we observe a good overlap of the boundaries of the plastic zones. This may be explained by the small difference (<2%) between the deflections of the shell at these times.

In conclusion, when elastic–plastic deformation can occur, one needs to use more sophisticated models, because simple models give results that are both qualitatively and quantitatively wrong.

We now analyse the dynamic behaviour of a shallow rectangular shell with the same material parameters (8.34) and the same dimensions, but with curvatures $k_{x_1} = k_{x_2} = 48$. Curves characterizing the variation with time of the deflection at the centre of the shell given by the nonlinear elastic model (model 2) for one loading level and by the elastic–plastic model (model 3) for two loading levels are shown in Fig. 8.63. Graphs of stresses and curvatures for the shell with a load $q = 925$ given

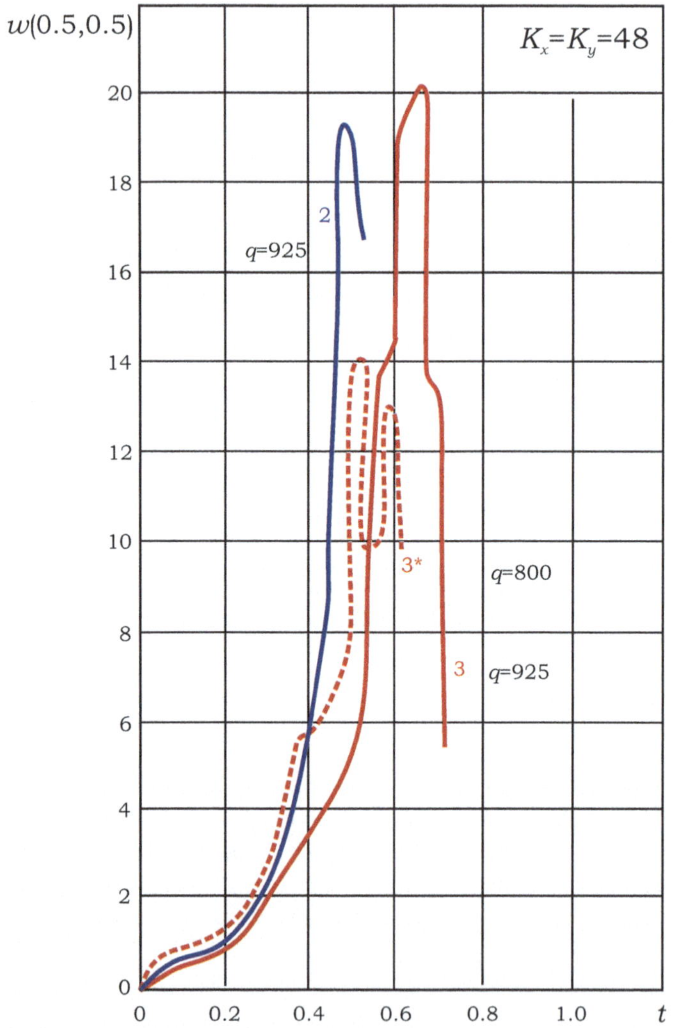

Fig. 8.63 Deflection of the centre of the shell $w(0.5, 0.5)$ as a function of time for different loads and materials: 2, nonlinear elastic material; 3, elastic–plastic material

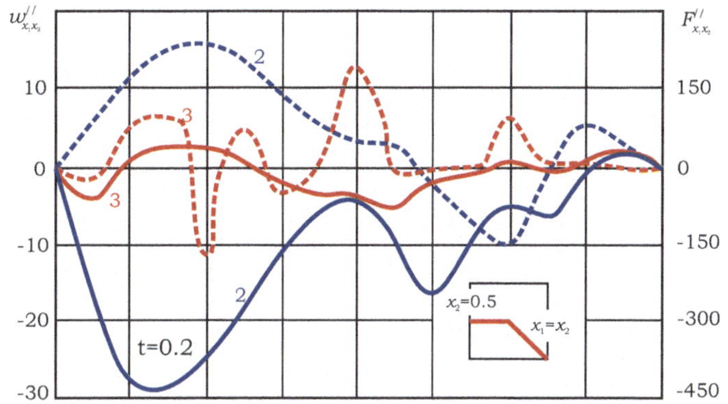

Fig. 8.64 Distribution of $F''_{x_1x_1}$ and $w''_{x_1x_1}$ for load $q = 925$ at $t = 0.2$

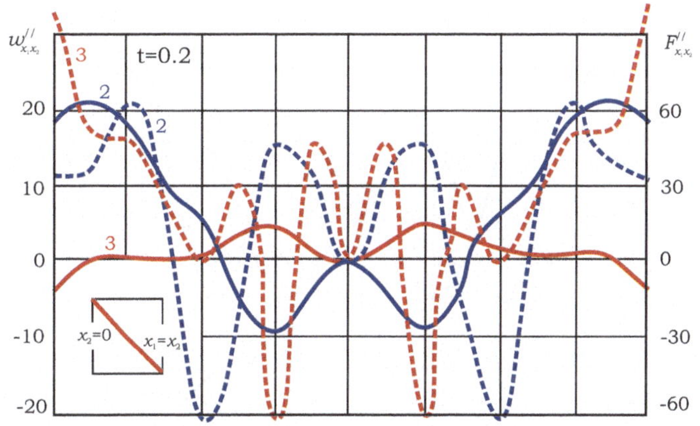

Fig. 8.65 Distribution of $F''_{x_1x_2}$ and $w''_{x_1x_2}$ for load $q = 925$ at $t = 0.2$

by models 2 and 3 are presented in Figs. 8.64, 8.65, 8.66, 8.67, 8.68, 8.69, 8.70 and 8.71. Analysis of those figures indicates that the stress–strain state of the shell for $k_{x_1} = k_{x_2} = 48$ is much more variable than that for $k_{x_1} = k_{x_2} = 24$.

In Figs. 8.72, 8.73, 8.74 and 8.75, the distributions of plastic zones for models 2 and 3 at different time moments are presented. The development of plastic zones shown here for the elastic–plastic shell (model 3) implies that strong plastic deformation occurs with the passage of time, although elastic properties persist in some regions of the shell. Examining the behaviour of the plastic zones for model

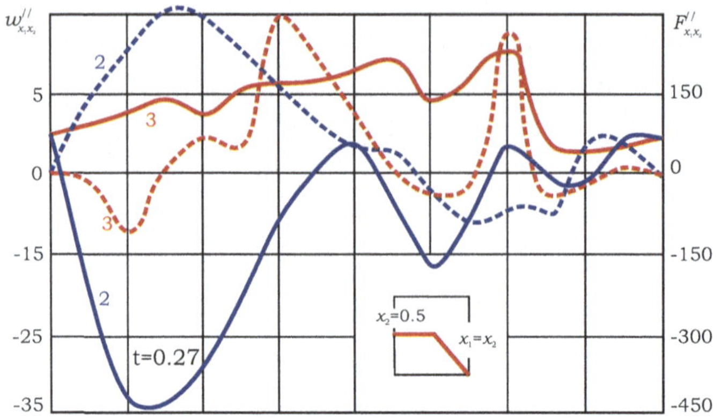

Fig. 8.66 Distribution of $F''_{x_1x_1}$ and $w''_{x_1x_1}$ for load $q = 925$ at $t = 0.27$

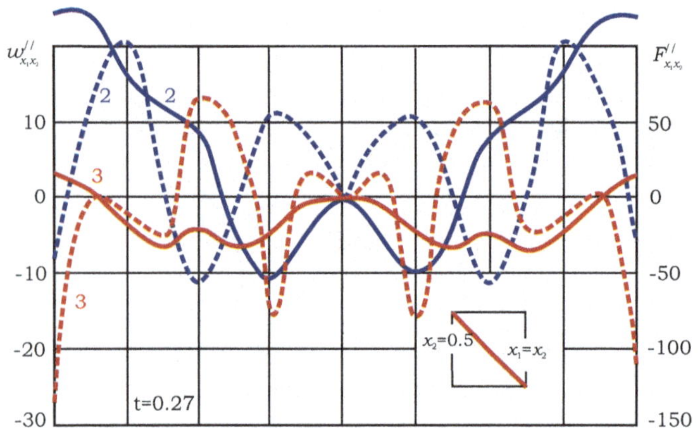

Fig. 8.67 Distribution of $F''_{x_1x_2}$ and $w''_{x_1x_2}$ for load $q = 925$ at $t = 0.27$

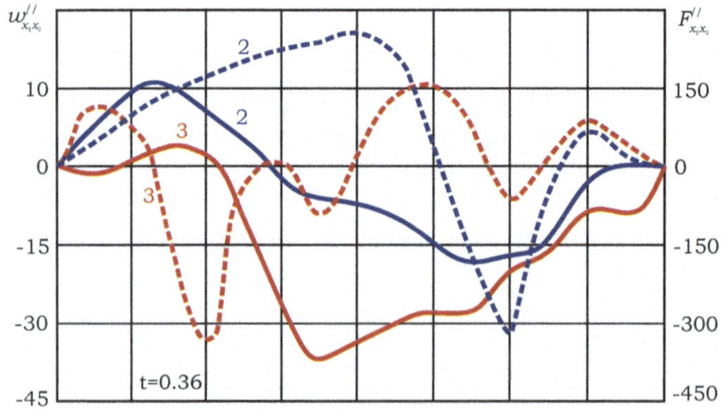

Fig. 8.68 Distribution of $F''_{x_1x_1}$ and $w''_{x_1x_1}$ for load $q = 925$ at $t = 0.36$

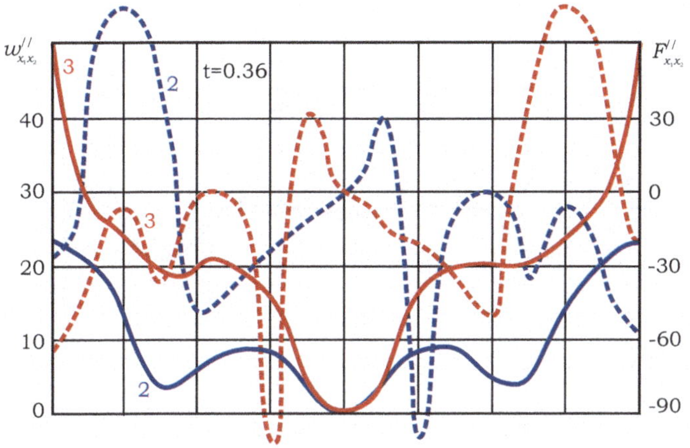

Fig. 8.69 Distribution of $F''_{x_1x_2}$ and $w''_{x_1x_2}$ for load $q = 925$ at $t = 0.36$

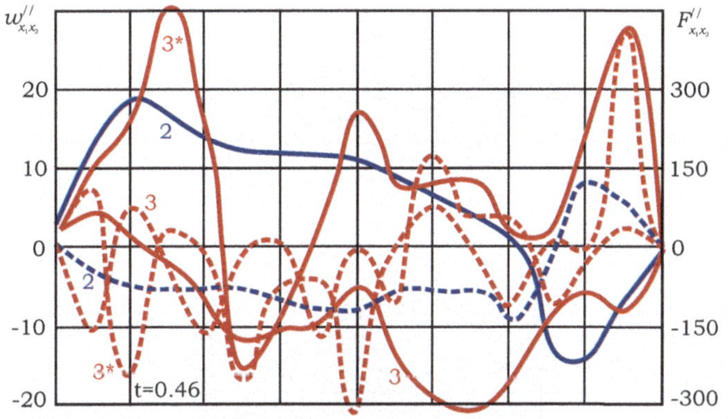

Fig. 8.70 Distribution of $F''_{x_1x_1}$ and $w''_{x_1x_1}$ for load $q = 925$ and $q = 800$ (asterisk) at $t = 0.46$

2, we observe only small changes with time, which indicates a high level of stress in the shell. The distributions of plastic zones models 2 and 3 are similar at some time moments. However, for model 3, a more complicated picture of plastic deformation is seen.

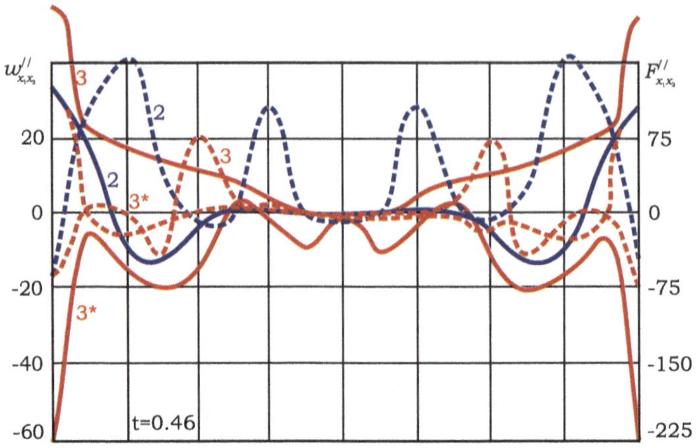

Fig. 8.71 Distribution of $F''_{x_1x_2}$ and $w''_{x_1x_2}$ for load $q = 925$ and $q = 800$ (asterisk) at $t = 0.46$

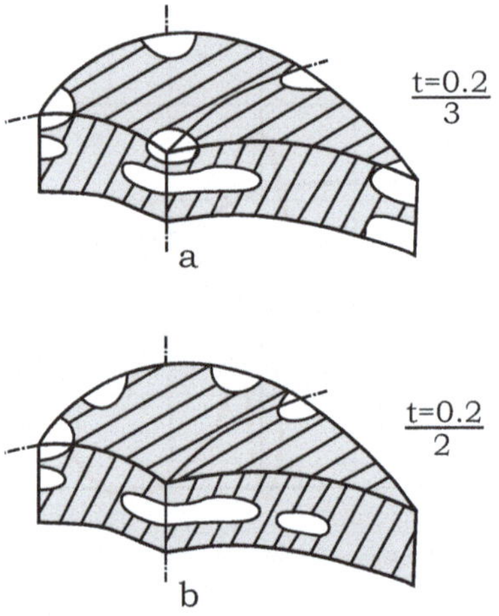

Fig. 8.72 (a, b) Distribution of plastic zones for models 2 and 3 at $t = 0.2$

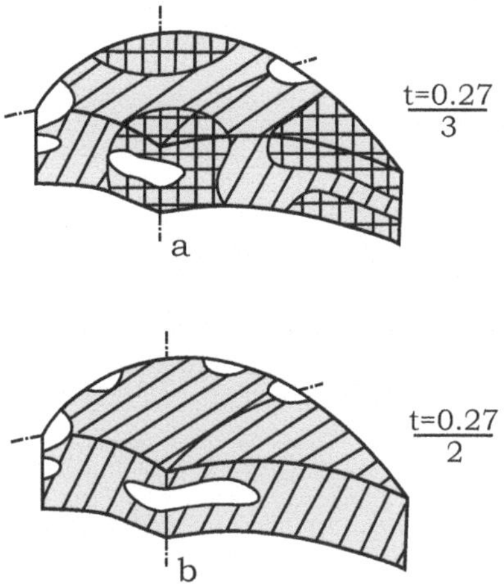

Fig. 8.73 (**a, b**) Distribution of plastic zones for models 2 and 3 at $t = 0.27$

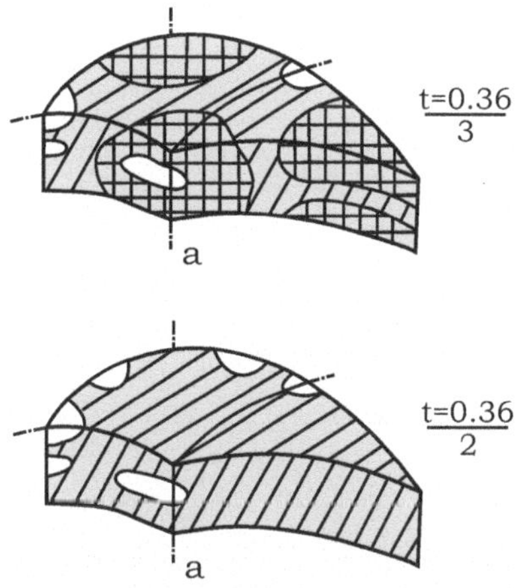

Fig. 8.74 (**a, b**) Distribution of plastic zones for models 2 and 3 at $t = 0.36$

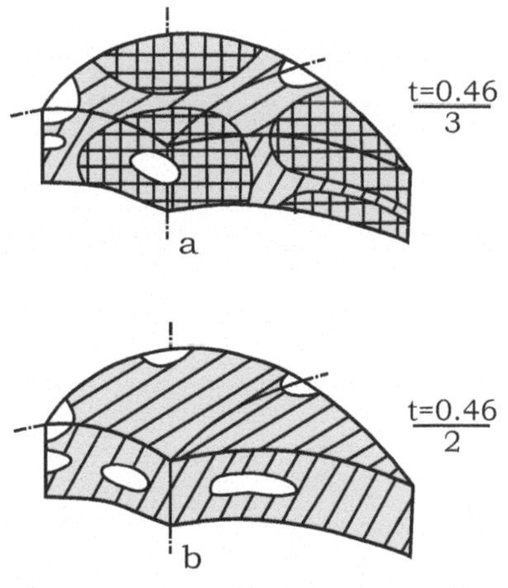

Fig. 8.75 (**a, b**) Distribution of plastic zones for models 2 and 3 at $t = 0.46$

Chapter 9
Mathematical Model of Cylindrical/Spherical Shell Vibrations

In this chapter a mathematical model of flexible shallow shells with an account of geometric non-linearity is derived. The results reliability is discussed. The stiff stability loss versus the cylindrical panels curvature k_y and versus the spherical shell with rectangular planform curvatures k_x, k_y under static load and an impulse of infinite length is analysed (conservative systems).

It has been found that the dynamics stability loss occurs earlier, i.e. for lower value of the load than in the static case.

Finally, in the case of the closed cylindrical shells the dynamical stability loss versus the width of the external transversal pressure has been investigated in both static and dynamic cases.

9.1 Fundamental Relations and Assumptions

We are aimed on analysis of a shallow shell as a closed 3D subspace of the space R^3 in the curvilinear system of coordinates x, y, z introduced in the following way. In the shell body its middle surface is fixed for $z = 0$, the axes ox and oy are directed along the main curvatures of the middle shell surface, whereas the axis oz tends into the curvature centre (Fig. 9.1). In the given coordinates the shell 3D space is defined in the following way:

$$\Omega = \left\{ x,y,z \,/\, (x, y, z) \in [0, \alpha] \times [0, \beta] \times \left[-\frac{h}{2}, \frac{h}{2} \right] \right\}.$$

It is assumed that the Lamé parameters A, B and the radii R'_1, R_{12}, R'_2 of the middle surface and their derivatives are continuous functions with regard to x, y. In contrary to the main radii R_1, R_2, by R'_1 and R'_2 are denoted the radii of the

J. Awrejcewicz, V. A. Krysko, *Elastic and Thermoelastic Problems in Nonlinear Dynamics of Structural Members*, Scientific Computation,
https://doi.org/10.1007/978-3-030-37663-5_9

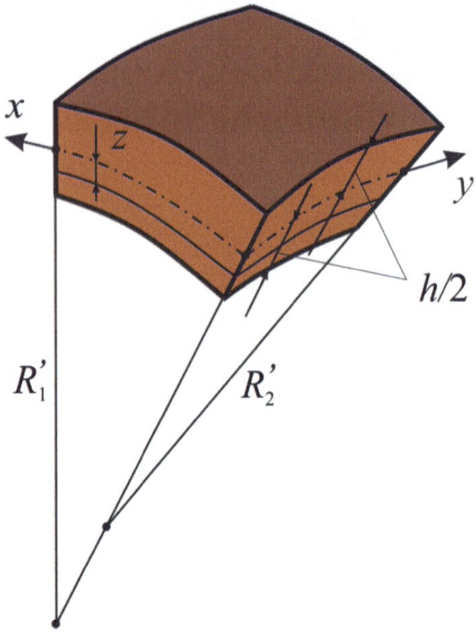

Fig. 9.1 The studied shallow shell (see the text for detailed description)

middle shell surface curvature in directions x, y. A coordinate normal to the middle shell surface is denoted by z (see Fig. 9.1).

It is assumed that $h(x, y)$ does not exhibit the first order discontinuities, and that the shell maximum thickness $h_{\max} \equiv h_0$ is essentially less than the smallest main curvature radius $R_{\min}$. We also assume that ${}^{h_0}/_{R_{\min}}$ is less than 1, and hence it is further omitted. Shells with an account of the latter assumption are called thin, and they obey the Kirchhoff-Love hypothesis of straight lines.

Shell deformation in its arbitrary point are defined as follows

$$e_1 = \varepsilon_1 + \gamma\chi_1 \quad e_2 = \varepsilon_2 + \gamma\chi_2 \quad e_{12} = \varepsilon_{12} + 2\gamma\chi_{12}, \tag{9.1}$$

where the middle shell surface deformations are governed by the following relations

$$\begin{aligned} \varepsilon_1 &= \frac{\partial u}{\partial x} + \frac{1}{2}\left(\frac{\partial w}{\partial x}\right)^2 - k_x w \,, \\ \varepsilon_2 &= \frac{\partial v}{\partial y} + \frac{1}{2}\left(\frac{\partial w}{\partial y}\right)^2 - k_y w \,, \\ \varepsilon_{12} &= \frac{\partial u}{\partial y} + \frac{\partial v}{\partial x} + \frac{\partial w}{\partial x}\frac{\partial w}{\partial y}, \end{aligned} \tag{9.2}$$

where u, v stand for displacements with regard to x and y, respectively. Observe the terms of power two appear in formula (9.2), which are associated with occurrence of a geometrical nonlinearity (rotation angles of power two of the normal to the middle shell surface are of the same order as the shell shear and extension effects). Formulas (9.2) can be obtained from more general relations reported in reference [337] ($A = B = 1$). The parameters responsible for the variations of the middle shell curvatures are introduced in the following way:

$$\chi_1 = -\frac{\partial^2 w}{\partial x^2},\ \chi_2 = -\frac{\partial^2 w}{\partial y^2},\ \chi_{12} = -\frac{\partial^2 w}{\partial x \partial y}. \tag{9.3}$$

The being sought relation, between shell stresses and deformations is as follows

$$\begin{aligned}\sigma_1 &= \frac{E}{1-\mu^2}\left[\varepsilon_1 + \nu\varepsilon_2 + \gamma\left(\chi_1 + \nu\chi_2\right)\right],\\ \sigma_1 &= \frac{E}{1-\mu^2}\left[\varepsilon_2 + \nu\varepsilon_1 + \gamma\left(\chi_2 + \nu\chi_1\right)\right],\\ \sigma_{12} &= \frac{E}{2\left(1+\mu\right)}\left(\varepsilon_{11} + 2\gamma\chi_{12}\right),\end{aligned} \tag{9.4}$$

where: E—elasticity modulus, μ—Poisson's coefficient. We carry out integration along shell thickness with a help of formulas (9.4), and in result the forces T_1, T_2, T_{12} are obtained. Multiplying both sides of formulas (9.4) by z and carrying out the integration we get the moments. The computation yields the following results

$$\begin{aligned}M_1 &= \frac{Eh^3}{12(1-\mu^2)}\left(\chi_1 + \mu\chi_2\right), \quad M_2 = \frac{Eh^3}{12(1-\mu^2)}\left(\mu\chi + \chi_2\right),\\ M_{12} &= \frac{Eh^3}{24(1+\mu)}\chi_{12}\end{aligned} \tag{9.5}$$

$$T_1 = \frac{Eh}{1-\mu^2}\left(\varepsilon_1 + \mu\varepsilon_2\right), \quad T_2 = \frac{Eh}{1-\mu^2}\left(\mu\varepsilon_1 + \varepsilon_2\right), \quad T_{12} = \frac{Eh}{2(1+\mu)}\varepsilon_{12}, \tag{9.6}$$

where M_1, M_2, are bending torques (moments) (M_{12}), χ_1, χ_2, χ_{12} stand for the parameters responsible for the change of the middle shell surface curvatures, T_1, T_2, T_{12} are the stress in the middle shell surface, and ε_1, ε_2, ε_{12} stand for deformations in the middle shell surface. Let us consider trajectories of motion of the shell points between their initial (t_0) and final (t_1) states. The real trajectories satisfy the following condition:

$$\int_{t_0}^{t_1} \left(\delta K - \delta V + \delta' W\right) dt = 0, \tag{9.7}$$

where K stands for the shell kinetic energy, V is the shell potential energy, and $\delta' W$ presents a sum of elementary works of the external forces. Based on the general theory of elasticity with an account of the hypothesis of a non-deformed normal, we get

$$V = \frac{1}{2} \iint_{\Omega} \int_{-h/2}^{h/2} (\sigma_1 \varepsilon_1 + \sigma_2 \varepsilon_2 + \sigma_{12} \varepsilon_{12}) dz ds = 0. \tag{9.8}$$

Now, taking into account of (9.2) and (9.4) and carrying out integration with regard to z, we obtain

$$V = V_1 + V_2,$$

$$V_1 = \frac{1}{2} \iint_{\Omega} \frac{2hE}{1-\mu^2} \left[\varepsilon_1^2 + \varepsilon_2^2 + 2\mu \varepsilon_1 \varepsilon_2 + \frac{1-\mu}{2} \varepsilon_{12}^2 \right] ds,$$

$$V_2 = \frac{1}{2} \iint_{\Omega} \frac{2h^3 E}{3(1-\mu^2)} \left[\left(\frac{\partial^2 w}{\partial x^2} \right)^2 + \left(\frac{\partial^2 w}{\partial y^2} \right)^2 + 2(1-\mu) \left(\frac{\partial^2 w}{\partial x \partial y} \right)^2 + 2\mu \frac{\partial^2 w}{\partial x^2} \frac{\partial^2 w}{\partial y^2} \right] ds. \tag{9.9}$$

Introducing variation of V_1 and taking into account of (9.6), we get

$$\delta V_1 = \iint_{\Omega} (T_1 \delta \varepsilon_1 + T_2 \delta \varepsilon_2 + T_{12} \delta \varepsilon_{12}) \, ds,$$

$$\delta V_1 = \iint_{\Omega} \frac{1}{2hE} \left[T_1 \delta (T_1 - \mu T_2) + T_2 \delta (T_2 - \mu T_1) + 2(1+\mu) T_{12} \delta (T_{12}) \right] ds =$$

$$= \iint_{\Omega} \frac{1}{2hE} \left[(T_1 - \mu T_2) \, \delta (T_1) + (T_2 - \mu T_1) \, \delta (T_2) + 2(1+\mu) T_{12} \delta (T_{12}) \right] ds. \tag{9.10}$$

In the next step, we introduce $T_1 = \frac{\partial^2 F}{\partial y^2}$, $T_2 = \frac{\partial^2 F}{\partial x^2}$, $T_{12} = -\frac{\partial^2 F}{\partial x \partial y}$ into formulas (9.10).

Deformations in the middle shell surfaces are defined based on the stresses (9.6) which are required to get a functional in the mixed form. Substituting deformations of the middle shell surface defined through the forces, moments and curvatures into (9.9), (9.10) and (9.7) and carrying out the standard operations (integration by parts and introduction of the Airy's function), one gets

$$\delta_F V_1 = \iint_{\Omega} \frac{1}{2hE} \left[\left(\frac{\partial^2 F}{\partial y^2} - \mu \frac{\partial^2 F}{\partial x^2} \right) \frac{\partial^2 (\cdot)}{\partial y^2} + \left(\frac{\partial^2 F}{\partial x^2} - \mu \frac{\partial^2 F}{\partial y^2} \right) \frac{\partial^2 (\cdot)}{\partial x^2} + \right.$$

$$\left. +2(1+\mu) \frac{\partial^2 F}{\partial x \partial y} \frac{\partial^2 (\cdot)}{\partial x \partial y} + \nabla_k^2 w + \frac{1}{2} L(w, w) \right] \delta(F) ds, \tag{9.11}$$

where $(\cdot)$ denotes derivative of the variation of function F, and

$$L(w,F)=\frac{\partial^2 w}{\partial x^2}\frac{\partial^2 F}{\partial y^2}+\frac{\partial^2 F}{\partial x^2}\frac{\partial^2 w}{\partial y^2}-2\frac{\partial^2 w}{\partial x\partial y}\frac{\partial^2 F}{\partial x\partial y},$$
$$\nabla_k^2=k_y\frac{\partial^2}{\partial x^2}+k_x\frac{\partial^2}{\partial y^2}. \tag{9.12}$$

We consider V_2 (see (9.9)) and we separate the variation of w. The kinetic energy of the shell (without an account of the shell element rotation) is

$$K=\frac{1}{2}\iint\limits_{\Omega}\frac{2h\rho}{g}\left[\left(\frac{\partial u}{\partial t}\right)^2+\left(\frac{\partial v}{\partial t}\right)^2+\left(\frac{\partial w}{\partial t}\right)^2\right]ds. \tag{9.13}$$

In what follows we compute integrals with regard to time of the variation of the shell kinetic energy. Carrying out the mentioned integration from t_0 up t_1, one obtains

$$\int\limits_{t_0}^{t_1}\delta_u K dt=\iint\limits_{\Omega}\frac{2h\rho}{g}\left[\frac{\partial u}{\partial t}\delta(u)\right]\Bigg|_{t_0}^{t_1}ds-\int\limits_{t_0}^{t_1}\iint\limits_{\Omega}\frac{2h\rho}{g}\frac{\partial^2 u}{\partial t^2}\delta(u)dsdt,$$
$$\int\limits_{t_0}^{t_1}\delta_v K dt=\iint\limits_{\Omega}\frac{2h\rho}{g}\left[\frac{\partial v}{\partial t}\delta(v)\right]\Bigg|_{t_0}^{t_1}ds-\int\limits_{t_0}^{t_1}\iint\limits_{\Omega}\frac{2h\rho}{g}\frac{\partial^2 v}{\partial t^2}\delta(v)dsdt, \tag{9.14}$$
$$\int\limits_{t_0}^{t_1}\delta_w K dt=\iint\limits_{\Omega}\frac{2h\rho}{g}\left[\frac{\partial w}{\partial t}\delta(w)\right]\Bigg|_{t_0}^{t_1}ds-\int\limits_{t_0}^{t_1}\iint\limits_{\Omega}\frac{2h\rho}{g}\frac{\partial^2 w}{\partial t^2}\delta(w)dsdt.$$

The elementary work of the external forces is as follows

$$\delta' W=\iint\limits_{\Omega}\left[p_x\delta(u)+p_y\delta(v)+\left(q-\varepsilon\frac{2h\rho}{g}\frac{\partial w}{\partial t}\right)\delta(w)\right]ds. \tag{9.15}$$

Substituting the all obtained expressions into (9.7) based on the Hamilton principle, and taking into account that variations $\delta u, \delta v, \delta w, \delta F$ are considered as the functions on time t and they are chosen in an arbitrary way, the following PDEs governing shell motion and the deformation compatibility are obtained:

$$\frac{2h^3E}{3(1-\mu^2)}\left[\frac{\partial^2 w}{\partial x^2}\frac{\partial^2(\cdot)}{\partial x^2}+\frac{\partial^2 w}{\partial y^2}\frac{\partial^2(\cdot)}{\partial y^2}+2(1-\mu)\frac{\partial^2 w}{\partial x\partial y}\frac{\partial^2(\cdot)}{\partial x\partial y}+\right.$$
$$\left.+\mu\left(\frac{\partial^2 w}{\partial x^2}\frac{\partial^2(\cdot)}{\partial y^2}+\frac{\partial^2 w}{\partial y^2}\frac{\partial^2(\cdot)}{\partial x^2}\right)\right]-\nabla_k^2 F+p_x(t)\frac{\partial^2 w}{\partial y^2}+$$
$$+p_y(t)\frac{\partial^2 w}{\partial x^2}-L(w,F)+q(t)-\frac{2h\rho}{g}\left(\frac{\partial^2 w}{\partial t^2}+\varepsilon\frac{\partial w}{\partial t}\right)=0, \tag{9.16}$$
$$\frac{1}{2hE}\left[\left(\frac{\partial^2 F}{\partial y^2}-\mu\frac{\partial^2 F}{\partial x^2}\right)\frac{\partial^2(\cdot)}{\partial y^2}+\left(\frac{\partial^2 F}{\partial x^2}-\mu\frac{\partial^2 F}{\partial y^2}\right)\frac{\partial^2(\cdot)}{\partial x^2}+\right.$$
$$\left.+2(1+\mu)\frac{\partial^2 F}{\partial x\partial y}\frac{\partial^2(\cdot)}{\partial x\partial y}\right]+\nabla_k^2 w+\frac{1}{2}L(w,w)=0.$$

In order to recast the obtained PDEs into their counterpart nondimensional forms, the following nondimensional quantities are introduced: $w = h\bar{w}$, $F = Eh^2\bar{F}$, $t = t_0\bar{t}$, $\varepsilon = \bar{\varepsilon}/\tau$ (bars stand for nondimensional quantities). Furthermore, in the case of a spherical panel with the rectangular planform and for the cylindrical panel we take: $x = a\bar{x}$, $y = a\bar{y}$; $k_x = \bar{k}_x\frac{h}{b^2}$, $k_y = \bar{k}_y\frac{h}{a^2}$, $q = \bar{q}\frac{Eh^4}{a^2b^2}$, $p_x = \bar{p}_x\frac{Eh^3}{b^2}$, $p_y = \bar{p}_y\frac{Eh^3}{a^2}$, $\tau = \frac{ab}{h}\sqrt{\frac{\rho}{Eg}}$, $M = 1$, $\lambda = \frac{a}{b}$, where a, b denote the shell dimension with regard to x and y, respectively. In the case of the closed cylindrical shell: $x = L\bar{x}$, $y = R\bar{y}$, $k_y = \bar{k}_y\frac{h}{R^2}$ ($k_x = 0$), $q = \bar{q}\frac{Eh^4}{L^2R^2}$, $p_x = \bar{p}_x\frac{Eh^3}{R^2}$, $p_y = \bar{p}_y\frac{Eh^3}{L^2}$, $\tau = \frac{LR}{h}\sqrt{\frac{\rho}{Eg}}$, $M = {k_y}^2$, $\lambda = \frac{L}{R}$, where L and $R = R_y$ denote the shell length and radius, respectively. In addition, t stand for time, ε is the surrounding medium damping, $\mu = 0.3$, $p_x(t)$, $p_y(t)$ are the longitudinal loads, and $q(x, y, t)$ present the transversal load. The governing nondimensional PDEs (bars are omitted) take the following form

$$\left[\frac{1}{\lambda^2}\frac{\partial^2 w}{\partial x^2}\frac{\partial^2(\cdot)}{\partial x^2}+\lambda^2\frac{\partial^2 w}{\partial y^2}\frac{\partial^2(\cdot)}{\partial y^2}+2(1-\mu)\frac{\partial^2 w}{\partial x\partial y}\frac{\partial^2(\cdot)}{\partial x\partial y}+\right.$$
$$\left.+\mu\left(\frac{\partial^2 w}{\partial x^2}\frac{\partial^2(\cdot)}{\partial y^2}+\frac{\partial^2 w}{\partial y^2}\frac{\partial^2(\cdot)}{\partial x^2}\right)\right]-\nabla_k^2 F+p_x(t)\frac{\partial^2 w}{\partial y^2}+$$
$$+p_y(t)\frac{\partial^2 w}{\partial x^2}-L(w,F)+M\cdot q(t)-\left(\frac{\partial^2 w}{\partial t^2}+\varepsilon\frac{\partial w}{\partial t}\right)=0, \tag{9.17}$$
$$\left[\left(\lambda^2\frac{\partial^2 F}{\partial y^2}-\mu\frac{\partial^2 F}{\partial x^2}\right)\frac{\partial^2(\cdot)}{\partial y^2}+\left(\frac{1}{\lambda^2}\frac{\partial^2 F}{\partial x^2}-\mu\frac{\partial^2 F}{\partial y^2}\right)\frac{\partial^2(\cdot)}{\partial x^2}+\right.$$
$$\left.+2(1+\mu)\frac{\partial^2 F}{\partial x\partial y}\frac{\partial^2(\cdot)}{\partial x\partial y}\right]+\nabla_k^2 w+\frac{1}{2}L(w,w)=0.$$

The following boundary conditions are considered.

1. Moving clamping on the shell faces

$$
\begin{aligned}
& w=0, \quad \frac{\partial w}{\partial x}=0, \quad F=0, \quad \frac{\partial F}{\partial x}=0 \quad \text{for} \quad x=0;\,1; \\
& w=g(x,y,t), \quad \frac{\partial w}{\partial y}=p(x,y,t), \quad F=u(x,y,t), \\
& \frac{\partial F}{\partial y}=v(x,y,t) \quad \text{for} \quad y=0;\,\xi.
\end{aligned} \tag{9.18}
$$

2. Simple support on the shell faces

$$
\begin{aligned}
& w=0, \quad \frac{\partial w}{\partial x}=0, \quad F=0, \quad \frac{\partial^2 F}{\partial x^2}=0 \quad \text{for} \quad x=0;\,1; \\
& w=g(x,y,t), \quad \frac{\partial w}{\partial y}=p(x,y,t), \quad F=u(x,y,t), \\
& \frac{\partial F}{\partial y}=v(x,y,t) \quad \text{for} \quad y=0;\,\xi.
\end{aligned} \tag{9.19}
$$

3. Movable clamping on the shell faces with ribs

$$
\begin{aligned}
& w=0, \quad \frac{\partial^2 w}{\partial x^2}=0, \quad F=0, \quad \frac{\partial F}{\partial x}=0 \quad \text{for} \quad x=0;\,1; \\
& w=g(x,y,t), \quad \frac{\partial w}{\partial y}=p(x,y,t), \quad F=u(x,y,t), \\
& \frac{\partial F}{\partial y}=v(x,y,t) \quad \text{for} \quad y=0;\,\xi.
\end{aligned} \tag{9.20}
$$

4. Movable clamping on the shell faces with flexible ribs

$$
\begin{aligned}
& w=0, \quad \frac{\partial^2 w}{\partial x^2}=0, \quad F=0, \quad \frac{\partial^2 F}{\partial x^2}=0 \quad \text{for} \quad x=0;\,1; \\
& w=g(x,y,t), \quad \frac{\partial^2 w}{\partial y^2}=r(x,y,t), \quad F=u(x,y,t), \\
& \frac{\partial^2 F}{\partial y^2}=z(x,y,t) \quad \text{for} \quad y=0;\,\xi.
\end{aligned} \tag{9.21}
$$

Here and further $\xi = 2\pi$ for the case of the closed cylindrical shell, and $\xi = 1$ for the case of either a cylindrical or a spherical shell. Non-homogeneity of the boundary conditions for $y = 0;\,\xi$ imply occurrence of imperfections and stresses in the shell. Equations (9.17), in spite of the boundary conditions, require a definition of the initial conditions (9.18)–(9.21), which are taken as follows

$$w|_{t=0} = w_0, \qquad \dot{w}|_{t=0} = 0. \tag{9.22}$$

9.2 The Bubnov-Galerkin Method

In order to solve PDEs (9.17), the functions w and F are approximated by analytical expressions containing the finite number of arbitrary parameters in the form of a product of the functions depending on time and coordinates, i.e. we take

$$w = \sum_{i=0}^{N_1}\sum_{j=0}^{N_2} A_{ij}(t)\varphi_{ij}(x, y), \quad F = \sum_{i=0}^{N_1}\sum_{j=0}^{N_2} B_{ij}(t)\psi_{ij}(x, y). \tag{9.23}$$

In order to find the approximate value of the functions w and F, we choose the coordinate systems of the functions $\{\varphi_{ij}(x, y), \psi_{ij}(x, y)\}$ $(i, j = 0, 1, 2\ldots)$ in (9.23), which satisfy the following requirements:

1. $\varphi_{ij}(x, y) \in H_A$, $\psi_{ij}(x, y) \in H_A$, where H_A stands for the Hilbert space which is further called energetic space;
2. $\forall i, j$ functions $\varphi_{ij}(x, y)$ and $\psi_{ij}(x, y)$ are linearly independent, continuous with their partial derivatives up to the fourth order in the space Ω;
3. $\varphi_{ij}(x, y)$ and $\psi_{ij}(x, y)$ satisfy rigorously the main boundary conditions (and possibly boundary conditions);
4. $\varphi_{ij}(x, y)$ and $\psi_{ij}(x, y)$ are complete in H_A;
5. $\varphi_{ij}(x, y)$ and $\psi_{ij}(x, y)$ should represent N first elements of the full system of functions.

The coefficients $A_{ij}(t)$ and $B_{ij}(t)$ are being searched functions on time. Denoting the left hand sides of Eqs. (9.17) by Φ_1 and Φ_2, we get

$$\Phi_1(w, F, \frac{\partial^2 w}{\partial x^2}, \frac{\partial^2 F}{\partial x^2}, \ldots) + M \cdot q(x, y, t) = 0, \quad \Phi_2(w, F, \frac{\partial^2 w}{\partial x^2}, \frac{\partial^2 F}{\partial x^2}, \ldots) = 0. \tag{9.24}$$

Employing the Bubnov-Galerkin procedure, PDEs (9.24) yield

$$\begin{gathered}\int_0^1\int_0^{\xi} \Phi_1\varphi_{rs}(x, y)dxdy + \int_{x_1}^{x_2}\int_{y_1}^{y_2} Mq(x, y, t)\varphi_{rs}(x, y)dxdy = 0,\\ \int_0^1\int_0^{\xi} \Phi_2\psi_{rs}(x, y)dxdy = 0, \; r = 0, 1, \ldots, N_1; s = 0, 1, \ldots N_2.\end{gathered} \tag{9.25}$$

Equations (9.24), with an account of (9.23) take the following form

$$\sum_{rs}\left[\sum_{ij}A_{ij}\sum_{kl}H_{ijklrs}+\sum_{ij}B_{ij}C_{1,ijrs}+\sum_{ij}A_{ij}W_{ijrs}+MqQ_{rs}+\right.$$
$$\left.+\sum_{ij}A_{ij}\sum_{kl}B_{kl}D_{1,ijklrs}+\sum_{ij}\left[\frac{d^2A_{ij}}{dt^2}+\varepsilon\frac{dA_{ij}}{dt}\right]G_{ijrs}\right]=0,$$
$$\sum_{rs}\left[\sum_{ij}A_{ij}C_{2,ijrs}+\sum_{ij}B_{ij}\sum_{kl}P_{ijklrs}+\sum_{ij}A_{ij}\sum_{kl}A_{rs}D_{2,ijklrs}\right]=0. \tag{9.26}$$

Here the symbol $\sum_{rs}[*]$ standing before each of equations of the system (9.26) means abbreviation of the rs equations, and the associated integrals of the Bubnov-Galerkin procedure take the following form

$$H_{ijklrs}=\int_0^1\int_0^\xi\frac{1}{12(1-\mu^2)}\left[\frac{1}{\lambda^2}\frac{\partial^2\varphi_{ij}}{\partial x^2}\frac{\partial^2\varphi_{kl}}{\partial x^2}+\lambda^2\frac{\partial^2\varphi_{ij}}{\partial y^2}\frac{\partial^2\varphi_{kl}}{\partial y^2}+\right.$$
$$\left.+2\frac{\partial^2\varphi_{ij}}{\partial x\partial y}\frac{\partial^2\varphi_{kl}}{\partial x\partial x}\right]\varphi_{rs}dxdy,$$
$$C_{1,ijrs}=\int_0^1\int_0^\xi\left[-k_y\frac{\partial^2\psi_{ij}}{\partial x^2}-k_x\frac{\partial^2\psi_{ij}}{\partial y^2}\right]\varphi_{rs}dxdy,$$
$$C_{2,ijrs}=\int_0^1\int_0^\xi\left[k_y\frac{\partial^2\varphi_{ij}}{\partial x^2}+k_x\frac{\partial^2\varphi_{ij}}{\partial y^2}\right]\psi_{rs}dxdy, \tag{9.27}$$
$$D_{1,ijklrs}=\int_0^1\int_0^\xi[-L(\varphi_{ij},\psi_{kl})]\varphi_{rs}dxdy,$$
$$D_{2,ijklrs}=\int_0^1\int_0^\xi\frac{1}{2}L(\varphi_{ij},\varphi_{kl})\psi_{rs}dxdy,$$

$$G_{ijrs} = \int_0^1 \int_0^{\xi} [-\varphi_{ij}\varphi_{rs}] dxdy,$$

$$Q_{rs} = \int_{x_1}^{x_2} \int_{y_1}^{y_2} Mq(x, y, t)\varphi_{rs} dxdy,$$

$$W_{ijrs} = \int_0^1 \int_0^{\xi} \left[p_y(x, t)\frac{\partial^2 \varphi_{ij}}{\partial x^2} + p_x(y, t)\frac{\partial^2 \varphi_{ij}}{\partial y^2} \right] \varphi_{rs} dxdy,$$

$$P_{ijklrs} = \int_0^1 \int_0^{\xi} \left[\frac{1}{\lambda^2}\frac{\partial^2 \psi_{ij}}{\partial x^2}\frac{\partial^2 \psi_{kl}}{\partial x^2} + \lambda^2 \frac{\partial^2 \psi_{ij}}{\partial y^2}\frac{\partial^2 \psi_{kl}}{\partial y^2} + 2\frac{\partial^2 \psi_{ij}}{\partial x \partial y}\frac{\partial^2 \psi_{kl}}{\partial x \partial x} \right] \psi_{rs} dxdy.$$

Integrals (9.27), with a possible exception of Q_{rs}, are computed with regard to the whole shell surface. The Bubnov-Galerkin method yields the system of ODEs with respect to the functions $A_{ij}(t)$ and $B_{ij}(t)$, which has the following matrix form representation

$$\begin{aligned} &\mathbf{G}(\ddot{\mathbf{A}} + \varepsilon\dot{\mathbf{A}}) + \mathbf{HA} + \mathbf{WA} + \mathbf{C}_1\mathbf{B} + \mathbf{D}_1\mathbf{AB} = \mathbf{Q}q(t), \\ &\mathbf{C}_2\mathbf{A} + \mathbf{PB} + \mathbf{D}_2\mathbf{AA} = 0, \end{aligned} \tag{9.28}$$

where: $\mathbf{H} = \|H_{ijrs}\|$, $\mathbf{G} = \|G_{ijrs}\|$, $\mathbf{C}_1 = \|C_{1ijrs}\|$, $\mathbf{C}_2 = \|C_{2ijrs}\|$, $\mathbf{D}_1 = \|D_{1ijklrs}\|$, $\mathbf{D}_2 = \|D_{2ijklrs}\|$, $\mathbf{W} = \|W_{ijrs}\|$, $\mathbf{P} = \|P_{ijrs}\|$ stand for the square matrices of dimension $2 \cdot N_1 \cdot N_2 \times 2 \cdot N_1 \cdot N_2$, whereas $\mathbf{A} = \|A_{ij}\|$, $\mathbf{B} = \|B_{ij}\|$, $\mathbf{Q} = \|Q_{ij}\|$ are the matrices of dimension $2 \cdot N_1 \cdot N_2 \times 1$.

Observe that the second equation of system (9.28) is solvable with respect to **B**, and in fact it is solved at each time step as follows (9.29):

$$\mathbf{B} = \left[-\mathbf{P}^{-1}\mathbf{D}_2\mathbf{A} - \mathbf{P}^{-1}\mathbf{C}_2 \right] \mathbf{A}. \tag{9.29}$$

Multiplying the first equation of the systems (9.28) by $\mathbf{G}^{-1}$, and denoting $\dot{\mathbf{A}} = \mathbf{R}$, the following Cauchy problem of the first order ODEs is formulated

$$\begin{cases} \dot{\mathbf{A}} = \mathbf{R}, \\ \dot{\mathbf{R}} = -\bar{\varepsilon}\mathbf{R} - \left[\mathbf{G}^{-1}\mathbf{C}_1 + \mathbf{G}^{-1}\mathbf{D}_1\mathbf{A} \right] \cdot \mathbf{B} - \mathbf{G}^{-1}\mathbf{HA} - \mathbf{G}^{-1}\mathbf{WA} + \mathbf{G}^{-1}\mathbf{Q}q(\bar{t}). \end{cases} \tag{9.30}$$

The carried out procedure is allowed since the inverse matrices $\mathbf{G}^{-1}$ and $\mathbf{P}^{-1}$ exist assuming that the coordinate functions are linearly independent.

Equations (9.30) are supplemented by one of the boundary conditions (9.18)–(9.21) and the initial condition (9.22) and the obtained Cauchy problem is solved by the Runge-Kutta fourth order. The time integrations step is chosen based on the Runge's rule.

9.2.1 Closed Cylindrical Shell

We investigate variations of a simply supported along the faces (circles) of the closed cylindrical shell with homogeneous boundary conditions (9.21) and zeroth initial conditions (9.22). Namely, the functions φ_{ij}, ψ_{ij} in (9.23) are presented as a product of two functions where each of them depend only on one argument, and they satisfy the boundary conditions (9.21) with respect to x:

$$w = \sum_{i=1}^{N_1}\sum_{j=0}^{N_2} A_{ij}(t)\sin(i\pi x)\cos(jy), \quad F = \sum_{i=1}^{N_1}\sum_{j=0}^{N_2} B_{ij}(t)\sin(i\pi x)\cos(jy). \tag{9.31}$$

The integrals of the Bubnov-Galerkin method are computed as follows

$$I_{1,r} = \int_{x_1}^{x_2} \sin(r\pi x)dx = \frac{\cos(r\pi x_1) - \cos(r\pi x_2)}{r\pi},$$

$$I_{2,s} = \int_{y_1}^{y_2} \cos(s\pi y)dy = \frac{\sin(s\pi y_2) - \sin(s\pi y_1)}{s\pi}$$

$$I_{3,ir} = \int_0^1 \sin(i\pi x)\sin(r\pi x)dx = \begin{cases} \frac{1}{2}, i = r, \\ 0, i \neq r, \end{cases}$$

$$I_{4,js} = \int_0^{2\pi} \cos(j\pi y)\cos(s\pi y)dy = \begin{cases} 2\pi, j = s = 0, \\ \pi, j = s \neq 0, \\ 0, , j \neq s, \end{cases}$$

$$I_{5,ikr} = \int_0^1 \sin(i\pi x)\sin(r\pi x)\sin(k\pi x)dx =$$

$$= \begin{cases} \frac{1}{4\pi}\Bigg[-\frac{\cos(\alpha_1\pi)}{\alpha_1} - \frac{\cos(\alpha_2\pi)}{\alpha_2} - \frac{\cos(\alpha_3\pi)}{\alpha_3} - \frac{\cos(\alpha_4\pi)}{\alpha_4} + \\ + \frac{1}{\alpha_1} + \frac{1}{\alpha_2} + \frac{1}{\alpha_3} - \frac{1}{\alpha_4}\Bigg], \quad \alpha_\upsilon \neq 0; \\ \left[\frac{\cos(\alpha_\upsilon\pi)}{\alpha_\upsilon} \approx 0, \frac{1}{\alpha_\upsilon} \approx 0\right], \upsilon = 1, 2, 3, \alpha_\upsilon = 0; \end{cases}$$

$$I_{7,ikr} = \int_0^1 \cos(i\pi x)\cos(k\pi x)\sin(r\pi x)dx =$$

$$= \begin{cases} \frac{1}{4\pi}\left[\frac{\cos(\alpha_1\pi)}{\alpha_1} - \frac{\cos(\alpha_2\pi)}{\alpha_2} - \frac{\cos(\alpha_3\pi)}{\alpha_3} - \frac{\cos(\alpha_4\pi)}{\alpha_4} - \right. \\ \left. -\frac{1}{\alpha_1} + \frac{1}{\alpha_2} + \frac{1}{\alpha_3} + \frac{1}{\alpha_4}\right], \quad \alpha_v \neq 0; \\ \left[\frac{\cos(\alpha_v\pi)}{\alpha_v} \approx 0, \frac{1}{\alpha_v} \approx 0\right], \quad v = 1, 2, 3, \alpha_v = 0; \end{cases}$$

$$I_{8,jls} = \int_0^{2\pi} \sin(jy)\sin(ly)\cos(sy)dy =$$

$$= \begin{cases} \frac{1}{4}\left[-\frac{\sin(\beta_1\pi)}{\beta_1} + \frac{\sin(\beta_2\pi)}{\beta_2} + \frac{\sin(\beta_3\pi)}{\beta_3} - \frac{\sin(\beta_4\pi)}{\beta_4} + \right. \\ \left. +\frac{1}{\beta_1} - \frac{1}{\beta_2} - \frac{1}{\beta_3} + \frac{1}{\beta_4}\right], \quad \beta_v \neq 0; \\ \left[\frac{\sin(\beta_v\pi)}{\beta_v} \approx 0, \frac{1}{\beta_v} \approx 0\right], \quad v = 1, 2, 3, 4, \beta_v = 0; \end{cases}$$

$$I_{6,jls} = \int_0^{2\pi} \cos(jy)\cos(ly)\sin(sy)dy = 0,$$

where

$$\alpha_1 = j + k - r, \quad \alpha_2 = k + r - i, \quad \alpha_3 = r + i - k, \quad \alpha_4 = i + k + r,$$
$$\beta_1 = j + l - s, \quad \beta_2 = l + s - j, \quad \beta_3 = s + j - l, \quad \beta_4 = j + l + s.$$

The following notation is introduced

$$I_Q^{rs} = M \cdot I_{1r}I_{2s}, \quad I_P^{rs} = \left(s^2 p_x(t) + r^2 p_y(t)\right)\pi^2 I_{3,ir}I_{4,js},$$

$$I_{AB}^{rs} = \left(s^2 k_x + r^2 k_y\right)\pi^2 I_{3,ir}I_{4,js}, \quad I_{rs}^t = I_{3,ir}I_{4,js},$$

$$I_{ijklrs} = \pi^2\left[\left(i^2l^2 + j^2k^2\right)I_{5ikr}I_{6jls} - 2ijlkI_{7ikr}I_{8jls}\right], \quad M = k_y{}^2,$$

$$J_{1,ijkl}^{rs} = \frac{\pi^2}{12(1-\mu^2)}\left[\frac{r^4}{\lambda^2} + 2r^2s^2 + \lambda^2 s^4\right] \cdot I_{3,ir} I_{4,js},$$

$$J_{2,ijkl}^{rs} = \left[\frac{r^4}{\lambda^2} + 2r^2s^2 + \lambda^2 s^4\right] \cdot \pi^2 I_{3,ir} I_{4,js}.$$

Finally, with an account of the above integrals, the system (9.30) takes the following form

$$\sum_{rs}\left\{\sum_{ij}\sum_{kl}\left[J_{1,ijkl}^{rs} A_{ij} + I_{AB}^{rs} B_{rs} + I_P^{rs} A_{ij} + I_Q^{rs} q(t) + A_{ij} B_{kl} I_{ijklrs} + \right.\right.$$
$$\left.\left. + \left(\frac{d^2 A_{ij}}{dt^2} + \varepsilon \frac{dA_{ij}}{dt}\right) I_{rs}^t\right]\right\} = 0,$$
$$\sum_{rs}\left\{\sum_{ij}\sum_{kl}\left[J_{2,ijkl}^{rs} B_{ij} + I_{AB}^{rs} A_{rs} + \frac{1}{2} A_{ij} A_{kl} I_{ijklrs}\right]\right\} = 0.$$

(9.32)

9.2.2 Cylindrical Panel

Investigation of the vibrations of the simply supported cylindrical panel and spherical panel with rectangular planform and boundary conditions (9.21) is carried out with a help of the following approximating functions:

$$w = \sum_{i=1}^{N}\sum_{j=1}^{N} A_{ij}(t)\sin(i\pi x)\sin(j\pi y), \quad F = \sum_{i=1}^{N}\sum_{j=1}^{N} B_{ij}(t)\sin(i\pi x)\sin(j\pi y).$$

(9.33)

Integrals of the Bubnov-Galerkin procedure differ from the previous case by integration with regard to coordinate y, and they are computed in the following way

$$I_{2,s} = \int_0^1 \sin(s\pi y)dy = \begin{cases} 0, & s = 0, 2, 4\ldots \\ \frac{2}{s\pi}, & s = 1, 3, 5, \ldots \end{cases}$$

$$I_{4,js} = \int_0^1 \sin(j\pi y)\sin(s\pi y)dy = \begin{cases} \frac{1}{2}, & j = s, \\ 0, & j \neq s, \end{cases}$$

$$I_{6,jls} = \int_0^1 \sin(j\pi y)\sin(l\pi y)\sin(s\pi y)dy =$$

$$= \begin{cases} \dfrac{1}{4\pi}\left[-\dfrac{\cos(\beta_1\pi)}{\beta_1} - \dfrac{\cos(\beta_2\pi)}{\beta_2} - \dfrac{\cos(\beta_3\pi)}{\beta_3} - \dfrac{\cos(\beta_4\pi)}{\beta_4} + \right. \\ \left. + \dfrac{1}{\beta_1} + \dfrac{1}{\beta_2} + \dfrac{1}{\beta_3} - \dfrac{1}{\beta_4}\right], \quad \beta_v \neq 0; \\ \left[\dfrac{\cos(\beta_v\pi)}{\beta_v} \approx 0, \dfrac{1}{\beta_v} \approx 0\right], \quad v = 1, 2, 3, \quad \beta_v = 0; \end{cases}$$

$$I_{8,jls} = \int_0^1 \cos(j\pi y)\cos(l\pi y)\sin(s\pi y)dy =$$

$$= \begin{cases} \dfrac{1}{4\pi}\left[\dfrac{\cos(\beta_1\pi)}{\beta_1} - \dfrac{\cos(\beta_2\pi)}{\beta_2} - \dfrac{\cos(\beta_3\pi)}{\beta_3} - \dfrac{\cos(\beta_4\pi)}{\beta_4} + \right. \\ \left. - \dfrac{1}{\beta_1} + \dfrac{1}{\beta_2} + \dfrac{1}{\beta_3} + \dfrac{1}{\beta_4}\right], \quad \beta_v \neq 0; \\ \left[\dfrac{\cos(\beta_v\pi)}{\beta_v} \approx 0, \dfrac{1}{\beta_v} \approx 0\right], \quad v = 1, 2, 3, \quad \beta_v = 0; \end{cases}$$

where

$$\alpha_1 = j + k - r, \quad \alpha_2 = k + r - i, \quad \alpha_3 = r + i - k, \quad \alpha_4 = i + k + r,$$
$$\beta_1 = j + l - s, \quad \beta_2 = l + s - j, \quad \beta_3 = s + j - l, \quad \beta_4 = j + l + s.$$

In what follows, taking into account (9.33), we finally get

$$I_Q^{rs} = I_{1r}I_{2s}, \quad I_P^{rs} = \left(s^2 p_x(t) + r^2 p_y(t)\right)\pi^4 I_{3,ir}I_{4,js},$$

$$I_{AB}^{rs} = \left(s^2 k_x + r^2 k_y\right)\pi^4 I_{3,ir}I_{4,js},$$

$$I_{ijklrs} = \pi^4\left[\left(i^2l^2 + j^2k^2\right)I_{5ikr}I_{6jls} - 2ijlkI_{7ikr}I_{8jls}\right],$$

$$J_{1,ijkl}^{rs} = \frac{\pi^4}{12(1-\mu^2)}\left[\frac{r^4}{\lambda^2} + 2r^2s^2 + \lambda^2 s^4\right] \cdot I_{3,ir}I_{4,js},$$

$$J_{2,ijkl}^{rs} = \left[\frac{r^4}{\lambda^2} + 2r^2s^2 + \lambda^2 s^4\right] \cdot \pi^4 I_{3,ir}I_{4,js}.$$

In words, the governing PDEs for the case of the cylindrical panel and spherical panel with the rectangular planform are obtained from (9.32) after substitution of the corresponding integrals.

9.3 Reliability of the Obtained Results

The so far briefly described and employed the Bubnov-Galerkin algorithm allows to solve a wide class of static and dynamic problems of structural members. The static problems can be solved using dynamic approach based on the set up method [111]. It is well known and documented that in order to solve static problems of theory of plates and shells there exist numerous traditional approaches mainly based on reduction of the governing nonlinear PDEs to a system of nonlinear algebraic equations, which then are solved and/or linearized. In contrary to the approaches in the case of the set up method, the evolutionary PDEs are reduced to consideration of nonlinear ODEs.

We briefly exhibit benefits of the mentioned approach from the mathematical point of view. The set up method can be understood as an iterational method of solution to a system of nonlinear algebraic equations where in each computational step we approach the exact solution of the problem. In this sense the set up method is characterized by its high accuracy. However, its drawback relies on a large sensitivity to the choice of the initial approximation. In addition, the set up method has one more strong advantage. It allows to find non-unique solutions of the static problems.

Indeed, introduction of different initial conditions may yield different steady state solutions. Besides, while solving homogeneous problems using traditional methods, one needs to introduce a certain initial imperfection, i.e. either small transversal load or a small curvature or a small initial deflection.

However, the introduction of the mentioned initial imperfection may violate the obtained results. In contrary, in the case of searching for solution of the analogous problem using the set up method, a role of the initial imperfections are undertaken by the nonhomogeneous initial conditions, and small changes of the initial conditions do not influence the obtained static solutions. Solving the Cauchy problem for $\varepsilon = \varepsilon_{cr}$ for the series $\{q_i\} \rightarrow \{w_i\}$ of the parameter of the transversal constant load allows to construct the dependence $q(w)$, and hence to investigate the stress-strain state of a studied construction.

Employing the mentioned approach for the closed cylindrical shells, we consider a dissipative linear system. We take the transversal external pressure $q(t) = q_0$ uniquely distributed along the sectorial zone characterized by an angle φ_0, $0 \leq x \leq 0.5$.

We study the cylindrical shell behaviour depending on the width of the pressure zone characterized by the angle φ_0. Shagivalev [294] solved the mentioned problem in an analytical way. Since we are going to compare the obtained by him results with ours, we remove nonlinear terms in Eqs. (9.32), i.e. we take $I_{ijklrs} = 0$.

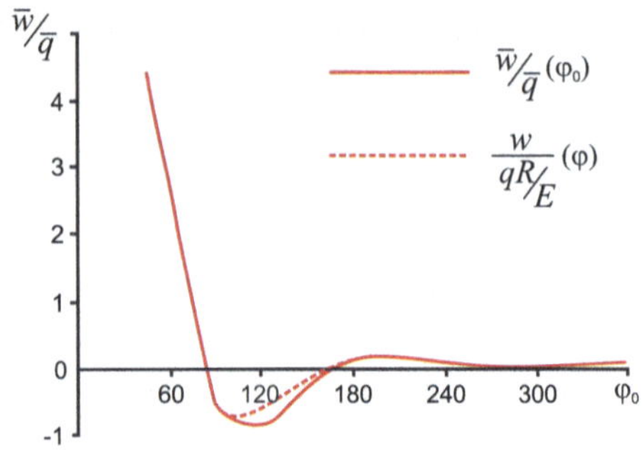

Fig. 9.2 The dependencies $\overline{w}/\overline{q}\,(\varphi_0)$ and $\frac{w}{qR/E}\,(\varphi)$

Table 9.1 Analytical versus numerical solution

φ_0	Analytical solution $\frac{w}{qR/E}$	Numerical solution $\overline{w}/\overline{q}\cdot(R/2h)\cdot\lambda^2$	Ratio $\overline{w}/\overline{q}$	Error
2π	50	52.8417	0.058713	5%
$\frac{3\pi}{2}$	6.97	7.4808	0.008312	7%
π	101.07	108.9054	0.121006	7%
$\frac{2\pi}{3}$	−524.2	−744.2469	−0.826941	28%
$\frac{\pi}{2}$	−479.47	−436.212	0.48468	8%
$\frac{\pi}{4}$	3886.27	3948.354	4.38706	1.5%

Now, having in hand the obtained simple mathematical model, we constructed the dependencies $w_{\max}(q_0)$ for various angles characterizing sectorial loads which are linear, i.e. $y = kx$, where $y = \overline{w}$, $x = \overline{q}_0$. Therefore, we have obtained $y = \overline{w}/\overline{q}_0 x$. Now, the mentioned analytical solution $\frac{w}{qR/E}\,(\varphi)$ can be compared with our numerical one $\overline{w}/\overline{q}\,(\varphi)$ by using the non-dimensional parameters in the following way: $\bar{w} = \frac{w}{2h}$, $\bar{q} = \frac{L^2R^2}{E(2h)^4\bar{k}_y^2}q$, and hence $\frac{\bar{w}}{\bar{q}} = \frac{w}{qR/E}\left(\frac{2h}{R}\right)\frac{1}{\lambda^2}$.

Figure 9.2 reports the dependencies $\frac{w}{qR/E}\,(\varphi)$ (analytical solution) and $\bar{w}/\bar{q}\,(\varphi_0)$ (numerical solution), where $\bar{w}$ and $\bar{q}$ are nondimensional deflection and load of the central point of the cylindrical shell (0.5; 0.0). The numerical solution obtained with a help of the higher order Bubnov-Galerkin method and the analytical solution are shown In Table 9.1 for the certain control angles of loading, where also the relative errors of the numerical results are presented.

Based on the results shown in Fig. 9.2, one may conclude that the numerical results practically coincide with the analytical solution which means that the numerical results are reliable. The relative error includes the errors introduced by numerical algorithm (the method of inverse matrix, the Runge-Kutta method, the set up method) and the computational errors. Therefore, the results obtained with a help of the Bubnov-Galerkin method in higher approximations while solving difference equations for the static linear problems are validated and they are reliable.

Now we approach the static nonlinear problems of the theory of shallow shells. The so far described method is employed to study closed cylindrical shells with

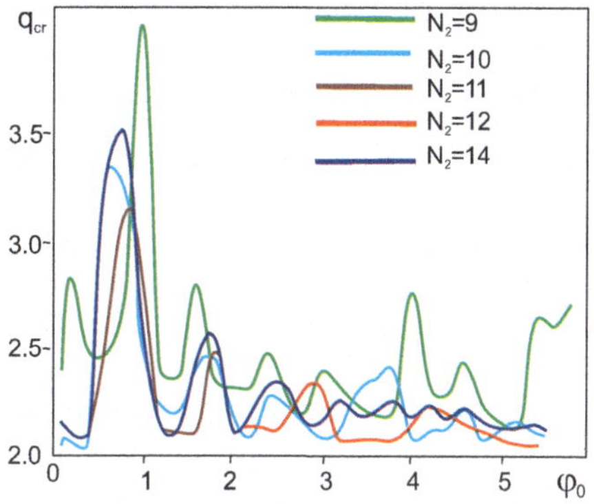

Fig. 9.3 The dependence of the critical loads on the pressure width measured by angle φ_0

$\lambda = 2$, and our results are compared with those reported by Obodan [5] in the case of static nonlinear problem. We consider the case of the transverse external pressure localized within sectorial band with the central angle φ_0.

In order to get $q_{cr}(\varphi_0)$ we need to construct a set $\forall \varphi_0 \in [0; 2\pi]$ for $\{q_i, w_i\}$, which serves for the critical load q_{cr} estimation. Let us investigate the dependence of the critical loads versus the width of the pressure sector φ_0 (Fig. 9.3). We study the obtained results versus various approximations. Since the load is distributed along the whole length of the cylindrical shell, a number of the series terms with regard to the coordinate x can be limited to only one series term, i.e. $N_1 = 1$. We investigate the dependence of the obtained results versus series terms number with regard to the circle coordinate y, i.e. on N_2. The mentioned dependencies for $N_2 \geq 9$ are shown in Fig. 9.3 and they exhibit non-monotonous and oscillating character. Though for different values of N the curves shape is similar, the increase of the number of approximations improves accuracy of the obtained results (Fig. 9.3).

This observation allows to conclude that in the case of the nonhomogeneous load the employment of only a few terms of the series (9.31) implies large errors, and the results depend on the number of used approximations. However, beginning with $N_2 = 13$, the system dynamic properties begin to stabilize, i.e. further increase of the equations does not introduce any changes of the results. Therefore, beginning from $N_2 = 13$ the convergent sequence is observed. All further reported results are given for $N_2 = 13$.

Consequently, we report the dependence of the critical loads versus the pressure sector width $\overline{q}_{cr}(\varphi_0)$ for $N = 13$ in Fig. 9.4. Here $\overline{q}_{cr} = q_{cr}/\tilde{q}_{cr}$, where $\tilde{q}_{cr}$ refers to the classical critical value in the case of the uniformly distributed external pressure which is estimated by the Mises formula [270] $\overline{q}_{cr} = 0.92\frac{R}{L}\left(\frac{h}{R}\right)^{5/2}$.

The dependencies $\overline{q}_{cr}(\varphi_0)$ reported by Obodan [5] for the case of closed cylindrical shell for $\lambda = 2$, fully coincide with the results obtained by our method which are denoted by solid curve (Fig. 9.4). The obtained results validate the high efficiency of the proposed method aimed on solving problems of statics.

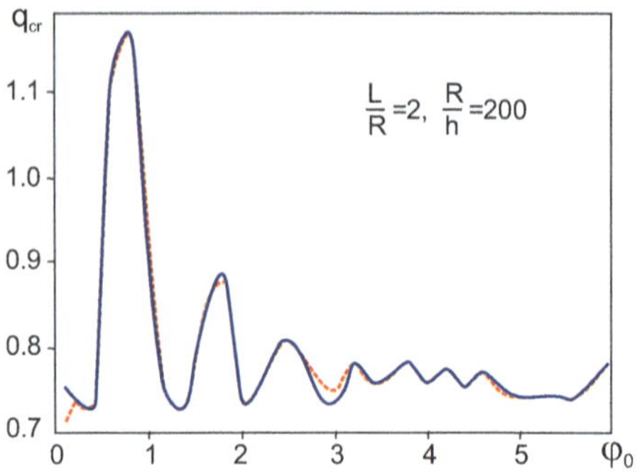

Fig. 9.4 The dependence of the critical loads against the width φ_0 of the pressure band (solid/dashed curve correspond to our/[5] results)

9.4 On the Set Up Method in Theory of Flexible Shallow Shells

We use again the approach presented in the previous section for the spherical panels with the rectangular planform with $k_x = k_y = 12$, and we report the results obtained with a help of the set up methods. In this case the problem is reduced to that of the hyperbolic (9.30) and parabolic type which is yielded by Eq. (9.30) while neglecting the time derivative:

$$\dot{\mathbf{A}} = -\frac{1}{\varepsilon} \cdot \Big[\Big[\mathbf{G}^{-1}\mathbf{C}_1 + \mathbf{G}^{-1}\mathbf{D}_1\mathbf{A}\Big] \cdot \Big[-\mathbf{P}^{-1}\mathbf{D}_2\mathbf{A} - \mathbf{P}^{-1}\mathbf{C}_2\Big] \cdot \mathbf{A} - \\ - \mathbf{G}^{-1}\mathbf{H}\mathbf{A} - \mathbf{G}^{-1}\mathbf{W}\mathbf{A} + \mathbf{G}^{-1}\mathbf{Q}q(\bar{t})\Big]. \tag{9.34}$$

The time series $w(t)$ for various damping coefficients $\varepsilon = 1,\ 10,\ 100$ obtained using the set up method for the hyperbolic problem (9.30) (Fig. 9.5a) and for the parabolic problem (9.34) are shown in Fig. 9.5. The key problem while using

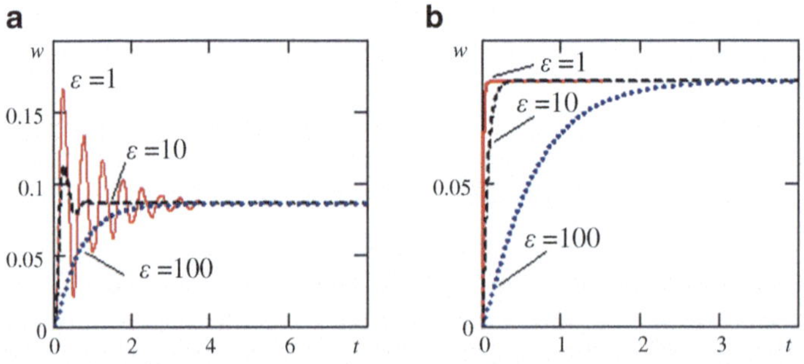

Fig. 9.5 The time histories obtained using the set up method: hyperbolic (**a**) and parabolic (**b**) problem

Table 9.2 Number of iterations required for obtaining reliable stationary solutions

$\bar{\varepsilon}$	Hyperbolic system	Parabolic system
1	5632	113
10	1722	568
100	6144	5677

the set up method for solving stationary problems concerns an estimation of its convergence.

Table 9.2 shows a number of iterations required for getting the solution with the absolute error $\bar{\varepsilon} = |w(t_i) - w(t_{i-1})| \leq 10^{-5}$. It should be emphasized that the damping coefficient $\bar{\varepsilon}$ essentially influences the iterations number and the character of its convergence for both the hyperbolic and parabolic problems.

In the case of the parabolic (hyperbolic) system the optimal value of the damping coefficient is $\bar{\varepsilon} = 1$ ($\bar{\varepsilon} = 10$). Observe that the number of iterations for the hyperbolic problem is of an order higher than the parabolic problem which gives preference for the choice of the parabolic type of the set up method.

As it has been earlier mentioned the worked out algorithm allows to solve the wide class of the problems of statics and dynamics. In what follows we demonstrate advantages of the so far described approach to solve the geometrically nonlinear problems of statics and dynamics. We investigate the mechanical system subjected to action of the uniformly distributed and constant load along the shell surface as well as the load of impulse type. Here a key role plays estimation of the critical values of the static and dynamic loads.

We begin with a brief description of the stability criterions proposed by various authors.

Volmir [337] proposed a sudden increase of deflection under a infinitely small load increase or an occurrence of the deflection point of the function $q(w)$ ($\frac{\partial^2 q}{\partial w^2} = 0$) as the dynamics stability criterion. On the other hand Shiau et al. [300] showed that the load responsible for the inversed process of the time change is necessary to achieve the first maximum of the dependence "load-time" is understood as critical one.

Kantor [160] employed the Ritz method in higher approximations while studying axially symmetric spherical shells and suggested that first buckling occurs if deflection in the shell centre achieves the value larger than the relation shell height $K \cong 2\bar{f}$, where $\bar{f} = \frac{f}{h}$ stands for the nondimensional height of the shell with regard to its planforms may serve as the dynamical stability loss.

Another criterion has been introduced in reference [212] owing to which the dynamic load is defined based on the observation that the system achieves its new equilibrium configuration with zero velocity. The analogous criterion has been employed in reference [59]. Namely, it has been observed that in the beginning of the loading process the inertial forces act against the external load, and next, after transition through zero and after changing their sign, they amplify the external load action. In this event the acceleration also changes its sign. It means that in a certain time instant a sudden increase of the deflection is observed. Therefore, the time instant associated with zero velocity shell occurrence is called critical.

On the other hand in reference [94] as the stability loss criterion it is suggested the time instant where deformation of the elastic body are unstable, i.e. the critical time instant t_0 is linked with such excitation where the displacement field is changed but the fields of velocity and accelerations remain unchanged.

In some works the process of defining the dynamic stability loss is associated with the reduction of the dynamic problem to quasi-static one. In the latter approach the pre-critical stresses in the middle shell surface are defined with an account of inertial forces obtained from dynamic relations while the buckling process is estimated using the static approaches. This freezing techniques in time is often practically employed in the case of investigation of complex constructions.

There are also researchers who take the occurrence of the plastic deformations of the shell type constructions as the dynamic criterion.

In reference [228], in the case of the arcs, it has been shown that the buckling process is characterized by two different mechanisms: the straight and coupling buckling process. In the first case the buckling is realized when the unstable state is achieved via symmetric form, and non-symmetric forms are unstable. In the second case the stability loss is realized through the complex interaction of the symmetric and non-symmetric forms, i.e. the system looses its stability via the non-symmetric form. It means that the character of stability loss with regard to symmetric and non-symmetric forms is qualitatively different, and consequently two different dynamic stability loss criterions should be taken into account.

In this chapter under the critical load we understand bifurcation load or a point of deflection in the dependence $w_{\max}(q)$. Let us investigate the critical loads for the case of the closed cylindrical shell, as well as the cylindrical panel and spherical shell with the rectangular planform. In Fig. 9.6 there are reported dependencies of the critical loads versus the parameters defining the shell curvature k_y. Solid curves concern the cylindrical panel ($k_x = 0$), whereas dashed curve characterises the spherical shell with two curvatures ($k_x = k_y$). As it is seen in Fig. 9.6, in the case

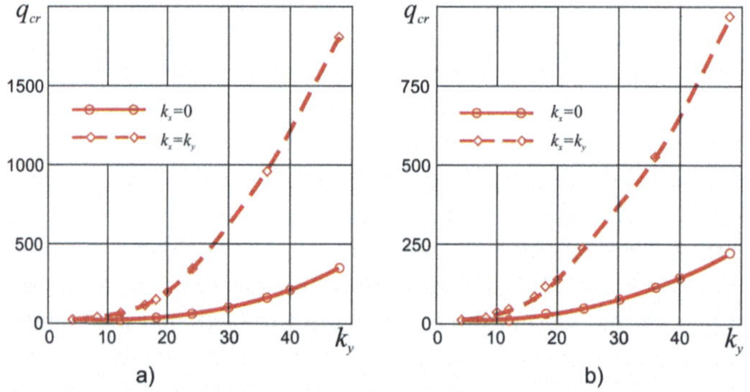

Fig. 9.6 The critical load values versus k_y: (**a**) statics and (**b**) step-type load of infinite length

of the static problem (a) and in the case of the load of infinite length (b) the curves corresponding to the spherical shell are located above the curve associated with the cylindrical panel, i.e. stiff stability loss takes case for large values of the load q than in the case of the cylindrical panel.

9.5 Dynamic Stability Loss of the Shells Under the Step-Type Function of Infinite Length

In this section we consider the influence of the type of the load on the critical load for the studied mechanical system. The dependencies of the critical loads for the cylindrical panel and spherical shell with the rectangular planform are under the static load and the step type load for $\varepsilon = 0$. Solid curves refer to static problem whereas dashed curves refer to dynamic problem.

As it can be seen from Fig. 9.7 the curves associated with the systems and step type load of infinite length are located below the curves corresponding to static problem. Analysis of the critical loads reported in Fig. 9.7 shows that the loading type had an essential influence on the critical load value.

Dynamic stability loss generated by the step type load occurs for the small value of the load than in the case of the static problems. Hence, one may conclude that in the case of cylindrical panels and spherical shell with the rectangular planform an increase of the curvature k_y implies construction resistance to higher loads before suffering for stability loss.

Let us study dynamics and static critical loads for the case of the closed cylindrical shell.

In Fig. 9.8 there are reported dependencies of the critical load versus the width of the sectorial band of the external pressure characterized by the angle φ_0 for the cylindrical shells subjected to step type load for $\varepsilon = 0$ and for the case of the transversal external constant pressure.

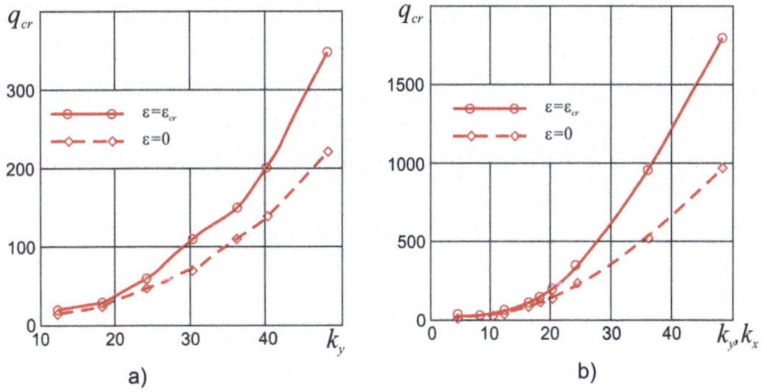

Fig. 9.7 Dependence of the critical loads on k_y for the static (**a**) and dynamic (**b**) problem

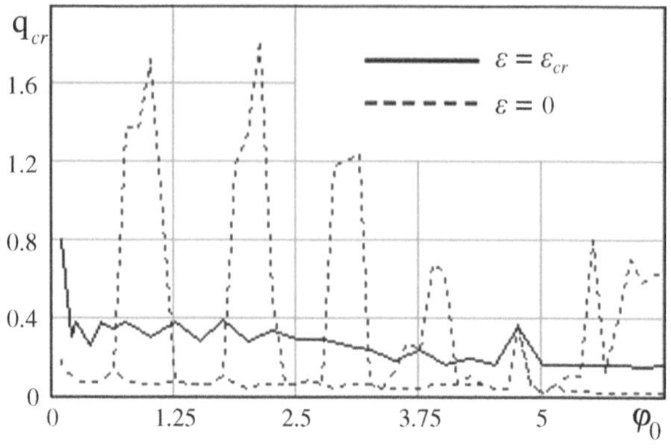

Fig. 9.8 Dependence of the critical load on the loading angle φ_0 for the closed cylindrical shell

Solid curves present results of the static problems whereas dashed curves refer to the step type load. In the case of the cylindrical shells with $\lambda = 2$ being under actions of the step type load of infinite length for the particular angles of the load two dynamic critical loads are found. This non-classical feature is associated with the occurrence of the local stability loss, i.e. we deal in fact with the global stability loss matching the local and general stability loss (Fig. 9.8).

Analysis of the critical loads reported in Fig. 9.8 allows to conclude that the loading character essentially influences the critical load values. The curves corresponding to the systems and step type load are located below the curves corresponding to static problem in the case if one deals with the local stability loss. If there are two critical loads than the curves corresponding to that type of loading are located below the static load cases. Therefore, the width of the external loading band (either static or dynamic) has the essential influence on the value of the critical loads on the dynamic stability loss.

Figure 9.9 presents the dependence $w_{\max}$ (0.5; 0.0) versus q_0 for the cylindrical shells under the step type load acting on the sectorial band of the width $\varphi_0 = 4.75$. In this case two stiff stability loss are observed, i.e. local and global one.

Let investigate the spatial-temporal characteristics of the cylindrical shell subjected to step type load acting on the sectorial band of the width $\varphi_0 = 4.75$ for two loads: pre-critical (point A in Fig. 9.9) and post-critical (point A in Fig. 9.9). In Fig. 9.10 there are presented time series $w(0.5, 0.0, t)$ for the precritical (Fig. 9.10a) and postcritical (Fig. 9.10b) loads. Figure 9.11 reports circle cross section of the shell surface for $x = 0.5$ in certain time instants for the pre-critical (Fig. 9.11a) and post-critical (Fig. 9.11b) loads.

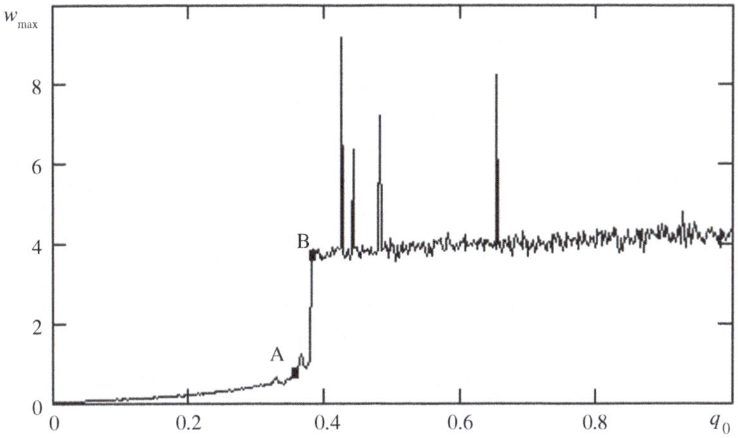

Fig. 9.9 Dependencies w_{max} (0.5; 0.0) versus q_0 for the cylindrical shell

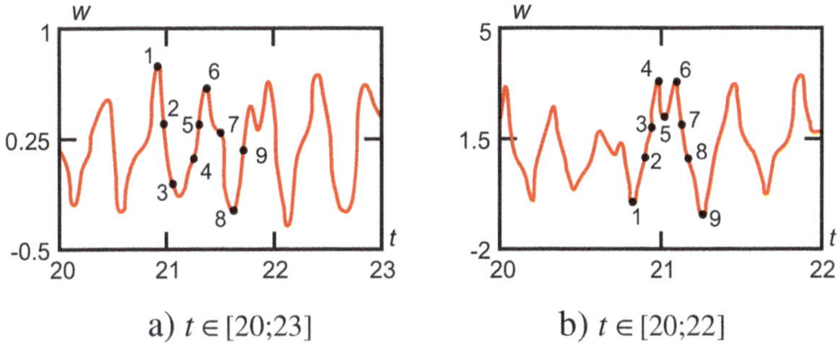

Fig. 9.10 Signals $w(0.5, 0.0, t)$ for pre-critical (**a**) and post-critical (**b**) loads

A simultaneous consideration of both graphs allows to analyse signal changes in the space-time plane. The number of a point in Fig. 9.10a, b corresponds to the curve with the same number in Fig. 9.11a, b. We analyse the shell deflection in time separately for the points A and B. The mentioned two points differ in the numbers of half waves along a circle coordinate while transiting through the critical load value. Namely, the number of half waves is changed with regard to different time instants in contrary to the pre-critical state where the number of half waves was fixed and equal to 6. The stiff stability loss in all shell points along the circle co-ordinate $y \in [0; 2\pi]$ takes place for the same load value and differs only with respect to the deflection magnitude.

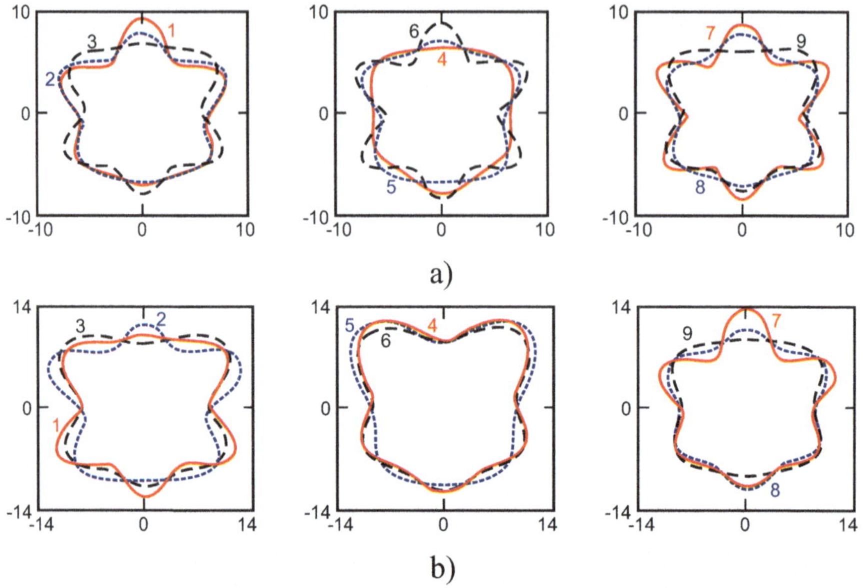

Fig. 9.11 Space-timing characteristics of the cylindrical shell cross sections

Chapter 10
Scenarios of Transition from Periodic to Chaotic Shell Vibrations

In this chapter we carry out numerical analysis of the scenarios of transition from periodic to chaotic vibrations exhibited by shells. Namely, we have detected and studied the so called modified Ruelle-Takens-Newhouse (RTN) scenario the modified Ruelle-Takens-Feigenbaum (RTF) scenario as well as the modified Pomeau-Manneville scenario found in our analysed mechanical continuous systems. We have numerically investigated the Sharkovsky's orders of periodicity for PDEs governing dynamics of shallow shells. In the case of analysis of vibrations of the closed cylindrical shell with $k_y = 112.5$ and $\lambda = 3$ under transversal uniformly distributed harmonic load $q = q_0 \sin(\omega_p t)$, localized in the shell zone characterized by the angle $\varphi_0 = 6$, we have detected windows of periodicity which coincide with the order predicted by Sharkovsky's theorem, i.e. $3;\ 5;\ 2^2 \cdot 5;\ 9$. We have also investigated a cylindrical panel under action of the longitudinal two-parametric load $p_x = p_0 \sin(\omega_p t)$ and $p_y = \alpha \cdot p_x$ for $k_y = 24$ and $\alpha = 3$ where we have found also the periodicity windows following the Sharkovsky's order $3;\ 2 \cdot 3;\ 5;\ 2 \cdot 5;\ 7;\ 2 \cdot 7;\ 11;\ 13$. The carried out analysis of the phase and model portraits constructed for both periodic and chaotic shell vibrations revealed that the space and temporal (timing) chaos appear simultaneously, i.e. in the case of shells it refers to occurrence of the space-temporal chaos.

10.1 Novel Models of Scenarios of Transition from Periodic to Chaotic Orbits

In the studied problems we have detected the novel scenario which is further called the modified Ruelle-Takens-Newhouse (RTN) scenario. It was exhibited by the cylindrical shells and spherical shells with the rectangular planform. We briefly describe this scenario using an example of the spherical shell with $k_x =$

J. Awrejcewicz, V. A. Krysko, *Elastic and Thermoelastic Problems in Nonlinear Dynamics of Structural Members*, Scientific Computation,
https://doi.org/10.1007/978-3-030-37663-5_10

$k_y = 12$ subjected to action of the uniformly distributed transverse harmonic load $q = q_0 \sin(\omega_p t)$. The fundamental vibration characteristics, i.e. time history $w\,(0.5, 0.5, t)$, phase portrait $w(\dot{w})$, power spectrum $S(\omega)$, Poincaré map $w_t(w_{t+T})$, where $T = \frac{2\pi}{\omega_p}$, versus excitation amplitude values q_0 are shown in Table 10.1.

Between the chosen values q_0 the mentioned characteristics are almost unchanged. A study of the results given in Table 10.1 allows to distinguish the key features of the RTN scenario.

1. First, the shell vibrations are periodic and exhibiting excitation frequency ω_p $(q_0 = 7.9)$.
2. Second, increase of the excitation amplitude up to $q_0 = 7.94$, results in occurrence of the linearly non-dependent frequency a_1 and the linearly dependent frequency $b_1 = \omega_p - a_1$. The system dynamics is transited to quasi-periodic vibrations with a_1 and ω_p. The phase orbit lies on a torus since the ratio $\left({}^{\omega_p}/{}_{a_1} = {}^{m}/{}_{n} = 9.696\ldots\ldots\right)$ is irrational.
3. Further increase of the load amplitude implies appearance of the new but dependent frequencies $a_n = n \cdot a_1,\ b_n = \omega_p - n \cdot a_1$. This is manifested by increase of the thickness of the trajectories in the phase space and creation of an attractor having the shape of a ring in the Poincaré map.
4. Then, for $q_0 = 8.05$ one more type of the dependent frequencies $c_1 = b_n - a_n$ is exhibited in the frequency spectrum. The Poincaré section demonstrates increase of the attractor in the phase space.
5. Therefore, the series of occurrence of the dependent frequencies shifts the shell dynamics into chaos $(q_0 = 8.21)$.

Consequently, a transition into chaos under the modified RTN scenario takes place in the following way. There is a birth of a new linearly independent frequency and the transition into chaos is realized via a series of linear combination between the latter frequency and ω_p.

Investigation of chaotic vibrations of the closed cylindrical shells of the circle cross section with $\lambda = 2$ under sinusoidal transversal load $q = q_0 \sin(\omega_p t)$ acting along the sectorial band $\varphi_0 = 343°$, yielded one more novel scenarios which will be further called the Ruelle-Takens-Feigenbaum (RTF) scenario. It is worthy to mention that the latter is only exhibited by cylindrical shells. It is based on combination of two known Ruelle-Takens and Feigenbaum scenarios. Its main dynamical characteristics are reported in Table 10.2, and it exhibits the following nonlinear features.

1. First the shell vibrations are periodic with one (excitation) frequency. Phase portrait presents a closed orbit, whereas the Poincaré map consists of one point $(q_0 = 0.44)$.
2. Increase of the excitation load $(q_0 = 0.48)$ implies a birth of the new independent frequency ω_1, and the system exhibits two-frequency quasi-periodic vibrations $(\omega_p/\omega_1 = m/n = 6.148\ldots$ is irrational).
3. Further increase of the excitation amplitude $(q_0 = 0.4990637)$ yields the sequence of the fourth successive period doubling bifurcations. Observe that the

Table 10.1 Vibration characteristics of the RTN scenario

q_0	Signal $w\,(0.5, 0.5, t)$	Phaseportrait $w(\dot{w})$	Powerspectrum $S\,(\omega)$	Poincarémap $w_t\,(w_{i+T})$
7.9	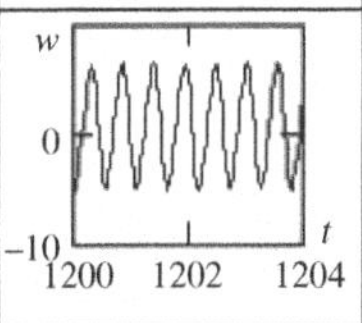	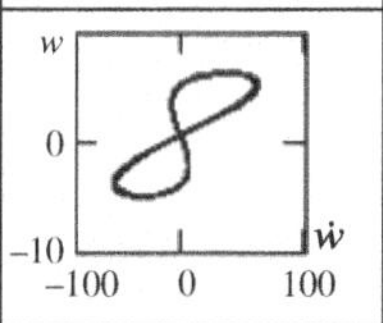	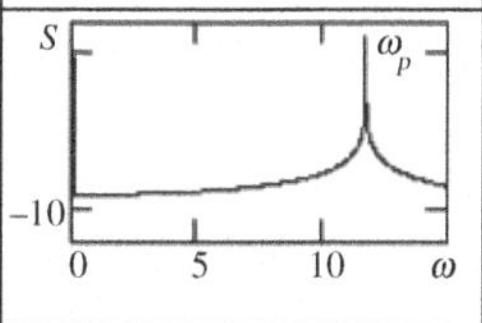	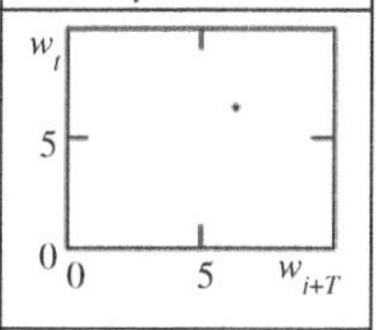
7.94	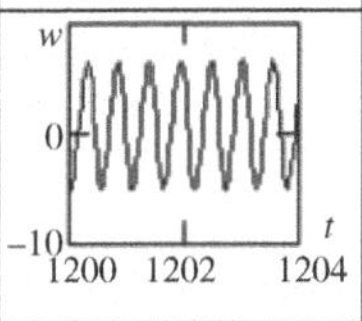	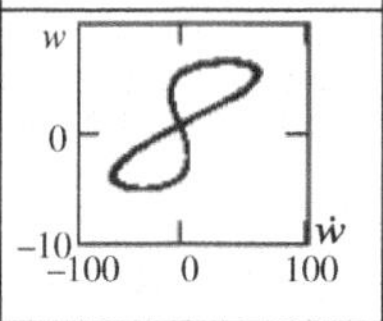	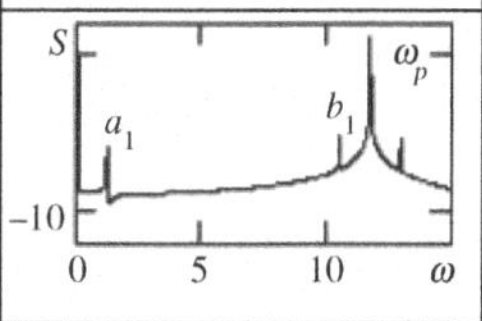	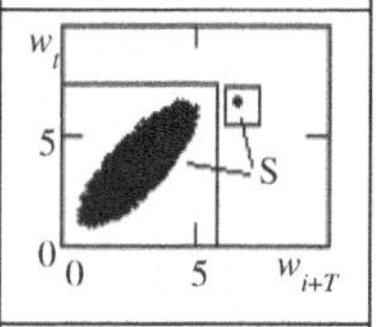
7.95	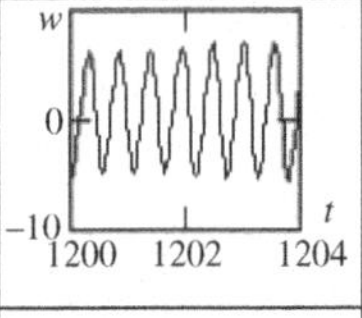	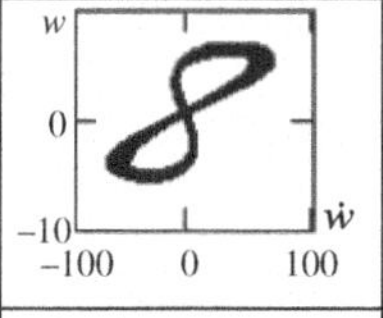	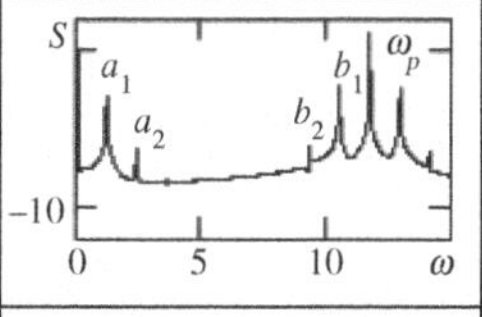	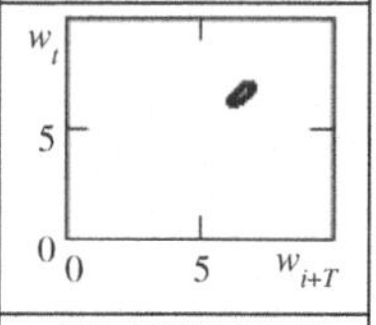
7.99	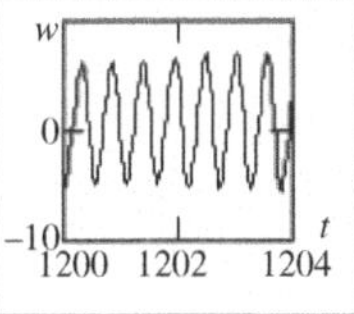	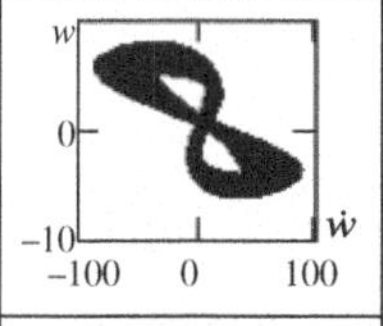	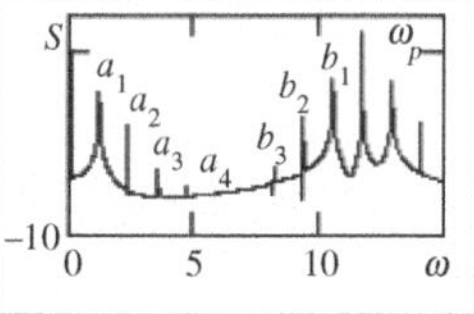	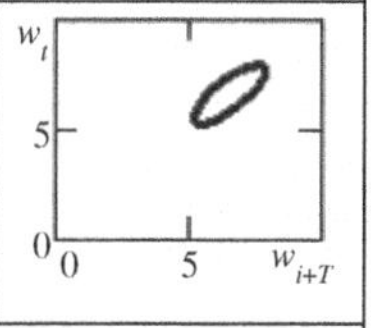
8.0	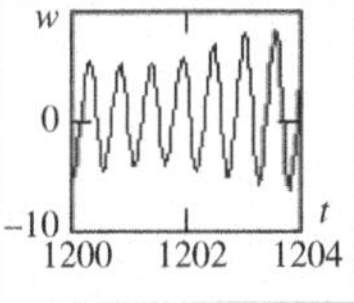	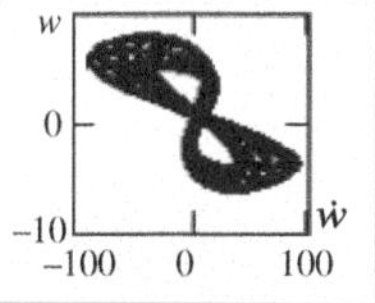	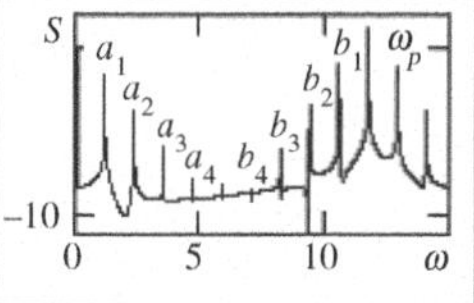	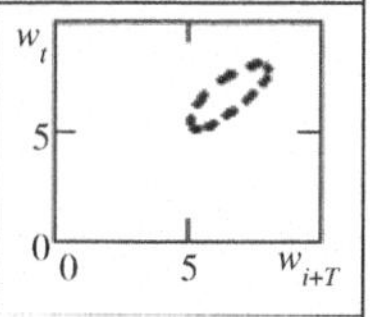
8.05	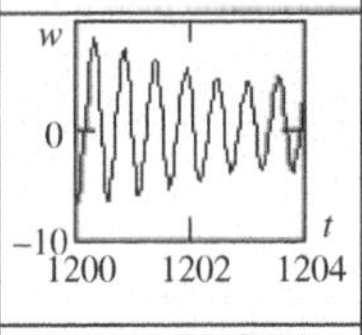	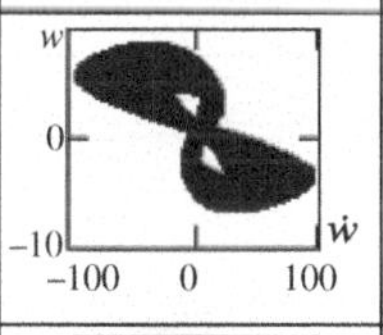	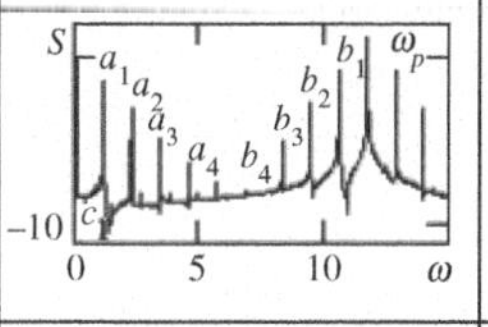	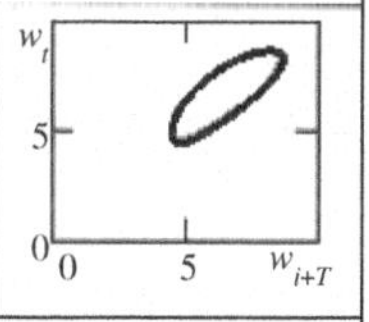
8.11	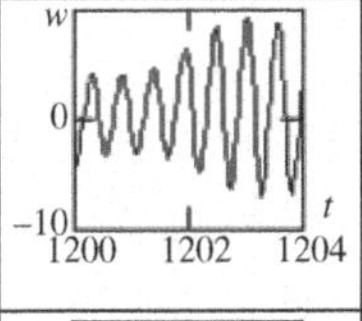	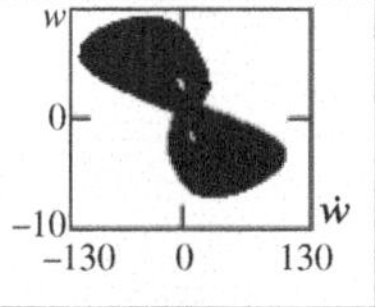	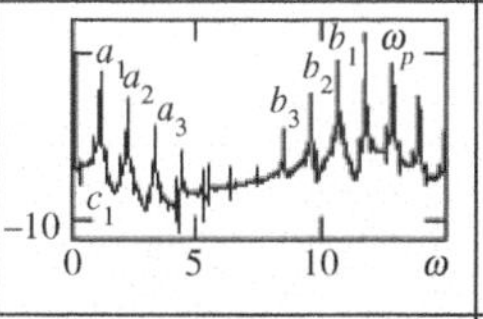	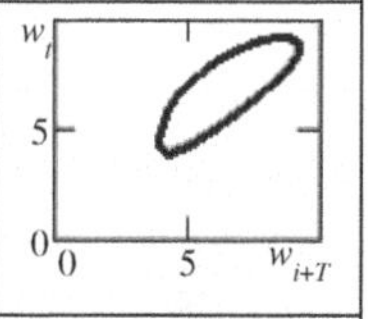
8.21	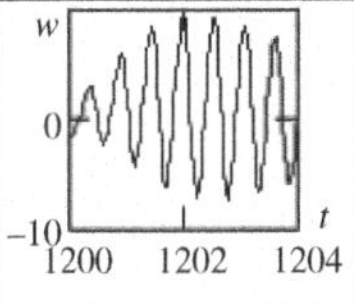	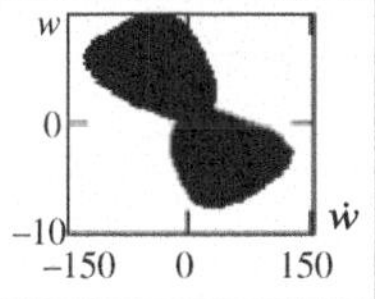	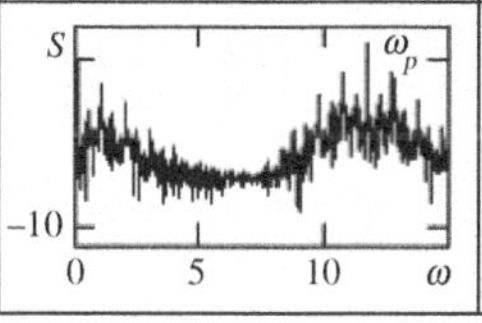	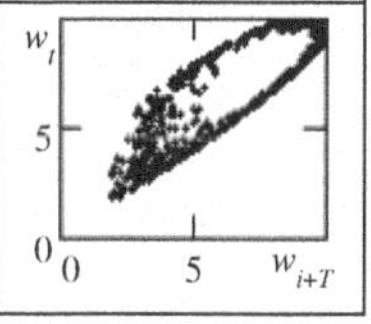

Table 10.2 Vibration characteristics of the RTF scenario

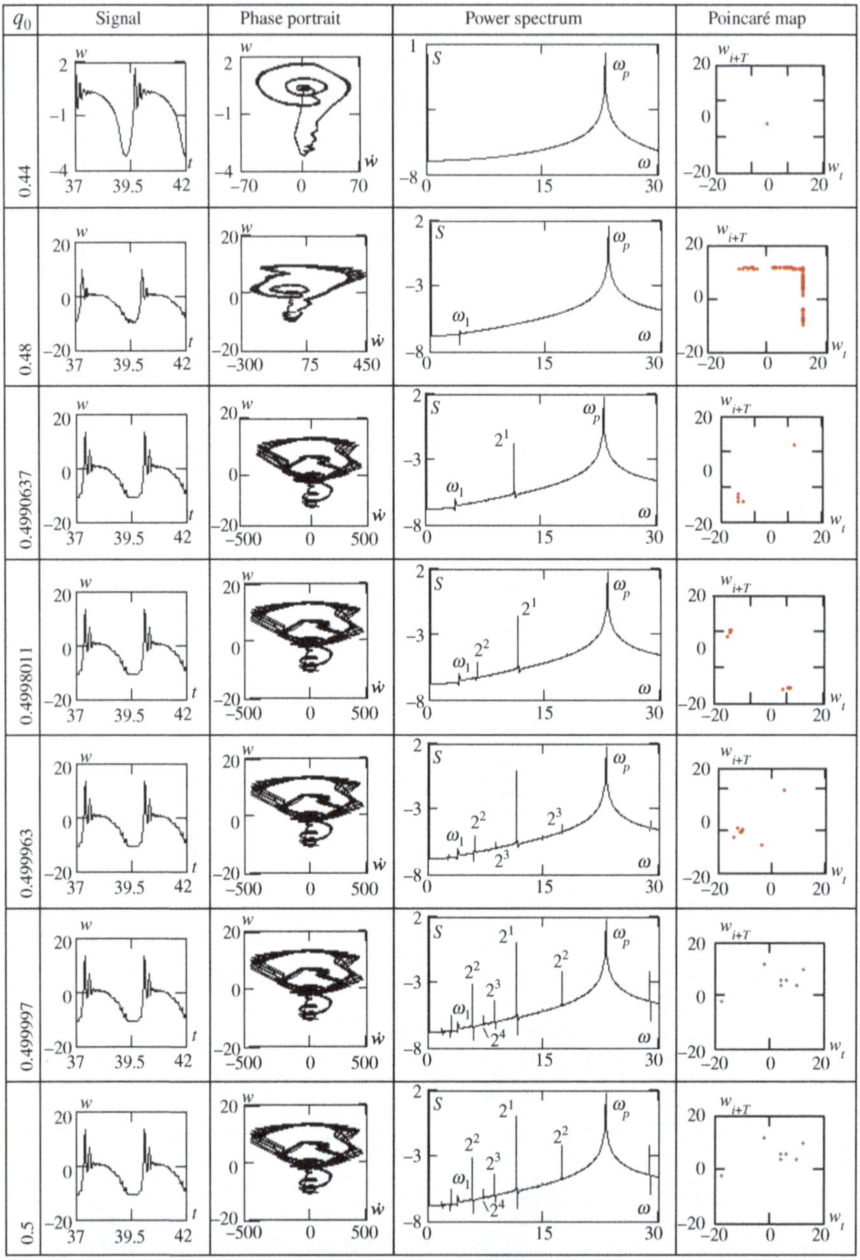

Table 10.3 Estimating the Feigenbaum constant in the modified RTF scenario

i	1	2	3	4
$q_{0,i}$	0.4990637	0.4998011	0.499963	0.499997
d_∞		4.557665	4.668532	

Hopf bifurcations do not introduce the qualitative changes in the signals and phase portraits though the Poincaré section is essentially changed.

4. Increase of the excitation amplitude on amount of 10^{-3} ($q_0 = 0.5$) shifts the system into chaos spanned on the excitation frequency.

In spite of that the transition into chaos has not been clearly exhibited by the classical Feigenbaum scenario but rather by its modification, we have estimated the Feigenbum constant (Table 10.3). The difference between our results in estimating, the Feigenbaum constant and its known theoretical value achieves 0.02%.

Therefore, the modified RTF scenario is realized in the following way. First, a new linearly independent frequency is born, and then a sequence of the successive four Hopf bifurcations are detected. The transition into chaos is realized via combination of the independent frequency and a series of the Hopf bifurcations.

It should be emphasized that while investigating of chaotic vibrations of the closed cylindrical shells under the transverse harmonic local pressure distributed along the sectorial band $\varphi_0 = 90°$, we have detected one more novel scenario. The latter stands for a mixture of the modified Ruelle-Takens and the classical Pomeau-Maneville scenarios. Here, as in the previous case, after the periodic vibrations with ω_p increase of q_0 yields occurrence of a new independent frequency and frequency linear combinations, which under further increase of q_0 are collapsed through a bifurcations associated with the stiff stability loss. The time history/signal presents then intermittency, and the vibrations become chaotic. This scenario will be described in more details in the next section.

10.2 Sharkovsky's Periodicity Exhibited by PDEs Governing Dynamics of Flexible Shells

One of the key problems in theory of turbulence is associated with a forecast of its occurrence based on the stability and equilibrium configurations (laminar flows). Feigenbaum studied intervals between period doubling bifurcation in the diagrams of the orbits for a square function, i.e. the so called logistic map. It originates from the ecosystems consideration and has been first investigated by Verhulst in 1845, and hence the bifurcation diagram of the map $y = x^2 + c$ is called the Verhulst diagram. The main feature of the Feigenbaum scenario stands for its universality. The described by Feigenbaum period doubling route to chaos are also exhibited by the following maps into itself: $cx\,(1-x)\,,\ x^2+c,\ c\sin(\pi x)$ and $cx^2\sin(\pi x)$.

The diagram of the orbits shown in Fig. 10.1 presents the attractive periodic orbits of the function $f_c(x) = x^2 + c$. In some intervals its dynamics is reach.

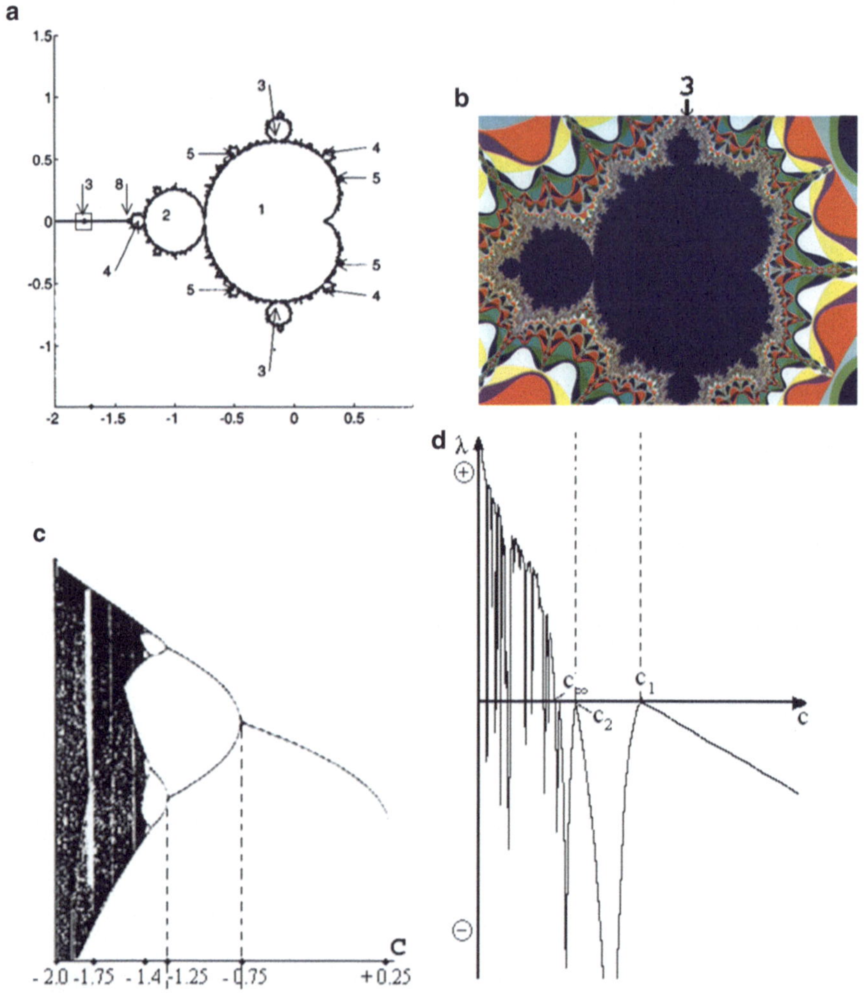

Fig. 10.1 The Mandelbrodt sets: (**a**) borders of the maps $f_c(z) = z^2 +$ and their periods; (**b**) window taken from figure (**a**) referring to period-3 set; (**c**) bifurcation tree for the map $f_c(z) = x^2 + C$; (**d**) Lyapunov exponent vs. C for the map $f_c(z) = x^2 + C$

For instance, for $c \approx 1.75$ there is a white zone where stable orbits are of period 3. The following question arises: are there other periodic orbits? It is expected that they should be repliers since the studied diagram exhibits only attracting orbits. It occurs that occurrence of period 3 orbits implies occurrence of the orbits with the periods $n = 1, 2, 3, \ldots$. Though in 1975 Li and Yorke [221] observed and reported the mentioned behaviour, but in fact their results belong to a particular case of the Sharkovsky's theorem [296].

Theorem ([296]) *Let I stands for a finite or infinite interval in R. Assume that the map* $f : Z \to I$ *is continuous. If there exists a point f of period n, then there exists* *the point f of period k for each integer k, $k > n$ which belong to the following Sharkovsky's order*

$$3, \quad 5, \quad 7, \quad 9, \ldots$$

$$2 \cdot 3, \quad 2 \cdot 5, \quad 2 \cdot 7, \quad 2 \cdot 9, \ldots$$

$$2^2 \cdot 3, 2^2 \cdot 5, 2^2 \cdot 7, 2^2 \cdot 9, \ldots$$

$$\ldots\ldots\ldots$$

$$\ldots 2^n, \ldots, \quad 2^2, \quad 2^1, \quad 1$$

Remark If there is an orbit of odd period larger than 1, then a number of different periods tends to infinity.

However, the Sharkovsky's theorem holds only for a real function defined on a real interval. For example, if a function f is given by rotation of each point of a circle on angle $2/n$, then orbits of all points have one and same period n. It means that in this case any other periods do not exist, and consequently, the Sharkovsky's theorem cannot be applied.

Let us compare two logistic maps $f_c(x) = x^2 + c$ and $f_c(x) = z^2 + c$ in real and complex planes and the Lyapunov exponent $\lambda_1(c)$ is constructed for the map $f_c(x) = x^2 + c$. Diagrams of the orbits of two functions exhibit attractors, repliers and neutral points, which correspond to stable, unstable and infinitely many of equilibrium states. If one starts in the vicinity of a fixed point, and it will be approached in the iterational process, than it is called an attractor. Similarly, starting in the vicinity of a replier, we away from that point. Finally, a neutral fixed point is defined as follows: starting in efficiently small neighbourhood of this point we will stay all the time in this neighbourhood neither approaching nor leaving that point. In order to define a character of the fixed point $\bar{z}$ of the map $f(z)$ one needs to compute $f'(z)$: if $|f'(\bar{z})| < 1$, then $\bar{z}$ is an attractor and if $|f'(\bar{z})| > 1$ it stands for replier. The case $|f'(\bar{z})| = 1$ defines a neutral point.

In the case of rational maps, as it has been already mentioned, owing to Sharkovsky's theorem, all cycles are of orders $n = 2, 3, 4, \ldots$.

By an attractor in complex plane we understand a point (or points) into which tends the iterational process $z_{n+1} = f(z_n)$, $n \to \infty$. It is already well known that sometimes such set can be consisted of more than one attractor, it may consists of infinitely many points or it may present a continuous curve/line or other set like the Kantor set. Figure 10.1a presents some zones (borders) of the Mandelbrodt set $f_c(z) = z^2 + > 0$ [235], corresponding to a set attractive periodic points of different periods. Figure 10.1b presents a window exhibiting period-3 area of the Mandelbrodt set. The orbital diagram $f_c(x) = x^2 +$ (Fig. 10.1c) reports the behaviour of the Mendelbrodt set on a real axis. Each of the bifurcations corresponds to a new zone, which intersects the axis x, and a period in this case corresponds a number of branches of the orbits. The point marked as C_∞ is often referred to as the Feigenbaum point. In the diagram in between $C = 1.4$ and C_∞ the period doubling bifurcation is associated with $C \to C_\infty$. The other zone where $C > C_\infty$ is called chaos zone which contains period-3 window in the neighbourhood of $C = -1.7548777\ldots$ (white zone in Fig. 10.1c). Occurrence of the period-3 orbit implies occurrence of orbits with periods $n = 1, 2, 3, \ldots$. Figure 10.1d presents the Lyapunov exponent λ_1 for the logistic map $f_C(x) = x^2 + C$. Analysis of the figures a, c, d reveals that the chaotic regime is interrupted where the sequence $\left|f_C^n(x)\right|$ again is located in the periodic interval which corresponds to $\lambda_1 < 0$.

The so far reported analysis shows that a construction (for simple dynamical systems) of one-dimensional maps allows to get qualitatively similar bifurcation mechanisms of transition of mechanical systems into chaotic vibrations. In the case of complex mechanical objects like shells one may obtain a similar like behaviour while using a strong reduced models up to one degree of freedom (1DoF) system. Though in the case of continuous mechanical system its dynamics is more rich and complicated, but in the charts $\left\{q_0, \omega_p\right\}$ one may find zones with periods $3;\ 2 \cdot 3;\ 5;\ 2 \cdot 5;\ 7;\ 9;\ 11;\ 13 \ldots$. It allows to extend the Sharkovsky's theorem to fit more complicated class of functions i.e. this governing nonlinear dynamics of shells.

Under investigation of the cylindrical shell with $k_y = 112.5$ and $\lambda = 3$, and subjected to transverse harmonic load $q = q_0 \sin(\omega_p t)$ acting along the band $\varphi_0 = 6$, there are detected windows of periodic vibrations following the Sharkovsky's order within chaotic zone. In what follows we report the time series (signals) computed in the shell centre, phase portraits, power spectra, Poincaré maps, and modal characteristics $w(w_{yy})$ for the orbits predicted by the Sharkovsky's theorem.

Sharkovsky's orders $3;\ 5;\ 2^2 \cdot 5;\ 9$ are shown in Table 10.4. It should be mentioned that the so called Sharkovsky's orders in our continuous system do not follow subsequently each other but they are rather randomly distributed in the whole plane $\{q_0, \omega_0\}$ of the control parameters. The following features have been found: in the case of period-3 bifurcation the fundamental period has been divided into three equal parts; and the same holds for period-5 bifurcation resulted in 5 equal period parts, etc.

The Poincaré section consists of $3;\ 5;\ 2^2 \cdot 5;\ 9$ points. The phase portraits present period doubling bifurcations. The mentioned orbits belong to periodic windows in the ocean of chaos, and their geometric structure is the same as in the whole analysed set.

Investigation of chaotic vibration of cylindrical panels under the longitudinal harmonic two-parametric load $p_x = p_0 \sin(\omega_p t)$ for $p_y = \alpha \cdot p_x$ and $k_y = 24$ and $\alpha = 3$ also showed the periodic windows belonging to the Sharkovsky's orders. We report power spectra for the orbits given in the Sharkovsky's theorem. Table 10.5 reports the Sharkovsky's order $3;\ 2 \cdot 3;\ 5;\ 2 \cdot 5;\ 7;\ 2 \cdot 7;\ 11;\ 13$.

In the case of the cylindrical panel we have detected the same qualitative dynamic behaviour as in the case of the closed cylindrical shell.

Table 10.4 Dynamic characteristics of the cylindrical shell

№	q_0 ω_p	Signal $w(0.5;0.0;t)$	Phase portrait $w(\dot{w})$	Power spectrum $S(\omega)$	Modal portrait $w(\partial^2 w/\partial y^2)$	Poincaré map $w_t(w_{t+T})$
3	0.284 18.68	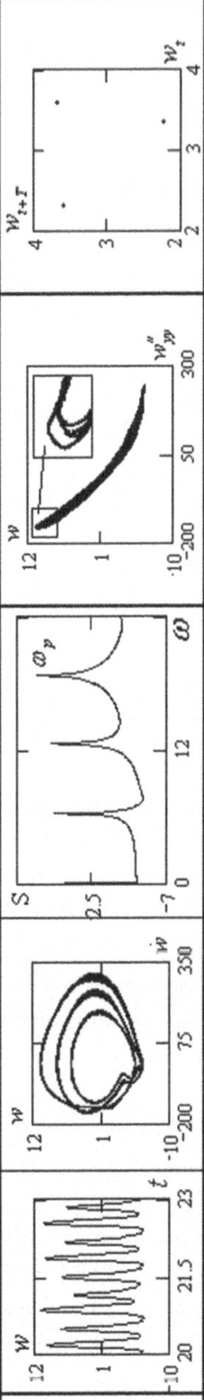				
5	0.293 18.24	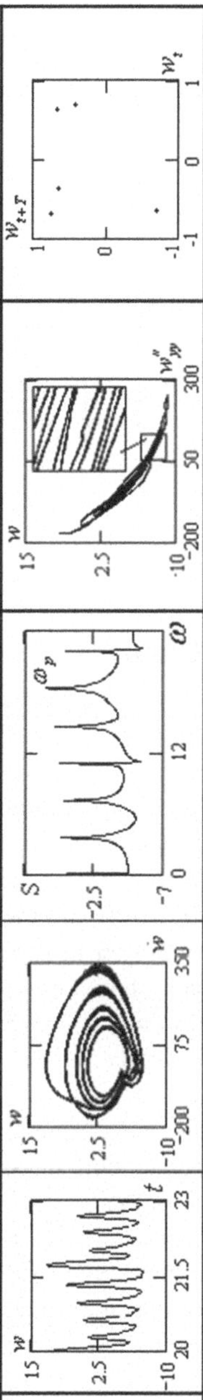				
$2^2\cdot 5$	0.323 26.642	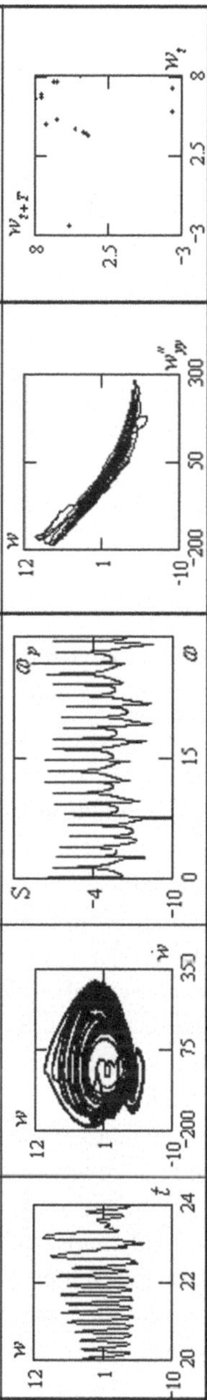				
9	0.315 26.642	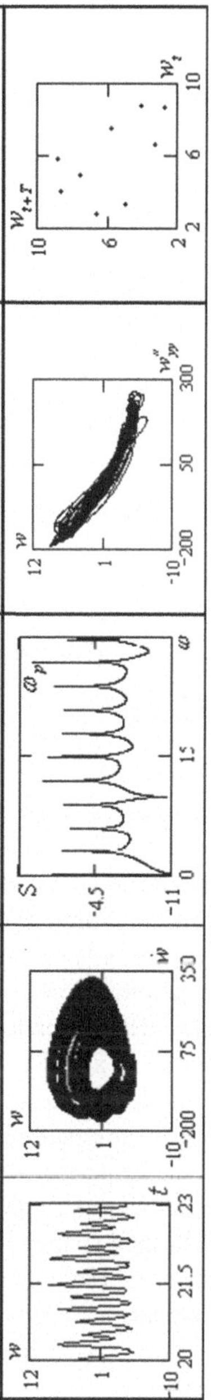				

Table 10.5 Sharkovsky's order based on the power spectrum analysis (cylindrical panel longitudinally loaded)

3	5	7	11
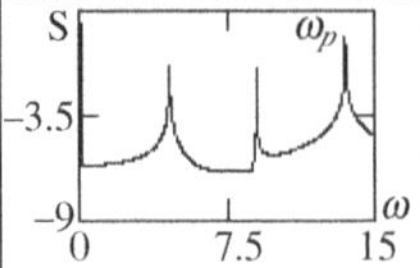	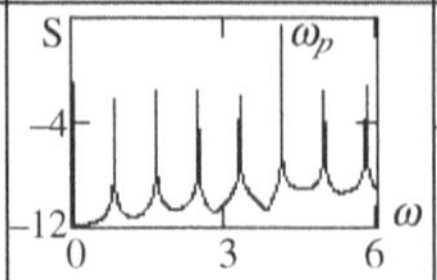	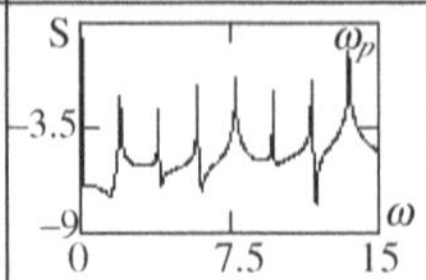	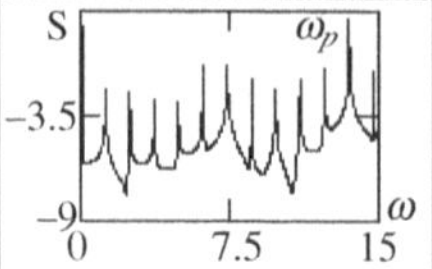
$p_0 = 12.16$, $\omega_p = 13.41$	$p_0 = 3.02$, $\omega_p = 4.11$	$p_0 = 8.68$, $\omega_p = 13.41$	$p_0 = 9.337$, $\omega_p = 13.41$
$2 \cdot 3$	$2 \cdot 5$	$2 \cdot 7$	13
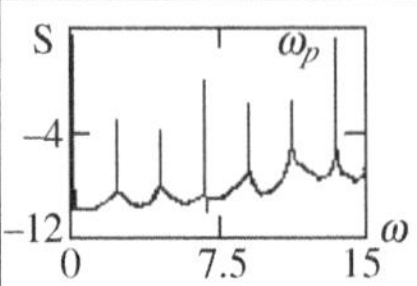	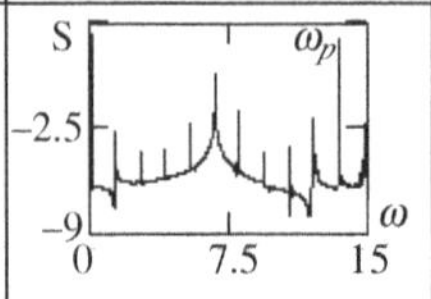	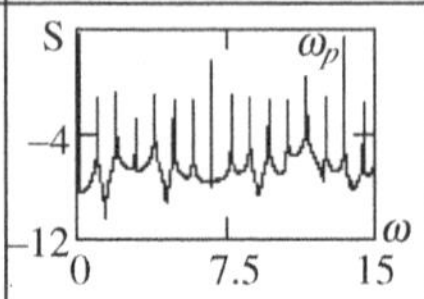	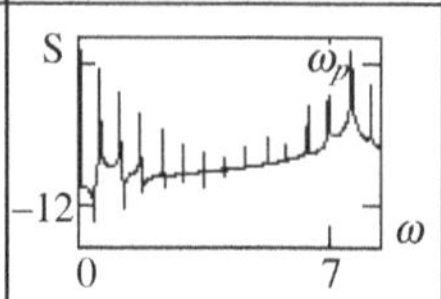
$p_0 = 3.5$, $\omega_p = 13.41$	$p_0 = 15.33$, $\omega_p = 13.41$	$p_0 = 10.25$, $\omega_p = 13.41$	$p_0 = 6.192$, $\omega_p = 7.12$

10.3 On the Space-Temporal Chaos

In order to get a picture on the system behaviour in time the following characteristics are constructed and analysed: time series $w(t)$, velocity $\dot{w}(t)$ and acceleration $\ddot{w}(t)$. Let us construct the function $f(w, \dot{w}, \ddot{w})$ further referred as the 3D phase portrait. Analogously, in order to study the bending shell surface the following characteristics of the deflection function $w(t, x, y)$, angle of inclination of and tangent measured $w'_x(t, x, y)$ and $w'_y(t, x, y)$, and the surface curvature defined by $w''_{xx}(t, x, y)$ and $w''_{yy}(t, x, y)$ are reported in a point (x, y). The chosen set of functions allows to achieve information about character of bending of the shell surface. The dependencies $f(w, w'_x w''_{xx})$ and $f(w, w'_y w''_{yy})$ allow to withdraw information about features of the shell surface, and also about a transition from periodic to chaotic shell vibrations.

The mentioned dependencies $f(w, w'_x w''_{xx})$ and $f(w, w'_y w''_{yy})$ are referred to as modal portraits and hence we investigate both phase and modal portraits in the case of either periodic or chaotic shell vibrations.

A birth of space-temporal chaos in time is studied using the example of vibrations of the closed cylindrical shell under the band type load ($k_y = 112.5$ and $\lambda = 3$). The signals $w(0.5, \frac{\pi}{2}; t)$ and the forms of the circle shell cross sections $w(0.5, y, t)$ in certain chosen time instant for the periodic (Fig. 10.2) and chaotic (Fig. 10.3) vibrations for $\omega_p = 26.642$ are analysed. Besides, there is presented phase portrait $w(\dot{w}, \ddot{w})$ in 3D space (Figs. 10.2b, 10.3b) and the modal portrait $w(w'_y, w''_{yy})$ in 3D space (Figs. 10.2c, 10.3c) as well as the power spectrum (Figs. 10.2d, 10.3d) in the quarter of shell along the circle coordinate ($y_0 = \frac{\pi}{2}$).

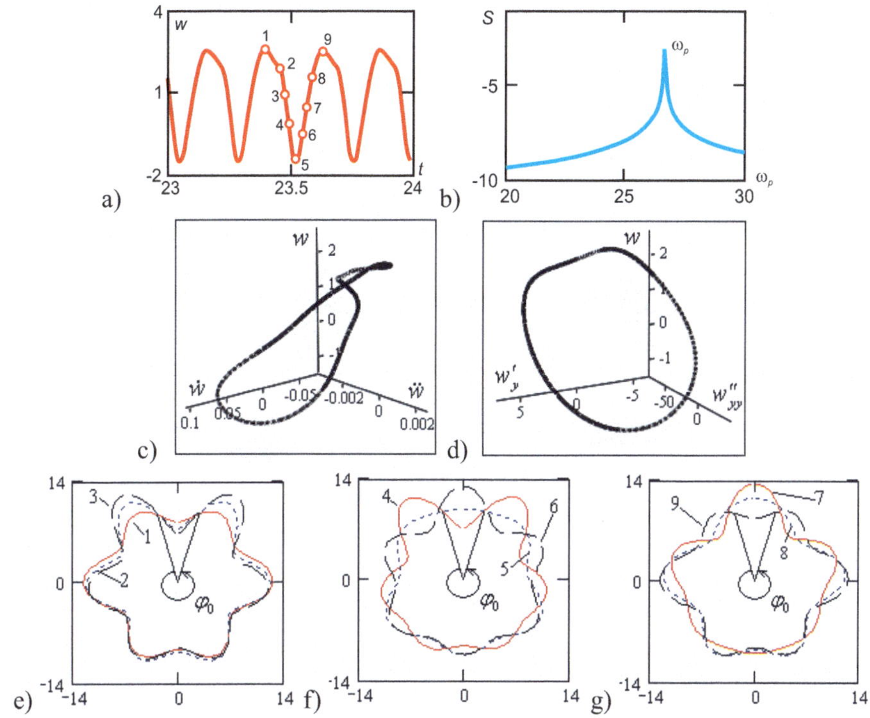

Fig. 10.2 Space-temporal shell characteristics in a periodic regime: (**a**) signal $w(0.5, \frac{\pi}{2}; t)$, $t \in [23; 24]$; (**b**) power spectrum $S(\omega)$; (**c**) phase portrait; (**d**) modal portrait

Simultaneous consideration of the all mentioned characteristics allows to carry out the reliable signal analysis in the space-time plane. The number of a point of Figs. 10.2a, 10.3a corresponds to the curve number with the same period reported in Figs. 10.2e–g, 10.3e–g. Let us analyse the shell deflection in time separately for the cases of periodic and chaotic shell vibrations. Analyzing the curves 1, 2, 3 in Fig. 10.2e one may observe movement of the shell points being unloaded into the shell centre whereas the loading points move into opposite direction (Fig. 10.2e, curve 3).

Next the inversed movement takes place and the periodic vibrations are localized around the neutral equilibrium state (curves 4–9 in Fig. 10.2f, g). It is clear and expected that the maximum shell deflections are located on the borders of the loading zone.

The so far carried out analysis of both phase and modal portraits of the shell chaotic vibrations based on Fig. 10.3c, d allows to conclude that both pictures present the washed out set of points being characteristic for the chaotic regimes. Another important observation worthy to be mentioned is that the temporal and space chaos occur simultaneously.

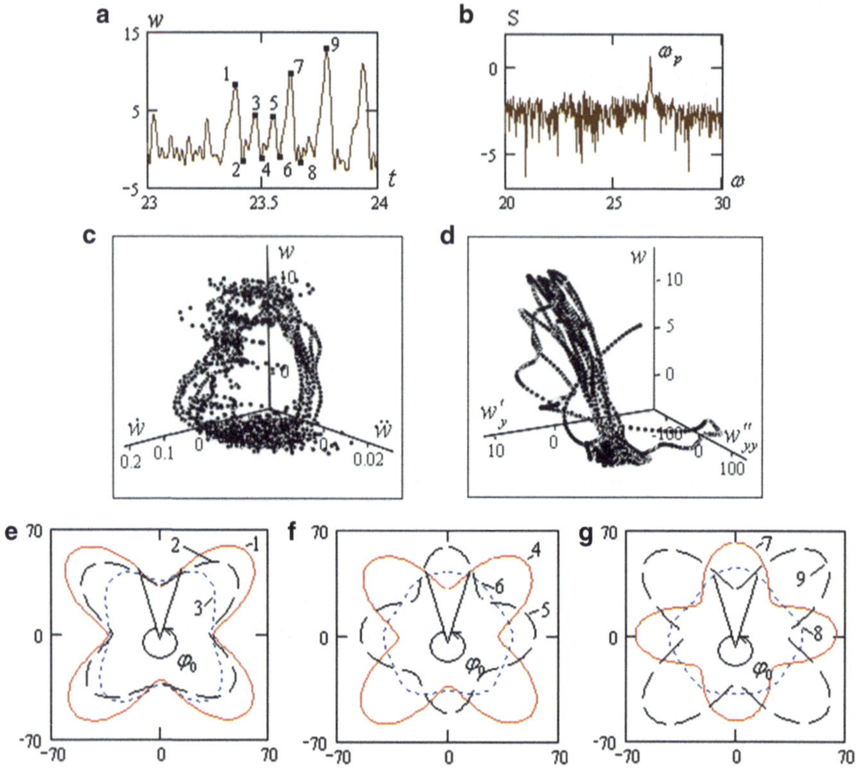

Fig. 10.3 Space-temporal characteristics within shell chaotic vibrations: (**a**) signal $w(0.5, \frac{\pi}{2}; t)$, $t \in [23; 24]$; (**b**) power spectrum $S(\omega)$; (**c**) phase portrait; (**d**) modal portrait

Chapter 11
Mathematical Models of Chaotic Vibrations of Closed Cylindrical Shells with Circle Cross Section

This chapter presents investigation of convergence of the Bubnov-Galerkin method for the cylindrical shells under action of the transverse harmonic local load on an example of a shell with $\lambda = 2;\ 6$ and under the width of the external pressure band measured by the central angle $\varphi_0 = 343$.

We demonstrate results obtained by our simulation package devoted to study chaotic vibrations of closed cylindrical shells with the circle cross section based on the Bubnov-Galerkin method in higher approximations.

The dependencies $w_{\max}(q_0)$ and the vibration scales of the dynamics regimes of the closed cylindrical shells with $\lambda = 2$ and for $\varphi_0 = 90°; 180°; 343°$, and with $\lambda = 1;\ 2;\ 3;\ 4;\ 5;\ 6;\ 7$ and for $\varphi_0 = 343°$ are presented. In particular, a novel criterion of the stiff stability loss of the cylindrical shells is formulated and illustrated.

Charts of the shell vibration character versus the control parameters $\{q_0, \omega_p\}$ corresponding to the amplitude and frequency of excitation for the mentioned shells for $\lambda = 2$ and for $\varphi_0 = 90°;\ 180°;\ 343°$, as well as for $\lambda = 1;\ 2;\ 3;\ 4;\ 5;\ 6;\ 7$ and for $\varphi_0 = 343°$, are constructed and an influence of the geometric shell dimensions and area of the band load localization of the vibration character are reported.

We have detected zones with the classical Feigenbaum scenario on the charts $\{q_0, \omega_p\}$ for the closed cylindrical shells with seven Hopf period doubling bifurcations; the latter allowed to estimate the Feigenbaum constant with the accuracy of 0.01%.

We have also detected a novel scenario of transition of the shell periodic vibrations to chaotic vibration referred by us as the modified Pomeau-Manneville scenario as well as the phenomenon of intermittency in the shell vibrations. The latter scenario has been investigated and illustrated by a case study of the closed cylindrical shell with $\lambda = 2$ and for $\varphi_0 = 90°$.

Influence of the area of the transverse local load acting in direction x and y as well as of the coordinates of their localization on the shell vibration character

J. Awrejcewicz, V. A. Krysko, *Elastic and Thermoelastic Problems in Nonlinear Dynamics of Structural Members*, Scientific Computation,
https://doi.org/10.1007/978-3-030-37663-5_11

are analysed. Finally, the features of the spatial-temporal chaos are illustrated and discussed.

11.1 On the Convergence of the Bubnov-Galerkin (BG) Method in the Vase of Chaotic Vibrations of Closed Cylindrical Shells

In order to quantify the whole palette of dynamic behaviour of continuous systems within the same factor, we use the same approach to study their dynamics in the phase space. A key role here plays reduction of the systems of infinite dimension to those of finite dimension and a proper interpretation of the obtained numerical results. It is hypothesized that an increase of ODEs implies achievement of a certain limit where the dynamic behaviour of the studied system is stabilized in the sense that a further increase of equation does not produce any additional dynamic effects. Here a particular role play dimension of the system attractor. However, even if dimension of the studied attractors is bounded an important role may play effects introduced by truncation of the approximated ODEs. In the case of an unsuitable choice of the basis responsible from transition of PDEs to ODEs the resulted truncated system may exhibit false attractors different from those hidden in the input studied system. This problem was observed already in the case of a 2D PDE governing heat convection of a fluid. The obtained Lorenz system [230] based on three modes approximation exhibits chaotic dynamics while increasing the number of ODEs first yields increase of the chaotic dynamics but than its remarkable decrease.

On the other hand, it has been shown in reference [92] that in the case of large Prandtl numbers δ in the studied 2D Bousinesque convection there are critical values of the Rayleigh number *Ra* responsible for occurrence of one- and two-mode vibrations whereas a further increase of *Ra* implies occurrence of the one frequency periodic convection. The mentioned example shows that in order to obtain the qualitatively reliable correspondence between dynamics of the input system (PDEs) and the truncated systems (ODEs) based on the employment of the BG or the Ritz procedure there is a need to take into account also higher modes. Owing to our knowledge the problem regarding the best approximation of the BG method while analysing closed cylindrical shells subjected to harmonic load has been not investigated.

In this chapter we present the case studied dealing with convergence of the BG method for various types of shallow shells.

Let us begin with a study of the number of modes required by the BG procedure to get reliable results using an example of vibrations of closed cylindrical and geometrically nonlinear shells with boundary conditions (9.21) and zero initial

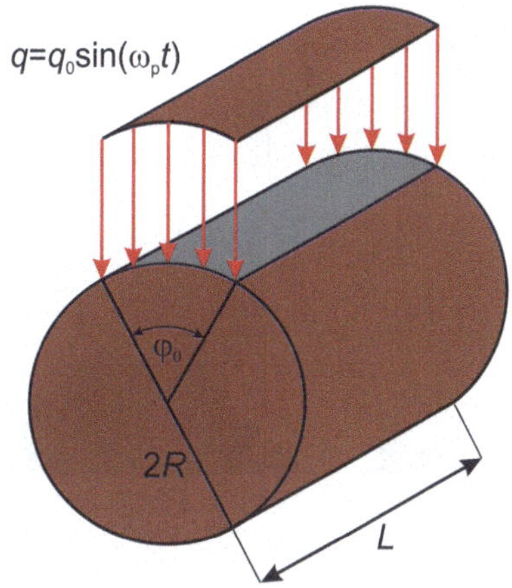

Fig. 11.1 Computational scheme of the circle cylindrical shell

conditions (9.22). The used local load is placed in the sectorial part of the shell defined by angle φ_0 (Fig. 11.1) and has the following harmonic form

$$q = q_0 \sin(\omega_p t). \tag{11.1}$$

In order to study shell behavior we have worked out the programs package allowing to construct the charts presenting dynamics of the shell versus the control parameters $\{q_0, \omega_p\}$. Those charts allow to study all possible dynamic shell behavior. The used algorithm allows also to distinguish zones with the purely exhibited Feigenbaum scenario as well zones of the modified RTN (Ruelle-Takens-Newhouse) scenario.

Let us first investigate the problem of the solution convergence while increasing the number of partition of the plane $\{q_0, \omega_p\}$ (Fig. 11.2). In order to construct the mentioned plane charts we have put a mesh consisting of the nodes where the vibrations character was identified. Consequently, the dynamic problem for each of the mesh node should be solved and for each node of the mesh $\{q_0, \ \omega_p\}$ the associated power spectrum should be also analyzed. The carried out computations have shown that the mesh resolution $N \times N \geq 350 \times 350$ is optimal with regard to the computational time in both period and chaotic regimes in the investigated plane $\{q_0, \omega_p\}$.

Let us analyze the problem regarding a choice of the modes while employing the BG procedure using the case study of closed cylindrical and geometrically nonlinear shells. Since we employ the local load along the shell surface zone defined by (11.1) then in order to achieve the reliable results it is sufficient to take only one series term along x in the development of the series of the functions w and F, i.e. we take $N_1 = 1$.

Let us construct charts characterizing the behavior of the cylindrical shells with $\lambda = 2$ and the width $\varphi_0 = 6$ of the pressure band (Fig. 11.3) versus the control parameters $\{q_0, \omega_p\}$ for various numbers of terms in (9.31), i.e. for $N_2 = 9 \div 14$.

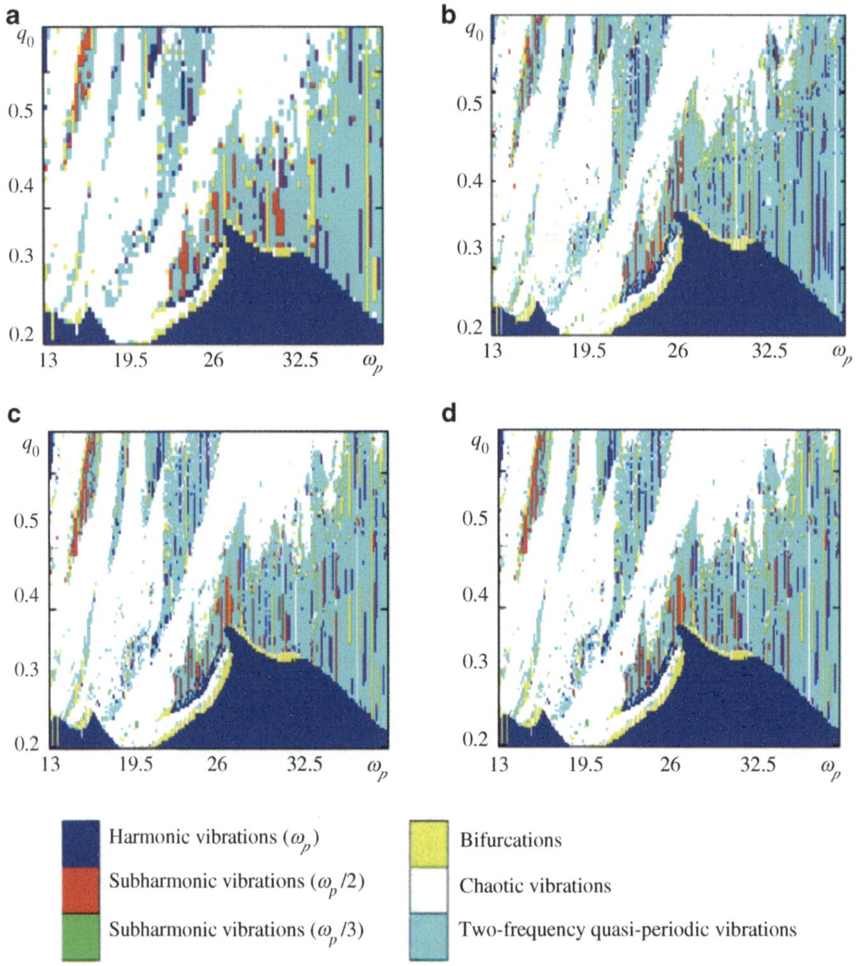

Fig. 11.2 Charts of the dynamic regimes vs. control parameters $\{q_0, \omega_p\}$ for the cylindrical shell with $\lambda = 2$. (**a**) 100×100 (**b**) 200×200 (**c**) 350×350 (**d**) 500×500

Further increase of N_2 in (9.31) did not introduce any qualitatively change of the chart $\{q_0, \omega_p\}$.

The largest value of the load amplitude q_0 is bounded by a limit given by a linear theory. The parameter ω_p is changed from $\frac{\omega_0}{2}$ to $\frac{3}{2}\omega_0$, where ω_0 stands the shell fundamental frequency.

Since the charts construction is based on analysis of the frequency power spectra and the Lyapunov exponents then the mentioned charts illustrate process of convergence of the power spectrum and the vibrations character for each of the employed mesh approximation. Therefore, for small values of $N_2 \leq 9$ mainly periodic vibrations are presented and only small zones of bifurcations are visible

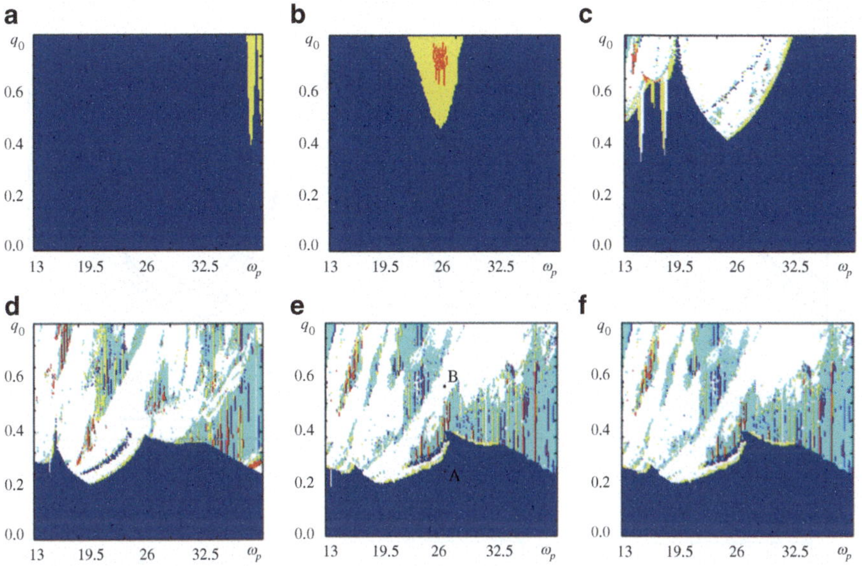

Fig. 11.3 Charts of the dynamic regimes depending on the control parameters $\{q_0, \omega_p\}$ for the cylindrical shell ($\lambda = 2$). (**a**) $N_2 = 9$. (**b**) $N_2 = 10$. (**c**) $N_2 = 11$. (**d**) $N_2 = 12$. (**e**) $N_2 = 13$. (**f**) $N_2 = 14$

with a lack of chaotic zones. For $N_2 = 10$ the chart undergoes a qualitative change; the bifurcation zones are vanished for the high frequencies but they appear on the frequencies close to the fundamental shell frequency. Further increase of N_2 up to $N_2 = 11$ generates occurrence of large zones of chaotic vibrations on low frequencies and being close to the fundamental frequency; the previous bifurcation areas observed for $N_2 = 9$ are vanished. Further increase of $N_2 = 12$ implies changes on high frequencies within chaotic zones whereas zones associated with low frequencies are practically unchanged. Finally, the constructed charts for $N_2 = 13$ and $N_2 = 14$ are very similar to each other; hence one may conclude that for $N_2 = 13$ the process of convergence is terminated and the obtained results are reliable. Observe that the process of convergence is more fast for the low frequencies than for high frequencies and frequencies being close to the fundamental frequency ($\omega_0 \approx 26.1$).

Let us consider two points shown in Fig. 11.3 in the bifurcation zone (point A in the chart $\{q_0, \omega_p\}$ for $N_2 = 13$) and in the chaotic zone (point ĐŠ in the chart $\{q_0, \omega_p\}$ for $N_2 = 13$). Figures 11.4 and 11.5 report the signals w (0.5; 0.0; t) and power spectra $S(\omega)$ in those points. Analysis of the results based on consideration of Fig. 11.4 shows that beginning with $N_2 \geq 12$ the power spectra coincide in full and time series are very close to each other. However, the results obtained for $N_2 = 9, 10, 11$ differ essentially from those of $N_2 \geq 12$. Based on that observation one may conclude that beginning with $N_2 \geq 12$ the process of bifurcation is described

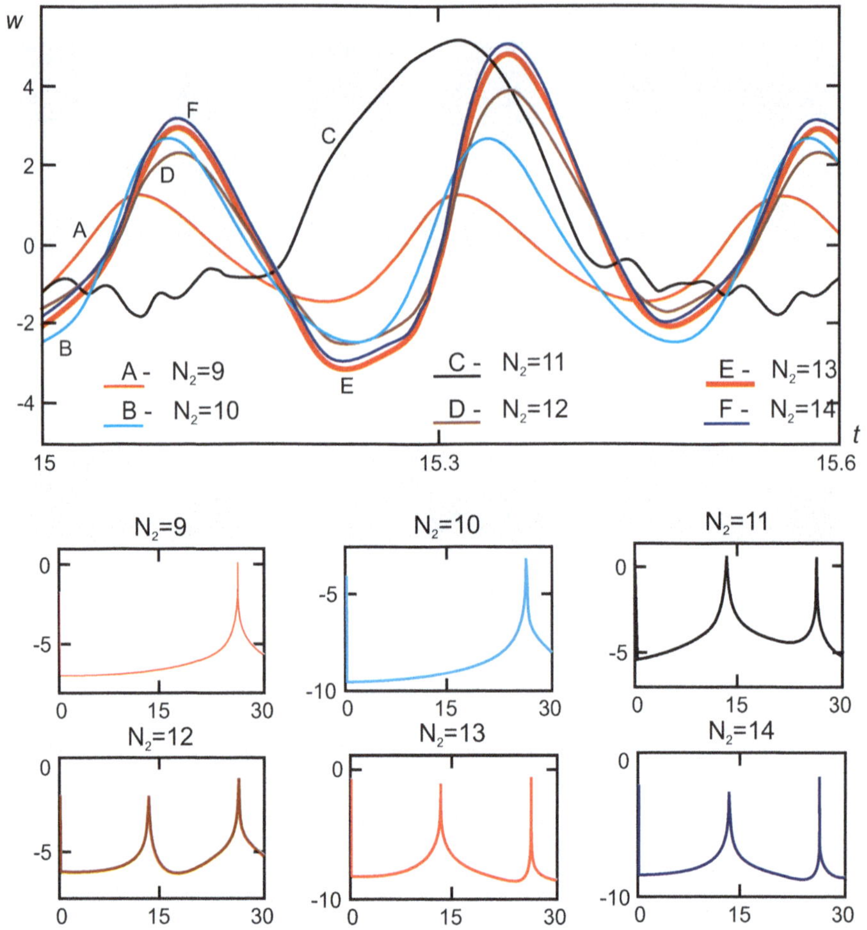

Fig. 11.4 Dependencies $w(0.5, 0.0, t)$, $t \in [15; 15.7]$ and $S(\omega)$ vs. N_2 in (1.31) for the cylindrical shell ($\lambda = 2$)

properly and the results are reliable, i.e. there exist a convergent subsequence, i.e. one may formally define the condition

$$\left[w - \sum_{i=0}^{N_1} \sum_{j=0}^{N_2} A_{ij}(t)\varphi_{ij}(x, y) \right] = \min_{t \in [15, 15.7]}$$

$$\left[F - \sum_{i=0}^{N_1} \sum_{j=0}^{N_2} B_{ij}(t)\psi_{ij}(x, y) \right] = \min_{t \in [15, 15.7]}$$

which allows to obtain the best approximation to the functions w and F.

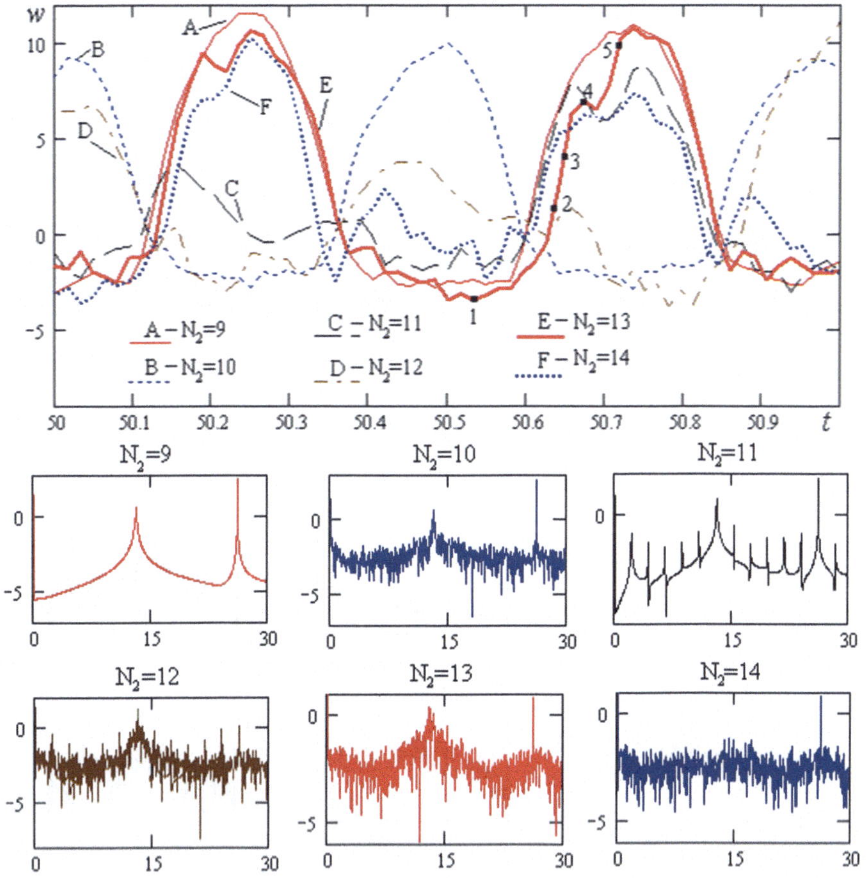

Fig. 11.5 Dependencies $w(0.5, 0.0, t)$, $t \in [50; 51]$ and $S(\omega)$ vs. N_2 in (1.31) for the cylindrical shell ($\lambda = 2$)

In Fig. 11.5 there are reported the same characteristics as in Fig. 11.4. However, in this case we do not see the uniform results convergence as it was in the previous case. For $N_2 = 9$ we have period doubling bifurcation (orbit 2); $N_2 = 11$ is associated with the period doubling and tripling bifurcation; for $N_2 = 10,\ 12,\ 13,\ 14$ we deal with chaotic vibrations but the character of chaos is principally different; for $N_2 = 14$ chaos is spanned on the fundamental frequency, whereas for $N_2 = 10,\ 13$ chaos is spanned of on $\omega_p/3$, and for $N_2 = 12$ on $\omega_p/12$; in words, each term of the series introduces own peculiarities to the system dynamics. On may also conclude that in chaos there exist regularity conditions defined by a density of periodic points.

Let us investigate convergence of the BG method regarding the shell geometry using an example of the cylindrical shell with $\lambda = 6$. We consider the point which for $N_2 = 13$ is located in chaotic zone for the following fixed parameters of the load: $q_0 = 0.65$, $\omega_p = 26.114$. The associated signals $w(0.5, 0.0, t)$, $t \in [15; 16]$

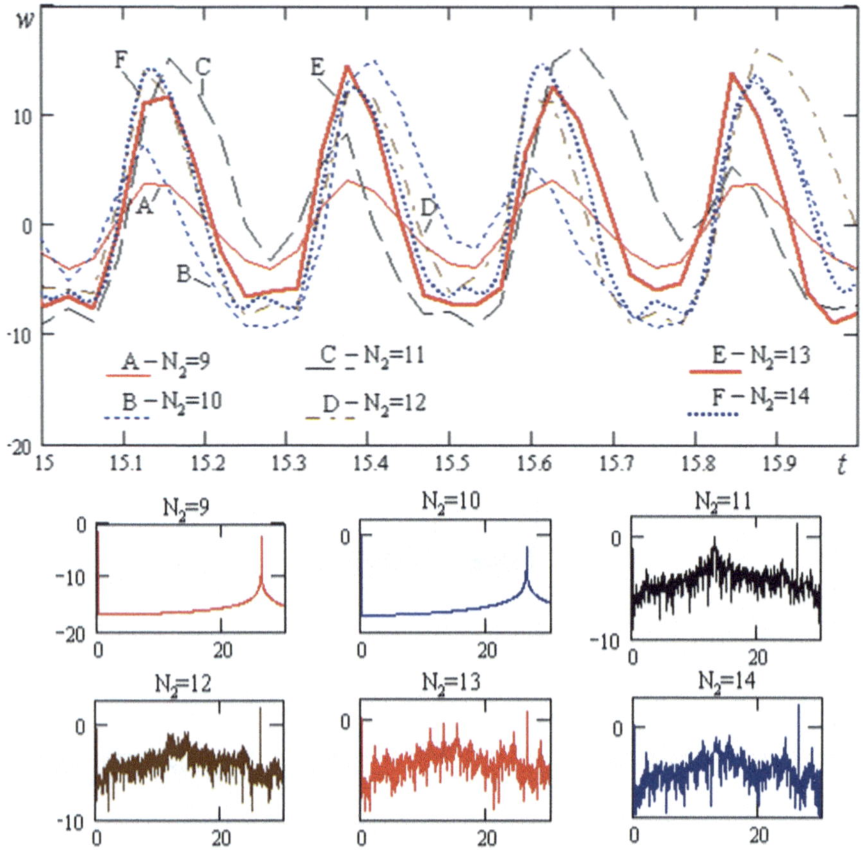

Fig. 11.6 Dependencies $w(0.5, 0.0, t)$, $t \in [50; 51]$ and $S(\omega)$ vs. N_2 in (1.31) for cylindrical shell ($\lambda = 6$)

and frequency power spectra are presented in Fig. 11.6. For $N_2 = 9,\ 10$ periodic vibrations with ω_p are observed; $N_2 = 11,\ 12$ describes chaos spanned on $\omega_p/2$, whereas $N_2 = 13, 14$ are associated with chaos spanned on $\omega_p/13$. Here the convergence of the signal is better than in the previous cases; if we consider the convergence in average, i.e. based on the computed power spectrum then beginning with $N_2 = 12$ a convergent subseries is visible.

The so far carried out investigation and description yields the following main conclusion: a convergence of the BG method versus the number of terms in the series (9.31) essentially depends on the geometry of the cylindrical shell and on its dynamical regime.

In order to estimate convergence of the method we need to analyze the form of the shell bending along the circle coordinate with fixed load and fixed width of the pressure for various approximations. In Table 11.1 there are reported cross sections of the cylindrical shell with $\lambda = 2$ in the defined time instants under fixed

Table 11.1 Cross section of the circle cylindrical shell for different N_2

N_2	Points in the signal $w(0.5;0.0;t)$				
	1	2	3	4	5
11	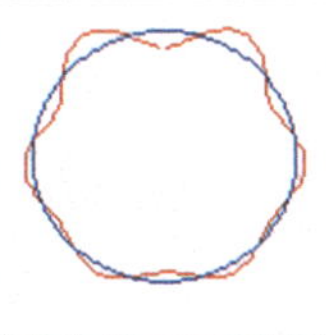	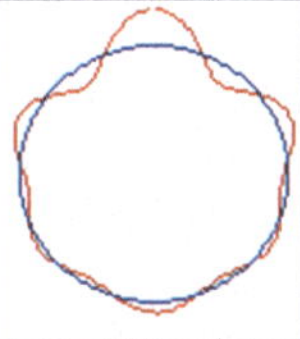	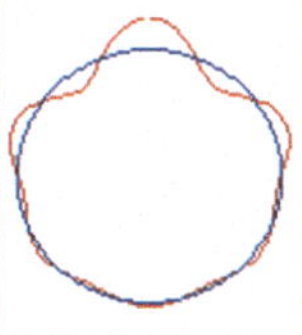	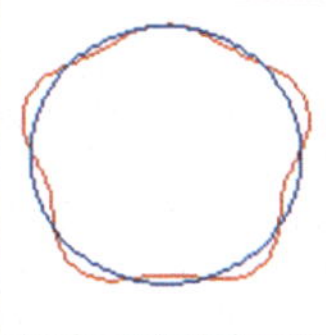	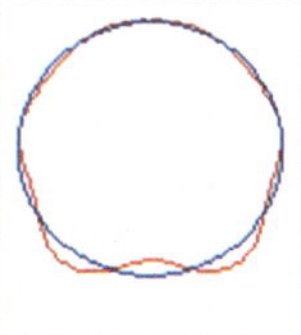
12	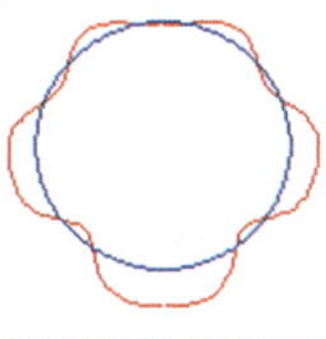	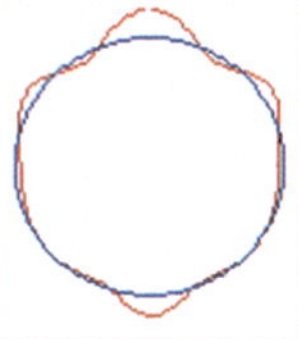	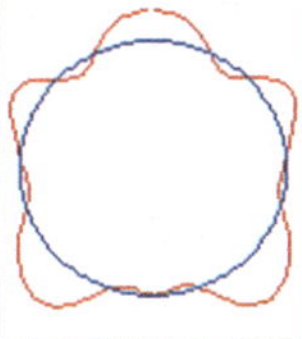	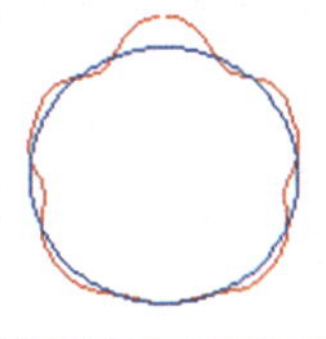	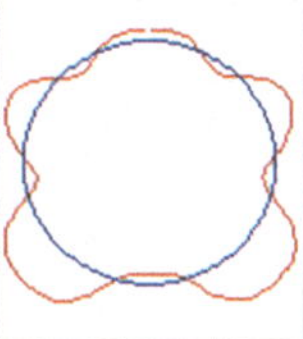
13	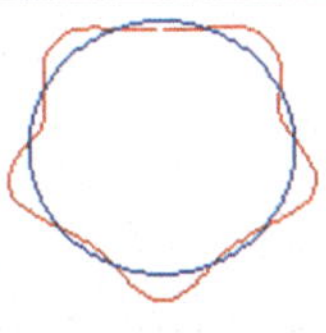	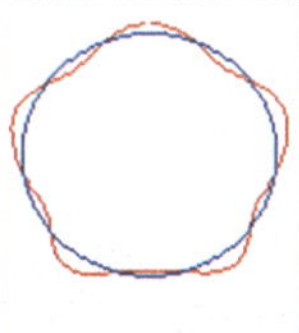	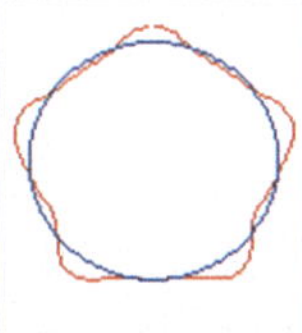	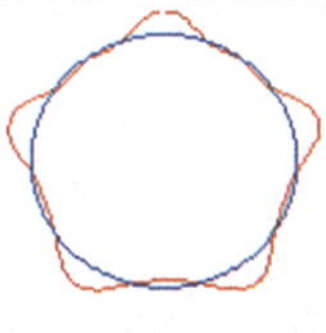	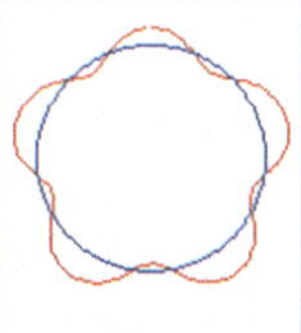
14	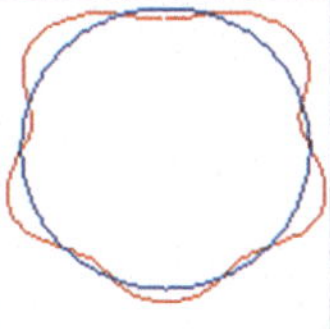	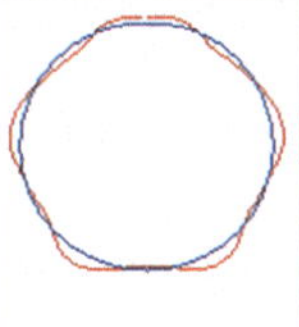	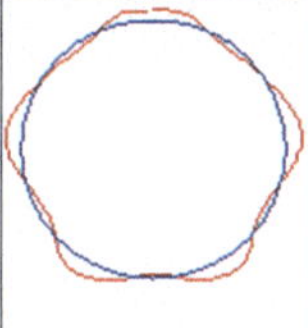	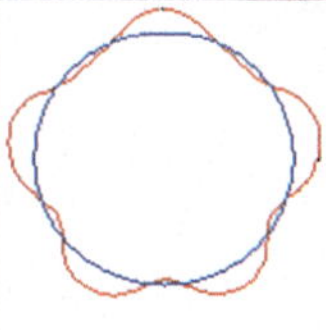	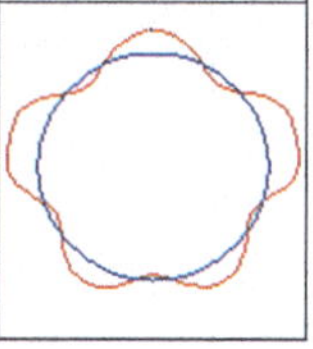

$x = 0.5$ and $0 \leq y \leq 2\pi$ for the point B in the chart $\{q_0, \omega_p\}$ and for $N_2 = 13$ (Fig. 11.3), i.e. in the chaotic zone. The points 1–5 correspond to those depicted the corresponding time histories (Fig. 11.4). Beginning with $N_2 = 13$ one may observe the same chart in the form of the transverse cross section as in the signal $w(0.5; 0.0; t)$. Namely, we have achieved convergence with respect to time history as well as with respect to the bending shell forms in space. Increase of the terms of the series (9.31) does not improve accuracy of the obtained results.

It is worthy to mention that in the considered problem an increase of the number of degree of freedom of the approximating lumped mass systems does not simplify the problem, which stand in contrary to the archetype Lorenz model. The shell instead of regular behaviour may present chaotic one for the same control parameters q_0 and ω_p and different N_2.

Let us study an input of each of the coefficients A_{1j} in the series (9.31) onto the final results. Keeping fixed $k_y = 112.5$, $\lambda = 2$, $\varphi_0 = 3.0$ we consider the dependencies $A_{1j}(t)$ ($j = 0, \ldots, 13$) for $N_2 = 13$ (Fig. 11.7) in a periodic regime. Figure 11.7a presents the first coefficients of the series (9.31) ($j = 0, \ldots, 7$), whereas Fig. 11.7b reports higher coefficients ($j = 8, \ldots, 13$).

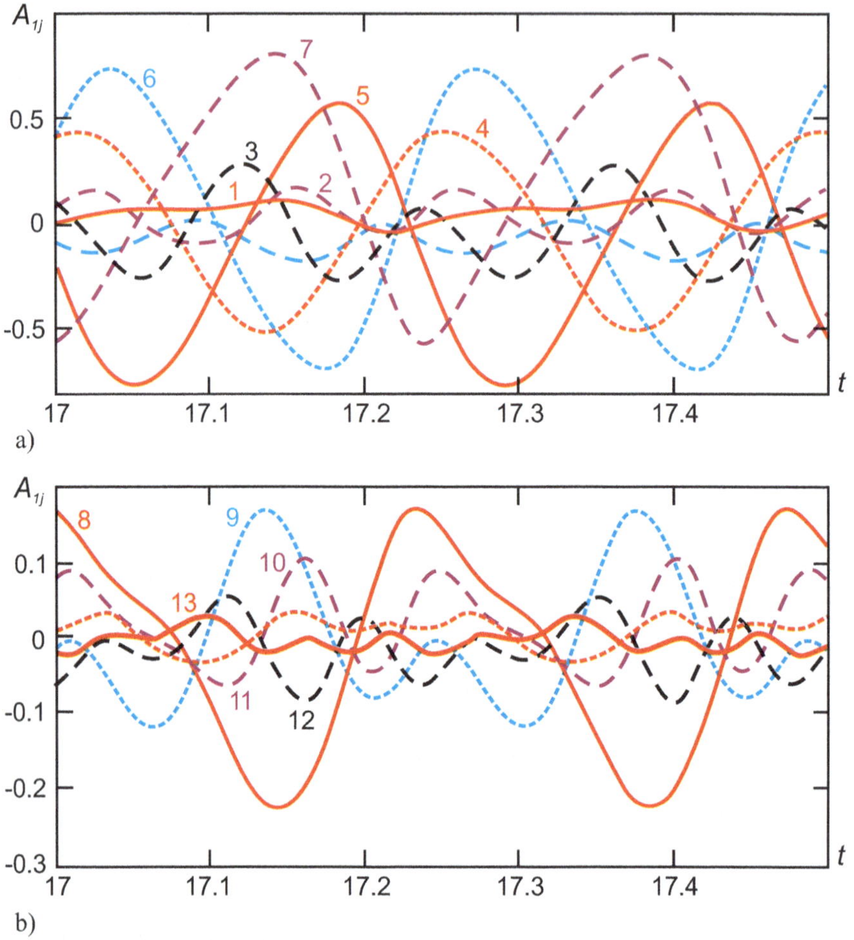

Fig. 11.7 Dependencies $A_{1j}(t)$ for $N_2 = 13$ and for different series (9.31) coefficients $A_{1j}(t)$: (**a**) $j = 0, \ldots, 7$; (**b**) $j = 8, \ldots, 13$

Analysis of the dependencies $A_{1j}(t)$ show that for a given value $N_2 = 13$ the fundamental contribution to the general chart of vibration are introduced by average coefficients of the series. It should be also mentioned that the character of the dependencies $A_{1j}(t)$ for each j is different not only with regard to the amplitude of vibrations but also because with respect to frequency.

Figure 11.8 presents the dependencies $\max(A_{1j}(q_0))$ versus the load amplitude for some values of $j = 0, 2, 4, 7, 9, 11, 13$. Now changing the load amplitude implies change the contribution of each of the coefficients into the chart of vibration. For small values of q_0 the fundamental contribution is given by the first coefficients of the solution approximation whereas an increase of the load amplitude q_0 the fundamental contribution is associated with the average coefficients of the series.

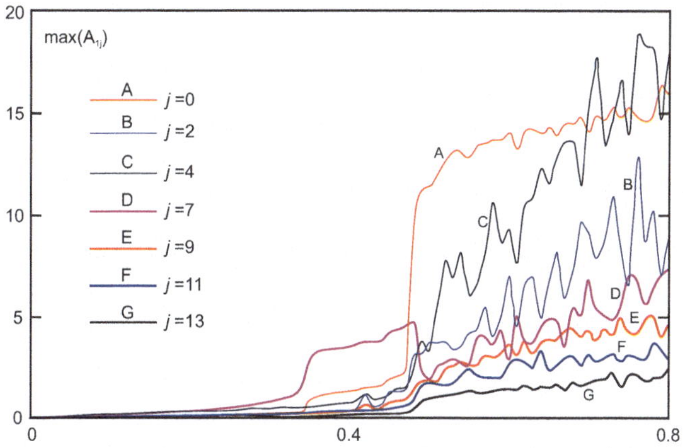

Fig. 11.8 Dependencies $\max(A_{1j}(q_0))$ for $N_2 = 13$ and for different j

The higher coefficients contribute essentially to the general chart of the system vibrations, i.e. increase of N_2 implies essential improvement in achieving reliable results; neglection of higher coefficients in the series (9.31) introduces errors in approximation to the reliable solution, and the obtained results are far from the reliable ones.

Let us investigate the BG method convergence while studying dynamics of the cylindrical shell under the local load employed in both x and y direction. In this case in approximation of the solution we should take $N_1 > 1$, $N_2 > 1$. Let $N_1 = N_2$ and $N_1 = N_2 = N$. Figures 11.9 and 11.10 presents the signals $w(0.4, 2.0, t)$, $t \in [30; 40]$ in the centre of the load area and the power spectra in both regular and chaotic regimes in the case of the load localization: $(x_1, x_2) = (0.3; 0.5)$, $(y_1, y_2) = (1.0; 3.0)$.

Analysis of the results shows that beginning with $N \geq 7$ the power spectra coincide and the time series are a very close to each other. The results obtained for $N = 3; 5$ essentially differ from those for $N \geq 7$. It means that beginning with $N \geq 7$ the process of periodic vibrations is reliable, i.e. there exists a convergent subsequence. In Fig. 11.10 the same characteristics are reported as in Fig. 11.9. Here we do not see the same unique convergence as in the previous case. Namely, for $N = 3$ we have harmonic vibration; for $N = 5, 7, 9, 11$ we have chaotic vibrations with different features depended on N; for $N = 7, 11$ chaos is spanned on period doubling frequency, whereas for $N = 5, 9$ chaos is spanned on the fundamental frequency, i.e. each of the series term contributes to the full picture of chaos in a different way.

We have omitted here a construction of the dynamic charts in the case of chaotic vibrations of the cylindrical shell subjected to the local load action in x and y directions because the time needed to construct one chart with increasing number of the employed series terms increased in an exponential way. In order to construct one power spectrum we need to solve $N \cdot (N + 1)$ linear algebraic equations (AEs)

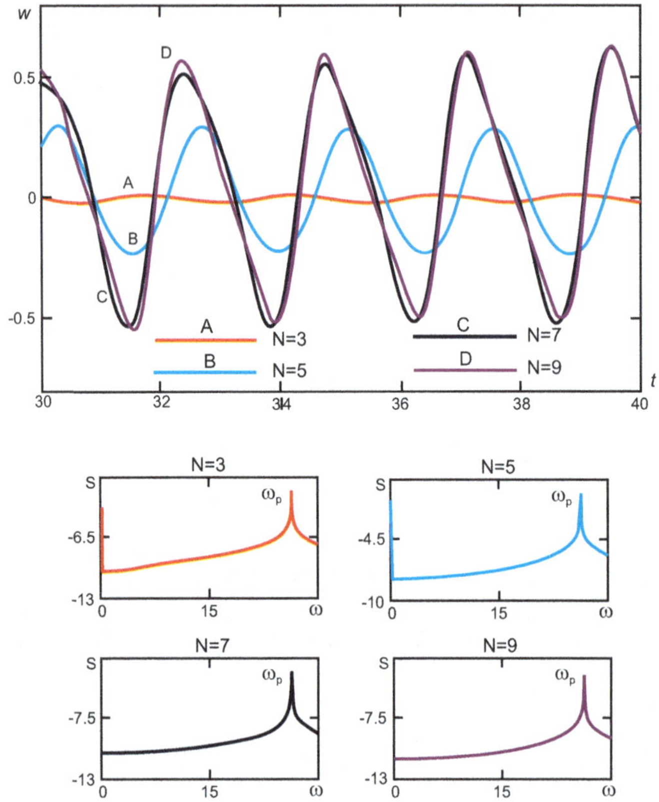

Fig. 11.9 Dependence $w(0.4, 2.0, t)$, $t \in [30; 40]$ and $S(\omega)$ for the cylindrical shell with $\lambda = 2$ subjected to the local load

and $2 \cdot N \cdot (N + 1)$ ODEs. Consequently, in order to compute one chart with the mesh resolution 200×200 of the plane $\{q_0, \omega_p\}$ and for $N = 7$ approximation for each mesh node one must solve 56 AEs and 112 ODEs. It corresponds to the work of a few months of a computer.

11.2 Chaotic Vibrations of Closed Cylindrical Shells Versus Their Geometric Parameters and Area of the External Load Action

Here and further we present results for $N_2 = 13$ since further increase of N_2 of the series terms approximating solution (9.31) does not introduced any observed changes in the obtained numerically results. We investigate vibrations of the cylindrical shells of the circle cross section under load harmonic load (11.1) within

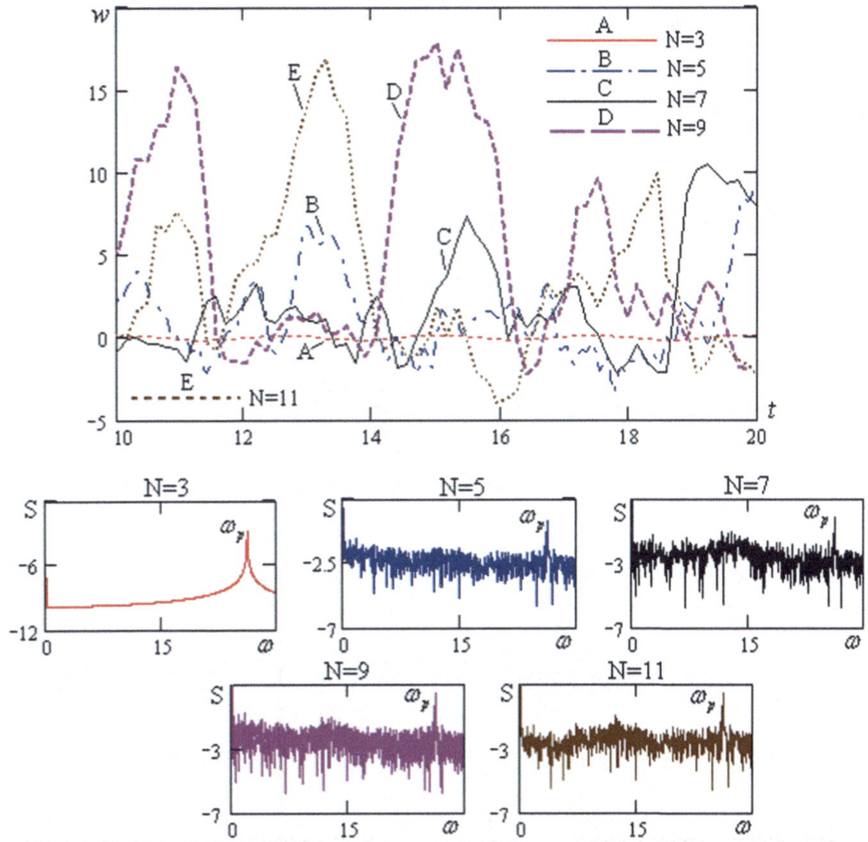

Fig. 11.10 Dependencies $w(0.4, 2.0, t)$, $t \in [10; 20]$ and $S(\omega)$ for the cylindrical shell ($\lambda = 2$)

interval of a zone associated with the central angle φ_0 (Fig. 11.1). The coordinates of the load location are: $0 \leq x \leq 1$; $-\varphi_0/2 \leq y \leq \varphi_0/2$. In the case of very small loading angles where the pressure distribution is close to the line loading, there are observed large shell deflections in the zone action neighbourhood. However, the bending form of the shell does not exhibit remarkable changes in the loading process (Fig. 11.11).

In the case of pre-critical shell bending with the mentioned band load the largest deflections are observed in zones located close to the borders of the loading area and inside of it. Small loading angles φ_0 are associated with development of one cavity under the pressure band (Fig. 11.11a), whereas further increase of φ_0 yields occurrence of two cavities along the borders of the load are (Fig. 11.11b). For $\varphi_0 \to 0$ and $\varphi_0 \to 2\pi$ on the contour parts lying far from the edges of the loading band the bending is weak (Fig. 11.11a, c). For some discrete values of φ_0 while increasing the load, a qualitative change of the bending form is observed with a simultaneous increase of the number of half waves. Large loading angles φ_0 imply

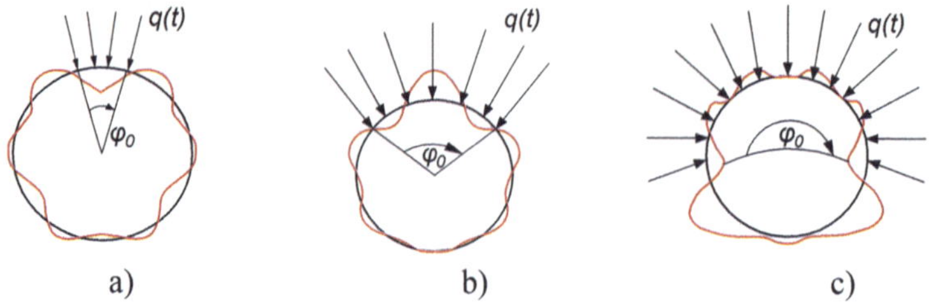

Fig. 11.11 Bending forms of the shell under different loading angles φ_0

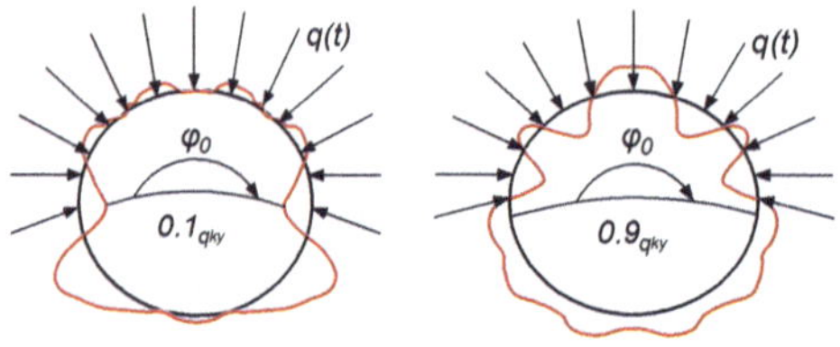

Fig. 11.12 Bending forms of the shell under different angles of the load

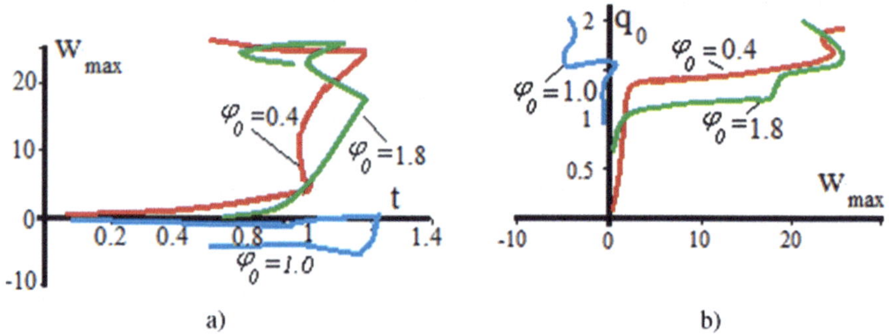

Fig. 11.13 Dependence maximum deflection versus time for different loading angles φ_0 (**a**) and maximum deflection versus q_0 (**b**) for different loading angles φ_0

also the qualitative change of the bending form but it has a local character and it is observed in the centre of the pressure band (see Fig. 11.12 here and further q_{cr} means critical load).

Stability loss of the shell is observed for different angles of the load, i.e. for $\varphi_0 \geq \varphi_{\min}$ where $\varphi_{\min}$ corresponds to the limiting point in the diagram load-deflection. The mentioned diagrams are presented in Fig. 11.13 for various loading angles. In order to define the critical load we employ two earlier described criteria, i.e. Shio et al. criterion (Fig. 11.13a) and the Volmir dynamic criterion (Fig. 11.13b). Analysis of the obtained results allows to conclude that the critical load estimated through two mentioned criterions coincides with the accuracy of 10^{-3}.

Table 11.2 Characteristics of the shell dynamics for different q_0

φ_0	Pre-critical state		Post-critical state	
0.4 rad = 23°	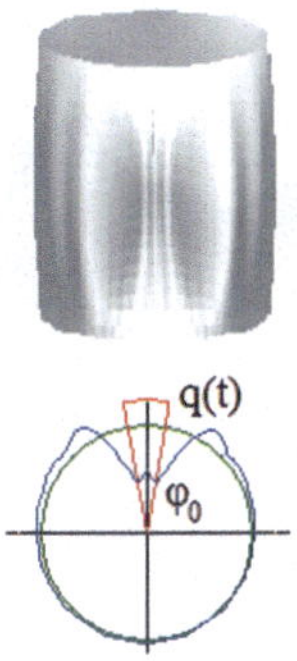	$q_0 = 1.3$ 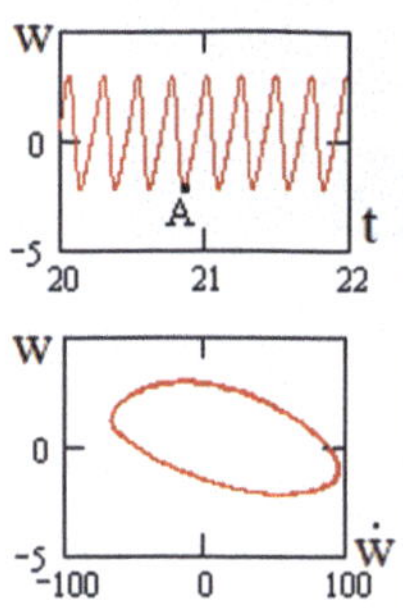	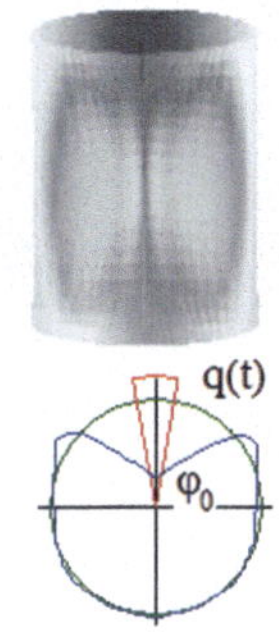	$q_0 = 1.5$ 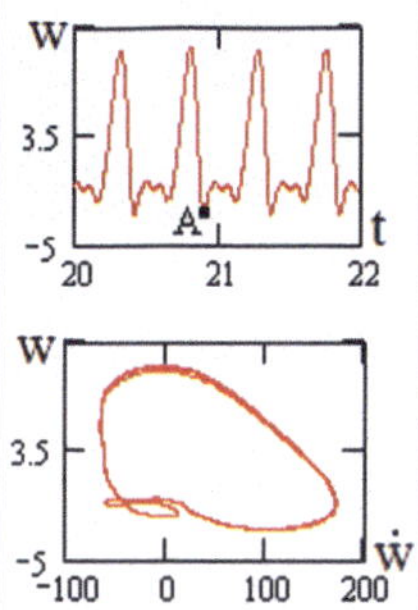
1.0 rad = 57°	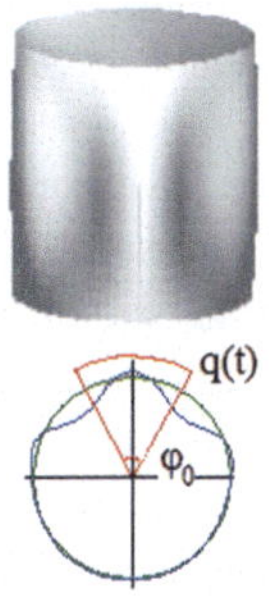	$q_0 = 1.4$ 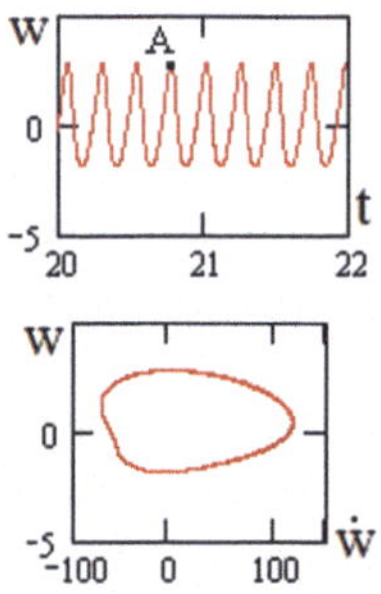	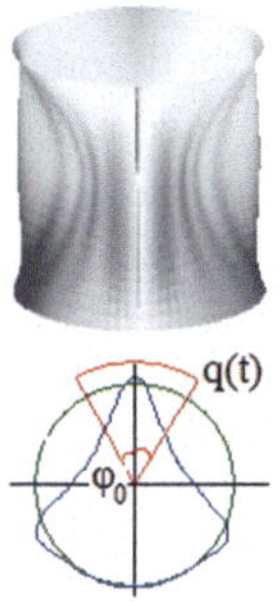	$q_0 = 1.5$ 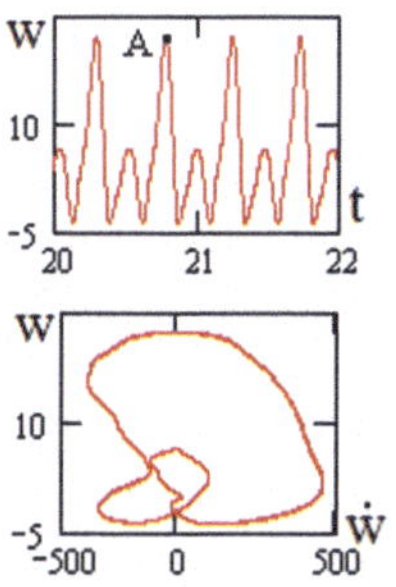
1.8 rad = 103°	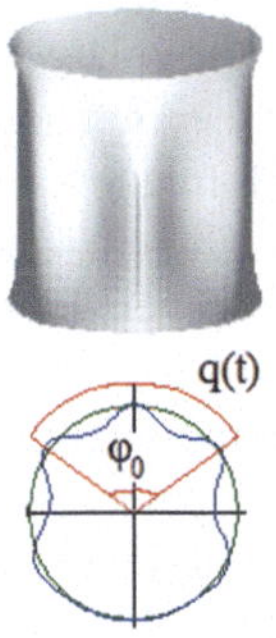	$q_0 = 1.0$ 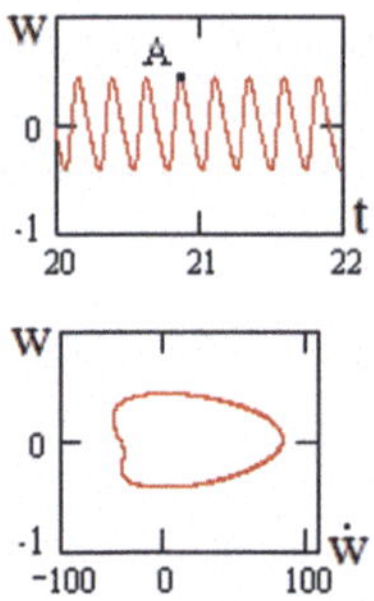	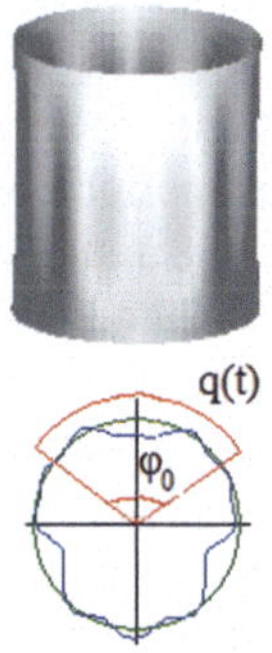	$q_0 = 1.1$ 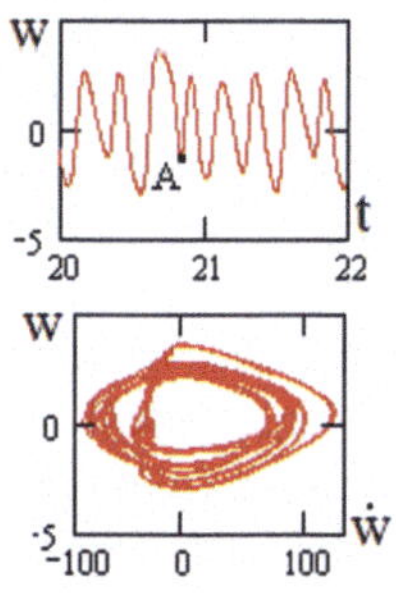

We have also carried out a similar analysis of the shell bending forms and the forms of the transverse shell cross sections in the shell pre- and post-critical states for the series of the values of φ_0 (Table 11.2). The forms of the transverse shell cross sections ($x = 0.5$; $y \in [0; 2\pi]$) and the characteristic wave forms ($x \in [0; 1]$, $y \in [0; 2\pi]$) of the cylinder correspond to the marked point A in the signal. In the pre-critical state the largest deflection is observed in the area of

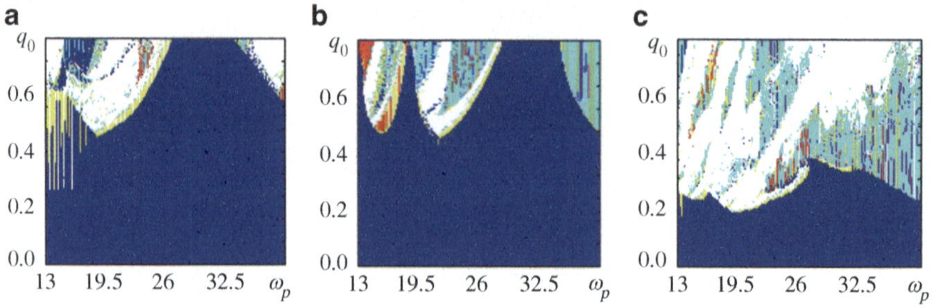

Fig. 11.14 Charts of dynamic regimes depending on the control parameters $\{q_0, \omega_p\}$ for the cylindrical shells with $\lambda = 2$ and for different loading angles φ_0. (**a**) $\varphi = 90°$. (**b**) $\varphi_0 = 180°$. (**c**) $\varphi_0 = 343°$

the load location. Transition to the post-critical state implies propagation of the large displacements into shell zones previously free of loading, and the character of vibrations becomes more complex.

For the certain values of the width of the pressure band φ_0 we have constructed charts of the dynamic regimes depending on the control parameters $\{q_0, \omega_p\}$ (Fig. 11.14). The mentioned charts allow to study all possible shell dynamic behaviour. In particular, we have detected that the vibrations character depends strongly on the loading angle. For small values of φ_0 the summed area of chaotic zone is large and consists of two subareas corresponding to the frequencies $\omega_p < \omega_0$ and $\omega_p > \omega_0$. Increase of the loading zone of the cylindrical shell implies decrease of the chaotic zone and its transition into low and average frequencies. This phenomena is accompanied by increase of the two-frequency quasi-periodic vibrations. Large area of the external pressure is associated with the enlargement of chaotic zone. Namely, chaotic zones are distributed along the whole vibration chart, there are also visible quasi-periodic vibrations in the case of high frequencies.

Let us now give a few remarks concerning turbulence. Though the latter phenomenon is known for hundreds of years, the first attempt devoted to give a rigorous mathematical description of transition from laminar to chaotic (turbulent) motion has been proposed by Landau. In the case of the solid mechanics objects like the studied here cylindrical shells we did not detect any purely exhibited transition from regular to chaotic dynamics of the shells for the considered loading angles and geometric parameters. For each of the fixed set of the mentioned parameters a key role play amounts of the amplitude and frequency of the external sinusoidal excitation. In the case of cylindrical shells we have detected a novel mechanism of transition from regular to chaotic vibrations which is matched with the modified Ruelle-Takens scenario and the classical Pomeau-Manneville scenario. The mentioned scenario has been also found while analysing vibrations of the spherical panels. We consider this scenario based on the case study of the cylindrical shell with $\lambda = 2$ and the width of the transversal pressure band $\varphi_0 = 90°$. We fix the excitation frequency $\omega_p = 24.74$, and we use q_0 as a control parameter. The following characteristics are analysed: signal $w\,(0.5; 0.0; t)$, phase portrait $w(\dot{w})$, power spectrum $S(\omega)$ and the Poincaré map $w_{t+T}(w_t)$ versus the limiting values of q_0, and they are shown in Table 11.3 (limiting values of q_0 means that the chart of dynamics is almost unchanged between the successive values of q_0).

In what follows we briefly describe the system main dynamic features based on the results shown in Table 11.3.

1. The shell vibrations are harmonic with the excitation frequency ω_0, and the phase portrait exhibits one rotational cycle ($q_0 = 0.58$).
2. Further increase of q_0 ($q_0 = 0.6$) generates a new linearly independent frequency a_1, i.e. a quasi-periodic orbit is spanned on the frequencies ω_0 and a_1 ($\omega_0/a_1 = m/n = 8.756\ldots$). The phase portrait presents two-rotational cycle. Besides, there is a third dependent frequency $b_1 = \omega_0 - a_1$. In the latter case the phase portrait exhibits thick phase trajectory.
3. Increasing q_0 up to $q_0 = 0.615$ yields the series of the linearly dependent frequencies $a_n = n \cdot a_1$ and $b_n = \omega_0 - (n-1)\,a_1$, and this process takes place until both frequencies a_k and $b_k \in [b_1, a_1]$ approach each other; then a qualitative change takes place where the independent frequency and the series of the dependent frequencies vanish followed by the period-7 bifurcation ($q_0 = 0.76$).
4. Further increase of q_0 ($q_0 = 0.769$) implies the occurrence of the third frequency c_1 accompanied by a series of the successive linear combinations $c_n = x_n \pm c_1$, $c_1 = a_4 - b_3$, $x_n = a_n, b_n$. Further increase of the parameter q_0 results in the system reconstruction; the frequency c_1 is substituted by the period doubling bifurcation frequency ($q_0 = 0.781$) with a birth of the new dependent frequency $d_1 = a_1/2 - (a_4 - b_3)$, i.e. resonance effects are exhibited. Further increase of q_0 up to the value of $q_0 = 0.793$ implies occurrence of chaos.

The reported in Table 11.2 scenario of transition from the periodic to chaotic vibrations belongs to the first investigation of the problems of evolution of one-, two- and three-frequency vibrations, i.e. the transition into chaos is realized by a collapse of the quasi-periodic regime with two independent frequencies. The carried out numerical analysis indicates and existence of a robust structure of the two-frequency regime (2D torus T^2) and the occurrence of a strange chaotic attractor in the neighbourhood of T^2.

Besides the mentioned scenario of transition into chaotic state, the set $\{q_0, \omega_p\}$ includes zones where a transition of our mechanical system is realized in the form of the classical Feigenbaum scenario [114] associated with the period doubling bifurcations. The Feigenbaum model has been well validated by the numerical experiments devoted to simple physical models.

The signal $w\,(0.5; 0.0; t)$, $100 \le t \le 103$, phase portrait $w(\dot{w})$, power spectrum $S(\omega)$ and the Poincaré map $w_t(w_{t+T})$ for the cylindrical shell with $\lambda = 2$ subjected to the harmonical load localized on the band with $\varphi_0 = 180°$ and with the excitation frequency $\omega_p = 13.25$ are shown in Table 11.4.

The values of the subseries $q_{0,n}$ and subseries d are shown in Table 11.5 which allows to estimate the Feigenbaum constant

$$D_N = \frac{Q_{0,N} - Q_{0,N-1}}{Q_{0,N+1} - Q_{0,N}} = 4.669444\ldots$$

Table 11.3 Dynamic characteristics of the cylindrical shell ($\lambda = 2$)

q_0	Signal $w(0.5;0.0;t)$	Phase portrait $w(\dot{w})$,	Power spectrum $S(\omega)$	Poincaré section $w_{t+T}(w_t)$
0.58	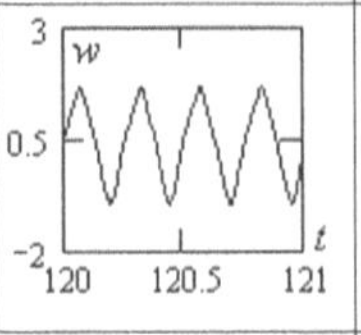	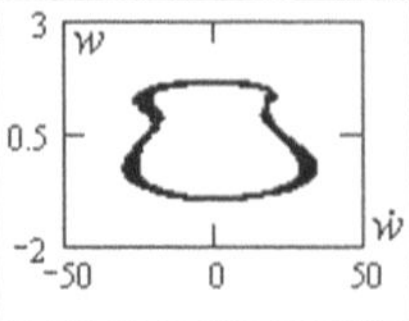	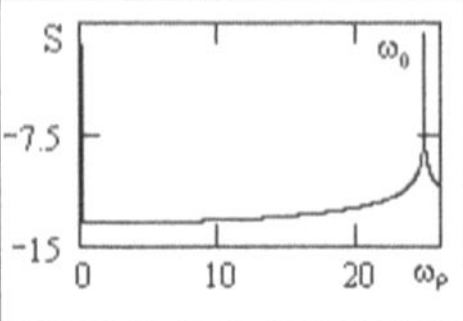	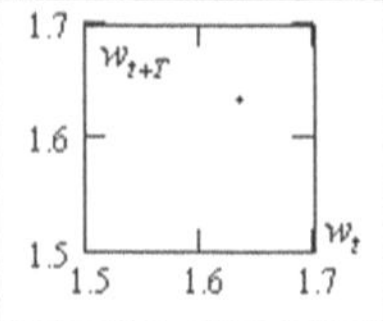
0.6	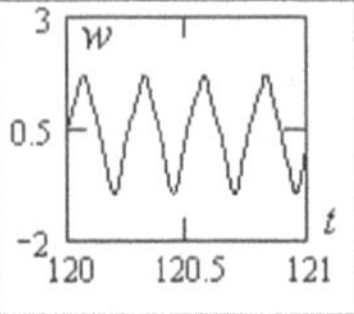	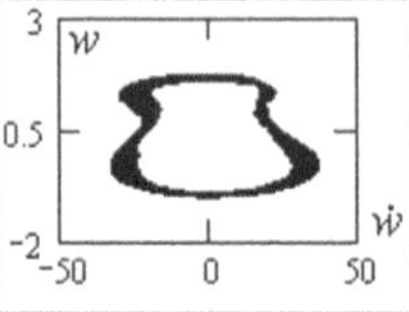	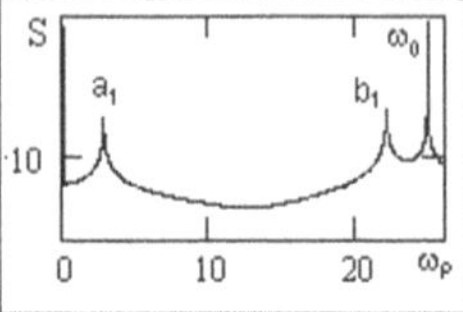	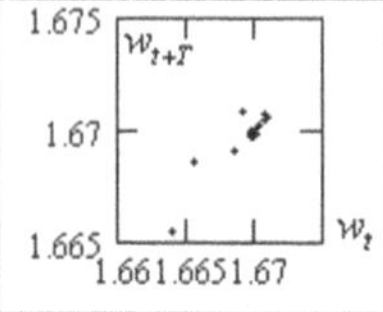
0.61	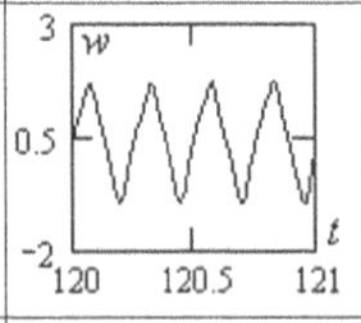	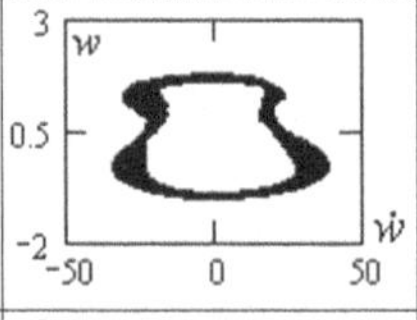	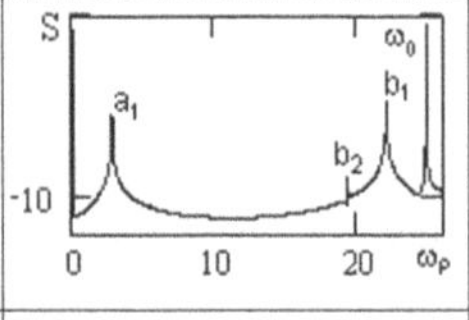	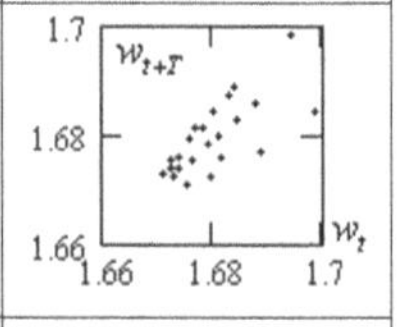
0.615	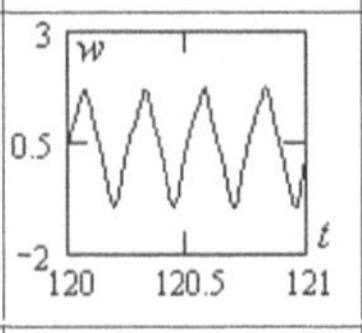	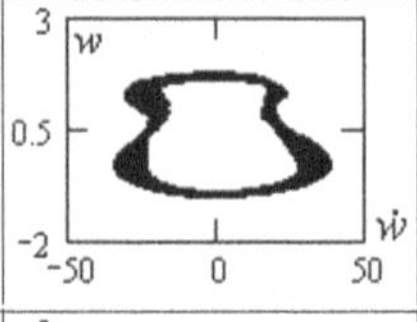	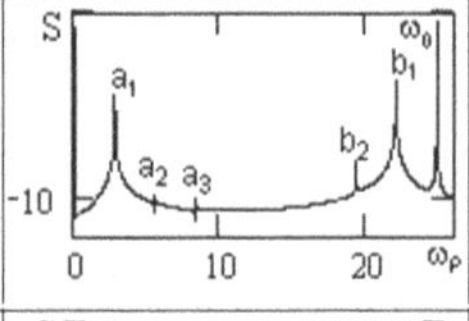	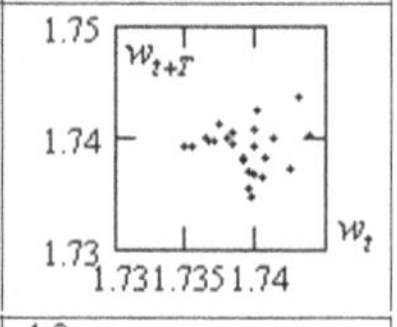
0.625	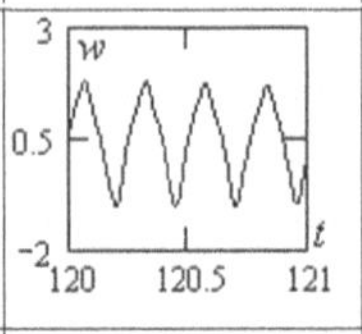	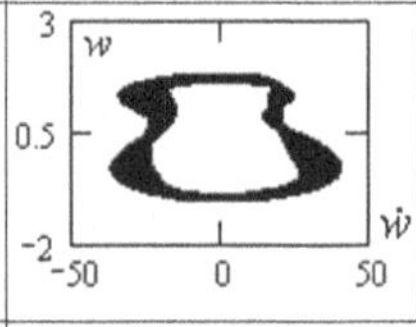	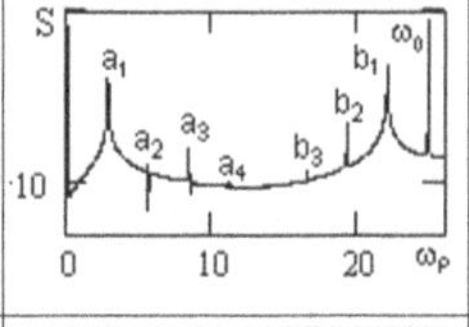	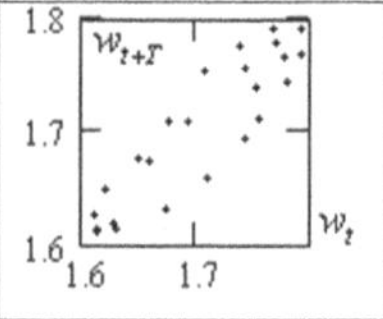
0.75	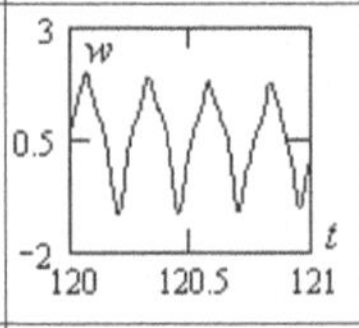	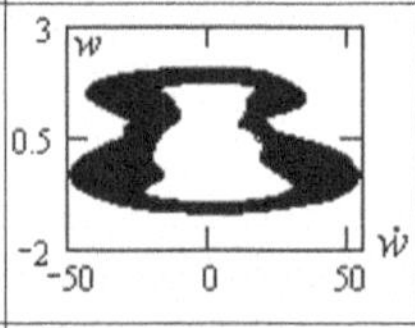	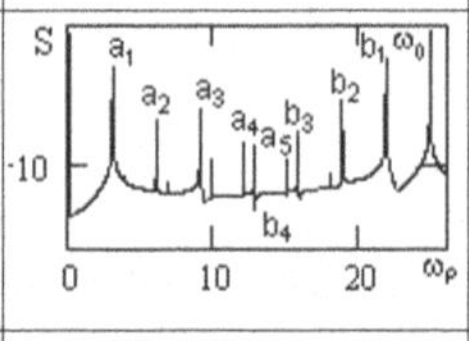	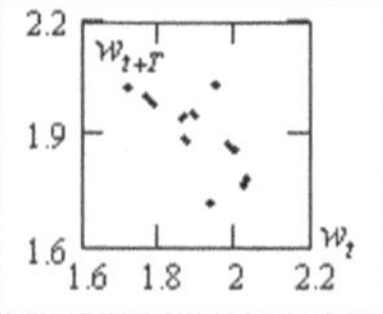
0.76	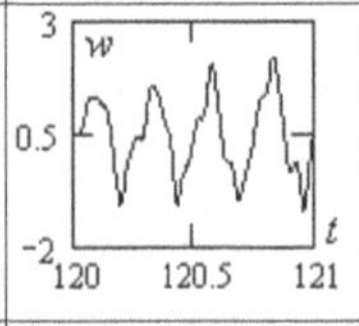	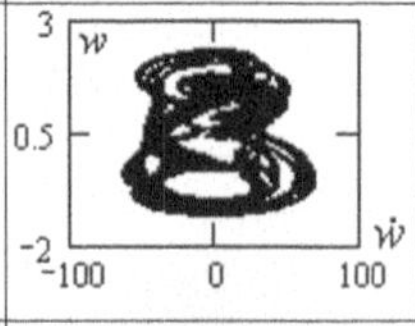	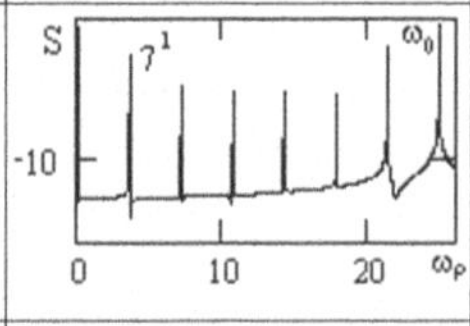	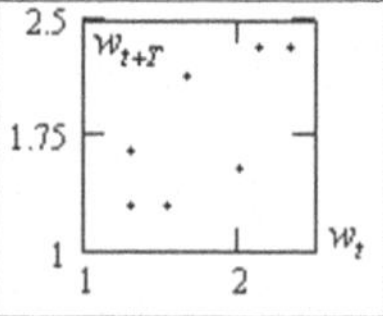
0.769	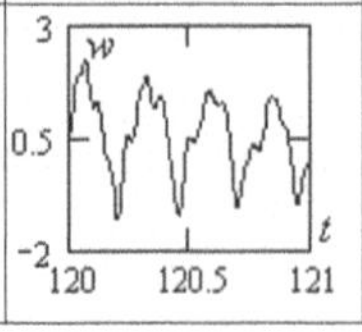	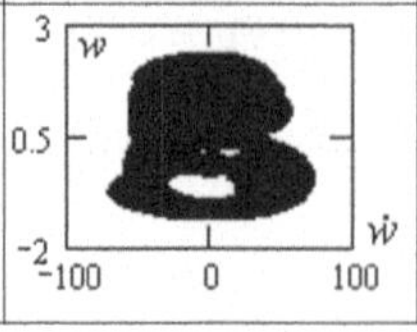	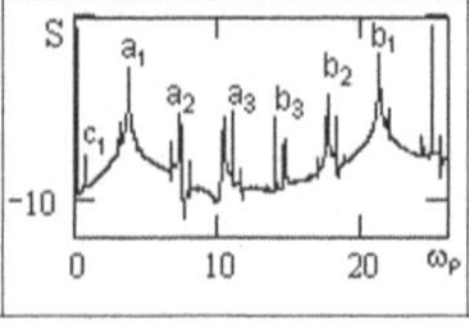	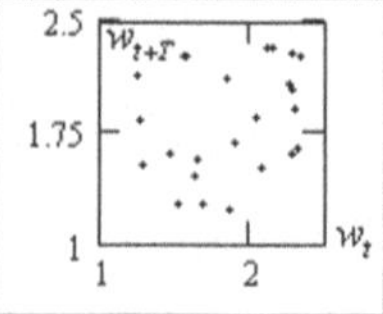

(continued)

Table 11.3 (continued)

0.781	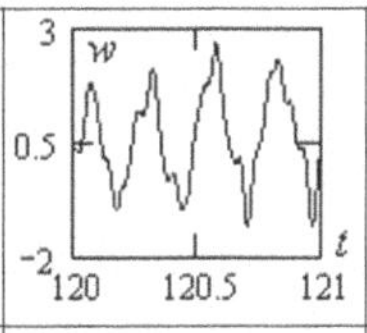	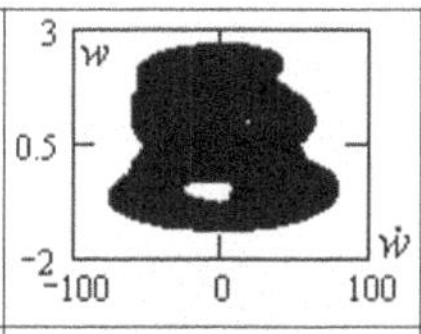	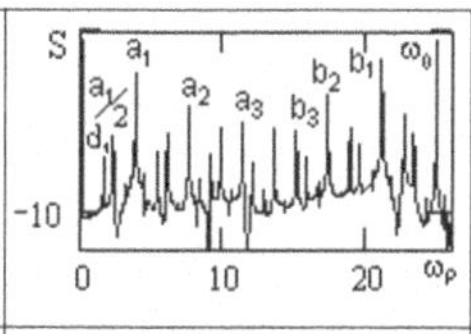	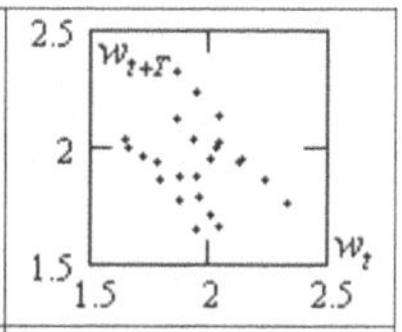
0.793	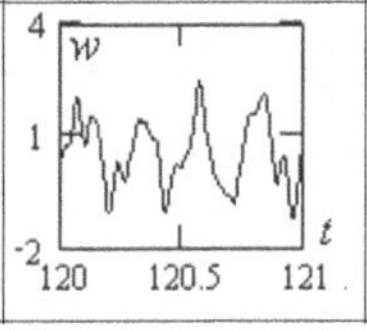	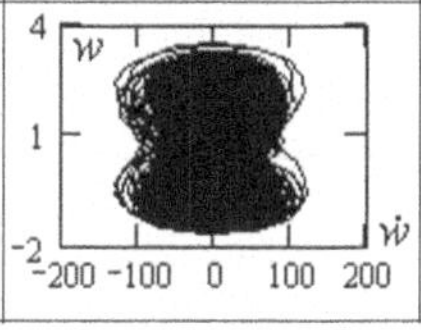	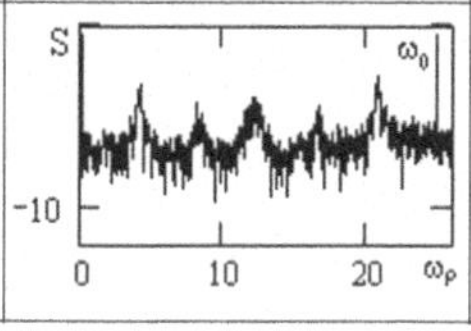	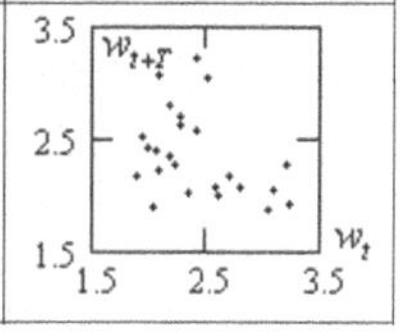

Theoretical value of d obtained for the function $f = \left(1 - cx^2\right)$ [90, Tab. 1.25] is equal to $d = 4.66916224\ldots$. The difference between theoretical computations and numerical result is 0.01%.

In order to investigate the critical loads we refer to the dependence $w_{\max}(q_0)$ in the shell centre with $\omega_p = 26.642$ and bifurcation scale for the cylindrical shells with the width of the external pressure band $\varphi_0 = 90°; 180°; 343°$, as well as the powers spectra before and after stability loss (Fig. 11.15).

Analysis of the dependencies $w_{\max}(q_0)$ and comparing them with the results presented in the scales of the dynamic regimes yields the following conclusion valid for all values φ_0: increase of the load amplitude allows to find the value where further even very small increase of the q_0 causes the shell stability loss (deflections increase suddenly which yields qualitative change of the vibration character).

For all investigated values of φ_0 the system dynamics is transited from the chaotic vibrations to periodic with ω_p or subharmonic with $\omega_p/2$ vibrations. The described process can be viewed as the dynamic stability loss of the cylindrical shells under periodic transverse local loads. This can be treated as a novel criterion of the stiff stability loss when the system dynamics is transited from one regime to qualitatively different another regime.

It is well known that the Lyapunov exponents play a crucial role while analyzing the dissipative mechanical systems. They can measure strength of chaos. The proposed and developed idea of construction of the scales of the vibration character of the dynamical systems [50] is based on the analysis of the power spectrum $S(\omega)$ which well coincides with the evolution of the largest Lyapunov exponent (LLE) $\lambda_1(q_0)$. Positive value of the LLE corresponds to occurrence of chaos, which is also confirmed by the scale of bifurcations and $w_{\max}(q_0)$. In the latter case the series of stiff bifurcations is observed associated with dynamic stability loss. This observation can be treated as the dynamic criterion of the stability loss.

As it has been already mentioned, we have not constructed charts of the dynamic regimes in the case of the local load action in two directions x and y (Fig. 11.16), but we have used scales of dynamic regimes and the dependencies $w_{\max}(q_0)$. Let

Table 11.4 Dynamic characteristics of the cylindrical shell with $\lambda = 2$ and the loading band angle $\varphi_0 = 180°$

q_0	Signal $w(0.5;0.0;t)$	Phase portrait $w(\dot{w})$,	Power spectrum $S(\omega)$	Poincaré map $w_{t+T}(w_t)$
0.302	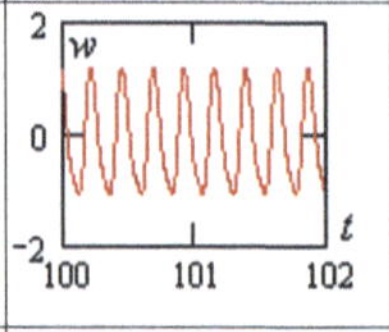	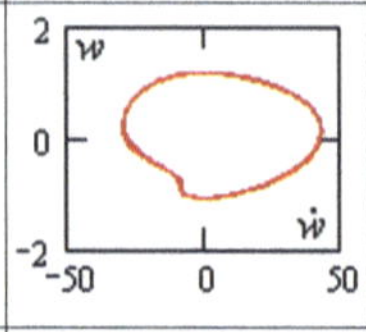	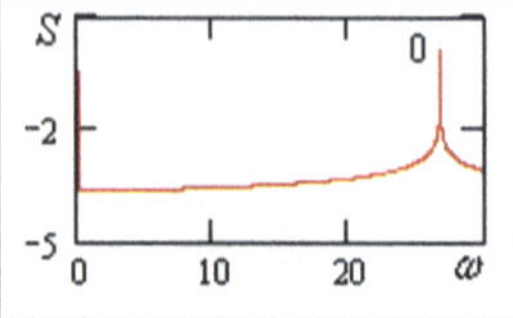	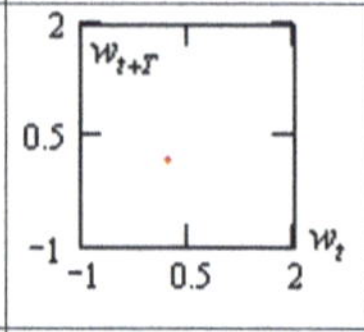
0.589764	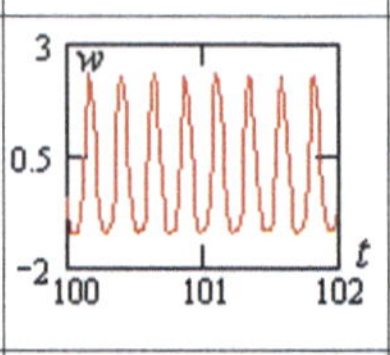	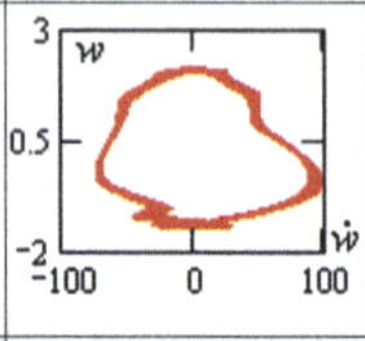	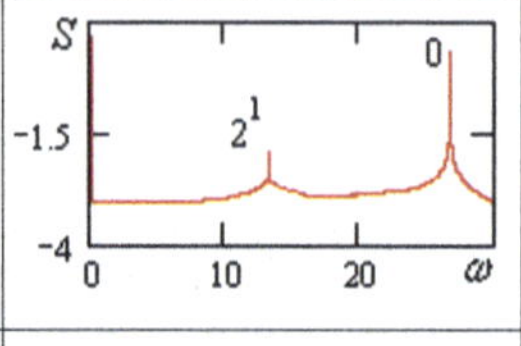	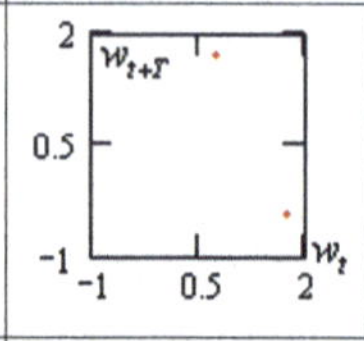
0.58977	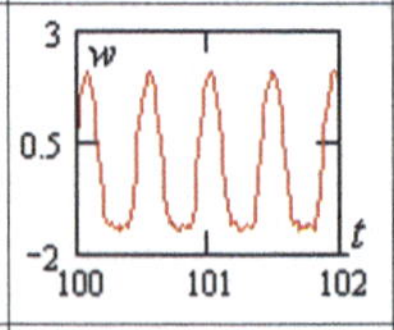	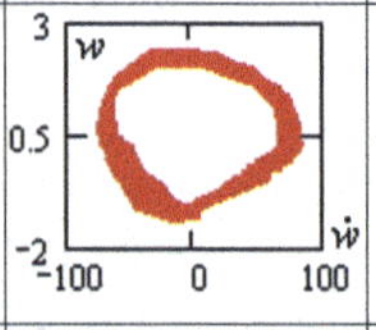	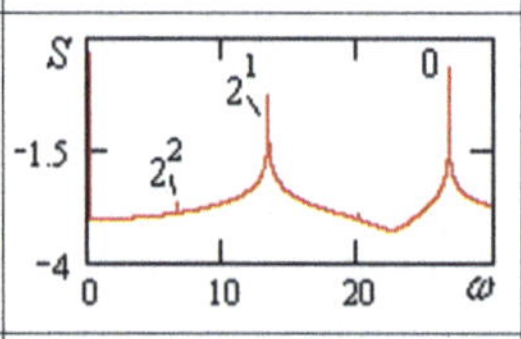	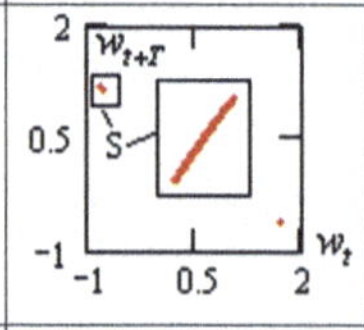
0.589778	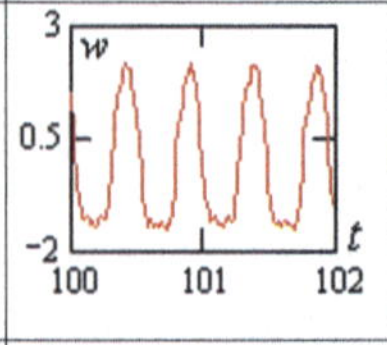	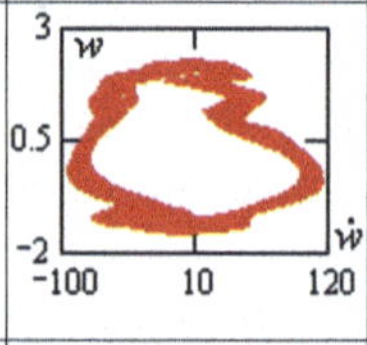	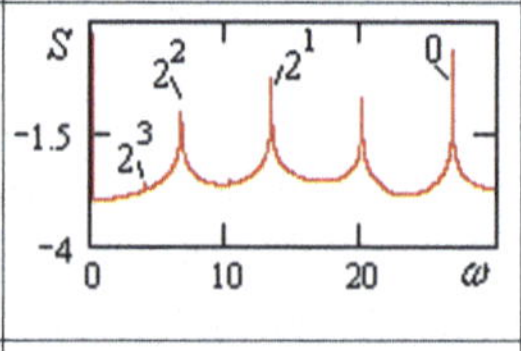	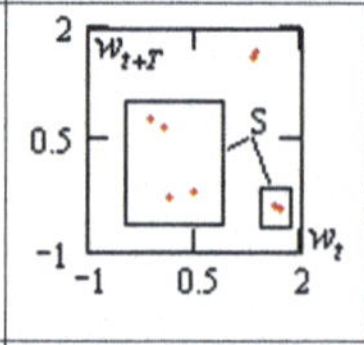
0.58978	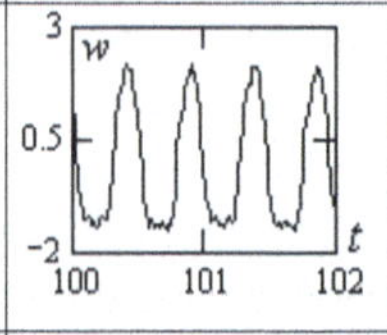	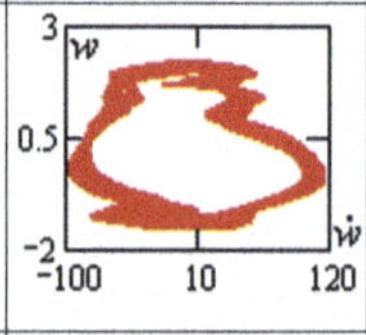	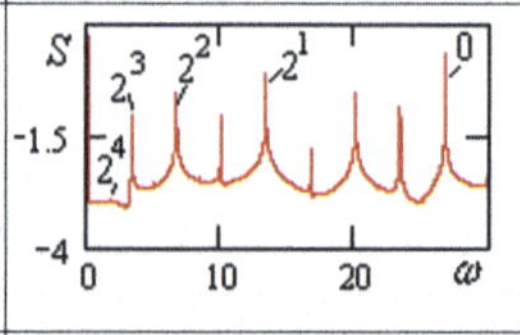	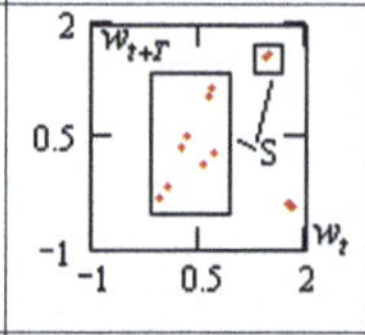
0.589785	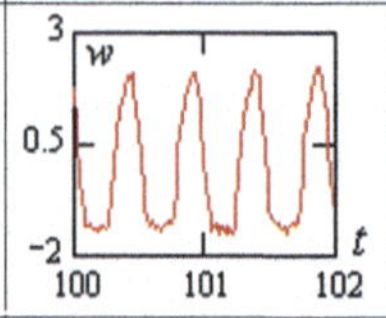	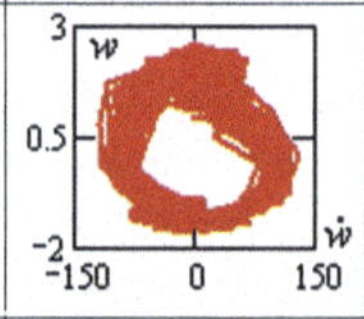	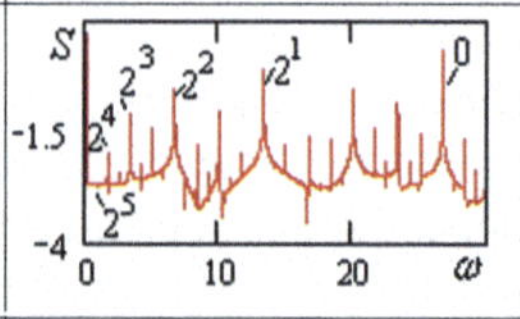	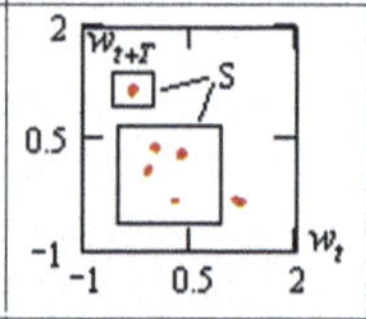
0.589803	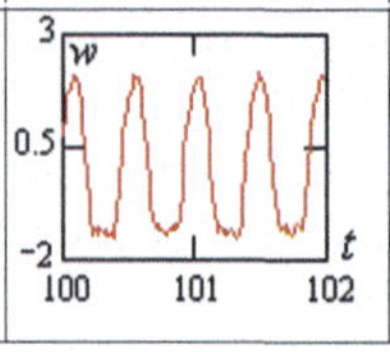	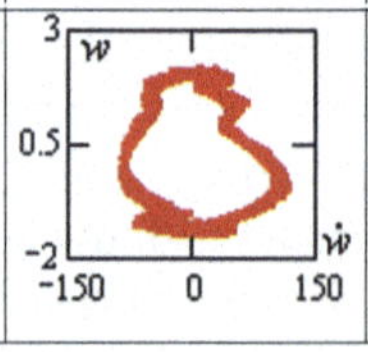	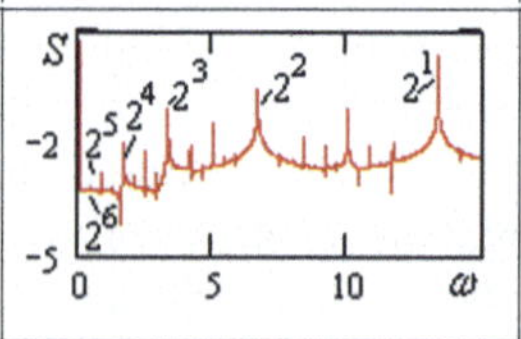	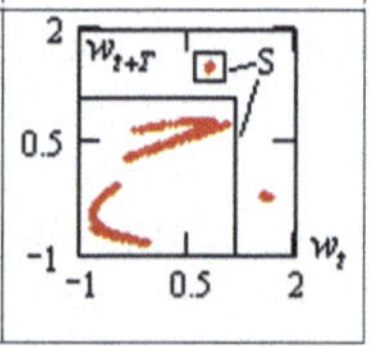

(continued)

Table 11.4 (continued)

0.589785	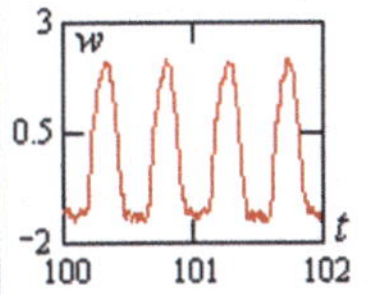	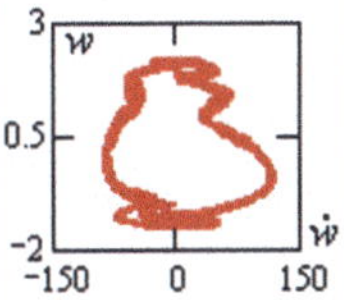	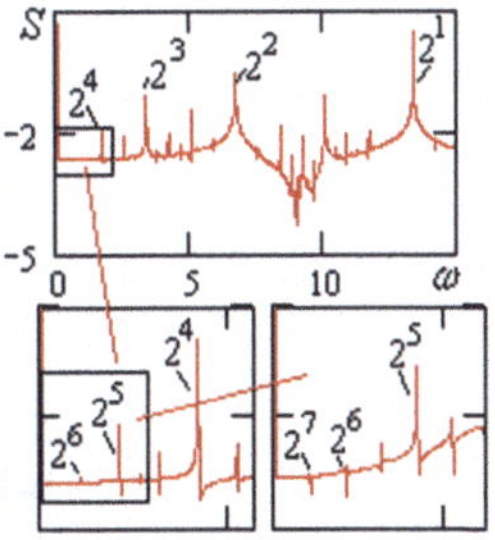	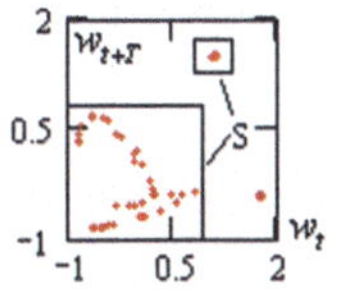
0.6	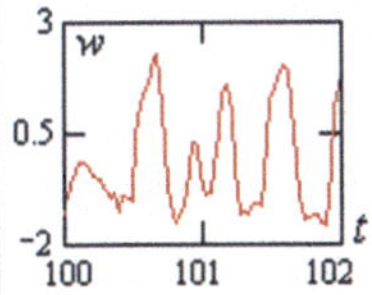	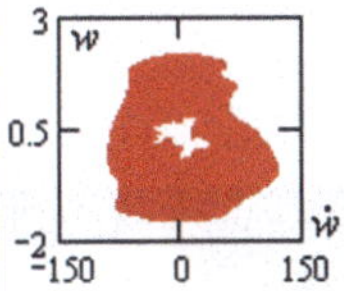	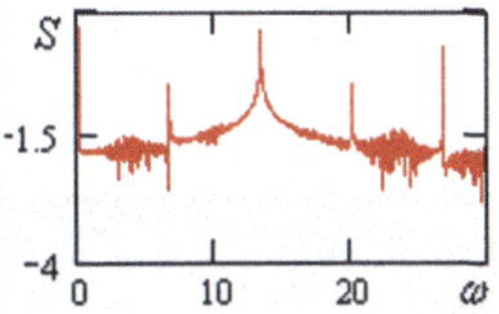	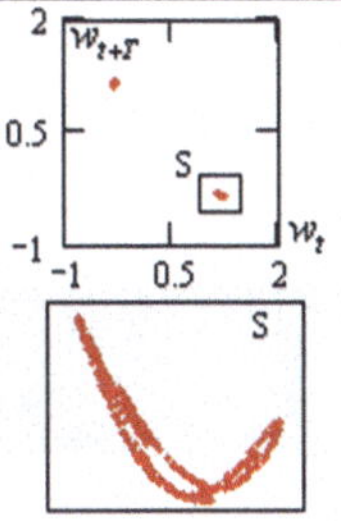

Table 11.5 Estimating the Feigenbaum constant

n	1	2	3	4	5	6	7
$q_{0,n}$	0.589764	0.58977	0.589778	0.58978	0.589785	0.589803	0.589785
d_n		1.66667	2.0023	2.5091	3.6118	4.669444	

us analyze an influence of the area of the load and the coordinates of its location on the cylindrical shell vibrations.

Figure 11.17 presents the dependencies $w_{\max}$ (q_0) and scales of dynamic regimes for various cases of loading. Schemes of loading along the longitudinal coordinate x are shown in Fig. 11.17. Area of the load location is a square, in spite of the case in Fig. 11.17e, where the load along y coordinate is localized on the closed ring. Analysis of the obtained results shows that large load area is associated with a large chaotic zone. Localization of the load also plays a crucial role on the vibrations character. Localization of the load along longitudinal coordinate close to the centre implies large chaotic zones. However, localization of the load along the shell edge (Fig. 11.17a) allows to remove chaotic vibrations due to the so called boundary effect. Let us study the space-temporal characteristics of the cylindrical shell behaviour in a zone of two-frequency vibrations and in a chaotic zone when the transversal local load is located at $(x_1, x_2) = (0.2; 0.4)$. Figures 11.18 and 11.19 presents changes of the wave forms in time $w(x; y; t)$, $0 \leq x \leq 1$, $0 \leq y \leq 2\pi$, signal $w(0.3; 0.0; t)$ in the center of the load location, phase portrait $w(\dot{w})$, Poincaré section $w_t(w_{t+T})$ and power spectrum $S(\omega)$. In the two-frequency vibration a_1 and ω_p all shell points move regularly. The points in the load area move into

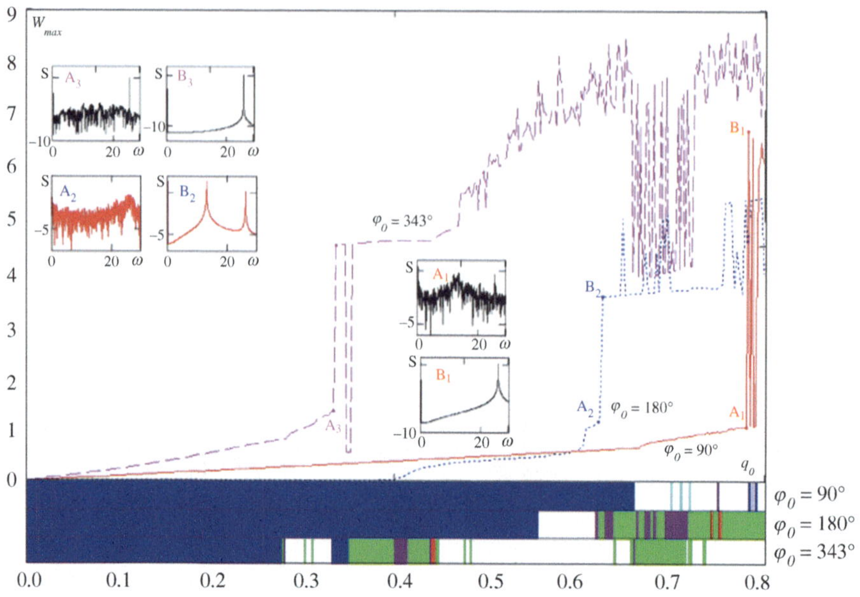

Fig. 11.15 Dependencies $w_{max}(q_0)$ and scales of the dynamic regimes for the cylindrical shell for $\lambda = 2$

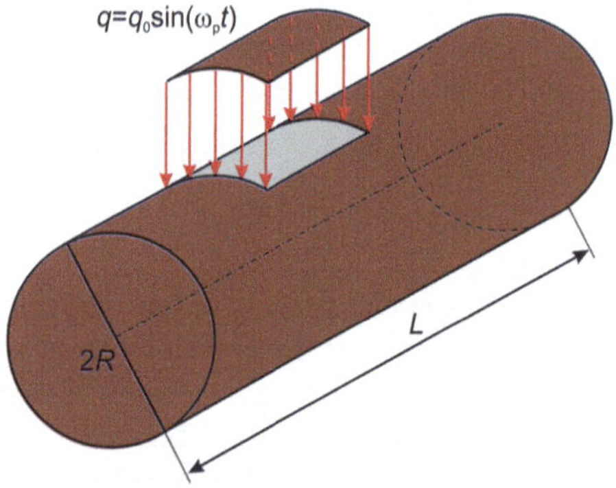

Fig. 11.16 Computational scheme of the cylindrical shell with the marked loading zone

the shell curvature center, whereas the points free from the load move up. In the circle coordinate the maximum deflections are concentrated in the loading zone. In the chaotic regime all shell points exhibit complex motion. Maximum deflections propagate in the whole shell surface; points under the load and points free of load may move in one direction (Fig. 11.19e, $t = 3, 7, 10$) or in the reversed one (Fig. 11.19e, $t = 2, 4$). Sudden changes of the signs of shell deflections are observed, i.e. buckling effects are exhibited in contrary to the quasi-periodic vibrations where the change of the sign of deflection was carried out smoothly.

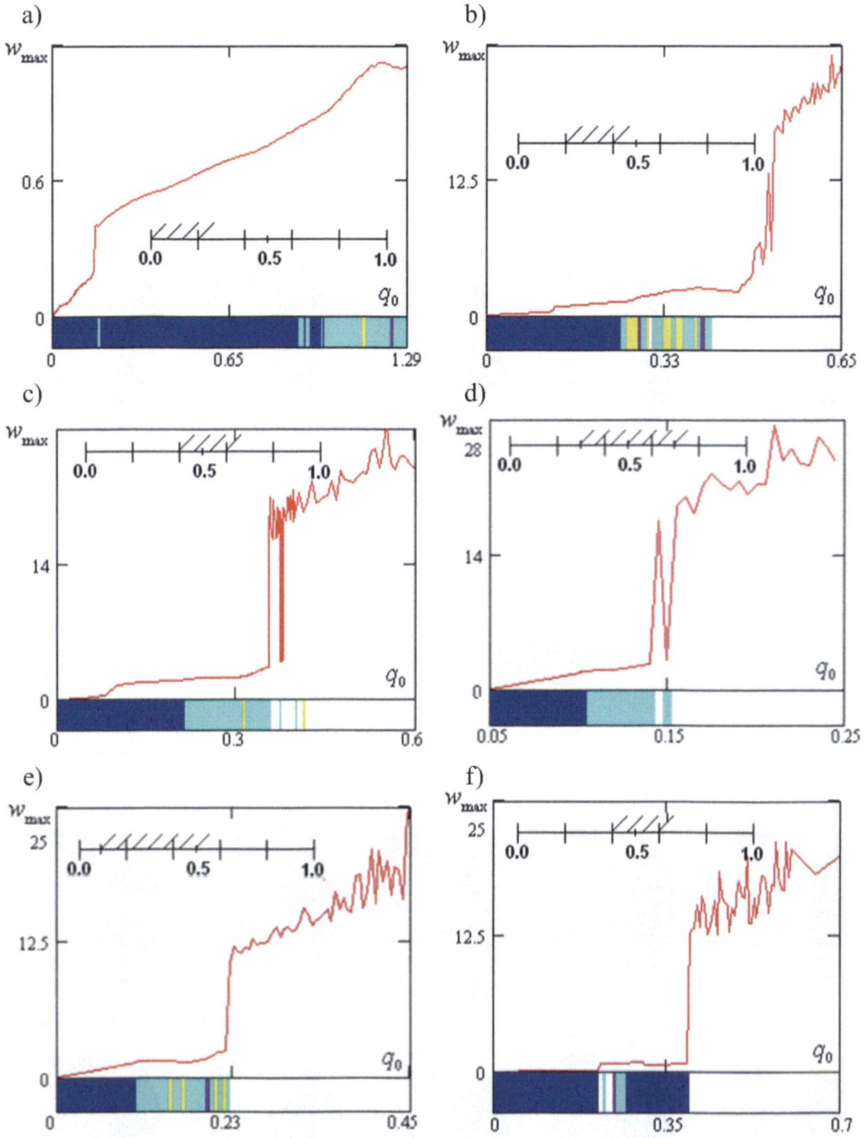

Fig. 11.17 Dependencies $w_{\max}(q_0)$, scales of dynamical regimes and schemes of the load location of the cylindrical shell. (**a**) $(x_1, x_2) = (0.0; 0.2)$. (**b**) $(x_1, x_2) = (0.2; 0.4)$. (**c**) $(x_1, x_2) = (0.4; 0.6)$. (**d**) $(x_1, x_2) = (0.3; 0.7)$. (**e**) $(x_1, x_2) = (0.1; 0.5)$. (**f**) $(x_1, x_2) = (0.4; 0.6)$

All of the reported characteristics allow to conclude that spatial and timing chaos appear simultaneously which support the conclusion about system transition into the space-temporal (timing) chaos.

In what follows we investigate influence of the geometric parameters of the cylindrical shell on its vibration characteristics. We consider action of the harmonic local

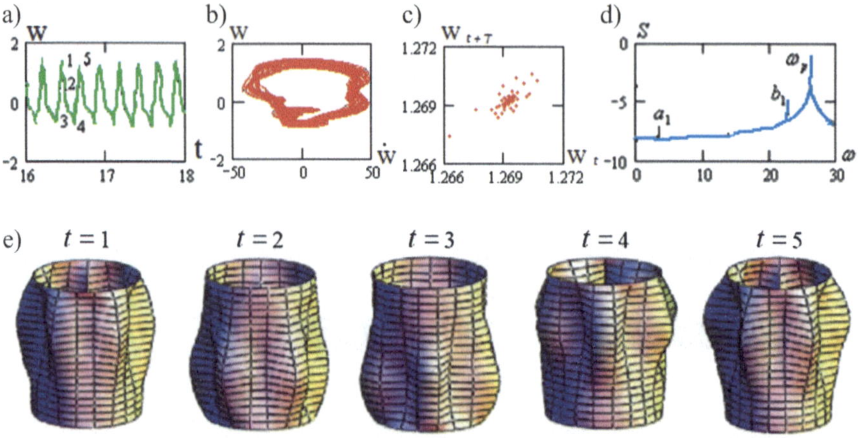

Fig. 11.18 Spatial-temporal characteristics of the cylindrical shell in the quasi-periodic regime. (**a**) $w(0.3; 0.0; t)$. (**b**) $w(\dot{w})$. (**c**) $w_t(w_{t+T})$. (**d**) $S(\omega)$. (**e**) $w(x; y; t)$, $0 \le x \le 1$, $0 \le y \le 2\pi$

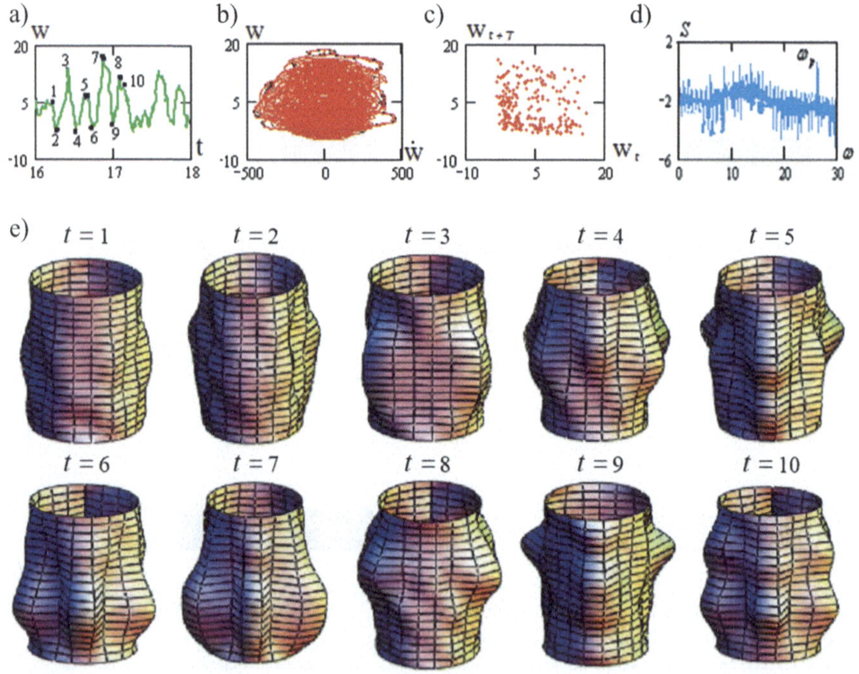

Fig. 11.19 Spatial temporal characteristics of the cylindrical shell in the chaotic regime. (**a**) $w(0.3; 0.0; t)$. (**b**) $w(\dot{w})$. (**c**) $w_t(w_{t+T})$. (**d**) $S(\omega)$. (**e**) $w(x; y; t)$, $0 \le x \le 1$, $0 \le y \le 2\pi$

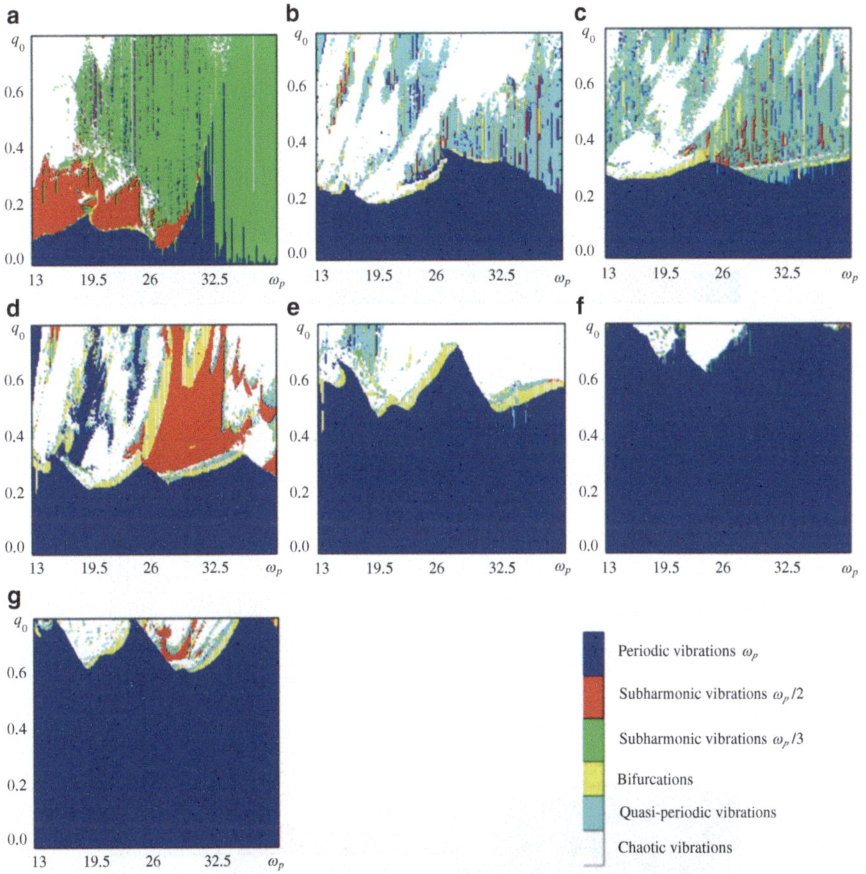

Fig. 11.20 Charts of the vibration regimes depending on the control parameters $\{q_0, \omega_p\}$ for the cylindrical shell for $\varphi_0 = 6$. **(a)** $\lambda = 1$. **(b)** $\lambda = 2$. **(c)** $\lambda = 3$. **(d)** $\lambda = 4$. **(e)** $\lambda = 5$. **(f)** $\lambda = 6$. **(g)** $\lambda = 7$

load employed on a band of the width defined by the central angle $\varphi_0 = 6.0\,\text{rad} = 343°$. We consider a few values of the fixed parameter $\lambda = \frac{L}{R}$: 1, 2, 3, 4, 5, 6, 7. The λ parameter is non-dimensional and it is proportional to the cylindrical shell length along its longitudinal coordinate and inversely proportional to the shell radius; in words, it characterizes the relative shell length. Charts of the shell dynamic regimes are constructed for fixed values of λ (Fig. 11.20). In addition we have also investigated the dependencies $w_{\max}(q_0)$ under fixed values of the excitation frequency $\omega_p = \omega_0$, where ω_0 stands for the fundamental frequency of the linear shell vibrations, as well as the scales of dynamic regimes. A scale presents a vertical part of the chart of the system dynamic regimes cut out for a given frequency. In order to analyse the space vibrations, the wave forms of the cylindrical shell for $0 \leq x \leq 1;\ 0 \leq y \leq 2\pi$ and the forms of the transverse shell cross section $x = 0.5;\ 0 \leq y \leq 2\pi$ are studied in both pre-critical and post-critical shell states.

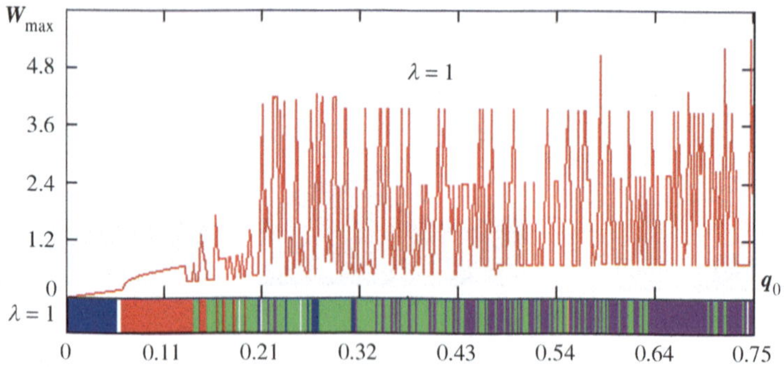

Fig. 11.21 Dependence $w_{\max}(q_0)$ and scale of dynamic regimes for $\lambda = 1$

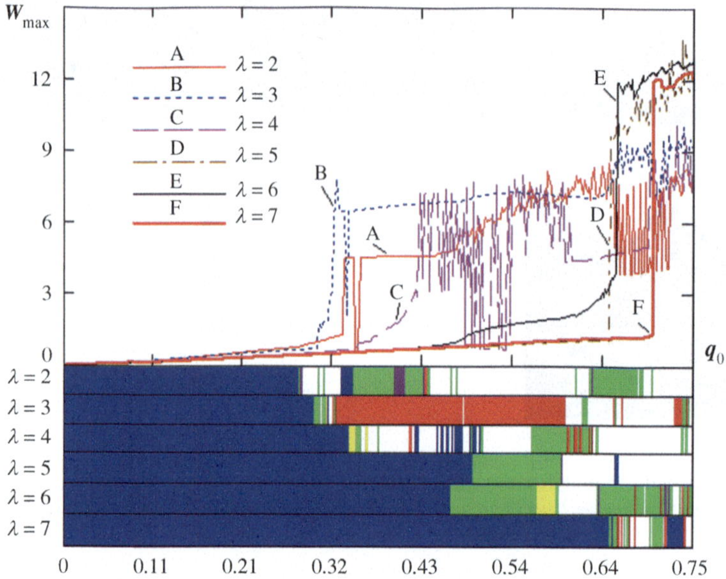

Fig. 11.22 Dependence $w_{\max}(q_0)$ and scales of dynamic regimes for different λ

Analysis of the relations $w_{\max}(q_0)$ for each λ allows to define zones of the stiff stability loss (Figs. 11.21, 11.22) and hence formulate criterions of the dynamical stability loss for the studied shells. It is worthy to mention that for all λ we may separate certain general properties. The stiff stability loss is characterized by a qualitative change of the vibration character and "trembling" of the relations $w_{\max}(q_0)$ and occurrence of chaos exhibited on the vibration scales. On the other hand, increase of q_0 from 0 implies a large zone of periodic vibrations which are associated with smooth increase of deflections reported on the dependence $w_{\max}(q_0)$.

Analyzing scales of the dynamic regimes together with the dependencies $w_{\max}(q_0)$, one may conclude that the stiff stability loss is characterized by a transition from chaotic vibrations to periodic vibrations either spanned on ω_0 or on $\omega_0/2$, i.e. the mechanism of stability loss is the same for all investigated interval of the parameter λ.

Let us analyze an occurrence of stability loos of the cylindrical shell versus the parameter λ (Fig. 11.23). Recall, that the dynamic stability loss we understand a sudden increase of deflections under very small increase of the exciting load amplitude. In order to define those values of λ associated with the occurrence of chaotic vibrations of the cylindrical shell, we construct the dependencies $\ell(\lambda)$ characterizing a length of a chaotic zone versus the parameter λ, i.e. ℓ stands for the length of chaotic zone for each of λ, where the other conditions are fixed. The length of a chaotic zone is estimated numerically using the constructed scale for each of the considered λ values.

Figures 11.23 and 11.24 report the mentioned graphs. The dependencies $q_{cr}(\lambda)$ and $\ell(\lambda)$ are non-monotonous. Movement of λ from 0 allows to observe two local minimum and two local maximum in the dependence $\ell(\lambda)$, whereas in the dependence $q_{cr}(\lambda)$ we see one local minimum and one local maximum. For $\lambda \geq 4$ we observe monotonous decrease of the dependence $\ell(\lambda)$, which corresponds to the monotonous increase of the counterpart segment in the graph $q_{cr}(\lambda)$. Therefore, for long shells ($\lambda \geq 4$) increase of critical load implies decrease of chaotic zones, i.e. decrease of the critical load yields increase of zones of chaotic vibrations. In the

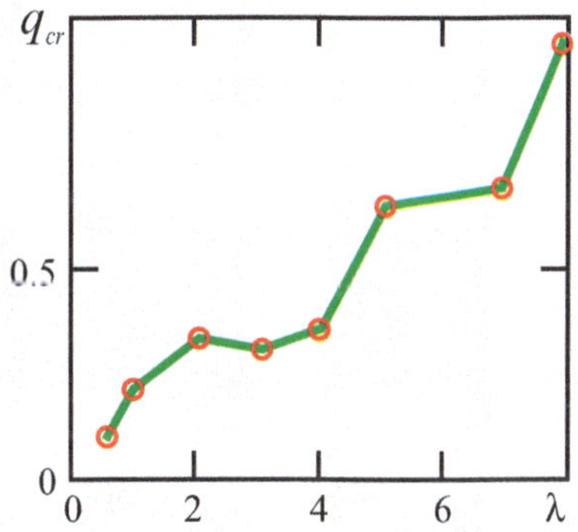

Fig. 11.23 Dependency $q_{cr}(\lambda)$

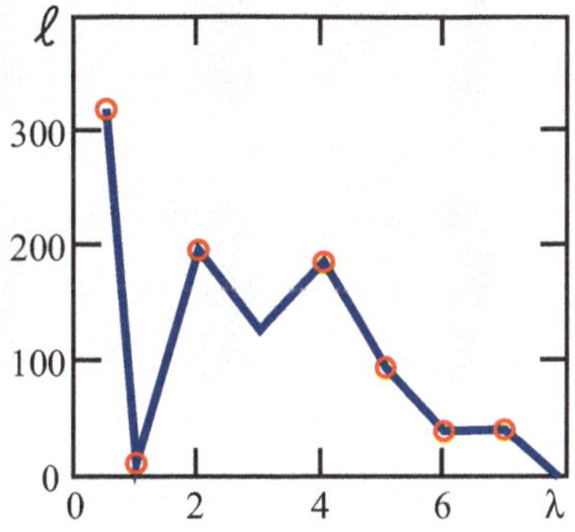

Fig. 11.24 Dependency $\ell(\lambda)$

case of the average and short length shells the dependencies $q_{cr}(\lambda)$ and $\ell(\lambda)$ are not monotonous.

Consequently, the dynamic critical load essentially depends on the relative length of a shell and it has different behavior for the short ($\lambda < 4$) and long ($\lambda \geq 4$) shells. It means that one may control shell vibrations and allow to increase dynamic critical load by the change of the shell geometric dimensions keeping fixed other conditions. The shell wave forms $0 \leq x \leq 1$; $0 \leq y \leq 2\pi$ for some fixed values of λ and for different time instants defined by number 1, 2, 3, 4 in the signal $w(0.5; 0.0; t)$ are shown in Table 11.6. It is seen the wave forms have complex character with a large number of half waves along the circle coordinate for short shells.

Table 11.6 Shell wave forms for different λ and time histories with marked different time instants

λ	1	2	3	4	Signal $w(t;0.5;3.0)$
1	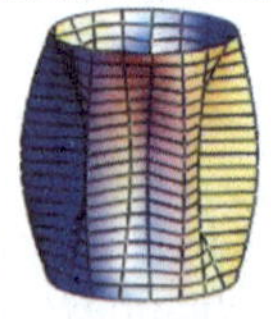	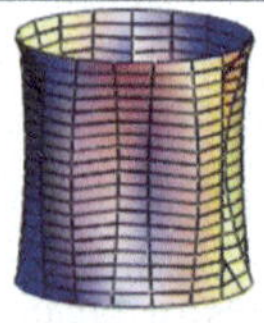	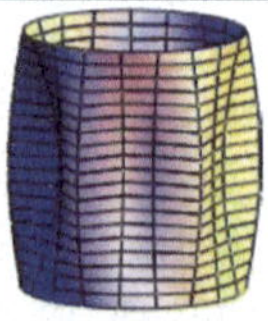	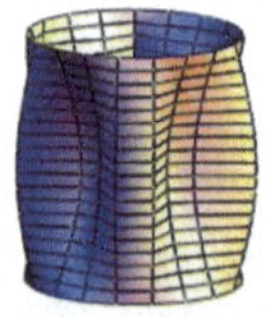	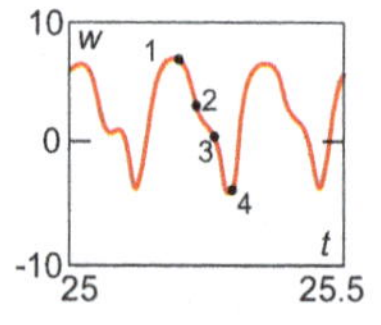
2			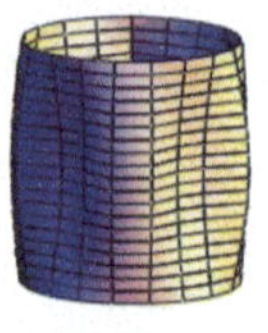	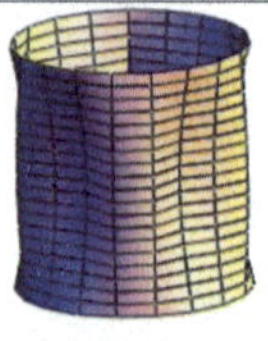	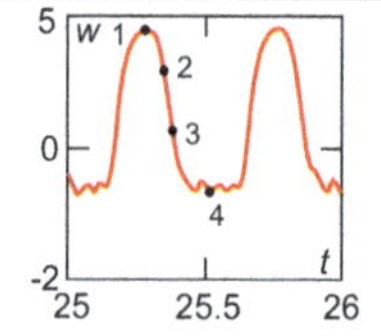
3	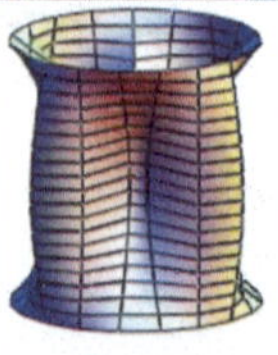	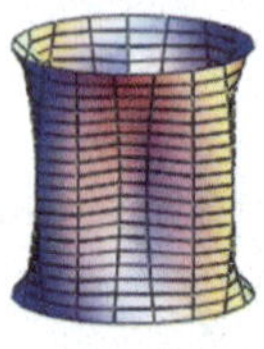	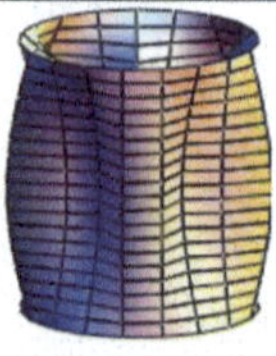	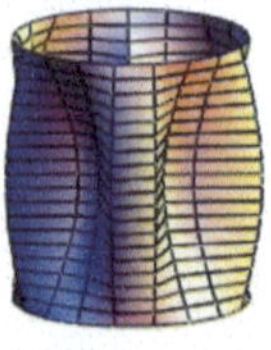	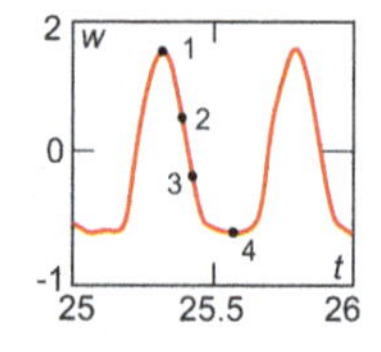
5		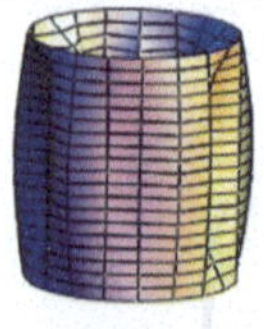	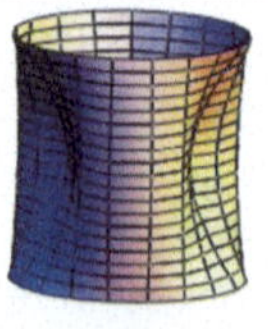	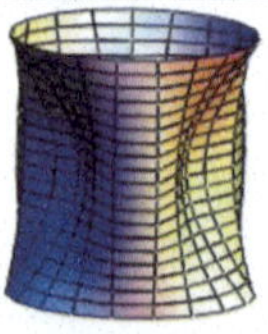	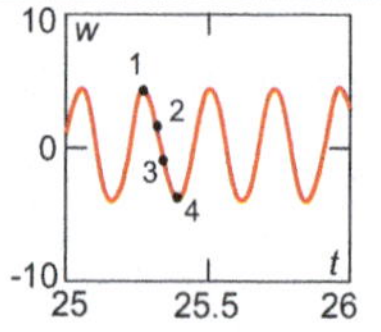
7	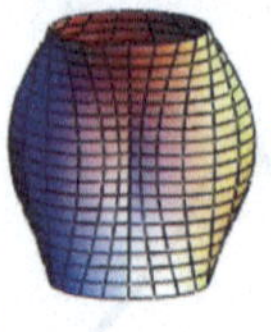				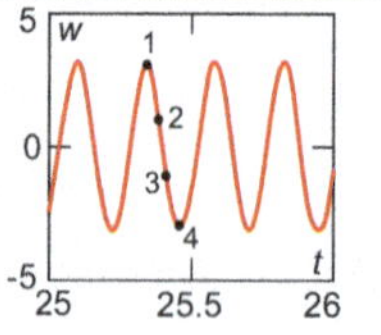

Chapter 12
Chaotic Dynamics of Flexible Closed Cylindrical Nano-Shells Under Local Load

In this chapter the mathematical model of a geometrically nonlinear closed cylindrical nano-shell is constructed. The solution convergence depending on the used series number terms in the Bubnov-Galerkin method is investigated. In particular, scenarios of transition into chaos of the nano-shell versus a number of the mentioned series terms N and the size dependent parameter l are analyzed. We have also detected, illustrated and discussed the scenarios of the Feigenbaum, Ruelle-Takens-Newhouse and their combination for differential values of N and l.

The mathematical model of a flexible closed cylindrical nano-shell obeying the first order approximation hypothesis (Kirchhoff-Love) is derived. The von Kármán geometrical nonlinearity is taken into account. The resolving PDEs, boundary and initial conditions with a help of the modified couple stress theory of elasticity are introduced on a basis of the energetic Hamilton principle. The nano-shell is subjected to sinusoidal load. The nonlinear PDEs are reduced to ODEs through the Bubnov-Galerkin method in higher approximations. In addition, the nonlinear vibrations of the closed cylindrical nano-shells under local load for a few size dependent parameters are studied. It is detected and illustrated that a scenario of transition from regular to chaotic vibrations essentially depends on a value of the size dependent parameter and the number of DoFs of the studied problem.

12.1 Statement of the Problem

We study a closed cylindrical shell occupying in the space R^3 the subspace $\Omega = \{x, y, z | (x, y, z) \in [0; L] \times [0; 2\pi] \times [-h; h]\}$. The studied nano-shell as well as the employed system of coordinates are shown in Fig. 12.1.

The Kirchhoff-Love hypothesis of the straight normal is applied. Owing to the modified couple stress theory [251], shell deformation energy U is a function of the

J. Awrejcewicz, V. A. Krysko, *Elastic and Thermoelastic Problems in Nonlinear Dynamics of Structural Members*, Scientific Computation,
https://doi.org/10.1007/978-3-030-37663-5_12

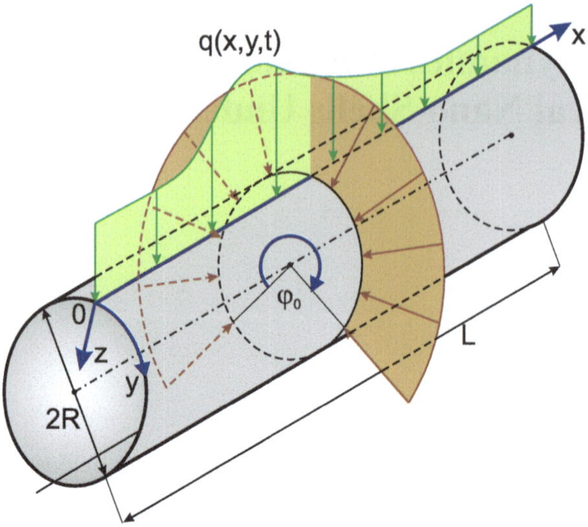

Fig. 12.1 Scheme of the closed cylindrical nano-shell with marked kinematic parameters, the system of coordinates and the shell geometry

deformation and shell curvature $U = \frac{1}{2}\int\limits_V (\sigma_{ij}\varepsilon_{ij} + m_{ij}\chi_{ij})\,dV$, where σ_{ij} stands for the stress tensor components, ε_{ij} are deformation tensors components and χ_{ij} are symmetric tensors of the curvature gradient components defined as follows: $\varepsilon_{ij} = \frac{1}{2}(u_{i,j} + u_{j,i} + u_{m,i}u_{m,j})$, $\chi_{ij} = \frac{1}{2}(\theta_{i,j} + \theta_{j,i})$, $\theta_i = \frac{1}{2}(rot(u))_i$. Here u_i stands for the component of the vector displacements $\mathbf{u}$. $\boldsymbol{\theta}$ stands for the infinitely small rotation vector with components θ_i, and δ_{ij} is the Kronecker's symbol. In the case of a linear isotropic elastic material the stresses generated by the kinematic parameters occurred in the relation governing density of the deformation energy are defined by the following state equations: $\sigma_{ij} = \lambda_1\varepsilon_{mm}\delta_{ij} + 2\mu\varepsilon_{ij}$, $m_{ij} = 2\mu l^2\chi_{ij}$, where σ_{ij}, ε_{ij}, m_{ij}, χ_{ij} denote components of the classical tensor of stresses $\boldsymbol{\sigma}$, tensor of deformation $\boldsymbol{\varepsilon}$, deviatory part of the symmetric tensor $\mathbf{m}$, and the symmetric part of the curvature tensor of higher $\boldsymbol{\chi}$, respectively. In addition, $\lambda_1 = \frac{E\nu}{(1+\nu)(1-2\nu)}$, $\mu = \frac{E}{2(1+\nu)}$ are Lamé parameters, E, ν stand for the Young modulus and the Poisson's coefficient, respectively. ρ is the material density and l presents the scale parameter of the material length taking into account effect of the moments of the higher orders.

The PDEs governing shell motion, the boundary and initial conditions are yielded by the Hamilton variational principle. Owing to that principle a comparison of the neighborhood motion carrying out the system of material points from its initial position at time instants t_0 to its terminal position in time interval t_1 is considered. In the case of real motion the following condition is satisfied

$$\int\limits_{t_0}^{t_1} (\delta K - \delta U + \delta W)dt = 0, \tag{12.1}$$

where: K—kinetic energy, U—potential energy, W—external work coupled with the space-distributed load and energy dissipation.

Using the standard procedure of the variational computation, the following governing motion PDEs and deformation compatibility equations are obtained

$$\begin{aligned}
&\left(-\frac{1}{12\left(1-\nu^2\right)}+\frac{l^2}{2\left(1+\nu\right)}\right)\cdot\left[\frac{1}{\lambda^2}\frac{\partial^2 w}{\partial x^2}\frac{\partial^2(\cdot)}{\partial x^2}+\lambda^2\frac{\partial^2 w}{\partial y^2}\frac{\partial^2(\cdot)}{\partial y^2}+\right.\\
&\left.+2(1-\nu)\frac{\partial^2 w}{\partial x\partial y}\frac{\partial^2(\cdot)}{\partial x\partial y}+\nu\left(\frac{\partial^2 w}{\partial x^2}\frac{\partial^2(\cdot)}{\partial y^2}+\frac{\partial^2 w}{\partial y^2}\frac{\partial^2(\cdot)}{\partial x^2}\right)\right]-\\
&-\nabla_k^2 F+p_x(t)\frac{\partial^2 w}{\partial y^2}+p_y(t)\frac{\partial^2 w}{\partial x^2}-\\
&-L(w,F)+M\cdot q(t)-\left(\frac{\partial^2 w}{\partial t^2}+\varepsilon\frac{\partial w}{\partial t}\right)=0,\\
&\left[\left(\lambda^2\frac{\partial^2 F}{\partial y^2}-\nu\frac{\partial^2 F}{\partial x^2}\right)\frac{\partial^2(\cdot)}{\partial y^2}+\left(\frac{1}{\lambda^2}\frac{\partial^2 F}{\partial x^2}-\nu\frac{\partial^2 F}{\partial y^2}\right)\frac{\partial^2(\cdot)}{\partial x^2}+\right.\\
&\left.+2(1+\nu)\frac{\partial^2 F}{\partial x\partial y}\frac{\partial^2(\cdot)}{\partial x\partial y}\right]+\nabla_k^2 w+\frac{1}{2}L(w,w)=0.
\end{aligned} \tag{12.2}$$

The following boundary conditions are supplemented to PDEs (12.2):

1. Movable clamping along the shell faces

$$\begin{aligned}
&w=0,\quad \frac{\partial w}{\partial x}=0,\quad F=0;\quad \frac{\partial F}{\partial x}=0\quad \text{for}\quad x=0;\,1.\\
&w=g(x,y,t),\quad \frac{\partial w}{\partial y}=p(x,y,t),\quad F=u(x,y,t),\\
&\frac{\partial F}{\partial y}=v(x,y,t)\quad \text{for}\quad y=0;\,\xi.
\end{aligned} \tag{12.3}$$

2. Simple support along the shell faces

$$\begin{aligned}
&w=0,\quad \frac{\partial w}{\partial x}=0,\quad F=0,\quad \frac{\partial^2 F}{\partial x^2}=0\quad \text{for}\quad x=0;\,1.\\
&w=g(x,y,t),\quad \frac{\partial w}{\partial y}=p(x,y,t),\quad F=u(x,y,t),\\
&\frac{\partial F}{\partial y}=v(x,y,t)\quad \text{for}\quad y=0;\,\xi.
\end{aligned} \tag{12.4}$$

3. Movable clamping along the shell faces with ribs

$$w = 0, \quad \frac{\partial^2 w}{\partial x^2} = 0, \quad F = 0, \quad \frac{\partial F}{\partial x} = 0 \quad \text{for} \quad x = 0; 1.$$
$$w = g(x, y, t), \quad \frac{\partial w}{\partial y} = p(x, y, t), \quad F = u(x, y, t), \tag{12.5}$$
$$\frac{\partial F}{\partial y} = v(x, y, t) \quad \text{for} \quad y = 0; \xi.$$

4. Simple support along shell faces with flexible ribs

$$w = 0, \quad \frac{\partial^2 w}{\partial x^2} = 0, \quad F = 0, \quad \frac{\partial^2 F}{\partial x^2} = 0 \quad \text{for} \quad x = 0; 1.$$
$$w = g(x, y, t), \quad \frac{\partial^2 w}{\partial y^2} = r(x, y, t), \quad F = u(x, y, t), \tag{12.6}$$
$$\frac{\partial^2 F}{\partial y^2} = z(x, y, t) \quad \text{for} \quad y = 0; \xi.$$

where $\xi = 2\pi$. Beside of the boundary conditions (12.3)–(12.6) the following initial conditions are attached

$$w|_{t=0} = w_0, \qquad \dot{w}|_{t=0} = 0. \tag{12.7}$$

Equations (12.2)–(12.7) are transformed to their non-dimensional counterpart form with a help of the following parameters: $w = h\bar{w}$, $F = Eh^2\overline{F}$, $t = t_0\bar{t}$, $\varepsilon = \overline{\varepsilon}/\tau$, $x = L\bar{x}$, $y = R\bar{y}$, $k_y = \bar{k}_y\frac{h}{R^2}$ ($k_x = 0$), $q = \bar{q}\frac{Eh^4}{L^2R^2}$, $\tau = \frac{LR}{h}\sqrt{\frac{\rho}{Eg}}$, $M = {k_y}^2$, $\lambda = \frac{L}{R}$, where L and $R = R_y$ stand for the length and radius, respectively. The following notation is used: t—time, ε—external damping coefficient, $\nu = 0.3$, $q(x, y, t)$—transversal load. In the governing PDEs bars are already omitted.

12.2 Algorithm of the Bubnov–Galerkin Method

In order to solve Eqs. (12.2) the functions w and F standing for their solution are approximated by analytical expressions containing the finite number of arbitrary parameters in the form of the dot product of the functions dependent on time and coordinates:

$$w = \sum_{i=0}^{N_1}\sum_{j=0}^{N_2} A_{ij}(t)\varphi_{ij}(x, y), \qquad F = \sum_{i=0}^{N_1}\sum_{j=0}^{N_2} B_{ij}(t)\psi_{ij}(x, y). \tag{12.8}$$

In order to find approximation to w and F, we choose the coordinate system of functions $\{\varphi_{ij}(x, y), \psi_{ij}(x, y)\}$ $(i, j = 0, 1, 2 \ldots)$ in (12.8), satisfying following five requirements:

1. $\varphi_{ij}(x, y) \in H_A$, $\psi_{ij}(x, y) \in H_A$, where H_A stands for a Hilbert space being called the energetic space;
2. $\forall i, j$ functions $\varphi_{ij}(x, y)$ and $\psi_{ij}(x, y)$ are linearly independent with their partial derivatives up to the fourth order in the space Ω;
3. $\varphi_{ij}(x, y)$ and $\psi_{ij}(x, y)$ satisfy main boundary conditions (and initial conditions, if they exist) exactly;
4. $\varphi_{ij}(x, y)$ and $\psi_{ij}(x, y)$ obey property of density in H_A;
5. $\varphi_{ij}(x, y)$ and $\psi_{ij}(x, y)$ should present N first elements of the full system of functions.

Coefficients $A_{ij}(t)$ and $B_{ij}(t)$ depend on time. For simplicity, the left hand sides of PDEs (12.2) are denoted by Φ_1 and Φ_2, respectively, and hence we get

$$\Phi_1(w, F, \frac{\partial^2 w}{\partial x^2}, \frac{\partial^2 F}{\partial x^2}, \ldots) + M \cdot q(x, y, t) = 0, \quad \Phi_2(w, F, \frac{\partial^2 w}{\partial x^2}, \frac{\partial^2 F}{\partial x^2}, \ldots) = 0. \tag{12.9}$$

Employing the Bubnov-Galerkin procedure to (12.2), one obtains

$$\begin{aligned}
&\int_0^1\int_0^\xi \Phi_1\varphi_{rs}(x, y)dxdy + \int_{x_1}^{x_2}\int_{y_1}^{y_2} Mq(x, y, t)\varphi_{rs}(x, y)dxdy = 0,\\
&\int_0^1\int_0^\xi \Phi_2\psi_{rs}(x, y)dxdy = 0, \quad r = 0, 1, \ldots, N_1; \quad s = 0, 1, \ldots N_2.
\end{aligned} \tag{12.10}$$

Equations (12.9), taking into account (12.8), can be presented in the following form:

$$\begin{aligned}
&\sum_{rs}\Bigg[\sum_{ij} A_{ij}\sum_{kl} H_{ijklrs} + \sum_{ij} B_{ij}C_{1,ijrs} + \sum_{ij} A_{ij}W_{ijrs} + MqQ_{rs} +\\
&+\sum_{ij} A_{ij}\sum_{kl} B_{kl}D_{1,ijklrs} + \sum_{ij}\left[\frac{d^2A_{ij}}{dt^2} + \varepsilon\frac{dA_{ij}}{dt}\right]G_{ijrs}\Bigg] = 0,\\
&\sum_{rs}\left[\sum_{ij} A_{ij}C_{2,ijrs} + \sum_{ij} B_{ij}\sum_{kl} P_{ijklrs} + \sum_{ij} A_{ij}\sum_{kl} A_{rs}D_{2,ijklrs}\right] = 0.
\end{aligned} \tag{12.11}$$

Here the sign $\sum\limits_{rs}[*]$ standing before each equation of the system (12.11) points out that a given equation corresponds to a set of rs equations of the similar form, and integrals of Bubnov-Galerkin procedure take the following form

$$H_{ijklrs} = \int\limits_0^1\int\limits_0^\xi \left(\frac{1}{12(1-\nu^2)} + \frac{l^2}{2(1+\nu)}\right)\left[\frac{1}{\lambda^2}\frac{\partial^2\varphi_{ij}}{\partial x^2}\frac{\partial^2\varphi_{kl}}{\partial x^2} + \right.$$

$$\left. + \lambda^2\frac{\partial^2\varphi_{ij}}{\partial y^2}\frac{\partial^2\varphi_{kl}}{\partial y^2} + 2\frac{\partial^2\varphi_{ij}}{\partial x\partial y}\frac{\partial^2\varphi_{kl}}{\partial x\partial x}\right]\varphi_{rs}dxdy,$$

$$C_{1,ijrs} = \int\limits_0^1\int\limits_0^\xi \left[-k_y\frac{\partial^2\psi_{ij}}{\partial x^2} - k_x\frac{\partial^2\psi_{ij}}{\partial y^2}\right]\varphi_{rs}dxdy,$$

$$C_{2,ijrs} = \int\limits_0^1\int\limits_0^\xi \left[k_y\frac{\partial^2\varphi_{ij}}{\partial x^2} + k_x\frac{\partial^2\varphi_{ij}}{\partial y^2}\right]\psi_{rs}dxdy,$$

$$D_{1,ijklrs} = \int\limits_0^1\int\limits_0^\xi [-L(\varphi_{ij}, \psi_{kl})]\varphi_{rs}dxdy,$$

$$D_{2,ijklrs} = \int\limits_0^1\int\limits_0^\xi \frac{1}{2}L(\varphi_{ij}, \varphi_{kl})\psi_{rs}dxdy, \tag{12.12}$$

$$P_{ijklrs} = \int\limits_0^1\int\limits_0^\xi \left[\frac{1}{\lambda^2}\frac{\partial^2\psi_{ij}}{\partial x^2}\frac{\partial^2\psi_{kl}}{\partial x^2} + \lambda^2\frac{\partial^2\psi_{ij}}{\partial y^2}\frac{\partial^2\psi_{kl}}{\partial y^2} + 2\frac{\partial^2\psi_{ij}}{\partial x\partial y}\frac{\partial^2\psi_{kl}}{\partial x\partial x}\right]\psi_{rs}dxdy,$$

$$G_{ijrs} = \int\limits_0^1\int\limits_0^\xi [-\varphi_{ij}\varphi_{rs}]dxdy, \qquad Q_{rs} = \int\limits_{x_1}^{x_2}\int\limits_{y_1}^{y_2} Mq(x,y,t)\varphi_{rs}dxdy,$$

$$W_{ijrs} = \int\limits_0^1\int\limits_0^\xi \left[p_y(x,t)\frac{\partial^2\varphi_{ij}}{\partial x^2} + p_x(y,t)\frac{\partial^2\varphi_{ij}}{\partial y^2}\right]\varphi_{rs}dxdy.$$

Integrals (12.12), perhaps with exception of Q_{rs}, when the transversal load acts only on a part of the shell surface, are computed with regard to the whole middle shell surface. The Bubnov-Galerkin procedure yields a system of ODEs with regard to $A_{ij}(t)$ and $B_{ij}(t)$, which has the following matrix representation

$$\begin{aligned} &\mathbf{G}(\ddot{\mathbf{A}} + \varepsilon\dot{\mathbf{A}}) + \mathbf{HA} + \mathbf{WA} + \mathbf{C}_1\mathbf{B} + \mathbf{D}_1\mathbf{AB} = \mathbf{Q}q(t), \\ &\mathbf{C}_2\mathbf{A} + \mathbf{PB} + \mathbf{D}_2\mathbf{AA} = 0, \end{aligned} \tag{12.13}$$

where: $\mathbf{H} = \|H_{ijrs}\|$, $\mathbf{G} = \|G_{ijrs}\|$, $\mathbf{C}_1 = \|C_{1ijrs}\|$, $\mathbf{C}_2 = \|C_{2ijrs}\|$, $\mathbf{D}_1 = \|D_{1ijklrs}\|$, $\mathbf{D}_2 = \|D_{2ijklrs}\|$, $\mathbf{W} = \|W_{ijrs}\|$, $\mathbf{P} = \|P_{ijrs}\|$—square matrices of dimension $2 \cdot N_1 \cdot N_2 \times 2 \cdot N_1 \cdot N_2$, $\mathbf{A} = \|A_{ij}\|$, $\mathbf{B} = \|B_{ij}\|$, $\mathbf{Q} = \|Q_{ij}\|$—matrices of dimension $2 \cdot N_1 \cdot N_2 \times 1$.

Furthermore, the second equation of system (12.13) is solved with respect to $\mathbf{B}$ on each time step:

$$\mathbf{B} = \left[-\mathbf{P}^{-1}\mathbf{D}_2\mathbf{A} - \mathbf{P}^{-1}\mathbf{C}_2\right]\mathbf{A}. \tag{12.14}$$

Multiplying the first equation of system (12.13) by $\mathbf{G}^{-1}$, and denoting $\dot{\mathbf{A}} = \mathbf{R}$, the problem is reduced to the Cauchy problem for nonlinear first order ODEs of the following form

$$\begin{cases} \dot{\mathbf{A}} = \mathbf{R}, \\ \dot{\mathbf{R}} = -\bar{\varepsilon}\mathbf{R} - \left[\mathbf{G}^{-1}\mathbf{C}_1 + \mathbf{G}^{-1}\mathbf{D}_1\mathbf{A}\right] \cdot \mathbf{B} - \mathbf{G}^{-1}\mathbf{HA} - \mathbf{G}^{-1}\mathbf{WA} + \mathbf{G}^{-1}\mathbf{Q}q(\bar{t}). \end{cases} \tag{12.15}$$

The introduced transformation is allowed, since the inverse matrices $\mathbf{G}^{-1}$ and $\mathbf{P}^{-1}$ exist if the coordinate functions are linearly independent.

Equations (12.15) are supplemented by the boundary conditions (12.4) and initial conditions (12.7), and the obtained Cauchy problem is solved by the 4th order Runge-Kutta method. The integration step is chosen by the Runge's rule.

Let us present φ_{ij}, ψ_{ij} in (12.9) in the form of a multiplication of two functions, where each of them depends only on one parameter and satisfy boundary conditions (12.4) with regard to x :

$$w = \sum_{i=1}^{N_1}\sum_{j=0}^{N_2} A_{ij}(t)\sin(i\pi x)\cos(jy), \; F = \sum_{i=1}^{N_1}\sum_{j=0}^{N_2} B_{ij}(t)\sin(i\pi x)\cos(jy). \tag{12.16}$$

Integrals of the Bubnov-Galerkin procedure are computed through the following integrals

$$I_{1,r} = \int_{x_1}^{x_2} \sin(r\pi x)dx = \frac{\cos(r\pi x_1) - \cos(r\pi x_2)}{r\pi},$$

$$I_{2,s} = \int_{y_1}^{y_2} \cos(s\pi y)dy = \frac{\sin(s\pi y_2) - \sin(s\pi y_1)}{s\pi},$$

$$I_{3,ir} = \int_{0}^{1} \sin(i\pi x)\sin(r\pi x)dx = \begin{cases} \frac{1}{2}, i = r, \\ 0, i \neq r, \end{cases}$$

$$I_{4,js} = \int_{0}^{2\pi} \cos(j\pi y)\cos(s\pi y)dy = \begin{cases} 2\pi, j = s = 0, \\ \pi, j = s \neq 0, \\ 0, , j \neq s, \end{cases}$$

$$I_{5,ikr} = \int_{0}^{1} \sin(i\pi x)\sin(r\pi x)\sin(k\pi x)dx =$$

$$= \begin{cases} \frac{1}{4\pi}\left[-\frac{\cos(\alpha_1\pi)}{\alpha_1} - \frac{\cos(\alpha_2\pi)}{\alpha_2} - \frac{\cos(\alpha_3\pi)}{\alpha_3} - \frac{\cos(\alpha_4\pi)}{\alpha_4} + \right. \\ \left. + \frac{1}{\alpha_1} + \frac{1}{\alpha_2} + \frac{1}{\alpha_3} - \frac{1}{\alpha_4}\right], \quad \alpha_\upsilon \neq 0; \\ \left[\frac{\cos(\alpha_\upsilon\pi)}{\alpha_\upsilon} \approx 0, \frac{1}{\alpha_\upsilon} \approx 0\right], \upsilon = 1, 2, 3, \quad \alpha_\upsilon = 0; \end{cases}$$

$$I_{7,ikr} = \int_{0}^{1} \cos(i\pi x)\cos(k\pi x)\sin(r\pi x)dx =$$

$$= \begin{cases} \frac{1}{4\pi}\left[\frac{\cos(\alpha_1\pi)}{\alpha_1} - \frac{\cos(\alpha_2\pi)}{\alpha_2} - \frac{\cos(\alpha_3\pi)}{\alpha_3} - \frac{\cos(\alpha_4\pi)}{\alpha_4} - \right. \\ \left. - \frac{1}{\alpha_1} + \frac{1}{\alpha_2} + \frac{1}{\alpha_3} + \frac{1}{\alpha_4}\right], \quad \alpha_\upsilon \neq 0; \\ \left[\frac{\cos(\alpha_\upsilon\pi)}{\alpha_\upsilon} \approx 0, \frac{1}{\alpha_\upsilon} \approx 0\right], \quad \upsilon = 1, 2, 3, \quad \alpha_\upsilon = 0; \end{cases}$$

$$I_{8,jls} = \int_0^{2\pi} \sin(jy)\sin(ly)\cos(sy)dy =$$

$$= \begin{cases} \frac{1}{4}\left[-\frac{\sin(\beta_1\pi)}{\beta_1} + \frac{\sin(\beta_2\pi)}{\beta_2} + \frac{\sin(\beta_3\pi)}{\beta_3} - \frac{\sin(\beta_4\pi)}{\beta_4} + \right. \\ \left. + \frac{1}{\beta_1} - \frac{1}{\beta_2} - \frac{1}{\beta_3} + \frac{1}{\beta_4}\right], \quad \beta_\upsilon \neq 0; \\ \left[\frac{\sin(\beta_\upsilon\pi)}{\beta_\upsilon} \approx 0, \frac{1}{\beta_\upsilon} \approx 0\right], \; \upsilon = 1, 2, 3, 4, \qquad \beta_\upsilon = 0; \end{cases}$$

$$I_{6,jls} = \int_0^{2\pi} \cos(jy)\cos(ly)\sin(sy)dy = 0,$$

where

$$\alpha_1 = j + k - r, \quad \alpha_2 = k + r - i, \quad \alpha_3 = r + i - k, \quad \alpha_4 = i + k + r,$$
$$\beta_1 = j + l - s, \quad \beta_2 = l + s - j, \quad \beta_3 = s + j - l, \quad \beta_4 = j + l + s.$$

In what follows, the following notation is employed

$$I_Q^{rs} = M \cdot I_{1r} I_{2s}, \quad I_P^{rs} = \left(s^2 p_x(t) + r^2 p_y(t)\right)\pi^2 I_{3,ir} I_{4,js},$$

$$I_{AB}^{rs} = \left(s^2 k_x + r^2 k_y\right)\pi^2 I_{3,ir} I_{4,js}, \quad I_{rs}^t = I_{3,ir} I_{4,js}, \quad M = {k_y}^2,$$

$$I_{ijklrs} = \pi^2\left[\left(i^2 l^2 + j^2 k^2\right) I_{5ikr} I_{6jls} - 2ijlk I_{7ikr} I_{8jls}\right],$$

$$J_{1,ijkl}^{rs} = \frac{\pi^2}{12(1-\mu^2)}\left[\frac{r^4}{\lambda^2} + 2r^2 s^2 + \lambda^2 s^4\right] \cdot I_{3,ir} I_{4,js},$$

$$J_{2,ijkl}^{rs} = \left[\frac{r^4}{\lambda^2} + 2r^2 s^2 + \lambda^2 s^4\right] \cdot \pi^2 I_{3,ir} I_{4,js}.$$

Finally, taking into account the integrals, the system (12.15) can be recast to the following form

$$\sum_{rs}\left\{\sum_{ij}\sum_{kl}\left[J^{rs}_{1,ijkl}A_{ij} + I^{rs}_{AB}B_{rs} + I^{rs}_{P}A_{ij} + I^{rs}_{Q}q(t) + \right.\right.$$
$$\left.\left. + A_{ij}B_{kl}I_{ijklrs} + \left(\frac{d^2A_{ij}}{dt^2} + \varepsilon\frac{dA_{ij}}{dt}\right)I^{t}_{rs}\right]\right\} = 0, \tag{12.17}$$

$$\sum_{rs}\left\{\sum_{ij}\sum_{kl}\left[J^{rs}_{2,ijkl}B_{ij} + I^{rs}_{AB}A_{rs} + \frac{1}{2}A_{ij}A_{kl}I_{ijklrs}\right]\right\} = 0.$$

The so far described algorithm allows to optimize the solution since the solution to the Cauchy problem is reduced to a solution of the algebraic system of equations on each time step.

12.3 Numerical Experiment

We report the results of numerical experiment for the closed cylindrical size dependent shell with simply supported edges. The shell is subjected to action of the transversal compress sinusoidal load with the amplitude $q_0 = 0.323$, $\omega_p = 20.3$, $\varepsilon = 1$, $\lambda = 2$ and with the same loading angle $\varphi_0 = 5.98$ along the entire length of the cylinder (Fig. 12.2).

Table 12.1 reports the signals obtained using different number of the series terms ($N = 5-15$), the Fourier power spectra in the function development of the Bubnov-Galerkin method for $l = 0$. In the beginning, for $N = 5, 7$ the cylinder vibrates with the frequency of the excitation load. However, increase of the number of the series terms implies a qualitative change of vibrations what can be followed by the dynamic characteristics. For $N = 9$ three frequencies are born with the values

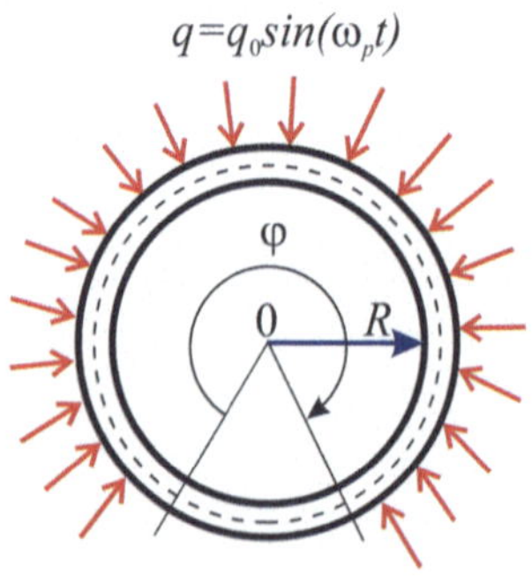

Fig. 12.2 Scheme of the shell loading

Table 12.1 Convergence of the numerical results for $N = 5 - 15, \quad l = 0$

Signal for $t \in \left[10^2,\ 1.1 \cdot 10^2\right], \quad (N = 5 - 15)$	Power spectrum $(N = 5 - 7)$
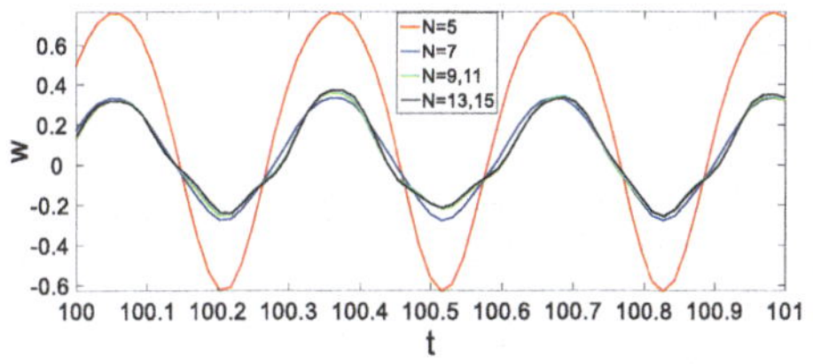	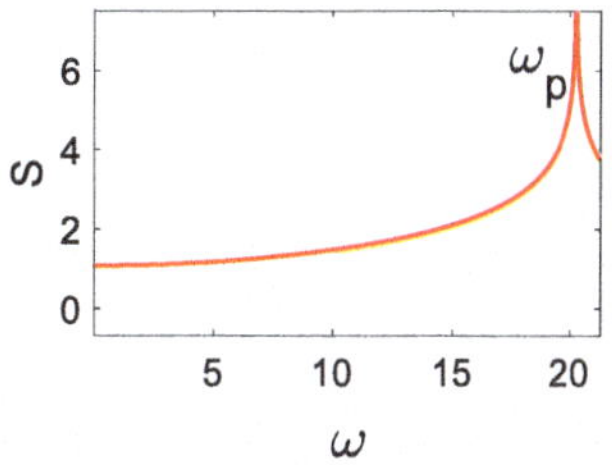
Power spectrum $(N = 11)$	Power spectrum $(N = 13 - 15)$
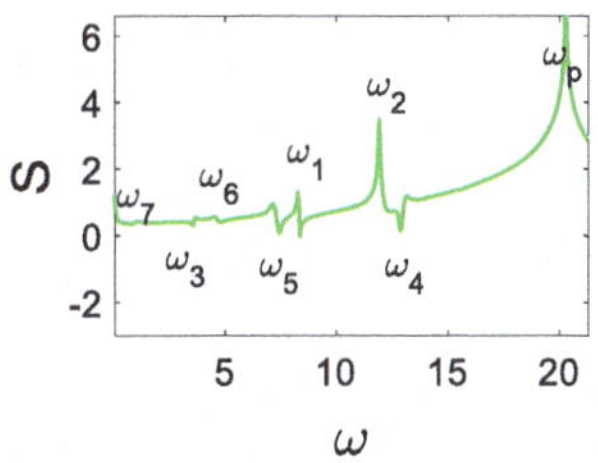	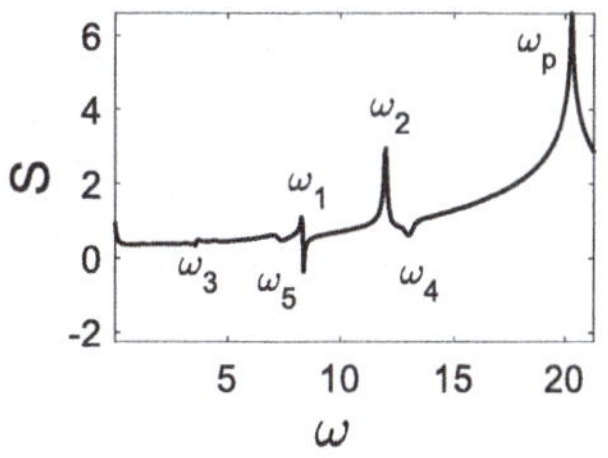

linearly dependent on the excitation frequency ω_p. Further increase of N up to 11 implies occurrence of more frequencies.

Therefore, the Fourier power spectrum consists of the following frequencies: $\omega_p = 20.3$, $\omega_1 = 8.3$, $\omega_2 = \omega_p - \omega_1 = 12$, $\omega_3 = \omega_p - 2\omega_1 = 3.7$, $\omega_4 = \frac{21}{20}\omega_p - \omega_1 = 13.015$, $\omega_5 = \omega_p - \omega_4 = 7.2$, $\omega_6 = \frac{\omega_p}{4} = 5.07$, $\omega_7 = \frac{\omega_p}{20} = 1.015$. For $N = 13, 15$ the dynamic characteristics coincide.

For the case of $N = 5$ the amplitude and vibration form is essentially different in comparison to other N. The signals for $N = 13$ and $N = 15$ coincide. Therefore, one may conclude that the obtained numerical solutions coincide in full for $N = 13$.

We investigate scenarios of transition from the shell regular vibrations to chaotic one depending on the terms number in the Bubnov-Galerkin method and depending on the size dependent parameter l. In Table 12.1 there are reported scenarios for $N = 5$ and for $N = 11 - 13$, respectively, for $l = 0$.

We have detected one period doubling bifurcation for $N = 5$, and a further increase of the load amplitude was impossible due to the large amplitude of vibrations. For $N = 11, 12$ another scenario of transition from regular to chaotic vibrations is observed. For $q = 0.32$, in spite of the excitation frequency ω_p, there is a birth of linearly non-dependent frequency $\omega_1 = 8.3$ as well as of the linear combination of frequencies $\omega_2 = \omega_p - \omega_1 = 12$ and $\omega_3 = \omega_p - 2\omega_1 = 3.7$. For $q = 0.3245$, new linear combinations of the frequencies ω_p and ω_1 appear. Therefore, a scenario of transition into chaos is in agreement with the classical Ruelle-Takens-Newhouse scenario via two linearly independent frequencies and their linear combinations.

In Table 12.3 a scenario of transition of vibrations of the closed size-dependent cylindrical shell from the regular to chaotic vibrations is shown for the fixed size dependent parameter $l = 0.3$ and for $N = 5$ and $N = 11 - 13$.

The account of the size-dependent parameter l has an essential influence on the scenario of transition from regular to chaotic shell vibrations. For large q the amplitude of the shell vibrations becomes smaller ($N = 5$) and for $q = 0.927834$ and for $N = 5$ the linearly independent frequency (in comparison to ω_p) is born. For $q = 0.927835$ the frequency ω_2 achieves the value $\omega_2 = 5.8$, and a new linearly dependent frequency $\omega_p - \omega_2 = 14.5$ occurs. In the given case, we deal with the Ruelle-Takens-Newhouse scenario.

In the case $l = 0$ (lack of the size dependent parameter) for $N = 11, 13$ (Table 12.2) the amplitude of the excited vibrations is changed within the interval (0.2; 0.324). An account of the size dependent parameter yielded increase of amplitude up to 0.46. For $q = 0.3$ the power spectrum exhibits only the excitation frequency though the signal and phase portrait correspond to the quasi-periodic vibrations. This is because we have occurrence of the large magnitude frequency comparing to the excitation frequency. Increase of q implies occurrence of the linearly independent frequency $\omega_1 = 8.3$ (the same holds for $l = 0$ in Table 12.2).

Table 12.2 Scenarios of transition of regular to chaotic shell vibrations, $l = 0, N = 5$ and $N = 11 - 13$

	Power spectra ($N = 5$)		Power spectra ($N = 11 - 13$)
q=0.7	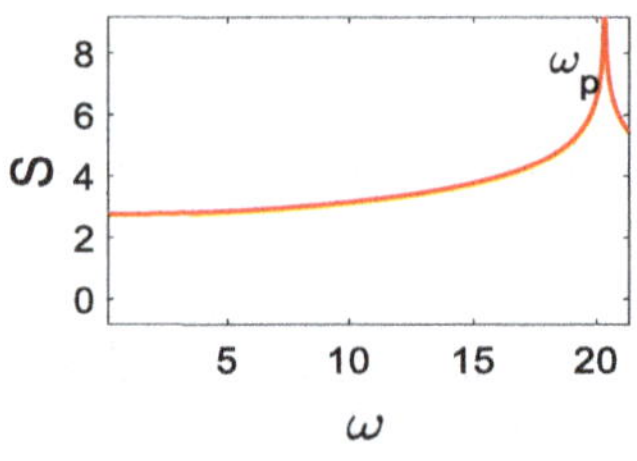	q=0.323	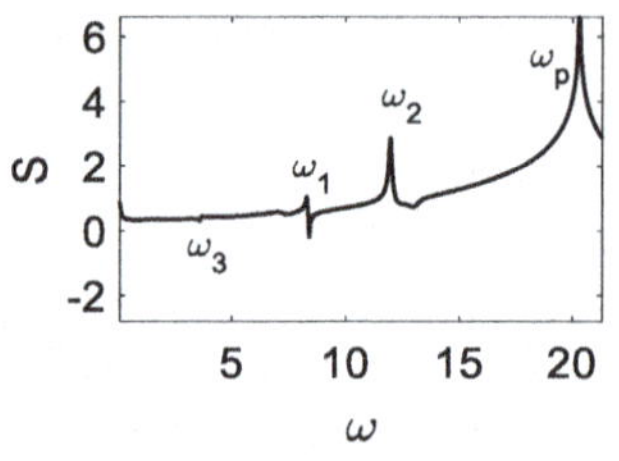
q=0.774	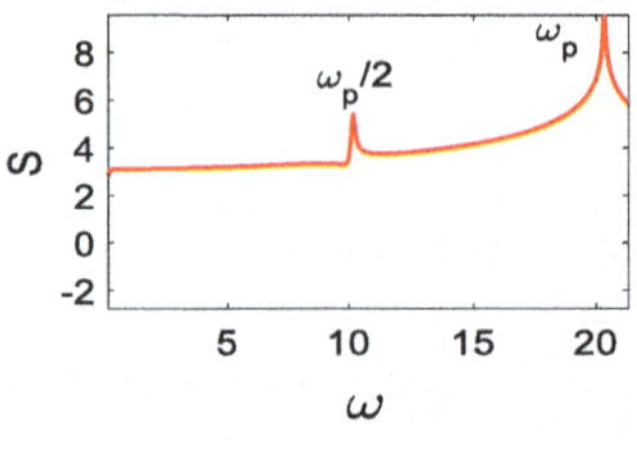	q=0.3245	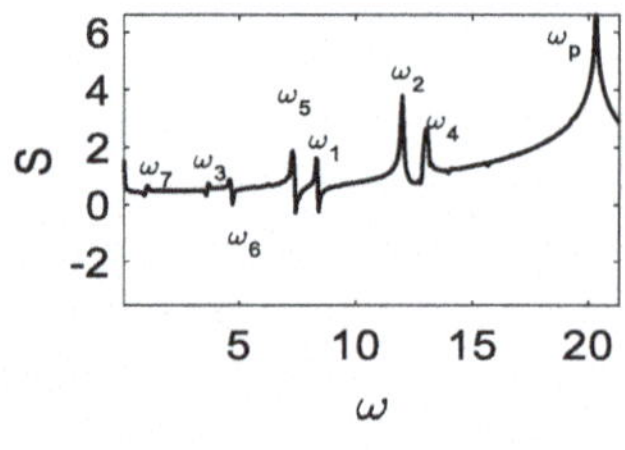

Further, we have period doubling bifurcation ($\omega_p/2$) and a linear dependence of ω_p and ω_1 is exhibited ($\omega_p - \omega_1 = 12$).

For we have not observed the subharmonic frequency $\omega_p/2$. Occurrence of the periodic doubling bifurcation has been also exhibited in the phase portrait (for instance $\omega_p/11 = 1.845$). For $q = 0.46$ we have detected one more period doubling bifurcation $\omega_p/4$; $3\omega_p/4$, and a new linear combination of the frequencies.

Hence, it is clear that an account of the size-dependent parameter l implies the change of transition from regular to chaotic shell vibrations. In the case of $l = 0.3$ the existence of two scenarios, i.e. Feigenbaum in the form of the period doubling bifurcation and the Ruelle-Takens-Newhouse scenario associated with two independent frequencies and their linear combinations, are detected (Table 12.3).

Table 12.3 Scenarios of transition of regular to chaotic shell vibrations, $l = 0.3$, $N = 5$ and $N = 11 - 13$

	Power spectra ($N = 5$)		Power spectra ($N = 11 - 13$)
q=0.9	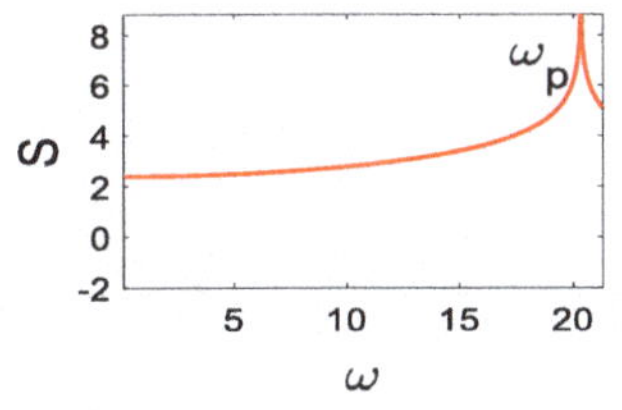	q=0.42	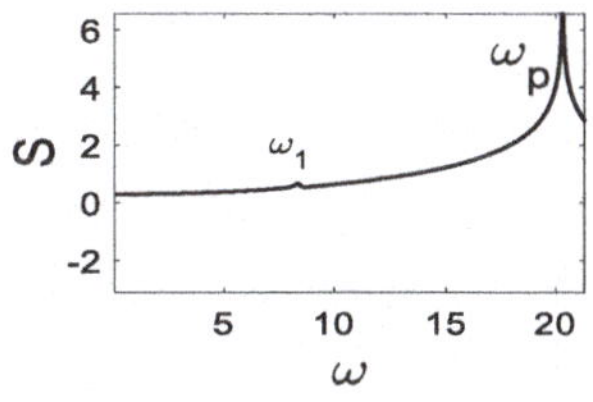
q=0.927834	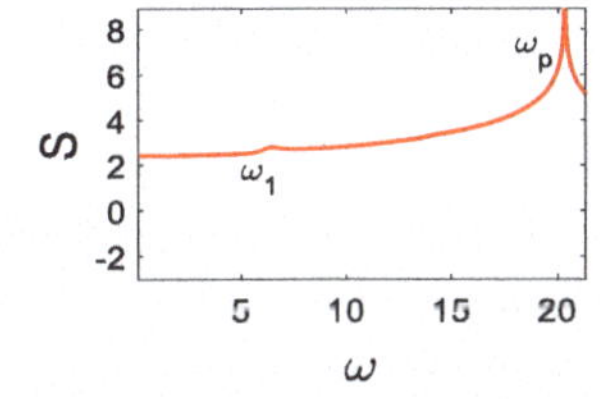	q=0.45	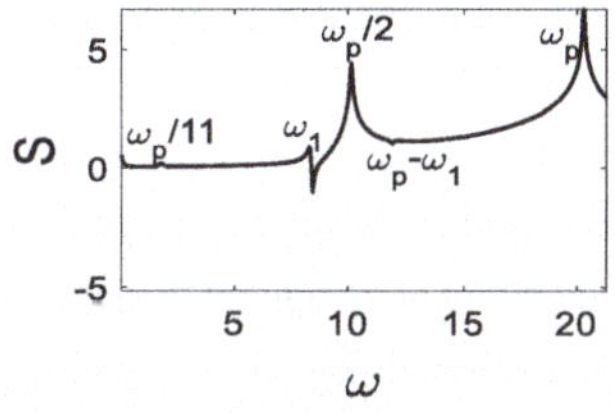
q=0.927835	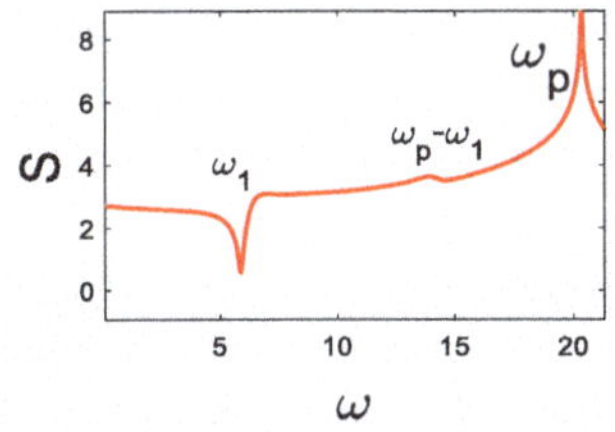	q=0.46	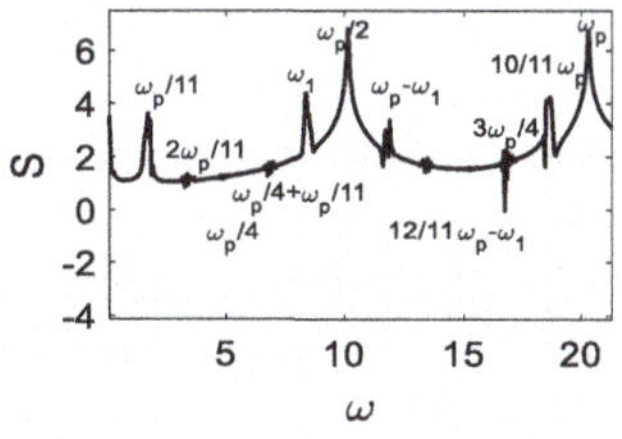

Table 12.4 Scenarios of transition from periodic to chaotic vibration for $l = 0.5$, $N = 5$ and $N = 11 - 13$

	Fourier power spectra ($N = 5$)		Fourier power spectra ($N = 11 - 13$)
q=0.6	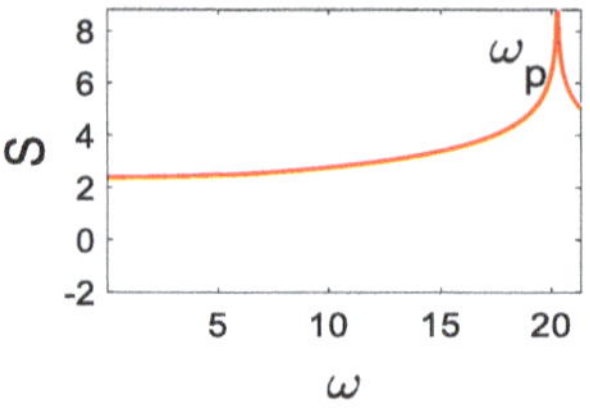	q=0.51	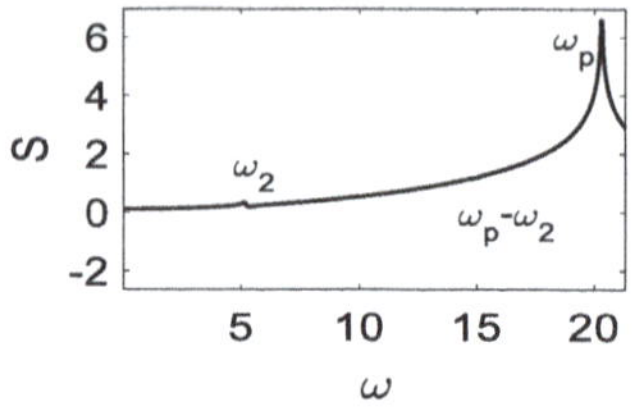
q=0.62	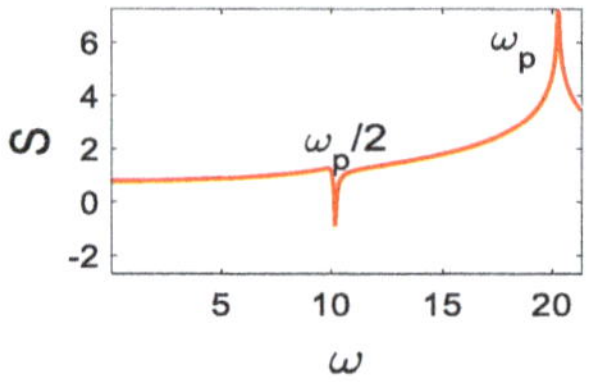	q=0.53	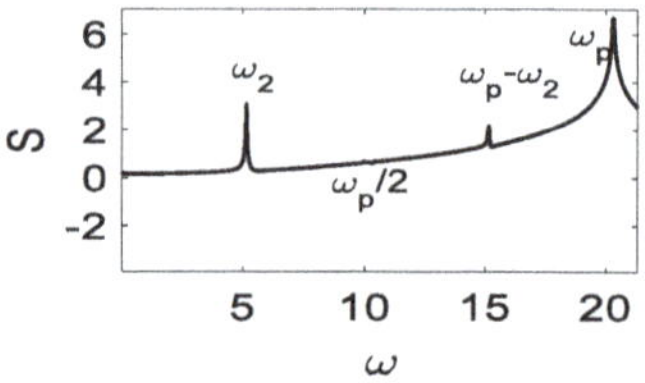
q=0.71	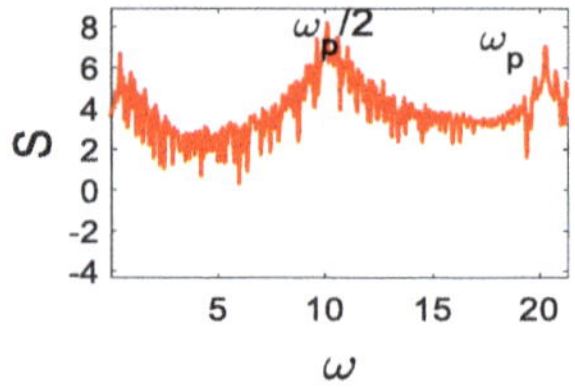	q=0.59	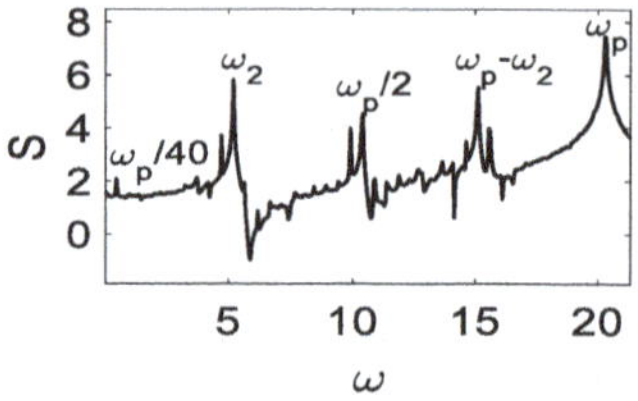

We increase the size dependent parameter up to 0.5 and investigate a scenario of transition into chaos for different numbers of terms used in the Bubnov-Galerkin series. Scenarios for $N = 5$ and $N = 11, 13$ are shown in Table 12.4.

In the case of $N = 5, l = 0.5$ a transition from regular to chaotic vibrations is realized for large values of the excitation amplitude q via the period doubling bifurcation. Increase of the parameter l, for $N = 11, 13$ yielded increase of the vibration amplitude which was observed also for $N = 5$. However, the scenario of transition from regular to chaotic vibrations has not been changed.

For $l = 0.3$ (Table 12.3) first a birth of the frequency $\omega_1 = 8.3$ is observed, and then (for $l = 0.5$) the frequency $\omega_2 = 5.16$ appears being linearly dependent $\omega_p - \omega_2 = 15.1$ for $q = 0.51$. Next, the period doubling bifurcation takes place (for $q = 0.53$), and the system regular vibrations move to chaotic vibration via occurrence of new linear combinations of the frequencies. The phase portrait consists of two rings for $q = 0.4$; 0.51, which demonstrates the occurrence of a high frequency. The signal is quasi-periodic with two frequencies.

Table 12.5 Scenarios of transition from regular to chaotic shell dynamics for $l = 0.7$, $N = 5$ and $N = 11 - 13$

	Fourier power spectra ($N = 5$)		Fourier power spectra ($N = 11 - 13$)
q=1.35	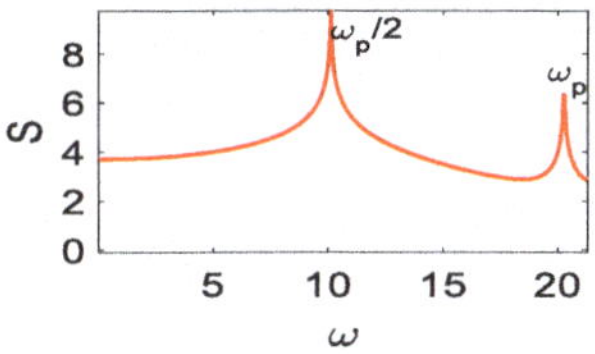	q=0.7	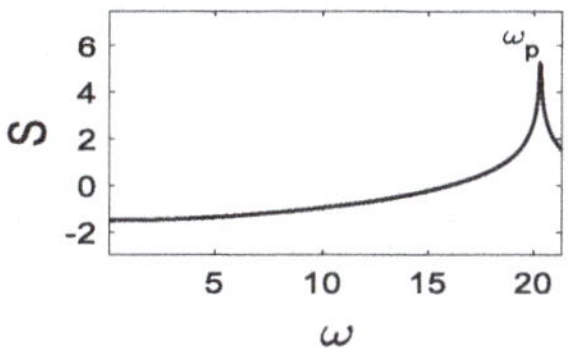
q=1.374	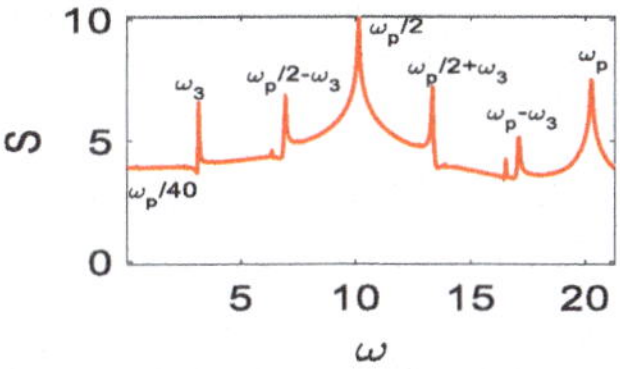	q=0.755	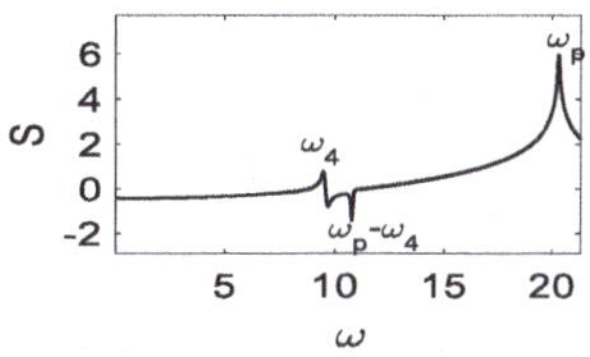
q=1.378	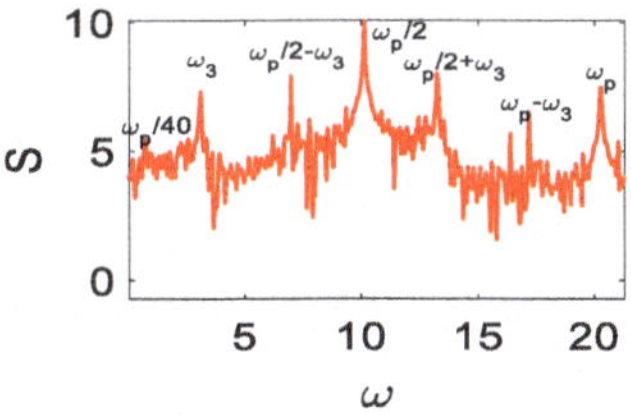	q=0.75722	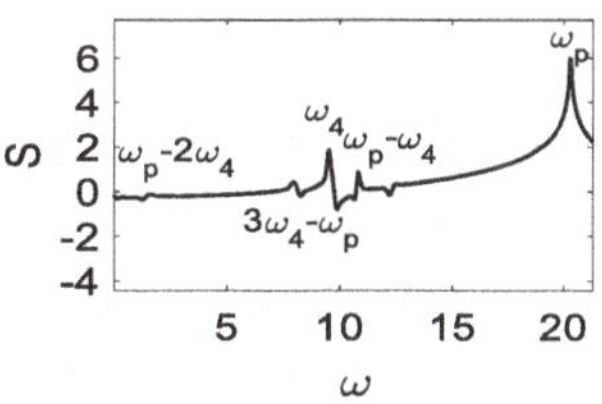

Finally, we report the transition scenario from periodic to chaotic shell vibrations where the size dependent parameter l plays a role of the control parameter and we consider $N = 5$ and $N = 11, 13$ (see Table 12.5).

For $l = 0$, $N = 5$ the first bifurcation is fixed for $q = 0.74$ (Table 12.2), whereas for $l = 0.5$, $N = 5$ (Table 12.4) the first period doubling bifurcation takes place for $q = 0.62$. In the studied so far case the first period doubling occurs for $q = 1.35$. Further, an independent frequency $\omega_3 = 3.17$ appears and combination of the frequencies ω_p, ω_3 increases thickness of the ring in the corresponding phase portrait.

For $q = 0.7$ we have two-frequency shell vibrations where the second frequency is twice larger than the excitation frequency ω_p, ω_3. Then a linearly independent frequency $\omega_4 = 9.5$, in comparison to ω_p, ω_3 appears, and the linear combination of the two mentioned frequencies takes place. In this case, the scenario of transition into chaotic vibrations follow the Ruelle-Takens-Newhouse scenario.

Chapter 13
Contact Interaction of Two Rectangular Plates Made from Different Materials with an Account of Physical Non-Linearity

In this chapter an approach convenient for analysis of stress-strain state (SSS) of multi-layer plates and beams with welded layers and with the occurrence of separated contacting zones due to the technological and exploitation reasons is proposed. The method offers a possibility of studying the constructions with contacts between layers, when the contact condition may depend on coordinates and all non-homogenity properties of the studied contact.

Important results yielded by our theoretical investigations concern removal of the function describing the contact interaction between the structural members from a number of unknowns.

The given algorithm allows one to study SSS of the structures, in which the dependencies $\sigma_i^{(i)}(e_i^{(i)})$ can be given either in digital forms or can be formulated analytically. In the case of a two-layer plate-plate package with design non-linearity, phase synchronization takes place in the frequency interval $\omega \in [4, 8]$ where there is lack of synchronization (bright zones in the graphs of the phase difference) in certain time intervals, but amplitude locking phenomena appear instead.

Finally, we present modification of the methods of reduction of elliptic types PDEs to ODEs. The mentioned methods possess many advantages in comparison to the Bubnov-Galerkin and Ritz methods since they do not require an a priori knowledge of the system of approximating functions, and the obtained solution satisfies the governing equations. The so far mentioned methods can be also successively applied to solve hyperbolic and parabolic PDEs including 2D and 3D problems.

J. Awrejcewicz, V. A. Krysko, *Elastic and Thermoelastic Problems in Nonlinear Dynamics of Structural Members*, Scientific Computation,
https://doi.org/10.1007/978-3-030-37663-5_13

13.1 Statement of the Problem

We consider a problem dealing with one-constrained mechanical interaction between two rectangular plates of variable thickness. The plates are treated as thin and their stress-strain state (SSS) can be described by the classical Kirchhoff theory supplemented with physical non-linearity and considered in the frame of the theory of small elastic-plastic deformations. We assume that the contact pressure (normal to the surface of generated stress) is much less in comparison to the normal stress occurred in the plates cross-sections, and hence a free slip of the contacting plates is allowed. A choice of the classical theory of plates (the Kirchhoff model) is motivated by an observation that, in comparison to the transversal crushing in a contact zone, the deformation of the transverse shear has a negligible influence on both the SSS and the distribution of the contact pressure [161]. This validated observation stands for us as the main purpose of developing and further employing our theoretical background.

Let us present a system of the input PDEs for two contacting plates in the following form [161]:

$$(A\,u_N - q, \psi_j(y))_{H(c,d)} = 0, \tag{13.1}$$

where: $w_i(x, y)$—vector functions; $q_i(x, y)$—vector of the external load; $i = 1, 2$—number of a plate coinciding with a positive sense of the normal; $\Omega_1 = \{a_1(x, y) \leq z \leq b_1(x, y)\}$, $\Omega_2 = \{a_2(x, y) \leq z \leq b_2(x, y)\}$—volumes occupied by plates; $z = a_i(x, y)$ and $z = b_i(x, y)$, $(x, y) \in \Omega_i$ are the functions governing forms of external surfaces of the plates, which allow to take into account changes in thickness of each plate (Fig. 13.1).

The influence of different plate moduli on the plate behaviour is investigated based on Zhao and Ye [358] as well as He et al. [146] proposals. Each plate is divided into two subspaces by means of employing a concept of the so-called elastic neutral layer $z = c_i(x, y)$: $\Omega_i^+ = \{a_i(x, y) \leq z \leq c_i(x, y)\}$, $\Omega_i^- = \{c_i(x, y) \leq z \leq b_i(x, y)\}$. The material in the space Ω_i^+ is defined by Young's modulus $E_i^+(x, y, z, e_i^+)$ and Poisson's coefficient $\nu_i^+(x, y, z, e_i^+)$ corresponding to extension (+), whereas the space Ω_i^- is defined by Young's modulus $E_i^-(x, y, z, e_i^-)$

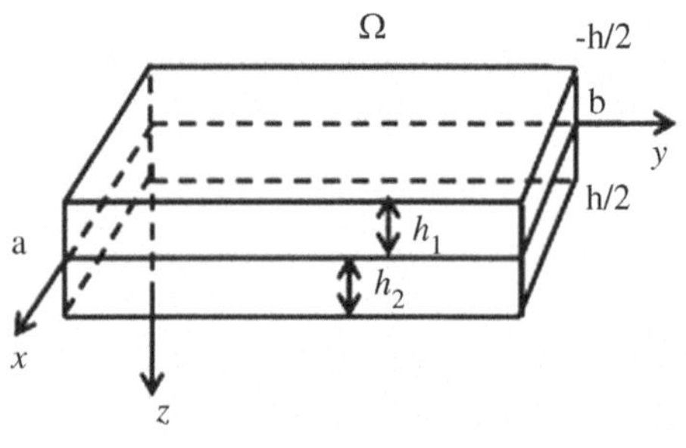

Fig. 13.1 Geometric parameters of the plates

and Poisson's coefficient $\nu_i^-(x, y, z, e_i^-)$ corresponding to compression (−). Furthermore, $e_i^{+/-}$ stands for the deformation intensity. The stress-strain state, owing to reference [146], is defined by non-constant elasticity modulus and Poisson's coefficient, and the following relations hold

$$\sigma_x^{+/-} = \frac{E_i^{+/-}}{1-\left(\nu_i^{+/-}\right)^2}\left[\varepsilon_x + \nu_i^{+/-}\varepsilon_y\right], \quad \sigma_y^{+/-} = \frac{E_i^{+/-}}{1-\left(\nu_i^{+/-}\right)^2}\left[\varepsilon_y + \nu_i^{+/-}\varepsilon_x\right],$$

$$\tau_{xy}^{+/-} = -\frac{E_i^{+/-}z}{1+\left(\nu_i^{+/-}\right)}\varepsilon_{xy}.$$

In physically linear problems, based on the introduced assumptions, the operator A_i $(i = 1, 2)$ takes the following form

$$A_i w_i(x, y) = D_i\left[\frac{\partial^2}{\partial x^2}\left(\frac{\partial^2 w_i}{\partial x^2} + \frac{\partial^2 w_i}{\partial y^2}\right) + \frac{\partial^2}{\partial y^2}\left(\frac{\partial^2 w_i}{\partial x^2} + \frac{\partial^2 w_i}{\partial y^2}\right)\right], \tag{13.2}$$

where

$$D_i = \frac{E_i^+\int_{a_i(x,y)}^{c_i(x,y)} z^2 dz}{1-\left(\nu_i^+\right)^2} + \frac{E_i^-\int_{c_i(x,y)}^{b_i(x,y)} z^2 dz}{1-\left(\nu_i^-\right)^2}.$$

On the other hand, for physically non-linear stress-strain problem of plates with non-constant thickness and made from a material with different moduli, the operator A_i $(i = 1, 2)$ has the following form

$$A_i w_i(x, y) = \frac{\partial^2}{\partial x^2}\left(B_{11,i}\frac{\partial^2 w_i}{\partial x^2} + B_{10,i}\frac{\partial^2 w_i}{\partial y^2}\right) + \frac{\partial^2}{\partial y^2}\left(B_{10,i}\frac{\partial^2 w_i}{\partial x^2} + B_{11,i}\frac{\partial^2 w_i}{\partial y^2}\right) + \\ + 2\frac{\partial^2}{\partial x\,\partial y}\left(\left(B_{11,i} - B_{10,i}\right)\frac{\partial^2 w_i}{\partial x\,\partial y}\right), \tag{13.3}$$

where

$$B_{mn,i} = \frac{1}{2}\left[\frac{E_{11,i}}{E_{01,i}} + (-1)^m\frac{E_{10,i}}{E_{00,i}} - E_{21,i} + (-1)^n E_{20,i}\right],$$

$$E_{mn,i} = \int\limits_{a_i(x,y)}^{c_i(x,y)} \frac{E_i^+ z^m}{1+(-1)^n\nu_i^+}\,dz + \int\limits_{c_i(x,y)}^{b_i(x,y)} \frac{E_i^- z^m}{1+(-1)^n\nu_i^-}\,dz,$$

$$n = 0,\ 1; \quad m = i = 1,\ 2.$$

The operators A_i $(i = 1, 2)$ are defined as follows (linear case)

$$A_i w_i\,(x\,,y) = D_i\left[\frac{\partial^2}{\partial x^2}\left(\frac{\partial^2 w_i}{\partial x^2}+\frac{\partial^2 w_i}{\partial y^2}\right)+\frac{\partial^2}{\partial y^2}\left(\frac{\partial^2 w_i}{\partial x^2}+\frac{\partial^2 w_i}{\partial y^2}\right)\right], \tag{13.4}$$

where $D_i = \frac{E_i h^3}{12(1-\nu_i{}^2)}$ stands for a cylindrical stiffness, E_i is Young modulus, and ν_i is Poisson's coefficient.

In a non-linear problem governing the SSS of plates of variable thickness and made from bi-modulus material, the operators A_i $(i = 1, 2)$ have the following form

$$A_i w_i\,(x\,,y) = \frac{\partial^2}{\partial x^2}\left(B_{11,i}\frac{\partial^2 w_i}{\partial x^2}+B_{10,i}\frac{\partial^2 w_i}{\partial y^2}\right)+\frac{\partial^2}{\partial y^2}\left(B_{10,i}\frac{\partial^2 w_i}{\partial x^2}+B_{11,i}\frac{\partial^2 w_i}{\partial y^2}\right)+ \\ +2\frac{\partial^2}{\partial x\,\partial y}\left(\left(B_{11,i}-B_{10,i}\right)\frac{\partial^2 w_i}{\partial x\,\partial y}\right), \tag{13.5}$$

where

$$B_{mn,i} = \frac{1}{2}\left[\frac{E_{11,i}}{E_{01,i}}+(-1)^{n+1}\frac{E_{10,i}}{E_{00,i}}-E_{21,i}+(-1)^n E_{20,i}\right],$$

$$E_{mn,i} = \int\limits_{a_i(x,y)}^{b_i(x,y)}\frac{E_i z^m}{1+(-1)^n\nu_i}\,dz, \qquad n = 0,\ 1, \qquad m = i = 1,\ 2.$$

The following notation is employed: $z = ai(x, y)$ and/or $z = bi(x, y)$, $(x, y) \in \Omega_i$ denote equations of external surfaces of the plates, which allow for inclusion of variable thickness of each plate; $E_i\,(x, y, z, e_i^{(i)})$, $\nu_i(x, y, z, e_i^{(i)})$ describe different stiffness modulus and Poisson's coefficients, respectively; $e_i^{(i)}$ stands for intensity of deformation of i-th plate.

It should be emphasized that the method of variable parameters of stiffness [70], employed in our physically non-linear problem, is based on observation that $E_i(x, y, z, e_i^{(i)})$, $\upsilon_i(x, y, z, e_i^{(i)})$ are non-constant parameters dependent on the deformation state in a plate point, and they are governed by the following formulas

$$E_i = \frac{9K_{1i}G_i}{3K_{1i}+G_i}, \qquad \nu_i = \frac{1}{2}\frac{3K_{1i}-2G_i}{3K_{1i}+G_i}, \tag{13.6}$$

where $K_{1i} = const$. The theory of deformation yields the following shear modulus

$$G_i = \frac{1}{3}\frac{\sigma_i^{(i)}(e_i^{(i)})}{e_i^{(i)}}, \tag{13.7}$$

where $\sigma_i^{(i)}$, $e_i^{(i)}$ stand for intensity of stresses and deformations of a plate, respectively, and

$$e_k^{(i)} = \frac{\sqrt{2}}{3}\left[\left(e_{xx}^{(i)} - e_{yy}^{(i)}\right)^2 + \left(e_{yy}^{(i)} - e_{zz}^{(i)}\right)^2 + \left(e_{xx}^{(i)} - e_{zz}^{(i)}\right)^2 + 1.5\left(e_{xy}^{i}\right)^2\right]^{1/2},$$

$$e_{xx}^{(i)} = -z\frac{\partial^2 w_i}{\partial x^2}, \quad e_{yy}^{(i)} = -z\frac{\partial^2 w_i}{\partial y^2}, \quad e_{xy}^{(i)} = -2z\frac{\partial^2 w_i}{\partial x \partial y}.$$

The value of $e_{zz}^{(i)}$ is defined through the condition of a plane stress state ($\sigma_{zz} = 0$):

$$e_{zz}^{(i)} = -\frac{\nu_i}{1-\nu_i}(e_{xx}^{(i)} + e_{yy}^{(i)}). \tag{13.8}$$

Two types of boundary conditions are analysed, i.e.

(1) simple support

$$w_i \mid_{\partial\Omega} = \frac{\partial^2 w_i}{\partial n^2} \mid_{\partial\Omega} = 0, \quad i = 1, 2; \tag{13.9}$$

(2) rigid clamping

$$w_i \mid_{\partial\Omega} = \frac{\partial w_i}{\partial n} \mid_{\partial\Omega} = 0, \quad i = 1, 2. \tag{13.10}$$

Note that in mathematical models (13.1), (13.4), (13.9), (13.10) and (13.1), (13.5), (13.9), (13.10), the contact pressure is proportional to the transversal crushing $w_1 - h_1 - w_2$ in a zone of contact between plates, i.e. we have

$$q_k(x, y) = k\frac{E}{h}(w_1 - h_1 - w_2), \tag{13.11}$$

and the function ψ has the following form

$$\psi = [1 + sign(w_1 - h_1 - w_2)]/2. \tag{13.12}$$

Here h_1 stands for a clearance between plates, k is the stiffness coefficient of the transverse crushing of the plates.

Due to occurrence of the multiplier ψ in (13.1), the system of equations becomes non-linear, and consequently the entire problem becomes non-linear. The design non-linearity governs the possibility of switching on/off of the one-sided constraint. Furthermore, inclusion of the physical non-linearity in the considered problem makes it impossible to solve the problem via classical analytical, semi-analytical or numerical approaches. Therefore, we employ here the method of successive approximations matched with a simultaneous process of improvements of the borders of the contact zones.

Observe that formula (13.11) holds for the case of a contact between plates having the same thickness h and the same parameters E and k. In the case of contact problems of Kirchhoff's plates, Winkler's type coupling between the crushing zone and the contact pressure is utilised. If there is a thin layer between plates, it can be also taken into account by means of changing the parameter k.

If the initial plates configuration (function h_1) and the employed load correspond to lack of contact between the deformed plates, then $\psi \equiv 0$, and the system (13.1) is splitted into two separate subsystems. Otherwise, two subsystems (plates) are coupled. Substituting (13.11) into (13.1) yields

$$\begin{cases} A_1 w_1(x, y) = q_1(x, y) - k\dfrac{E_1}{h}(w_1 - h_1 - w_2)\psi, \\ A_2 w_2(x, y) = q_2(x, y) + k\dfrac{E_2}{h}(w_1 - h_1 - w_2)\psi. \end{cases} \tag{13.13}$$

Observe that the system (13.13) is of the eight order and hence it should be studied together with the boundary conditions (13.9), (13.10). It includes both design and physical non-linearity.

In order to solve the derived design-non-linear problem (13.13), one has to construct an iterational process allowing (for each step of loading) for solving only one of the equations of the system (13.13). This procedure yields a decrease in the systems order twice for a two-layer package and consequently, into n times while dealing with a package of n layers. The iterational procedure has the following form

$$\begin{cases} A_1(w_1^{(n)}) + k\dfrac{E}{h}\,\psi_{(n-1)} w_1^{(n)} = q_1(x, y) + k\dfrac{E_1}{h}(h_1 + w_2^{(n-1)})\,\psi_{(n-1)}, \\ A_2(w_2^{(n)}) + k\dfrac{E}{h}\,\psi_{(n-1)} w_2^{(n)} = q_2(x, y) + k\dfrac{E_2}{h}(w_1^{(n-1)} - h_1)\,\psi_{(n-1)}. \end{cases} \tag{13.14}$$

The system (13.14) should be supplemented with the corresponding boundary conditions (13.6) or (13.7) for the i-th plate. If the values of E and h are different, then for the upper/lower plate we use notation $E_1, h_1/E_2, h_2$, and we have

$$q_k(x, y) = 2k\frac{E_1}{h_1}\left(1 + \frac{E_1 h_2}{E_2 h_1}\right)(w_1 - h_j - w_2). \tag{13.15}$$

In the case of fixed/frozen contact zone, Eqs. (13.14) can be solved using one of the methods devoted to reduction of PDEs to ODEs. Those methods are briefly discussed in the next sections of the paper.

13.2 Reduction of PDEs to ODEs

Nowadays, there exist a few methods aiming at reducing the PDEs to ODEs. The principal concepts and ideas of those methods will be discussed using the operator-type equation of the form

$$Aw\,(x\,,\,y) = q\,(x\,y)\,, \quad (x\,,\,y) \in \Omega = (a\,,\,b) \times (c\,,\,d)\,, \tag{13.16}$$

where: A—operator of a boundary problem including both differential operator and boundary conditions; $q(x, y)$—known function (vector function of the system of equations); $w(x, y)$—function being searched (vector-function of the system of equations).

13.2.1 *The Method of Kantorovich-Vlasov (MKV)*

Owing to the MKV, the following form of solution to (13.16) can be assumed [163, 332]:

$$w_N\,(x\,,\,y) = \sum_{i=1}^{N} \varphi_i\,(x)\;\psi_i\,(y)\,, \tag{13.17}$$

where: $\varphi_i(x)$ are given functions satisfying the boundary conditions (13.9) or (13.10), whereas $\psi_i(y)$ are the being searched functions defined by the following system of projected equations

$$(Aw_N - q, \varphi_j(x))_{H(a,b)} = 0, \qquad j = 1, 2, \ldots, N. \tag{13.18}$$

Here $(A\,u_N - q, \varphi_j(x))$ stands for notation of a scalar product in the space $H(a, b)$, $H = H(a, b) \times H(c, d)$ is the space in which the exact solution w_0 to Eq. (13.16) exists. The procedure (13.18) reduces the problem of PDEs to ODEs by employing the Bubnov-Galerkin method regarding one of the coordinates. This procedure, in fact, generalises the classical Bubnov-Galerin method [118].

Disadvantages If the expected solution (13.17) should obey the same symmetry property with regard to all variables, then the MKV method sometimes cannot exhibit this property due to the limited number of terms used in the series. In other words, in case of taking a limited number of the series terms, the mentioned symmetry can be lost.

Advantages Independence of the basic functions versus one of the variables (here y) on the form of the boundary conditions.

13.2.2 Method of Vaindiner (MV)

In this case a solution to Eq. (13.15) is assumed to be as follows [323]:

$$w_N(x, y) = \sum_{i=1}^{N} \varphi_i(x)\, \psi_i(y) + u_i(x)\, v_i(y), \tag{13.19}$$

where $\psi_i(y)$ and $u_i(x)$ are given functions, $\varphi_i(x)$ and $v_i(x)$ are the being searched functions satisfying the system of projected equations of the following form

$$(A\, w_N - q, \psi_j(y))_{H(c,d)} = 0, \quad (A\, w_N - q, u_j(x))_{H(a,b)} = 0, \quad j = 1, 2, \ldots, N, \tag{13.20}$$

where $H((a, b) \times (c, d)) = H(a, b) \times H(c, d)$ is the space, where the exact solution w_0 exists.

Drawbacks There is a need to choose a full coordinate system of equations; there exists a dependence of accuracy of the obtained solutions on a number of equations in the system of approximating equations $2N$, i.e. the number of equations is twice increased in comparison to the MKV.

Advantages In spite of the same advantages as those exhibited by MKV, there is a possibility of obtaining a symmetric solution with respect to all variables, if the being searched exact solutions have the mentioned symmetry.

13.2.3 Method of Variational Iteration (MVI)

This method presents a modification of the MKV [179]. Owing to MVI, a solution to Eq. (13.16) can be searched in the form of (13.17), where the functions $\varphi_i(x)$ and $\psi_i(y)$ are defined through the following equations

$$(A\, u_N - q, \varphi_j(x))_{H(a,b)} = 0, \tag{13.21}$$

$$(A\, u_N - q, \psi_j(y))_{H(c,d)} = 0, \qquad j = 1, 2, \ldots, N, \tag{13.22}$$

in the following way: a certain system of N functions with regard to one of the variables, let us say $\varphi_j^0(x)$ $j = 1, 2, \ldots, N$, is given. Then, the system (13.21) defines a system of functions $\psi_j^{(1)}(y)$, which are substituted into the system (13.22), and a new choice with regard to the variable x is defined as $\varphi_j^{(2)}(x)$. Then, using the chosen functions, the system (13.21) yields a new function $\psi_j^{(3)}(y)$ with respect to y, and so on. Truncating the process of searching for the functions $\varphi_j(x)$ and $\psi_j(y)$

on the k-th step, we define the following function

$$w_N = \sum_{i=1}^{N} \varphi_i^{(k-1)}(x)\psi_j^{(k)}(y), \tag{13.23}$$

which serves as an approximate solution to Eq. (13.14).

Disadvantages The accuracy of the obtained approximate solution depends on a number of the series terms of (13.17).

Advantages There is no need to define the initial approximations satisfying, for instance, the boundary conditions; symmetry of the approximate solutions, if it is a priori known, holds; it is possible to achieve the maximum allowed accuracy of the approximate solutions even for a bounded number of terms approximating the solutions.

13.2.4 Method of Arganovskiy-Baglay-Smirnov (MABS)

The so far discussed method of variational iterations (MVI) coincides with the first term of the series of the method developed by Arganovskiy and Baglay [13], Baglay and Smirnov [58]. The formal scheme of the MABS can be briefly summarised in the following way.

A solution to Eq. (13.16) in the first approximation ($N = 1$) is searched in a way similar to the MVI, i.e. we take

$$w_1 = \varphi_1^{(k-1)}(x)\ \psi_1^{(k)}(y). \tag{13.24}$$

The new equation is defined as follows

$$Aw_2(x, y) = q(x, y) - Aw_1(x, y), \tag{13.25}$$

i.e. we have changed the right hand side of equation (13.16). Equation (13.16) is solved again with the help of MVI, and its first approximation yields

$$w_2(x, y) = \varphi_2^{(k-1)}(x)\ \psi_2^{(k)}(y). \tag{13.26}$$

The next new equation follows: $Aw_3(x, y) = q(x, y) - Aw_1(x, y) - Aw_2(x, y)$, and then one employs the MVI again in the first approximation, and so on.

Finally, the following series is used as the input solution:

$$w(x, y) = \sum_{n=1}^{N} w_n(x, y). \tag{13.27}$$

Drawbacks This method cannot be used to solve a class of problems (for instance, if the boundary conditions are changed in a rectangular plate).

Advantages The advantages are analogous to those of MVI. Furthermore, each of the defined right-hand sides of the type of Eq. (13.25) can be solved by the MVI only in the first approximation, which generates algebraic systems of a small dimension.

13.2.5 Combined Method (MC)

This method is a combination of the MV, MABS, and MVI. Owing to the MC, the approximate solution is searched in the following form

$$w^{(k)}(x, y) = \varphi^{(k)}(x)\ \psi^{(k-1)}(y) + u^{(k-1)}(x)\ v^{(k)}(y), \tag{13.28}$$

where: $\psi^0(y)$ and $u^0(x)$ are assumed as known functions; $\varphi^{(1)}(x)$ ad $v^{(1)}(y)$ are the being searched functions defined by the projected equations (13.16). However, on the contrary to the MV, where the searched functions $\varphi^{(1)}(x)$ and $v^{(1)}(y)$ have been treated as the final ones, here they are taken as known functions used to define the new $\psi^{(2)}(y)$ and $u^{(2)}(x)$ unless the accuracy between $u^{(k)}(x, y)$ and $u^{(k-1)}(x, y)$ exceeds the assumed magnitude. We are looking for a solution of the equations

$$Aw_2(x, y) = q(x, y) - Aw_1(x, y) \tag{13.29}$$

defined through the so far described way, i.e. $w_2(x, y) = \varphi_2^{(k)}(x)\,\psi_2^{(k-1)}(y) + u_2^{(k-1)}(x)\,v^{(k)}(y)$.

If the error between u_1 and (u_1+u_2) is less than a given quantity, then the process is terminated. If not, then the following equation is constructed

$$Aw_3(x, y) = q(x, y) - Aw_1(x, y) - Aw_2(x, y), \tag{13.30}$$

and $w_3(x,\ y)$ is defined, and so on. The final solution to Eq. (13.16) is defined by the following series

$$w_N(x, y) = \sum_{j=1}^{N} w_j(x, y). \tag{13.31}$$

Advantages This method includes all advantages of MV and MABS; furthermore, it may be used for a wide class of problems (for example, for the problems of rectangular variable boundary conditions along their contours).

13.2.6 Matching of the Methods of Kantorovich-Vlasov and Arganovskiy-Baglay-Smirnov (MKV+MABS)

The key idea of this matching is the application of the procedure of MABS to the MKV without the employment of the method of variational iterations (MVI).

Owing to the method, an approximate solution of equation (13.16) is found based on the Kantorovich-Vlasov method, assuming its following form

$$w_N(x, y) = \sum_{i=1}^{N} \varphi_i(x)\, \psi_i(y), \tag{13.32}$$

where $\varphi_i(x)$ are the given functions satisfying the corresponding boundary conditions (13.9) or (13.10), $\psi_i(y)$ are the searched functions defined by the system of projection equations (13.18). However, if the functions $w_1(x, y)$ found in the MKV are the final ones, they can be employed to construct the operator $Aw_1(x, y)$ on the right-hand side of equation

$$Aw_2(x, y) = q(x, y) - Aw_1(x, y), \tag{13.33}$$

which is solved using the MKV.

If the error between w_1 and $w_1 + w_2$ is smaller than a given quantity, then the iterative process is stopped. If this is not the case, than we construct the following equation

$$Aw_3(x, y) = q(x, y) - Aw_1(x, y) - Aw_2(x, y), \tag{13.34}$$

and we define $w_3(x, y)$ with the help of the MKV, and so on. The terminal/final solution to Eq. (13.16) is given by the following formula

$$w_N(x, y) = \sum_{j=1}^{N} w_j(x, y). \tag{13.35}$$

13.2.7 Matching of the Methods of Vaindiner and the Arganovskiy-Baglay-Smirnov (MV+MABS)

In this case the method of Vaindiner (MV) is supplemented with the procedure of the MABS but without a procedure of MVI.

A solution to Eq. (13.15) is searched in the following form:

$$w_N(x, y) = \sum_{i=1}^{N} \varphi_i(x)\, \psi_i(y) + u_i(x)\, v_i(y), \tag{13.36}$$

where $\psi_i(y)$ and $u_i(x)$ are the given functions, $\varphi_i(x)$ and $v_i(x)$ are the searched functions defined by the system of projection equations (13.18).

However, while the found function $w_1(x, y)$ is treated as the final one in the MV, it is used to define the operator $Aw_1(x, y)$ on the right-hand side of the following equation

$$Aw_2(x, y) = q(x, y) - Aw_1(x, y), \tag{13.37}$$

which is solved by the MV.

If the error between w_1 and $w_1 + w_2$ is smaller than a given quantity, then the iterative process is stopped. Otherwise, we construct the following equations

$$Aw_3(x, y) = q(x, y) - Aw_1(x, y) - Aw_2(x, y), \tag{13.38}$$

and we define $w_3(x, y)$ with the help of the MKV. As the final solution of equation (13.16) we take the following

$$w_N(x, y) = \sum_{j=1}^{N} w_j(x, y). \tag{13.39}$$

13.2.8 Matching of the Methods of Vaindiner with the Method of Variational Iterations (MV+MVI)

In this case the method of Vaindiner (MV) is extended to include a procedure of the variational iterations.

The approximate solution to the problem is searched in the following form

$$w^{(k)}(x, y) = \varphi^{(k)}(x)\ \psi^{(k-1)}(y) + u^{(k-1)}(x)\ v^{(k)}(y), \tag{13.40}$$

where $\psi^0(y)$ and $u^0(x)$ are known functions, $\varphi^{(1)}(x)$ and $v^{(1)}(y)$ are the searched functions defined by the system of projection equations (13.16). However, if the functions $\varphi^{(1)}(x)$ and $v^{(1)}(y)$ found with the help of MV are the final ones, they are taken as known functions, and the being searched solution takes the form

$$w^{(2)}(x, y) = \varphi^{(1)}(x)\ \psi^{(1)}(y) + u^{(1)}(x)\ v^{(1)}(y). \tag{13.41}$$

Employment of the MVI yields new functions $\psi^{(2)}(y)$ and, and so on. The iterative procedure is carried out until the error between $w^{(k)}(x, y)$ and $w^{(k-1)}(x, y)$ is smaller than a priori given value.

13.2.9 Numerical Example

In order to compare the accuracy of each of the discussed methods, we consider the problem of a linear bending of the Germain-Lagrange plate governed by the following PDE

$$w(x,y) = D\nabla^2\nabla^2 w(x,y) = q(x,y), \tag{13.42}$$

which has known exact solution.

The following boundary conditions are applied:

(1) simple support

$$w\Big|_{\partial\Omega} = \frac{\partial^2 w}{\partial n^2}\Big|_{\partial\Omega} = 0; \tag{13.43}$$

(2) stiff clamping

$$w\Big|_{\partial\Omega} = \frac{\partial w}{\partial n}\Big|_{\partial\Omega} = 0. \tag{13.44}$$

The space $\Omega = (0,a)\times(0,b)$ has the boundary $\partial\Omega$. The differential operator (13.42) and the employed boundary conditions are presented in their counterpart non-dimensional form

$$x = \bar{x}a, \quad y = \bar{y}b, \quad w = \bar{w}h, \quad \lambda = a/b, \quad \bar{q}(x,y) = \frac{q(x,y)}{12(1-\nu^2)}\frac{a^4}{Eh^4}, \tag{13.45}$$

$$\bar{\nabla}^4\bar{w} = \frac{\partial^4 w}{\partial \bar{x}^4} + 2\lambda^2\frac{\partial^4\bar{w}}{\partial\bar{x}^2\partial\bar{y}^2} + \lambda^4\frac{\partial^4\bar{w}}{\partial\bar{y}^4},$$

and the bars over the non-dimensional quantities are further omitted.

Note that a system of integral-differential equations yielded by the Kantorovich-Vlasov procedure can be solved through the method of finite differences (FDM) with the successive employment of the Gauss method while solving a system of algebraic equations produced during searching for the solutions. The solution to the stated problems has been solved using unitary and doubled accuracy. The interval [0,1] has been divided into 30 nodes. Further increase in nodes number has not changed the obtained results. The relative error of computations has been traced using the formula $\varepsilon = \left|\frac{w^k(x,y)-w^{k-1}(x,y)}{w^k(x,y)}\right| < 10^{-3}$. The obtained results are reported in Table 13.1 for $\bar{q}(x,y) = 50$.

The so far described methods as well as the results of the numerical experiment allow one to conclude that the MC has the highest accuracy in comparison to other methods and it yields the results being close to the exact solution. However, a drawback of this method is associated with the increase in the number of differential

Table 13.1 Comparison of the results obtained with different methods

		Deflection $w(0.5, 0.5)$/absolute error	
		Boundary condition	
No.	Employed method	(13.9)	(13.10)
1	MKV (3.1)	0.2056/1.381%	0.0654/3.645%
2	MVI (3.3)	0.2030/0.099%	0.0648/2.694%
3	MV (3.2)	0.2029/0.049%	0.0674/6.815%
4	MV + MVI (3.8)	0.2033/0.247%	0.0668/5.864%
5	MABS (3.4)	0.2030/0.099%	0.0648/2.694%
6	MV + MABS (3.7)	0.2029/0.049%	0.0672/6.498%
7	MC (3.5)	**0.2029**/0.049%	**0.0644**/2.06%
8	MKV + MABS (3.6)	0.2056/1.381%	0.0674/6.815%
9	Exact solution	**0.2028**	**0.0631**

equations, and hence one more iteration should be involved to get reliable results, which implies an increase in the computational time. However, the method of variational iterations is free from this drawback, and the yielded results are also close to the exact ones. Therefore, particularly in the case of complex problems, the MVI is recommended.

13.3 Mathematical Background

13.3.1 Theorems on Convergence of MVI

In the case of a fixed contact zone, the procedure of the MVI [179] is constructed (see the comments given in Sect. 3.3) together with the method of variable parameters of stiffness (MVPS) [70] unless the required accuracy is achieved. A successive improvement of a contact zone is achieved due to the method of direct iteration. Then the solution procedure is repeated, i.e. we employ three iterational procedures nested in each other. Scheme of MVI can be formally described in the following way. We are aimed at finding a solution to equation

$$Aw(x, y) = q(x, y), \qquad x, y \in \Omega(x, y), \tag{13.46}$$

where: A stands for a certain operator defined on the manifold $D(A)$ of the Hilbert space $L_2(\Omega)$; $q(x, y)$ stands for a given function of two variables x, y; $w(x, y)$ is a searched function; $\Omega(x, y)$ is a space associated with variations of x and y.

If $\Omega(x, y) = X \times Y$ (X—a certain bounded set of variables x; Y—a bounded set of y), then a solution to Eq. (13.46) has the following form

$$w_N(x, y) = \sum_{i=1}^{N} u_i(x)v_i(y), \tag{13.47}$$

where the functions $u_i(x)$ and $v_i(y)$ are defined by the following system of equations

$$
\begin{aligned}
&\int_X (Aw_N - q)\, u_1(x)\, dx = 0, \qquad \int_Y (Aw_N - q)\, v_1(y)\, dy = 0, \\
&\dots\dots\dots\dots\dots\dots\dots\dots\dots\dots\dots\dots\dots\dots\dots\dots \qquad (13.48)\\
&\int_X (Aw_N - q)\, u_N(x)\, dx = 0, \qquad \int_Y (Aw_N - q)\, v_N(y)\, dy = 0,
\end{aligned}
$$

in the following way. A certain system composed of N functions with respect to one of the variables, for instance, $u_1^0(x),\ u_2^0(x), \dots,\ u_N^0(x)$ is given. Then, the first N equations of the system (13.48) yield N functions $v_1^1(x),\ v_2^1(x), \dots,\ v_N^1(x)$. Next, the obtained functions are employed to create a new set of functions x – $u_1^2(x),\ u_2^2(x), \dots,\ u_N^2(x)$, which is further used to construct a set of new functions with respect to the variable y, i.e. $v_1^3(x),\ v_2^3(x), \dots,\ v_N^3(x)$, and so on.

In the case of the iterational procedure MVI [179], proves of the theorems constituting the theoretical background of the MVI convergence were given for the problems of the theory of plates.

Theorem 13.1 *If A is a positively defined operator with its action space $D(A) \subset H_A$, then the sequence of elements $\alpha_k = \left\| w_1^k(x, y) - w_0 \right\|_{H_A}$ is monotonously decreasing, i.e. for arbitrary i and j, if only $i \geq j$, then*

$$
\left\| w_1^i - w_0 \right\|_{H_T} \leq \left\| w_1^j - w_0 \right\|_{H_T}. \tag{13.49}
$$

Theorem 13.2 *Let each element of the basis system of the space $\overset{0}{W_2^m}(X \times Y)$ has the form*

$$
\theta_i(x, y) = \varphi_i(x)\psi_i(y), \tag{13.50}
$$

where $\{\varphi_i(x)\}$ is a basis system in the space $\overset{0}{W_2^m}(X)$, and respectively $\{\psi_i(y)\}$ in the space $\overset{0}{W_2^m}(Y)$, and in order to get an arbitrary N-th approximation regarding MVI, the components of the elements of the basis system $\{\theta_i(x, y)\}$ are taken as initial conditions. Then, for sufficiently large N, the MVI gives a unique approximate solution ω_N, and the sequence $\{\omega_N\}$ is convergent with regard to the norm of the space $\overset{0}{W_2^m}(X \times Y)$, and tends to the exact solution ω_0 independently of a number of steps k, which can be defined for each of the N-th approximation, i.e. $\left\| w_N^k - w_0 \right\|_{\overset{0}{W_2^m}} \to 0, \quad N \to \infty$.

The ordinary differential equations obtained on each step of the iterational procedure by the method of variational iterations can be solved with numerous methods such as reduction of the problem to the algebraic equations by FDM with the successive solutions of the obtained algebraic equations by the Gauss method.

The method of variational iterations (MVI) removes the necessity of constructing the system of approximating functions, on the contrary to the widely employed Bubnov-Galerkin and Ritz methods [206, 248]. The initially given arbitrary functions (perhaps satisfying the well-known conditions of smoothness) are modified in the process of computations through MVI beginning with a solution of the input system of PDEs governing the plates behaviour. Therefore, in a natural way, the MVI constructs the approximating functions in the form of a product of two functions, where one of them depends on one argument in a way similar to the Fourier method. This implies the necessity of solving two ODEs instead of one. In fact, the method of variational iterations generalises the Fourier method.

The most fascinating result is that the procedure of the MVI is converged even if functions not satisfying the boundary condition (13.9), (13.10) in the form: $w_1(x) = 1$, $w_2(y) = 1$ are taken as the initial approximations.

The MVI procedure was stopped already on the 4-th step of the variational iterations. Remarkably, almost accurate coincidence with the exact solution has been obtained, i.e. the boundary conditions (13.6)/(13.7) correspond to the difference between the approximate and exact solution in percentage of 0.1%/0.6%.

13.3.2 Convergence Theorem

In what follows, we formulate a theorem (without a proof) regarding convergence of the iterational procedure (13.14) associated with solving our formulated problem aimed at designing a non-linear system composed of two plates made from materials of different moduli.

Let R^2 be an Euclidean plane with a Descartes basis: $\Omega_i \in R^2$—area on a plane with the boundary $\partial\Omega_i$ $(i = 1, 2)$, $\bar{\Omega}_i = \Omega_i \bigcup \partial\Omega_i$ $(x, y) \in \Omega_I$; Ω^*—sub-area of Ω_i $\forall i$, $\Omega^* \subseteq \Omega_I$; n_i—external normal to $\partial\Omega_i$.

We consider a space plate-type object composed of two contacting plates (Fig. 13.2) obeying Kirchhoff's hypotheses and governed by the following PDEs:

$$A_1(w_1(x, y)) + k\frac{E_1}{h}w_1\,\psi(x, y) = q_1 + k\frac{E_1}{h}(w_2 + h_1)\,\psi(x, y),$$
$$A_2(w_2(x, y)) + k\frac{E_2}{h}w_2\,\psi(x, y) = q_2(1 - \psi(x, y)) + k\frac{E_2}{h}(w_1 - h_1)\,\psi(x, y), \tag{13.51}$$

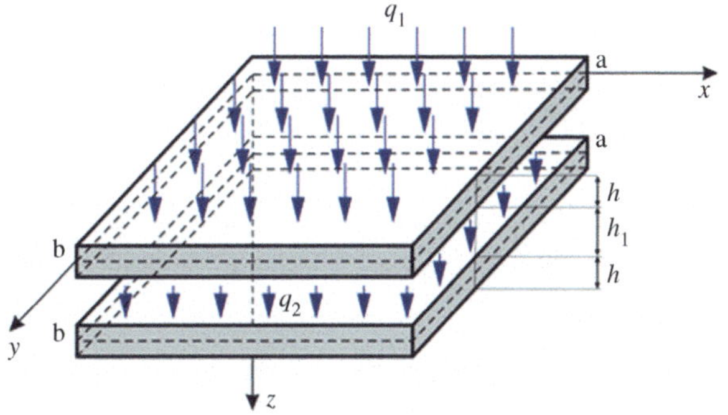

Fig. 13.2 Two plates with a clearance

with the boundary conditions (13.10). The function defining the contact zone Ω^* of the plates is as follows

$$\psi(x, y) = \begin{cases} 1, & (x, y) \in \Omega^*, \\ 0, & (x, y) \notin \Omega*, \end{cases} \tag{13.52}$$

and $q_1(x, y), q_2(x, y)$ are the functions of external load acting on the first and the second plate, respectively; the operators $A_i(w_i)$ have the form (13.5), which in the case of elastic problems is governed by (13.4); $\| \|_A$—norm in a normalized space A; $(,)$—scalar product in the Hilbert space L_2. The employed notation of the functional spaces coincides with that given in reference [215], whereas $E_1,\ E_2$ stand for different moduli in tension and compression.

In order to solve the problem (13.51), (13.10), the following iterational algorithm is employed:

$$A_1(w_1^{(n+1)}(x, y)) + k\frac{E_1}{h} w_1^{(n+1)} \psi(x, y) = q_1 + k\frac{E_1}{h}(w_2^{(n)} + h_1)\psi(x, y),$$

$$A_2(w_2^{(n+1)}(x, y)) + k\frac{E_2}{h} w_2^{(n)} \psi(x, y) = q_2(1 - \psi(x, y)) + k\frac{E_2}{h}(w_1^{(n)} - h_1)\psi(x, y), \tag{13.53}$$

$$w_1^{(n+1)}\Big|_{\partial\Omega_1} = \frac{\partial w_1^{(n+1)}}{\partial n_1}\Big|_{\partial\Omega_1} = 0, \qquad w_2^{(n+1)}\Big|_{\partial\Omega_2} = \frac{\partial w_2^{(n+1)}}{\partial n_2}\Big|_{\partial\Omega_2} = 0. \tag{13.54}$$

Then, the following theorem holds.

Theorem 13.3 *Let Ω_i, $(i = 1, 2)$ are bounded spaces and let their boundaries $\partial\Omega_i$ satisfy conditions of the Sobolev embedding theorems [215], where Ω^* stands for a measurable space, $q_i(x, y) \subseteq L_2(\Omega_i)$ and besides, there exist real constants $c_i > 0$,*

$D_i > 0$ *such that*

$$D_1 \parallel \Delta (w_i) \parallel^2_{L_2(\Omega^*)} \leq \left(B_{11,i} \frac{\partial^2 w_i}{\partial x^2} + B_{10,i} \frac{\partial^2 w_i}{\partial y^2}, \frac{\partial^2 w_i}{\partial x^2} \right)_{L_2(\Omega_2)} +$$

$$+ \left(B_{10,i} \frac{\partial^2 w_i}{\partial x^2} + B_{11,i} \frac{\partial^2 w_i}{\partial y^2}, \frac{\partial^2 w_i}{\partial y^2} \right)_{L_2(\Omega_2)} + \Bigg[\ (B_{11,i} +$$

$$+ B_{10,i}) \frac{\partial^2 w_i}{\partial x \partial y}, \frac{\partial^2 w_i}{\partial x \partial y} \Bigg]_{L_2(\Omega_2)} \leq c_i \parallel \Delta (w_i) \parallel^2_{L_2(\Omega_i)} .$$

Then:

(1) $\forall n, w_i^{(n)} \in W_2^4(\Omega_i) \bigcap \dot{W}_2^2(\Omega_i), i = 1, 2;$

(2) There exist functions $w_i^*(x, y) \in \dot{W}_2^2(\Omega_i)$, $i = 1, 2$, being solutions to the problems (13.14), (13.10) and $\lim_{n\to\infty} \left\| w_i^{(n)} - w_i^* \right\|_{W_2^2(\Omega_i)} = 0$.

Remark 13.1 The method of the variable stiffness parameters allows one to study SSS of the plates and plate-type constructions made from materials of different moduli, i.e. materials having different values of the elastic modulus and the Poisson's coefficients regarding tension and compression.

Remark 13.2 The given theorem implies the theorem reported in reference [206] valid for a material with one modulus.

13.4 Contact Interaction of Two Square Plates

In this section we consider bending of two squared plates made from materials having different moduli in tension and compression, taking the physical non-linearity into account (see Fig. 13.2).

If values of the material parameters E^+, E^-, ν^+, ν^- are slightly changed, the change in the elastic neutral layer (surfaces of the zeroth order) can be neglected [146, 358]. This observation has been taken into account in our further considered computational examples.

In order to investigate the stress-strain state of the plate-type structures made from materials of different moduli, one needs to use two different dependencies of the stress intensity on deformation. Let $\sigma_i^{(i)+}(e_i^{(i)+})$ and $\sigma_i^{(i)-}(e_i^{(i)-})$ be dependencies of the stress on deformation for the i-th plate ("+" refers to tension, whereas "−" refers to compression). Depending on a sign of the averaged stress of i-th plate $\sigma_v^{(i)} = \frac{1}{3}\left(\sigma_X^{(i)} + \sigma_Y^{(i)} + \sigma_Z^{(i)}\right)$, one may construct a diagram $\sigma_i^{(i)}(e_i^{(i)})$ corresponding to either tension or compression for each of the plates. In this case, the previously implemented algorithm, including the method of variable elasticity parameters, method of variational iterations, iterational method of variational iterations, and iterational method of contact interaction, is employed.

13.4.1 Computational Examples

We consider an example of a computation of the contact interaction between the square plates made from materials of different tension/compression moduli.

The plate material is a pure aluminium, and its non-linear diagram is governed by the following formula

$$\sigma_i = \sigma_s \left[1 - \exp\left(-\frac{e_i}{e_{is}} \right) \right], \tag{13.55}$$

where $e_{iS} = \frac{\sigma_S}{3\sigma_0}$ is a conventional intensity of plasticity deformation, and hence

$$\sigma = \sigma_1 \frac{e_{is}}{e_i} \left[1 - \exp\left(-\frac{e_i}{e_{is}} \right) \right]. \tag{13.56}$$

While approximating the $\sigma_i(e_i)$ dependence shown in Fig. 13.3, one may use the following formula

$$\sigma_i = \sigma_s \left[1 - \exp\left(-\frac{e_i}{e_{is}} \left\{ 1 + \frac{e_i}{e_{is}} \right\} \right) \right]. \tag{13.57}$$

The coefficient b can be identified assuming that the curve L goes through the point $(\sigma_i^{(*)},\ e_i^{(*)}) \in L$, and hence

$$b = -\frac{e_i}{e_{is}} \left[1 + \frac{e_{iS}}{e_i^*} \ln\left(1 - \frac{\sigma_i^*}{\sigma_s} \right) \right]. \tag{13.58}$$

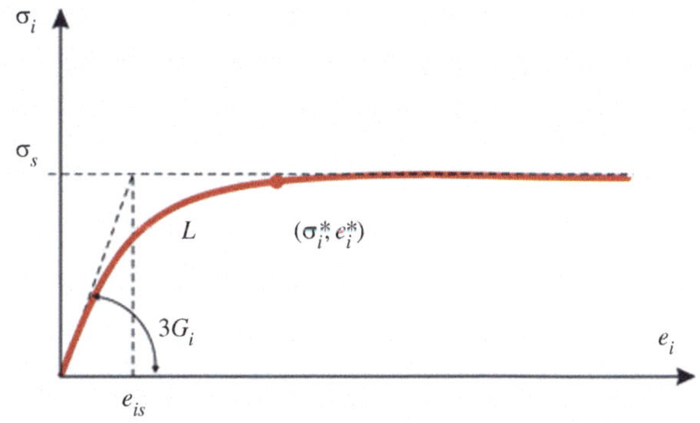

Fig. 13.3 The diagram $\sigma_i(e_i)$

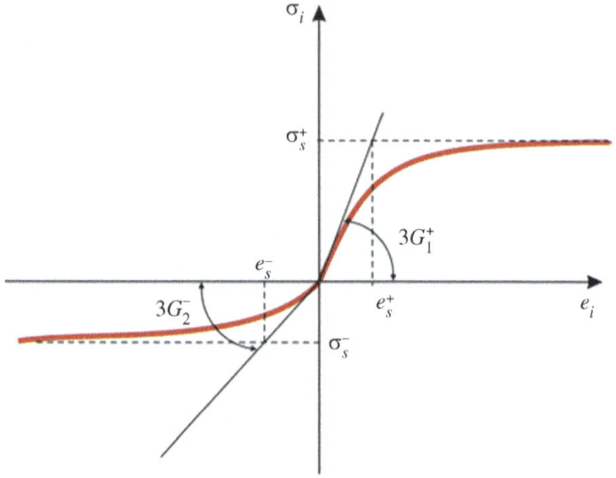

Fig. 13.4 Diagram $\sigma_i\left(e_i\right)$ for the material of different moduli and with physical non-linearity

This allows for more feasible tracing of the velocity of approximating σ_s by σ_i^* with an increase in e_i. In the case of a material with different moduli (Fig. 13.4) and with an account of $\sigma_i\left(e_i\right)$ in the case of physical non-linearity, we have

$$\sigma_i^{\pm} = \alpha\sigma_s^{\pm}\left[1 - \exp\left(-\frac{e_i^{\pm}}{e_{is}^{\pm}}\right)\right]. \tag{13.59}$$

The conventionally recognized notation $\pm$ for σ_i, e_i, e_{is} denote the values of the stress, deformation and plasticity deformation, respectively, for the case of tension (+) and compression (−). The coefficient α may have values larger or smaller than 1 in both cases of deformation, i.e. tension or compression.

As an example, we consider two square plates simply supported along their contours (see (13.10)) for $\lambda = \frac{a}{b} = 1$ and for their relative thickness $a/h = 12.5$. Each plate is made from aluminium and exhibits the stress plasticity threshold $\sigma_S = 1023 \cdot 10^5$ Pa, $G_1 = G_2 = 0.383 \cdot 10^{11}$ Pa (here stresses are measured with respect to $G_1(h/a)^2$), the non-dimensional deformations $\overline{e} = e_i(a/h)^2$, whereas the plasticity deformations are described by the formula $\overline{e_S} = e_{iS}(a/h)^2$. In reference [268], it has been shown that the properties of aluminium are well approximated by the dependencies $\sigma_i\left(e_i\right)$ through the formulas (13.55), (13.56) and they are validated experimentally. Owing to the data reported in this reference, we take $k = k_1 = 2G_1$, and the $\sigma_i^{(i)}(e_i^{(i)})$ function is governed by the following formula

$$\sigma_i^{(i)\pm} = \alpha^{\pm}\sigma_s^{(i)}\left[1 - \exp\left(-\frac{e_i^{(i)\pm}}{e_s^{(i)\pm}}\right)\right], \tag{13.60}$$

the algorithms developed by us for estimating the dependence $\sigma_i^{(i)}(e_i^{(i)})$ can be given in an arbitrary way, i.e. either in a tabular form (digits) or in an analytical way.

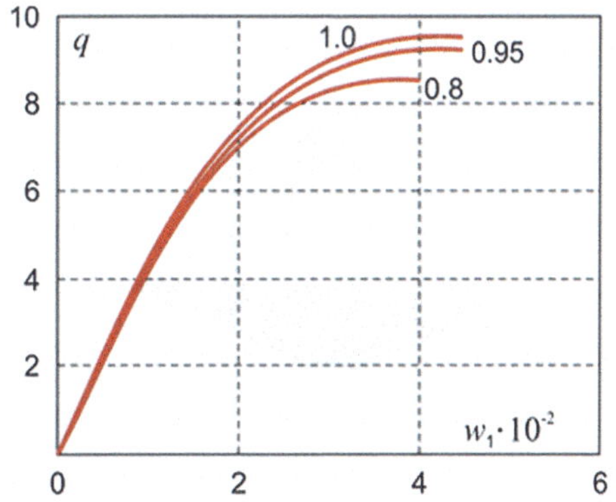

Fig. 13.5 The load -deflection dependence of the upper plate (both plates are made from material of different moduli)

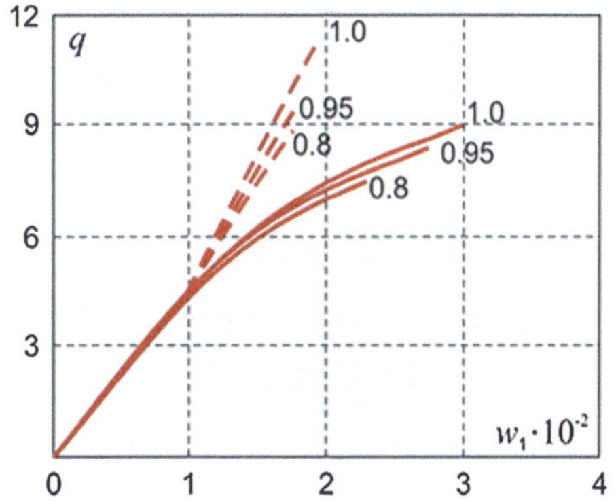

Fig. 13.6 The load -deflection dependence of the upper plate (only upper plate is made from material of different moduli)

The volume of each plate is covered by a mesh B_i with respect to x, y, z of the resolution $30 \times 30 \times 8$. In each point of two volumes, the following quantities have been computed σ_{av}, E_i, ν_i $(i = 1, 2)$ as a result of employment of the formulas reported in Sect. 13.1, and hence the iterational procedures have been constructed.

The load-deflection dependence in the centre plate of the upper plate $\overline{q}(w_1(0.5,\ 0.5))$ is reported in Fig. 13.5, where $\left(\bar{q}(x, y) = \frac{q(x,y)}{12(1-\nu_1{}^2)} \frac{a^4}{E_1 h^4}\right)$. The values of α^- are shown in the figure, whereas $\alpha^+ = 1.0$ for all investigated variants. Observe that an increase in the loading yields essential differences in SSS in comparison to decreasing α^-. The value of the threshold load decreases on amount of 2.1% with a decrease in σ_S on amount of 5%.

Now we consider the SSS of the contact two-layer construction simply supported along the contour (13.9). Here the upper plate is made from a material of different moduli and physically non-linear (solid curves), whereas the lower plate is made from one-modulus material (dashed curves). Relations $q(w_1(0.5,\ 0.5))$ are reported in Fig. 13.6.

The value of deflection, for which a contact occurred, is denoted by a dot ($w_1 = 0.01$). In the case of contact problems, the differences exhibited by the curves $q(w_1)$ are essentially less than for the one-layer plates for $\alpha^- = 0.8$; 0.95; 1.0. Analysis of the SSS of the plates implies that although the qualitative behaviour of

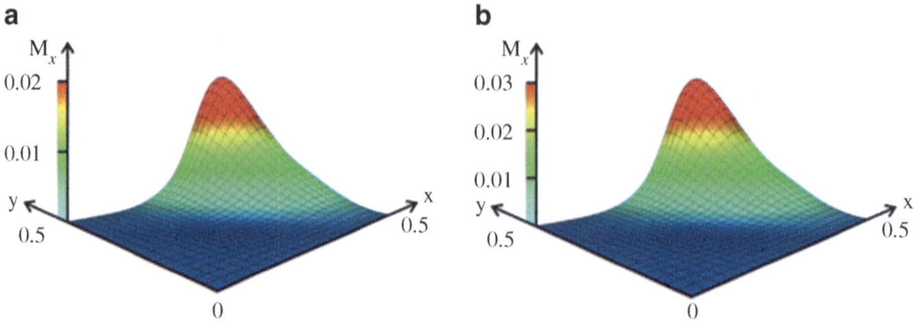

Fig. 13.7 Surfaces $M_x(x, y)$ for different (**a**) and the same (**b**) modulus corresponding to the upper and lower plate, respectively

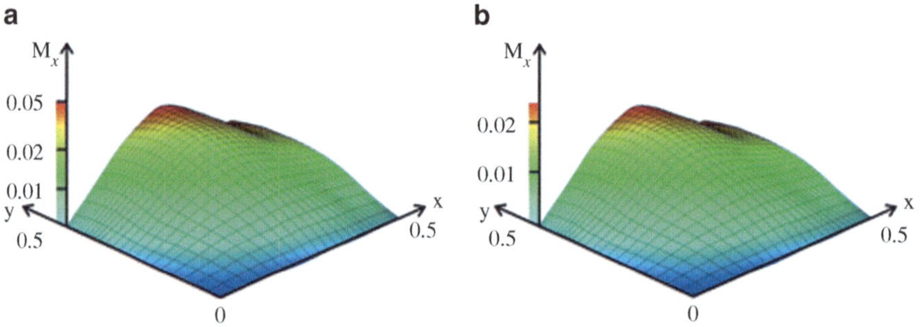

Fig. 13.8 Surfaces $M_x(x, y)$ for the upper plate with $\alpha^- = 0.8$ (**a**) and $\alpha^- = 1.0$ (**b**), respectively

the one-modulus ($\alpha^- = 1$) and different moduli ($\alpha^- \neq 1$) materials is the same, there is an essential quantitative difference.

The surface $\bar{M}_x(x, y) = -\frac{\partial}{\partial x}\left(\overline{D_1 \frac{\partial w}{\partial x}}\right)$ for different modulus $\alpha^- = 0.8$ of the first plate and one modulus of the second (lower) plate are shown in Fig. 13.7a and b, respectively. Figure 13.8a and b show the surfaces $\bar{M}_x(x, y)$ for the upper plate for the same parameters $\alpha^- = 0.8$, and 1.0, respectively. Analysis of the plates SSS implies that their qualitative behaviour is similar in both cases ($\alpha^- = 1$), and ($\alpha^- \neq 1$), but there is an essential quantitative difference.

13.5 Dynamics of a Contact Interaction

It should be emphasized that the phenomenon of contact interplay exhibited by mechanical systems composed of multi-layer plates coupled by boundary conditions has been reported for the first time in [56].

The further employed wavelet-based analysis allows one to study different kinds of vibration character in a given time interval. It allows also for detection of a birth/death of a frequency and time of its occurrence/vanishment. Despite there exist different types of wavelets [82, 220], the Morlet wavelet is the most suitable for our purpose. The analysis of advantages and disadvantages of different wavelet types to be employed in the study of shells can be found in reference [54].

A phase synchronization means phase locking of chaotic vibrational signals in time, although amplitudes of the mentioned signals remain uncoupled with each other and exhibit chaotic behaviour. Furthermore, phase locking implies frequency locking (resonance phenomenon). The wavelet surface $W(s, t_0) = \|W(s, t_0)\| \exp[j\varphi_S(t_0)]$ characterizes the system behaviour in each time interval s at an arbitrary time instant t_0. The quantity $\|W(s, t_0)\|$ describes the occurrence and intensity of the corresponding time interval s at the time instant t_0. The integral energy distribution of the wavelet spectrum with respect to time intervals $E(s) = \int \|W(s, t_0)\|^2 dt_0$ is introduced. The studied phase is defined as $\varphi_S(t_0) = \arg W(s, t)$ for each time interval s, i.e. one can characterize behaviour of each time interval s with the help of the associated phase $\varphi_s(t)$. In particular, we have employed the Morlet wavelet while investigating the synchronization phenomenon of the considered mechanical structures.

Chaotic phase synchronization plays a key role in modern theory of non-linear vibrations. The occurrence of the chaotic phase synchronization has been experimentally validated in radio-frequency generators, electromechanical oscillators, lasers, heart beating, and many others. It has also a significant influence on robustness of information transmission with the help of deterministic chaotic vibration.

Equations governing the dynamics of two-layer package of physically linear elastic plates have the following form (see Fig. 13.2)

$$\begin{aligned}
&\left[\frac{E_1^+ \int_{a_i(x,y)}^{c_i(x,y)} z^2 dz}{1-\left(\nu_i^+\right)^2} + \frac{E_1^- \int_{c_i(x,y)}^{b_i(x,y)} z^2 dz}{1-\left(\nu_i^-\right)^2}\right] \nabla_\lambda{}^4 w_1 + \\
&\quad + \frac{\gamma}{g} h \frac{\partial^2 w_1}{\partial t^2} + \varepsilon \frac{\gamma}{g} h \frac{\partial w_1}{\partial t} - \frac{k}{h} E_1^+ (w_1 - w_2 - h*)\psi = q(t), \\
&\left[\frac{E_2^+ \int_{a_i(x,y)}^{c_i(x,y)} z^2 dz}{1-\left(\nu_i^+\right)^2} + \frac{E_2^- \int_{c_i(x,y)}^{b_i(x,y)} z^2 dz}{1-\left(\nu_i^-\right)^2}\right] \nabla_\lambda{}^4 w_2 + \\
&\quad + \frac{\gamma}{g} h \frac{\partial^2 w_2}{\partial t^2} + \varepsilon \frac{\gamma}{g} h \frac{\partial w_2}{\partial t} + \frac{k}{h} E_2^- (w_1 - w_2 - h*)\psi = 0.
\end{aligned} \tag{13.61}$$

In a particular case of one-modulus ($E_1^{+/-} = E_2^{+/-} = E$ and $\nu_1^{+/-} = \nu_2^{+/-} = \nu$) physically linear material ($E = const, = const$), PDEs (13.61) can be reduced to the counterpart non-dimensional form as follows

$$\begin{cases} \frac{1}{12(1-\nu^2)}\left(\nabla_\lambda{}^4 w_1\right) + \frac{\partial^2 w_1}{\partial t^2} + \varepsilon\frac{\partial w_1}{\partial t} - K(w_1 - w_2 - h*)\Psi = q(t), \\ \frac{1}{12(1-\nu^2)}\left(\nabla_\lambda{}^4 w_2\right) + \frac{\partial^2 w_2}{\partial t^2} + \varepsilon\frac{\partial w_2}{\partial t} + K(w_1 - w_2 - h*)\Psi = 0. \end{cases} \tag{13.62}$$

System of equations (13.62) has been obtained by the following introduced relations between dimensional and non-dimensional parameters: $x = a\overline{x}$, $y = a\bar{y}$; $q = \overline{q}\frac{E(h)^4}{a^2b^2}$, $\tau = \frac{ab}{h}\sqrt{\frac{\gamma}{Eg}}$, $\lambda = \frac{a}{b}$, where: a, b—plate dimension with respect to x and y, respectively; t—time; ε—damping coefficient; w—deflection; h—plate thickness; $\nu = 0.3$—Poisson's coefficient; g—the gravity of Earth; E—Young's modulus; $q(x, y, t)$—transverse load; γ—specific material weight. Note that bars over non-dimensional quantities have been omitted in (13.62) for simplicity reasons.

As boundary conditions, simple support along a contour is taken:

$$\begin{aligned} &w_m = 0, w''_m|_x = 0, \quad for \quad x = 0; 1, \\ &w_m = 0, w''_m|_y = 0, \quad \text{for} \quad y = 0; 1, \quad m = 1, 2. \end{aligned} \tag{13.63}$$

The initial conditions are as follows

$$w_m(x, y)|_{t=0} = 0, \dot{w}_m|_{t=0} = 0, \quad m = 1, 2. \tag{13.64}$$

We take the control function $\psi = \frac{1}{2}[1 + sign\,(w_1 - h * - w_2)]$; $\Psi = 1$, if $\mathrm{w}_1 > w_2 + h_k$, i.e. there is contact between plates; otherwise $\Psi = 0$; w_1, w_2—function of deflection of upper and lower plates, respectively; K—stiffness coefficient of transverse clamping in the contact zone; h^*—clearance between plates, and $q(t) = q_0 \sin(\omega_0 t)$.

A solution to system (13.60)–(13.62) is being searched by the Faedo-Galerkin method, and the problem is reduced to investigating non-linear ODEs (the Cauchy problem). The function w_i, $i = 1, 2$ is approximated by a function being a product of two functions (one depending on time and the other depending on coordinates) of the following form

$$w_m = \sum_{i=1}^{N}\sum_{j=1}^{N} A^{(m)}{}_{ij}(t)\sin(i\pi x)\sin(j\pi y), \quad m = 1, 2. \tag{13.65}$$

The derived Cauchy problem has been solved using the 4th-order Runge-Kutta method.

As an example, let us consider vibration of the plates under the sinusoidal and uniformly distributed load for the following fixed parameters: $q_0 = 2$, $K = 5000$, $h^* = 0.01$, $\varepsilon = 0.5$, $\omega_0 = 5.0887$ (the last one stands for the fundamental frequency of linear plate vibration).

It has been shown that to verify convergence of the numerical results obtained by the Faedo-Galerkin method for $N = 1, 3, 5$, it is sufficient to take only three first terms of the series in formula (13.65). It has been found that once a contact between two plates takes place, vibration become chaotic. The character of plates vibration has been investigated with the help of qualitative methods of non-linear dynamical systems, i.e. we have employed Fourier and wavelet based analysis, studied contact pressure between plates, and monitored the phase difference of plate vibration.

Graphs of simultaneous vibration of the centres of both plates $w(t)$, their contact pressure $Pq(t)$, time history of phases difference $\Delta\varphi_s(t)$ as well as the Fourier power spectrum $F(\omega)$ and 2D/3D wavelet-spectrum for each of the plate, respectively, are presented in Table 13.2.

Time histories of plate deflections (a) and time histories of contact pressures (b) exhibit complex chaotic interaction of the plates. A dark colour used in the plots showing the phase difference (c) exhibits the frequencies zone $\omega \in [4, 8]$ where the phase synchronization has been detected. Vibrations exhibit clearly the frequency ω_0, i.e. the vibrations take place on the fundamental frequency of free vibrations of the two-layer structural package.

This is validated by the Fourier spectrum (d, g) and the Morlet spectrum (e, h, f, i). The latter exhibits energetic component of each of the frequencies at a given time instant.

Remark 13.3 It should be mentioned that the so far solved dynamical problems and used methodology can be successfully employed also to solve static counterpart problems. The method is known as the relaxation method and has been used by us earlier to solve different non-linear problems [35]. From the mathematical point of view, the relaxation method belongs to a class of iterational methods devoted to solve a system of algebraic non-linear equations where each time step creates a new approximation to a being searched solution. This method is of high accuracy, and it is robust to the choice of the initial approximation. This observation is motivated by physical correspondence of Eqs. (13.50) to vibrations of plates embedded into a viscous environment.

Remark 13.4 In the case of the studied non-linear dynamical problems, after transition from PDEs to ODEs, one can (on each time interval) improve the system of approximating functions $w_n(x, y)$, where a solution to the dynamical problem can be presented in the following form

$$w(x, y, t) = A(t)w_n(x, y), \tag{13.66}$$

by using one of the earlier described method. This approach can be used particularly to the problems where the boundary conditions are changed due to vibrations.

Table 13.2 Characteristics of the plate contact interactions (see text for more details)

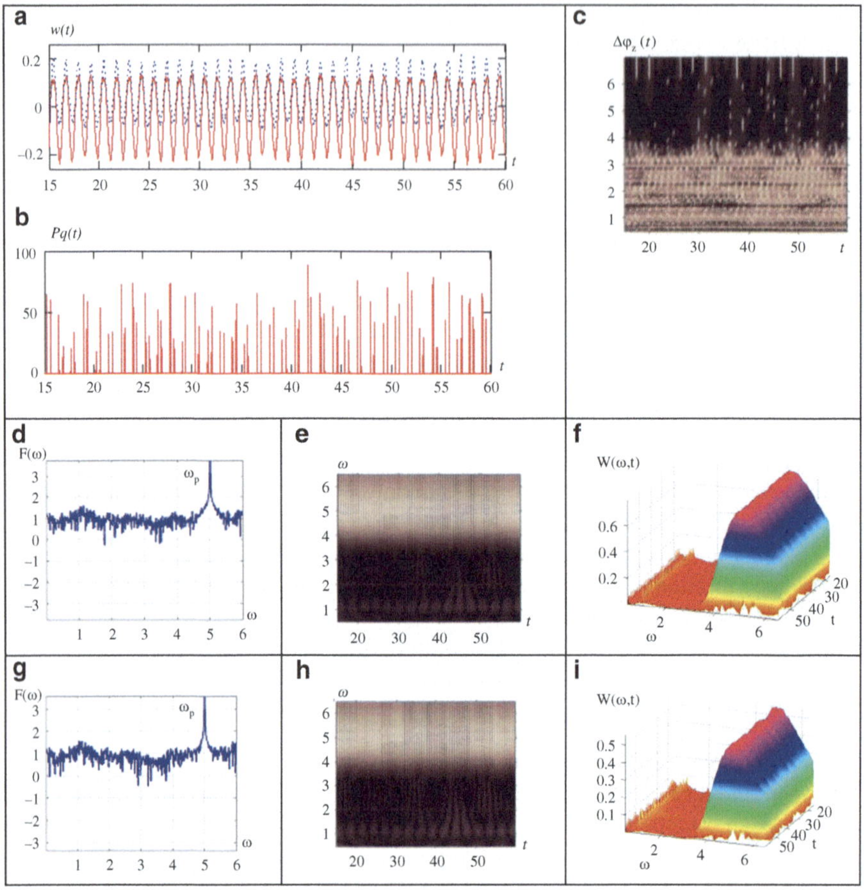

Chapter 14
Chaotic Vibrations of Flexible Shallow Axially Symmetric Shells vs. Different Boundary Conditions

In the present chapter, novel results of the study of chaotic vibration of flexible circular axially symmetric shallow shells subjected to sinusoidal transverse load are presented for four different boundary conditions are presented. The study is conducted with the use of the finite difference method (FDM), which differentiates the study from the majority of research conducted with the finite element method (FEM).

Chaos reliability is verified and validated by solving the implemented mathematical model represented by partial nonlinear equations (PDEs) governing the dynamics of flexible spherical shells and by estimating the signs of the largest Lyapunov exponents (LLEs) with the help of qualitatively different approaches. It is shown how the scenario of transition of the investigated shells from regular to chaotic vibrations depends on the boundary conditions. The following cases are considered: (1) movable and fixed simple supports along the shell contours, taking into account shell stiffness (Feigenbaum scenario) and shell damping (Ruelle-Takens-Newhouse scenario), and (2) movable clamping (regular shell vibrations).

It is detected, illustrated, and discussed that the employed boundary conditions significantly influence the obtained solutions. For simple nonmovable and movable supports of the shell, for $\omega_p = \omega_0$ (ω_p stands for the harmonic excitation frequency, and ω_0 refers to the shell fundamental frequency) a transition from regular to chaotic shell vibration is realized following the Feghenbaum scenario, whereas for a rigidly clamped shell, a transition to chaos follows the Ruelle-Takens-Newhouse scenario. Finally, in the case of the movable clamped shell, a series of Sharkovsky's windows is reported.

It is found that a number of ribs exhibited by the shell surface strongly depends on the boundary conditions. The maximum number of ribs is detected for the shell with movable simple support, one rib is found for the shell with nonmovable simple support and the shell with movable clamping, whereas three ribs are detected in the case of rigid clamping.

J. Awrejcewicz, V. A. Krysko, *Elastic and Thermoelastic Problems in Nonlinear Dynamics of Structural Members*, Scientific Computation,
https://doi.org/10.1007/978-3-030-37663-5_14

This chapter is based on the reference [208], where one may find also a transition from dimensional to nondimensional PDEs (here omitted).

14.1 Problem Statement and the Method of Solution

Let a spherical shallow shell be considered in the polar system of coordinates introduced in the following way: $\Omega = \{ (r, z) \mid r \in [0, b], -h/2 \leq z \leq h/2 \}$ (see Fig. 14.1a). Owing to the earlier work of Novozhilov [17], a shell is called thin if $h/R << 1$ (see Fig. 14.1b), and the inequality $h/R \leq 1/40$ should be employed in such a case, as suggested by Novozhilov. If this inequality is not satisfied, then the shell is called thick. However, the majority of practically applied shells satisfy the inequalities $1/2000 \leq h/R \leq 1/50$, i.e., they belong to the class of thin shells. On the other hand, the threshold used to identify shallow shells is defined by the Reissner [283] ratio $f/2c << 1/8$, where f is the distance between the shell vertex and a plane f and the distance $2c$ is measured in plane (see Fig. 14.1b). Vlasov [335] suggested $f/2c \leq 1/5$. Employing the Reissner criterion for shallowness ($\alpha < \alpha^* = 60°$), one gets $2c \leq R$.

The system of nonlinear PDEs governing dynamics of the axially symmetric shells has the following form [342]:

$$\frac{\partial^2 w}{\partial t^2} + \varepsilon \frac{\partial w}{\partial t} = -\frac{\partial^4 w}{\partial r^4} - \frac{2}{r}\frac{\partial^3 w}{\partial r^3} + \frac{1}{r^2}\frac{\partial^2 w}{\partial r^2} -$$

$$-\frac{1}{r^3}\frac{\partial w}{\partial r} - \frac{-}{r}\left(1 - \frac{\partial^2 w}{\partial r^2}\right) - \frac{\partial}{\partial r}\left(1 - \frac{1}{r}\frac{\partial w}{\partial r}\right) + 4q, \tag{14.1}$$

$$\frac{\partial^2}{\partial r^2} + \frac{1}{r}\frac{\partial}{\partial r} - \frac{1}{r^2}\frac{\partial}{\partial r} = \frac{\partial w}{\partial r}\left(1 - \frac{1}{2r}\frac{\partial w}{\partial r}\right), \quad = \frac{\partial F}{\partial r}.$$

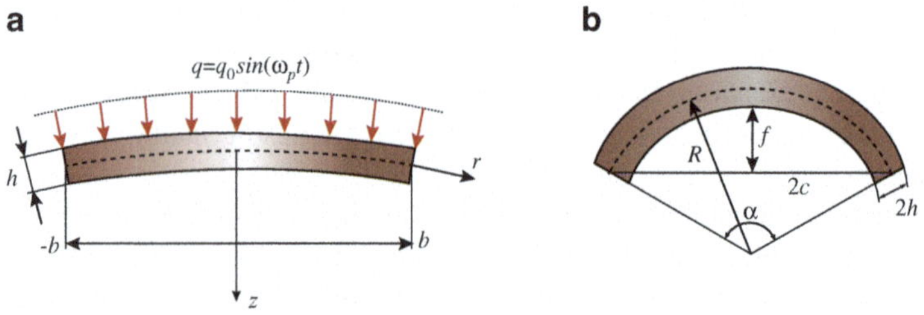

Fig. 14.1 The studied axially symmetric shallow shell with the uniformly distributed periodic load (**a**) and the scheme of a thin shell (**b**)

The below nondimensional quantities (with bars) are introduced:

$$\bar{t} = \omega_0 t, \quad \omega_0 = \sqrt{Eg/\gamma R^2}, \quad b = \left(c/\sqrt{Rh}\right), \quad \bar{\varepsilon} = \varepsilon R/h\,\sqrt{g/(\gamma E)},$$

$$\bar{F} = \eta F/Eh^3, \quad \bar{w} = w\sqrt{\eta}/h, \quad \bar{r} = br/c, \quad \eta = 12(1-\nu^2),$$

$$\bar{q} = \bar{q}_3 = (R/h)^2 q_3\sqrt{\eta}/(4E),$$

where: t [s]—time; ε [1/s]—coefficient of viscous-type external damping; F [Nm]—stress function; w[m]—deflection function; R [m], c [m]—main radius of the shell curvature and the radius of the shell contour, respectively; h [m]—shell thickness; b—shell parameter; ν—Poisson's coefficient; r [m]—distance between the rotation axis and the point of the shell middle surface; q [N/m^2]—amplitude of the external load; ω_0 [rad/s]—natural frequency of the small shell vibrations; E [N/m^2]—elasticity modulus, g [m/s^2]—gravity constant, γ [N/m^3]—unit weight (henceforward, bars over the nondimensional quantities are omitted). Dimensional PDEs can be found by means of substituting the above-mentioned nondimensional quantities to PDE (14.1) (see also [15, 208, 291]). In the present investigations, we are aimed at studying the boundary conditions and two excitation parameters q_0 and ω_p of the sinusoidally distributed $q = q_0 \sin \omega_p t$.

The following boundary conditions are considered:

(1) simple contour movable in the meridional direction

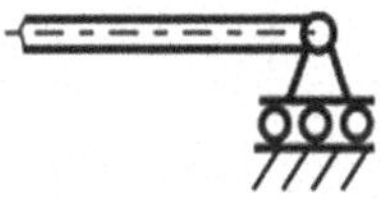

$$\Phi = w = 0, \quad \frac{\partial^2 w}{\partial r^2} + \nu\frac{\partial w}{\partial r} = 0, \quad \text{for} \quad r = b; \tag{14.2}$$

(2) rigidly clamped contour

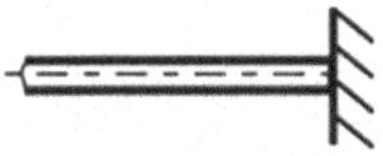

$$\frac{\partial \Phi}{\partial r} - \nu\frac{\Phi}{b} = 0, \; w = 0, \; \frac{\partial w}{\partial r} = 0, \quad \text{for } r = b; \tag{14.3}$$

(3) sliding clamping of the contour

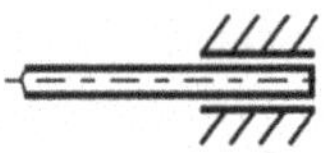

$$\Phi = w = 0, \quad \frac{\partial w}{\partial r} = 0, \quad \text{for} \quad r = b; \tag{14.4}$$

(4) simple nonmovable contour

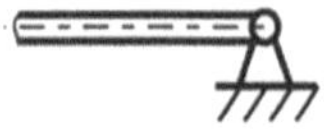

$$\frac{\partial \Phi}{\partial r} - \nu \frac{\Phi}{b} = 0, \quad w = 0, \quad \frac{\partial^2 w}{\partial r^2} + \frac{\nu}{r} \frac{\partial w}{\partial r} = 0, \quad \text{for} \quad r = b; \tag{14.5}$$

Moreover, the following initial conditions $\left(' = \frac{d}{dt} \right)$

$$w = 0, \quad w' = 0, \tag{14.6}$$

as well as the below conditions in the vicinity of the shallow top are employed

$$_{l} \approx Ar, \quad ' \approx A, \quad w \approx B + Cr^2, \quad w' \approx 2Cr, \quad w'' \approx 2C, \quad w''' \approx 0. \tag{14.7}$$

In order to reduce the problem (14.1)–(14.7) governing dynamics of the considered continuous system to a system with lumped parameters, the finite difference method (FDM) with approximation $O\left(\Delta^2\right)$ is used. PDEs as well as boundary and initial conditions (14.1)–(14.7) are recast to the following finite difference formulas with respect to the spatial coordinate r and time:

$$\begin{gathered}
\ddot{w}^i + \varepsilon \dot{w}^i = -\frac{w^{i+1} - w^{i-1}}{2\Delta} \left(\frac{1}{r_i^3} - \frac{\Phi^{i+1} - \Phi^{i-1}}{2 r_i \Delta} \right) + \\
+ \frac{w^{i+1} - 2w^i + w^{i-1}}{r_i \Delta^2} \left(\Phi^i + \frac{1}{r_i} \right) - \frac{\Phi^{i+1} - \Phi^{i-1}}{2\Delta} - \frac{\Phi^i}{r_i} - \\
- \frac{w^{i+2} - 4w^{i+1} + 6w^i - 4w^{i-1} + w^{i-2}}{\Delta^4} - \\
- \frac{w^{i+2} - 2w^{i+1} + 2w^{i-1} - w^{i-2}}{2 r_i \Delta^3} + 4q^i, \\
\Phi^{i+1} \left(-\frac{1}{\Delta^2} - \frac{1}{2 r_i \Delta} \right) + \Phi^i \left(\frac{2}{\Delta^2} + \frac{1}{r_i^2} \right) + \Phi^{i-1} \left(-\frac{1}{\Delta^2} + \frac{1}{2 r_i \Delta} \right) = \\
= -\frac{w^{i+1} - w^{i-1}}{2\Delta} \left(1 - \frac{w^{i+1} - w^{i-1}}{4 r_i \Delta} \right),
\end{gathered} \tag{14.8}$$

where: $\Delta = b/n$ and n denotes the number of nodes of the shell radius.

The finite difference method (FDM) is employed only to spatial coordinates. It reduces the problem governed by PDEs (14.1) of an infinite dimension to the system of ODEs (14.8) with respect to time (the Cauchy problem). The Cauchy

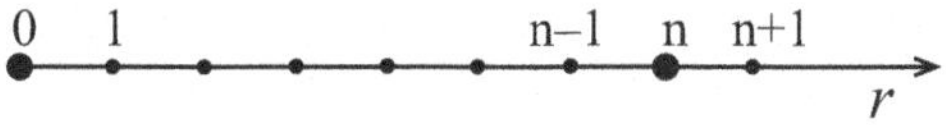

Fig. 14.2 Location of nodes

problem is solved using a few methods of the Runge-Kutta types, i.e. the 4th-order Runge-Kutta method (RK4), the 2nd-order Runge-Kutta method (RK2), the 4th-order Runge-Kutta-Fehlberg method (rkf45), the 4th-order Cash-Karp method (RKCK), the 8th-order Runge-Kutta-Prince-Dormand method (rk8pd), the 2nd-order and the 4th-order implicit Runge-Kutta method (rk2imp and rk4imp). After conducting numerous case studies, we have finally chosen the 4th-order Runge-Kutta method. The reason is that the accuracy of the this method is the same as of the Runge-Kutta methods of higher orders, but the computational time is shorter. Location of nodes along the shell radius r is shown in Fig. 14.2.

The counterpart difference forms of the boundary conditions are as follows:

1. Simple contour movable in the meridional direction

$$\Phi^n = 0, \quad w^{n+1} = \frac{\nu\,\Delta - 2b}{2b + \nu\,\Delta} w^{n-1}, \quad w^n = 0 \quad \text{for} \quad r^n = b; \qquad (14.9)$$

2. Rigidly clamped contour

$$\Phi^{n+1} = \Phi^{n-1} + \frac{2\Delta\nu}{b}\Phi^n, \quad w^{n+1} = \frac{\nu\,\Delta - 2b}{2b + \nu\,\Delta} w^{n-1}, \quad w^n = 0 \quad \text{for} \quad r^n = b; \qquad (14.10)$$

3. Sliding clamping of the shell contour

$$\Phi^n = 0, \quad w^{n+1} = w^{n-1}, \quad w^n = 0 \quad \text{for} \quad r^n = b; \qquad (14.11)$$

4. Simple nonmovable contour

$$\Phi^{n+1} = \Phi^{n-1} + \frac{2\Delta\nu}{b}\Phi^n, \quad w^{n+1} = w^{n-1}, \quad w^n = 0 \quad \text{for} \quad r^n = b; \qquad (14.12)$$

and the below initial conditions are taken

$$w^k = 0, \quad w'^k = 0, \quad 0 \le k \le m, \quad 0 \le t < \infty. \qquad (14.13)$$

If small terms are neglected and differential operators are substituted by the central finite differences for $r = \Delta$, the following conditions are obtained at the top of the shell:

$$\Phi^0 = \Phi^2 - 2\Phi^1, \quad w^0 = \frac{4}{3}w^1 - \frac{1}{3}w^2, \quad w^{-1} = \frac{8}{3}w^1 - \frac{8}{3}w^2 + w^3. \qquad (14.14)$$

The FDM (the finite difference method) is an alternative to the Faedo-Galerkin, the Ritz and the finite element methods. Let us cite a few recently published papers, in which the FDM has been employed to study dynamics of structural members.

Medina et al. [239] studied axially symmetric nonlinear behavior of electrostatically actuated and initially curved bistable microplates. Chen et al. [83] considered post-buckling of size-dependent microplate, focusing on damage effects due to the transverse load. Mehditabar et al. [238] carried out the thermoelastic study of a functionally graded piezoelectric rotating hollow cylindrical shell subjected to dynamic loads.

Aranda-Iglesias et al. [11] also investigated spherical shells. The solution was found using both the finite element method and the finite difference method. The same results were obtained for both methods, which validates the reliability of the FDM.

The transverse load can be changed arbitrarily with respect to the spatial coordinate and time. In this work, the harmonic transverse load of the form $q = q_0 \sin(\omega_p t)$, where q_0 stands for an amplitude and ω_p is a frequency of the excitation, is used.

14.2 Quantification of True Chaotic Vibrations

The problem related to reliability of chaotic vibration is extremely important, since in many cases errors introduced by numerical computations can be interpreted as chaotic behavior, what has been illustrated and discussed in reference [231]. It is known that one of the fundamental properties of chaotic dynamics is associated with its essential dependence on the initial conditions. On the other hand, more rigorous definition of chaos, given by Duavaney in 1989 [99], aside from the already mentioned property, emphasizes the importance of two other characteristics of chaos, i.e. intermittency/transitivity and regularity/periodicity conditions manifested by the density of periodic points. In 1992, Banks et al. [60] proved that the transitivity and periodicity conditions imply sensitivity to the initial conditions. On the other hand, Knudsen [183] defined chaos with the help of a function given on the bounded metric space, which is understood as a chaotic one if it exhibits a dense orbit and presents an essential dependence on the initial conditions.

Owing to Gulick results [138], chaos exists if sensitivity to the initial conditions is exhibited or a positive LE (Lyapunov exponent) is identified in each point of the space on which the chaotic function is defined, and the latter does not tend to periodicity. In this work, we employ the Gulick definition.

To solve the problem of reliability of chaotic vibrations while numerically solving nonlinear PDEs governing dynamics of spherical axially symmetric shells subjected to harmonic transverse load, the following steps are carried out:

1. Since PDEs are reduced to a Cauchy problem by means of the second-order FDM, the solution strongly depends on the number of nodes involved in the shell radius $r \in [0; b]$ partition. In the present paper, nodes in which the solutions for n and $n + 1$ nodes coincide. In particular, convergence of the employed FDM is analyzed for different shallowness parameters and boundary conditions.
2. Since the Cauchy problem is also solved numerically, the solution essentially depends on both the used method and the employed time step. Therefore, the following methods are used: 4th- and 2nd-order Runge-Kutta (RK4 and RK2) methods, the 4th-order Runge-Kutta-Fehlberg (rkf45) method, the 4th-order Cash-Karp (RKCK) method, combination of the 8th-order Runge-Kutta and Dormand method (rk8pd) as well as 2nd- (rk2imp) and the 4th-order (rk4imp) implicit Runge-Kutta methods.
3. The numerically obtained data are used to construct the time series (signals), phase portraits, Fourier power spectra (FFT), Morlet wavelets, and shell deflection surfaces. The following wavelets are used: Meer wavelets, Shennon-Kotelnikov and Meier wavelets, Daubechies wavelets from db2 to db4, coiflets, Morlet wavelets, complex Morlet wavelets as well as the wavelets based on a derivative of the Gauss function of higher than the 8th order. It should be emphasized that it was not feasible to investigate the considered problem with Meer and Shennon-Kotelnikov wavelets, i.e., the first one did not identify the frequency zones properly, whereas the second one did not yield the expected localization in time. The carried out analysis of the wavelet spectra obtained by the Daubechies wavelets, coiflets, and simlets has shown that an increase in the order of the employed filter improves proper frequency identification. Furthermore, in spite of differences between the employed wavelets and their filters, the wavelet spectra yielded by the Daubechies wavelets, simlets, and coiflets of the same order give almost the same results. However, the frequency localization yielded by them is not accurate enough, what makes them unsuitable for analysis of vibration of the investigated mechanical system.

   Considering the results obtained by means of derivatives of the Gauss functions, the accuracy of the frequency localizations increased with the increase in the order of the derivative. It should be noted that the frequency spectra obtained by the Meer wavelet, which is a smooth variant of the Shennon-Kotelnikov wavelet, has been localized better than in the case of Morlet wavelets only for a low frequencies band. In what follows we give results for Morlet wavelets that were specially chosen and adopted to study the stated problem.

In these wavelets, there is no scale function φ, and ψ is given explicitly, i.e., there is no compact carrier. Therefore, to study complex vibration of continuous mechanical system, like a plate/shell, one can employ either complex or real Morlet wavelets or wavelets based on either real or complex derivatives of the Gauss functions of the order higher than 16.

4. Since the definition of chaos given by Gulick [138] is used in this work, one needs to compute Lyapunov exponents (LEs) and define their signs. In this paper, a spectrum of the LEs for each partition of the interval $[0; b]$ was obtained by three methods: Wolf's [351], Rosenstein's [285], and Kantz's [164]. Furthermore, convergence of the above-mentioned methods was also investigated.

The final solution was accepted and treated as reliable/true if coincidence up to four digits after the decimal point was achieved. According to the already given steps of investigation, convergence of the solution depending on the number n of partitions of the shell radius was considered at first. The study was carried out for all types of boundary conditions: movable simple support (14.9), rigid clamping (14.10), movable clamping (14.11), and nonmovable simple support (14.12) for $b \in [3; 8]$. Then, results for $n = 10, 20, 40, 80, 100, 120, 140$ were compared.

After reducing the problem (14.1)–(14.5) to its counterpart normal form, the first equation of the Cauchy problem was solved by the Runge-Kutta methods with respect to the deflection function w. At the same time, algebraic equations were solved (on each time step) with respect to the stress function $\Phi(r, t)$ by the method of inverse function. The time step was yielded by the Runge principle. The computational results in the form of frequency power spectra, wavelet spectra, and the surfaces of shell deflection for fixed time instances $t = 300$, $t = 304$, $t = 305$ are reported in Table 14.3. The power spectrum obtained for $n = 10$ yields chaos spanned on five frequencies. In addition, the associated wavelet spectrum exhibits intermittency for $n = 10$, whereas for $n = 20$, chaos spanned on ω_p and $\omega_p/2$ occurs (for $n = 40$, one can detect ω_p, $\omega_p/2$,. $\omega_p/4$ and $3\omega_p/4$). Since the latter behavior is exhibited by both FFT and wavelet spectrum, where intermittency does not appear, the number of partitions of the radius plays a crucial role in identifying the scenario of transition from periodic to chaotic vibrations.

Furthermore, convergence of the deflection surfaces was investigated at the same time instances. As the reported graphs show, the surfaces coincide only for $n = 120, 140$. To make the results more clear, each of the 3D surfaces is supplemented with the deflection isoclines. Figure 14.3a presents the time histories/signals in relation to the number n of partitions of the shell radius (Fig. 14.3b reports a window of the signal in the interval $t \in [290; 330]$). Remarkably, for $n = 120$ and $n = 140$, the graphs coincide.

Integration with respect to time was also carried out by both explicit and implicit Runge-Kutta methods of different orders, as it is shown in Table 14.1. it is worth mentioning that triangular or arbitrary matrices are used in explicit and implicit methods, respectively.

The methods RKF45, RKCK, RK8pd allow for an automatic change of the integration steps as well as control of the integration errors. The carried out inves-

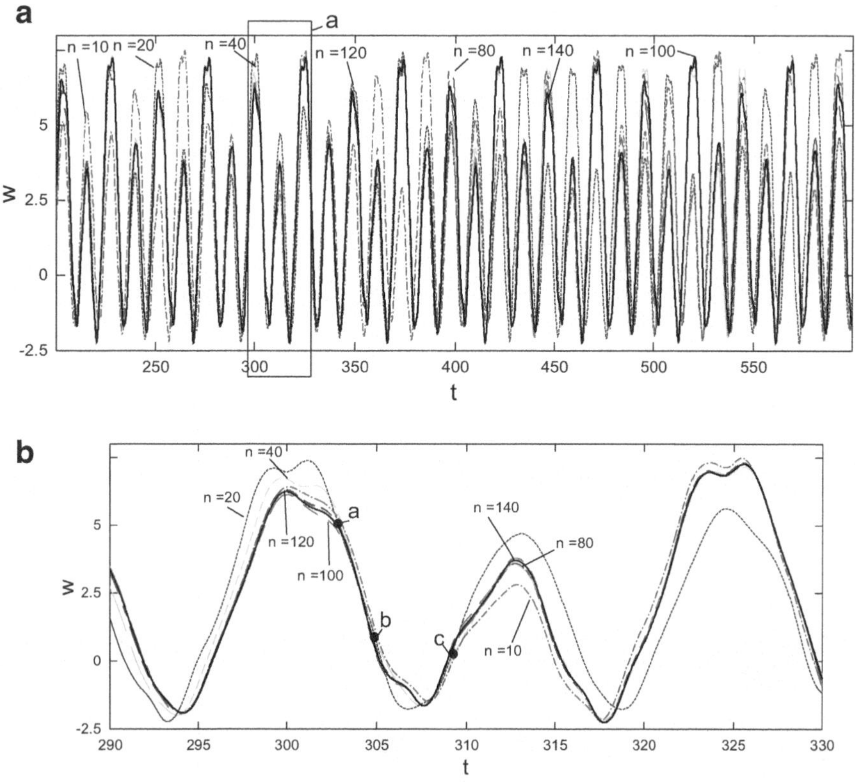

Fig. 14.3 Convergence of FDM regarding time history $w(t)$ versus the number of partitions of the radius n (**a**) and the window "a" for $t \in [290; 330]$ (**b**)

Table 14.1 The methods used in numerical simulations

Explicit methods		Implicit methods
Variable mesh	Constant mesh	
RKF45, RKCK, RK8pd	RK2, RK4	RK2inf, RK4inf

tigations showed that the 4th-order Runge-Kutta method is optimal to investigate chaotic vibration of the considered flexible spherical shell subjected to harmonic load.

To discuss how the scenario to chaos will be changed while a number of nodes will be increased in the FDM, let us begin with the results reported in Table 14.2.

In this table the frequency power spectra are reported for the vibration signals obtained with the FDM for different nodes numbers $n = 20; 40; 80; 120$. For $q_0 = 0.9$, all frequency spectra exhibit ω_p and $\omega_p/2$. Amplitudes of these harmonics are the same for all employed n. For $q_0 = 0.1335$, the regular bifurcations are exhibited on the frequencies ω_p, $\omega_p/2$, $\omega_p/4$ also for all $n = 20; 40; 80; 120$ nodes, but the values of harmonic amplitudes associated with $\omega_p/4$ are different

Table 14.2 Frequency power spectra for different n and different q_0

n=20	n=40	n=80	n=120, 140; 160
q_0=0.09			
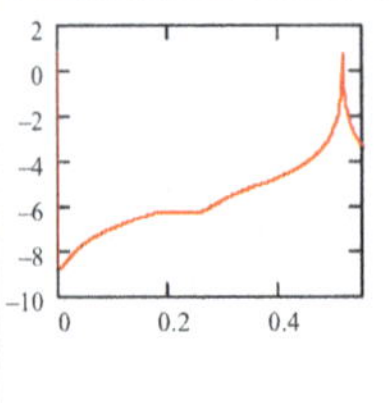	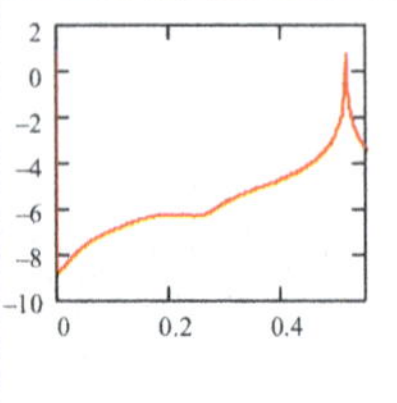	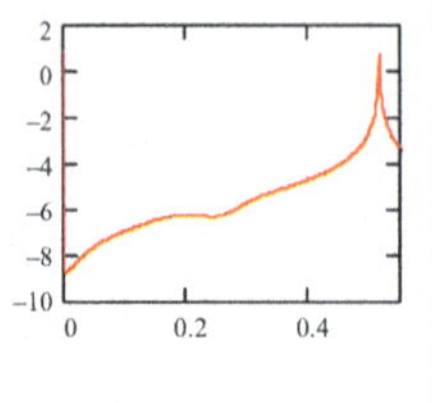	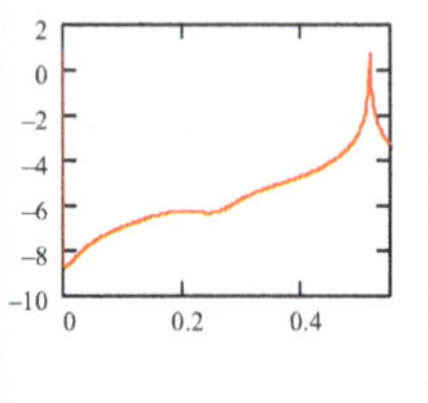
q_0 =0.1335			
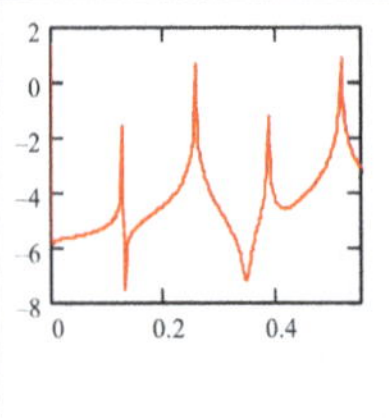	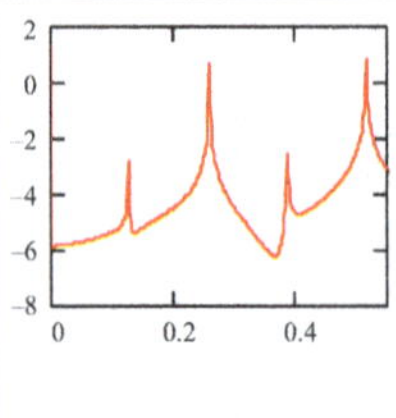	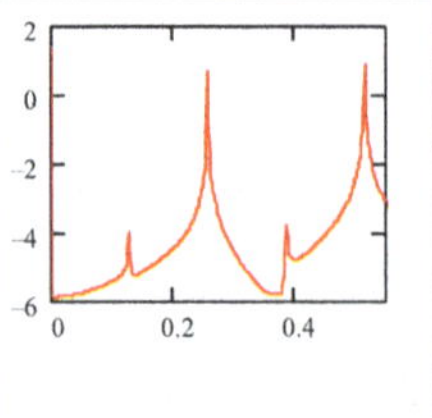	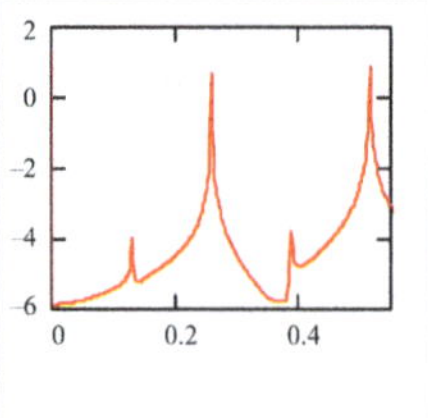
q_0=0.1395			
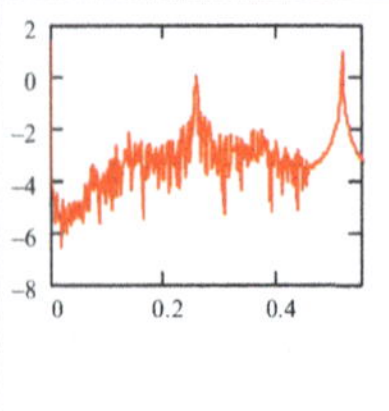	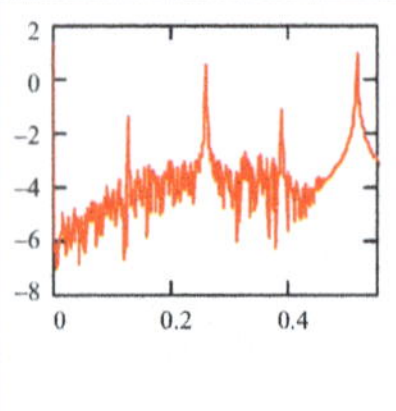	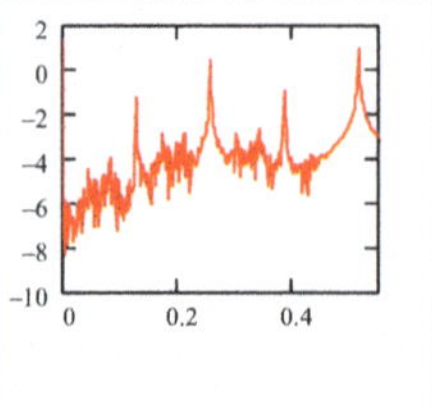	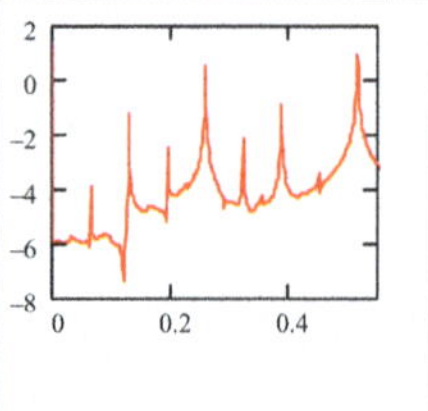
q_0=0.14			
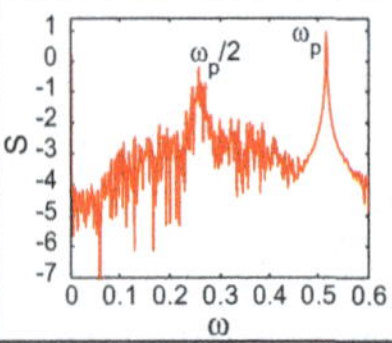	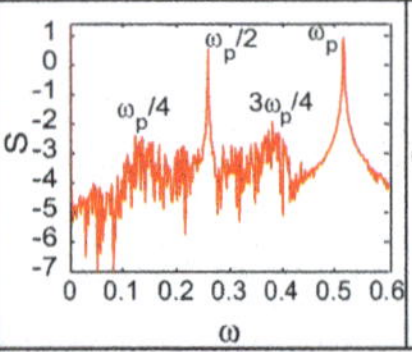	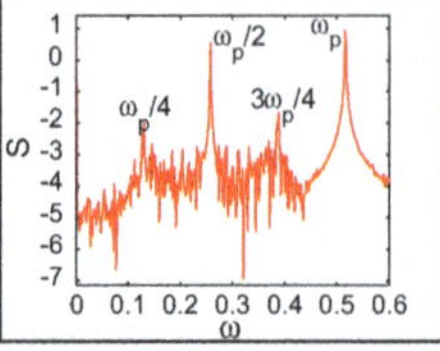	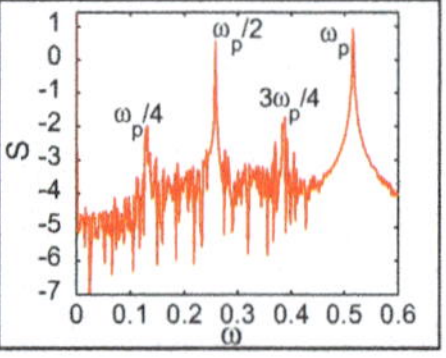

for different n. This requires the use of higher nodes number, i.e. the convergence of the results is achieved for $n = 80$; 120; 140; 160. In the case of $q_0 = 0.1395$, the vibration signals and their frequency characteristics strongly differ from each other: for $n = 20$, we have chaos with the distinguished ω_p and $\omega_p/2$; for $n = 40$, we have also chaos, but with dominating ω_p and $\omega_p/4$, whereas for $n = 120$; 140; 160, one deal with quasi-periodic vibration associated with the frequencies ω_p, $\omega_p/2$, $\omega_p/4$, $\omega_p/8$ and $\omega_p/16$. For $q_0 = 0.14$, beginning already from $n = 20$, the obtained results coincide, exhibiting the system chaotic vibrations.

Thus, the following conclusion can be yielded. A number of nodes employed in the FDM should be chosen differently for different vibrational regimes. Namely, periodic regimes require $n = 20$ nodes, whereas quasi-periodic/chaotic regimes require $n = 80$ or $n = 120$. It means that if one investigates a mechanical system with an infinite number of degrees-of-freedom, one has to investigate convergence of numerical solutions obtained by the finite difference method in order to get reliable/true results.

Table 14.3 contains frequency power spectra (FFT), 2D wavelet spectra, and deflection forms of the shell surfaces at three different time instants, put one next to another for better clarity of the obtained results. Table 14.4 reports the shell cross-sections obtained for different n at three chosen time instants. The LEs computed by Wolf's, Rosenstein's, and Kantz's methods are given in Table 14.5. It should be emphasized that to get reliable chaotic dynamics, a few different methods should be employed to compute LEs. In the present work, $n = 100; 120; 140$ were used in the FDM, and the obtained LEs coincide to three decimal places for all three considered methods. In the case of LEs computed for $n = 140$ by both Rosenstein's and Kantz's methods, the results coincide up to three decimal places, whereas in the case of Wolf's method, the values of LEs are three times larger than those yielded by the remaining methods. Finally, $n = 1\,20$ was used in further computations.

The carried out preliminary investigations allow one to conduct reliable analysis of mechanical systems with infinitely many degrees of freedom.

14.3 Modes of Vibrations (Simple Support)

In 1978, Feigenbaum [115] proposed a universal mechanism of period doubling bifurcation route to chaos. The Feigenbaum scenario was validated for the simply supported (14.2) axially symmetric spherical shells subjected to transverse harmonic load in reference [51], where the following parameters were used: $b = 4$, $\varepsilon = 0.01$ (damping coefficient), $\nu = 0.3$ (Poisson's ratio), $t \in [0, 1400]$, $\omega_p = \omega_0 = 0.516$, $q_0 \in [0.01, 0.15]$ (amplitude of harmonic excitation). As the excitation frequency ω_p was equal to the shell frequency ω_0, the case of the fundamental resonance was considered.

Six Hopf bifurcations were detected and served later for estimation of the Feigenbaum constant yielding $\alpha_6 = 4.65608466$ (the exact numerical value obtained by Feigenbaum equals 4.66916224 and differs by 0.28% from the above-mentioned numerical results).

The signals obtained in the present numerical experiments have the amplitude-time representation. However, in order to understand the behavior of any dynamical system under action of an arbitrary external load, and to control the system, one needs to have an insight into information hidden in the frequency space of the signal. For this reason, the wavelet transform approach was employed. Recall that the wavelet-based analysis enables one to detect and trace the localized particular features of the signal, whereas the Fourier-based approach (FFT) makes it possible

Table 14.3 Number of radius partitions n, power spectra, 2D wavelet spectra, and the shell deflection surfaces at three different time instants

	Power spectrum	Wavelet-analysis	Deflection surface t = 300	Deflection surface t = 304	Deflection surface t = 305
0	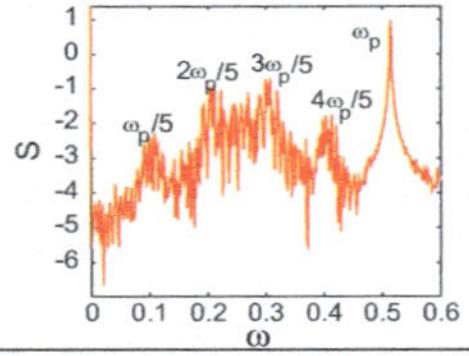	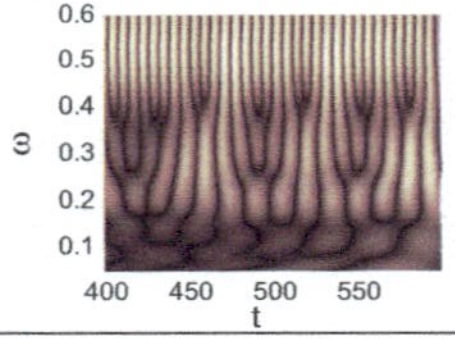		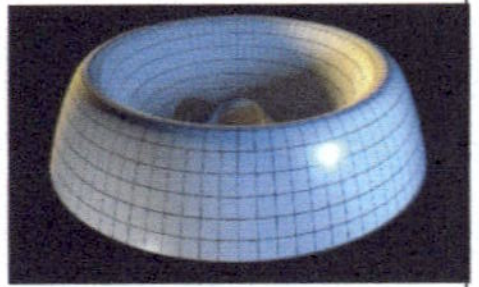	
0	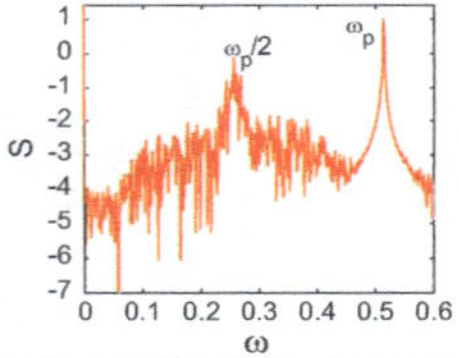	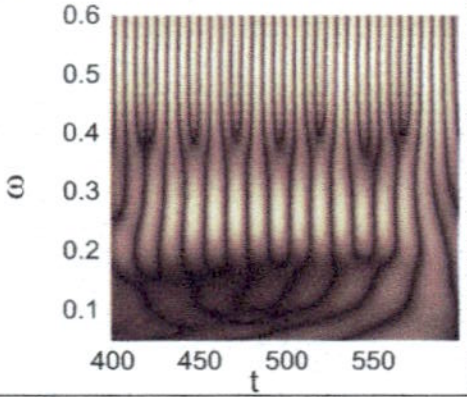			
0	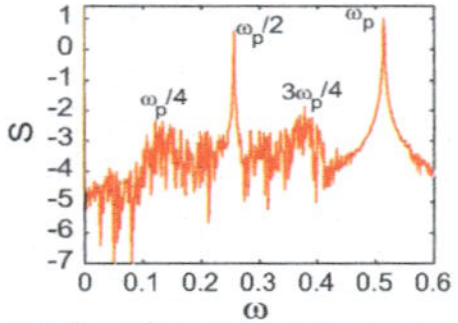	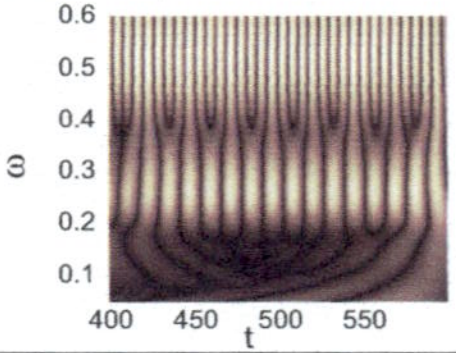			
0	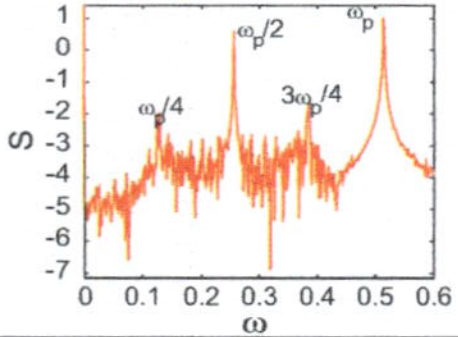	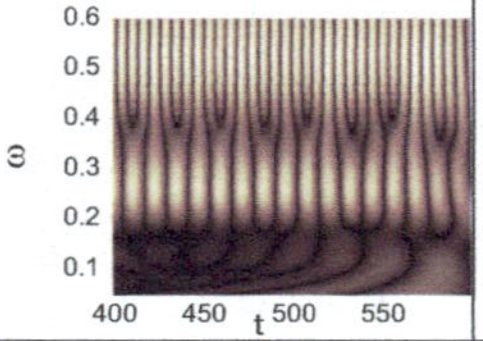	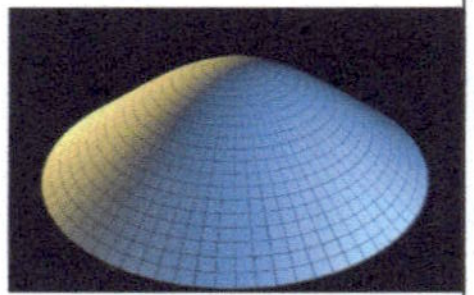		

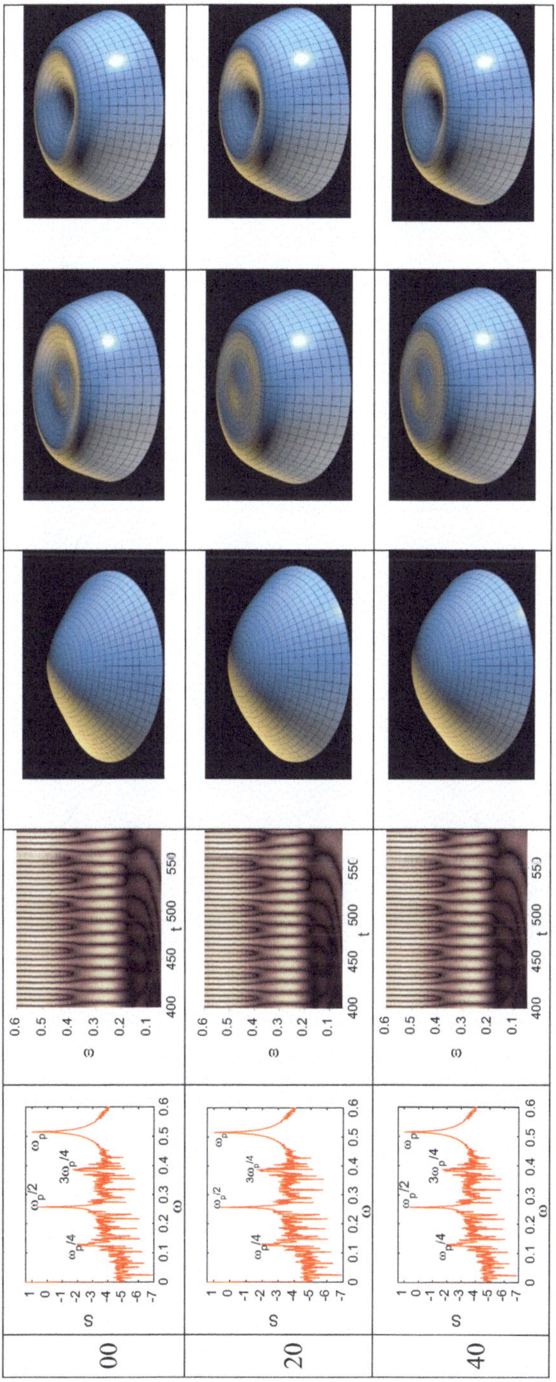

00
20
40
S
ω
t
ω_p/4
ω_p/2
3ω_p/4
ω_p

Table 14.4 Shell cross sections for different n at three time instants

$t = 300$	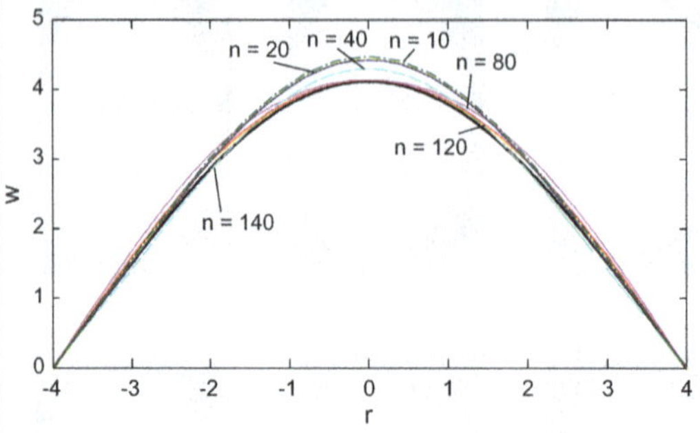
$t = 304$	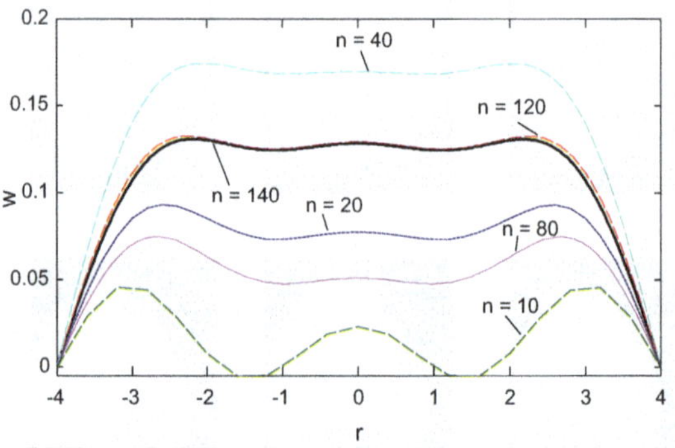
$t = 310$	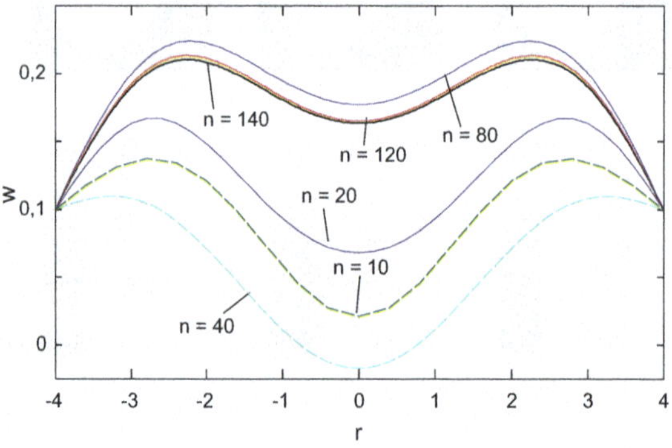

to follow the signal properties on the whole investigated time interval (see the description given in Sect. 14.2).

An increase in the number of new frequencies in the spectrum of vibration of the system yields a decrease in the informative properties of the employed wavelets, since harmonics with larger power shadow those with less power. Aside from this, it should be noted that the wavelet spectrum significantly depends also on the employed wavelet (each wavelet exhibits its own properties and

Table 14.5 Calculated values of the LEs versus the number n of partitions of the shell radius

Number n	Wolf	Rosenstein	Kantz
10	0.02894	0.06167	0.01969
20	0.01543	0.02832	0.01427
40	0.02776	0.01091	0.00510
80	0.03691	0.01553	0.01307
100	0.03959	0.01671	0.01013
120	0.03754	0.01666	0.01075
140	0.03642	0.01698	0.01071

peculiarities). Therefore, to obtain reliable results, it is recommended to use both FFT and wavelet spectra simultaneously. The so far presented remarks concerning investigation of continuous structural systems were taken into account while constructing Table 14.5, where the Fourier frequency power spectra, 2D wavelet spectra, $(w, \dot{w}, \ddot{w})$ phase plots, time history with marked time instants a, b, c, and the snapshots of the shell surface cross sections and the shell 3D shapes at a, b, c are reported for different values of q_0 (a similar table was constructed when studying the problem with boundary conditions (14.3)).

In the considered case, six period doubling bifurcations were numerically detected and the Feigenbaum constant was estimated. The investigations showed that for the implemented boundary condition, a transition from periodic to chaotic vibrations is carried out following the Feigenbaum scenario. Remarkably, owing to the employed boundary conditions, one of the ribs at the surface of the shell deflections occurred regardless of the amplitude of the excitation load, i.e., the occurrence of the mentioned rib was not influenced by the number of the Hopf bifurcations. Furthermore, the mentioned rib appeared at the time instants when the shell top deflection was close to zero.

14.4 Modes of Vibrations (Rigid Clamping)

In this section, vibrations of rigidly supported spherical shell ($b = 8$, boundary conditions (14.3)) subjected to the harmonic load $q_0 = \{0.01;\ 0.17;\ 0.21\}$ and $\omega_p = 1.34$ close to ω_0 is studied. Namely, the case of the fundamental resonance is investigated. The associated dynamical characteristic, similar to those reported in Table 14.6, are shown in Table 14.7. In this case, a transition from regular to chaotic vibrations follows the Ruelle-Takens-Newhouse route, i.e., the increase in the excitation amplitudes yields the birth of the independent frequency, which is linearly dependent on the excitation frequency. For $q_0 = 0.01$, the shell exhibits a periodic regime (phase portraits present ellipses, whereas the LLE is less than zero). The shell surface deflections exhibit two ribs, which occur at time instants when the shell center deflection is close to zero. An increase in the excitation load amplitude up to $q_0 = 0.17$ yields the occurrence of an independent frequency.

Table 14.6 Fourier frequency power spectra, 2D wavelet spectra, $(w, \dot{w}, \ddot{w})$ phase plots, time history with marked time instants a, b, c, shell surface cross sections, and shell 3D shapes at a, b, c for different values of q_0 (simple support)

$q_0 = 0.08$; $LE < 0$

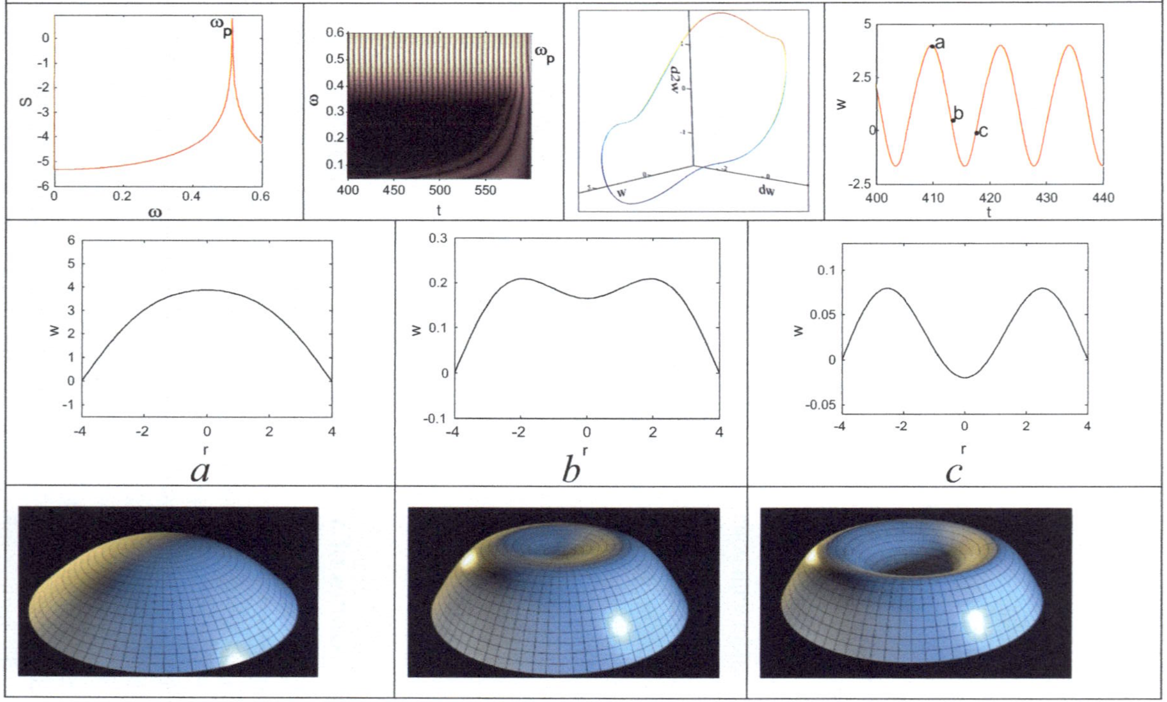

$q_0 = 0.1335$; LE > 0

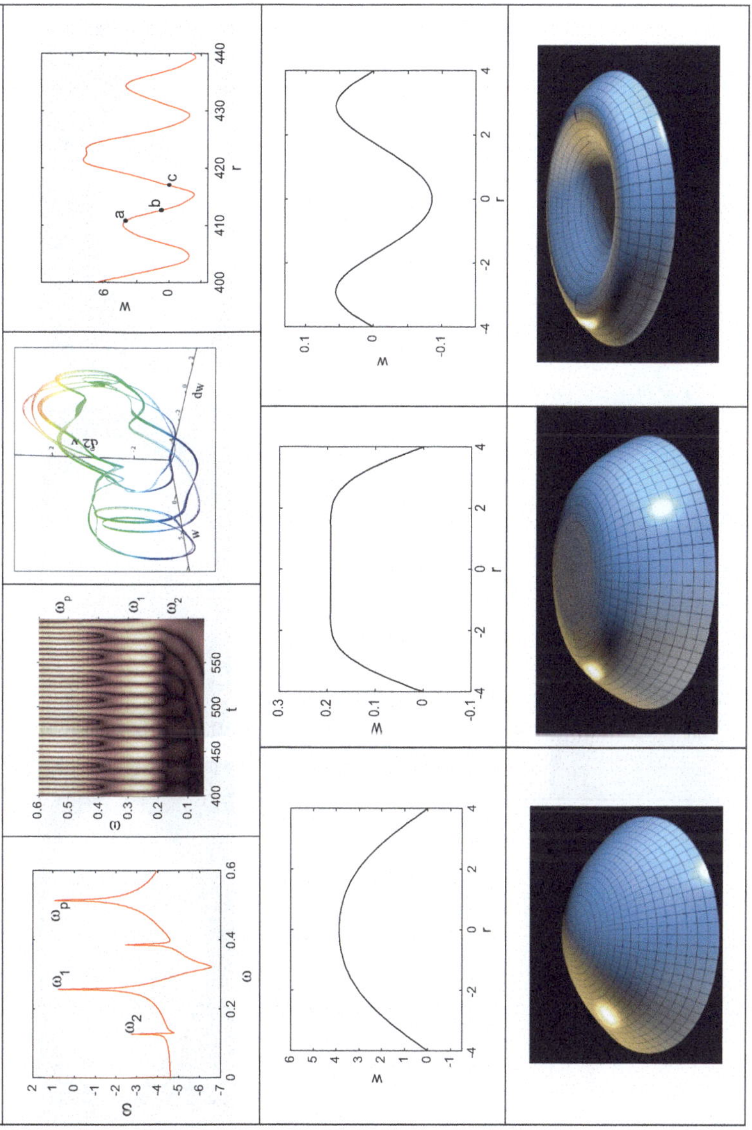

(continued)

Table 14.6 (continued)

$q_0 = 0.14$; LE > 0

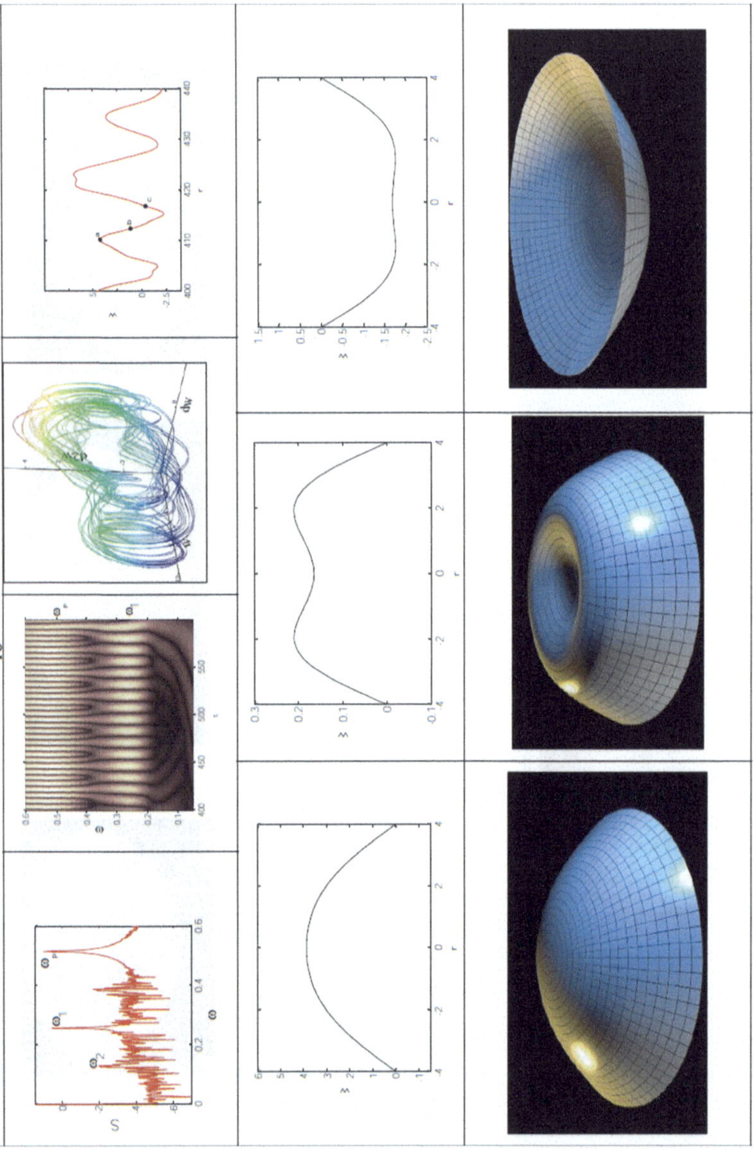

Table 14.7 Fourier frequency power spectra, 2D wavelet spectra, $(w, \dot{w}, \ddot{w})$ phase plots, time history with marked time instants a, b, c, shell surface cross sections, and the shell 3D shapes at a, b, c for different values of q_0 (rigid clamping)

$q_0 = 0.01$, LE < 0

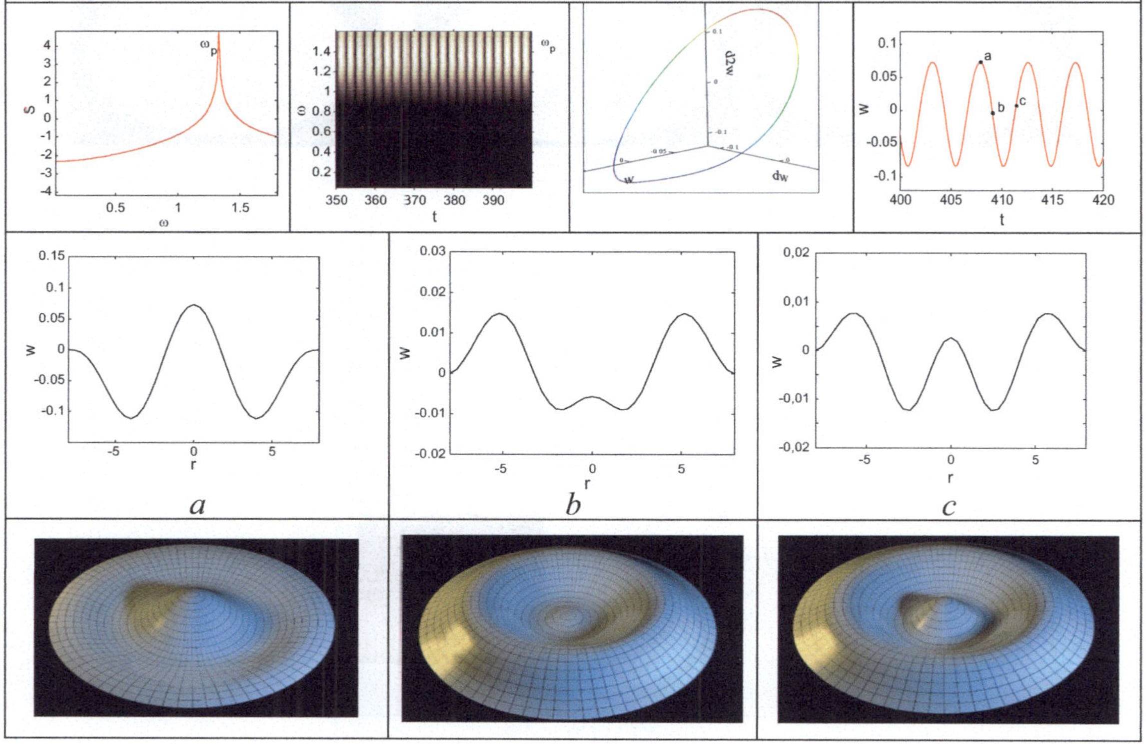

(continued)

Table 14.7 (continued)

$q_0 = 0.17$, LE > 0

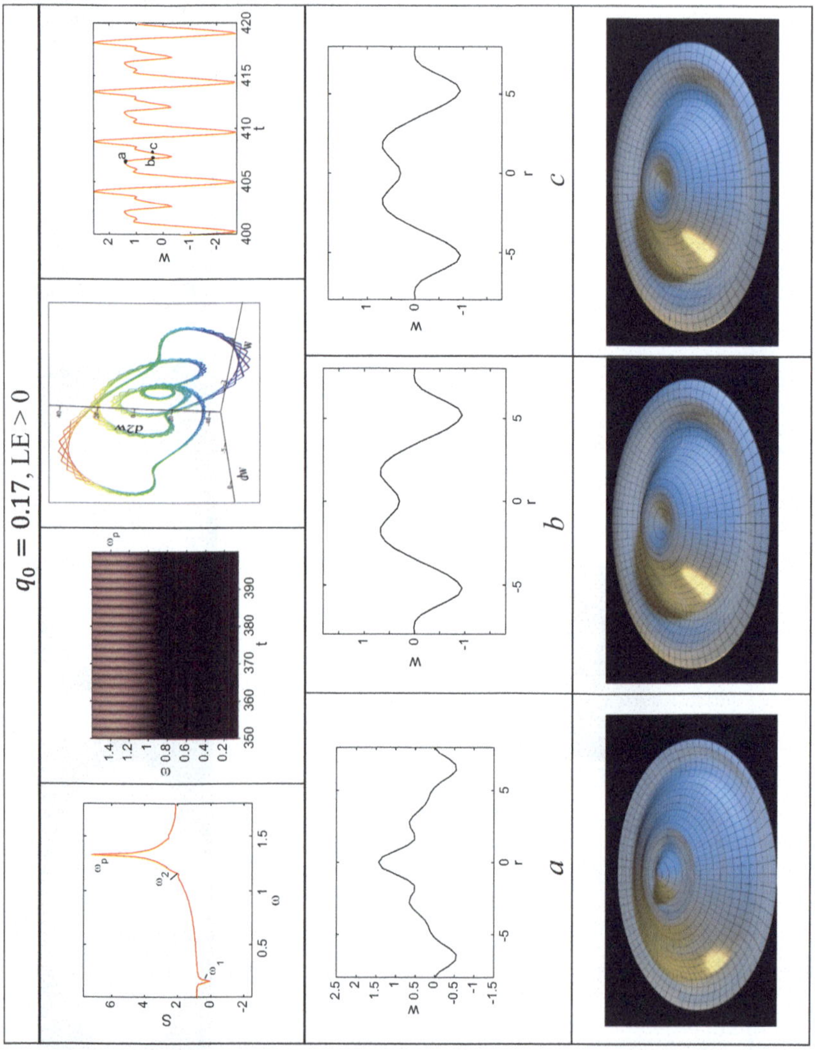

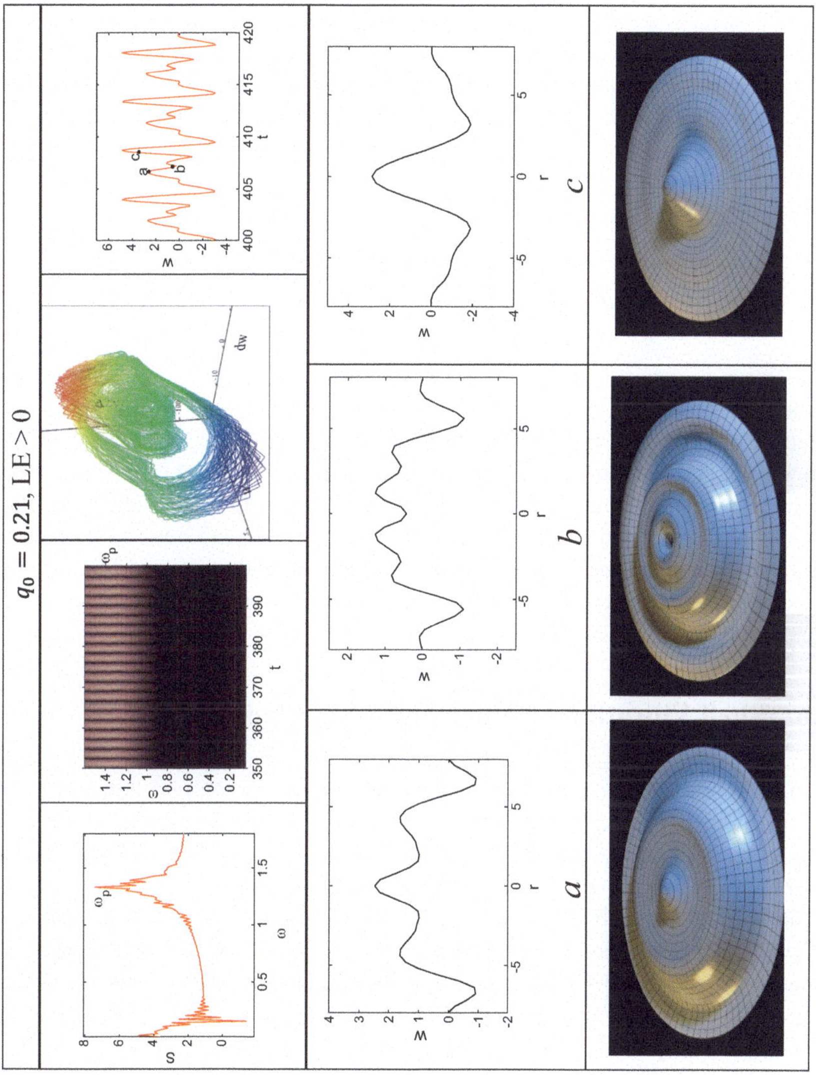

$q_0 = 0.21$, LE > 0
S
ω
ω_p
t
w
dw
r
a
b
c

Furthermore, the ribs are exhibited by the shell surface, and the LLE> 0. The wavelet spectrum and the FFT spectrum also reports the independent frequency, and a linear combination of the excitation and independent frequencies is observed. The investigations imply that in the case of the rigidly clamped axially symmetric spherical shell, a transition from regular to chaotic vibrations is realized via the Ruelle-Takens-Newhouse scenario along the shell contour. Moreover, deflection of the shell surfaces exhibit from one up to four ribs. It is worth mentioning that the Ruelle-Takens-Newhouse scenario plays a crucial role in the ribs occurrence. Namely, the presence of only one rib is associated with regular shell vibration, two ribs are associated with the occurrence of the independent frequency, whereas the occurrence of chaos yields the existence of up to three ribs.

14.5 Investigation of Occurrence of Ribs (Simple Nonmovable Shell Support)

This section is devoted to vibrations of the axially symmetric shell with simple nonmovable support of its contours and the shallow parameter $b = 4$. If the shell is subjected to transverse and uniformly disturbed periodic load, a period doubling route to chaos following the Feigenbaum scenario is detected and seven Hopf bifurcations can be estimated numerically. On the contrary to the previous examples, a number of the detected ribs is not fixed, but changes in time (Table 14.8). Changes in the vibration character do not influence the number of the detected ribs, i.e., the Feigenbaum scenario does not influence the number of ribs.

14.6 Shell Vibration Modes (Movable Clamping)

In what follows, vibration of a spherical axially symmetric shell are considered for fixed parameters $b = 5, \quad \varepsilon = 0.1$. The presence of the Sharkovsky's orbits is detected for the transverse uniform harmonic load $q = q_0 sin\omega_0 t$ for the case of movable clamping (14.4) and for zero initial conditions (14.6).

The shell is made from an elastic, homogenous and isotropic material with $\nu = 0.3$, and it is excited by the frequency ω_p being equal to the fundamental shell frequency $\omega_0 = 0.773$ (linear vibration). Phase portraits (3D), frequency power spectra, wavelet-based analysis results as well as the shapes of the shell surface deflections are reported. Table 14.9 shows the Sharkovsky's ordering $3, 3^2$. The detected Sharkovsky's ordering occurred successively, and this novel result, up to the knowledge of authors, is reported for the first time in the literature. It should be mentioned that Sharkovsky's windows have been already detected and illustrated when studying vibration of beams.

Table 14.8 Fourier frequency power spectra, 2D wavelet spectra, $(w, \dot{w}, \ddot{w})$ phase plots, time history with marked time instants a, b, c, and the snapshots of the shell surface cross sections and the shell 3D shapes at a, b, c for different values of q_0 (simple nonmovable shell support)

$q_0 = 1; \quad LE < 0$

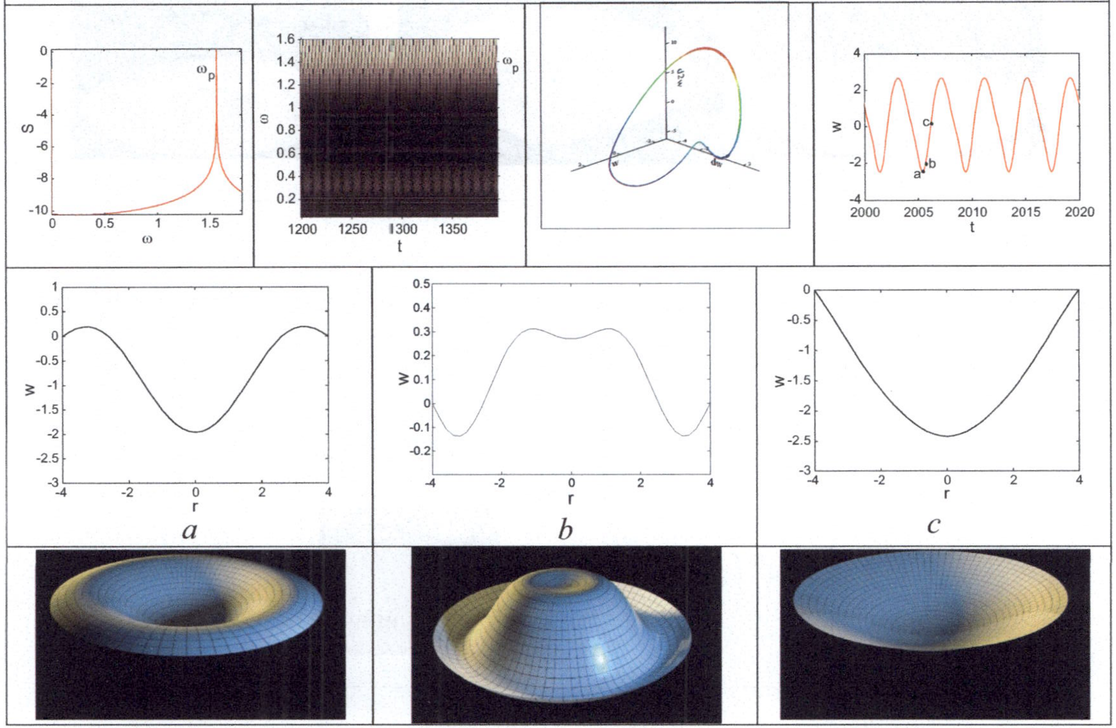

(continued)

Table 14.8 (continued)

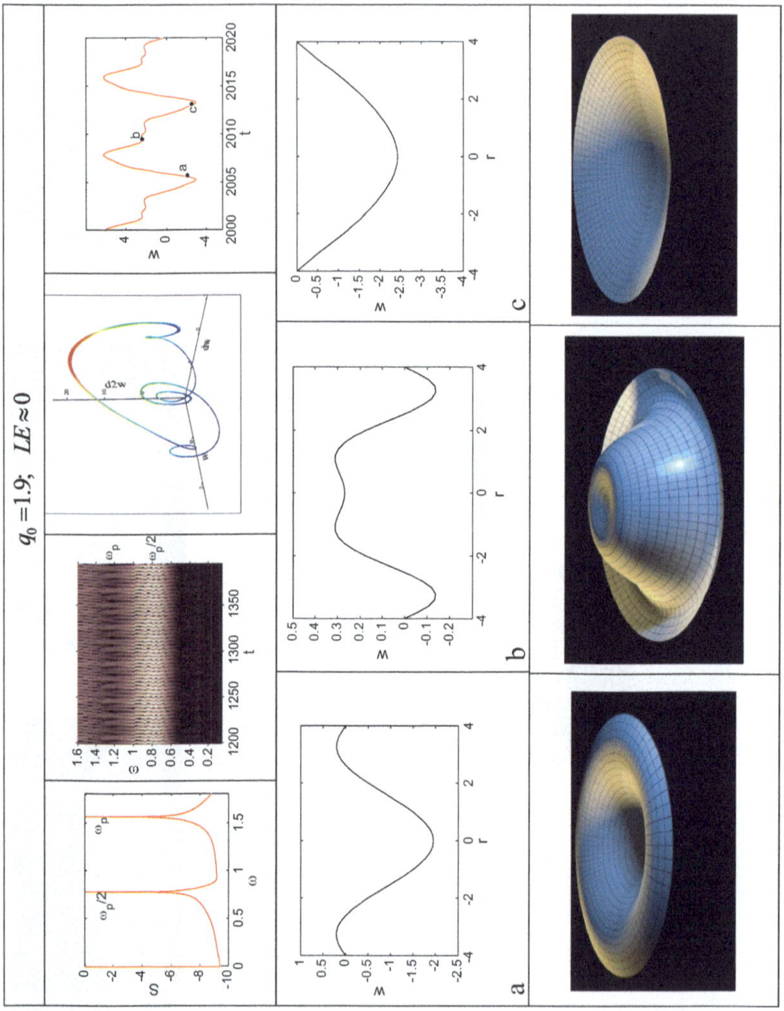

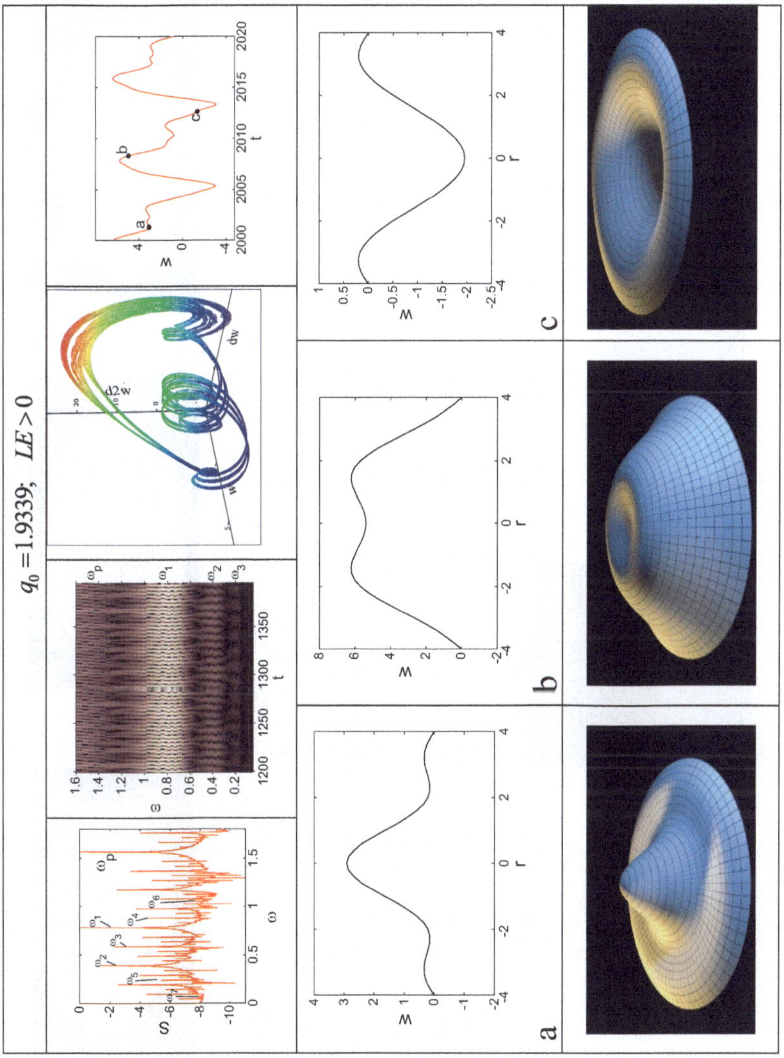

q0 = 1.9339; LE > 0
a
b
c

Table 14.9 Fourier frequency power spectra, 2D wavelet spectra, $(w, \dot{w}, \ddot{w})$ phase plots, time history with marked time instants a, b, c, and the snapshots of the shell surface cross sections and the shell 3D shapes at a, b, c for different values of q_0 (movable clamping)

$q_0 = 0.02; \quad LE < 0$

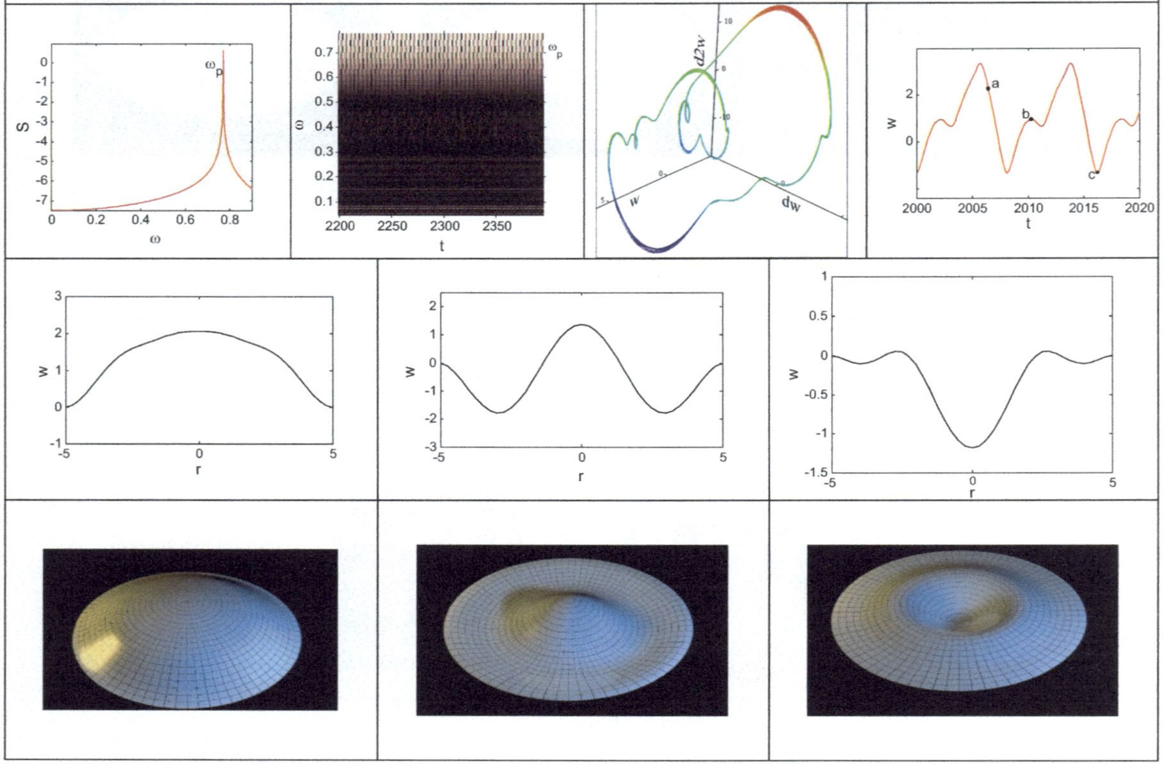

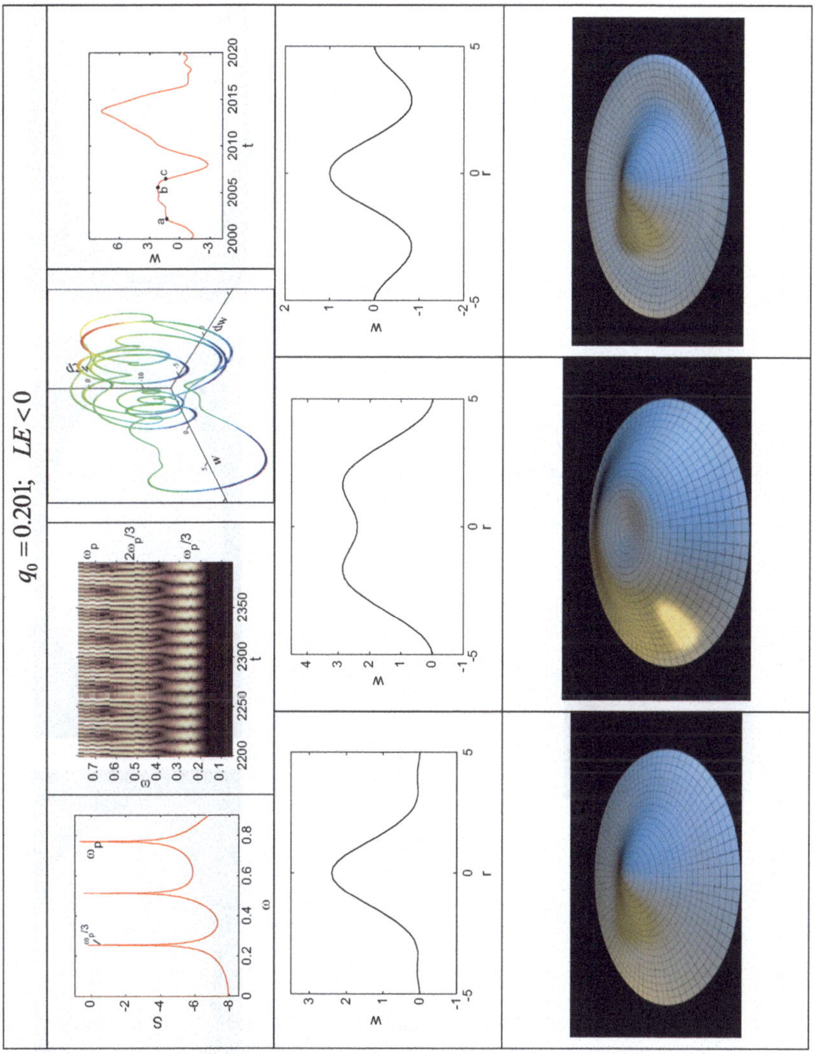

(continued)

Table 14.9 (continued)

$q_0 = 0.21;\quad LE \approx 0$

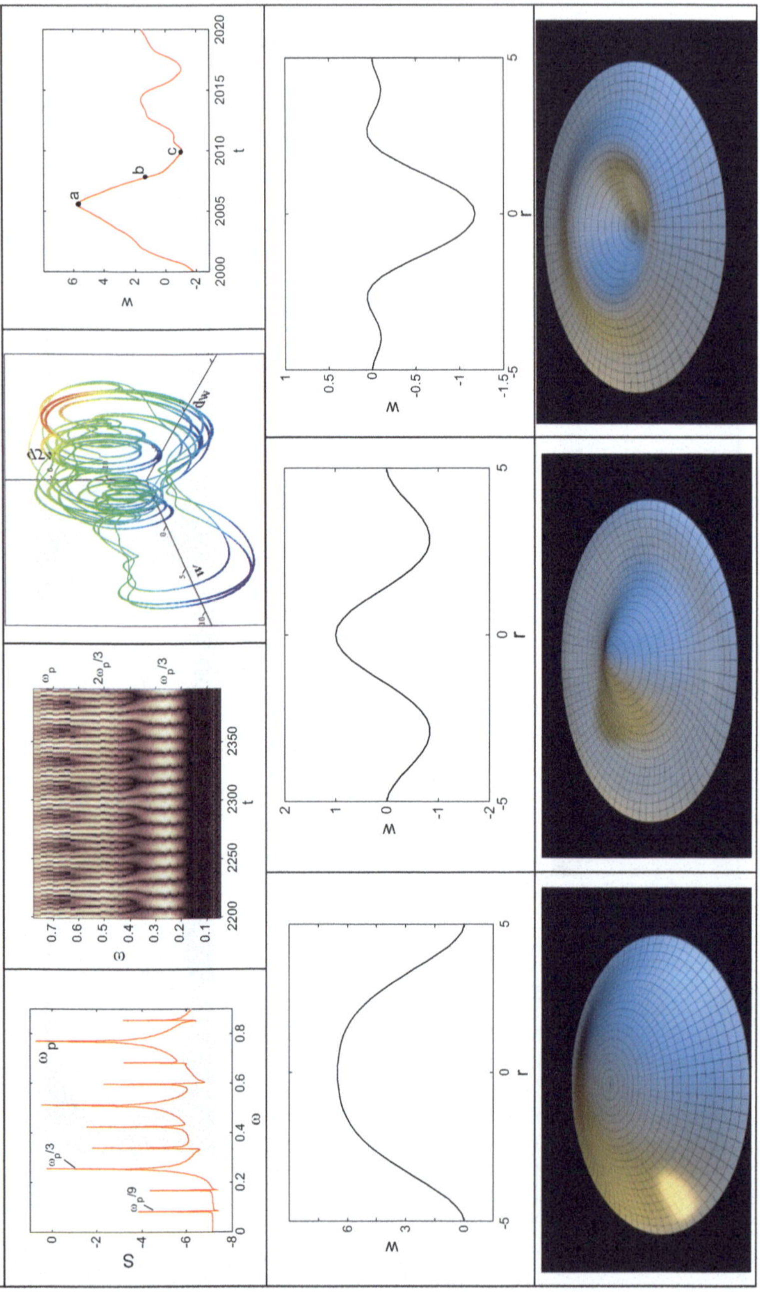

For the applied boundary conditions, one time-dependent rib was detected in the snapshots of the shell main cross sections. Here, no relationship between the vibration regimes and the number of ribs was found.

Chapter 15
Chaotic Vibrations of Two Euler-Bernoulli Beams With a Small Clearance

In this chapter a methodology to detect true/reliable chaos (in terms of non-linear dynamics) is developed on an example of a structure composed of two beams with a small clearance. The Euler-Bernoulli hypothesis is employed, and the contact interaction between beams follows the Kantor model. The complex non-linearity results from the von Kármán geometric non-linearity as well as the non-linearity implied by the contact interaction. The governing PDEs are reduced to ODEs by the second-order Finite Difference Method (FDM). The obtained system of equations is solved by Runge-Kutta methods of different accuracy. To purify the signal from errors introduced by numerical methods, the principal component analysis is employed and the sign of the first Lyapunov exponent is estimated by the Kantz, Wolf and Rosenstein methods and the method of neural networks. In the latter case a spectrum of the Lyapunov exponents is estimated. It has been illustrated how the number of nodes in the employed FDM influences the numerical results regarding chaotic vibrations. We have also shown how increase of the beams distance implies stronger action of the geometric nonlinearity, and hence influence of convergence of the used numerical algorithm for FDM has been demonstrated. Effect of essential dependence of the initial conditions choice on the numerical results of the studied contact problem is presented and discussed.

We have used the methods which guarantee reliability and truth of the obtained solutions. In particular, we have demonstrated that the increase in the number of beam partitions (nodes) implies regularization of the obtained vibrations. After comparing the results obtained within linear and non-linear problems and taking into account contact between beams, we have also illustrated that the account of the spatial coordinate x in $u(x, t)$, i.e. the increase in the number of equations, decreases the chaotic components in the analysed signals.

The results of this chapter are based on reference [209].

J. Awrejcewicz, V. A. Krysko, *Elastic and Thermoelastic Problems in Nonlinear Dynamics of Structural Members*, Scientific Computation,
https://doi.org/10.1007/978-3-030-37663-5_15

15.1 Mathematical Model

In this chapter, the following main steps of investigations are carried out to numerically solve non-linear problems of structures composed of two beams with a clearance and subjected to the harmonic load, and thus aimed on detection of chaotic vibrations:

1. Since non-linear PDEs are reduced to ODEs due to the employed FDM of the 2nd order accuracy, the obtained solutions essentially depend on the number of partitions of the beam. In other words, we need to find the number of n beam partitions, for which the solutions obtained for $2n$ and n partitions coincide. We require convergence with respect to the signal/time history also in the case of chaotic vibrations. It should be mentioned that in the previous investigations [54], the signal convergence was required rather only in the case of periodic vibrations, whereas for chaos, the integral convergence was accepted.
2. The Cauchy problem is also solved numerically, and hence solutions essentially depend on both the chosen method and the time step integration. Therefore, in order to achieve reliable results, the Cauchy problem is solved using the Runge-Kutta of the 4th (RK4) and the 2nd (2RK) orders [313], the Runge-Kutta-Fehlberg of the 4th order (rkf45) [112, 113], the Cash-Karp of the 4th order (RKCK) [80], the Runge-Kutta-Prince-Dormand of the 8th order (rk8pd) [100] as well as the implicit Runge-Kutta methods of the 2nd (rk2imp) and the 4th (rk4imp) orders. The explicit method is characterised by the triangular form of the matrix coefficients (including zeroth main diagonal). On the contrary, in the implicit method, the governing matrix has an arbitrary shape. Moreover, control of integration errors can be implemented in a relatively easy manner.
3. For each spatial coordinate partition as well as for the chosen method, the signals/time histories, 2D/3D phase portraits, Fourier power spectra, the Morlet wavelets, snapshots of beam deflections, the Poincaré sections and the 2D wavelet-spectra are constructed while solving the Cauchy problem. Haar [141], Shannon-Kotelnikov, Meyer [241], and Daubechies (from db2 to db16) [95] wavelets, coiflets and simlets, Morlet wavelet as well as the complex arbitrary Gauss function of the order higher than 8th (see also [137, 240, 242]) are implemented. The Haar and Shannon-Kotelnikov wavelets are rather not feasible for investigation of the beam constructions, i.e. the first one is badly localised in time. The carried out analysis of the wavelet-spectra obtained with the help of the Daubechies wavelets, coiflets and simlets yielded a conclusion that an increase in the used wavelets order implies the improvement of the distinction of the frequency localisation. However, no matter how different wavelets and different filters are employed, the wavelet spectra obtained by the Daubechies wavelets, simlets and coiflets are practically the same, but still not enough accurate for investigation of the vibration character of the studied continuous mechanical systems. On the other hand, employment of the arbitrary Gauss function shows that an increase in the derivative order implies better resolution of the detected frequencies. It should be emphasized that the spectra obtained on a basis of

the Meyer wavelets [240, 242] (smoothened variant of the Shannon-Kotelnikov wavelets) are localised better in the case of the low frequency band in comparison to the Morlet wavelets. However, the higher spectrum part is better identified using the Morlet wavelets. In what follows we report results obtained based on the Morlet wavelets. Observe that real wavelets exhibit a lack of the scaling function φ, whereas the function ψ does not have a compact carrier and is given explicitly. The complex Morlet and Gauss wavelets exhibit better localisation with respect to frequency than their real counterparts. Therefore, in order to analyse complex vibrations of continuous systems composed of beams, one may employ either complex or real Morlet wavelets as well as wavelets based on the derivatives of the Gauss function of the order higher than 16.

4. Since we would like to use chaos definition given by Gulick [138], we need to compute and estimate the sign of the spectrum of the Lyapunov exponents. In this work, the spectrum of the Lyapunov exponents for all beam partitions has been estimated using three methods based on Kantz [164], Wolf [351] and Rosenstein [285] algorithms. The final choice was made after obtaining four-digit accuracy.

   As it has been already mentioned, this chapter is focused on the non-linear dynamics and the contact interaction of two beams with a small clearance, for the case when one of the beams (beam 1) is subjected to the action of the transverse harmonic load, causing movement of the second beam. The contact interaction between beams follows the Kantor model [162].

   In the literature, one can find numerous works devoted to investigation of beams, plates and shells in the frame of the Euler-Bernoulli [107] and Kirchhoff-Love [204, 205] models. The review of the state-of-the-art in this field has been given by Alijani and Amabili [4]. In the majority of considered cases, the solutions were found using simple models including a few degrees of freedom, and eventually yielded unreliable results [290]. This has been pointed out in the recent monograph [54], where the problem of chaotic vibrations of beams, plates and shells has been studied treating the mentioned structural members as those of an infinite number of degrees of freedom.

The considered structure composed of two beams occupies a 2D space within the R^2 space with the rectangular system of coordinates given in the following way: a reference line, further called the middle line $z = 0$, is fixed in the beam 2, the axis OX is directed from the left to the right of the middle line, and the axis OZ is directed downwards ($OZ \perp OX$). In the given system of coordinates, the space Ω is defined in the following way (see Fig. 15.1): $\Omega = \{x \in [0, a]\,;\ -h \leq z \leq h_k + 3h\}$, $0 \leq t \leq \infty$.

In order to construct the mathematical model of contact interaction of two beams, we introduce the following assumptions and hypotheses:

(1) each beam is composed of one layer;
(2) the beams are isotropic, elastic, and obey Hook's law;
(3) the longitudinal dimensions of beams are larger than their transverse dimensions and the beams have the unit thickness;
(4) the axis of each of the beams is a straight line;

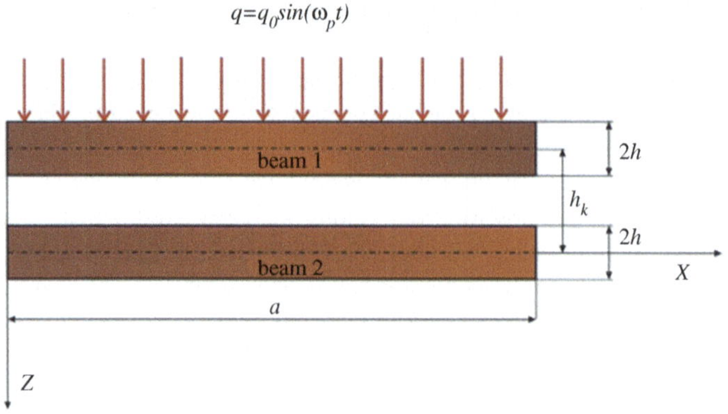

Fig. 15.1 The computation scheme of two beams with a clearance

(5) the load acts in the direction of the OZ axis and the external forces do not change their direction during the beam deformation;
(6) contact pressure is estimated within the Kantor model [162];
(7) normal stresses regarding the zones being parallel to the axis are small and negligible;
(8) geometric non-linearity is taken in the von Kármán form [169].

In order to model the contact interaction of the beam within the Kantor model, we introduce the term $(-1)^i K(w_1 - w_2 - h_k)\Psi$ into the equation governing the beams vibrations. In the above term, "i" stands for the beam number, and the function Ψ is defined by the formula $\Psi = \frac{1}{2}[1 + sign(w_1 - h_k - w_2)]$, i.e. if $\Psi = 1$, then the beams are in contact $w_1 > w_2 + h_k$, otherwise there is no contact between beams [162]. The coefficient K describes the beam transverse stiffness in a contact zone, whereas h_k represents the clearance between beams (see Fig. 15.1).

In order to keep clarity, in the further parts of the paper, by "beam 1" we understand the externally beam loaded, whereas "beam 2" stands for the unloaded beam.

Equations of motion of beams with the account of hypotheses of the first order as well as boundary and initial conditions are yielded from the Hamilton energetic principle.

It should be noted that neglecting of the von Kármán geometric non-linearity does not reduce the investigated problem to the linear one, since the design non-linearity implied by contact of two beams is also taken into account. We consider a small clearance between beams, i.e. contact of beams occurs even for small deformations of the beam 1, and the vibrations can be treated as linear only for $w_1 \leq 0.25$. One of our aims is to verify if the geometric non-linearity needs to be taken into account due to small amplitudes of vibrations.

Equations of motion of a beam element with an account of contact interaction in displacements have the following form [54]:

$$\begin{cases} E(2h)\left\{L_2(w,w)+L_1(u_i,w_i)-\dfrac{(2h)^2}{12}\dfrac{\partial^4 w_i}{\partial x^4}\right\}+q_i(t)-(2h)\dfrac{\gamma}{g}\dfrac{\partial^2 w_i}{\partial t^2}- \\ -\varepsilon(2h)\dfrac{\gamma}{g}\dfrac{\partial w_i}{\partial t}+(-1)^i K\dfrac{E}{(2h)}(w_1-w_2-h_k)\Psi=0, \\ E(2h)\left\{\dfrac{\partial^2 u_i}{\partial x^2}+L_3(w_i,w_i)\right\}-(2h)\dfrac{\gamma}{g}\dfrac{\partial^2 u_i}{\partial t^2}=0, \end{cases} \tag{15.1}$$

where: $L_1(u_i,w_i) = \frac{\partial^2 u_i}{\partial x^2}\frac{\partial w_i}{\partial x} + \frac{\partial u_i}{\partial x}\frac{\partial^2 w_i}{\partial x^2}$, $L_2(w_i,w_i) = \frac{3}{2}\frac{\partial^2 w_i}{\partial x^2}\left(\frac{\partial w_i}{\partial x}\right)^2$, $L_3(w_i,w_i) = \frac{\partial^2 w_i}{\partial x^2}\frac{\partial w_i}{\partial x}$ are non-linear operators; w_i, u_i are functions describing deflections and displacements of the upper and the lower beam, respectively; K stands for the coefficient of transverse stiffness of the contact zone, and h_k is the clearance between the beams; E—Young's modulus; g—gravity of Earth; ν—Poisson's ratio; γ—volume weight of the beam material; ε—damping coefficient of the surrounding matter.

The system of governing PDEs supplemented by boundary and initial conditions is reduced to the counterpart dimensionless form using the following variables and parameters (quantities with bars are non-dimensional)

$$\begin{gathered} \bar{w}=\frac{w}{2h},\quad \bar{u}=\frac{ua}{(2h)^2},\quad \bar{x}=\frac{x}{a},\quad \lambda=\frac{a}{(2h)},\quad \bar{q}=q\frac{a^4}{(2h)^4 E}, \\ \bar{t}=\frac{t}{\tau},\quad \tau=a\sqrt{\frac{\gamma}{Eg}},\quad \bar{\varepsilon}=\varepsilon a\sqrt{\frac{\gamma}{Eg}}. \end{gathered} \tag{15.2}$$

Equations governing the dynamics of two Euler-Bernoulli beams with respect to displacements and taking into account frictional energy loss (dissipation) are governed by the following PDEs:

$$\begin{cases} \dfrac{1}{\lambda^2}\left\{L_2(w,w)+L_1(u_i,w_i)-\dfrac{1}{12}\dfrac{\partial^4 w_i}{\partial x^4}+q_i(t)\right\}-\dfrac{\partial^2 w_i}{\partial t^2}-\varepsilon\dfrac{\partial w_i}{\partial t}+ \\ +(-1)^i K(w_1-w_2-h_k)\Psi=0, \\ \dfrac{\partial^2 u_i}{\partial x^2}+L_3(w_i,w_i)-\dfrac{\partial^2 u_i}{\partial t^2}=0. \end{cases} \tag{15.3}$$

PDEs (15.3) are non-dimensional, i.e. bars are omitted. We need to supplement the equations (15.1) by the boundary and initial conditions. Furthermore, in the case when the geometric non-linearity is neglected, we need to take $L_1 = 0,\ L_2 = 0,\ L_3 = 0$.

The system of non-linear PDEs is reduced to the one of non-linear ODEs by means of the FDM (Finite Difference Method) of the second-order approximation $O(c^2)$, where c is a step regarding the spatial coordinate.

In each node of the mesh, one obtains the following set of ODEs:

$$\begin{cases} L_{1,c}(w_{i,j}(t), u_{i,j}(t)) + q_{i,j}(t) = \varepsilon_1 \dot{w}_{i,j}(t) + \ddot{w}_{i,j}(t); \\ L_{2,c}(w_{i,j}(t), u_{i,j}(t)) = \varepsilon_2 \dot{u}_{i,j}(t) + \ddot{u}_{i,j}(t); \quad (i = 1,\ 2;\ j = 0, \ldots, n), \end{cases} \tag{15.4}$$

where: i—beam number; n—number of beam partitions; c—step regarding the spatial coordinate; $L_{1,c}(w_{i,j}(t), u_{i,j}(t))$, $L_{2,c}(w_j(t), u_j(t))$ stand for the finite difference operators of the following form

$$L_{1,c}(w_{i,j}, u_{i,j}) = \frac{1}{\lambda^2}\left(-\frac{1}{12}\frac{1}{c^4}\left(w_{i,j+2} - 4w_{i,j+1} + 6w_{i,j} - 4w_{i,j-1} + w_{i,j-2}\right) + \right.$$

$$+ \frac{1}{2c}(w_{i,j+1} - w_{i,j-1})\frac{1}{2}\left(u_{i,j+1} - 2u_{i,j} + u_{i,j-1}\right) +$$

$$+ \frac{1}{2c}(w_{i,j+1} - w_{i,j-1})\frac{1}{2}\left(u_{i,j+1} - 2u_{i,j} + u_{i,j-1}\right) +$$

$$+ \left(\frac{1}{2c}(w_{i,j+1} - w_{i,j-1})\right)^2 \frac{1}{c^2}(w_{i,j+1} - 2w_{i,j} + w_{i,j-1}) +$$

$$+ \frac{1}{c^2}(w_{i,j+1} - 2w_{i,j} + w_{i,j-1})\left(\frac{1}{2}\left(u_{i,j+1} - u_{i,j-1}\right) + \right.$$

$$\left.\left. + \frac{1}{8^2}(w_{i,j+1} - w_{i,j-1})(w_{i,j+1} - w_{i,j-1})\right) + q_{i,j}\right) +$$

$$+ (-1)^i K(w_{1,j} - w_{2,j} - h_k)\Psi_j,$$

$$L_{2,c}(w_{i,j}, u_{i,j}) = \frac{1}{2}\left(u_{i,j+1} - 2u_{i,j} + u_{i,j-1}\right) +$$

$$+ \frac{1}{2c}(w_{i,j+1} - w_{i,j-1})\frac{1}{c^2}(w_{i,j+1} - 2w_{i,j} + w_{i,j-1}); \quad j = 0,\ \ldots,\ n.$$

In Eqs. (15.4), for $j = 1,\ j = n - 1$, we need to employ the values of w the contour points, which are defined by the boundary conditions.

In order to show, how the procedure works, let us take five points pattern $(x - 2c,\ x - c, x, x + c, x + 2c)$, and let us define

$$L_C w = w_{\bar{x}x\bar{x}x} = \frac{1}{c^2}(w_{\bar{x}x}(x + c) - 2w_{\bar{x}x}(x) + w_{\bar{x}x}(x - c)) =$$

$$= \frac{1}{c^4}(w(x + 2c) - 4w(x + c) + 6w(x) - 4w(x - c) + w(x - 2c)).$$

L_C approximates L with a second order, i.e. $w_{\bar{x}x\bar{x}x} = w^{(4)} = \frac{c^2}{6}w^{(6)} + O\left(c^4\right)$. Employing the Taylor formula [79] we obtain $w(x \pm kc) = w(x) + \sum_{s=1}^{7} \frac{(-1)^s k^s c^s}{s!} \frac{d^s w(x)}{dx^s} + O\left(c^8\right)$ for $k = 1, 2$. Since the sum $w(x + kc) + w(x - kc)$ contains only even powers, one obtains the earlier reported approximation for $w_{\bar{x}x\bar{x}x}$.

The obtained Cauchy problem is solved by the 4th- and the 2nd-order Runge-Kutta, the 4th-order Runge-Kutta-Fehlberg, the 4th-order Cash-Karp, the 8th-order Runge-Kutta-Prince-Dormand and the implicit 2nd- and 4th-order (rk4imp) Runge-Kutta methods. On the basis of the described algorithms, the program package has been developed, which allows one to solve the given problem with respect to the control parameters $\left\{q_0, \omega_p\right\}$. The main attention has been paid to control and avoid the occurrence of penetration of the structural elements. As it has been already pointed out, the studied problems are strongly non-linear, and hence an important question regarding reliability of the obtained results arises.

While studying the problems of contact interaction, one needs to detect the fundamental frequencies of the vibrational processes first, because, as it will be shown later, a transition into chaotic vibrations takes place after the beam contact. One of the methods devoted to studying localisation of the fundamental frequencies is the principal component analysis [151, 222, 315].

15.2 Principal Component Analysis (PCA)

It should be emphasized that, formally, the principal component analysis, aimed at estimation of the principal components with respect to the dissipative problems, can be always employed. However, in the case when the "signal" cannot be distinguished from "noise", any earlier given accuracy is helpful. Typically, the "signal" exhibits a relatively small dimension while keeping the relatively large amplitude. On the other hand, "noise" exhibits a large dimension for a relatively small amplitude. This observations allow us to understand the main features of the principal component analysis. Namely, the method works as a filter, i.e. the signal is mainly kept in projections onto the first principal components, whereas the remaining components include mainly noise (for more details see [207]).

As a result of solving PDEs by FDM, a matrix W composed of the values of the time-dependent beam deflections measured in the nodes are obtained. Let us present a matrix $\tilde{W}$ in the form of its linear splitting $\tilde{W} = TP^t + E$. The matrix W contains all elements, whereas the matrix $\tilde{W}$ is yielded by the series with the account of k principal components. In order to compute the score matrix and the matrix of loading, the singular series development of the matrix W is carried out through the so-called auto-scaling process. The auto-scaled matrix $\bar{W} = USV^t$, where $T = US$, $P = Vt$, is obtained. The main diagonal of the matrix S is composed of eigenvalues, whereas the remaining matrix elements are equal to zero.

The values of the principal components k are chosen in a way to keep the eigenvalues of the matrix S larger than 1. Equivalently, the dependence of dispersion on the number of principal components k $TRV(k)$ and the dependence of the dispersion on the number of principal components k $ERV(k)$ should rapidly change their behaviour, i.e. the truncated singular series are employed. After definition of the required number of principal components, the matrix $T = [t_{ij}]$; $t_{ij} = (x_i, a_j)$ has a dimension $m \times k$, and its each row presents a projection of the data vector onto k principal components (the number of columns k corresponds to the number of vectors of the principal components chosen for the purpose of projection). The loading matrix $P = \{a_1, \ldots, a_k\}$ and each of its columns correspond to the vector of principal components, whereas the number of rows n corresponds to the dimension of the data space chosen for the purpose of projection. Each row of the computation matrix stands for the projection of the data vector onto k principal components. The matrices U and V are orthogonal. The characteristics of dispersion TRV and of dispersion ERV show the percentage of noise remaining after projection onto the multidimensional space $PC1 - PCk$. In other words, the whole characteristics shows what number of principal components is necessary for purification of the signal from noise (more description of PCA is given in Sect. 15.4).

15.3 Numerical Experiment

It is assumed that two beams satisfy the hypotheses of the first-order Euler-Bernoulli approximation. The system (15.1) is solved taking into account the boundary (15.4), (15.5) and initial conditions (15.6).

Both ends of the beams are rigidly clamped:

$$w_i(0,t) = w_i(1,t) = u_i(0,t) = u_i(1,t) = \\ = \frac{\partial w_i\,(0,t)}{\partial x} = \frac{\partial w_i\,(1,t)}{\partial x} = 0, \;\; i = 1,\;2, \tag{15.5}$$

or they are simply supported

$$w_i(0,t) = w_i(1,t) = u_i(0,t) = u_i(1,t) = \\ = \frac{\partial^2 w_i(0,t)}{\partial x^2} = \frac{\partial^2 w_i(1,t)}{\partial x^2} = 0, \quad i = 1, 2, \tag{15.6}$$

and the initial conditions are taken in the following form

$$w_i(x)_{|t=0} = 0, \;\; u_i(x)_{|t=0} = 0, \;\; \frac{\partial w_i\,(x)}{\partial t}_{|t=0} = 0, \\ \frac{\partial u_i\,(x)}{\partial t}_{|t=0} = 0, \;\; i = 1,\;2. \tag{15.7}$$

The beam 1 is subjected to the uniformly distributed transverse harmonic excitation of the following form

$$q = q_0 \sin(\omega_p t), \tag{15.8}$$

where q_0 stands for the amplitude and ω_p for the frequency of excitation.

In order to carry out the numerical experiments we fix $\omega_p = 5.1$, and the chosen excitation frequency is equal to the beam fundamental frequency (primary resonance). Observe that $q_0 = 5000$ (amplitude of vibrations) corresponds to the case where the beam 1 deflection exceeds the clearance h_k, and the magnitude of clamping $\varepsilon = 1$.

The obtained system of non-linear PDEs is reduced to ODEs using the FDM with the approximation $O(c^2)$. The beam clearance equals $h_k = 0.1$. First, estimation of convergence of the FDM is conducted. For this purpose, the signals of both beams were compared employing different kinds of Runge-Kutta methods: 4th- and 2nd-order Runge-Kutta methods, Runge-Kutta-Fehlberg method, the 4th-order Cash-Karp method, the 8th-order Runge-Kutta Prince-Dormand method as well as the implicit 4th- and 2nd-order Runge-Kutta methods. According to the obtained results, all the above-mentioned methods ensure good convergence. In further computations, the 8th-order Runge-Kutta method keeping the Prince-Dormand accuracy is used.

Table 15.1 reports signals/time histories $w(0.5, t)$ obtained for different number of beam length partitions: $n = 40; 80; 100; 120; 140; 160$.

For $n = 40; 80$, beam deflections strongly differ from each other. Beginning from $n = 100; 120; 140; 160$, deflections of the first beam coincide. In the case of the second beam, the convergence with respect to the number of beam partitions nodes is significantly worse compared to the previous case and occurs only for $n = 120; 140; 160$.

Owing to the already introduced methodology, Table 15.2 represents results of convergence of the FDM regarding the time series $w(0.5, t)$ for $500 \leq t \leq 506$.

Table 15.1 Time histories for different number of nodes $n = 40; 80; 100; 120; 140; 160$

Beam 1	Beam 2
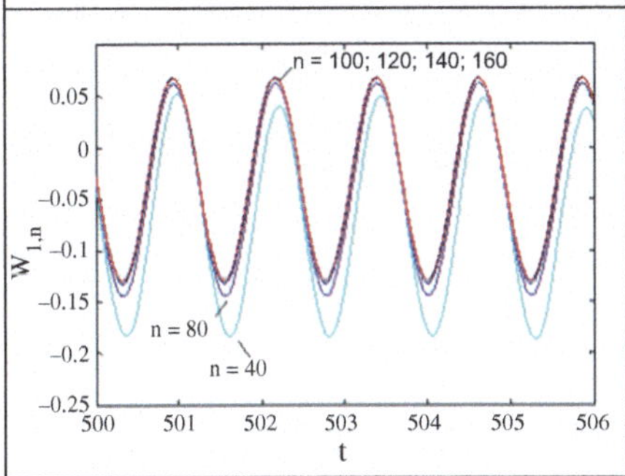	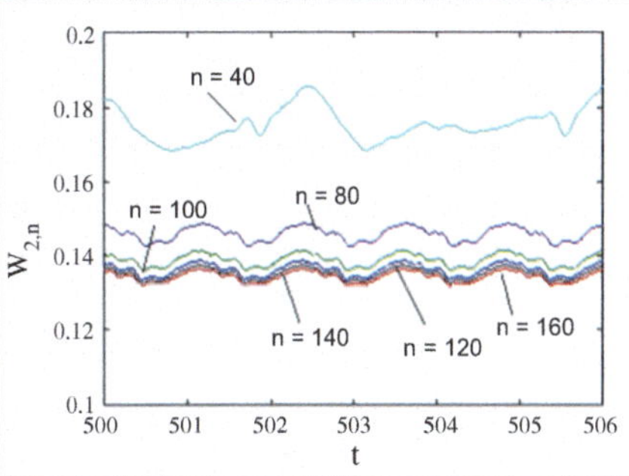

Table 15.2 Convergence of the signals $w_(i, n)(0.5, t)$ for $n = 40; 80; 120$

n	Beam 1	Beam 2
40	20.447%	24.882%
80	7.221%	9.071%
120	0.00166%	0.00521%

In the case of the test/reference signal, the calculations were made for $n = 160$. It can be observed in Table 15.2 that an increase in the number of beam nodes for $x \in [0; 1]$ implies an increase in the convergence. In order to validate convergence of the numerical results, a difference between maximum value of the pattern/reference signal for $n = 160$ and the maximum value of the signals for $n = 40; 80; 120$ obtained in the same time instance have been compared.

Though the beam deflection shapes have been analysed for each time instant, we present here only their exemplary shapes for $t = 501; 501.3; 501.6$ for $x \in [0; 1]$, and for $n = 40; 80; 120; 160$. For $t = 501.3$, $w(x = 0.5, t = 501.3)$ deflection is around zero. The number of maxima and minima for $n = 40$ is fixed, whereas an increase in $n = 80; 120; 160$ implies decrease of the beam deflection to four and five times, respectively. Increase in $n = 120; 160$ causes a sinusoidal shape of the beam 2 snapshot, i.e. the maximum is achieved only for $x = 0.5$. The deflection snapshots of the first and second beams coincide for $n = 120; 160$. Snapshots of beams configurations for $t = 501.3$ and for $n = 40; 80; 120; 160$ are reported in Table 15.3.

In Table 15.4 the frequency power spectra, the 2D and 3D phase portraits, the wavelet spectra for $\omega_p \in [0.5; 5.1]$ and $\omega_p \in [0.5; 2]$, and the Poincaré pseudo maps are reported for different numbers of the beam spatial coordinates.

In Tables 15.4 and 15.7, the number 1 corresponds to the number of spatial coordinate partition, while 2 corresponds to the beam number.

For $n = 40$, the frequency power spectrum exhibits linearly dependent frequencies ω_p: $\omega_{p/2}$, $\omega_1 = (2\omega_p)/5$, $\omega_2 = (3\omega_p)/5$, a linear combination of those

Table 15.3 Snapshots of the beam shape (for $w_{i,n}(x)$, $x \in [0; 1]$), computed for different $n = 40; 80; 100; 120; 140; 160$ and $t = 501.3$

Beam 1	Beam 2
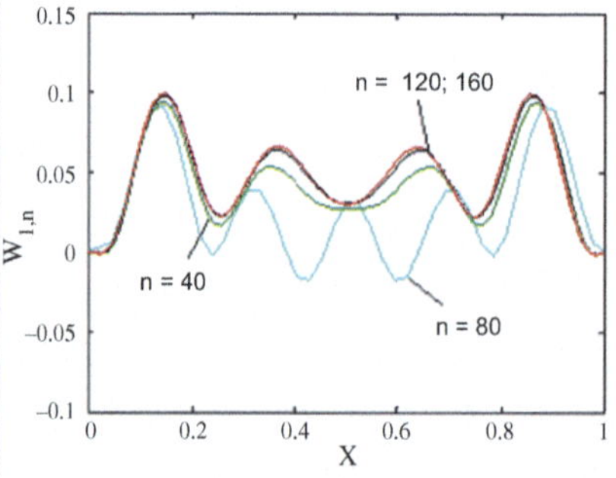	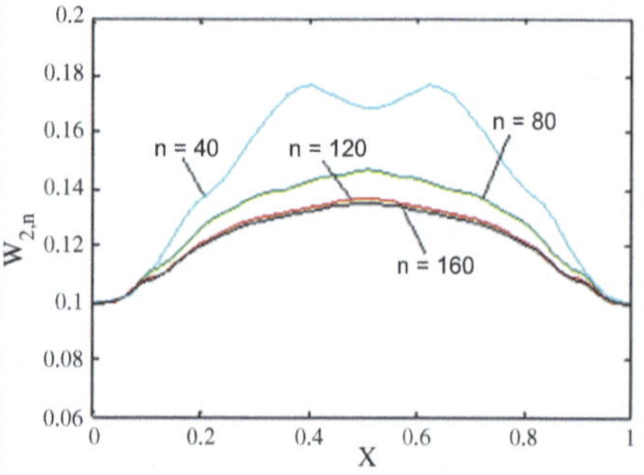

Table 15.4 Dynamical characteristics of the beams 1 and 2 for $n = 40; 80; 120; 160$

40		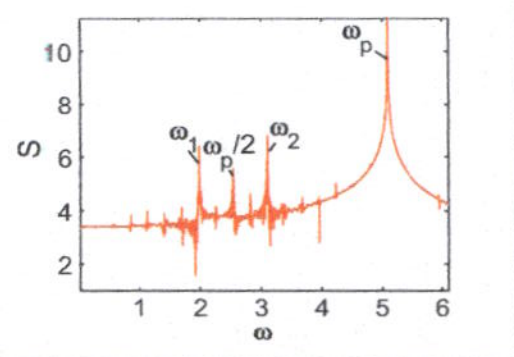	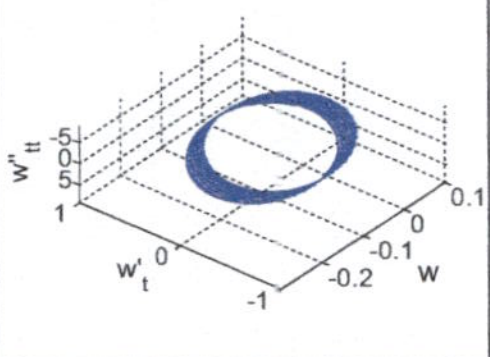	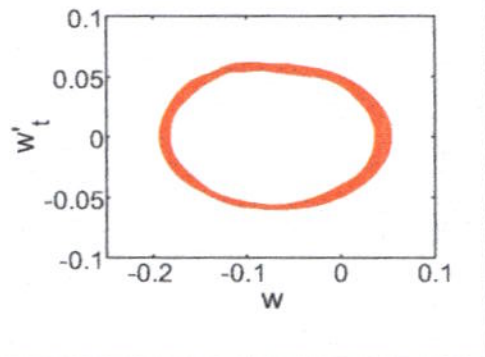	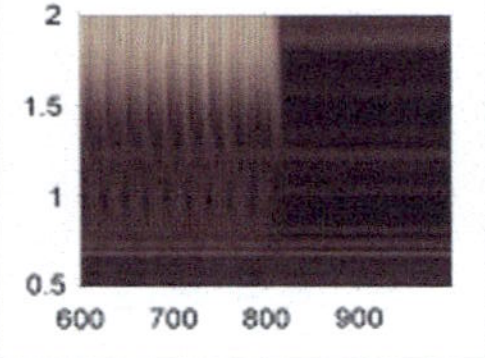	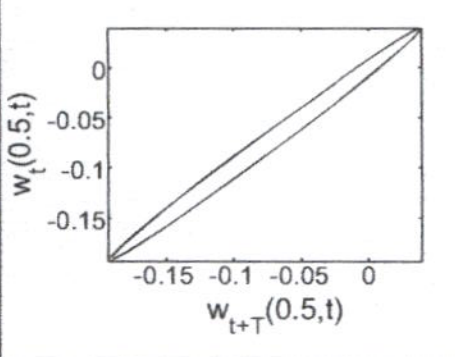
		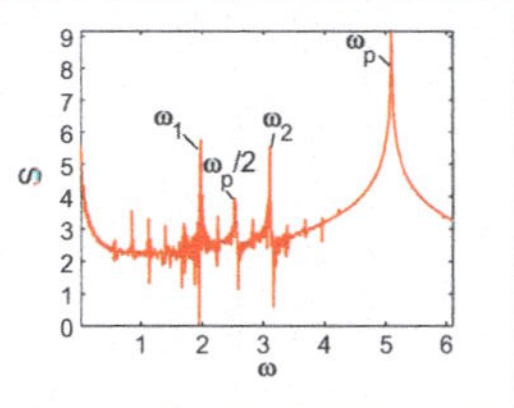	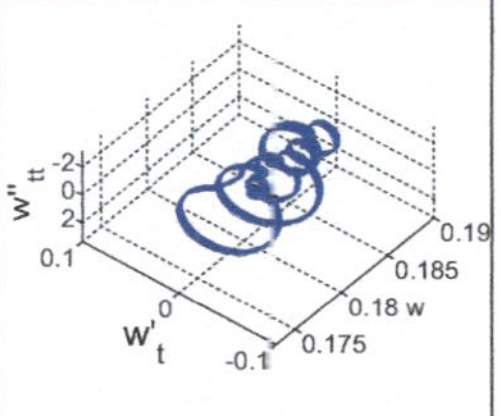	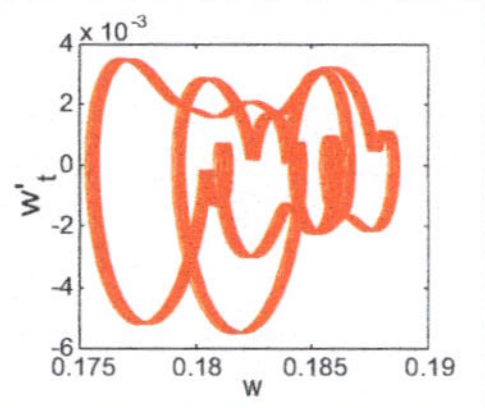	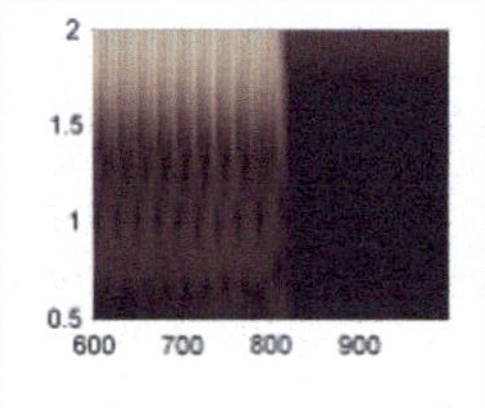	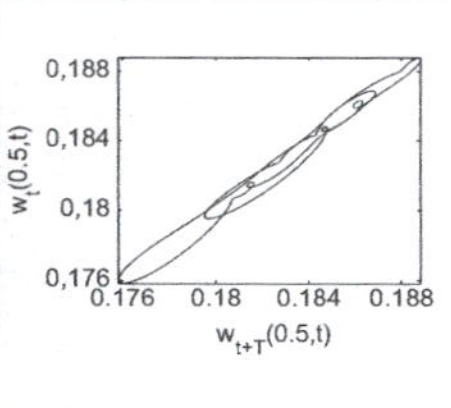
120, 160		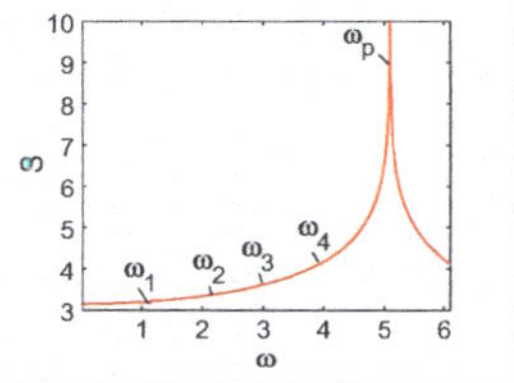	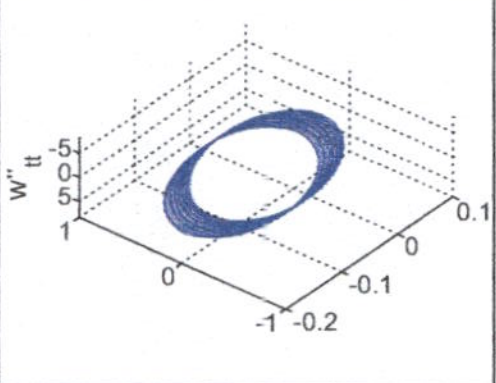	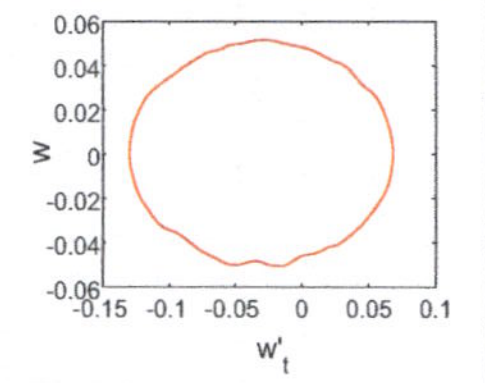	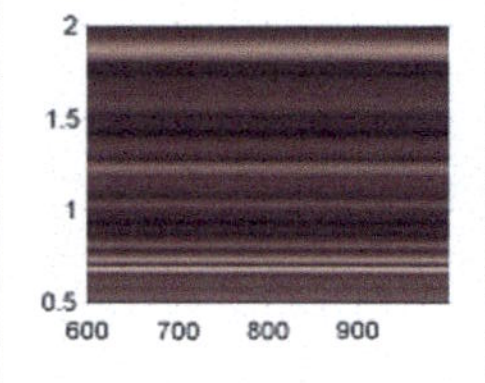	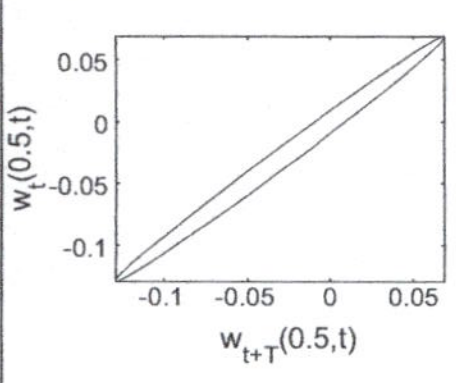
		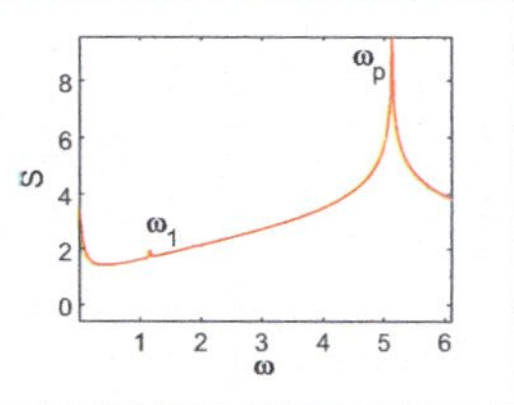	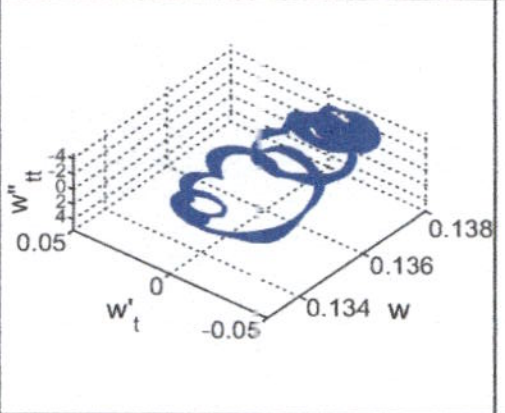	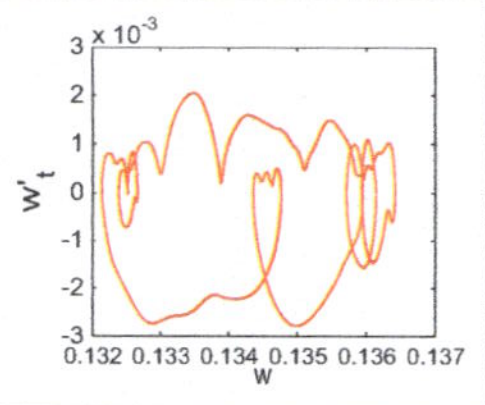	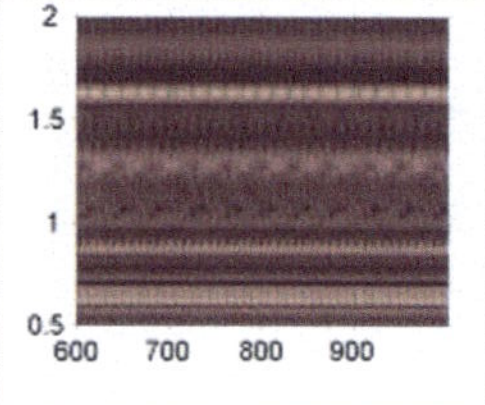	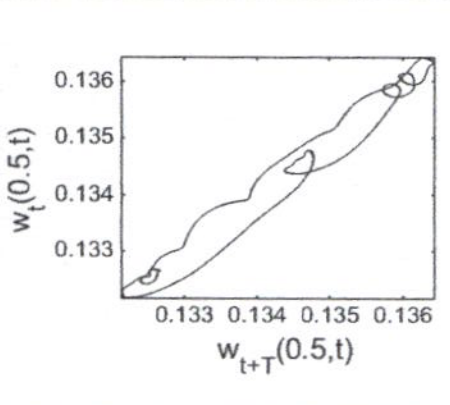

frequencies ω_5, $\omega_{(7,)}$ ω_8, and a linearly independent frequency. Power spectra of both beams are the same. However, the 2D and 3D phase portraits are different for both beams. The wavelet spectrum exhibits $\omega_{p/15}$, $(4\omega_p)/15$, $\omega_{p/3}$ for $\omega \in [0.2; 2]$. The remaining frequencies ω_5, $\omega_{(7,)}$ ω_8 present in the frequency spectrum are not exhibited by the wavelet spectrum (the amplitude of harmonics associated with these frequencies is very small in comparison to the frequency ω_2).

The Poincaré pseudo-map presents a simple closed curve for the first beam, whereas a closed curve with loops can be observed for the second beam. This corresponds to a stable quasi-periodic orbit being simple for the beam 1, and its period-doubling bifurcated counterpart exhibited by the beam 2.

A similar situation takes place for $n = 120$ and $n = 160$. The power spectra of both beams comprise the following frequencies ω_p, $\omega_1 = \omega_{p/5}$, $\omega_2 = 2\omega_{p/5}$, $\omega_3 = 3\omega_{p/5}$, $\omega_4 = 4\omega_{p/5}$, and a small component of the independent (irrational) frequency. This means that we have detected a quasi-periodic stable orbit, with a qualitatively similar interpretations as in the previous case ($n = 40$), i.e. the beam 2 exhibits period-doubled quasi-periodic orbit. The wavelet spectrum for $\omega \in [0.2; 2]$ exhibits the same frequencies as the Fourier spectrum. Pseudo-Poincaré maps confirm the results obtained by phase portraits. On the graph, the dark brown phase difference corresponds to synchronization of vibrations. Synchronization for $n = 40$ takes place for $\omega \in [3.5; 5]$. For $\omega \in [1; 2]$ and $t \in [315; 340]$, synchronization takes place for five frequencies. For $n = 160$, synchronization of vibrations has been detected for $\omega \in [3; 5]$ on the whole considered time interval, whereas for $\omega \in [0.5; 2]$, synchronization is associated with nine frequencies for $t \in [315; 340]$. Both graphs present also synchronization on the frequency $\omega_{p/2}$ at certain time instants.

An increase in the number of nodes yields a decrease in the frequencies number to five, i.e. the so-called regularisation of beam vibrations takes place. Finally, let us emphasize the occurrence of the chaotic phase synchronisation of vibrations of beams.

In Table 15.5, we report surfaces of the beams deflections for $t \in [95; 100]]$ and for $n = 40; 120; 160$. The horizontal axis denotes the beam length, whereas the vertical axis stands for the beam deflection, and the remaining axis denotes time.

We have achieved convergence of the deflection surfaces of the middle lines of the beams with respect to time for $n = 120$ and $n = 160$. Therefore, convergence of the numerical results is guaranteed on the whole time and space intervals.

Figure 15.2a, b presents the curves obtained based on estimation of the largest Lyapunov exponents by the Wolf, Rosenstein and Kantz methods versus n, i.e. nodes number. The obtained values result from the computation using the 8th-order Runge-Kutta method. It should be emphasized that different methods of computation of the Lyapunov exponent are needed to obtain reliable/true value and, consequently, reliable estimation of chaos. When using the FDM, for the number of beams nodes $n = 40, 80, 120; 160$ for any mentioned method, the Lyapunov exponents coincide with an accuracy up to the third decimal digit. Furthermore, all of the largest Lyapunov exponents (LLEs) are positive.

Table 15.5 Surfaces of the middle beam line deflections

$n = 40$	Beam 1	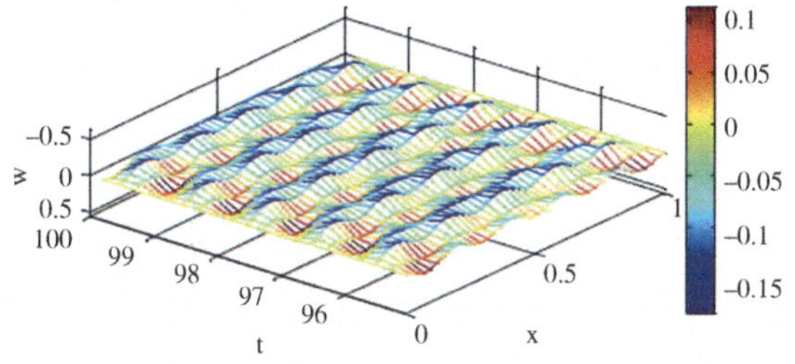
	Beam 2	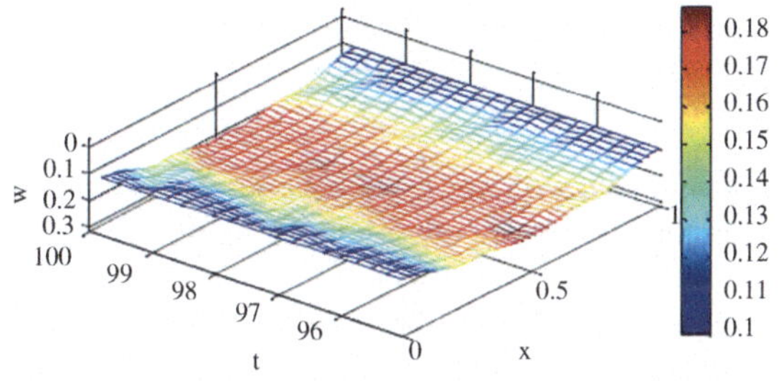
$n = 120; 160$	Beam 1	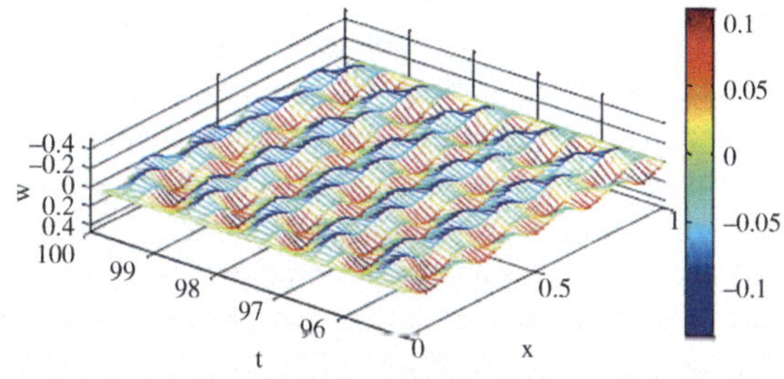
	Beam 2	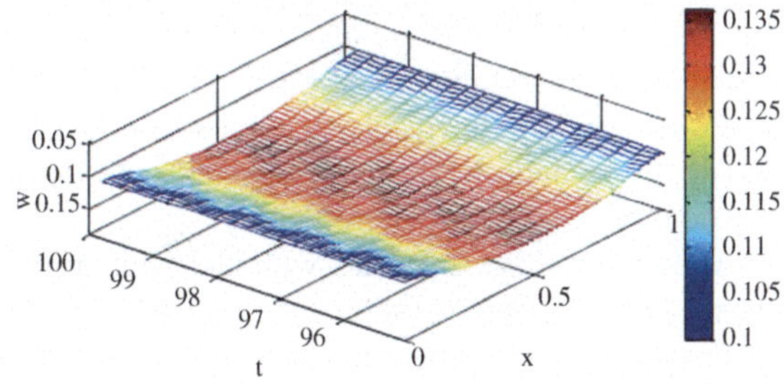

In what follows we define how the LLEs depend on the method employed for solution of the Cauchy problem. For this purpose, we have computed the Lyapunov exponents using the Wolf, Kantz and Rosenstein algorithms for $n = 160$. On the

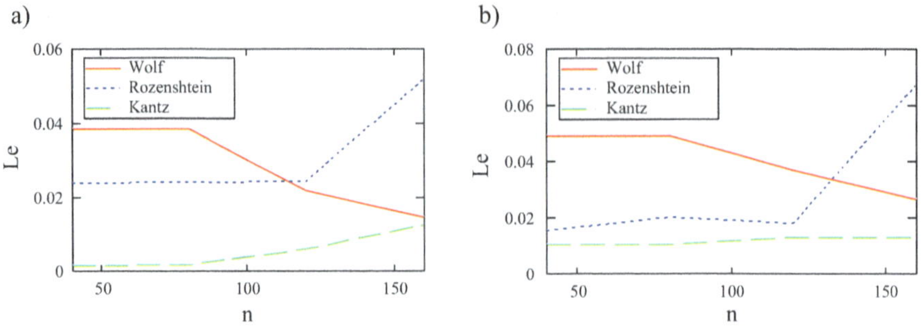

Fig. 15.2 Largest Lyapunov exponent (LLE) versus n, obtained using the Wolf, Rosenstein and Kantz methods. (**a**) Beam 1. (**b**) Beam 2

contrary to the dynamic characteristics, we have not observed full coincidence of the results. However, the difference between the minimum and the maximum of the exponents computed for different Runge-Kutta methods using the Wolf algorithm for the beam 1 is about 0.07, whereas the same done by the Rosenstein and Kantz methods yields the difference of 0.008. Convergence up to the second decimal digit has been observed for each of the computational methods of the LEs computation.

It should be noted that all values of the LLE, independently of the employed method of the solution of the Cauchy problem, the beam partition number, and the employed LLE method, are positive. Since the amplitude of the beams vibrations is small, one can assume that it is sufficient to use a linear theory of beam vibrations.

This can be done using our developed approach based on the neural networks (see references [52, 53], where the spectrum of the Lyapunov exponent while employing FDM for $n = 40$ and $n = 160$ has been computed for both beams). For $n = 40$ the second beam exhibited hyper-chaos (two positive LEs), whereas the first beam exhibited classical chaos (one positive LE). Increase of n yielded change of vibration regime of the second beam. For $n = 160$ we have found only one positive LE for both beams (chaos). We emphasize again that the number of chosen nodes essentially influences the obtained numerical results, i.e. if the beam is modelled only by a few degrees-of-freedom, the results may be not correct.

In order to validate this remark, let us compare signals of other beam characteristics, taking into account the geometric non-linearity.

A comparison of the beams time histories with and without the account of the geometric non-linearity are reported in Table 15.6.

Table 15.7 contains dynamic characteristics of the geometrically linear beam for $n = 160$ (rk8pd).

For the same parameters, we have obtained the results for geometrically linear beams taking into account their contact interactions (Table 15.7).

While comparing vibration signals of geometrically linear and non-linear problems, we detected coincidence of the results in the case of the first beam. For the second beam, the vibration signals regarding linear and non-linear cases are similar with respect to their forms, but differ with respect to amplitude magnitudes.

Table 15.6 Comparison of the beams signals with/without the geometric non-linearity

Signal $w(t)$ of beam 1	Signal $w(t)$ beam 2
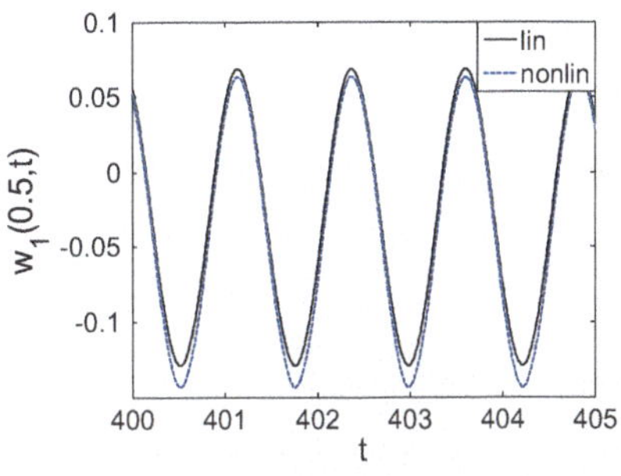	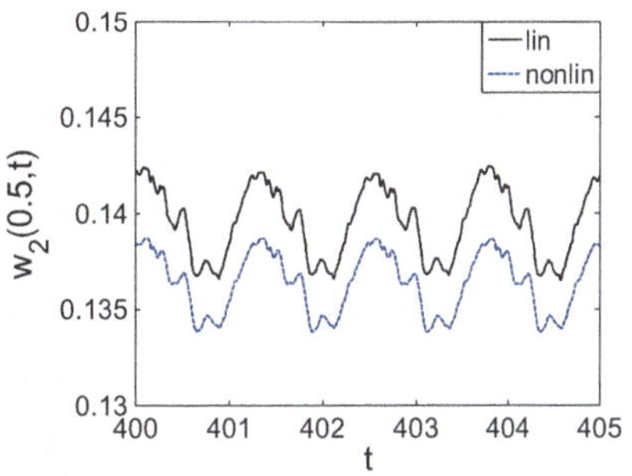

Therefore, we demonstrated the influence of the geometric non-linearity even for small deflections of beams.

A comparison of the phase portraits and Poincaré pseudo-maps for geometrically linear (Table 15.7) and non-linear (Table 15.4) cases indicates that convergence of results obtained for both beams should be investigated to validate their reliability.

The frequency power spectra differ from each other in the linear and non-linear case. Namely, in the linear case, the frequencies ω_1 and ω_2 are not exhibited. This can be caused by their small power. The Lyapunov exponents for both linearly geometric beams are positive.

Therefore, carrying out the complex analysis of solutions to the problem, one can conclude that vibrations of the studied two-layer beam with a small clearance between the layers are chaotic. The efficient beam partition while using the FDM is $n = 160$. Since the solutions for the Cauchy problem coincide for all employed modifications of the Runge-Kutta methods, the 2nd-order Runge-Kutta method can be used. The comparison of the geometrically linear and non-linear problems implies the occurrence of non-linear terms in the solution, despite the amplitude of beam vibrations coincides with the linear approximation.

Let us consider how the clearance (h_k) influences vibrations of the package composed of two beams. We report here the numerical results obtained for $h_k = 0.5$. Table 15.8 presents vibration signals $w(0.5, t)$ obtained for the interval $x \in [0; 1]$ divided into different numbers of partitions: $n = 40; 120; 240; 360$. It has been found that convergence with respect to the vibrational signal $w(x, t)$ takes place for both beams already in the initial time interval $t \in [0; 908]$. However, it is lost for $t \in (908; +\infty]$ and the considered n. Furthermore, convergence with respect to the initial time interval is achieved by employing the FDM procedure for $n = 240$ only. It means that an increase in deflection requires an increase in the number of partitions (nodes) used in the FDM. Table 15.9 presents frequency power spectra, phase portraits, and wavelet spectra. Observe that for $n = 40$ the power spectrum consists of the following frequencies: $\omega_1 = 0.939$, $\omega_2 = 2.1419$, $\omega_3 = 3.4737$,

Table 15.7 Dynamical characteristics of the geometrically linear beams 1 and 2 for $n = 160$ (rk8pd method)

		Power spectrum	Phase portrait 3D	Phase portrait 2D	Wavelet spectrum $\omega(t)$, $\omega \in [0.2; 2]$	Poincaré pseudo-map
160	1	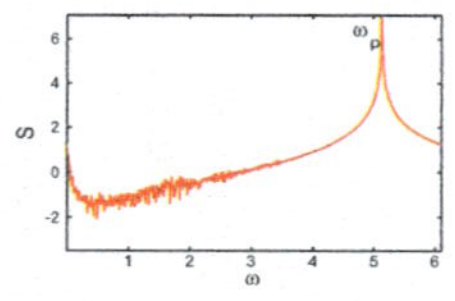	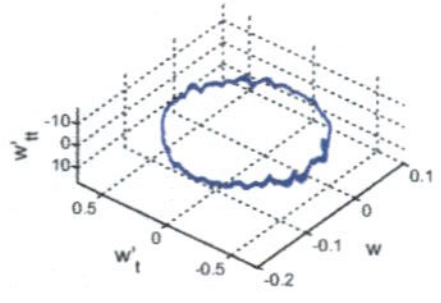	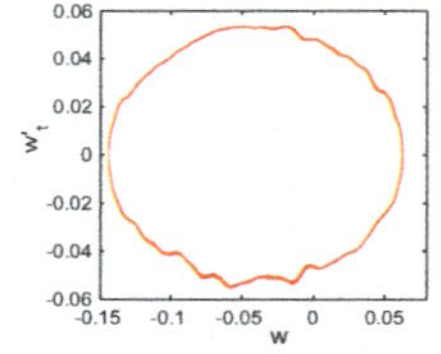	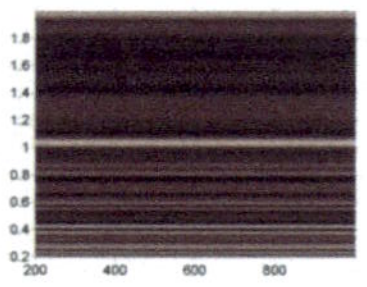	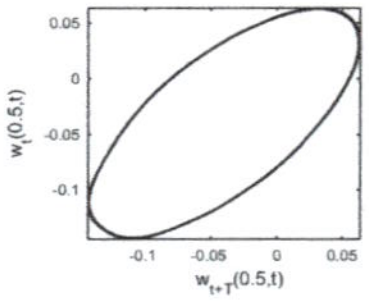
160	2	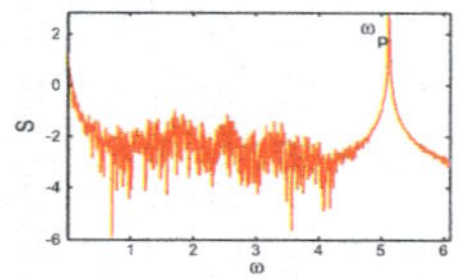	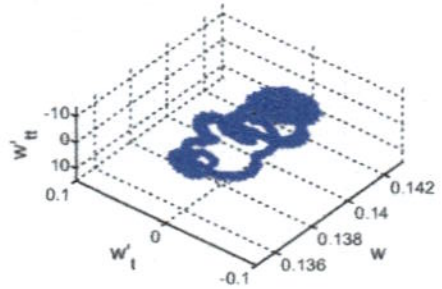	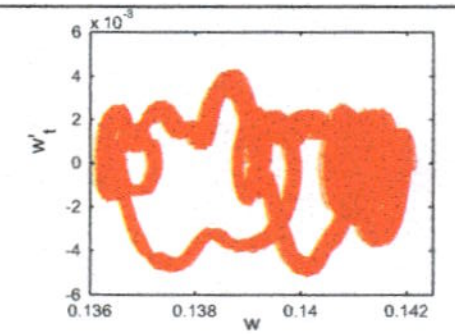	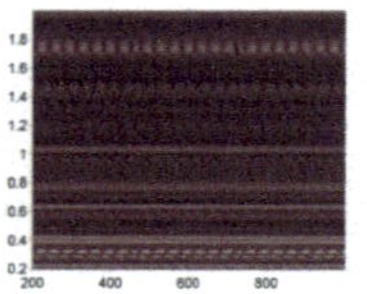	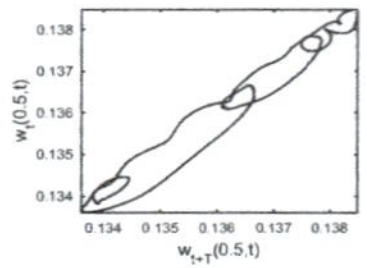

Table 15.8 Time histories for different number of nodes $n = 40; 120; 240; 360$

Signal $w(t)$ of beam 1	Signal $w(t)$ of beam 2
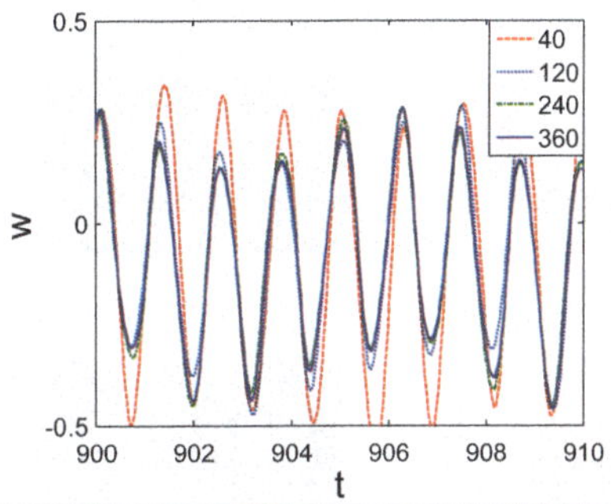	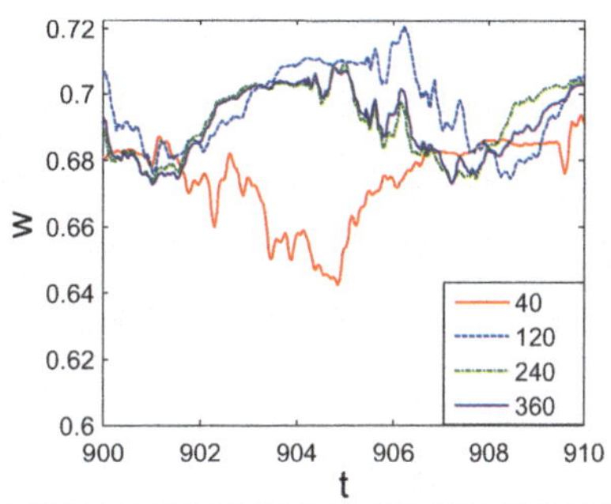

$\omega_4 = 4.2899$, ω_p for both beams, i.e. the frequencies are synchronized. However, for $n = 240; 360$ the frequencies changes and the power spectrum exhibits $\omega_1 = 0.81625$, $\omega_2 = 1.878$, $\omega_3 = 3.222$, $\omega_4 = 4.161$, ω_p for both beams. Therefore, convergence of the results for the PDEs with the account of geometric non-linearity in the chaotic regime can be only achieved based on the frequency characteristics, i.e. in the averaged sense (this has been pointed out in monograph [54]). However, in spite of the essential influence of the geometric non-linearity, strong dependence on the initial conditions is also observed here.

15.4 Application of the Principal Component Analysis

As an example, let us consider a matrix of deflections for the beam 1, obtained as a solution to the system of geometrically non-linear equation with the account of contact interaction. The matrix $\tilde{\mathbf{W}} = \left(\tilde{w}_{ij}\right)_{i=1,j=1}^{n=73535,m=160}$ is linearly split into $\tilde{\mathbf{W}} = \mathbf{TP}^t + \mathbf{E}$. To compute the matrix $\mathbf{T}$ and the matrix of loads $\mathbf{P}$, we employ a singular matrix $\tilde{\mathbf{W}}$ in the form $\tilde{\mathbf{W}} = \mathbf{USV}^t$, where $\mathbf{T} = \mathbf{US}$ $\left(\mathbf{S} = \left(s_{ij}\right)_{i=1,j=1}^{n=160,m=160}\right)$, $\mathbf{P} = \mathbf{V}^t$. Matrices $\mathbf{U} = \left(u_{ij}\right)_{i=1,j=1}^{n=73535,m=160}$ and $\mathbf{V} = \left(v_{ij}\right)_{i=1,j=1}^{n=160,m=160}$ are orthogonal. The main diagonal of the matrix $\mathbf{S}$ contains the eigenvalues, whereas the remaining elements of the matrix $\mathbf{S}$ are equal to zero. The eigenvalues s_{ij} of the matrix $\tilde{\mathbf{W}}$ follow: $s_{11} = 989$; $s_{22} = 724$; $s_{33} = 176$; $s_{44} = 41$; $s_{55} = 14$; $s_{66} = 11.76$; $s_{77} = 3.96$; $s_{88} = 2.86$; $s_{99} = 1.9$; $s_{1010} = 1.85$; $s_{1111} = 1.59$; $s_{1212} = 1.41$; $s_{1313} = 1.04$; $s_{1414} = 0.61$; $s_{1616} < 1$; $s_{3131} = \cdots = s_{160160} = 0$.

The value of main components A has been chosen so as to make sure that the dependence of dispersion on the number of main components A TRV(A) and the dependence of the dispersion on the number of main components ERV(A) will rapidly change their behaviour.

Table 15.9 Dynamical characteristics of the beams 1 and 2 for $n = 40; 240; 360$

	n=40		
	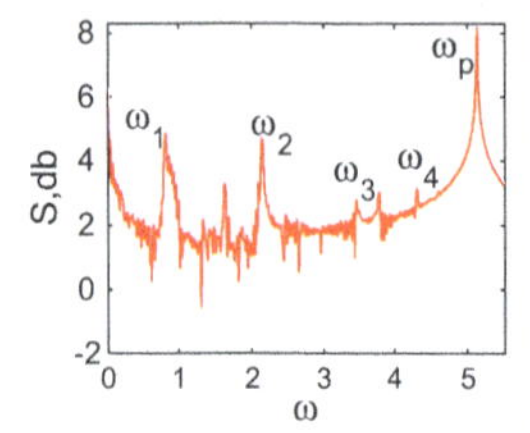	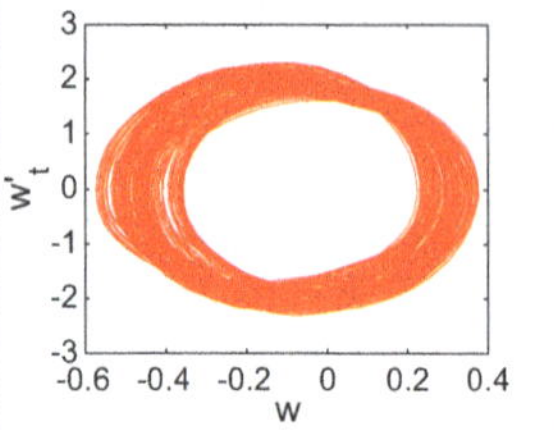	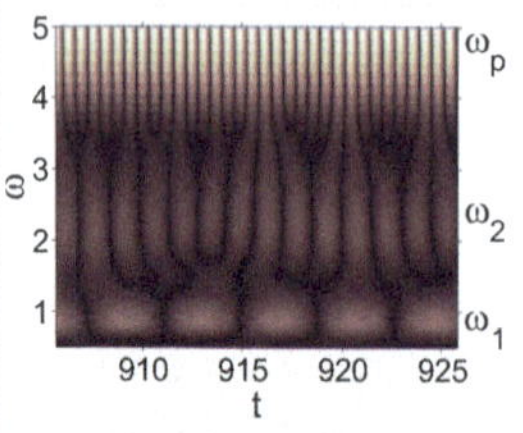
	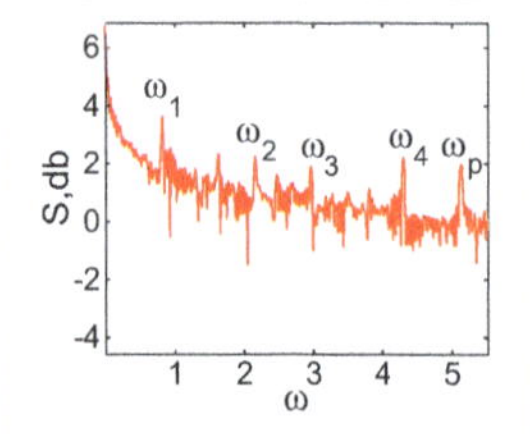	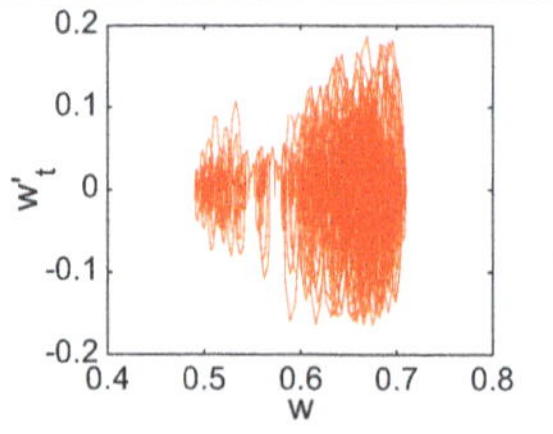	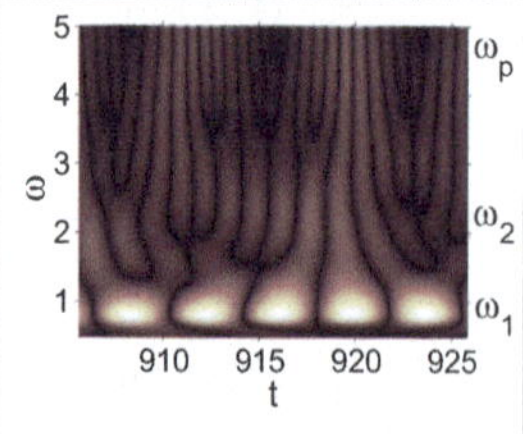
	n=240, 360		
	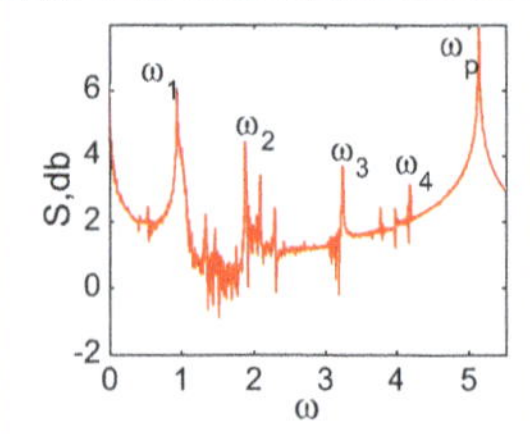	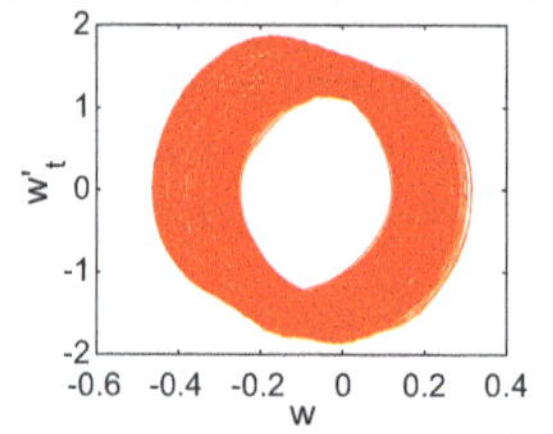	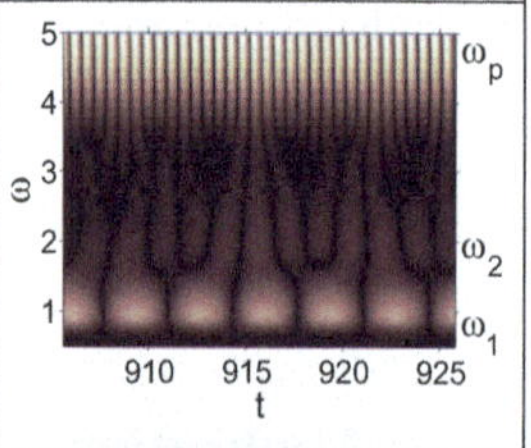
	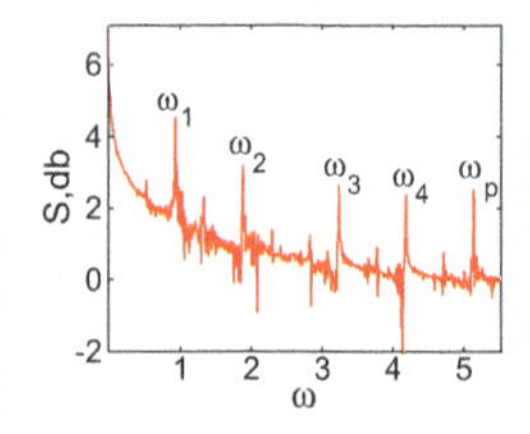	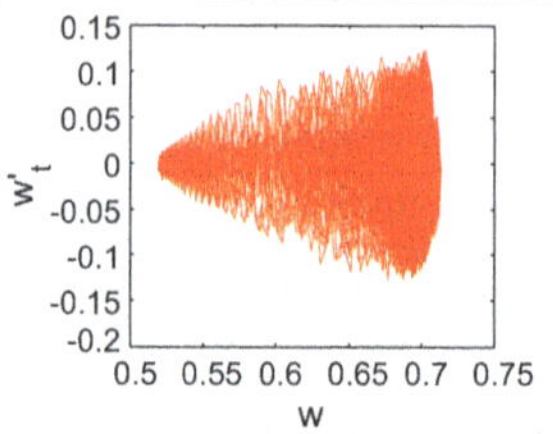	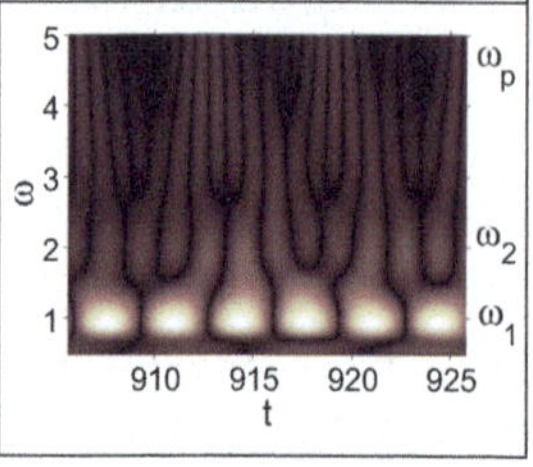

If one takes the number of main components, then the matrix of residuals $\mathbf{E}_A = \left(\mathbf{e}_{Aij}\right)_{i=1,j=1}^{n=73535,m=160}$ and it is computed as $\mathbf{E}_A = \tilde{\mathbf{W}} - \mathbf{T}_A\mathbf{P}_A^t$. The matrices have different dimensions: $\mathbf{T}_A = \left(\mathbf{t}_{Aij}\right)_{i=1,j=1}^{n=73535,m=A}$ and $\mathbf{P}_A = \left(\mathbf{p}_{Aij}\right)_{i=1,j=1}^{n=160,m=A}$. One of the important properties of the Principal Component Method is the orthogonality (independency) of main components. This is why $\mathbf{T}_A$ does not change with increasing number of components, i.e. only one more column, corresponding to a new direction, is added. The same holds for the loading matrix $\mathbf{P}_A$.

In order to compute exponents TRV and ERV, one can obtain the residuals matrix $\mathbf{E}_A = \left(\mathbf{e}_{Aij}\right)_{i=1,j=1}^{n=73535,m=160}$ for each number of the main components A, and hence compute the required exponents. The following steps are needed.

1. Compute $v_i = \sum_{j=1}^{j=160} e_{Aij}$. This quantity defines a square of the distance between the input vector $\mathbf{w}_i$ and its projection onto the space PC. Its small magnitude defines better approximation to the i-th pattern.
2. Averaged (for all patterns) square of the distance is yielded by the formula $v_0 = \frac{1}{160}\sum_{i=1}^{n} v_i$.
3. Find the full dispersion TRV $= \frac{v_0}{73535}$ and the explaining dispersion ERV $= 1 - \frac{160 * v_0}{\sum_{ij} \tilde{w}_{ij}^2}$.

In order to separate the signal from the noise, one should employ the linear decomposition of the matrix $\mathbf{W}_A = \mathbf{T}_A\mathbf{P}_A^t + \mathbf{E}_A$, where the matrix $\mathbf{W}_A = \left(\mathbf{w}_{Aij}\right)_{i=1,j=1}^{n=73535,m=160}$ represents the signal purified from noise, and A stands for the amount of main components. The matrix $\mathbf{T}_A = \left(\mathbf{t}_{Aij}\right)_{i=1,j=1}^{n=73535,m=A}$ is called a score matrix. The matrix $\mathbf{P}_A = \left(\mathbf{p}_{Aij}\right)_{i=1,j=1}^{n=160,m=A}$ is known as a matrix of loadings. $\mathbf{E}_A = \left(\mathbf{e}_{Aij}\right)_{i=1,j=1}^{n=73535,m=160}$ is referred to as a residual matrix. In the considered example, the exponents TRV(A) and ERV(A) essentially change the direction for $A = 3; 4$. This means that in order to get the matrix $\tilde{W}$, the value of main components was chosen to be three or four.

As it can be seen from the reported characteristics (Table 15.10), for $n = 160$ and for the contact interaction problem, the frequency spectra for both beams exhibit five linearly dependent frequencies ω_p, ω_1, ω_2, ω_3, ω_4, whereas for geometrically nonlinear and contact interaction problem, the noisy components of the frequency power spectra are exhibited.

After the time series reconstruction by means of the principal component analysis, the number of frequencies has been reduced for the beam 1 in the case of the geometrically linear problem and the locations of four frequencies have been detected (they coincide with the frequency amplitudes of the geometrically non-linear problem). In the case of the signal reconstruction obtained for the contact interaction problem, all frequencies are exhibited. The beam 2, with no account of the geometric non-linearity, exhibits chaotic vibrations.

After time series reconstruction using the principal component analysis, the frequencies localization takes place. The frequencies are linearly dependent regarding the geometrically non-linear problem. It should be emphasized that since we considered the contact interactions of the beams, the first contact of the beams exhibited for small beam deflections, i.e. where geometric non-linearity can be neglected, implies a rapid occurrence of the chaotic vibrations of the beams (Table 15.7). Therefore, the contact interaction implies chaos occurrence.

Table 15.10 Comparison of the power spectra $S(\omega)$ for the geometrically linear and non-linear beams

		n =160	n =160
	After PCA	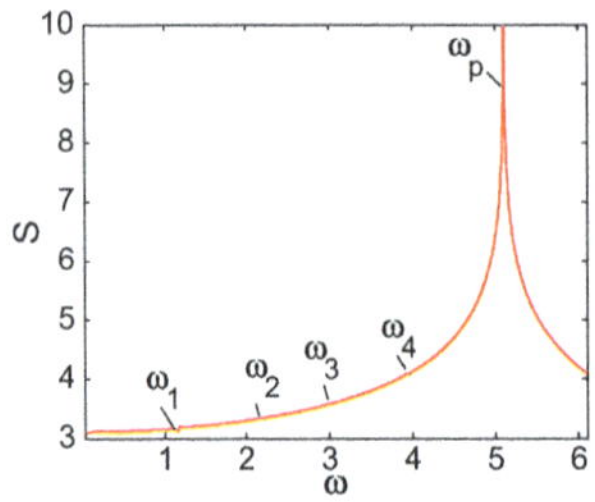	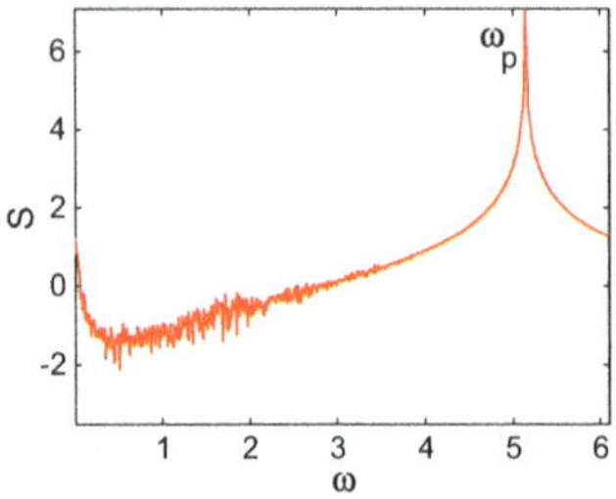
	Before PCA	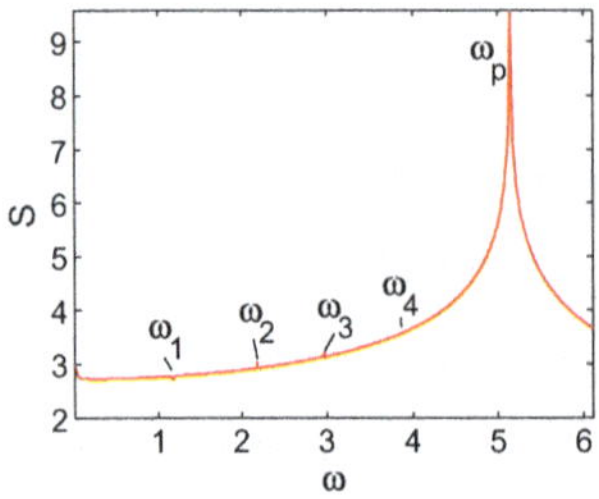	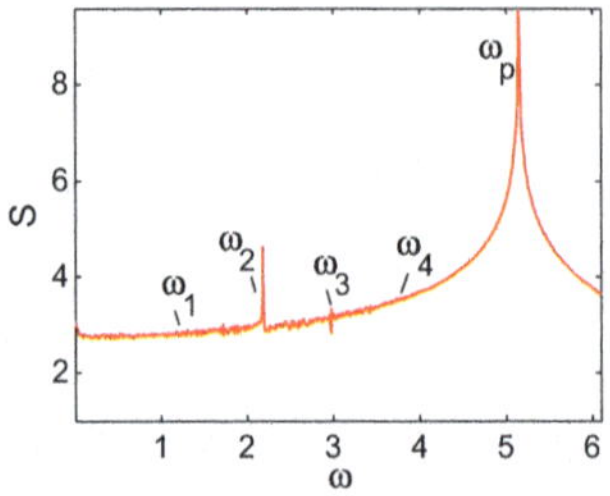
	After PCA	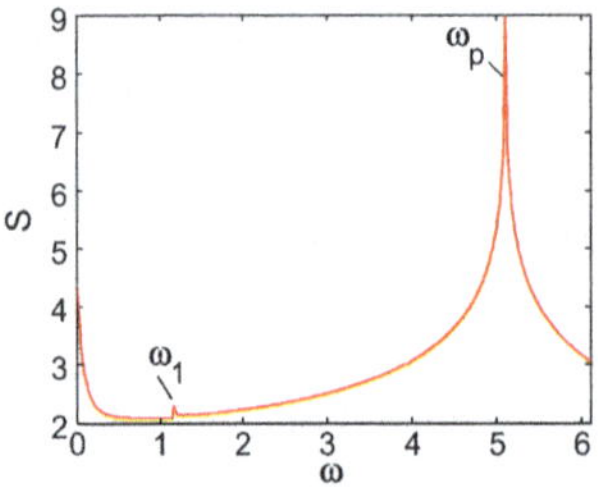	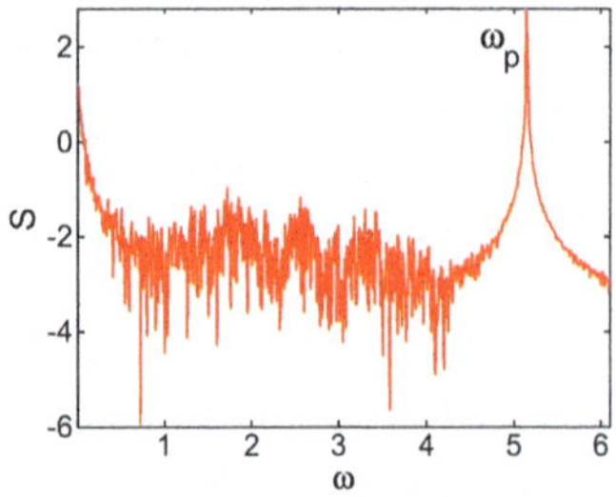
	Before PCA	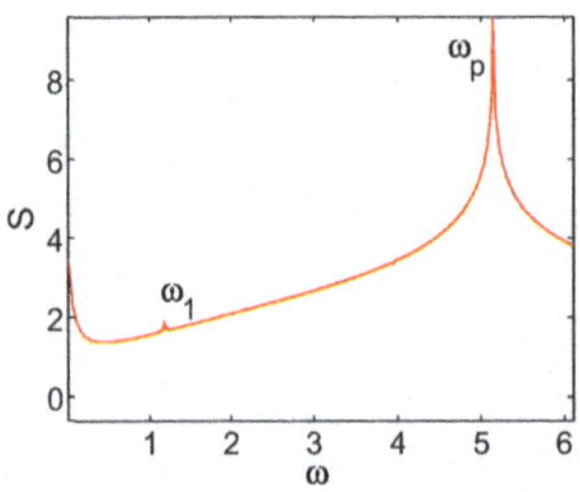	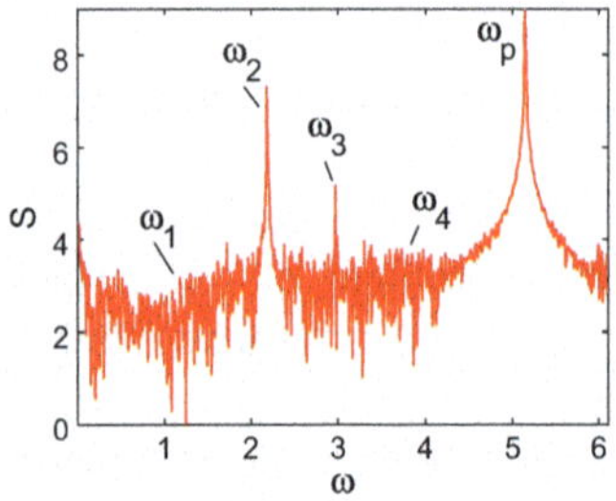

In addition, we have studied how the choice of boundary conditions influences non-linear vibrations of simply supported beams (15.6) and we have detected a different type of non-linear dynamics in comparison to the earlier studied boundary conditions (15.5).

In particular, we have illustrated how the number of nodes, i.e. number of partitions of the interval [0; 1] and the magnitude of the clearance between beams affects convergence of numerical results.

To conclude the proposed methodology for detection of true chaos is based on the complex investigation of a signal/time history from the point of view of non-linear dynamics, focusing on construction and analysis of the signals, power frequency spectra, Poincaré pseudo maps, wavelet spectra, Lyapunov exponents as well as 2D and 3D phase portraits.

Owing to the introduced methodology, the convergence of FDM has been achieved while solving the system of non-linear PDEs/ODEs with the account of the geometric von Kármán non-linearity and the contact interaction for the mathematical model established in the frame of the kinematic first-order hypotheses.

The employed methodology has proved that the chaotic vibrations of two-layer beams can be detected based on the used methods and computational algorithms. Although the magnitude of the clearance between the beams is small, it has been found that chaotic vibrations of beams appear even for small amplitudes. However, we have illustrated that in spite of small amplitudes, there is a need to take the geometric non-linearity into account while constructing a feasible mathematical model. We have detected, the occurrence of the phase synchronization of beams vibrations for the investigated system, among others. In addition, all frequencies exhibited by beam vibrations are in resonance relation with the excitation frequency.

In order to confirm the occurrence of true chaotic vibration, we have demonstrated convergence of the numerical results regarding the signal, not only with respect to the Fourier frequency power spectrum, which is recognized and often employed. Furthermore, we obtained convergence of the numerical results when the computation step (with respect to the spatial coordinate) was increased twice.

Furthermore, we have shown how applications of the PCA for purification of the chaotic signal of the beam allows to localize the fundamental frequency of vibrations of the studied structure composed of two beams.

Finally, it should be mentioned that investigation of vibrations character should be carried out with the help of both Fourier- and wavelet-type analyses. The wavelet analysis can be viewed as a microscope allowing for vibration analysis on the given time interval. It yields the system vibration types and the associated frequency characteristics, which cannot be detected by using only standard FFT (Fast Fourier Transform). Also, according to the Gulick definition of chaos, there is a need to estimate a sign of at least the largest Lyapunov exponent while defining either regular or chaotic dynamics.

Chapter 16
Unsolved Problems in Nonlinear Dynamics of Thin Structural Members

The problems in the field of nonlinear dynamics of shells that remain to be solved can be summarized as follows (this item has been extended in comparison to the first book editions).

1. Formulating the important boundary problems for the nonlinear dynamics of shells without assuming slenderness and central bending (i.e. for arbitrary rotations).
2. Formulating the important boundary problems for the coupled theory of thermoelasticity of shells without assuming slenderness and central bending, which means formulating the boundary problem for an initial set of hyperbolic and parabolic equations or of hyperbolic equations.
3. Obtaining numerical results for coupled problems of thermoelasticity for a wider class of boundary conditions than that investigated in this book.
4. Formulating a mathematical theory of the initial–boundary conditions within the Timoshenko theory of shells, taking into account different types of nonlinearity (geometric and physical nonlinearities, nonlinear elastic–plastic properties, and structural nonlinearities).
5. Formulating the important initial–boundary conditions for the coupled theory of thermoelasticity for polymer shells.
6. Investigation of the coupling of deflection and temperature fields for polymers in the framework of nonlinear geometry.
7. Research on obtaining an energy solution for nonlinear initial–boundary conditions in the vicinity of corners and of points or curves where the type of boundary conditions changes.
8. Detailed analysis of the regions of applicability of various nonlinear initial–boundary conditions in the theory of shallow shells rectangular in plan.
9. Investigation of the effect of local thermal shock in a three-dimensional theory, without the hypothesis of a linear distribution of the thermal field through the shell thickness.

J. Awrejcewicz, V. A. Krysko, *Elastic and Thermoelastic Problems in Nonlinear Dynamics of Structural Members*, Scientific Computation,
https://doi.org/10.1007/978-3-030-37663-5_16

10. Detailed investigation of the initial–boundary conditions in the theory of multilayer shells with coupled temperature and strain fields, in geometrically linear and nonlinear cases.
11. Investigation of the influence of local nonideal thermal contact on the stress–strain state in the theory of multilayer plates and shells for different kinematic models (Kirchhoff–Love, Timoshenko, etc.).
12. Derivation of the important initial–boundary problems in the nonlinear theory of shallow shells, including coupling of strain and temperature fields, from initial–boundary conditions of the three-dimensional coupled theory of thermoelasticity (for hyperbolic–parabolic and hyperbolic sets of equations).
13. Research on the initial–boundary conditions for nonlinear coupled thermoelastic problems of shells applying the quantitative theory of differential equations.
14. Finding a class of nonlinear initial–boundary problems of mathematical physics for which prior estimates of the solution can be obtained using the methods presented in this book.
15. Investigation of spatial–temporal chaos in the theory of plates and shells including coupling of strain and temperature fields with geometric, physical and structural nonlinearities.
16. Problems of aeroelasticity of shells interacting with transonic and supersonic flows of an ideal gas, including coupling of the strain, temperature and gas fields.
17. Research on nonsymmetric buckling of shells subjected to longitudinal periodic loads.
18. Novel structural modelling and systems identification.
19. Buckling and postbuckling behaviour of composite shells vs. their load-carrying capability.
20. Development of optimal control problems of minimizing the dynamic response of composite shells.
21. Prediction of the fatigue and cracks in industrial machineries composed of classical structural members (beams, plates and shells) and vibration monitoring.
22. Modelling of cracks including transverse/longitudinal, open, breathing, slant, surface/subsurface cracks.
23. Modelling and investigation of the effects of multi-layer symmetric/asymmetric laminates with regard to the angles of fibre orientation.
24. Modelling and analysis of failure and damage of structural members subjected to various loads.
25. Investigation of fluid/gas structure interaction for composite shells and panels.
26. Mathematical non-classical modelling of non-linear vibrations of functionally graded shells and panels under thermo-mechanical loads.
27. Development of mathematical models governing non-linear dynamics of hyper-elastic shells based on the modified neo-Hookean, Mooney-Rivlin and Ogden-Fung models.
28. Modelling and experimental investigations of the influence of the particles size of micro/nano state on elastic properties of matrix composites.

29. Study of effects of neutral surface concept for fundamentally graded piezoelectric plates/shells.
30. Development of the micromechanics theories of multiple inclusions into homogenmeous/heterogeneous matrices.
31. Research devoted to modelling theoretical and numerical studies of the influence of porosity within functionally graded nanostructures under hygro-thermo-mechanical loading.
32. Development of theories to fit the size-dependent dynamic/static phenomena of micro/nano-structural members (so far the strain gradient and modified couple stress theories are available) with application to MEMS, NEMS and atomic force microscope micro/nano-cantilevers.
33. Modelling and investigation of bifurcation and chaos of micro/nano-structural members and their interaction in thermal magnetic and environment.

Some of the mentioned items have been analysed in this book. To conclude this book, a prior estimate of the error of the Bubnov–Galerkin method for an abstract hyperbolic equation with variable operator coefficients will be presented.

In Sect. 3.1, prior estimates of the error of the Bubnov–Galerkin method in the case of the Cauchy problems (3.59), for an abstract equation with constant operator coefficients, were presented. This problem generalizes class of initial–boundary problems in the theory of elasticity (for three-dimensional bodies and thin-walled structures). Here the Cauchy problem for an abstract, hyperbolic equation more general than (3.59) will be discussed. This equation has variable operator coefficients, containing non-adjoint operators of small modulus. Prior estimations of the error of the Bubnov–Galerkin method of type (3.60) and (3.62) will be presented for this case. An application of the results to a Cauchy–Dirichlet problem of partial differential equations of the second order, of hyperbolic type with variable coefficients and small elements, is also presented.

Let H be a Hilbert space, with a scalar product $(\cdot,\cdot)_H$ and a norm $\|\cdot\|_H$. Let us investigate the Cauchy problem

$$u''(t) + A(t)u(t) + B(t)u(t) = f(t), \tag{16.1}$$

$$u(0) = u'(0) = 0, \tag{16.2}$$

where u is a sought function that projects a segment $[0, T]$ of the number line into the space H, and $\{A(t)|t \in [0, T]\}$ is a family of unbounded self-adjoint, positive definite operators acting in H and determined on any linear set $D(A)$ included in H. $\{B(t)|t \in [0, T]\}$ is a family of linear (possibly unbounded) operators, defined on the same linear set $D(B)$, $D(A^{1/2}) \subset D(B) \subset H$; f is a given rational function that maps the segment $[0, T]$ into the space H. It is assumed that for all $t \in [0, T]$, the generalized derivatives $A'(t)$, $A''(t)$, $B'(t)$ are linear operators determined on $D(A)$, $D(A)$, $D(B)$, respectively. Furthermore, A_0 is a self-adjoint positive definite operator, having a completely continuous reverse operator A_0^{-1}. We assume also that the operator A_0 is convergent and forms an acute angle with all of the operators $A(t)$. The norms $\|A(t)A_0^{-1}\|_{H\to H}$ and $\|A_0^{1/2}(A(t))^{1/2}\|_{H\to H}$ are bounded uniformly from above in $t \in [0, T]$, and

$$\gamma[A] \equiv \inf_{0\le t\le T} \gamma(A(t),\ A_0) > 0, \tag{16.3}$$

where $\gamma(A(t),\ A_0)$ is the positive constant used in the acute-angle inequality (2.107) for the operators $A(t)$ and A_0. The following constrains are imposed on the operator coefficients in (16.1) and the operator A_0:

$$|(A'(t)u_1u_2)_H| \le C_1\|A_0^{1/2}u_1\|_H\|A_0^{1/2}u_2\|_H, \tag{16.4}$$

$$|(A''(t)u_1u_2)_H| \le C_2\|A_0^{1/2}u_1\|_H\|A_0^{1/2}u_2\|_H, \tag{16.5}$$

$$\|B(t)u_2\|_H \le C_3\|A_0^{1/2}u_2\|_H, \tag{16.6}$$

$$\|B'(t)u_2\|_H \le C_4\|A_0^{1/2}u_2\|_H, \tag{16.7}$$

where u_1 is any element from $D(A)$, u_2 is any element from $D(A^{1/2})$, and t is any number from $[0, T]$. Here and in what follows, the elements c_i $(i = 1, 2, \dots, 13)$ are various positive constants which can depend only on the operators $A(t)$, $B(t)$, A_0 and T. Conditions for boundness and differentiability are formulated in the inequalities (3.4)–(3.7). These conditions are used in the theorems given below about the uniqueness of the solution of the problem (16.1), (16.2), and also in obtaining the prior estimates of the error of the Bubnov–Galerkin method. The inequalities (16.5), (16.7) are used only for proving the existence of a solution of the problem (16.1), (16.2) for $f \in W_1^1(0, T; H)$. They are not used in obtaining the estimates of the error.

Definition 16.1 The function $u : [0, T] \to H$ is called the generalized solution of the Cauchy problem (16.1), (16.2) if $u \in W_\infty^1(0, T; H) \cap L_\infty(0,\ T; D(A_0^{1/2}))$, $u(0) = 0$ and, for any function $v \in W_\infty^1(0,\ T; H) \cap L_1(0,\ T; D(A_0^{1/2}))$ satisfying conditions $v(T) = 0$, the following condition is satisfied:

$$\int_0^T \Big[-(u'(t),\ v'(t))_H + ((A(t))^{1/2}u(t),\ (A(t))^{1/2}v(t))_H$$

$$+(B(t)u(t),\ v(t))_H\Big]\,dt = \int_0^T (f(t),\ v(t))_H dt. \tag{16.8}$$

Definition 16.2 The function $u : [0, T] \to H$ is called the "classical" solution of the Cauchy problem (16.1), (16.2) if $u \in W_\infty^2(0,\ T; H) \cap W_\infty^1(0,\ T; D(A_0^{1/2})) \cap L_\infty(0,\ T; D(A_0))$, and the condition (16.1) (almost always $(t \in [0, T])$) and conditions (16.2) are satisfied.

Remark 16.1 In Definitions 16.1 and 16.2 we use a function with maximum smoothness $u : [0, T] \to H$, which can be estimated from the conditions on the operators $A(t)$, $B(t)$ and A_0 presented above for $f \in L_2(0, T; H)$ or for $f \in W_2^1(0, T; H)$. The integral condition (16.8) was obtained, following common assumptions used in defining generalized solutions in mathematical physics, by formally multiplying the differential equation (16.1) by a test function $v(t)$ and following it by integrating with respect to time, taking into account the initial condition $u'(0) = 0$. Hence, one can conclude that the "classical" solution (if it exists) of the problem (16.1), (16.2) is also its generalized solution.

Let $\{v_i\}_{i=1}^{\infty}$ be a basis in H, constructed from eigenelements of the operator A_0, and let $\{\lambda\}_{i=1}^{\infty}$ be the corresponding series of eigenvalues of the operator $A_0 : A_0 v_i = \lambda_i v_i$, $i = 1, 2. \dots$ The eigenelements and eigenvalues of the operator A_0 are treated as known.

Let H_n be a linear environment of the bounded system of basis elements $\{v_i\}_{i=1}^{\infty}$, let P_n be an orthogonal transformation H on H_n, and let R_n be an operator $E - P_H$, where E is an identity operator on H. The system of (16.1) and (16.2) is to be solved by means of the following equations given by the Bubnov–Galerkin method:

$$\begin{cases} P_n(u_n''(t) + A(t)u_n(t) + B(t)u_n(t)) = P_n f(t), \\ u_n(t) \in H_n, \ \forall t \in [0, T], \\ u_n(0) = u_n'(0) = 0. \end{cases} \tag{16.9}$$

The exact solution of the problem (16.9) is called the approximate solution of the problem (16.1), (16.2) obtained by the Bubnov–Galerkin method. For $f \in L_2(0, T; H)$ the problem (16.9) is equivalent to the Cauchy problem for a linear system of ordinary differential equations and consequently has an equivalent solution for any natural number n such that $u_n \in W_2^2(0, T; D(A_0))$, for $f \in L_\infty(0, T; H)$ (and for $f \in W_2^1(0, T; H)$, $u_n \in W_\infty^2(0, T; D(A_0))$). Other necessary conditions for the existence and uniqueness of a solution to the problem (16.1), (16.2) can be obtained from the following considerations: for $f \in L_2(0, T; H)$, a generalized solution of the problem (16.1) and (16.2) exists and is unique, and for $f \in W_2^1(0, T; H)$ a "classical" solution of the problem (16.1), (16.2) exists and is unique. The proof of the relevant theorems is based on the compactness method [226], and is analogous to that of Theorems 2.1 and 2.2 (Sect. 2.1.5) on the existence and uniqueness of the solution of the abstract problem (2.54)–(2.56). The solutions of the problems (16.1), (16.2), (16.9) satisfy the following prior estimates for $f \in L_2(0, T; H)$:

$$\left[\|u_n'(t)\|_H^2 + \|A_0^{1/2}u_2\|_H^2\right]^{1/2} \le C_5\|f\|_{L_2(0,T;H)}, \tag{16.10}$$

$$\left[\|u'(t)\|_H^2 + \|A_0^{1/2}u\|_H^2\right]^{1/2} \le C_5\|f\|_{L_2(0,T;H)} \tag{16.11}$$

(almost always ($t \in [0, T]$)). If $f \in W_2^1(0,\ T;H)$ then, in addition to (16.10) and (16.11),

$$\|u_n''(t)\|_H^2 + \|A_0^{1/2} u_n'\|_H^2 \leq C_6 \|f\|_{L_2(0,T;H)}^2 + C_7 \|f'\|_{L_2(0,T;H)}^2 + C_8 \|f(0)\|_H^2, \tag{16.12}$$

$$\|u''(t)\|_H^2 + \|A_0^{1/2} u'\|_H^2 \leq C_6 \|f\|_{L_2(0,T;H)}^2 + C_7 \|f'\|_{L_2(0,T;H)}^2 + C_8 \|f(0)\|_H^2 \tag{16.13}$$

(almost always ($t \in [0, T]$)). In the inequalities (16.10)–(16.13), $u(t)$ is an exact (generalized or "classical") solution of the problem (16.1), (16.2), and $u_n(t)$ is an approximate solution obtained with the Bubnov–Galerkin method from (16.9). The estimates (16.10)–(16.13) are obtained in the process of proving the theorems about the existence of the generalized and "classical" solutions of the problem (16.1), and (16.2) with the compactness method.

We shall now obtain prior estimates of the error of the Bubnov–Galerkin method for the problem (16.1), (16.2). We introduce

$$\Delta u_n(t) = u_n(t) - u(t),$$

where $u(t)$ is the exact solution and $u_n(t)$ is the approximate solution of the problem (16.1), (16.2).

Theorem 16.1 *For $f \in W_2^1(0,\ T;H)$, the accuracy of the Bubnov–Galerkin method using (16.9) to solve the problem (16.1), (16.2) can be estimated as follows:*

$$\begin{aligned} \max_{0 \leq t \leq T} & \left[\|\Delta u_n'(t)\|_H^2 + \|A_0^{1/2} \Delta u_n\|_H^2 \right] \\ & \leq C_9 (C_6 \|f\|_{L_2(0,T;H)}^2 + C_7 \|f'\|_{L_2(0,T;H)}^2 \\ & + C_8 \|f(0)\|_H^2)^{1/2} \Big[\|f\|_{L_2(0,T;H)} \\ & + C_{11} (C_6 \|f\|_{L_2(0,T;H)}^2 + C_7 \|f'\|_{L_2(0,T;H)}^2 + \\ & + C_8 \|f(0)\|_H^2)^{1/2} \Big] \tilde{\lambda}_{n+1}^{-1/2} \equiv 0(\tilde{\lambda}_{n+1}^{-1/2}), \end{aligned} \tag{16.14}$$

where

$$\tilde{\lambda}_R = \min_{i \geq R} \lambda_i . \tag{16.15}$$

Proof We rewrite the first equation of (16.9) in the following form:

$$P_n A(t) u_n(t) = P_n (f(t) - u_n''(t) - B(t) u_n(t)) .$$

We then multiply both sides scalarly by $A_0 u_n(t)$. From the equation obtained and the definition (16.3) of the constant $\gamma[A]$, we obtain

$$\gamma[A]\|A_0 u_n(t)\|_H^2 \leq (A(t)u_n(t)A_0 u_n(t))_H$$
$$= (f(t) - u_n''(t) - B(t)u_n(t),\ A_0 u_n(t))_H$$
$$\leq \|f(t) - u_n''(t) - B(t)u_n(t)\|_H \times \|A_0 u_n(t)\|_H.$$

Hence, from the inequality (16.3), we obtain that

$$\|A_0 u_n(t)\|_H \leq \frac{1}{\gamma[A]}\|f(t) - u_n''(t) - B(t)u_n(t)\|_H.$$

Applying the above inequality and the condition of uniform boundedness of the norm $\|A(t)A_0^{-1}\|_{H\to H}$, we obtain

$$\|A(t)u_n(t)\|_H \leq \frac{\|A(t)A_0^{-1}\|_{H\to H}}{\gamma[A]}\|f(t) - u_n''(t) - B(t)u_n(t)\|_H \tag{16.16}$$
$$\leq C_{12}\|f(t) - u_n''(t) - B(t)u_n(t)\|_H.$$

We introduce the notation

$$\delta_n(t) = u_n''(t) + A(t)u_n(t) + B(t)u_n(t) - f(t).$$

Applying the inequalities (16.16), (16.6), (16.10) and (16.12), we obtain

$$\|\delta_n(t)\|_H \leq (1 + C_{12})\left(\|u_n''(t)\|_H + \|B(t)u_n(t)\|_H + \|f(t)\|_H\right)$$
$$\leq (1 + C_{12})\left(\|u_n''(t)\|_H + C_3\left\|A_0^{1/2}u_n(t)\right\|_H + \|f(t)\|_H\right)$$
$$\leq (1 + C_{12})\left[\left(C_6\|f\|^2_{L_2(0,T;H)} + C_7\|f'\|^2_{L_2(0,T;H)} + C_8\|f(0)\|_H^2\right)^{1/2}\right.$$
$$\left. + C_3C_5\|f\|_{L_2(0,T;H)} + \|f(t)\|_H\right].$$

Transforming to the norm of the space $L_2(0,\ T;\ H)$, we obtain the estimate

$$\|\delta_n(t)\|_{L_2(0,T;H)} \leq C_{10}\|f\|_{L_2(0,T;H)} + C_{11}\left(C_6\|f\|^2_{L_2(0,T;H)}\right.$$
$$\left. + C_7\|f'\|^2_{L_2(0,T;H)} + C_8\|f(0)\|_H^2\right)^{1/2}, \tag{16.17}$$

where

$$C_{10} = (1 + C_{12})(1 + C_3 C_5 T^{1/2}), \quad C_{11} = (1 + C_{12})T^{1/2}.$$

Hence, from (16.1) and the first equation of (16.9), we obtain

$$\Delta u_n''(t) + A(t)\Delta u_n(t) + B(t)\Delta u_n(t) = R_n \delta_n(t) .$$

We multiply this equation scalarly by $2\Delta u_n'(t)$ and then integrate by parts over the interval $[0, T]$, where $t \le T$. Taking into account the initial conditions $\Delta u_n(0) = \Delta u_n'(0) = 0$, we obtain

$$\|\Delta u_n'(t)\|_H^2 + \|(A(t))^{1/2}\Delta u_n(t)\|_H^2] = 2\int_0^t (R_n\delta_n(\tau),$$
$$\Delta u_n'(\tau))_H d\tau - 2\int_0^t (B(\tau)\Delta u_n(\tau),\ \Delta u_n'(\tau))_H d\tau \tag{16.18}$$
$$+\int_0^t (A'(\tau)\Delta u_n(\tau),\ \Delta u_n(\tau))_H d\tau.$$

Because of the special choice of the basis $\{v_i\}_{i=1}^{\infty}$, we obtain

$$\|A_0^{1/2} R_n\|_{H\to G} = \tilde{\lambda}_{n+1}^{-1/2},$$

and consequently the first term of the right-hand side of (16.18) can be estimated in the following way:

$$2\int_0^t (R_n\delta_n(\tau),\ \Delta u_n'(\tau))_H d\tau = 2\int_0^t (A_0^{1/2} R_n\delta_n(\tau),\ A_0^{1/2}\Delta u_n'(\tau))_H d\tau$$
$$\le 2\left\|A_0^{-1/2} R_n\right\|_{H\to H} \|\delta_n\|_{L_2(0,T;H)} \left\|A_0^{1/2}\Delta u_n'\right\|_{L_2(0,T;H)}$$
$$\le 2\|\delta_n\|_{L_2(0,T;H)} \times \left(\left\|A_0^{1/2} u_n'\right\|_{L_2(0,T;H)} + \left\|A_0^{1/2} u\right\|_{L_2(0,T;H)}\right)\tilde{\lambda}_{n+1}^{-1/2}.$$

Hence (16.12), (16.13), (16.17) give

$$2\int_0^t (R_n\delta_n(\tau),\ \Delta u_n'(\tau))_H d\tau \le 4\Big(C_6\|f\|^2_{L_2(0,T;H)} + C_7\|f'\|^2_{L_2(0,T;H)} + C_8\|f(0)\|^2_H\Big)^{1/2}\Big[C_{10}\|f\|_{L_2(0,T;H)} + C_{11}(C_6\|f\|^2_{L_2(0,T;H)} + C_7\|f'\|^2_{L_2(0,T;H)} + C_8\|f(0)\|^2_H)^{1/2}\Big]\tilde{\lambda}_{n+1}^{-1/2}. \tag{16.19}$$

The sums of the second and third terms of the right-hand side of (16.18) can be estimated using the conditions (16.4), (16.6) and the Cauchy inequality. We obtain

$$-2\int_0^t (B(\tau)\Delta u_n(\tau),\ \Delta u_n'(\tau))_H d\tau + \int_0^t (A'(\tau)\Delta u_n(\tau),\ \Delta u_n(\tau))_H d\tau \le C_{13}\int_0^t (\|\Delta u_n'(\tau)\|^2_H + \|A_0^{1/2}\Delta u_n(\tau)\|^2_H)d\tau. \tag{16.20}$$

We now estimate the right-hand sides of (16.19), (16.20) using the inequalities (16.18). We obtain the following inequality:

$$\|\Delta u_n'(t)\|^2_H + \|(A(t))^{1/2}\Delta u_n(t)\|^2_H \le 4\Big(C_6\|f\|^2_{L_2(0,T;H)} + C_7\|f'\|^2_{L_2(0,T;H)} + C_8\|f(0)\|^2_H\Big)^{1/2}\Big[C_{10}\|f\|_{L_2(0,T;H)} + C_{11}(C_6\|f\|^2_{L_2(0,T;H)} + C_7\|f'\|^2_{L_2(0,T;H)} + C_8\|f(0)\|^2_H)^{1/2}\Big]\tilde{\lambda}_{n+1}^{-1/2} + C_{13}\int_0^t (\|\Delta u_n'(\tau)\|^2_H + \|A_0^{1/2}\Delta u_n(\tau)\|^2_H)d\tau.$$

Hence, applying the condition of uniform boundness of the norm

$$\|A_0^{-1/2}(A(t))^{-1/2}\|_{H\to H}$$

and the Gronwill lemma, we obtain the estimate (16.14).

Therefore Theorem 16.1 has been proved.

We now use the notation (3.21), (3.22), where H is the Hilbert space under consideration, f is the free element of (16.1).

Theorem 16.2 *For $f \in L_2(0, T; H)$, the accuracy of the Bubnov–Galerkin method using (16.9) to solve the problem (16.1), (16.2) can be estimated as follows:*

$$\sup_{0\le t\le T} ess \left(\|\Delta u_n'(t)\|_H^2 + \|A_0^{1/2}\Delta u_n(t)\|_H^2\right)^{1/2}$$

$$\le D_1\omega(f, \tilde{\lambda}_{n+1}^{-1/4}) + C_9^{1/2}\Big[C_6(e(f, \tilde{\lambda}_{n+1}^{-1/4}))^2\tilde{\lambda}_{n+1}^{-1/2}$$

$$+D_2(\omega(f, \tilde{\lambda}_{n+1}^{-1/4}))^2\Big]^{1/4}\Big\{C_{10}e(f, \tilde{\lambda}_{n+1}^{-1/4})\tilde{\lambda}_{n+1}^{-1/4}$$

$$+C_{11}\Big[C_6(e(f, \tilde{\lambda}_{n+1}^{-1/4}))^2\tilde{\lambda}_{n+1}^{-1/2} + D_2(\omega(f, \tilde{\lambda}_{n+1}^{-1/4}))^2\Big]^{1/2}\Big\}^{1/2}$$

$$\equiv 0(\max(\tilde{\lambda}_{n+1}^{-1/4}, \omega(f, \tilde{\lambda}_{n+1}^{-1/4}))), \tag{16.21}$$

where $\tilde{\lambda}_R$ is the series defined by (16.15), D_1 and D_2 are positive constants, defined below by (16.29), and $e(f, \delta)$ is a quantity bounded for $\delta \to 0$, defined below by (16.30).

Proof Let $\varphi(t)$ be any averaging kernel of class W_2^1 on the segment [–1, 1], i.e. an even function from the Sobolev space $W_2^1([-1, 1])$, equal to zero on the ends of the segment [–1, 1] and satisfying (3.25). Let $\varphi_\delta(t) = \frac{1}{\delta}\varphi\left(\frac{t}{\delta}\right)$, $\delta > 0$, $-\delta \le t \le \delta$. We introduce the notation

$$f_\delta(t) = \int_{t-\delta}^{t+\delta} \tilde{f}(\tau)\varphi_\delta(t-\tau)d\tau, \quad \Delta f_\delta(t) = f(t) - f_\delta(t),$$

where $\delta > 0$, $0 \le t \le T$, and $\tilde{f}$ is the extension of the function f from $[0, T]$ to the whole number line, defined by (3.21). Analogous, (3.28)–(3.30) lead to the inequalities:

$$\|\Delta f_\delta\|_{L_2(0,T;H)} \le C_{0,\varphi}\omega(f, \delta), \tag{16.22}$$

$$\|\Delta f_\delta'\|_{L_2(0,T;H)} \le \frac{1}{\delta}C_{1,\varphi}\omega(f, \delta), \tag{16.23}$$

where

$$C_{0,\varphi} = \sqrt{2}\left(\int_{-1}^{1}(\varphi(\tau))^2 d\tau\right)^{1/2}, \quad C_{1,\varphi} = \sqrt{2}\left(\int_{-1}^{1}(\varphi'(\tau))^2 d\tau\right)^{1/2}$$

(for example, for $\varphi(t) = 1 - |t|$, $C_{0,\varphi} = 2\sqrt{3}/3$, $C_{1,\varphi} = 2$).

We consider an investigated family of Cauchy problems with free elements $f_\delta \in W_2^1(0,\ T; H)$, indexed by the parameter $\delta > 0$:

$$\begin{cases} u_\delta''(t) + A(t)u_\delta(t) + B(t)u_\delta(t) = f_\delta(t), \\ u_\delta(0) = u_\delta'(0) = 0. \end{cases} \tag{16.24}$$

The approximate solution of (16.24) obtained by the Bubnov–Galerkin method is denoted by $u_{\delta,n}$. The following notation is also introduced:

$$\Delta u_\delta(t) = u(t) - u_\delta(t),$$

$$\Delta u_{\delta,n}(t) = u_n(t) - u_{\delta,n}(t),$$

$$\Delta u_{n,\delta}(t) = u_{\delta,n}(t) - u_\delta(t)\ .$$

Assuming that the function Δu_δ is the exact solution of the Cauchy problem of type (16.1) with free element Δf_δ, and that the function $\Delta u_{\delta,n}$ is the approximate solution of this problem obtained by the Bubnov–Galerkin method, we can write down prior estimates of type (16.10), (16.11) as follows:

$$\left[\|u_{\delta,n}'(t)\|_H^2 + \|A_0^{1/2}u_{\delta,n}(t)\|_H^2\right]^{1/2} \le C_5\|f_\delta\|_{L_2(0,T;H)}, \tag{16.25}$$

$$\left[\|u_\delta'(t)\|_H^2 + \|A_0^{1/2}u_\delta(t)\|_H^2\right]^{1/2} \le C_5\|f_\delta\|_{L_2(0,T;H)}. \tag{16.26}$$

(almost always ($t \in [0,\ T]$)).

By applying Theorem 16.1 to (16.24) and taking into account the fact that $f_\delta(0) = 0$, which is true in the case of an odd extension $\tilde{f}$ of the free element of (16.1), we obtain the following inequality:

$$\begin{aligned}\left[\|\Delta u_{n,\delta}'(t)\|_H^2 + \|A_0^{1/2}\Delta u_{n,\delta}(t)\|_H^2\right]^{1/2} &\le C_9^{1/2}\Big(C_6\|f_\delta\|_{L_2(0,T;H)}^2 \\ &+C_7\|f_\delta'\|_{L_2(0,T;H)}^2\Big)^{1/4}\Big[C_{10}\|f_\delta\|_{L_2(0,T;H)} + C_{11}(C_6\|f_\delta\|_{L_2(0,T;H)}^2 \\ &+C_7\|f_\delta'\|_{L_2(0,T;H)}^2)^{1/2}\Big]^{1/2}\tilde{\lambda}_{n+1}^{-1/4},\end{aligned} \tag{16.27}$$

where $0 \le t \le T$.

The error of the Bubnov–Galerkin method when applied to (16.1), (16.2) can be represented in the following form:

$$\begin{aligned}\left[\|\Delta u_n'(t)\|_H^2 + \|A_0^{1/2}\Delta u_n(t)\|_H^2\right]^{1/2} &\le \Big[\|\Delta u_\delta'(t)\|_H^2 \\ &+\|A_0^{1/2}\Delta u_\delta(t)\|_H^2\Big]^{1/2} + \left[\|\Delta u_{\delta,n}'(t)\|_H^2 + \|A_0^{1/2}\Delta u_{\delta,n}(t)\|_H^2\right]^{1/2} \\ &+\left[\|\Delta u_{n,\delta}'(t)\|_H^2 + \|A_0^{1/2}\Delta u_{n,\delta}(t)\|_H^2\right]^{1/2}.\end{aligned}$$

Hence, applying the inequalities (16.25)–(16.27), (16.22) and (16.23), we obtain

$$\left[\|\Delta u_n'(t)\|_H^2 + \|A_0^{1/2}\Delta u_n(t)\|_H^2\right]^{1/2} \le 2C_5\|\Delta f_\delta\|^2_{L_2(0,T;H)}$$
$$+C_9^{1/2}\Big(C_6\|f_\delta\|^2_{L_2(0,T;H)} + C_7\|f_\delta'\|^2_{L_2(0,T;H)}\Big)^{1/4}\Big[C_{10}\|f_\delta\|_{L_2(0,T;H)}$$
$$+C_{11}\Big(C_6\|f_\delta\|^2_{L_2(0,T;H)} + C_7\|f_\delta'\|^2_{L_2(0,T;H)}\Big)^{1/2}\Big]^{1/2}\tilde{\lambda}_{n+1}^{-1/4}$$
$$\le 2C_5C_{0,\varphi}\omega(f,\delta) + C_9^{1/2}\Big[\Big(C_6(\|f\|_{L_2(0,T;H)} + C_{0,\varphi}\omega(f,\delta)\Big)^2$$
$$+\frac{1}{\delta^2}C_7C_{1,\varphi}^2(\omega(f,\delta))^2\Big]^{1/4}\Big\{C_{10}(\|f\|_{L_2(0,T;H)} + C_{0,\varphi}\omega(f,\delta))$$
$$+C_{11}\Big[C_6(\|f_\delta\|_{L_2(0,T;H)} + C_{0,\varphi}\omega(f,\delta))^2 + \frac{1}{\delta^2}C_7C_{1,\varphi}^2$$
$$\times(\omega(f,\delta))^2\Big]^{1/2}\Big\}^{1/2}\tilde{\lambda}_{n+1}^{-1/4}.$$

In the above inequality, δ is any positive number. Let $\delta = \tilde{\lambda}_{n+1}^{-1/4}$; we now obtain

$$\left[\|\Delta u_n'(t)\|_H^2 + \|A_0^{1/2}\Delta u_n(t)\|_H^2\right]^{1/2} \le 2C_5C_{0,\varphi}\omega(f,\tilde{\lambda}_{n+1}^{-1/4})$$
$$+C_9^{1/2}\Big[C_6\Big(\|f\|_{L_2(0,T;H)} + C_{0,\varphi}\omega(f,\tilde{\lambda}_{n+1}^{-1/4})\Big)^2 + C_7C_{1,\varphi}^2$$
$$\times\Big(\omega(f,\tilde{\lambda}_{n+1}^{-1/4})\Big)^2\Big]^{1/4}\Big\{C_{10}\Big(\|f\|_{L_2(0,T;H)} + C_{0,\varphi}\omega(f,\tilde{\lambda}_{n+1}^{-1/4})\Big) \quad (16.28)$$
$$\times\tilde{\lambda}_{n+1}^{-1/4} + C_{11}\Big[C_6\|f\|_{L_2(0,T;H)} + (C_{0,\varphi}\omega(f,\tilde{\lambda}_{n+1}^{-1/4}))^2\tilde{\lambda}_{n+1}^{-1/2}$$
$$+C_7C_{1,\varphi}^2(\omega(f,\tilde{\lambda}_{n+1}^{-1/4}))^2\Big]^2\Big\}^{1/2}$$

(almost always ($t \in [0, T]$)). By introducing the notation

$$D_1 = 2C_7C_{1,\varphi}^2,\ D_2 = C_7C_{1,\varphi}^2, \quad (16.29)$$

$$e(f,\delta) = \|f\|_{L_2(0,T;H)} + C_{0,\varphi}\omega(d,\ \delta), \quad (16.30)$$

the (16.28) can be rewritten as (16.21). Therefore Theorem 16.2 has been proved.

Remark 16.2 The full continuity of the operator A_0^{-1}, $\lambda_i \to \infty$ for $i \to \infty$, implies that the estimates of error of the Bubnov–Galerkin method (16.14) and (16.21) are concised in the sense that the right-hand sides approach zero as $n \to \infty$.

As an example of the application of the above estimates, we discuss the Cauchy–Dirichlet problem for a hyperbolic equation of second order:

$$\frac{\partial^2 u}{\partial t^2} - \sum_{i,j=1}^{P} \frac{\partial}{\partial x}\left(a_{ij}(x,\ t)\frac{\partial u}{\partial x_j}\right) + \sum_{i=1}^{P} b_i(x,\ t)\frac{\partial u}{\partial x_i} + c(x,\ t)u = f(x,\ t),$$

$$u|_{t=0} = 0,\ \left.\frac{\partial u}{\partial t}\right|_{t=0} = 0,\ u|_{x\in S} = 0, \tag{16.31}$$

where $u = u(x,\ t)$ is a soughtfunction defined on $Q = \Omega \times [0, T]$; $x = (x_1,\ \ldots,\ x_p) \in \Omega$; $t \in [0, T]$, $T > 0$; and Ω is a p-metric space, surrounded by a surface S, such that the boundary functions $y_p = \Psi(y_1,\ \ldots,\ y_{p-1})$ describing the equations of the surface S in the local coordinate systems have uniformly bounded partial derivatives over all y_k up to the second order inclusive. The eigenfunctions of the Laplace operator ∇^2, equal to zero on S, are written for the space Ω in the explicit form (see, for example [249]); and $a_{ij},\ b_i,\ c$ are given functions, satisfying the following conditions:

$$a_{ij},\ \frac{\partial a_{ij}}{\partial x_k},\ \frac{\partial a_{ij}}{\partial t},\ \frac{\partial^2 a_{ij}}{\partial t^2},\ b_i,\ \frac{\partial b_i}{\partial t} \in L_\infty(Q),$$

$$\frac{\partial c}{\partial t} \in L_\infty(0,\ T; \in L_2(\Omega)), a_{ij} = a_{ji}(i,\ j,\ k = 1, 2,\ \ldots, p),$$

$$\sum_{i,j=1}^{P} a_{ij}(x,\ t)\xi_i\xi_j \geq \alpha \sum_{i=1}^{P} \xi_i^2. \tag{16.32}$$

Here $\xi = (\xi_1,\ \ldots,\ \xi_p)$ is any set of p real numbers; α is a positive constant, not depending on t or ξ, $f = f(x,\ t)$ is a given rational function defined on Q.

Let us convert the problem (16.31) into the form (16.1), (16.2). We introduce the following notation:

$$A(t)u = -\sum_{i,j=1}^{P} \frac{\partial}{\partial x_i}\left(a_{ij}(x,\ t)\frac{\partial u}{\partial x_j}\right),\ D(A) = W_2^2(\Omega) \cap \overset{\circ}{W}{}_2^1(\Omega),$$

$$B(t)u = \sum_{i=1}^{P} b_i(x,\ t)\frac{\partial u}{\partial x_i} + c(x,\ t)u,\ \ D(B) = \overset{\circ}{W}{}_2^1(\Omega), H = L_2(\Omega).$$

Here $\overset{\circ}{W}{}_2^1(\Omega)$ depicts the subspace of the Sobolev space $W_2^1(\Omega)$ of functions equal to zero on S. In [213] it is stated that when the conditions on the space Ω and the coefficients a_{ij} are recalculated, the following inequality can be derived for an arbitrary function $u \in W_2^2(\Omega) \cap \mathring{W}_2^1(\Omega)$:

$$m\|u\|^2_{W_2^2(\Omega)} \leq -\int_\Omega A(t)u\, \nabla^2\, u\, dx + \|u\|^2_{\mathring{W}_2^2(\Omega)}, \tag{16.33}$$

where the positive constant m depends only on α and $\max\left\{|a_{ij}|, \left|\frac{\partial a_{ij}}{\partial x_k}\right|, \left|\frac{\partial^2 \Psi}{\partial y_l \partial y_m}\right|\right\}$. Consequently, this constant can be chosen set independently of t. From (16.33), using (16.32), we obtain the inequality

$$\int_\Omega A(t)uA_0 u dx \geq m\|u\|^2_{W_2^2(\Omega)},$$

where $A_0 u = -\nabla^2 u + \dfrac{1}{\alpha} u$.

The last inequality and the conditions on the space Ω and coefficients a_{ij}, b_i, c given above ensure a uniform upper limit of the norms $\|A(t)A_0^{-1}\|_{H\to H}$, $\|A_0^{1/2}(A(t))^{-1/2}\|_{H\to H}$ and also ensure that the inequalities (16.3)–(16.7) are satisfied. We can consider the application of the Bubnov–Galerkin method to the problem (16.31), using the eigenfunctions of the Laplace operator for the space Ω as the basis functions $v_i(x)$. If $f \in L_2(0,\ T; L_2(\Omega)) = L_2(Q)$, an estimate of the form (16.21) exists for the problem (16.31) with $\lambda_n = -\mu_n + \frac{1}{\alpha}$, where $\{\mu_n\}_{n=1}^\infty$ is a series of positive eigenvalues of the Laplace operator for the space Ω. If $f \in W_2^1(0,\ T; L(\Omega))$ (this means also if $\dfrac{\partial f}{\partial t} \in L_2(Q)$), an estimate of the form (16.14) also exists.

References

1. Abali, B.E., Müler, W.H., Eremeyev, V.A.: Strain gradient elasticity with geometric nonlinearities and its computational evaluation. Mech. Adv. Mater. Moder. Process. **1**, 1–4 (2015)
2. Akgöz, B., Civalek, Ö.: Buckling analysis of functionally graded microbeams based on the strain gradient theory. Acta Mech. **224**(9), 2185–2201 (2013)
3. Alibeigi, B., Tadi Beni, Y., Mehralian, F.: On the thermal buckling of magneto-electro-elastic piezoelectric nanobeams. Eur. Phys. J. Plus **133**, 133 (2018)
4. Alijani, F., Amabili, M.: Non-linear vibrations of shells: a literature review from 2003 to 2013. Int. J. Non Linear Mech. **58**, 233–257 (2014)
5. Andreev, L.V., Obodan, N.I., Lebedev, A.G.: Shell Stability Under Non-Axisymmetric Deformation. Nauka, Moscow (1988) (in Russian)
6. Andrianov, I.V., Awrejcewicz, J.: Numbers or understanding: analytical and numerical methods in the theory of plates and shells. Facta Universitis, Mech. Automat. Control Robot. **2**(10), 1319–1327 (2000)
7. Andrianov, I.V., Awrejcewicz, J.: New trends in asymptotic approaches: summation and interpolation methods. Appl. Mech. Rev. **54**(1), 69–92 (2001)
8. Andronov, A.A., Vitt, A.A., Khaikin, S.E.: Theory of Oscillations. Dover, New York (1987)
9. Ansari, R., Pourashraf, T., Gholami, R., Sahmani, S., Ashrafi, M.A.: Size-dependent resonant frequency and flexural sensitivity of atomic force microscope microcantilevers based on the modified strain gradient theory. Int. J. Optomechatr. **9**, 111–130 (2015)
10. Ansari, R., Shoajaei, M.F., Ebrahimi, F., Rouhi, H., Bazdid-Vahdati, M.: A novel size-dependent microbeam element based on Mindlin's strain gradient theory. Eng. Comput. **32**, 99–108 (2016)
11. Aranda-Iglesias, D., Vadillo, G., Rodriguez-Martinez, J.A.: Oscillatory behaviour of compressible hyperelastic shells subjected to dynamic inflation: a numerical study. Acta Mech. **228**(6), 2187–2205 (2017)
12. Arefi, M., Arani, A.H.S.: Higher order shear deformation bending results of a magneto-electrothermoelastic functionally graded nanobeam in thermal, mechanical, electrical, and magnetic environments. Mech. Based Des. Struct. Mach. **46**(6), 669–692 (2018)
13. Arganovskyy, M.L., Baglay, R.D.: On an expansion in Hilbert space and its applications. Comput. Math. Math. Phys. **17**(4), 871–878 (1977) (in Russian)
14. Awrejcewicz, J. (ed.): Bifurcation and Chaos: Theory and Application. Springer, Berlin/Heidelberg (1995)
15. Awrejcewicz, J.: Ordinary Differential Equations and Mechanical Systems. Springer, Berlin (2014)

J. Awrejcewicz, V. A. Krysko, *Elastic and Thermoelastic Problems in Nonlinear Dynamics of Structural Members*, Scientific Computation,
https://doi.org/10.1007/978-3-030-37663-5

16. Awrejcewicz, J., Andrianov, I.V.: Asymptotic Methods and Their Applications in Shell Theory. WNT, Fundacja Ksiazka Naukowo-Techniczna, Warsaw (2000) (in Polish)
17. Awrejcewicz, J., Andrianov, I.V.: Plates and Shells in Nature, Mechanics and Biomechanics. WNT, Fundacja Ksiazka Naukowo-Techniczna, Warsaw (2001) (in Polish)
18. Awrejcewicz, J., Krys'ko, V.A.: Techniques and Methods of Plate and Shell Analysis. Lodz Technical University Press, Lodz (1996) (in Polish)
19. Awrejcewicz, J., Krys'ko, V.A.: Dynamics of a shell with the added masses. In: Awrejcewicz, J., Grabski, J., Mrozowski, J. (eds.) Proceedings of the 4th Conference on Dynamical Systems–Theory and Applications, Lodz, December 8–9, 1997, pp. 163–168. Technical University Press, Lodz (1997)
20. Awrejcewicz, J., Krys'ko, V.A.: Three dimensional problem of the orthotropic plate vibrations with the added masses. In: Awrejcewicz, J., Grabski, J., Mrozowski, J. (eds.) Proceedings of the 4th Conference on Dynamical Systems–Theory and Applications, December 8–9, Lodz, Poland, 1997, pp. 169–176. Technical University Press, Lodz (1997)
21. Awrejcewicz, J., Krys'ko, V.A.: Numerical Analysis of Shells Oscillations With Thermal Load. WNT, Scientific Book Foundation, Warsaw (1998) (in Polish)
22. Awrejcewicz, J., Krys'ko, V.A.: Dynamics and Stability of Shells With Thermal Excitations. WNT, Scientific Book Foundation, Warsaw (1999) (in Polish)
23. Awrejcewicz, J., Krys'ko, V.A.: Dynamical stability of thin shells with thermal convection. In: Maruszewski, B.T., Muschik, W., Radowicz, A. (eds.) Proceedings of the International Symposium on Trends in Continuum Physics, Poznan, Poland, 17–20 August, 1998, pp. 35–45. World Scientific, Singapore (1999)
24. Awrejcewicz, J., Krys'ko, V.A.: Vibration analysis of the plates and shells of moderate thickness. J. Tech. Phys. **40**(3), 277–305 (1999)
25. Awrejcewicz, J., Krys'ko, V.A.: 3D theory versus 2D approximate theory of the free orthotropic (isotropic) plates and shells vibrations. Part 1, derivation of governing equations. J. Sound Vib. **226**(5), 807–829 (1999)
26. Awrejcewicz, J., Krys'ko, V.A.: 3D theory versus 2D approximate theory of the free orthotropic (isotropic) plates and shells vibrations. Part 2, numerical algorithms and analysis. J. Sound Vib. **226**(5), 831–871 (1999)
27. Awrejcewicz, J., Krys'ko, V.A.: Abstract thermoelasticity problems of shallow shells. In: Awrejcewicz, J., Grabski, J., Mrozowski, J. (eds.) Proceedings of the 5th Conference on Dynamical Systems–Theory and Applications, Lodz, Poland, December 6–8, 1999, pp. 89–94. Technical University Press, Lodz (1999)
28. Awrejcewicz, J., Krys'ko, V.A.: Coupled thermomechanical problems in the shell theory. In: Skrzypek, J.J., Hetnarski, R.B. (eds.) Proceedings of the 3rd International Congress on Thermal Stresses '99, Cracow, Poland, June 13–17, 1999, pp. 53–56
29. Awrejcewicz, J., Krys'ko, V.A.: Optimization of plate and shell surfaces. In: Proceedings of the ASME Design Engineering Technical Conferences, Las Vegas, Nevada, September 12–16, 1999, DETC99/VIB-8139, (published on CD-ROM)
30. Awrejcewicz, J., Krys'ko, V.A.: Optimization of plate and shell surfaces. In: Proceedings of the 10th World Congress on the Theory of Machines and Mechanisms, Oulu, Finland, June 20–24, 1999, pp. 2128–2133
31. Awrejcewicz, J., Krys'ko, V.A.: Oscillations of Lumped Systems. WNT, Warsaw (2000) (in Polish)
32. Awrejcewicz, J., Krys'ko, V.A.: Period doubling bifurcation and chaos exhibited by an isotropic plate. Z. Angew. Math. Mech. **80**, S267–S268 (2000)
33. Awrejcewicz, J., Krys'ko, V.A.: Feigenbaum scenario exhibited by thin plate dynamics. Nonlinear Dyn. **24**, 373–398 (2001)
34. Awrejcewicz, J., Krys'ko, V.A.: Complex parametric oscillations of flexible rectangular plates. In: Awrejcewicz, J., Grabski, J., Nowakowski, J. (eds.) Proceedings of the 6th Conference on Dynamical Systems–Theory and Applications, Lodz, Poland, December 10–12, 2001, pp. 153–164. Technical University Press, Lodz (2001)

35. Awrejcewicz, J., Krysko, V.A.: Nonclassical Thermoelastic Problems in Nonlinear Dynamics of Shells. Springer, Berlin (2003)
36. Awrejcewicz, J., Krys'ko, V.A.: Coupled thermoelasticity problems of shallow shells. J. Syst. Anal. Modell. Simulat. **43**(3), 269–286 (2003)
37. Awrejcewicz, J., Krys'ko, V.A., Krys'ko, A.V.: Spatial-temporal chaos and solitions exhibited by von Kármán mode. Int. J. Bifurcation Chaos. **12**(7), 1465–1513 (2002)
38. Awrejcewicz, J., Krys'ko V.A., Krys'ko A.V.: Period doubling Hopf bifurcation of thin flexible isotropic plates with an impact load. In: Awrejcewicz, J. Grabski, J. Mrozowski, J. (eds.) Proceedings of the 4th Conference on Dynamical Systems–Theory and Applications, Lodz, Poland, pp. 63–68. Technical University Press, Lodz (1997)
39. Awrejcewicz, J., Andrianov, I.V., Manevitch, L.I.: Asymptotic Approach in Nonlinear Dynamics: New Trends and Applications. Springer, Berlin/Heidelberg (1998)
40. Awrejcewicz, J., Krys'ko, V.A., Kutsemako, N.: Free vibrations of doubly curved in-plane non-homogeneous shells. J. Sound Vib. **225**(4), 701–722 (1999)
41. Awrejcewicz, J., Krys'ko, V.A., Krys'ko, A.V.: Non-symmetric oscillations and transition to chaos in freely supported flexible plate sinusoidally excited. In: Awrejcewicz, J., Grabski, J., Mrozowski, J. (eds.) Proceedings of the 5th Conference on Dynamical Systems–Theory and Applications, Lodz, Poland, December 6–8, 1999, pp. 95–102. Technical University Press, Lodz (1999)
42. Awrejcewicz, J., Krys'ko ,V.A., Krys'ko, A.V.: Symmetric oscillations and transition to chaos in a freely supported sinusoidally excited flexible shell. In: Awrejcewicz, J., Grabski, J., Mrozowski, J. (eds.) Proceedings of the 5th Conference on Dynamical Systems–Theory and Applications, Lodz, Poland, December 6–8, 1999, pp. 103–112. Technical University Press, Lodz (1999)
43. Awrejcewicz, J., Krys'ko, V.A., Lamarque, C.-H.: Optimal vibroisolation problem of plates and shells. In: Awrejcewicz, J., Grabski, J., Mrozowski, J. (eds.) Proceedings of the 5th Conference on Dynamical Systems-Theory and Applications, Lodz, Poland, December 6–8, 1999, pp. 113–120. Technical University Press, Lodz (1999)
44. Awrejcewicz, J., Krys'ko, V.A., Mrozowski, J.: Analysis of free vibrations of non-homogeneous plates using finite element method. In: Fryba, L., Naprstek, J. (eds.) Proceedings of EURODYN'99, Fourth Conference of the European Association for Structural Dynamics, Prague, June 7–10, 1999, Structural Dynamics, vol. I, pp. 415–420. Balkema, Rotterdam (1999)
45. Awrejcewicz, J., Krys'ko, V.A., Krys'ko, A.V.: Symmetric and non-symmetric oscillations and bifurcations of periodically excited plates with non-homogeneous boundary conditions. In: Papadrakakis, M., Samartin, A., Onat, E. (eds.) IASS-IACM 2000 Fourth International Colloquium on Computation of Shell & Spatial Structures, Chania, Grecee, June 4–7, 2000, pp. 1–9. ISASRNTUA, Athens (2000) (CD-ROM)
46. Awrejcewicz, J., Krys'ko, V.A., Krys'ko, A.V., Vakakis, A.: Solitons and chaos exhibited by flexible plates sinusoidally excited. In: Balthazar, J.M., Goncalvez, P.B., Brasil, R.M., Caldas, I.L., Rizatto, F.B. (eds.) Nonlinear Dynamics, Chaos, Control and Their Applications to Engineering Sciences. Chaos Control and Times Series, vol. 5, pp. 258–267. Brazilian Society of Mechanical Sciences, Rio de Janeiro (2000)
47. Awrejcewicz, J., Krys'ko, V.A., Krys'ko, A.V.: Regular and chaotic behaviour of flexible plates. In: Zaras, J., Kowal-Michalska, K., Rhodes, J. (eds.) Proceedings of the Third International Conference on Thin-Walled Structures, Cracow, Poland, June 5–7, 2001, pp. 349–356. Elsevier Science (2001)
48. Awrejcewicz, J., Krys'ko, V.A., Krys'ko, A.V.: Regular and chaotic behaviour of flexible plates. In: European Congress on Computational Methods in Applied Sciences and Engineering, ECCOMAS Computational Fluid Dynamics Conference 2001, Swansea, UK, September 4–7, 2001 (CD ROM)

49. Awrejcewicz, J., Krys'ko. V.A., Krys'ko, A.V.: Solitons exhibited by the von Kármán equations. In: Kittl, P., Diaz, G., Mook, D., Geer, J. (eds.) Applied Mechanics in the Americas, Proceedings of the Seventh Pan-American Congress on Applied Mechanics, Temuco, Chile, January 2–4, 2002, vol. 9, pp. 653–9 to 653–12
50. Awrejcewicz, J., Krysko, V.A., Narkaitis, G.G.: Bifurcations of a thin plate-strip excited transversally and axially. Nonlinear Dyn. **32**, 187–209 (2003)
51. Awrejcewicz, J., Krysko, V.A., Papkova, I.V.: Dynamics and statics of flexible axially-symmetric shallow shells. Math. Probl. Eng. **2006**, ID 35672 (2006)
52. Awrejcewicz, J., Krysko, A.V., Dobriyan, V., Papkova, I.V., Krysko, V.A.: On the Lyapunov exponents computation of coupled non-linear Euler-Bernoulli beams. In: Topping, B.H.V., Ivanyi, P. (eds.) Proceedings of the Fourteenth International Conference on Civil, Structural and Environmental Engineering Computing, pp. 53. Civil-Comp Press, Stirlingshire (2013)
53. Awrejcewicz, J., Krysko, V.A., Papkova, I.V., Yakovleva, T.V., Zagniboroda, N.A., Zhigalov, M.V., Krysko, A.V., Dobriyan, V., Krylova, E.Yu., Mitskevich, S.A.: Application of the Lyapunov exponents and wavelets to study and control of plates and shells. In: Awrejcewicz J. (ed.) Computational and Numerical Simulations, pp. 1–19. In:Tech (2014)
54. Awrejcewicz J., Krysko V.A., Papkova I.V., Krysko A.V.: Deterministic Chaos in One Dimensional Continuous Systems. World Scientific, Singapore (2016)
55. Awrejcewicz, J., Krysko-Jr., V.A., Yakovleva, T.V., Krysko, V.A.: Noisy contact interactions of multi-layer mechanical structures coupled by boundary conditions. J. Sound Vib. **369**, 77–86 (2016)
56. Awrejcewicz, J., Krysko, V.A., Zhigalov, M.V., Krysko, A.V.: Contact interaction of two rectangular plates made from different materials with an account of physical non-linearity. Nonlinear Dyn. **91**(2), 1191–1211 (2018)
57. Bacinov, C.D.: Nonlinear oscillations of plates under simultaneous static and vibrating loads. Appl. Mech. **7**(10), 126–130 (1971) (in Russian)
58. Baglay, R.D., Smirnov, K.K.: To the processing of two-dimensional signals on a computer. Comput. Math. Math. Phys. **15**(1), 241–248 (1975) (in Russian)
59. Bagteva, L.U.: In:vestigation of the stability of thin shells and plates under dynamic loads. PhD Thesis, Kazan (1981) (in Russian)
60. Banks, J., Brooks, J., Davis, G., Stacey, P.: On Devaney's definition of chaos. Amer. Math. Monthly **99**(4), 332–334 (1992)
61. Barati, M.R.: Magneto-hygro-thermal vibration behavior of elastically coupled nanoplate systems incorporating nonlocal and strain gradient effects. J. Braz. Soc. Mech. Sci. Eng. **39**, 4335–4352 (2017)
62. Bauer, H.: Nonlinear response of elastic plates to pulse excitation. Trans. ASME **E35**(1), 47–52 (1968)
63. Beam, R.M., Warming, R.F.: Numerical Calculating of Two-Dimensional Unsteady Transonic Flows with Circulation. NASA, TND-76–05 (1974)
64. Beam, R.M., Warming, R.F.: An implicit finite-difference algorithm for hyperbolic system in conservation law form. J. Comput. Phys. **22**(1), 87–110 (1976)
65. Bell, J.F.: Experiment Fundamentals of Mechanics of Elastic Solids, vol. 2. Small Deformations. Nauka, Moscow (1984) (in Russian)
66. Bellman, R., Angel, E.: Dynamic Programming and Partial Differential Equations. Academic, New York (1972)
67. Bernadou, M., Oden, J.T.: An existence theorem for a class of nonlinear shallow shell problem. J. Math. Pure Appl. **60**, 1–24 (1981)
68. Biot, M.A.: Thermoelasticity and irreversible thermodynamics. J. Appl. Phys. **27**(3), 240–253 (1956)
69. Birger, I.A.: Circular Plates and Shells of Revolution. Oborongiz, Moscow (1971) (in Russian)
70. Birger, I.A., Mavlyutov, R.R.: The Resistance of Materials. Nauka, Moscow (1986) (in Russian)

71. Bogaryan, K.O.: On convergence of errors of the Bubnov–Galerkin and Ritz methods. Trans. Acad. Sci. SSSR **191**(2), 267–269 (1961) (in Russian)
72. Bollhaus, U.F., Lomex, G.: Numerical modelling of low-frequency nonstationary transonic flows. In: Ishlinskiy, A.Yu., Tchiorny, G.G. (eds.) Numerical Solutions of Hydromechanics Problems. News in Foreign Sciences, Mekhanika, No. 14. Mir, Moscow (1977) (in Russian)
73. Bolotin, V.V.: Nonconservative Problems of the Theory of Elastic Stability. Pergamon, Oxford (1963)
74. Bolotin, V.V.: The Dynamic Stability of Elastic Systems. Holden-Day, San Francisco (1964)
75. Borisyuk, A.I., Motovilovitz, I.A.: On temperature fields in a variable thickness shell. Appl. Mech. **3**(12), 84–89 (1961) (in Russian)
76. Bradley M.E., Lasiecka I.: Global stabilization of a von Kármán plate without geometric conditions. In: Banks, H.T., Fabiano, R.H., Ito, K. (eds.) Proceedings in Applied Mathematics Identification and Control in Systems Governed by Partial Differential Equations, No. 68. SIAM, Philadelphia (1993)
77. Budiansky, B., Roth, B.S.: Axisymmetric Dynamic Buckling of Clamped Shallow Spherical Shells. NASA, TN D-1510, pp. 379–606 (1962)
78. Buslov, E.P.: Experimental investigations on shallow behaviour under dynamic load. In: Lur'e, L.I. (ed.) Theory of Plates and Shells, pp. 363–365. Sudostroenie, Leningrad (1995) (in Russian)
79. Butcher, J.C.: Numerical Methods for Ordinary Differential Equations. Wiley, New York (2016)
80. Cash, J.R., Karp, A.H.: A variable order Runge-Kutta method for initial value problems with rapidly varying right-hand sides. ACM Trans. Math. Soft. **16**, 201–222 (1990)
81. Chen, L.-W., Hwang, J.-R.: Axisymmetric dynamic stability of transversely isotropic Mindlin circular plates. J. Sound Vib. **121**(2), 307–315 (1988)
82. Chen, X., Xiang, J., Li, B., He, Z.: A study of multiscale wavelet-based elements for adaptive finite elements analysis. Adv. Engng. Soft. **41**, 196–205 (2010)
83. Chen, C., Yuan, J., Mao, Y.: Post-buckling of size-depend micro-plate considering damage effects. Nonlinear Dyn. **90**(2), 1301–1314 (2017)
84. Chrzeszczyk, A.: Generalized solutions of dynamical equations in nonlinear theory of thin elastic shells. Archiwum Mechaniki Stosowanej **35**(5–6), 555–566 (1983)
85. Chrzeszczyk, A.: On the regularity, uniqueness and continuous dependence for generalized solutions of some coupled problems in nonlinear theory of thermoelastic shells. Archiwum Mechaniki Stosowanej **38**(1–2), 97–102 (1986)
86. Ciarlet, P.G.: Plates and Junctions in Elastic Multi-Structures: An Asymptotic Analysis, vol. 14. Masson, Paris (1990)
87. Ciarlet P.G.. Mathematical Elasticity, vol. 2. Theory of Plates. Elsevier Science, Amsterdam (1997)
88. Ciarlet, P.G., Rabier, P.: Les Equations de von Kármán. Springer, Berlin/Heidelberg (1980)
89. Cole, I.D.: On a quasi-linear parabolic equation occurring in aerodynamics. Q. Appl. Math **9**, 226–238 (1951)
90. Collet, P., Ecrmann, J.P.: Iterated Maps on the Interval us Dynamical Systems. Birkhauser, Boston (1980)
91. Crocker, M.J.: Response of panels to oscillating and to moving shock waves. J. Sound Vib. **6**(1), 38–58 (1967)
92. Curry, J.H., Herring, J.R., Loncaric, J., Orszag, S.A.: Order disorder in two- and three-dimensional Benard convection. S. Fluid Mech. **147**(1), 1–38 (1984)
93. Dafermos, K.H., Quasi-linear hyperbolic systems derived from conservation principles. In: Leibovich, S., Seebass, A.R. (eds.) Nonlinear Waves, pp. 91–112. Cornell University Press, Ithaca and London (1974)
94. Darevskiy, V.M., Nonlinear equations of the theory of shells and their linearization in problems of stability. Works of the VI National Conference on the Theory of Shells and Plates, pp. 355–368. Nauka, Moscow (1968) (in Russian)

95. Daubechies, I., Grossmann, A., Meyer, Y.: Painless nonorthogonal expansions. J. Math. Phys. **27**, 1271–1283 (1986)
96. Day, W.A.: Cesaro means and recurrence in dynamic thermoelasticity. Mathematika (London) **28**(2), 211–230 (1981)
97. Day W.A.: On the status of the uncoupled approximations within quasi-static thermoelasticity. Mathematika (London) **28**(2), 286–294 (1981)
98. Destuynder, C.: An existence theorem for a nonlinear shell model in large displacements analysis. Math. Meth. Appl. Sci. **5**, 68–83 (1983)
99. Devaney, R.L.: An Introduction to Chaotic Dynamical Systems. Addison-Wesley, Reading, MA (1989)
100. Dormand, J.R., Prince, P.J.: A family of embedded Runge-Kutta formulae. J. Comput. Appl. Math. **6**(1), 19–26 (1980)
101. Dowell, E.H.: Aeroelasticity of Plates and Shells. Nordhoff, Leiden (1974)
102. Duhamel, J.: Second memoire sur les phénomenes thermo-méchanique-nes. J. Ecole Polytechn **15**, 1–15 (1937)
103. Dzhishkaryani, A.V.: Convergence speed of approximate Ritz method. J. Comput. Math. Math. Phys. **3**(4), 654–663 (1963) (in Russian)
104. Dzhishkaryani, A.V.: On the convergence speed of the Bubnov-Galerkin method. J. Comput. Math. Math. Phys. **4**(2), 343–348 (1964) (in Russian)
105. Ehlers, F.E., Weatherill, W.H., Sebastian, J.D.: The Practical Application of a Finite Difference Method for Analyzing Transonic Flow Over Oscillating Airfoils and Wings. National Aeronautics and Space Administration, Scientific and Technical Information Office, Springfield, VA (1978)
106. El-Ghazdly, H.A., Sherboune, A.N.: Deformation theory for elastic-plastic buckling analysis of plates under nonproportional planar loading. Comput. Struct. **22**(2), 131–149 (1986)
107. Euler, L.: Sur la force des colones. Mem. de L'Acad. Berlin **13**, 252–282 (1757)
108. Faedo, S.: Un nuovo metodo per l'analisi esistenziale e duantativa dei problemi di propagazione. Ann. Scuola Norm. Super Pisa No. 1, 1–40 (1949)
109. Fathalilou, M., Sadeghi, M., Razazadeh, G.: Nonlinear behavior of compactive micro-beams based on strain gradient theory. J. Mech. Sci. Technol. **28**(4), 1141–1151 (2014)
110. Fedos'ev, V.I.: On a certain method of solution to the nonlinear problem of stability of deformable systems. Appl. Math. Mech. **27**, 265–274 (1963) (in Russian)
111. Fedos'ev, V.I.: Geometrically nonlinear problems in the theory of plates and shells. In: Proceedings of the VI Conference on 'Theory of Plates and Shells', pp. 971–976. Nauka, Moscow (2012)
112. Fehlberg, E.: Low-order classical Runge-Kutta formulas with step size control and their application to some heat transfer problems. NASA Tech. Rep. **315** (1969)
113. Fehlberg, E.: Klassische Runge-Kutta-Formeln vierter und niedrigerer Ordnung mit Schrittweiten-Kontrolle und ihre Anwendung auf Wärmeleitungsprobleme. Computing **6**(1–2), 61–71 (1970)
114. Feigenbaum, M.J.: Quantitative universality for a class of nonlinear transformations. J. Sat. Phys. **19**(25), 61–84 (1978)
115. Feigenbaum, M.J.: The universal metric properties of nonlinear transformations. J. Stat. Phys. **21**(6), 669–706 (1979)
116. Filippov, A.P.: Vibrations of Deformable Bodies. Mashinostroenie, Moscow (1970) (in Russian)
117. Fischer, G.: Existence theorems in elasticity. In: Handbuch der Physik Bd. 6a/s2, 347–389 (1972)
118. Fletcher, C.A.J.: Computational Galerkin Methods. Springer, Berlin (1984)
119. Fomin V.G., Modification of Godunov method for asymptotic transonic flows. In: Khromov, A.P. (ed.) Applied Mathematics and Mechanics, issue 2, pp. 61–66. Saratov University Press, Saratov (1983) (in Russian)
120. Galerkin, B.G.: Rods and plates. In: Some Problems on Elastic Equilibrium of Rods and Plates. Eng. News (Vestnik Inzhinierov) **1**(19), 897–908 (1915) (in Russian)

121. Galimov, K.Z.: General theory of plates and shells with finite displacements and deformations. Appl. Math. Mech. **15**(6), 723–742 (1951) (in Russian)
122. Gawinecki, J.: Existence, uniqueness and regularity of the first boundary–initial value for thermal stress equations of classical and generalized thermomechanics. J. Techn. Phys. **24**(4), 467–479 (1983)
123. Gawinecki, J.: On the first initial–boundary value problem for the equations of thermal stresses. Bull. Acad. Pol. Sci. Ser. Sci. Technol. **33**(2), 17–34 (1984)
124. Gelos, R., Dominiques, H., Laura, P.A.A.: Application of the optimized Galerkin method to the determination of the fundamental frequency of a vibrating circular plate subjected to non-uniform in-plane loading. J. Sound Vib. **114**(3), 598–600 (1987)
125. Germain, P., Bader, R.: Unicite des ecoulement avec chocs dans la mecanique de Burgers, pp. 1–13. Office National d'Etudes et de Recherches Aeronautiques, Paris (1953)
126. Ghadiri, M., Hosseini, S.H.S., Shafiei, N.: A power series for vibration of a rotating nanobeam with considering thermal effect. Mech. Adv. Mater. Struct. **23**(12), 1414–1420 (2016)
127. Godunov, S.K., Riabenkiy, V.S.: Difference Schemes. Elsevier Science, Amsterdam (1987)
128. Godunov, S.K., Zabrodin, A.V., Prokopov, G.P.: Difference scheme for two-dimensional, nonstationary problems of gas dynamics and calculations of a shock wave flow. Comput. Math. Tech. Phys. **1**(6), 1020–1050 (1961) (in Russian)
129. Godunov, S.K., Zabrodin, A.V., Ivanov, M.Ya., Krayko, A.N., Prokopov, G.P.: Numerical Solutions of Multi-Dimensional Problems in Gas Dynamics. Nauka, Moscow (1976) (in Russian)
130. Goldenweiser, A.L.: Theory of Thin Elastic Shells. Gostekhizdat, Moscow (1953) (in Russian)
131. Gorshkov, A.G.: Nonstationary interaction of plates and shells with continuous systems. Proc. Acad. Sci. SSSR, MTT No. 4, 177–189 (1981) (in Russian)
132. Gould, P.L.: Analysis of Shells and Plates. Prentice Hall, Englewood Cliffs, NJ (1999)
133. Gould, P.L.: Introduction to Linear Elasticity, 2nd edn. Springer, Berlin/Heidelberg (1999)
134. Gribanov, V.F., Panichkin, N.G.: Coupled and Dynamical Problems of Thermoelasticity. Mashinostroenie, Moscow (1984) (in Russian)
135. Grigolyuk, E.I., Kabanov, V.V.: Stability of Shells. Nauka, Moscow (1978) (in Russian)
136. Grigolyuk, E.I., Shalashilin, V.I.. Problems of Nonlinear Deformation. Kluwer Academic, Dordrecht (1991)
137. Grossman, A., Morlet, S.: Decomposition of Hardy functions into square separable wavelets of constant shape. SIAM J. Math. Anal. **15**(4), 723–736 (1984)
138. Gulick, D.: Encounters with Chaos. McGraw-Hill, NewYork (1992)
139. Gurov, N.P.: Phenomenological Thermodynamics of Nonconvertible Processes. Nauka, Moscow (1978) (in Russian)
140. Gvozdev, A.A.: Calculations of Load Limit of Construction by Limited Equilibrium Method. Strojizdat, Moscow (1949) (in Russian)
141. Haar, A.: Zur Theorie der orthogonalen Funktionensysteme. Math. Ann. **69**(3), 331–371 (1910)
142. Han, Q., Hu, H., Yang, G.: A study of chaotic motion in elastic cylindrical shells. Eur. J. Mech. A/Sol. **18**, 351–360 (1999)
143. Hao, Y.X., Chen, L.H., Zhang, W., Lei, J.G.: Nonlinear oscillations, bifurcations and chaos of functionally graded materials plate. J. Sound. Vib. **312**, 862–892 (2008)
144. Hartmann, F.: Ordinary Differential Equations. Birkhäuser, Boston (1982)
145. Hashemi, M., Asghari, M.: A semi-analytical solution for flexural vibration of micro beams. J. Ther. Appl. Vib. Acoust. **1**(1), 48 61 (2015)
146. He, X.T., Chen, Q., Sun, J.Y., Zheng, Z.L., Chen, S.L.: Application of the Kirchhoff hypothesis to bending thin plates with different moduli in tension and compression. J. Mech. Mater. Struct. **5**, 755–769 (2010)

147. Hinton, E.. Numerical Methods and Software for Dynamics Analysis of Plates and Shells. Pineridge, Swansea (1987)
148. Hopf, E.: The partial differential equation. Commun. Pure Appl. Math. No. 3, 201–230 (1950)
149. Horn, M.A., Lasiecka I.: Nonlinear boundary stabilization of parallelly connected Kirchhoff plates. Dyn. Control **6**(3), 263–292 (1996)
150. Hosseini, M., Jamalpoor, A.: Analytical solution for thermomechanical vibration of double-viscoelastic nanoplate-systems made of functionally graded materials. J. Therm. Stress. **38**, 1428–1456 (2015)
151. Hotelling, H.: Analysis of a complex of statistical variables into principal components. J. Edu. Phychol. **24**, 417–441 (1933)
152. Hu, Y., Zhang, Z.: Bifurcation and chaos of thin circular functionally graded plate in thermal environment. Chaos Sol. Fract. **44**, 739–750 (2011)
153. Huang, H.-C.: Static and Dynamic Analyses of Plates and Shells. Theory, Software and Applications. Springer, Berlin/Heidelberg (1989)
154. Ignatiev, V.A., Sokolov, O.L.: Thin-Walled Cellular Structures. Methods for Their Analysis. Balkema, Rotterdam (1999)
155. Il'yushin, A.A.: Plasticity. Gostekhizdat, Moscow (1948) (in Russian)
156. Ivanov, G.V.: About solution of two-dimensional and spatial problems of transonic flow around bodies. Comput. Math. Tech. Phys. **15**(5), 1222–1240 (1975) (in Russian)
157. Ivanov, M.Ja., Ryl'ko, O.A.: Calculations of transonic flow in spatial nozzles. Comput. Math. Tech. Phys. **12**(5), 1280–1291 (1972) (in Russian)
158. Jawad, M.: Theory and Design of Plate and Shell Structures. Kluwer Academic, Dordrecht (1994)
159. Kaczkowski, Z.: On variational principles in thermoelasticity. Bull. Acad. Pol. Sci. Technol. **30**(5–6), 81–86 (1982)
160. Kantor, B.Ya.: On the Nonlinear Theory of Thin Shells. Kharkiv National University, Kharkiv (1967) (in Russian)
161. Kantor, B.Y.: Nonlinear Problems of the Theory of Nonhomogeneous Shallow Shells. Naukova Dumka, Kiev (1971) (in Russian)
162. Kantor, B.Y., Bohatyrenko, T.L.: The method for solving contact problems in the nonlinear theory of shells. Rep. Acad. Sci. Ukr. SSR **A**(1), 18–21 (1986)
163. Kantorovich, L.V.: One direct method of approximate solution of the problem of minimum of the double integral. News of Acad. Sci SSSR **5**, 647–652 (1933) (in Russian)
164. Kantz, H.: A robust method to estimate the maximum Lyapunov exponent of a time series. Phys. Lett. A **185**, 77–87 (1994)
165. Kaplunov, J.D., Kossovitch, L.Yu., Nolde, E.V.: Dynamics of Thin Walled Elastic Bodies. Academic, San Diego (1997)
166. Karličić, D., Kozić, P., Adhikari, S., Cajić, M., Murmu, T.: Nonlocal mass-nanosensor model based on the damped vibration of single-layer graphene sheet influenced by in-plane magnetic field. Int. J. Mech. Sci. **96–97**, 132–142 (2015)
167. Karličić, D., Jovanović, D., Kozić, P., Cajić, M.: Thermal and magnetic effects on the vibration of a cracked nanobeam embedded in an elastic medium. J. Mech. Mater. Struct. **10**(1), 43–62 (2015)
168. Karličić, D., Cajić, M., Adhikari, S., Kozić, P., Murmu, T.: Vibrating nonlocal multi-nanoplate system under inplane magnetic field. Eur. J. Mech. A/Sol. **64**, 29–45 (2017)
169. Kármán, T.: Festigkeitsprobleme in Maschinenbau. Encykle D Math. Wiss **4**(4), 311–385 (1910)
170. Karnaukhov, V.G., Kirichok, I.F.: Coupled Problems in Theory of ElastoPlastic Plates and Shells. Naukova Dumka, Kiev (1986) (in Russian)
171. Karsloy, G., Eger, D.: Heat Conductivity of Solids. Nauka, Moscow (1964) (in Russian)
172. Kauderer, H.: Nonlinear Mechanics. Springer, Berlin (1958) (in German)
173. Keldysh, M.V.: On the Galerkin method for solving boundary problems. Trans. Acad. Sci. SSSR Ser. Math. **6**(6), 309–330 (1942) (in Russian)

174. Kelly, J.M., Wierzbicki, T.: Motion of a circular viscoplastic plate subject to projectile impact. Z. Angew. Math. Phys. **18**, 236–246 (1967)
175. Kerid, R., Bourouina, H., Yahiaoui, R., Bounekhla, M., Aissat, A.: Magnetic field effect on nonlocal resonance frequencies of structure-based filter with periodic square holes network. Phys. E. **105**, 83–89 (2019)
176. Kiani, K.: Free vibration of conducting nanoplates exposed to unidirectional in-plane magnetic fields using nonlocal shear deformable plate theories. Phys. E **57**, 179–192 (2014)
177. Kikoin, I.K. (ed.): Tables of Physical Properties. Atomizdat, Moscow (1976) (in Russian)
178. Kil'chinskaya, G.A.: The smallest excitation force principle for generalized thermomechanics. Trans. Acad. Sci. USSR No. 2, 1092–1095 (1997) (in Russian)
179. Kirichenko, V.F., Krysko, V.A.: Substantiation of the variational iteration method in the theory of plates. Sov. Appl. Mech. **17**(4), 366–370 (1981) (in Russian)
180. Kirichenko, V.F., Krys'ko, V.A.: On the existence of solutions to a certain nonlinear coupled problem of thermoelasticity. Differ. Equ. **20**(9), 1583–1588 (1984) (in Russian)
181. Kirichenko, V.F., Krys'ko, V.A., Khametova, N.A.: On the influence of the coupling effect of temperature and deformation fields on the dynamic stability of shallow shells. Appl. Mech. **24**(11), 46–50 (1998) (in Russian)
182. Kleiber, M., Woźniak, C.: Nonlinear Mechanics of Structures, vol. 8. Kluwer Academic, Dordrecht (1991)
183. Knudsen, C.: Chaos without periodicity. Am. Math. Mon. **101**, 563–565 (1994)
184. Kochin, N.E., Kibel, I.A., Roze, N.V.: Theoretical Hydromechanics. Inter-Science, New York (1964)
185. Kolyano, Yu.M., Shter, E.I.: Thermoelasticity of nonhomogeneous bodies. Eng. Phys. J. **3**(6), 1111–1114 (1980) (in Russian)
186. Kong, S., Zhou, S., Nie, Z., Wang, K.: Static and dynamic analysis of micro beams based on strain gradient elasticity theory. Int. J. Eng. Sci. **47**, 487–498 (2009)
187. Kovalenko, A.D.: Thermoelasticity. Basic Theory and Applications. Wolters-Noordhoff, Groningen (1969)
188. Kovalenko, A.D.: Thermoelasticity. Vysshaia Shkola, Kiev (1975) (in Russian)
189. Kovalenko, A.D.: Collected Works. Naukova Dumka, Kiev (1976) (in Russian)
190. Kowalski, T., Piskorek, A.: Existence of solution of boundary-value problem in linear theory of thermoelasticity. Z. Angew. Math. Mech. **61**(5), T250–T252 (1981)
191. Kowalski, T., Litevska, K., Piskorek, A.: Uniqueness and regularity of the solution of the first initial–boundary value problem in linear thermoelasticity. Bull. Acad. Pol. Sci. Technol. **30**(3–4), 171–175 (1982)
192. Kozdoba, L.A.: Solution Methods for Nonlinear Problems of Thermoconductivity. Nauka, Moscow (1975) (in Russian)
193. Kozlov, V.I.: Coupling effect in a problem of thermal impact on rod surface. In: Kovalenko, A.D. (ed.) Thermal Stresses in Structures (Collection of Articles), Kiev, issue 11, pp. 97–102 (1971) (in Russian)
194. Kozlov, V.I.: Thermal impact on circular plate surface taking into account coupling of deformation fields. Trans. Acad. Sci. USSR Ser. A No. 10, 923–927 (1971) (in Russian)
195. Kozlov, V.I.: Temperature oscillations of rectangular plate. Appl. Mech. **8**(4), 123–127 (1972) (in Russian)
196. Krasnosel'skiy, M.A.: Convergence of Galerkin method for nonlinear equations. Trans. Acad. Sci. SSSR **23**(6), 1121–1124 (1950) (in Russian)
197. Krasnosel'skiy, M.A., Baynikko, G.M., Zabreyko, P.P.: Approximate Solution of Operator Equations. Nauka, Moscow (1969) (in Russian)
198. Krys'ko, V.A.: Nonlinear Statics and Dynamics of Nonhomogeneous Shells. Saratov University Press, Saratov (1976) (in Russian)
199. Krys'ko, V.A.: Dynamic buckling of shells, rectangular in plan, with finite deflection. Appl. Mech. **15**(11), 58–62 (1979) (in Russian)

200. Krylov, K.M., Bogolyubov, N.N.: On some theorems concerning the behaviour of integrals of partial differential equations of hyperbolic type. Trans. Acad. Sci. SSSR No. 3, 323–344 (1931) (in Russian)
201. Krys'ko, V.A., Kutsemako, A.N.: Stability and Oscillations of Nonuniform Shells. Saratov Technical University, Saratov (1999) (in Russian)
202. Krys'ko, V.A., Mishnik, M.P.: Calculation of coupled physically nonlinear three-dimensional plates in a temperature field. Trans. VUZ, Ser. Civ. Eng. Architect. No. 9, 33–37 (1984) (in Russian)
203. Krys'ko, V., Awrejcewicz, J., Bruk, V.: The existence and uniqueness of solution of one coupled plate thermomechanics problem. J. Appl. Anal. **8**(1), 129–139 (2002)
204. Krysko, A.V., Awrejcewicz, J., Kutepov, I.E., Zagniboroda, N.A., Dobriyan, V., Krysko, V.A.: Chaotic dynamics of flexible Euler-Bernoulli beams. Chaos **23**(4), 043130-1– 043130-25 (2013)
205. Krysko, A.V., Awrejcewicz, J., Saltykova, O.A., Zhigalov, M.V., Krysko, V.A.: Investigations of chaotic dynamics of multi-layer beams using taking into account rotational inertial effects. Commun. Nonlinear Sci. Num. Simul. **19**(8), 2568–2589 (2014)
206. Krysko, A.V., Awrejcewicz, J., Zhigalov, M.V., Krysko, V.A.: On the contact interaction between two rectangular plates. Nonlinear Dyn. **84**(4), 2729–2748 (2016)
207. Krysko, A.V., Awrejcewicz, J., Papkova, I.V., Szymanowska, O., Krysko, V.A.: Principal component analysis in the nonlinear dynamics of beams: purification of the signal from noise induced by the nonlinearity of beam vibrations. Adv. Math. Phys. **2017**, 1–9 (2017)
208. Krysko, A.V., Awrejcewicz, J., Zakharova, A.A., Papkova, I.V., Krysko, V.A., Chaotic vibrations of flexible shallow axially symmetric shells. Nonlinear Dyn. **91**(4), 2271–2291 (2018)
209. Krysko, V.A., Awrejcewicz, J., Papkova, I.V., Saltykova, O.A., Krysko, A.V.: On reliability of chaotic dynamics of two Euler-Bernoulli beams with a small clearance. Int. J. Non Linear Mech. **104**, 8–18 (2018)
210. Kupradze, V.D., Gegelija, T.G.: Three-Dimensional Problems of the Mathematical Theory of Elasticity and Thermoelasticity. North-Holland, Amsterdam (1979)
211. Kutateladze, S.S.: Fundamentals of Heat Transfer. Academic, New York (1963)
212. Kuznetzov, E.B., Kulakov, N.A., Shalashilin, V.I.: On the effect of dynamic loads on some systems with clicking. Selected Problems of Applied Mechanics, pp. 12–35. VINITI, Moscow (1974) (in Russian)
213. Ladyzhenskaya, O.A.: The Mathematical Theory of Viscous Incompressible Flow. Gordon and Breach, New York (1969)
214. Ladyzhenskaya, O.A.: The Boundary Value Problems of Mathematical Physics. Springer, New York (1985)
215. Ladyzhenskaya, O.A., Uraltseva, N.N.: Linear and Quasilinear Elliptic Equations. Academic, New York (1968)
216. Lam, D.C.C., Yang, F., Chong, A.C.M., Wang, J., Tong, P.: Experiments and theory in strain gradient elasticity. J. Mech. Phys. Sol. **51**, 1477–1508 (2003)
217. Lanczos, C.: Variational Principles of Mechanics, 2nd edn. Toronto University Press, Toronto (1964)
218. Landau, L.D., Lifshitz, E.M.: A Shorter Course of Theoretical Physics. Pergamon, Oxford (1972)
219. Lepik, Yu.A.: Dynamics of circular and toroidal plates made of rigid–plastic material sensitive to deformation speed. Appl. Mech. Acad. Sci. USSR **5**(1), 35–41 (1969) (in Russian)
220. Lepik, U.: Solving integral and differential equations by the aid of nonuniform Haar wavelets. Appl. Math. Comput. **198**, 326–332 (2008)
221. Li, T.Y., Yorke, L.A.: Period three implies chaos. Am. Math. Mon. **82**, 985–992 (1975)
222. Liang, Y.C., Lee, H.P., Lim, S.P., Lin, W.Z., Lee, K.H., Wu, C.G.: Proper orthogonal decomposition and its applications, Part I: Theory. J. Sound. Vib. **252**, 527–544 (2002)
223. Libai, A., Simmonds, J.G.: The Nonlinear Theory of Elastic Shells. Cambridge University Press, Cambridge (1998)

224. Liew, K.M., Wang, C.M., Xiang, Y., Kitipornchai, S.: Vibration of Mindlin Plates. Elsevier, Amsterdam (1998)
225. Lin, C.C., Reissner, E., Tsien, H.S.: On two-dimensional nonsteady motion of a slender body in a compressible fluid. J. Math. Phys. **27**(3), 220–231 (1948). Russian translation in Gas Dynamics. Moscow, 1955, pp. 183–196
226. Lions, J.L.: Problèmes aux limites dans les équations aux dérivées partielles. 2 éd. Presses de l'Université de Montréal, Montréal (1967) (in French)
227. Lions, J.L., Magenes, E.: Non-homogeneous Boundary Problems and Their Applications. Springer, Berlin (1972)
228. Lock, M.H.: Snapping of a shallow sinusoidal arch under a step pressure load. AIAA J. **4**(7), 31–41 (1966)
229. Lord, H.W., Shulman, Y.: A generalized dynamical theory of thermoelasticity. J. Mech. Phys. Sol. **15**(5), 299–309 (1967)
230. Lorenz, E.N.: Deterministic nonperiodic flow. Atmos. Sci. **20**(1), 130–141 (1962)
231. Lozi, R.: Can we trust in numerical computations of chaotic solutions of dynamical systems? In: World Scientific Series on Nonlinear Science. Topology and Dynamics of Chaos in Celebration of Robert Gilmore's 70th Birthday, vol. 84, pp. 63–98 (2013)
232. Lykov, A.V.: Analytical Heat Diffusion Theory. Academic, New York (1968)
233. Magnus, R., Yoshihara, H.: Calculations of transonic flow over an oscillating airfoil. AIAA J. No. 1, 75–98 (1975)
234. Malkin, Ja.F.: On problems of distribution of temperature in flat plates. Appl. Math. Mech. **2**(3), 317–330 (1939) (in Russian)
235. Mandelbrot, B.B.: The Fractal Geometry of Nature. Freemcen, San Francisco (1982)
236. Mansfield, E.H.: The Bending and Stretching of Plates. Cambridge University Press, Cambridge (1989)
237. Marguerre, K.: Temperature changes and temperature stresses in plates and shallow bodies. Ing. Arch. **8**(3), 216–228 (1937)
238. Mehditabar, A., Rahimi, G.H., Tarahhomi, M.H.: Thermo-elastic analysis of a functionally graded piezoelectric rotating hollow cylindrical shell subjected to dynamic loads. Mech. Adv. Mater. Struct. **0**, 1–12 (2017)
239. Medina, L., Gilat, R., Krylov, S.: Modeling strategies of electrostatically actuated initially curved bistable micro-plates. Int. J. Sol. Struct. **118–119**, 1339–1351 (2017)
240. Meyer, Y.: Ondelettes et functions splines. Ecole Polytech. Paris, Tech. Rep. Semin. EDP (1986)
241. Meyer, Y.: Ondelettes, fonctions splines et analyses graduees. Lecture Notes Univ. Torino (1986)
242. Meyer, Y.: Wavelets: Algorithms and Applications. SIAM, Philadelphia (1993)
243. Mikhaylov, V.P.: Partial Differential Equations. Mir, Moscow (1978) (in Russian)
244. Mikhaylovskaya, I.E., Novik, O.B.: Cauchy problems in a class of increasing functions for nonhyperbolic evolutionary systems which are not parabolic. Sib. Math. J. Novosibirsk (1979) (in Russian)
245. Mikhlin, S.G.: On the convergence of the Galerkin method. Trans. Acad. Sci. SSSR **61**(2), 197–199 (1948) (in Russian)
246. Mikhlin, S.G.: The Minimum Problem of a Square Functional. Gostekhizdat, Moscow (1952) (in Russian)
247. Mikhlin, S.G.: On the Ritz method. Trans. Acad. of Sci. SSSR **106**(3), 391–394 (1956) (in Russian)
248. Mikhlin, S.G.: Variational Methods in Mathematical Physics. Pergamon, Oxford (1964)
249. Mikhlin, S.G.: The Numerical Performance of Variational Methods. Wolters-Noordhoff, Groningen (1971)
250. Mikhlin, S.G.: Some Problems in Error Theory. Leningrad University Press, Leningrad (1988) (in Russian)
251. Mindlin, D., Tiersten, H.F.: Effects of couple-stresses in linear elasticity. Arch. Ration. Mech. Anal. **11**, 415–448 (1962)

252. Morozov, N.F.: On nonlinear oscillations of thin plates with consideration of moment of inertia. Trans. Acad. Sci. SSSR **176**(3), 522–525 (1967) (in Russian)
253. Morozov, N.F.: Collected Two-Dimensional Problems of Theory of Elasticity. Leningrad University Press, Leningrad (1978) (in Russian)
254. Moskvitin, V.V.: Plasticity Under Alternating Loads. MGU, Moscow (1965) (in Russian)
255. Moskvitin, V.V.: Periodic Loads on Structural Elements. Nauka, Moscow (1981) (in Russian)
256. Motovilovitz, I.A., Kozlov, V.I.: Mechanics of Coupled Fields in Structural Elements, vol. 1. Naukova Dumka, Kiev (1987) (in Russian)
257. Mukhopandhayay, M.: Free vibration of rectangular plates with edges having different degrees of rotational restraint. J. Sound Vib. **67**(4), 459–468 (1979)
258. Murman, E.M., Cole, I.D.: Calculation of plane steady transonic flows. AIAA J. **9**(1), 114–121 (1971)
259. Mushtari, Kh.M., Galimov, K.Z.: Non-Linear Theory of Elastic Shells. Tatknigoizdat, Academy of Sciences, USSR, Kazan' Branch (1957)
260. Nashed, M.Z.: The convergence of the method of steepest descents for nonlinear equations with variational or quasi-variational operators. J. Math. Mech. **13**, 765–794 (1964)
261. Nezlin, M.V., Snieshkin, E.N., Trubnikov, A.S.: Kelvin–Helmholtz instability and Great Red Spot of Jupiter. ZETF **32**, 190–193 (1982)
262. Nickell, R.E., Sackman, J.L.: Variational principles for linear coupled thermoelasticity. Q. Appl. Math. **26**(1), 11–26 (1968)
263. Nowacki, W.: Thermal stresses in orthotropic plates. Bull. Acad. Pol. Sci. Ser. Sci. Techol. **7**(1), 1–6 (1959)
264. Nowacki, W.: Some dynamic problems of thermoelasticity. Archiwum Mechaniki Stosowanej **11**(2), 259–283 (1959)
265. Nowacki, W.: Thermal Stresses in Shells. General Report on Non-Classical Shell Problem. IASS, Warsaw (1962)
266. Nowacki, W.: Thermoelasticity. Pergamon, London (1962)
267. Nowacki, W.: Dynamical Problems in Thermoelasticity. PWN, Warsaw (1966) (in Polish)
268. Ohashi, Y., Murakami, S.: The elasto-plastic bending of a clamped thin circular plate. In: Proceedings of the Eleventh International Congress of Applied Mechanics Munich, pp. 212–223 (1964)
269. Osaka, H., Fujita, M., Hanasaki, K., Fujinaka, R.: A numerical analysis of plates under transverse impact loading. J. Jpn. Soc. Technol. Plast. **27**(301), 288–294 (1986)
270. Papkovich, P.F.: Ship Building Mechanics. Sudpromgiz, Leningrad (1939) (in Russian)
271. Park, W.-T., Han, S.-Ch.: Buckling analysis of nano-scale magneto-electro-elastic plates using the nonlocal elasticity theory. Adv. Mech. Eng. **10**(8), 1–16 (2018)
272. Parton, V.Z., Perlin, P.I.: Mathematical Methods of the Theory of Elasticity. Mir, Moscow (1984) (in Russian)
273. Pelevic, N., van der Meer, T.: Numerical investigation of heat transfer enhancement by carbon nano fibres deposited on a flat plate. Comput. Math. Appl. **65**, 914–923 (2013)
274. Petrashvili, V.I.: Multidimensional solitons. In: Nonlinear Waves (ed. Gaponov), pp. 5–20. Nauka, Moscow (1979) (in Russian)
275. Petryshyn, W.V.: Direct and iterative methods for the solution of linear operator equations in Hilbert space. Trans. Am. Math. Soc. **105**(1), 136–175 (1962)
276. Pobedrya, B.E.: Numerical methods in viscoelasticity Polym. Mech. No. 3, 417–428 (1973) (in Russian)
277. Podstrigach, Ya.S., Kolyano, Yu.M.: Generalized Thermomechanics. Naukova Dumka, Kiev (1976) (in Russian)
278. Podstrigach, Ya.S., Shvetz, R.N.: Thermoelasticity of Thin Shells. Naukova Dumka, Kiev (1978) (in Russian)
279. Podstrigach, Ya.S., Lapakin, V.A., Kolyano, Yu.M.: Thermoelasticity of Nonhomogeneous Bodies. Nauka, Glav. red. fiziko-matematicheskoi lit-ry, Moscow (1984) (in Russian)
280. Pugno, N.: Non-linear statics and dynamics of nanoelectromechanical systems based on nanoplates and nanowires. Proc. Inst. Mech. Eng. Part N. **219**(1), 29–33 (2005)

281. Rabinovich, M.I., Trubetskov, D.I.: Oscillations and Waves in Linear and Nonlinear Systems. Kluwer Academic, Dordrecht (1989)
282. Rajabi, F., Ramezani, S.: A nonlinear microbeam model based on strain gradient elasticity theory with surface energy. Arch. Appl. Mech. **82**(3), 363–376 (2012)
283. Reissner, E.: Stress and small displacements of shallow spherical shells. J. Math. Phys. **25**, 279–300 (1946)
284. Ritz, W.: Uber eine neue Methode zur Losung gewisser Variationsprobleme der mathematischen Physik. J. Reine Angew. Math. **135**(1), 1–61 (1909)
285. Rosenstein, M.T., Collins, J.J., De Luca, C.J.: A practical method for calculating largest Lyapunov exponents from small data sets. Phys. D **65**, 117–134 (1993)
286. Rozhdestvenskiy, B.L., Janenko, N.N.: Systems of quasilinear equations and their applications to gas dynamics. Am. Math. Soc. **55**, 0065–9282 (1983)
287. Ryzhov, O.S.: Transonic Flow in Laval Nozzles. Computer Centre, Academy of Sciences of the SSSR, Moscow (1965) (in Russian)
288. Ryzhov, O.S., Shefter, G.M.: On the influence of viscosity and heat conductivity on the structure of compressible flow. Appl. Math. Mech. **28**(6), 996–1007 (1964) (in Russian)
289. Sathyamoorthy, M.: Nonlinear Analysis of Structures. CRC, Boca Raton, FL (1998)
290. Sedighi, H.M., Shirazi, K.H., Zare, J.: Novel equivalent function for deadzone nonlinearity: applied to analytical solution of beam vibration using He's parameter expanding method. Lat. Am. J. Sol. Struct. **9**(4), 443–452 (2012)
291. Sedov, L.: Similarity and Dimensional Methods in Mechanics. CRC Press, Boca Raton (1993)
292. Sedov, L.I.: Mechanics of Continuous Media. World Scientific, Singapore (1997)
293. Sevost'janov G.D., Fomin V.G., On nonstationary axisymmetric ideal gas flows. Aerodynamics (Saratov University Press, Saratov) **5**(8), 27–32 (1976) (in Russian)
294. Shagivalev, K.F.: Calculation of a Closed Cylindrical Shell on the Approximate Theory. Saratov State Technical University, Saratov (2001)
295. Shalov, V.M.: Solution of non-self-adjoint equations by variation method. Trans. Acad. Sci. SSSR **151**(3), 511–512 (1963) (in Russian)
296. Sharkovskiy, A.N.: The coexistence of cycles of continuous transformation of a straight line to itself. Ukr. Math. J. **26**(1), 6–71 (1964) (in Russian)
297. Shiau, A.C., Soong, I.T., Roth, R.S.: Dynamic buckling of conical shells with imperfections. AIAA J. **12**(6), 262–271 (1974)
298. Shvetz, R.N.: Variational theorem for mutually coupled thermoelasticity problems of thin shells. Math. Phys. No. 28, 104–108 (1980) (in Russian)
299. Shvetz, R.N., Flachok, V.M.: Equations of mechanical thermal diffusion for anisotropic shells with consideration of transverse deformations. Math. Meth. Phys. Mech. Fields No. 20, 54–61 (1984) (in Russian)
300. Shvetz, R.N., Lopat'ev, A.A.: On properties of dynamic processes in de- formable solids with consideration of finite speed of heat transfer. J. Eng. Phys. **25**(4), 705–712 (1978) (in Russian)
301. Shvetz, R.N., Lun', E.I.: Some aspects of the theory of thermoelasticity of an orthotropic shell with consideration of moments of inertia and transverse shear. Appl. Mech. **7**(10), 121–125 (1971) (in Russian)
302. Skrypnik, I.V.: Nonlinear Elliptic Equations of High Order. Naukova Dumka, Kiev (1973) (in Russian)
303. Skurlatov, E.D.: On the behaviour of cylindrical panels and shells subjected to an incoming pressure wave. In: Lur'e, L.I. (ed.) Theory of Plates and Shells, Nauka, Moscow (1971), pp. 256–261 (in Russian)
304. Skurlatov, E.D.: Experimental investigations of behaviour of shells under dynamic loads. Elast. Problems No. 9, 79–83 (1972) (in Russian)
305. Skurlatov, E.D., Solonenko, V.R.: Experimental and theoretical investigation of oscillations of cylindrical shells under moving, pulsing pressure jumps. In: State Symposium on Distribution of Elastic and Elasto-Plastic Waves. Nauka, Moscow (1978) (in Russian)
306. Smith, I.M., Griffiths, D.V.: Programming the Finite Element Method, 3rd edn. Wiley, Chichester (1998)

307. Sobolev, C.L.: Some Application of Functional Analysis in Mathematical Physics. Leningrad University Press, Leningrad (1950) (in Russian)
308. Sobolevskiy, P.E.: On the Bubnov–Galerkin method for parabolic equations in Hilbert space. Trans. Acad. Sci. SSSR **178**(3), 548–551 (1968) (in Russian)
309. Sobotka, Z.: Theory of Plasticity and Limit Design of Plates, vol. 18. Elsevier, Amsterdam (1989)
310. Soedel, W.: Vibrations of Shells and Plates, vol. 10. Marcel Dekker, New York (1981)
311. Spreiter, I.R.: Aerodynamics of wings and bodies at transonic speeds. J. Aero/Space Sci. **26**(8), 465–486 (1959). Russian translation in Mechanics, No. 3, 3–50 (1960)
312. Stepanov, G.B., Kovalenko, A.V.: Bending of slender plate loaded by short pressure impact. Strength Problems No. 3, 40–46 (1986) (in Russian)
313. Süli, E., Mayers, D.: An Introduction to Numerical Analysis, Cambridge University Press, Cambridge (2003)
314. Svirskiy, I.V.: Methods of the Bubnov–Galerkin Type and Successive Approximations. Nauka, Moscow (1968) (in Russian)
315. Sylvester, J.J.: On the reduction of a bilinear quantic of the nth order to the form of a sum of n products by a double orthogonal substitution. Mess. Math. **19**, 42–46 (1889)
316. Szilard, R.: Theory and Analysis of Plates, Classical and Numerical Methods. Prentice-Hall, Englewood Cliffs, NJ (1974)
317. Thompson, D.E.: Design Analysis. Mathematical Modeling of Nonlinear Systems. Cambridge University Press, Cambridge (1999)
318. Timashev, S.A.: Stability of Reinforced Shells. Stroyizdat, Moscow (1974) (in Russian)
319. Touzé, C., Thomas, O., Amabili, M.: Transition to chaotic vibrations for harmonically forced perfect and imperfect circular plates. Int. J. Non Linear Mech. **46**(1), 234–246 (2011)
320. Traci, R.M., Farr, I.L., Albano, E.: Perturbation method for transonic flow about oscillating airfoils. AIAA J. 75–877 (June 1975)
321. Ugural, A.C.: Stresses in Plates and Shells, 2nd edn. McGraw-Hill, New York (1999)
322. Ugural, A.C., Fenster, S.K.: Advanced Strength and Applied Elasticity. Prentice Hall, Upper Saddle River NJ (1995)
323. Vaindiner, A.I.: On a generalization of the Bubnov-Galerkin-Kantorovich method of approximate solution of boundary value problems. Bull. Mosc. St. Univ. Math. Mech. **2**, 1001–1003 (1967) (in Russian)
324. Vakhlaeva, L.F., Krys'ko, V.A.: Stability of thin shallow shells in a temperature field. Appl. Mech. **19**(1), 16–23 (1983) (in Russian)
325. Valishvilli, I.E.: Methods of Computer Calculation of Axisymmetric Shells. Mashinostroenie, Moscow (1976) (in Russian)
326. Vashakmadze, T.S.: The Theory of Anisotropic Elastic Plates. Kluwer Academic, Dordrecht (1999)
327. Vasil'kovskiy, S.N.: Theorem about uniqueness of solution of coupled thermoelasticity dynamics in stress equations. Trans. *VUZ*. Math. No. 9, 21–24 (1984) (in Russian)
328. Vatankhah, R., Kahrobaiyan, M.H.: In:vestigation of size-dependency in free-vibration of micro-resonators based on the strain gradient theory. Lat. Am. J. Sol. Struct. **13**(3), 498–515 (2016)
329. Vaynikko, G.M., Oya, P.E.: On convergence and convergence velocity of Galerkin method for abstract of evolutionary equations. Differ. Equ. **11**(7), 1269–1277 (1975) (in Russian)
330. Villagia, P.: Mathematical Models for Elastic Structures. Cambridge University Press, Cambridge (1997)
331. Vinson, J.R.: The Behavior of Thin Walled Structures: Beams, Plates and Shells, vol. 8. Nijhoff, Dordrecht (1988)
332. Vlasov, V.Z.: A new practical method for calculating folded coatings and shells. Construct. Indust. **11**, 33–37 (1932)

333. Vlasov, V.Z.: General Shell Theory and Its Technical Applications. Gostekhizdat, Moscow (1949) (in Russian)
334. Vlasov, V.Z.: Thin-Walled Structures. Gosstrojizdat, Moscow (1958) (in Russian)
335. Vlasov, V.Z.: General Theory of Shells and Its Application in Engineering. NASA-TT-F-99 (1964)
336. Volmir, A.S.: Survey of Investigations on the Theory of Flexible Plates and Shells (Covering the Period from 1941 to 1957). National Aeronautics and Space Administration, Washington, DC (1963)
337. Volmir, A.S.: Stability of Deformable Systems. Nauka, Moscow (1967)
338. Volmir, A.S.: Nonlinear Dynamics of Plates and Shells. Nauka, Moscow (1972) (in Russian)
339. Volmir, A.S.: Shells in Liquid and Gas Flow. Nauka, Moscow (1976) (in Russian)
340. Volmir, A.S.: Shells in Liquid and Gas Flow. Hydroelasticity Problems. Nauka, Moscow (1979) (in Russian)
341. Vorovich, I.I.: On some simple methods in the nonlinear theory of oscillations of elastic shells. Appl. Math. Mech. **21**(6), 747–784 (1957) (in Russian)
342. Vorovich, I.I.: Nonlinear Theory of Shallow Shells. Springer, New York (1998)
343. Vorovich, I.I.: Nonlinear Theory of Shallow Shells. Springer, New York (1999)
344. Vorovich, I.I., Lebedev, L.P.: On the existence of solutions in the nonlinear theory of elastic shells. Appl. Math. Mech. **36**(4), 691–704 (1972) (in Russian)
345. Vorovich, I.I., Lebedev, L.P., Shlafman, S.M.: On certain direct methods and existence of solutions in nonlinear theory of elastic non-shallow rotational shells. Appl. Math. Mech. **38**(2), 339–348 (1974)
346. Wang, Y.-G., Song, H.-F., Li, D., Wang, J.: Bifurcations and chaos in a periodic time-varying temperature-excited bimetallic shallow shell of revolution. Arch. Appl. Mech. **80**, 815–828 (2010)
347. Waszczyszyn, Z., Cichon, C., Radwanska, M.: Stability of Structures By Finite Element Method. Elsevier, Amsterdam (1994)
348. Weatherill, W.H., Sebastian, J.D., Ehlers, F.E.: The Practical Application of a Finite Difference Method for Analyzing Transonic Flow over Oscillating Airfoils and Wings. National Aeronautics and Space Administration, Scientific and Technical Information Office, Springfield, VA (1978)
349. Wierzbicki, T.: Impulsive loading of rigid viscoplastic plates. Int. J. Solids Struct. **3**, 635–647 (1967)
350. Wierzbicki, T.: Finite deflection of a circular viscoplastic plate subject to projectile impact. Int. J. Solids Struct. **4**, 1081–1092 (1968)
351. Wolf, A., Swift, J.B., Swinney, H.L., Vastano, J.A.: Determining Lyapunov exponents from a time series. Phys. D **16**, 285–317 (1985)
352. Yao, M., Zhang, W., Wang, D.M.: Modeling and chaotic dynamics of the laminated composite piezoelectric rectangular plate. Math. Probl. Eng. **2014**, 345072 (2014)
353. Zarubin, A.G.: On the convergence speed of the Faedo–Galerkin method for linear nonstationary equations. Differ. Equ. **8**(4), 639–645 (1982) (in Russian)
354. Zarubin, A.G.: Investigation of Galerkin–Petrov projection procedure by small-steps method. Trans. Acad. Sci. SSSR **297**(4), 780–784 (1987) (in Russian)
355. Zarubin, A.G., Tiunchik M.F.: On approximate solutions of a certain class of nonlinear nonstationary equations. Differ. Equ. **9**(11), 1966–1974 (1973) (in Russian)
356. Zhang, W., Yang, J., Hao, Y.: Chaotic vibrations of an orthotropic FGM rectangular plate based on third-order shear deformation theory. Nonlinear Dyn. **59**, 619–660 (2010)
357. Zhang, B., He, Y., Liu, D., Gan, Z., Shen, L.: Non-classical Timoshenko beam element based on the strain gradient elasticity theory. Fin. Elem. Anal. Desig. **79**, 22–39 (2014)
358. Zhao, H.L., Ye, Z.M.: Analytic elasticity solution of bi-modulus beams under combined loads. Appl. Math. Mech. **36**, 427–438 (2015)

359. Zhao, J., Zhou, S., Wang, B., Wang, X.: Nonlinear microbeam model based on strain gradient theory. Appl. Math. Model. **36**, 2674–2686 (2012)
360. Zino, I.E., Trepp, E.A.: Asymptotic Methods in Problems of Theory of Thermal Conductivity and Thermoelasticity. Leningrad University Press, Leningrad (1978) (in Russian)
361. Zubrikhin, O.A., Gribanov, V.F., Skurlatov, E.D.: Experimental investigation of oscillations of shells in a transonic gas flow. In: Volmir, A.S. (ed.) Oscillations of Elastic Structures in Liquids, pp. 109–115. Nauka, Moscow (1976) (in Russian)

The manufacturer's authorised representative in the EU is Springer Nature Customer Service Centre GmbH, Europaplatz 3, 69115 Heidelberg, Germany. If you have any concerns regarding our products, please contact ProductSafety@springernature.com

Printed and bound by CPI Group (UK) Ltd, Croydon, CR0 4YY
30/09/2026
02225647-0019